Advances in Cryogenic Engineering

VOLUME 37, PART B

A Continuation Order Plan is available for this series. A continuation order will bring delivery of each new volume immediately upon publication. Volumes are billed only upon actual shipment. For further information please contact the publisher.

A Cryogenic Engineering Conference Publication

Advances in Cryogenic Engineering

VOLUME 37, PART B

Edited by

R. W. Fast

Fermi National Accelerator Laboratory
Batavia, Illinois

SPRINGER SCIENCE+BUSINESS MEDIA, LLC

The Library of Congress cataloged the first volume of this title as follows:

Advances in cryogenic engineering. v. 1–
New York, Cryogenic Engineering Conference; distributed by Plenum Press, 1960–
v. illus., diagrs. 26 cm.
Vols. 1– are reprints of the Proceedings of the Cryogenic Engineering Conference, 1954–
Editor: 1960- K. D. Timmerhaus.

1. Low temperature engineering–Congresses. I. Timmerhaus, K. D., ed. II. Cryogenic Engineering Conference.

TP490.A3 660.29368 57-35598

Proceedings of the 1991 Cryogenic Engineering Conference,
held June 11–14, 1991, in Huntsville, Alabama

ISBN 978-1-4613-6486-3 ISBN978-1-4615-3368-9 (eBook)
DOI 10.1007/978-1-4615-3368-9

Softcover reprint of the hardcover 1st edition 1991
Originally published by Plenum Press in 1991

Printed in the United States of America

HIGH EFFICIENCY, HIGH RELIABILITY OIL INJECTION TYPE HELIUM SCREW COMPRESSOR

Takayuki Kishi, Mizuo Kudo, Hiromasa Iisaka and Nobumi Ino

Mayekawa Mfg.Co.,Ltd. MYCOM Advanced Technology Laboratory
Moriya, Ibaraki, Japan

ABSTRACT

Investigation of the optimum condition and the reliability of two-stage oil injection type helium screw compressor has been carried out, in order to determine ways of enhancing compressor performance with high reliability. It was found that the volumetric efficiency of the compressor is increased by reducing the mass flow ratio of the helium gas / lubricating oil while accelerating compressor revolution speed. And it was proved that compressor volumetric efficiency is significantly influenced by sound speed in the mixture of gas and lubricating oil. The authors were able to derive a formula for predicting the change of compressor volumetric efficiency, when the mass flow ratio of gas / lubricating oil and the revolution speed are changed. It was also found that shaft power decreases when the ratio of theoretical volumetric flow rates of the high- and low-stage machines is set at 1:4 or a low density lubricating oil is utilized. In result, the advanced experimental machine achieved an isothermal efficiency level of 60% without any trouble.

INTRODUCTION

Currently, screw compressors incorporating oil injection systems are widely used in helium refrigeration systems for superconducting equipment. Their popularity for such applications stems from the fact that they incorporate fewer moving parts and exhibit more stable operation over long period of time than the reciprocating compressor. Nevertheless, there remains scope for improvement of oil injection type helium screw compressors in the area of performance and reliability, particularly with superconducting generation systems which demand very high efficiency and reliability. When a large quantity of oil is injected into the screw compressor in order to improve performance, problems such as increased compression power and bearing failure occur due to oil compression phenomenon within the screw compressor. The authors carried out research into these aspects of compressor performance in order to determine ways of enhancing compressor performance and developing ways of avoiding problems. The research found that compressor performance can be improved by:

Advances in Cryogenic Engineering, Vol. 37, Part B
Edited by R.W. Fast, Plenum Press, New York, 1992

Table 1. Specifications of the experimental machines

Item		Low-stage machine	High-stage machine
Suction pressure	(MPa)	0.1	(0.44)
Discharge pressure	(MPa)	(0.44)	1.67
Suction temp.	(°C)	27.	
Mass flow rate	(g/s)	0 ~ 70	

(1) Increasing the lubricating oil quantity and decreasing the mass flow ratio of the helium gas / lubricating oil,
(2) Accelerating the revolution speed,
(3) Setting the ratio of theoretical volumetric flow rates of the high-and low-stage machines to 1:4,
(4) Using low density lubricating oil.

The above investigation results were verified using an experimental machine with the specifications shown in Table 1. The specifications in Table 1 are equivalent to those of a helium refrigeration compressor having 100(L/h) liquifaction capability.

ANALYSIS

Mass Flow Ratio of Helium Gas / Lubricating Oil and Volumetric Efficiency

(1) Relationship between internal leakage quantity "ΣL" and volumetric efficiency "E_v"

The volumetric efficiency of the screw compressor can be expressed by the next equation which contains two assumptions. One is that the suction resistance of the screw compressor is negligible, and another one is that heat transfer between the lubricating oil, the suction helium gas and the internal leakage are disregarded in the suction process.

$$E_v = 1 - \frac{\Sigma L}{V_{th}} \qquad (1)$$

where V_{th} = theoretical volumetric flow rate at suction condition.

The internal leakage is a characteristic of screw compressor operation; gas leaks from the higher-pressure side to the lower-pressure side through several sealing lines, as shown in Fig.1.

Regarding the amount of internal leakage at each sealing portion, some calculation methods utilizing nozzle flow formula have been announced [1,2]. The instantaneous quantity of internal leakage "ΔL" at a specific rotation angle can be calculated from the following equations:

i) When the pressure condition is : $\left(\frac{2}{\kappa+1}\right)^{\kappa/(\kappa-1)} \le \frac{P_j}{P_h} \le 1$

$$\Delta L = f*A*v_j*C \left\{ \frac{2}{\kappa-1} \left[\left(\frac{P_j}{P_h}\right)^{2/\kappa} - \left(\frac{P_j}{P_h}\right)^{(\kappa+1)/\kappa} \right] \right\}^{1/2} \qquad (2)$$

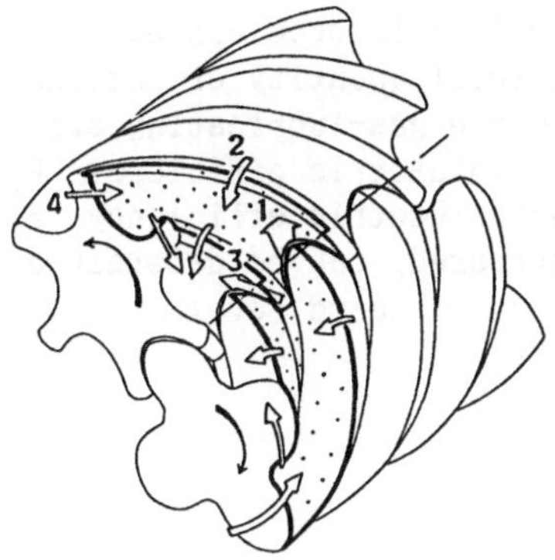

Fig. 1. Sealing lines and leakages of screw compressor

ii) When the pressure condition is : $0 \le \frac{P_j}{P_h} \le \left(\frac{2}{\kappa+1}\right)^{\kappa/(\kappa-1)}$

$$\Delta L = f*A*\left(\frac{v_j}{v_h}\right)*C\left[\left(\frac{2}{\kappa+1}\right)^{(\kappa+1)/(\kappa-1)}\right]^{1/2} \quad (3)$$

where f=coefficient of flow rate, A=leakage area, v=specific volume of gas, P=pressure of gas, C=sound speed, κ=specific heat ratio; subscript: j=lower pressure side at sealing line, h=higher pressure side at sealing line.

Since the amount of internal leakage at each sealing portion is proportional to the sound speed "C", if the sound speed is decelerated, the leakage volume is reduced and volumetric efficiency improved.

(2) Relationship between mass flow ratio "e" of helium gas / lubricating oil, and sound speed "C"

Keiffer indicated that sound speed in liquid-gas mixtures can be calculated from the following formula[3].

$$C = \left[e*R_{LA}\left(\frac{a}{P}\right)^{1/\kappa} + \exp\left(\frac{P_{LA}-P}{K}\right)\right]*\left\{\left[(1+e)R_{LA}\right]^{1/2} * \left[\frac{e*R_{LA}}{\kappa P}\left(\frac{a}{P}\right)^{1/\kappa} + \frac{1}{K}\exp\left(\frac{P_{LA}-P}{K}\right)\right]^{1/2}\right\}^{-1} \quad (4)$$

where $e = M_G/M_L$, $a = Pv^{\kappa}$, M=mass flow rate, R=density, K=bulk modulus of liquid ; subscript: L=liquid, G=gas, A=reference state.

Fig.2 shows the results of calculation with physical properties of the helium gas, the ammonia gas and lubricating oil substituted for symbols in the above equation. We can find it from Fig.2 that the sound speed in the helium gas-lubricating oil mixture fluid changes considerably according to the mass flow ratio of the helium gas / lubricating oil.

(3) Relationship between the mass flow ratio "e" of gas / lubricating oil and volumetric efficiency "E_v"

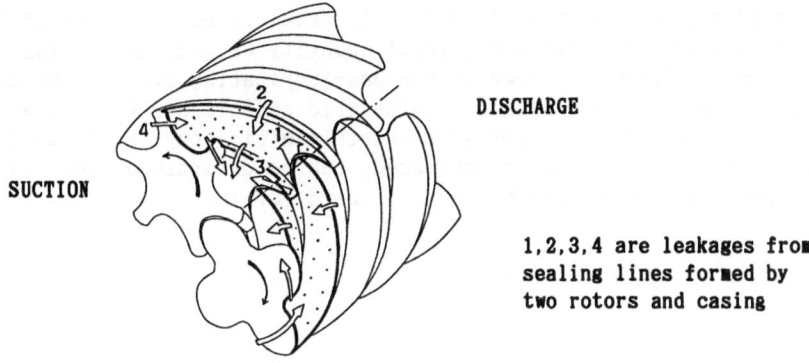

According to equation(2), (3) and Fig.2, if the mass flow ratio of gas / lubricating oil is reduced by supplying a larger quantity of oil than normal, sound speed at each sealing line is decelerated and volumetric efficiency is improved. Since the total quantity of internal leakage is proportionate to the sound speed in the gas-lubricating oil mixture fluid, the relationship between a standard volumetric efficiency "E_{vb}" at a certain quantity of oil and another volumetric efficiency "E_v", when the quantity of oil is increased or decreased, can be determined from the following equation. Subscript b means standard level.

$$E_v = 1-(1 - E_{vb}) \frac{C}{C_b} \tag{5}$$

Judging from Fig.2, if the mass flow ratio of gas / lubricating oil is set at 0.03~0.04, our purpose can be achieved. The oil quantity was therefore increased to 1.3~1.8 times the normal amount.

Revolution Speed and Volumetric Efficiency

The volumetric efficiency for a certain revolution speed "N" can be determined using the equation given below. Because suction resistance of screw compressor can be ignored and the theoretical volumetric flow rate at suction condition in formula(1) is proportional to the revolution speed and the sound speed changes according to changes of the mass flow ratio of gas / lubricating oil.

$$E_v = 1-(1 - E_{vb}) \frac{C}{C_b} * \frac{N_b}{N} \tag{6}$$

According to equation(6), a larger revolution speed results in a significant improvement in volumetric efficiency. But in the experiments, revolution speed was limited to a maximum of 4,500 rpm in consideration of increasing mechanical loss and going down the durabilities of the bearings etc.

Ratio of theoretical volumetric flow rates of high/low stage machines and compression power

In the case of two-stage compression, the optimum ratio of theoretical volumetric flow rates of them "ϕ" can be determined by the following equation:

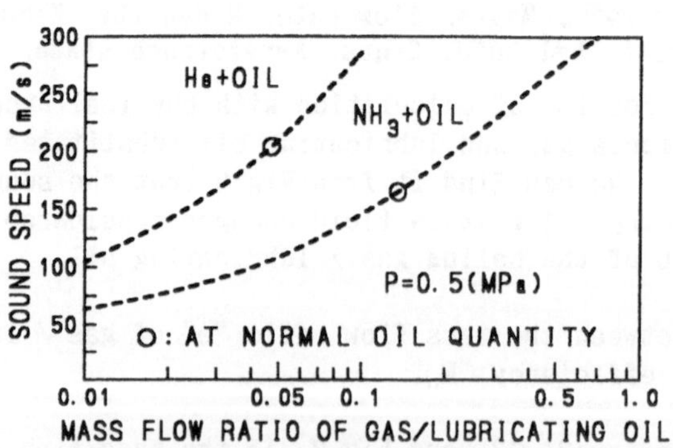

Fig.2. Sound speed in He + oil and NH3 + oil

Table 2 Comparison of Experimental and Conventional Machines

Item		Experimental machine	Conventional machine
Low-stage machine			
Rotor size	(mm)	ϕ 204x337	ϕ 204x337
Theo. vol. flow rate	(cm^3/rev)	6820	6820
Design volumetric ratio	(-)	3.6	3.6
Revolution speed	(rpm)	0～4500	0～3600
Oil quantity	(l/m)	130	85
High-stage machine			
Rotor size	(mm)	ϕ 127.5x210	ϕ 163.2x269.3
Theo. vol. flow rate	(cm^3/rev)	1660	2330
Design volumetric ratio	(-)	3.0	3.6
Revolution speed	(rpm)	0～4500	0～3600
Oil quantity	(l/m)	70	45

$$\Phi = \left(\frac{V_{th1}}{V_{th2}}\right) \geq \left\{\left(\frac{P_{d2}}{P_{s1}}\right)^{1/2}\right\}^{1/m} \qquad (7)$$

where m=polytropic index; subscript: s=suction condition, d=discharge condition, 1=low-stage machine, 2=high-stage machine.

With the experimental machine, the cooling effect of lubricating oil increases because of the increased quantity of it ,then the polytropic index of low-stage machine m_1 is estimated to be 1.1 or less. The following calculation can be carried out by substituting the pressure conditions and the polytropic index of the experimental machine in equation(7).

$$\Phi \geq \left\{\left(\frac{P_{d2}}{P_{s1}}\right)^{1/2}\right\}^{1/m_1} = \left\{\left(\frac{1.67}{0.1}\right)^{1/2}\right\}^{1/1.1} = 3.6$$

Consequently, the rated ratio of theoretical volumetric flow rates of the high- and low-stage machines was changed to 1:4 from the usual 1:3 ratio.

COMPARISON OF EXPERIMENTAL AND CONVENTIONAL MACHINES

Table 2 gives a comparison of the experimental machines used and a conventional machines. The experimental device is a closed loop consisting of high- and low-stage compressors, six separators, an after-cooler, buffer tank and oiling devices. The inverter motors are used to drive the compressors at variable revolution speeds of 0～4,500rpm.

RESULTS

Revolution speed and volumetric efficiency

Figure 3 shows the change in the volumetric efficiency of the low-stage machine when revolution speed is changed. The broken line in Figure 3 represents the change in volumetric efficiency calculated from equations (4)and(6). The calculated value is consistent with the experimental value. And the peak in volumetric efficiency does not appear in Figure 3, consequently the volumetric efficiency may be more improved with more accelerated revolution speed.

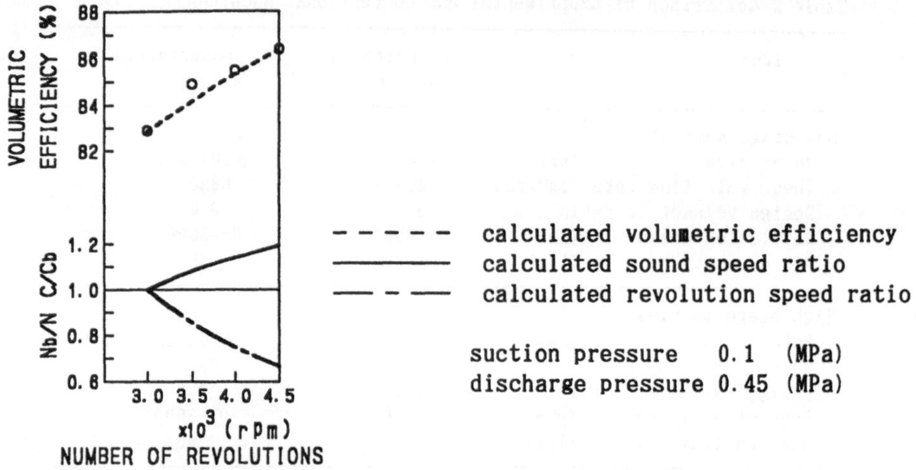

Fig. 3 Relationship between revolution speed ratio, sound speed ratio and volumetric efficiency of low-stage machine.

Total shaft power and efficiency

Figure 4 shows the change in total shaft power of the experimental machine in comparison with a conventional machine. The total shaft power of the experimental machine was reduced by 15~18(kW) compared to the conventional machine. 5~6(kW) of this reduction equivalent to the improvement due to use of low viscosity lubricating oil(18cSt at °40 C). The net 10~12(kW) represents the effective value of the ratio of theoretical volumetric flow rates of the high- and low-stage machines which were preset at 1:4.

Figure 5 parameterizes the total isothermal efficiency of an advanced experimental machine with a newly designed discharge side rotor shape. Regarding the redesign of the discharge side rotor shape, detailed information will be presented at a later date. Only the results are presented here. The advanced experimental machines displayed an improve-

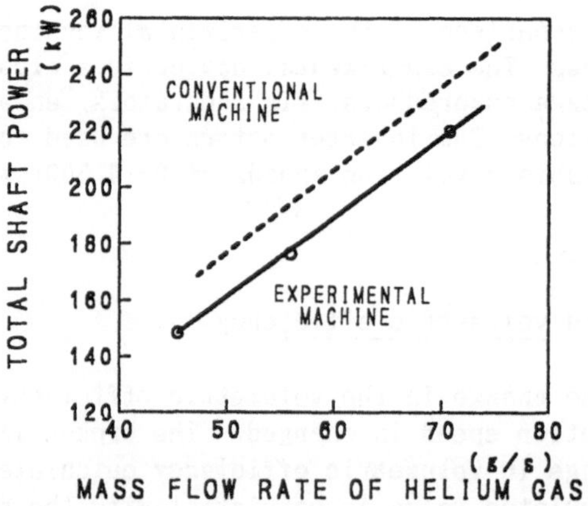

Fig. 4 Comparison of total shaft power
suction pressure of low-stage machine 0.1(MPa)
discharge pressure of high-stage machine 1.7(MPa).

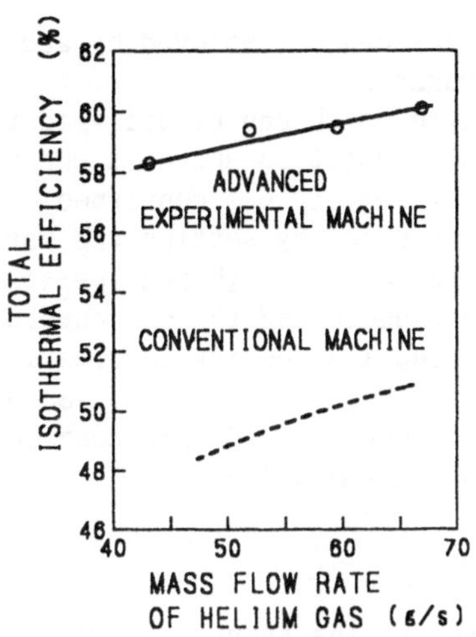

Fig. 5 Comparison of total isothermal efficiency
suction pressure of low-stage machine 0.1(MPa)
discharge pressure of high-stage machine 1.7(MPa).

ment of 3～4% in volumetric efficiency and a 13～18% improvement in shaft power. Maximum total isothermal efficiency of them achieved at 60%.

Oil compression phenomenon and 500 times start & stop running test

The measurements for the oil compression phenomenon in the low- and high-stage machines has been carried out. Oil compression pressure in the low-stage machine is 1.5(MPa) higher than discharge pressure of low-stage machine at 4000(rpm), and in the high-stage machine it is only 0.2(MPa) higher than discharge pressure of high-stage machine at same revolution speed. Both pressure values are lower level than 4~7(MPa) oil compression pressure which was measured in the conventional machine at 3000(rpm).

This reduction of oil compression pressure is also due to use utilizing low density lubricating oil.

The experimental machines performed without any problems through 500 start/ stop running test cycles. The vibration levels 3~5(μm peak to peak) were very low, and increased vibration levels 4~7(μm) after 500 times start and stop running test cycles were still low. In this way, the reliability of oil injection type screw compressor was improved to higher level.

But oil compression phenomenon, particularly in the low-stage machine, may increase the shaft power of the compressor, so authors are still investigating it.

CONCLUSIONS

Regarding the oil injection type helium screw compressor, the following points were clarified in the research:

① The sound speed of the gas-lubricating oil mixture fluid influences volumetric efficiency.
② Volumetric efficiency can be improved by reducing the mass flow ratio of the helium gas / lubricating oil.

③ Volumetric efficiency can be improved by accelerating the revolution speed of the compressor.

④ A formula was derived which can be utilized to estimate volumetric efficiency when the revolution speed is changed and the effectiveness of the formula was confirmed.

⑤ Shaft power can be reduced by setting the ratio of theoretical volumetric flow rates of the high-and low-stage machines to 1:4.

⑥ Shaft power can be reduced and the oil compression phenomenon decreased by utilizing low density lubricating oil.

⑦ Some 60% of total isothermal efficiency was obtained for the helium compressor, an improvement of 18～25% compared to the efficiency (48~51%) of a conventional machine.

ACKNOWLEDGMENTS

This project is sponsored by the Agency of Industrial Science and Tecnology of MITI, which commissioned NEDO to promote the work, and was carried out by Super-GM and its member companies. The authors would like to express their appreciation to individuals of the above organizations for their assistance in the preparation of this paper.

REFERENCE

1. T.Hirai et al,Performance Analysis of Oil Single Screw Compressor in "International Compressor Engineering Conference" (1986)
2. D.Xiao et al,The Computer Simulation of Oil-flooded Refrigeration Twin-screw Compressor in "International Compessor Engineering Coference" (1986)
3. S.W.Kieffer,Sound speed in Liquid-Gas Mixtures:Water-Air and Water-Steam,in "Journal of Geophysical Research"(July,1977)

A CLEAN HELIUM RECIPROCATING COMPRESSOR INCORPORATING ORBITAL SQUEEZE FILM BEARINGS

Victor Iannello, W. Dodd Stacy, and Herbert Sixsmith

Creare Incorporated
Hanover, New Hampshire

ABSTRACT

A contamination-free helium compressor for cryogenic refrigerators was designed, built, and tested. The machine incorporates a reciprocating shaft which floats on novel orbital squeeze film bearings to eliminate wear, sliding friction, and contaminating lubricants. The prototype compressor demonstrated pressure ratios greater than 4:1. At the design pressure ratio of 3:1, a 0.14 g/s flowrate was measured. A flowrate of 0.25 g/s was measured with the compressor unloaded. After 216 hours of endurance testing at design conditions, the compressor exhibited no visible signs of wear.

INTRODUCTION

Cryogenic refrigerators are in general intolerant of contaminants. Gaseous contaminants in the process gas stream condense and later freeze in the cold parts of the system. Frozen contaminants may block small passages, increase system pressure drops, and reduce the performance of heat exchangers. In J-T systems, the expansion orifices may become totally blocked. Most cryogenic refrigerators employ positive displacement compressors that are oil-flooded to provide lubrication, sealing, and cooling. To ensure adequate purity of the process gas, high effectiveness oil removal systems must be employed to reduce oil concentrations down to the ppm range or better.

An alternative approach would be the adoption of an oil-free compressor. There is a need for small cryogenic refrigerators which are capable of operating indefinitely without maintenance. A small oil-free compressor has been developed to meet this need. The machine incorporates a reciprocating shaft which floats on novel orbital squeeze film bearings which eliminate sliding friction, wear, and contaminating lubricants. The compressor was successfully designed, built, and tested.

DESCRIPTION OF COMPRESSOR

A drawing of the compressor is shown in Figure 1. The only moving component in the compressor is the shaft. The shaft is radially supported in the compressor by two orbital squeeze film bearings located at either end of the machine. The bearing journals at the ends of the shaft are of the simple sleeve type and function also as the cylinders and pistons of the compressor. The ends of the cylinders are provided with covers incorporating inlet and outlet reed valves. Reciprocation is effected by a pair of AC magnets which are alternately switched on and off. At the end of each stroke in either direction, the gas contained in the clearance space acts as a spring. During a

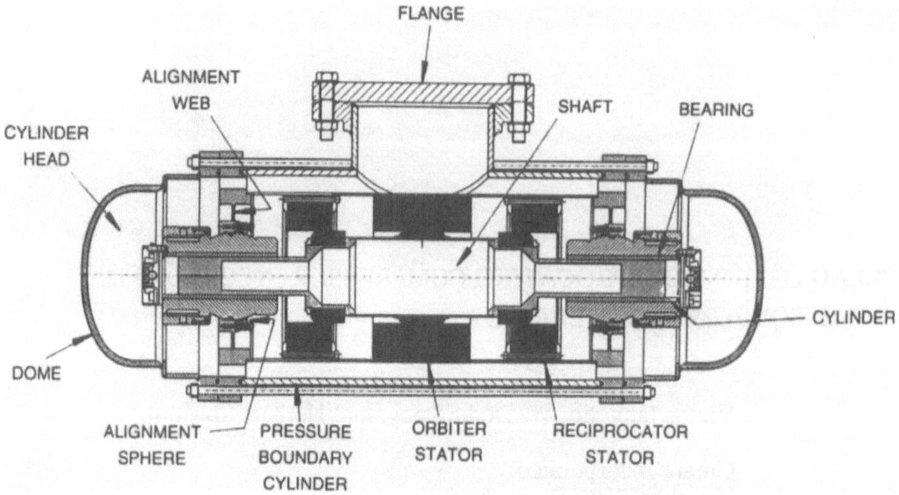

Fig. 1. Squeeze film bearing compressor.

displacement in one axial direction, gas compression and discharge occur in one cylinder while expansion of the cushion gas and intake occur in the other.

The gas springs together with the mass of the shaft form a resonant system. The compressor can be operated at the frequency of resonance for maximum transfer of energy to the process gas.

To demonstrate the technology, a design point of 3:1 pressure ratio at a helium flowrate of 0.2 g/s was chosen. However, the compressor was designed with the capability of individually exceeding the design point pressure ratio and flowrate. In this way, the achievable flowrate and pressure ratio of the concept could be separately demonstrated in the context of a well-proportioned machine.

DYNAMIC AND THERMODYNAMIC DESIGN

A dynamic analysis was performed to determine the important design parameters of the compressor. In particular, the analysis was used to determine the bore, stroke, and axial force necessary for resonant operation of the compressor given a design flowrate, pressure ratio, reciprocating mass, stroke frequency, and clearance volume.

The stroke frequency was chosen to be 6.86 Hz from valving considerations. The compressor was fabricated with reed valves for intake and discharge porting. However, it was desired to design the compressor with the flexibility to implement a valving scheme based on a rotating piston where ports on the wall of the piston rotate past stationary ports in the wall of the cylinder, and are designed to open and close at the proper times in the cycle. The axial and angular displacements of the piston must be synchronized for this implementation. The orbiter magnet is designed to rotate the shaft at 3.43 Hz, and as it is desirable to have pairs of ports at opposite ends of a diameter to ensure symmetry, the frequency of reciprocation must be twice this value, i.e., 6.86 Hz.

The basis for the dynamic analysis is the numerical integration of the equation of motion for the reciprocating mass, which is

$$M\ddot{x} = A_p[P_l(x)-P_r(x)] + F_{ax}(x) \quad (1)$$

where

M = reciprocating mass (kg)
x = displacement of piston
A_p = bore area of cylinder (m^2)
P_l = pressure in left cylinder (Pa)
P_r = pressure in right cylinder (Pa)
F_{ax} = reciprocator axial force (N)

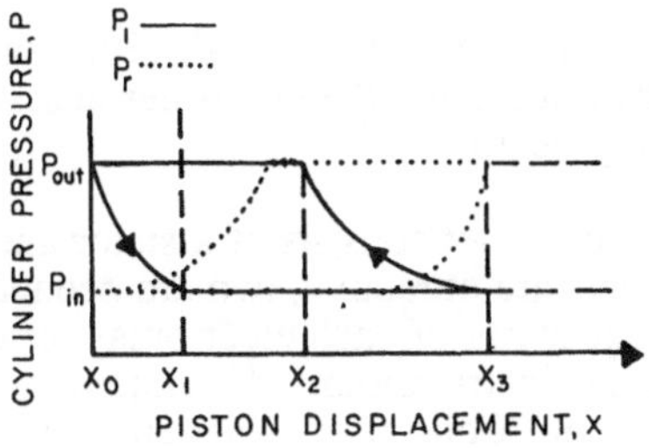

Fig. 2. Pressure history in cylinders as a function of piston displacement.

which is derived assuming friction forces, pressure drop due to porting, and leakage flow from the cylinder are all negligible.

The pressure histories in the cylinders as a function of piston displacement are shown in Figure 2 for a compressor operating with an inlet pressure P_{in} and outlet pressure P_{out}. At x_0, the piston is at TDC in the left cylinder, and $P_l = P_{out}$. As the piston moves to the right, the clearance volume gas is expanded, and the pressure drops to P_{in} at x_1. Isobaric intake occurs in the left cylinder until the piston reaches TDC in the right cylinder at x_3. The direction of the cylinder is then reversed. As the piston moves to the left, compression in the left cylinder occurs until the outlet pressure P_{out} is reached at x_2. Isobaric exhaust then occurs in the left cylinder as the piston reaches the TDC position in the left cylinder at x_o. The pressure in the right cylinder during this cycle is identical but phase shifted by 180 degrees.

The pressure in the cylinders as a function of shaft position was calculated for the limiting case of adiabatic compression/expansion. In the prototype, some heat is absorbed from the cylinders by coolant flowing in the cooling jackets, thereby reducing the required pumping power for a given flowrate and pressure ratio. We conservatively neglected this heat removal for the numerical analysis.

Based on these analyses, design specifications were chosen that would both satisfy design pressure ratio and flowrate requirements, as well as result in a well-proportioned and manufacturable machine. These specifications are summarize in Table 1.

ELECTROMAGNETIC AND ELECTRONIC DESIGN

The compressor includes the following electromagnetic and electronic subsystems: the reciprocator electromagnets (stators), which produce the axial force for

Table 1. Dynamic and Thermodynamic Design Specifications of Compressor

Number of cylinders	2
Inlet pressure	1 atm
Inlet temperature	300 K
Flowrate (2 cyl)	0.2 g/s He
Pressure ratio	3:1
Adiabatic Power (2 cyl)	173 W
Isothermal Power (2 cyl)	137 W
Piston diameter	5.72 cm (2.25 in)
Stroke length	4.06 cm (1.60 in)
Intake volume	95.0 cm^3 (5.80 in^3)
Clearance volume	9.5 cm^3 (0.58 in^3)
Clearance volume ratio	0.10
Reciprocating mass	16.5 kg (36.3 lb_m)
Maximum reciprocator force	440 N (98.8 lb_f)
Stroke frequency	6.86 Hz

compression; the orbiter stator and rotor, which induce the orbital motion of the shaft to generate the squeeze film lift forces; and the controller and switching circuits, which control the current to the reciprocators.

Each reciprocator electromagnet consists of a stator and rotor laminations on the shaft. Since the electromagnets are attractive, two are needed to effect the reciprocation of the shaft. The stator of each electromagnet was manufactured from the core of a conventional 3-phase electric motor. The internal diameter was 14 cm (5.5 in.) and there were 36 pole teeth. Coils at opposite ends of a diameter were connected in parallel to reduce any side pull due to eccentricity of the shaft. When supplied with 208 VAC, the current drawn was 22 A, and the force exceeded 446 N (100 lb_f). The force generated by the reciprocators was measured in separate effects experiments.

The function of the orbiter magnet is to generate a rotating radial force on the shaft which causes it to orbit in the bearings. This motion generates the squeeze film pressure forces in the bearing gap and produces the lift force that provides contact-free radial support of the shaft. The orbiter magnet also functions as a stepper motor to ensure synchronous rotation of the shaft.

The orbiter stator was constructed from the same laminations as the reciprocator stators, but wound to provide three groups of four AC magnets. The three groups are mutually spaced 120 degrees. Each group of magnets is connected to a phase of the 3-phase power supply. The magnets generate a radial force on the shaft that rotates at 120 rev/s. The magnitude of the force tends to be proportional to the square of the current in the coils. The radial force is designed to be about four times the weight of the shaft.

To rotate the shaft, 35 axial grooves are machined into the rotor. Because the stator has 36 teeth, the shaft will tend rotate one tooth spacing per rotation of force, or at 3.43 rev/s. The shaft rotation can be used for porting gas via porting holes in the cylinder walls and piston if reed valve-less porting is required. The rotation of the shaft also minimizes the chance of rubbing wear which could be caused by dirt particles between the shaft and cylinders.

Electrical and electronic subsystems were developed to perform the following functions:

- Supply the reciprocator stators with current of proper magnitude and switching frequency so that the stroke length and frequency may be controlled.
- Provide the proper voltage to the orbiter stator coils.
- Electrically isolate the compressor if a malfunction is detected.

The 3-phase voltage applied to the reciprocator stators is controlled by an AC current controller, which requires 3-phase 240 VAC input. Closed loop control of the current is achieved with the current transformers, which sense the currents flowing to the stators and thereby provide feedback signals to the controller. An error signal is generated by subtracting the sensed current value from the desired value. By varying the firing angle of the silicon controlled rectifiers (SCRs), the current delivered to the stator is controlled. The current establishes the axial force, and thus the stroke length of the compressor.

MECHANICAL DESIGN

The shaft is 75 cm (30 in.) long and is fabricated from carbon steel. The shaft is hollow along much of its length to reduce its total mass. As shown in Figure 1, the shaft includes the pistons, reciprocator rotors, and orbiter rotor.

The two cylinders function as journal bearings, compression chambers, and seals. The bearings generate their radial load capacity by means of squeeze film forces produced by the orbiting shaft. A small bearing clearance results in high load capacity and small leakage flow. However, fabrication constraints limit the achievable clearance. It was experimentally determined that a radial clearance of about 25 μm (0.001 in.) was suitable for this application. The leakage flow at this clearance was calculated to be 0.016 g/s for both cylinders at the design 3:1 pressure ratio. This represents 8% of the design flowrate of 0.2 g/s.

In order to align the bearings with the shaft, the bearings were structurally supported in the compressor housing by spherical surfaces. In addition, dynamic alignment is achieved by thin walled, annular webs which structurally connect the concave spherical surfaces with the compressor housing. This configuration maintains bearing alignment for small bending deflections of the shaft induced by its orbiting motion.

In order to reduce the pumping power required to achieve the 3:1 pressure ratio, the cylinder heads are surrounded by coolant jackets. A coiling coil also surrounds the structural shell which envelops the reciprocator stators.

A domed cylinder provides the pressure boundary for the compressor. All electrical, pneumatic, and hydraulic lines enter and exit the compressor through a 20 cm (8 in.) diameter flange. The free space in the compressor is maintained at the inlet pressure. The process gas enters the compressor through the flange, and flows in the annulus between the structural shell and a cylinder which forms the outer pressure boundary. The flow then passes through the compressor end plates and into the domed space. The compressed process gas exits the cylinder heads through the discharge ports and is sequentially ducted into the domed space, through the end plates, through the flow annulus, and finally exits the compressor by means of the flange.

TESTING

Initial tests were designed to measure the maximum flowrate attainable with the compressor fully unloaded, i.e., with $P_{in}=P_{out}$. With the reciprocator controller set for maximum current, the reciprocation frequency was decreased incrementally from 9.9 Hz. Because the gas spring forces are negligible for a unity pressure ratio, the stroke length should be inversely proportional to the square of the reciprocation frequency due to the shaft mass if the reciprocator force is independent of stroke length. However, the flowrate is proportional to the product of frequency and stroke length. Ideally one would expect the flowrate to be inversely proportional to the reciprocation frequency. The maximum flowrate would be obtained for an unloaded compressor in which the piston reciprocated with its maximum stroke length. Experimentally, this was determined to occur at a reciprocation frequency of 6.9 Hz. At this point of maximum stroke, a flowrate of 0.25 g/s (1.55 L/s) was measured.

Performance tests were then run to determine the flowrate versus pressure ratio characteristics of the compressor for reciprocation frequencies of 7.5, 8.2, and 9.4 Hz. The results of these tests are shown in Figure 3. At each frequency, three distinct regions can be observed:

- Inertially-dominated region
- Resonance region
- Pressure-dominated region

In the inertially-dominated region, the flowrate is independent of pressure ratio. Since the stroke length is determined only by the reciprocation frequency, the flowrate-pressure ratio curve is relatively flat at a given reciprocation frequency.

In the resonance region, the inertial and gas pressure forces acting on the shaft are of similar magnitude. As the pressure ratio is increased and resonance is approached, the flow remains approximately constant. Without resonance effects, the flowrate would decrease with increased pressure ratio as the pressure forces built and limited the stroke length. However, as the compressor approaches resonance, the

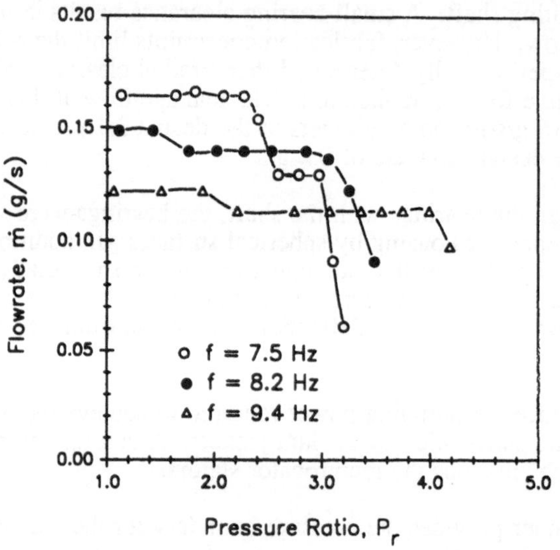

Fig. 3. Measured flowrate versus pressure ratio.

transfer between shaft kinetic energy and compression work increases, and the stroke length remains relatively constant. This is advantageous in that the flowrate is not very sensitive to system "tuning" about its resonance point.

For pressure ratios greater than the resonance condition, the flowrate falls off rapidly with increasing pressure ratio. This is the pressure-dominated region of operation, where gas compression forces exceed the inertial forces. As the pressure ratio increases, the compression work increases, and the efficiency of energy transfer between kinetic energy and gas compression work decreases. Both these effects result in a decrease of stroke and flowrate with pressure ratio in the region.

Table 2 summarized the resonance points for the three reciprocation frequencies for which data were obtained.

An endurance test lasting 216 hours (9 days) was performed at the design pressure ratio of 3:1 and at the resonant frequency of 8.2 Hz. This represents 6.4×10^6 cycles of shaft displacement. For most materials, the fatigue strength does not markedly decrease after about 10^6 cycles. The 216 hour endurance test therefore provides a good indication of the reliability of the compressor.

The compressor operated throughout the endurance test with no detectable change in performance. At the conclusion of the endurance test, the compressor domes and cylinder heads were removed for inspection of the bearing surfaces. No observable degradation of the surfaces was observed. This confirms the no wear aspect of the squeeze film bearings. No visible signs of wear or damage were found elsewhere in the machine.

Table 2. Summary of Resonance Conditions Measured

	Reciprocation Frequency		
	7.5 Hz	8.2 Hz	9.4 Hz
Pressure Ratio	2.40	3.06	3.99
Flowrate (g/s)	0.154	0.136	0.113
Isothermal Work (W)	85	96	100

CONCLUSION

A contamination-free reciprocating compressor has been developed for reliable service in helium cryogenic systems. The compressor shaft is supported by novel "orbital squeeze film bearing", which eliminate wear, sliding friction, and contaminating lubricants.

The performance and reliability of the concept were demonstrated by testing a prototype. A maximum pressure ratio of 4:1 was achieved with a helium flowrate of 0.113 g/s. At the design pressure ratio of 3:1, a flowrate of 0.139 g/s was measured. With the compressor unloaded, a flowrate of 0.254 g/s was measured.

An endurance test was performed to demonstrate the no wear and high reliability aspects of the compressor. After 216 hours of continuous operation at design conditions, the compressor exhibited no degradation of performance. Visual inspection of the bearings showed no signs of rubbing or wear.

Based on these successes, we conclude that the orbital squeeze film bearing compressor is ideally suited for cryogenic refrigerators. In particular, it's high reliability and contamination-free characteristics make it ideal for high pressure ratio cryogenic applications.

ACKNOWLEDGEMENTS

The work described in this paper was funded by the Office of High Energy Physics, U.S. Department of Energy, under contract DE-AC01-86-ER80335. The authors would like to thank DOE for their support of this work.

REFERENCES

1. H. Sixsmith, U.S. Patent No. 3,523,715 , U.K. Patent No. 1,232,999, Rotating Field Gas Bearings.

80 K CENTRIFUGAL COMPRESSOR FOR HELIUM REFRIGERATION SYSTEM

H. Asakura, D. Kato, N. Saji, and H. Ohya*

Ishikawajima Harima Heavy Industries Co., Ltd.,
Tokyo, Japan

* Super GM, Osaka, Japan

ABSTRACT

A centrifugal compressor for a completely lubricant free helium refrigeration system is now being developed using magnetic bearings. The compressor has four compression stages and each stage is independently driven by a 25-kW built-in motor at the speed of 100,000 rpm. The inlet of the compressor is maintained at a temperature of 80 K during operation by liquid nitrogen, and the motors are also maintained at the same temperature. Since the rotor of the motor has been fabricated from dissimilar materials to have a one piece construction using the hot isostatic press (HIP) technique, it is expected that it will deform when exposed to cryogenic temperature. The peripheral speed of the motor rotor will reach 340 m/s during operation. The peripheral speed of the impeller at the first stage will reach 568 m/s. This report summarizes the investigation results when the system was operated at 100,000 rpm using dummy impellers.

INTRODUCTION

We are participating in a Japanese national project which aims at the development of a 70-MW class superconducting generator. For this project, we are now developing a helium refrigeration system using a completely lubricant free centrifugal compressor for a power generating plant that is capable of operating continuously for prolonged periods of time. Figure 1 shows the time schedule for developing the compressor. The elementary design of the compressor is expected to be completed before the end of 1992, and the development of the refrigeration system is to follow thereafter.

The initial target for the compressor is to operate for 20,000 hours continuously using magnetic bearings which enables pure helium without containing any lubricant to be circulated inside the system. The motors used for driving the compressor are claw pole type synchronous motors, which have achieved the very high speed of 100,000 rpm at an output power of 25 kW. Since the motor has a sandwich structure in which non-magnetic material is diagonally placed between the magnetic materials, there exist a difference in the coefficients of thermal expansion between the two materials. It has thus been predicted that there would be a substantial deformation of the rotor when it is exposed to cryogenic temperature.

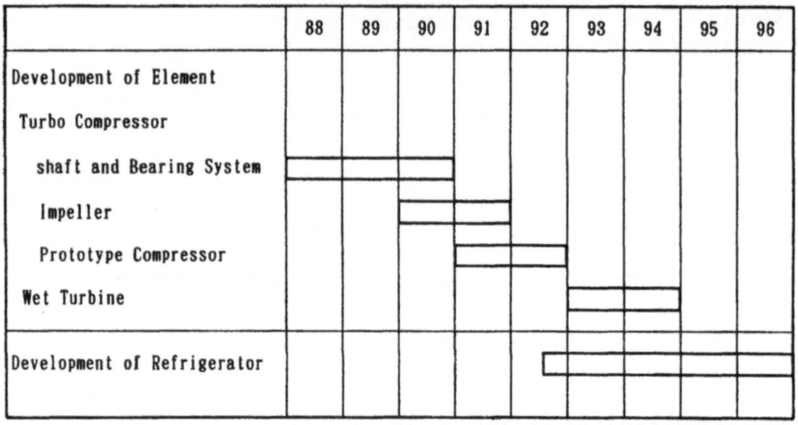

88
89
90
91
92
93
94
95
96
Development of Element
Turbo Compressor
shaft and Bearing System
Impeller
Prototype Compressor
Wet Turbine
Development of Refrigerator

Fig. 1 Schedule of turbo refrigerator development lower case

Determining the effect of the deformation on system operation was one of the objectives of this research.

DESIGN OF THE REFRIGERATION SYSTEM

Figure 2 shows the schematic diagram of the refrigeration system for the superconducting power generating system, which includes an 80 K nitrogen liquefier and two helium expansion turbines of Claude cycle. The design liquefying capacity is 120 liters of helium per hour. The compressor cooled by liquid nitrogen is of centrifugal type having four compression stages, and its total compression ratio is eight which is low compared with most helium refrigerators.

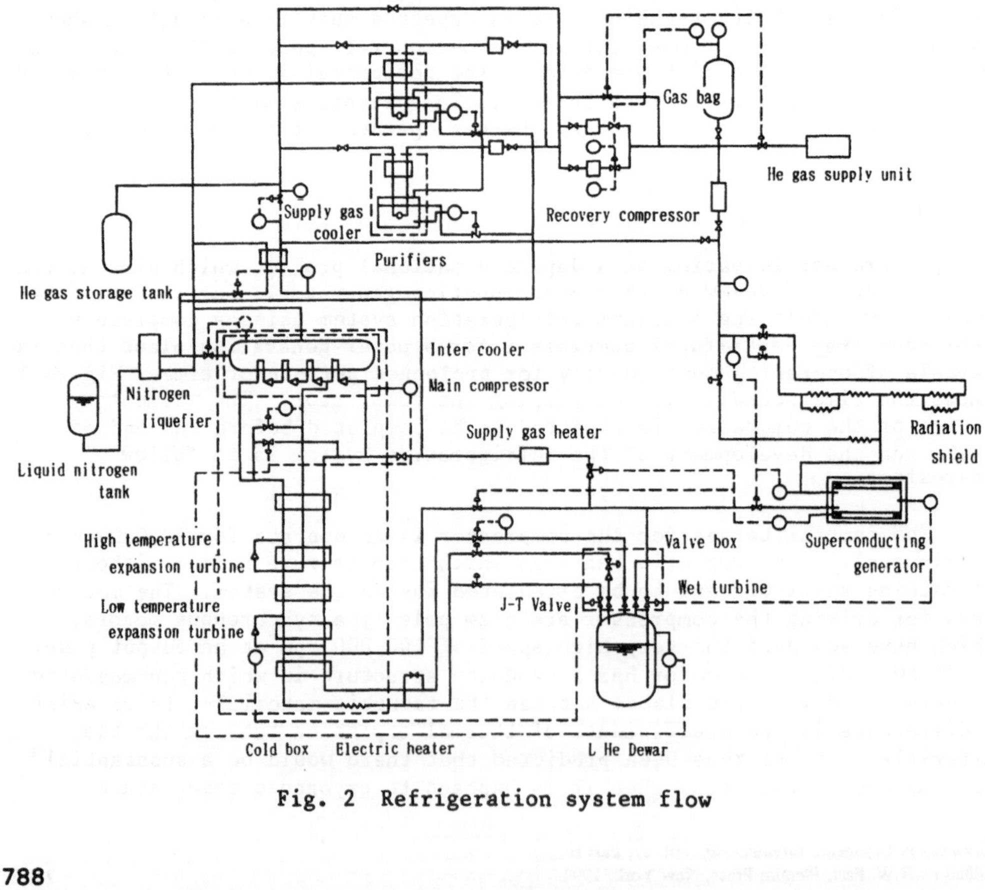

Fig. 2 Refrigeration system flow

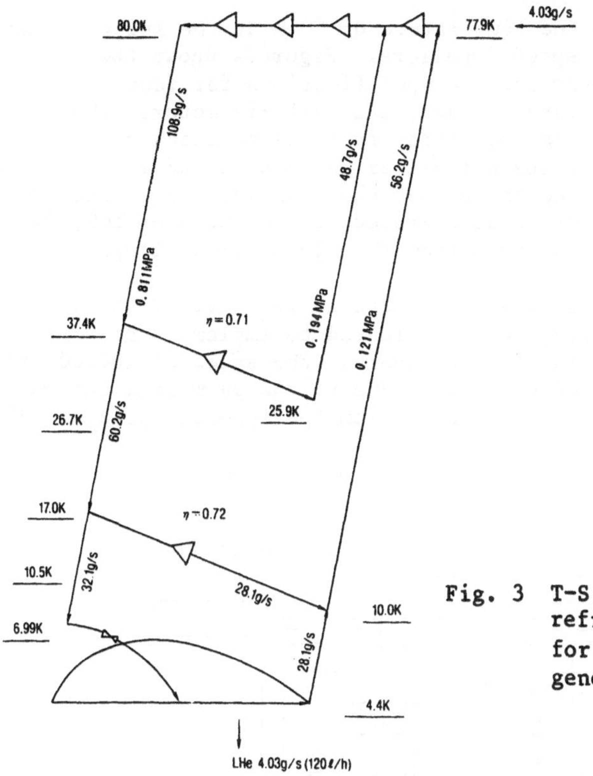

Fig. 3 T-S Diagram of helium refrigeration system for superconducting generator (with J-T valve)

The high temperature turbine, in which the gas enters at the temperature of 34.7 K, has an the expansion ratio of four, and the gas returns to the inlet of the second stage compressor. These conditions have been selected to assure good impeller efficiency for the centrifugal compressor since it will become difficult to improve the efficiency if the volumetric flow of the gas is further reduced at subsequent stages. Another reason was that we considered the efficiency of the high temperature turbine to be important. The gas enters the low temperature turbine at a temperature of 17 K, and the expansion ratio is 6.6. Since the overall compression ratio is low, development of a wet turbine is planned in order to improve the system efficiency.

During normal operation of the power generator, the gas bypasses the purifier and returns directly to the pre-cooler of the main compressor. The liquid helium for the power generator is supplied from the dewer having a capacity of 2,600 liters, and the refrigeration system is controlled by the liquid level inside the dewer. The amount of helium to be liquefied is controlled by the speed of the compressor. Bypass control is provided to avoid possible surging.

The helium gas storage tank has a capacity of 115 cubic meters. It is capable of accommodating all the returning helium supplied by the dewer to the generator upon the failure of the refrigeration system. The compressor for recovery of the helium is of a lubricant free reciprocating type, and has a capacity of 35 g/s at a discharge pressure of 2 MPa, which is large enough to cover the maximum flow during the pre-cooling operation.

COMPRESSOR DESIGN

Since the helium gas with a small molecular weight is to be compressed to 0.811 MPa from 0.101 MPa using the four stage impellers, it is necessary to maintain the temperature of 80 K at the inlet of each stage and to have very high speed impellers. Figure 4 shows the construction of the compressor and the specifications for four compressor. Each stage consists of identical built-in motors, though the profiles of the impeller and casing differ from one to another. The peripheral speed of the first stage impeller reaches at 568 m/s. Results of a stress analysis by FEM indicated that the disk would be tilted by 0.17 mm toward the shroud when it is operated at the speed of 100,000 rpm. The impeller has been machined from the Al-Zn-Mg-Cu alloy.

The calculated maximum stress is 354 MPa at the base of the shaft. The impeller for the first stage was fabricated to implement the spinning test. No permanent deformation was observed at the speed of 100,000 rpm. Its outside diameter increased by 0.15 mm due to the permanent deformation when it was operated at a speed of 146,000 rpm (peripheral speed of 830 m/s), however, it did not fracture.

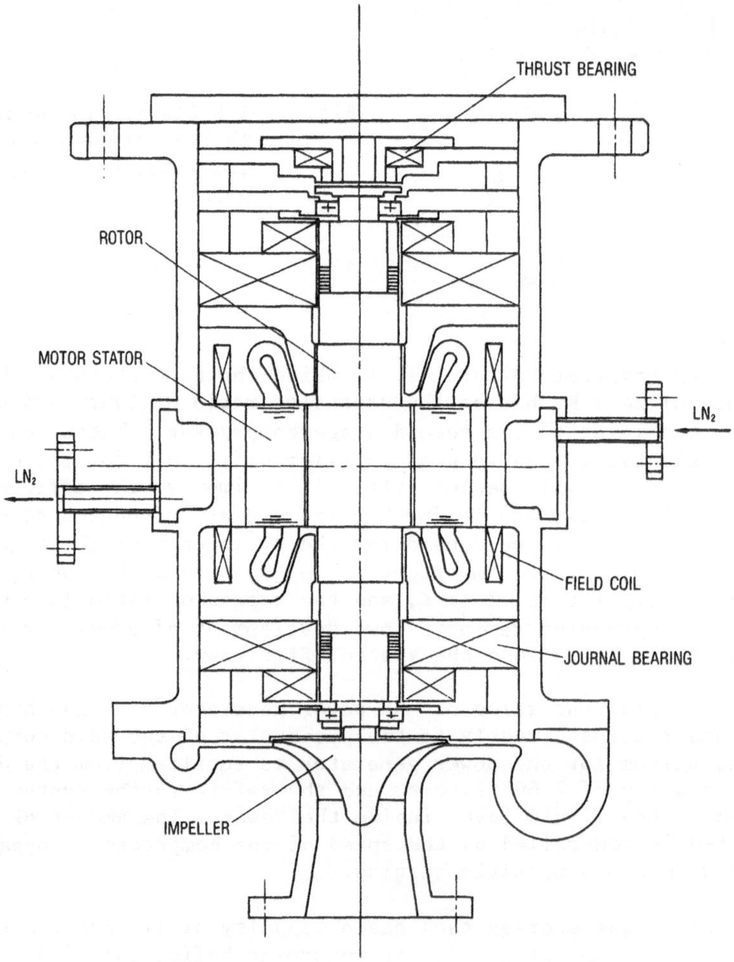

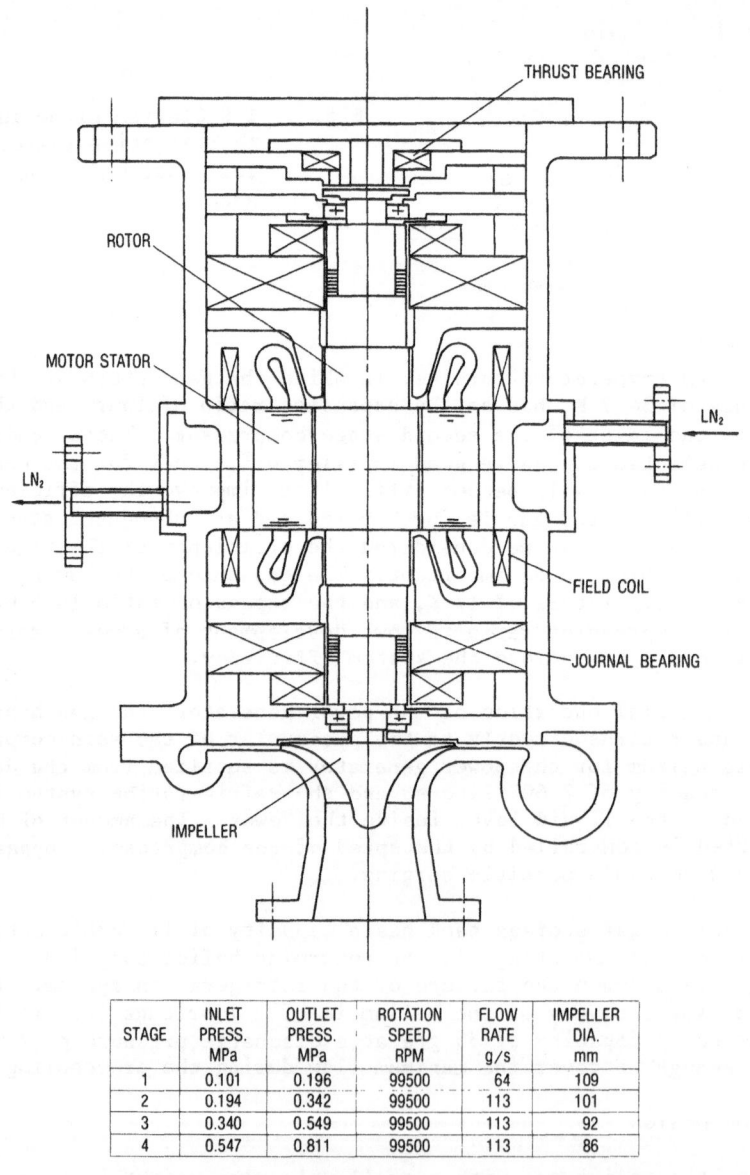

STAGE	INLET PRESS. MPa	OUTLET PRESS. MPa	ROTATION SPEED RPM	FLOW RATE g/s	IMPELLER DIA. mm
1	0.101	0.196	99500	64	109
2	0.194	0.342	99500	113	101
3	0.340	0.549	99500	113	92
4	0.547	0.811	99500	113	86

Fig. 4 Cross section of He cold compressor

Fig. 5 Impeller for spinning test

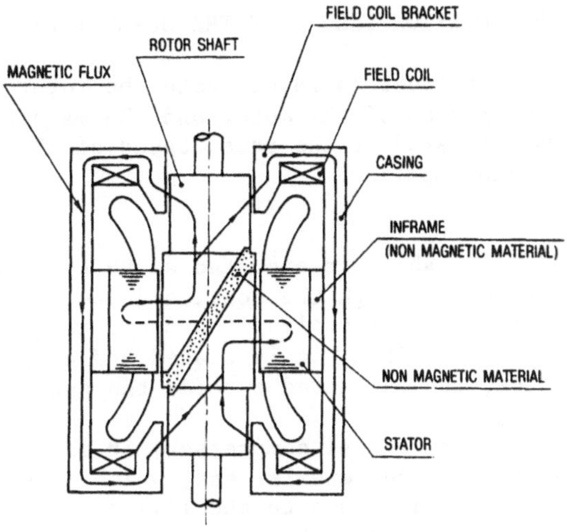

Fig. 6 Magnetic flux path of claw-pole type synchronous motor

The motor is a claw pole type synchronous motor, which generates a magnetic flux path by the field coils shown in Figure 6. The motor is cooled by flowing liquid nitrogen through the cooling jacket and the low temperature helium through the motor itself. It was necessary to use high strength material since the peripheral speed of the motor rotor was expected to reach 340 m/s at the speed of 100,000 rpm. For this reason, Inconel 718 was selected as the non-magnetic material to be placed between the magnetic pieces of 9% Nickel steel, which are bonded together to form a one piece structure using the HIP technique. Due to difference in the coefficients of thermal expansion between the two materials, it was expected that internal stresses would occur to cause deformation when they are exposed to cryogenic temperature.

Stress analysis predicted that the stress would not be too high even at cryogenic temperature. However, there exists a possibility of having a considerable deformation at cryogenic temperatures. The rotor revolves at a very high speed in a given direction around the center of inertia as determined by the bearing, however, the impeller might appear to be mounted with a certain inclination at cryogenic temperature if the shaft has been bent. When the bent shaft is running at cryogenic temperature, it is likely that there will be vibration. However, we expected that the vibration might be eliminated since the magnetic bearing is a non-contact type and the shaft rotates around the inertia center due to the automatic balancing function.

As a result of studies, we noted that it was necessary to reduce the weight of the impeller at the overhung portion in order to reduce vibration. In this respect, we have managed to achieve an impeller weight as light as approx. 200 g using an aluminum alloy. At present, there is not much margin left between the design speed of 100,000 rpm for the rotor and the critical speed of the primary bending mode for the first stage compressor. Therefore, serious attentions has been paid to balance it prior to assembly. The weight of the rotor was 6.5 kg and the datum value for design correction was set to 7.65×10^{-2} g-cm. The actual correction values before implementing the test were 1.4×10^{-2} g-cm for the impeller side and 1.35×10^{-2} g-cm for the thrust bearing side.

THE ROTATION TEST WITH THE DUMMY IMPELLER

In order to investigate the rotor dynamics behaviors for the combination of the motor and the magnetic bearings in advance, the system was operated at the rated speed of 100,000 rpm using a dummy disk instead of the impeller.

The shaft is operated at 80 K. The motor is cooled by the liquid nitrogen through the cooling jacket outside the stator, and the liquid nitrogen after cooling the motor is accumulated at the bottom of the cryostat. Then, it is heated and evaporated by a heater to cool the rotor, and exhausted outside the system.

On an actual compressor, it is not possible to install a vibration sensor on the impeller end, which is estimated to have the most potential for vibration, due to the limitations on the shaft dimensions and the obstruction of the flow at the compressor inlet. Therefore, it is more convenient if the rotation test can be implemented without using the impeller. Other vibration measurements were taken by obtaining signals on the magnetic bearing. The temperature and vibration signals were monitored and recorded using a data recorder, oscilloscope, and FFT analyzer.

Figure 8 shows the mounted conditions of the compressor. Figure 9 shows the vibration level analyzed by the FFT analyzer on the impeller side shaft end at a speed of 100,000 rpm, which represent data for every 30 seconds.

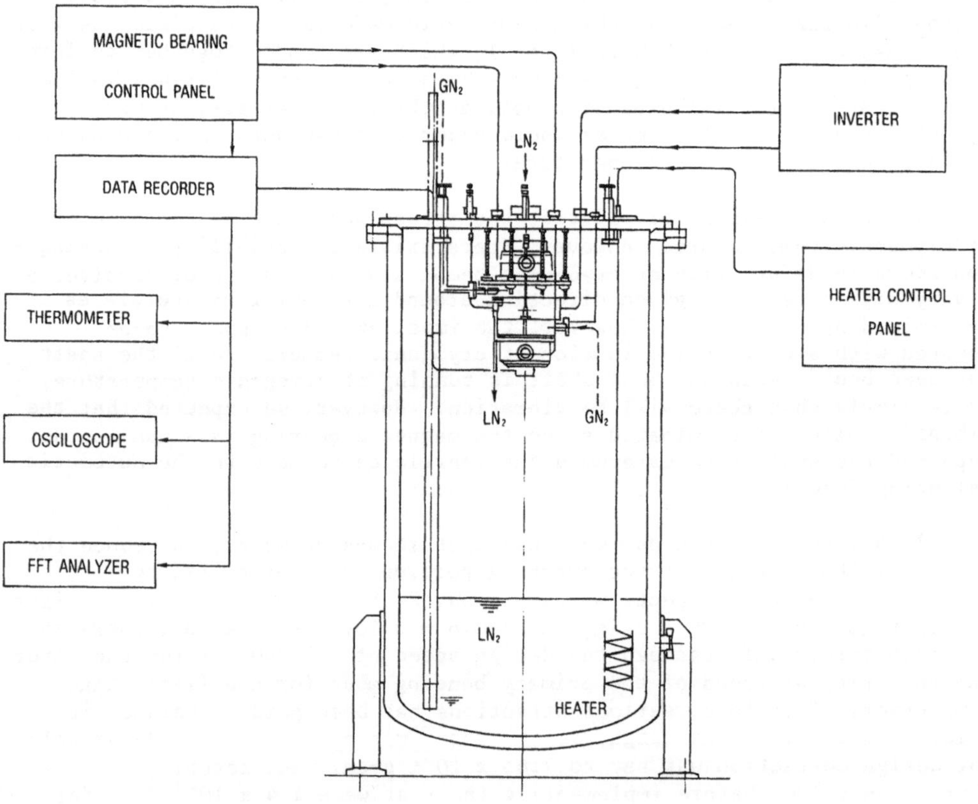

Fig. 7 The rotation testing apparatus

Fig. 8 He cold compressor

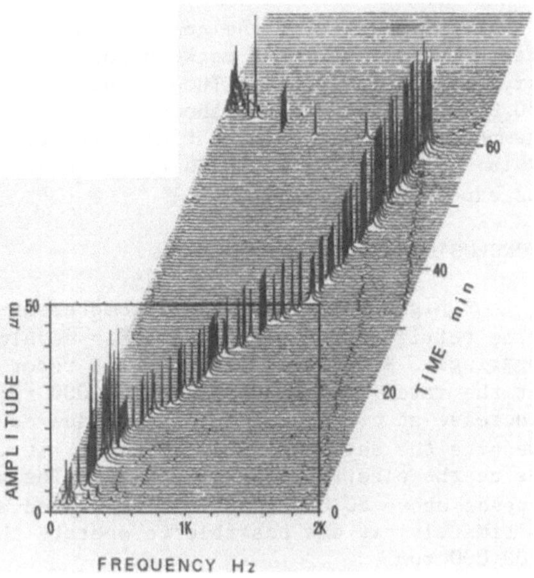

Fig. 9 3-dimensional record of FFT analysis of the shaft end vibration (at cryogenic temp.)

Since it is difficult to operate the motor synchronously at low revolution speeds, it was first accelerated up to 12,000 rpm as an induction motor, and then switched to synchronous operation. During the synchronous operation, increase in the vibration was recognized when compared with that of induction operation. The magnetic bearing switched into automatic balancing mode operation at the speed of approx. 18,000 rpm, and the vibration decreased accordingly since the magnetic bearings ceased to respond to the synchronous constituents of the revolution of the vibration. When the speed exceeded 80,000 rpm, the vibration amplitude of the shaft on the impeller side started to increase, and reached approx. 30 microns at a speed of 100,000 rpm (1,667 Hz).

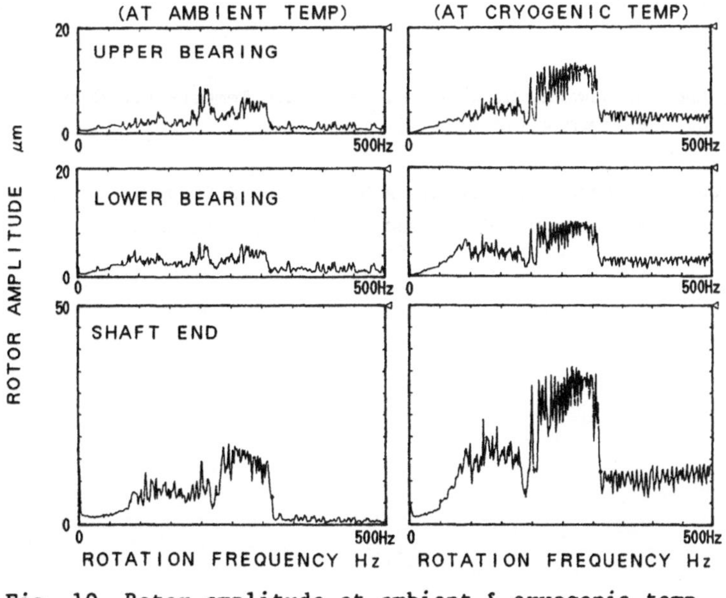

Fig. 10 Rotor amplitude at ambient & cryogenic temp.

Figure 10 shows the comparisons of the amplitude and vibration frequency of the rotor between the operation at room temperature and cryogenic temperature. Though the data represent rather low speeds (up to 30,000 rpm), it clearly shows that the vibration amplitude at cryogenic temperature would be many times greater than at room temperature. For this reason, it is believed that the deformation of the shaft is present as expected.

CONCLUSION

A high speed centrifugal compressor aiming at a completely lubricant free refrigeration system is under development. It uses electromagnetic bearings. A rotation test for the rotor confirms that it can be operated at the rated maximum speed of 100,000 rpm. Vibration was confirmed to increase at cryogenic temperature due to deformation of the rotor. Despite the deformed rotor, however, it was possible to continue operation using the electromagnetic bearing. The vibration further increased at speeds above 80,000 rpm since the speed approached the critical speed. Ultimately, it was possible to operate the system at its rated speed of 100,000 rpm.

Since the direction of the deformation has already been determined, it is possible to reduce the vibration by correcting the shaft balance. With this approach, the vibration at cryogenic temperature is expected to be reduced generally on a long term basis even though the vibration at room temperature is expected to increase by some degree.

From now on, the development work is to progress on the development of the impeller, fabrication and operation of the prototype system in order to achieve the development of a completely lubricant free centrifugal refrigeration system.

ACKNOWLEDGMENT

This work was performed as a part of "R&D on Superconducting Technology for Electric Power Apparatuses" as a subject of Super-GM under the Moonlight Project of Agency of Industrial Science and Technology, MITI, being consigned by New Energy and Industrial Technology Development Organization (NEDO).

REFERENCES

1. M. Okano, M. Kawada, S. Togo, N. Saji, Low Temperature Centrifugal Helium Compressors, ICEC 9,170 (1982).

2. S. Togo, M. Okano, Y. Akiyama, A New Helium Refrigerator System with Low-Temperature Centrigual Compressors, ICEC 9,174(1982).

DESIGN, CONSTRUCTION, AND OPERATION OF A TWO CYLINDER RECIPROCATING COLD COMPRESSOR

J.D. Fuerst

Fermi National Accelerator Laboratory*
Batavia, Illinois

ABSTRACT

In an effort to operate Fermilab's Tevatron synchrotron at higher beam energies, plans exist to reduce the working temperature of the superconducting components. A proposed upgrade will utilize cold compressors as an addition to each satellite refrigerator, lowering the pressure of the 2-phase helium returning to the refrigerator system from the magnets. Coil temperatures as low as 3.9K are anticipated.

Since initial tests were performed with CCI reciprocating compressors and turbo compressors from Creare, Inc. and IHI Co., Ltd., the scope of the upgrade has broadened such that these machines no longer meet the pressure ratio, throughput, and efficiency requirements. Therefore new cold compressor development has been funded. In parallel with the purchase of a new centrifugal machine, Fermilab has developed a reciprocating unit capable of meeting the new performance criteria. The development history and operating characteristics of this machine are presented.

INTRODUCTION

Fermi National Accelerator Laboratory intends to upgrade its satellite cryogenic refrigeration system in order to permit an increase in Tevatron particle beam energy from the current 900 GeV maximum up to perhaps 1100 GeV. Elevated beam energies require stronger fields in the bending and focussing magnets that make up the synchrotron lattice. Stronger magnetic fields imply higher current in the superconducting windings. Finally, peak (or critical) current in a superconductor increases as the conductor temperature is lowered below the transition temperature. Tevatron magnets were designed such that the latent heat of flowing two-phase helium is used to transport heat out of the system. Hence magnet operating temperature is fixed by the pressure of the two-phase helium stream. The cryogenic system upgrade revolves around the installation of cold compressors to reduce the pressure of the two-phase circuit. It is planned to lower this pressure from the current 126 kPa (1.2 atm) to about 51 kPa (0.5 atm).

Several cold compressors have been tested in the pursuit of this upgrade[1,2]. It is required that the machines pump saturated vapor returning from the magnet strings with acceptable pressure ratio, throughput, and efficiency. Although a phase separator is to be installed upstream of the compressor, it is likely that some portion of the liquid fraction returning via the two-phase circuit will occasionally enter the compressor. For this reason,

* Operated by Universities Research Association, Inc. under contract with the U.S. Department of Energy

Table 1. Cold Compressor Specification

Normal Operation:	
Inlet condition	51 kPa sat. vap.
Outlet condition	126 kPa
Flow rate	60 g/s
Adiabatic efficiency	>60%
(incl. static heat leak)	
Off-Design Operation:	
Inlet range	41-81 kPa
Flow rate range	40-70 g/s
Minimum efficiency	60%
Standby (compressor off) Operation:	
Inlet condition	126 kPa sat. vap.
Flow rate	55 g/s
Pressure drop	<6.9 kPa
Upset Conditions:	
• 275 kPa, 100 ms pressure spikes during operation	
• Unit must reliably pump or otherwise protect itself from two-phase flow	

the machines must tolerate two-phase flow as well as saturated vapor. Finally, a high degree of reliability is crucial to the smooth, uninterrupted operation of the cryogenic system and the physics program it supports.

Details concerning the specification have evolved, requiring equipment development beyond that described in (1) and (2). Current requirements are listed in Table 1.

DESIGN DESCRIPTION

Single cylinder reciprocating cold compressors supplied to Fermilab by Cryogenic Consultants, Inc. (CCI) were of insufficient capacity to meet the newest specifications. Otherwise, performance was acceptable. Given the units' straightforward design and uncomplicated operation, it was decided that some sort of upgrade or modification to the existing machine was appropriate to reach desired capacity. Tests of the CCI compressors showed that throughput was about half the required level at the new design pressure ratio and intake conditions. This information was passed on to CCI where they suggested that the machine's volumetric displacement be doubled. This option implied large pistons, valves, and associated seal difficulties. Concurrent with these efforts, an alternative concept was generated at Fermilab. Given the success of the single cylinder unit, sketches were prepared showing installation of two CCI units in a common cryostat. The resulting idea described a two cylinder reciprocating compressor based on the CCI cold end/linear bearing assembly. Benefits included acceptable throughput as well as proven performance and smoother operation.

Fermilab has considerable experience with the Koch Process Systems, Inc. (KPS) expansion engine, utilizing this platform as the "wet" and "dry" expander in all satellite refrigerators. However, dry expander operation is not required for normal operation due to the introduction of LHe into the refrigerator return stream from the Central Helium Liquefier[3]. Because of increased CHL reliability and space limitations imposed by other elements of the upgrade, tentative plans include removal of all satellite dry expanders. Given this imposed surplus of rotating machinery, it was suggested that the new reciprocating cold compressor platform make use of the obsolete dry expander cryostats, topworks, and motor/controllers. Compressor construction would amount to scavenging all useful components from KPS expanders, the purchase of CCI cold end assemblies, fabrication of a few new pieces, and assembly. Table 2 outlines the parts list. For each cold compressor required, there is a corresponding dry expander slated for removal.

CONSTRUCTION DETAILS

The unit is pictured in Figs. 1 through 3. Specific elements of the design will be discussed in this section. Adaptation of the CCI elements was straightforward, requiring

Table 2. Two Cylinder Reciprocating Cold Compressor Parts List

From CCI:	
(2)	Cold gas pump cold end assembly with linear bearing assembly
From Existing KPS Expander:	
(1)	7.5 HP electric motor with controller
(1)	Flywheel
(1)	Flywheel frame
(1)	Crankshaft with associated bearing assemblies
(1)	Jackshaft with bearing assembly
(1)	Outer vacuum vessel assembly
(1)	Bayonet assembly (or FNAL equivalent)
Designed and Fabricated by FNAL:	
(1)	Top plate
(2)	Housings for CCI cold end assemblies
(2)	Surge volumes for intake and exhaust, with associated piping
(1)	"Subatmospheric" bayonet assembly for intake bayonet
(1)	Frame riser
(2)	Connecting rod/clevis
(2)	Crank wheel
(1)	Motor mount with brackets

little work beyond accommodations for intake, exhaust, and warm end interfaces. The new cryostat top plate was machined to accept the CCI assemblies, as well as carry the flywheel, flywheel frame, and frame riser (required to allow for the increased height and stroke of the CCI assemblies relative to the KPS arrangement). While the motor/controller of an expander serves to extract work from the unit, a compressor consumes work. This switch was easily accommodated by the existing motor/controller package with a few

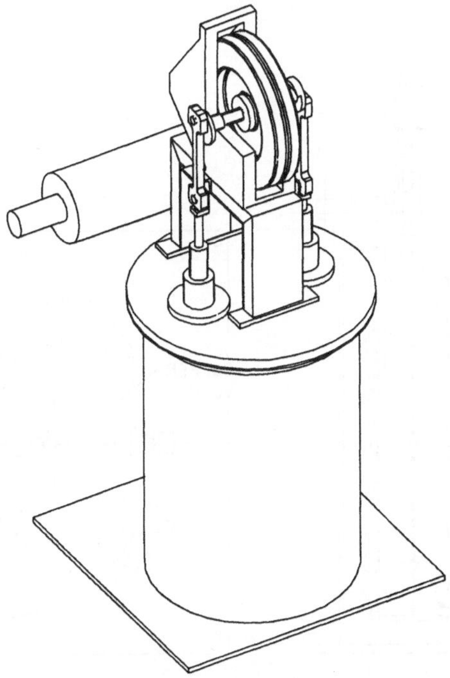

Fig. 1. I-DEAS Solid Model of the Prototype

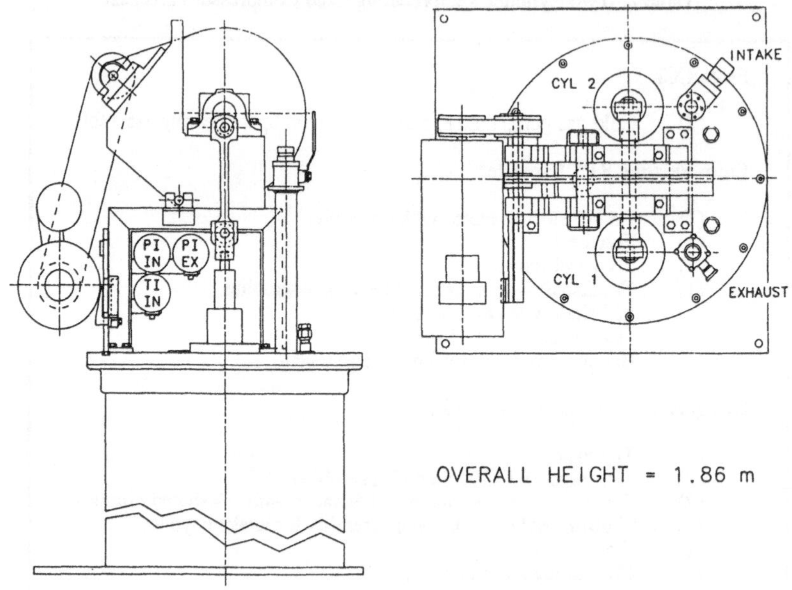

Fig. 2: Layout of the Prototype

electrical adjustments. The motor and instrument cluster were mounted on the frame riser to eliminate all connections to the cryostat. This simplifies maintenance by permitting unfettered removal of the top plate and attached cold end plumbing from the vacuum vessel. Warm end design is such that the cold end assemblies can be replaced by tilting back the flywheel and frame on the frame riser, a feature retained from the KPS design. New crank wheels and connecting rods were fabricated to mate the CCI piston shafts to the KPS flywheel. Finally, form-fitting aluminum covers with acrylic windows were fit to the flywheel, crank wheels/connecting rods, and jackshaft. The entire package was designed with the help of SDRC I-DEAS, a CAE program. This computer model was useful for overall component integration and as a visualization tool.

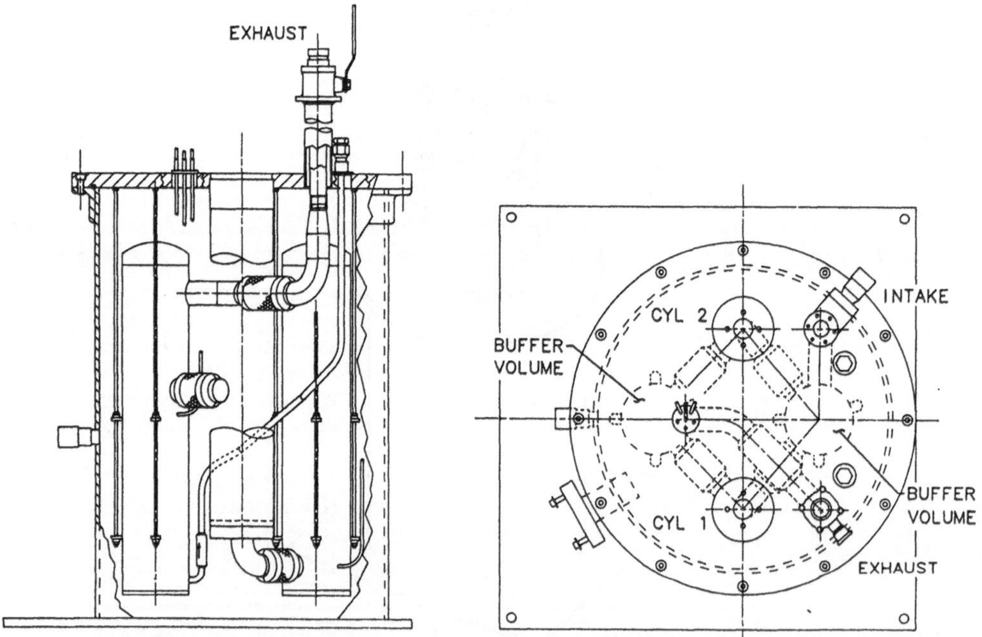

Fig. 3. Layout of the Internal Piping

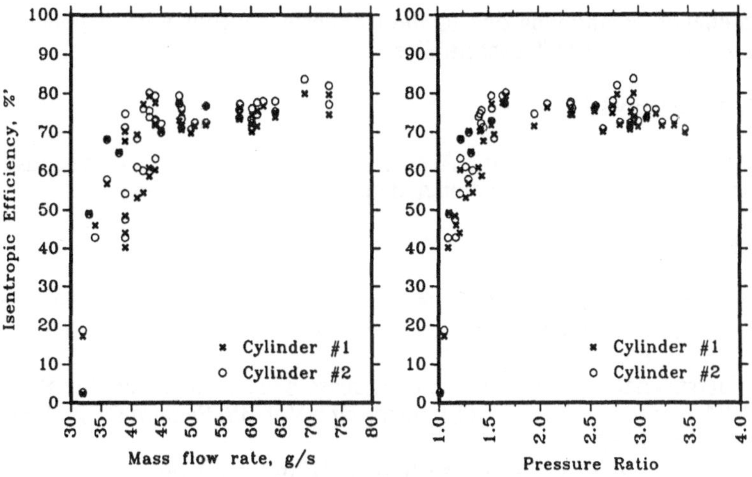

Fig. 4. Prototype Performance Data

Cold end piping was designed similar to that of the CCI single cylinder design. Braided flex-hose was used where appropriate to limit stresses induced by thermal contraction. Appropriately sized intake and exhaust buffer volumes serve to minimize pressure oscillations induced by the unit. Instrumentation consists of intake and exhaust pressure taps connected to compound test gauges, intake and exhaust carbon resistor thermometers, and a helium vapor pressure thermometer on the intake stream connected to a compound test gauge. There is an independent resistor at the exhaust of each cylinder to facilitate separate cylinder efficiency measurements. Because of the contamination risks associated with compressor operation at subatmospheric intake conditions, special precautions were taken to eliminate non-metallic helium-to-atmosphere seals on the intake side. As a consequence, gauges are welded or silver soldered directly to their sensing lines. The intake bayonet assembly is a special Fermilab design using copper gaskets instead of elastomer O-ring seals. Finally, rupture disks are used in favor of relief valves for overpressure protection. Given the inconvenience associated with the opening of a rupture disk, relief pressures are set well above the nominal system operating pressure.

PERFORMANCE

This prototype unit has performed as anticipated for over 500 hours of operation, both in a test refrigerator and in an actual Tevatron satellite. Efficiency is much the same as that of the original single cylinder CCI machines. As one would expect, throughput is double the CCI units. In Tevatron service, the prototype was more than capable of meeting the compressor specification. Flywheel speeds never exceeded 300 rpm, in part due to the limitations of the motor/controller and a somewhat inappropriate pulley ratio. CCI single cylinder units have operated at 400 rpm for extended lengths of time with acceptable performance; it is reasonable to expect the two cylinder prototype, with its improved balance, to meet or exceed this speed without difficulty. Therefore the prototype is most likely capable of well exceeding the specification with regard to throughput and pressure ratio. Fig. 4 shows performance data, taken independently for each cylinder. Maintenance requirements are proving to be a cross between those of our CCI compressors and our KPS expanders. Warm end maintenance is performed identically to our expander schedule, while cold end behavior is no different from our existing CCI machines.

CONCLUSIONS

A successful two cylinder reciprocating cold compressor was fabricated by combining two CCI compressor cold end assemblies with elements of a KPS dry expander. Performance is essentially identical to that of two CCI single cylinder compressors arranged in parallel, except for reduced static heat leak. The KPS platform is well suited to this modification, with appropriately sized cryostat and flywheel. In addition to meeting the FNAL cold compressor specification, the prototype is well understood by

FNAL maintenance technicians due to their experience with KPS expanders and the straightforward design of the CCI assemblies.

ACKNOWLEDGEMENTS

The author wishes to thank the Accelerator Division Cryogenic Systems Expander Group for their efforts in the construction and operation of this prototype.

REFERENCES

1. T.J. Peterson and J.D. Fuerst, Tests of cold helium compressors at Fermilab, in: "Advances in Cryogenic Engineering", Vol. 33, Plenum Press, New York (1988), p. 655
2. J.D. Fuerst, Trial operation of cold compressors in Fermilab Satellite Refrigerators, in: "Advances in Cryogenic Engineering", Vol. 35, Plenum Press, New York (1990), p. 1023
3. C.H. Rode, Tevatron cryogenic system, in "12th Inter. Conf. of High Energy Accel.", 1983

DEMONSTRATION OF A CRYOGENIC ISOTHERMAL EXPANDER

M.G. Norris, J.L. Smith Jr. and J.A. Crunkleton

Cryogenic Engineering Laboratory
Massachusetts Institute of Technology
Cambridge, Massachusetts

ABSTRACT

A new high-pressure-ratio steady-flow cryocooler employing an cryogenic isothermal expander has the potential for high reliability and efficiency at a modest cost. A demonstration test operated the isothermal expander in a test apparatus that simulated a high-pressure-ratio steady-flow cryocooler. The design of the expander is described including the expander surfaces, the recondenser and the valves. In these tests, the expander achieved produced 730 watts of net cooling at 78 K with an expander isothermal efficiency of 60%.

INTRODUCTION

This paper reports the results of a project[1] to demonstrate the feasibility of a cryogenic isothermal expander. This expander is the crucial component in a high-pressure-ratio steady-flow cryocooler that has the potential to combine high reliability, high efficiency and relatively low cost[2].

An isothermal expander was built and operated in a test apparatus (Figure 1). The test apparatus simulated a steady-flow cryocooler consisting of an compressor, counter-flow

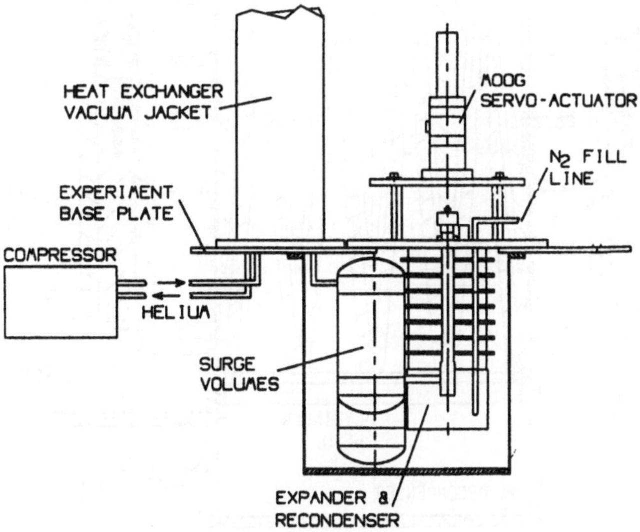

Figure 1: Isothermal Expander in Test Apparatus

Advances in Cryogenic Engineering, Vol. 37, Part B
Edited by R.W. Fast, Plenum Press, New York, 1992

heat exchanger and the isothermal expander. The test apparatus provided high pressure helium near 80 K and recovered low pressure helium. Enough flexibility was included in the experiment to operate the expander over a wide range of operating parameters: stroke length, speed, valve timing, pressure ratio, and mean pressure.

In the initial tests reported here, the expander was operated at three strokes, 2.5 cm, 3.8 cm and 5.1 cm, at a constant average piston speed of 5.1 cm/s and at 60 rpm for the same three strokes. The isothermal expander produced net cooling rates (Q'_{net}) from 276 to 773 watts at isothermal expander efficiencies from 60% to 55%. Large losses, mostly in the heat exchanger of the test apparatus, reduced cooling delivered to the load to less than 486 watts.

DEMONSTRATION EXPERIMENT

The isothermal expander assembly drawing (Figure 2) shows the expander integrated with the inlet and exhaust valves, the expander driver, and the recondenser. The major dimensions of expander were selected by parametric studies using a computer simulation

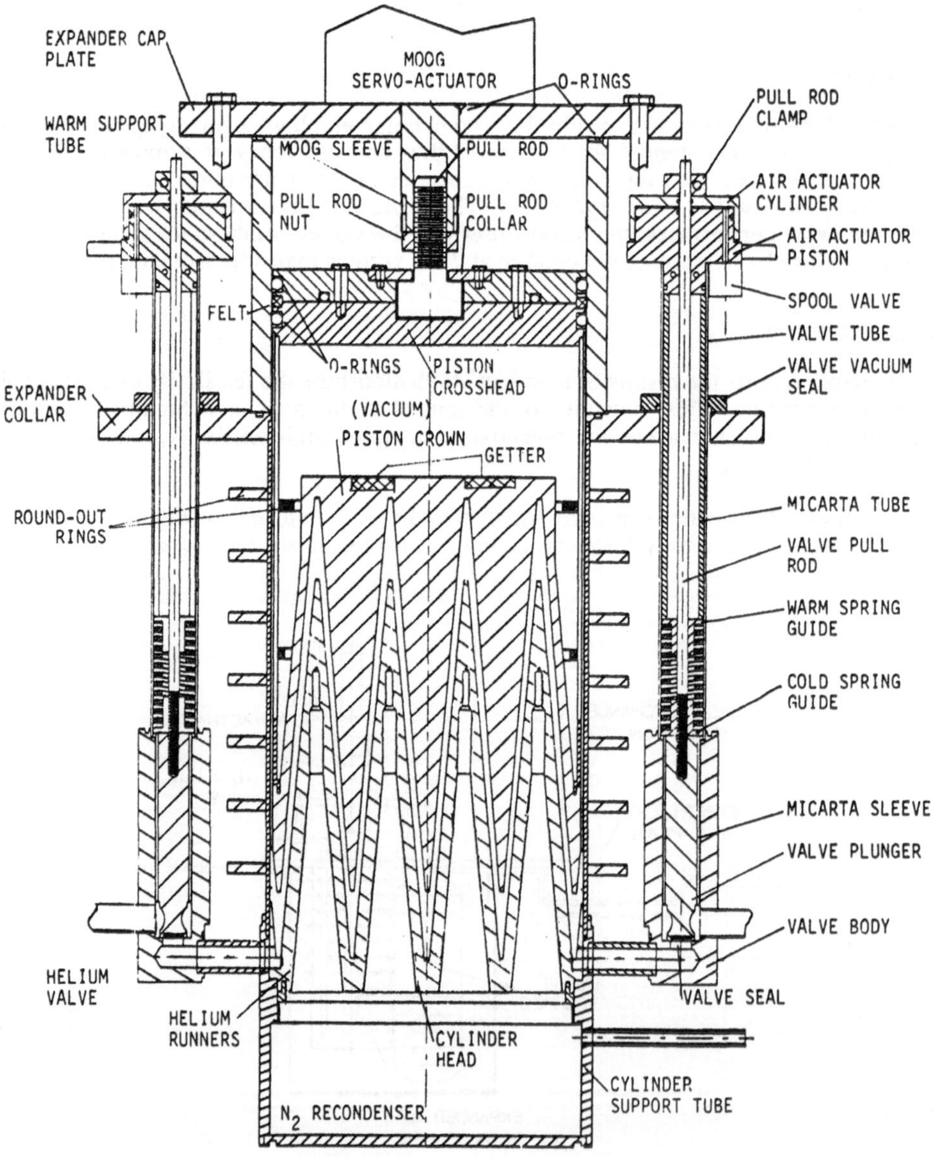

Figure 2: Isothermal Expander Assembly Drawing

Figure 3: Cylinder Head

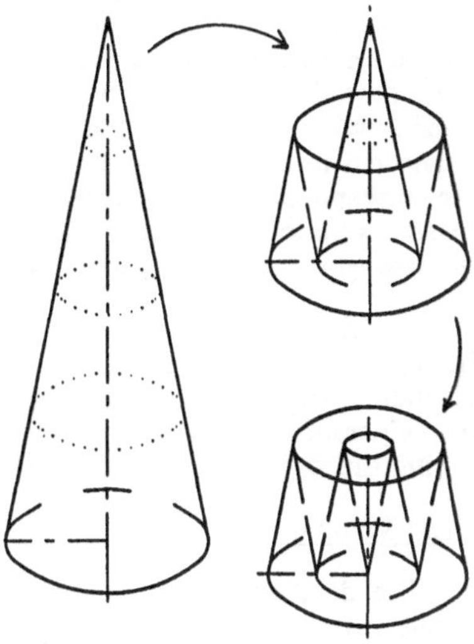

Figure 4: Folded Cone Structure

model of the expander[2]. The expander had a 19.1 cm bore and 22.7 cm tall fins. The piston crown was attached by a 18.7 cm OD tube to the top of the piston where O-rings seal the expansion working volume. The cylinder head was attached by the cylinder tube to the expander flange. A flow passage in the cylinder head (not shown) improved the distribution of gas from the inlet valve to the three concentric grooves in the cylinder head. Round-out rings on the outside of the cylinder tube and inside of the piston provided strength and rigidity to keep the tubes round. The recondenser was a volume under the cylinder head where nitrogen condensed in cavities on the underside of the cylinder head and was reboiled by a variable power heater at the bottom of the recondenser (not shown). The recondenser was charged with nitrogen gas by the single line shown.

The piston motion was stroked sinusoidally at 60 to 120 rpm by an Moog electro-hydraulic servo-actuator that was mounted on the expander cap plate. This cap plate was bolted to the expander flange. The inlet and outlet valves are shown with their air actuators. The connections from the valves to the surge volumes and the heat exchanger are not shown.

Expander surfaces

The tapered conical fins seen in Figure 3 were the heart of the isothermal expander. The triangular cross-sections (Figure 2) produced by the these conical fins provided two advantages. First, they allowed the expander to be operated with a very small minimum or dead volume and avoid the possibility of scraping during the rest of the cycle. A small minimum to maximum volume ratio is crucial to the performance of an isothermal expander, since larger ratios reduce both the capacity and efficiency. Second, the short stroke and long triangular-cross-section fins produced a uniform hydraulic diameter D_H, which minimized the losses due to mixing of gases at different temperature resulting from expansion at different D_H.

The tapered conical fins of cylinder head were hollowed out on the underside to provide a condensing surface for the nitrogen. The hollow fins minimized the thermal resistance from the load (condensing nitrogen) to gas in the expansion volume. The resulting shape of the cylinder head can be thought of as a folded cone (Figure 4). The folded cone was self supporting, so the only structural connections from the cylinder head and piston crown to the rest of the expander were at the outside diameter. The single point connection reduced the conduction heat leak to the expander and simplified construction. The folded cone structure was quite stiff. The 0.5 cm walls with 1.25 cm solid tips were analytically shown not to flex significantly during pressure cycling. The flex was important as small rotations of the fins would have led to interference between the piston and cylinder fins.

The manufacturing challenge of the piston and cylinder was making sure the fins of these two components fit together with a gap on the order of 0.012 cm. After the roughing out the brass cylinder head and piston crown, stainless steel inserts were brazed into the brass parts before machining to final dimensions. This allowed the brass to warp during brazing without affecting the final dimensions. The cylinder head and piston crown were connected to the rest of piston and cylinder by welding the steel inserts of each piece to the corresponding stainless steel tubes. The thin cross-sections and low conductivity of the inserts combined with the huge thermal capacitance of the brass head and crown isolated the fins from the high temperatures of the welding process.

The final fit between the assembled piston and cylinder was improved by hand fitting. Initially, contact between the sides of the piston and cylinder fins limited the minimum clearance length to 0.31 cm. The filing removed the high points on each piece, resulting in a clearance length (L_c) of 0.058 cm. Clearance length is the minimum axial distance during the stroke between the piston crown and cylinder head.

<u>Dead Volume</u>

The dead or minimum volume between the cylinder head and piston crown was estimated from dimensional measurements. The average gap was measured between the piston and cylinder fins with Plastigage, which is a plastic wire that deforms easily. Eight pieces of this plastic wire were taped to the piston: 4 on the fins and 4 on the fin tips. The piston was forced into the cylinder as far as possible with the hydraulic actuator, then removed. The deformed Plastigage indicated that the average fin to fin gap was 0.018 cm at a clearance length (L_c) of 0.058 cm.

The effective total dead volume is the sum of the individual dead volumes factored by their average temperatures. The piston/cylinder gap contained 59 cm^3, at an average temperature of 165 K, making the effective dead volume 30 cm^3. The helium runner in the cylinder head and the valves contributed 19 cm^3 of dead volume at 80 K. The dead volume contributed by the fin to fin gap is a function of the clearance length (L_C). At the minimum clearance of 0.58 cm, the fin to fin dead volume is 59 cm^3 for a total dead volume of 108 cm^3. The crucial dead volume parameter is the ratio of dead or minimum volume to maximum volume, which is 0.10 for a 3.8 cm stroke.

<u>Piston</u>

The cold piston crown was insulated by a vacuum cavity and connected by a long thin-walled tube to the ambient temperature top of piston. The cavity between the crown and the top was evacuated to at least 20 $\mu m Hg$ and insulated with aluminized mylar. To improve this static vacuum and to make up for small leaks, an absorbant was placed on the piston crown to cryopump the cavity. A warm seal system allowed standard buna-n O-rings to be used, which were inexpensive, reliable and any frictional heating from the seal occurs at the warm end. On the down side, the long piston/cylinder gap accounted for 28% of the total dead volume mass.

A micarta sleeve was placed on the piston to minimize friction and avoid galling between the piston and cylinder. This sleeve had a 0.025 cm radial clearance with respect to the cylinder over most of its length. The piston was guided in the cylinder at its cold end by a 2.5 cm long section of the sleeve that had a radial gap less than 0.01 cm. This cold end guide kept the cylinder and piston brass fins, which have a strong tendency to gall, from contacting during the stroke.

<u>Inlet and Exhaust Valves</u>

The inlet and exhaust valves, shown in Figures 2, were operated by air cylinder actuators at the warm end. Kel-f plastic washers provided a leak tight seal and the 1.27 cm diameter flow passages kept the pressure drop through the valves low. A micarta sleeve between the plunger and body minimized friction. The micarta sleeve was held in by the cold spring guide, which was threaded on the valve pull rod and was to keep the spring from rubbing on the tube wall. Larger guides should have been used since the spring flexed sideways and rubbed somewhat during the experiments. A warm spring guide transmitted the spring force through a micarta tube to the air actuator piston, which was pinned to the valve tube.

It was important to minimize the throttling associated with a slow opening valve to avoid losses and to make the experiment easier to run. In this design, an electrically operated spool valve (Numatics Mark 3) was mounted on the custom air actuator piston to minimize dead volume in the actuator thus maximizing the response time of the air actuator and the helium valve. The helium valve was opened by air pressure raising the air actuator cylinder which lifted the valve plunger via the valve pull rod and pull rod clamp. The air actuator piston and the warm spring guide were Delrin plastic to minimize friction.

Joy Tube Heat Exchanger

A Joy-tube counter-flow heat exchanger regeneratively cooled the high pressure helium from room temperature to near 80 K before it entered the expander. The 350 cm length was predicted to provide a 6 K difference between the high and low pressure flow at the cold end. The experimental temperature difference was closer to 15 K indicating that the heat transfer correlation needs to be corrected and that the heat exchanger should have been longer.

Instrumentation

The isothermal expander was instrumented to characterize its external parameters. The temperatures of the flow in and out of the surge volumes were measured with thermal diodes (Lakeshore DT-471) mounted on the stingers in the lines between the expander valves and the surge volumes. A thermocouple was also mounted on each stinger as a backup sensor. Schaevitz strain gauge pressure sensors (P5041) measured the pressure upstream of the inlet valve and downstream of the exhaust valve. The volume was calculated from the dead volume measurements described above and the position of the piston as measured by the LVDT on the Moog actuator.

The expander temperatures were measured with thermal diodes mounted on the piston crown, cylinder head and in the recondenser (Figure 2). A Dytran H2300C3 piezoelectric transducer measured the relative pressure in the working volume. The previously mentioned Schaevitz strain gauge pressure transducers (P5041) provide reference pressures for the piezoelectric sensor data. All the electrical sensor outputs for temperature, pressure, and piston location were recorded with the Acro-900 data system and transferred to a personal computer.

The pulsed nature of the low pressure flow made mass measurements with a rotometer impractical. In the following section a method to calculate the mass flow rate from temperature, pressure and volume data will be presented.

EXPERIMENTAL RESULTS

The testing had two main objectives. The first was to demonstrate that the isothermal expander could efficiently produce several hundred watts of cooling at liquid nitrogen temperatures. The second was to map out the performance of the expander for different speeds, and strokes. In addition, the static heat leak to the expander was determined to be 58 watts.

Performance Experiments at Liquid Nitrogen Temperatures

In these experiments the expander was operated between a high pressure of approximately 1.58 MPa and a low pressure of approximately 0.21 MPa producing 276 to 773 watts of net cooling at 78 K excluding counter-flow heat exchanger losses. The pressure ratio was held at 7.75:1, the blowdown pressure was approximately 0.26 P_I and the recompression pressure was approximately 0.40 P_I. The blowdown pressure is the expander pressure just before the exhaust valve is opened. The recompression pressure is the expander pressure just before the intake valve is opened. The coarseness of the electronic

Table 1: Measured and Calculated Results

Case	L_s	Speed	P_I	P_O	Q'_{Load}	$Q'_{Indicated}$	ΔT_{HE}	M'	η_I	η	Q'_{net}
-	cm	rpm	MPa	MPa	$watts$	$watts$	K	g/s	-	-	$watts$
1	2.55	120	1.53	0.199	441	957	14.9	3.74	0.78	0.597	732
2	2.55	60	1.66	0.213	215	397	8.6	1.36	0.855	0.595	276
3	3.85	90	1.55	0.203	451	924	13.9	3.88	0.725	0.575	732
4	3.85	60	1.61	0.206	347	724	12.8	2.98	0.731	0.552	547
5	5.04	60	1.55	0.199	486	981	13.9	3.95	0.751	0.591	773

valve timing adjustment limited the blowdown pressure to $0.26\,P_I$ for the 120 rpm case (Case 1). This blowdown pressure was maintained for the slower speed cases, so they could be compared. A lower blowdown pressure would have increased the efficiency of the expander with a slight decrease in cooling rate.

The first set of tests varied the stroke length and the speed to maintain a 1450 cm^3/s displacement rate while holding the blowdown and recompression pressure constant (cases 1,3,5). The second set of test varied the stroke (2.5, 3.8 and 5.1 cm) at a constant speed (60 rpm), blowdown and recompression pressure (cases 2,4,5).

Table 1 lists the measured data and calculated results for the 5 cases. Q'_{Load} was the cooling rate measured by the heater in the recondenser. The indicated cooling rate, $Q'_{Indicated}$, was the cyclic integral of pressure-volume over the cycle period ($\oint PdV\ /tau$). P_I and P_O were the inlet and outlet pressures. The ΔT_{HE} was based on the average exhaust and inlet temperatures while each valve was opened. For cases 3-5, electrical shorts in the data acquisition system caused by shock waves from the actuator absorbing the impulse of the inlet valve opening, cause the temperature measurement in both the inlet and exhaust temperatures to jump about 10 K. These spurious temperature spikes were ignored. The large differences between indicated cooling rates and the load cooling rates were primarily due to the very large temperature differences across the heat exchanger at the cold end.

The calculations for the mass rate (M'), the indicated and expander efficiencies (η_I and η) and the net cooling rate (Q'_{net}) are described in the following sub-sections. The mass rate was calculated from the pressure, volume and temperature data of the expander. The indicated efficiency is the isothermal efficiency of the expansion process not including any parasitic heat loads. The expander efficiency is the isothermal efficiency of the expander itself including the thermal conductance and shuttle heat leak losses. The net cooling rate (Q'_{net}) is the sum of the heater power (Q'_{Load}) and net enthalpy flow to and from the heat exchanger ($M'c_p\Delta T_{HE}$).

Mass Flow Rate Calculations

The mass rate was calculated from the volume, pressure and temperature of the expander at the beginning and end of the intake process. The equation for the mass flow rate was

$$M' = \Big[V_{ei}\,\rho(P_{ei},T_{ei}) - V_{bi}\,\rho(P_{bi},T_{bi})\Big]\frac{1}{\tau} + \Big[V_{CR}\,\rho(P_{ei},T_{CR}) - V_{CR}\,\rho(P_{bi},T_{CR})\Big]\frac{1}{\tau}$$

where V_{ei}, P_{ei} and T_{ei} were the volume, pressure and temperature of the expander at the end of the intake process and V_{bi}, P_{bi} and T_{bi} were the same quantities at the beginning of recompression, and V_{CR} and T_{CR} were the volume, and mass-averaged temperature of the piston/cylinder crevice. The density (ρ) was calculated from the ideal gas law.

The expander gas temperature at both the beginning (T_{bi}) and the end (T_{ei}) of the intake process, was assumed to be the average temperature of the cylinder head and the piston crown since the hydraulic diameter between the piston and cylinder fins was very small (.06 cm to .26 cm). Computer simulations of the experiments confirmed that the gas temperature is the average temperature of the piston crown and cylinder head at the beginning of intake, but drops 1 to 3 K by the end of intake. Since this temperature drop is small and not known without running a computer simulation for each case, it was ignored.

The recompression and end-of-intake pressures were measured by the piezoeletric pressure transducer using the Schaevitz sensor data for reference pressures. The volume measurement was the sum of the dead volume and stroke length as a function of time.

Efficiency Calculations

The isothermal efficiency of the expander was the ratio of actual cooling over the isothermal cooling potential of the gas entering and exiting the expander.

$$\eta = \frac{Q'}{M' \; R \; T_{CY} \; \ln \frac{P_I}{P_O}}.$$

Here P, V, and M' were pressure, volume and mass flow rate of the helium gas in the expander, R was the gas constant, P_I and P_O were the inlet and outlet pressures and T_{CY} was the temperature of the cylinder head, which was the structural temperature closest to the load temperature.

Two efficiencies were calculated for this experiment. The indicated efficiency defines the efficiency of the expansion process. It is calculated by substituting the indicated power ($\oint PdV \; /tau$) for Q'. The indicated efficiency includes only the losses in the expansion process, ignoring the conduction and shuttle heat leak of the expander.

The expander efficiency defines the efficiency of the expander itself. It is calculated by substituting the net cooling rate of the expander ($Q'_{Load} + M' \, c_p \, \Delta T_{HE}$) for Q'. The cooling provided to make up for the ΔT_{HE} was counted as part of the load since the ΔT_{HE} was due to inefficiencies in the heat exchanger.

Discussion of Results

The cooling capacity and efficiency of the expander are plotted here for both the constant displacement cases and the constant speed cases.

Figure 5 shows the indicated and expander efficiency vs. stroke and the indicated and net cooling vs. stroke for the constant displacement rate cases (1, 3, and 5). Both the cooling rate and efficiency were fairly constant at different stroke lengths. The computer simulations[3] predicted a 5% decrease in cooling rate and a 15% increase in efficiency with shorter strokes over this range of stroke lengths. With only three data points, the experimental data does not clearly show these trends.

Figure 6 shows the indicated and expander efficiencies vs. stroke and the indicated and net cooling vs. stroke for the constant expander speed cases (2, 4, and 5). The cooling rate increases linearly with the stroke length due mostly to larger mass flow rates. The simulation computer model[3] predicted a nearly linear rise in capacity and a linear decrease in efficiency at longer stroke lengths. Again with only three data points, the experimental data does not clearly show these trends in the indicated efficiency nor the expander efficiency.

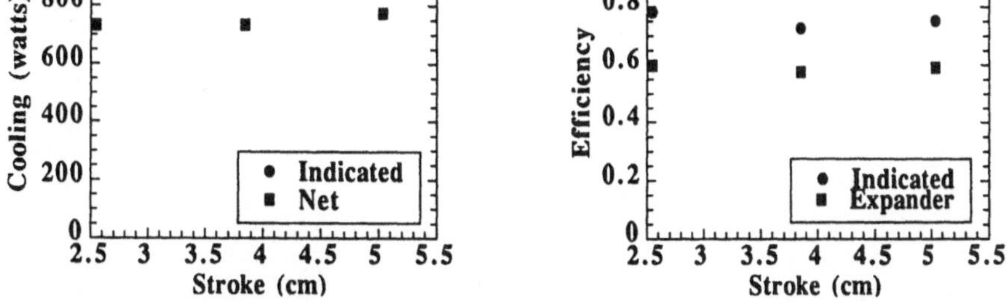

Figure 5: Cooling Rate and Efficiency vs. Stroke length at 1450 cm^3/s (Cases 1,3,5)

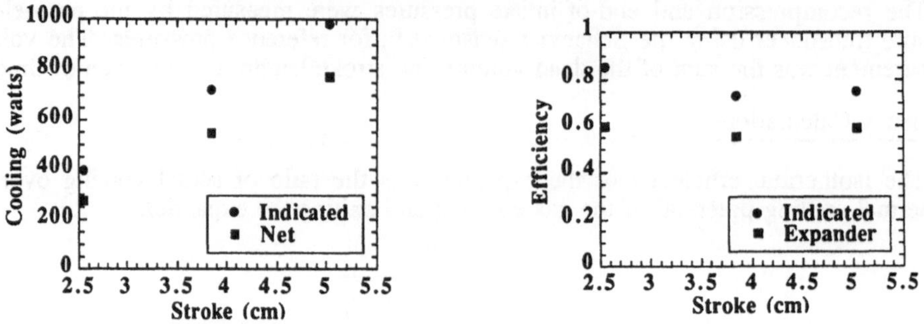

Figure 6: Cooling Rate and Efficiency vs. Stroke length at 60 rpm (Cases 2,4,5)

CONCLUSIONS AND RECOMMENDATIONS

The test demonstrated that this isothermal expander could efficiently provide several hundred watts of cooling at 78 K. The performance of the expander ranged from 215 watts of cooling at an expander efficiency of 62% to 773 watts of cooling at an efficiency of 59%. The tests also demonstrated the expansion process can reach indicated efficiencies better than 80% and produce approximately 950 watts of indicated cooling at efficiencies around 75%.

Additional testing of the isothermal expander with the existing test apparatus is recommended. Additional instrumentation is needed to accurately measure the mass flow rate of the expander. A dead volume test also needs to be completed. This data should be used to correct if needed and validate the simulation model of the expander.

Future work should focus on the efficient fabrication of the isothermal expander, especially the manufacturing of the expansion surfaces and the stroking mechanism. The experimental piston crown and cylinder head were machined, which was very expensive. Casting the parts would significantly lower the cost, if done on a sufficiently large scale. A drive mechanism needs to be developed that can handle the large forces of this expander while operating at a relatively low speeds from 60 to 120 rpm.

ACKNOWLEDGEMENT

This work was supported by the International Business Machines Corp.

REFERENCES

1. M.G. Norris, "Development of An Isothermal Expander for 80 K Cryocoolers", S.M. Thesis, MIT, 1991.
2. M.G. Norris, J.L. Smith Jr., J.A. Crunkleton, Near-Isothermal Expander for Cryocoolers in Superconducting Electronics, in: Advances in Superconducting Materials and Electronic Technologies, AES-Vol.22, ASME, 1990.
3. M.G. Norris, J.L. Smith Jr., J.A. Crunkleton, Simulation of a Near-Isothermal Expander, in: International Cryocooler Conference, Plymouth, Massachusetts 1990.

MAGNETIC BEARINGS FOR CRYOGENIC TURBOMACHINES

Victor Iannello and Herbert Sixsmith

Creare Incorporated
Hanover, New Hampshire

ABSTRACT

Magnetic bearings offer a number of advantages over gas bearings for the support of rotors in cryogenic turboexpanders and compressors. Their performance is relatively independent of the temperature or pressure of the process gas for a large range of conditions. Active magnetic bearing systems that use capacitive sensors have been developed for high speed compressors for use in cryogenic refrigerators. Here, the development of a magnetic bearing system for a miniature ultra high speed compressor is discussed. The magnetic bearing has demonstrated stability at rotational speeds exceeding 250,000 rpm. This paper describes the important features of the magnetic bearing and presents test results demonstrating its performance characteristics.

INTRODUCTION

This paper focuses on the development of high speed magnetic bearings for turbomachines used in cryogenic systems. The magnetic bearing work is an outgrowth of our experience with gas bearings. Over the years, we have developed turboexpanders, circulators, and compressors for cryogenic systems that vary in size from small cryocoolers for cooling of spaceborne sensors to large cryogenic refrigerators for cooling the superconducting magnets used in high energy particle accelerators. These machines use gas bearings to eliminate contamination of the process gas stream and to ensure reliable and stable high speed rotation with little or no maintenance.

Magnetic bearings offer the potential to increase the performance of present and future machines because they offer the following advantages over gas bearings: performance independence from process gas conditions, load capacity at low rotational speeds, and relative ease of fabrication. For medium sized applications, magnetic bearings in the size range of 2.5 cm (1.0 in.) to 5.0 cm (2.0 in.) have been developed for pumps and compressors and tested to rotational speeds exceeding 80,000 rpm. The development of a cryogenic vacuum compressor is currently underway. In this paper, we discuss the development of a miniature magnetic bearing for use in an ultra high speed compressor with a shaft diameter of 0.64 cm (0.25 in.) and rotational speed of about 500,000 rpm.

PRINCIPLE OF OPERATION

An active magnetic bearing is a type of position controlled servomechanism. A schematic drawing of one magnetic bearing configuration is shown in Figure 1. The bearing can be thought of as four electromagnets arranged circumferentially around the shaft. Each magnet controls the flux level for an individual quadrant of the bearing.

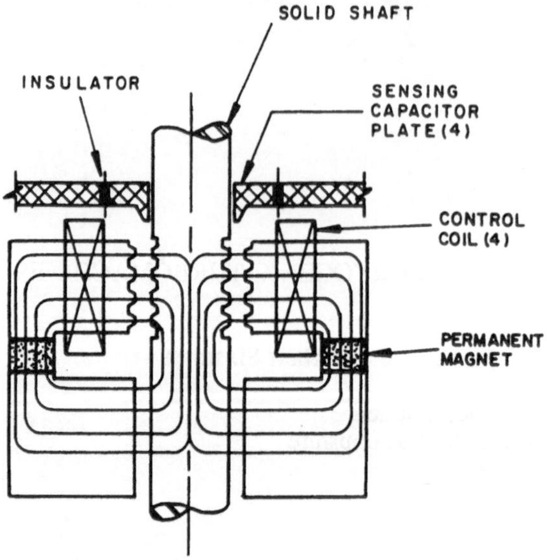

Fig. 1. Schematic drawing of side.

By individually adjusting the currents in the four coils, a radial force may be produced of a desired magnitude and direction. This magnetic force is used to control the position of the shaft. The flux emanating from a pole face is composed of two components--a DC or bias field B_o that remains relatively constant, and a control component B_c that varies with the shaft position.

An active control system is included for each radial axis. The system consists of the control electromagnets (actuator), a sensor for detecting radial displacements of the shaft, and control electronics which supply current to the coil in response to the error signal. One component of the control current is proportional to the shaft displacement and determines the stiffness of the bearing. Another component of the control current is proportional to the rate of change of position (velocity) of the shaft and determines the damping of the bearing. The stiffness and damping can be adjusted within a range of values via the controller electronics.

A bias flux B_o is included to linearize the relationship between the control current and force. The bias field can be produced by the control electromagnets, by a separate bias coil, or by one or more permanent magnets. For large applications where resistive heating loss in the bearings is not critical, the bias field may be supplied by the control electromagnets for simplicity. However, the use of permanent magnets to supply the bias field is preferred for small and miniature applications where electrical heating would adversely affect the performance of the machine. In the absence of external loads on the shaft (such as imbalance and acceleration forces), an active magnetic bearing employing a permanent magnet can be designed to dissipate almost zero power.

MINIATURE MAGNETIC BEARING FOR AN ULTRA HIGH SPEED COMPRESSOR

Creare is developing a miniature compressor that incorporates magnetic bearings as part of a program to develop the Single Stage Reverse Brayton (SSRB) cryocooler. This is a spaceborne mechanical cryocooler capable of 5 watts of cooling at 65 K[1,2]. The 0.64 cm (0.25 in.) shaft of this compressor spins at 500,000 rpm at its design point, and the 1.5 cm (0.60 in.) impeller produces about 100 watts of useful work. The nominal design parameters for this compressor are listed in Table 1.

The magnetic bearing that was conceived, built, and tested for this application is shown schematically in Figure 1 and photographically in Figure 2. The four

Table 1 Nominal Design Parameters for Miniature Compressor

Inlet pressure	1.1 atm
Inlet temperature	300 K
Pressure ratio	1.7
Mass flow rate	1.3 g/s
Design speed	480,000 rpm (8,000 rev/s)
Input shaft power	130 W
Impeller diameter	1.52 cm (0.60 in.)
Shaft diameter	0.635 cm (0.25 in.)

electromagnets that control the flux for each quadrant of the bearing can be seen in the photograph. In this configuration, the bias flux is supplied by a permanent magnet.

The miniature bearing is designed for low magnetic losses even at high rotational speeds. Unlike conventional magnetic bearings where the flux paths in the shaft contain an azimuthal component, the flux paths are purely radial and axial in the miniature bearing. As a result, the poles of the electromagnets surrounding the shaft are of all the same polarity, which substantially reduces the hysteresis and eddy current losses in the rotor due to shaft rotation.

Conventional magnetic bearings generally employ magnetic laminations in the rotor to reduce response lags due to eddy current effects. In this design, the miniature magnetic bearing employs a solid rotor with tapered poles. There are two advantages to this approach. By eliminating the rotor laminations, the entire shaft may be integrally machined. The bending stiffness is enhanced relative to a shaft that incorporates laminations, which generally lower the bending critical speeds of the shaft. The fabrication, assembly, and balancing processes are greatly simplified when compared with a composite shaft.

The other advantage of the tapered pole concept is that this configuration also provides passive stiffness in the axial direction. This passive restoring force eliminates the need for a separate thrust bearing for many applications. This is important for miniature turbomachines where the size envelopes for the bearings are small.

MINIATURE POSITION SENSOR

Magnetic bearings for large diameter shafts typically use commercially available proximity sensors of the capacitive, inductive, optical, or eddy current type. Unfortunately, there are no commercially available sensors suitable for miniature

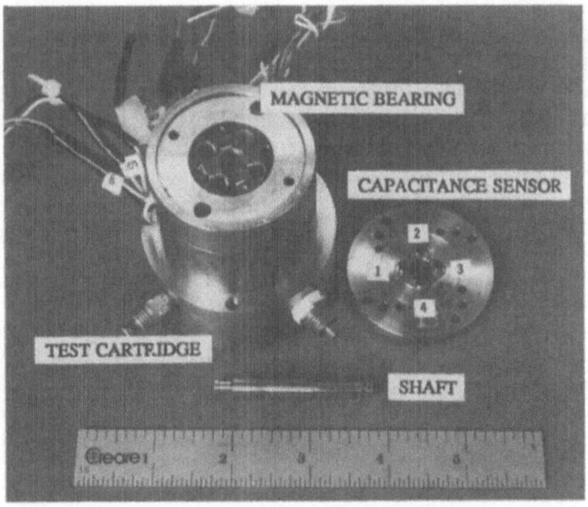

Fig. 2. Magnetic bearing components.

turbomachines in which high resolution and bandwidth are required from a miniature sensing element small enough to be mounted in close proximity to the stator of the bearing.

To satisfy these requirements, a compact capacitance sensor was developed to detect the shaft position. The shaft and sensing elements together comprise two plates of a capacitor, with the radial gap forming a portion of its capacitance, as shown in Figure 1. The sensor uses four electrically isolated electrodes which are supported in an annular plate. Because the magnetic fields from the bearing do not affect the sensor, the sensor may be placed as close to the bearing as mechanical constraints allow.

The signals from a pair of diametrically opposed quadrants determine the radial position of the shaft along that axis. The sensor circuits are designed to generate a signal which is proportional to shaft displacement along a diameter, and which is close to zero when the shaft is centered between the quadrants. Like other capacitance sensors, a high frequency signal is applied to the electrodes of the sensor to determine the shaft position. Typically, the output from the sensor circuit consists of a DC component proportional to the displacement of the shaft from its centered position, and an AC component, which represents the remnants of the high frequency signal supplied to the electrodes. The AC component is undesirable--it typically limits the achievable resolution of the sensor. Although the signal-to-noise ratio may be improved with filtering, this tends to reduce the bandwidth of the sensor, which must be kept high for high speed applications.

A novel feature of the sensor developed here is that high frequency signals 180 degrees out of phase are applied to diametrically opposed sensing plates. By using an "AC balanced" demodulation configuration for the sensor circuit, the AC components tend to cancel in the circuit, while the DC components are added. This is accomplished with passive components (resistor, diodes, and capacitors) for high reliability and relative simplicity. We are applying for a patent on this sensor.

ANALYSIS OF MAGNETIC BEARING

According to Earnshaw's theorem, a body having a relative permeability equal to or greater than one (i.e., is attracted to a magnet) cannot be stably suspended by a passive magnetic field. This theorem implies that the suspended body will be unstable along at least one axis of motion, and therefore active control techniques must be used. For the miniature magnetic bearing, passive stiffness is provided in the axial direction, and active support is provided in the radial directions. An summary of the analysis used to design the active feedback loop is presented here.

With no control currents and only bias currents supplied to the bearing coils, the centered shaft is in a state of unstable equilibrium, i.e., a displacement from the equilibrium position results in a force on the rotor in the same direction as the displacement. The unbalance stiffness associated by this effect is denoted K_u and can be empirically determined for a given bearing geometry and bias flux level. To stabilize the rotor, control currents are supplied to the stator quadrants. A current stiffness K_i is used to describe the ratio of control force for a given control current for the bearing. The equation that describes the dynamics of the bearing-shaft system is thus

$$F_d = m\dot{x} - K_u x + K_i I_c \tag{1}$$

where F_d is a disturbance force acting on the rotor, and the first, second, and third terms on the RHS of Eq. (1) are the respective inertial, magnetic unbalance, and magnetic control forces. In general, I_c will be a function of x. If this function is properly chosen, stability will be achieved, i.e., the bearing will exhibit positive stiffness and damping.

The basic concept for the feedback loop is shown in Figure 3, where $s=j\omega$ is the Laplace transform variable. The radial position signal(s) is produced by the sensor.

The signal is then conditioned in a phase compensation network, which is essentially a "proportional plus derivative" (PD) controller. The static stiffness is adjusted by adjusting the proportional gain K_p and the damping by adjusting K_d. Also included in the block diagram is a first-order model for the current amplifier and the current stiffness. The overall gain G_o is dimensionless and is related to the individual sensor, compensator, current stiffness, and amplifier gains as follows:

$$G_o = K_p K_s K_a K_i / K_u \quad (2)$$

where the constants have the following units: K_p(-), K_s(V/m), K_a(A/V), K_i(N/A), and K_u(N/m).

Stability analyses of the overall control system using classic open-loop techniques were performed. The analyses indicated that if the controller/amplifier bandwidth is sufficiently high, the phase of the open loop system is primarily determined by the damping constant K_d. As such, the stiffness and the damping of the bearing can be approximated by

$$K_b = (G_o-1)\, K_u \quad (3)$$

$$\zeta = 0.5\, K_d\, (K_b/m)^{0.5} \quad (4)$$

These equations allow the performance of the bearing to be easily characterized for a range of proportional and damping gain settings.

BEARING TEST RIG

A bearing cartridge was designed, fabricated, and tested to demonstrate the performance of the magnetic bearing system. A schematic drawing of the test rig is shown in Figure 4. The test rig includes a magnetic bearing, which provides radial and axial positioning of the shaft at one end, an externally pressurized (hydrostatic) gas bearing which provides radial positioning at the other end, and a nozzle and turbine for rotating the shaft.

The magnetic bearing consists of four tapered pole rings, each 0.127 mm (0.005 in.) wide at the stator-rotor gap. The bearing rotor diameter is 0.762 cm (0.3 in.) and the radial gap is 0.051 mm (0.002 in.), which are prototypical dimensions for the miniature compressor. A bias coil is also included in the test rig to produce a bias flux in the bearing air gap. This coil may be replaced by a permanent magnet in the actual compressor. The important design parameters for the bearing are included in Table 2.

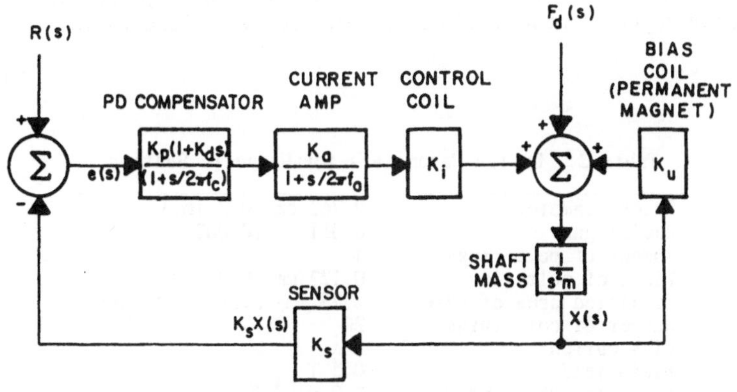

Fig. 3. Magnetic bearing control loop.

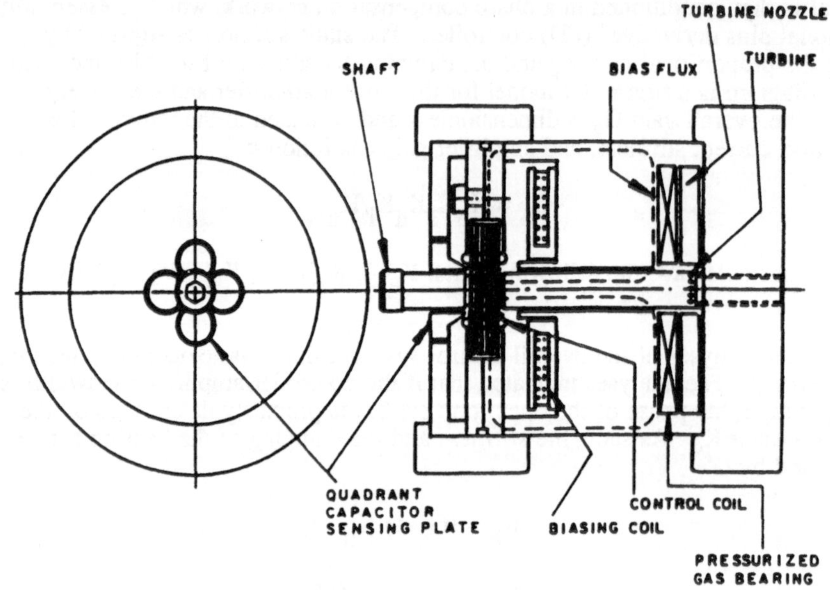

Fig. 4. Schematic drawing of test rig.

The sensor for the magnetic bearing consisted of four capacitance plates electrically isolated from the bearing cartridge. The radial clearance between the shaft and sensor was nominally 0.064 mm (0.0025 in.), and the axial length of the plates was 7.62 mm (0.3 in.). A 1.0 MHz sine wave was used as the high frequency signal applied to the electrodes.

An externally pressurized (hydrostatic) bearing was used to support and position one end of the shaft. To rotate the shaft, eight radial holes were drilled into this end of the shaft, which was counterbored to produce a turbine. A nozzle assembly in the bearing housing directed the gas flow towards the turbine. The supply gas for both the turbine and gas bearing was compressed air.

The controller electronics consisted of a high frequency oscillator to generate the carrier frequency for the sensor, the sensor conditioning electronics, and a proportional-derivative (PD) controller/amplifier, which supplied the current to the bearing control coils. A separate controller was used for each radial axis in the bearing. The PD controller was specifically built for magnetic bearing work at Creare and allowed us to easily vary the gain and damping settings via 10-turn potentiometers located on a panel. For this application, the gain G_o could be regulated between 0 and 15, and the damping constant K_d could be adjusted between 0 and 1.6 ms.

Table 2. Design Parameters for Magnetic Bearing

Rotor diameter	0.762 cm (0.3 in.)
Radial gap	0.051 mm (0.002 in.)
Number of pole rings	4
Width of pole ring	0.127 mm (0.005 in.)
Projected area of pole	2.73 mm^2 (4.24×10^{-3} in^2)
Number of coil turns	20
Bias current	1 A
Bias field	0.5 T
Effective shaft mass	7,2 g (0.25 oz.)

BEARING TESTS

The magnetic bearing was tested in an instrumented benchtop rig to demonstrate its performance. The testing consisted of sensor tests, impulse (shock) tests, and rotational tests. The results of these tests are described below.

Sensor Tests

Tests were initially performed to demonstrate the performance of the capacitance sensor. Because the positioning accuracy of the bearing is ultimately determined by the sensor performance, the sensor is an important part of the magnetic bearing system and must have sufficient sensitivity and signal-to-noise ratio for the miniature compressor application.

A high frequency signal of 1.0 MHz frequency and 6 V amplitude was applied to the sensor electrodes. The sensor detected the shaft position with an average sensitivity of about 100 V/mm (2.5 V/mil) over the range of allowable displacement range of the shaft. The circuit was balanced so the amplitude of the radio frequency signal superimposed on the output signal is only 10 mV. As a result, the sensor can detect displacements as small as 0.1 μm (4 μ-inch). By comparison, the radial gap in the bearing is 500 times this value.

Impulse (Shock) Tests

The purpose of these tests was to demonstrate the stiffness and damping characteristics of the magnetic bearing and to validate the simple models that were developed to predict the bearing performance. These tests were extensively performed for one radial axis, although several tests were performed for the second radial axis to confirm similar performance.

First, the unbalance stiffness was experimentally determined to be 1.32×10^4 N/m (75 lb_f/in). Next, the bearing stiffness was measured as the gain G_o was varied. As predicted by Eq. (3), the bearing stiffness was found to be linear with the gain for gains between 2 and 8.

The impulse testing was performed at a constant gain setting of G_o=3, which resulted in a bearing-shaft resonance at 300 Hz. Figure 5 compares the measured response with the theoretical response predicted by Eqs. (3) and (4) for two damping ratios. The agreement is excellent for ζ=0.12. For higher damping ratios, we measured a damping ratio that is higher than predicted due to dissipative forces not included in

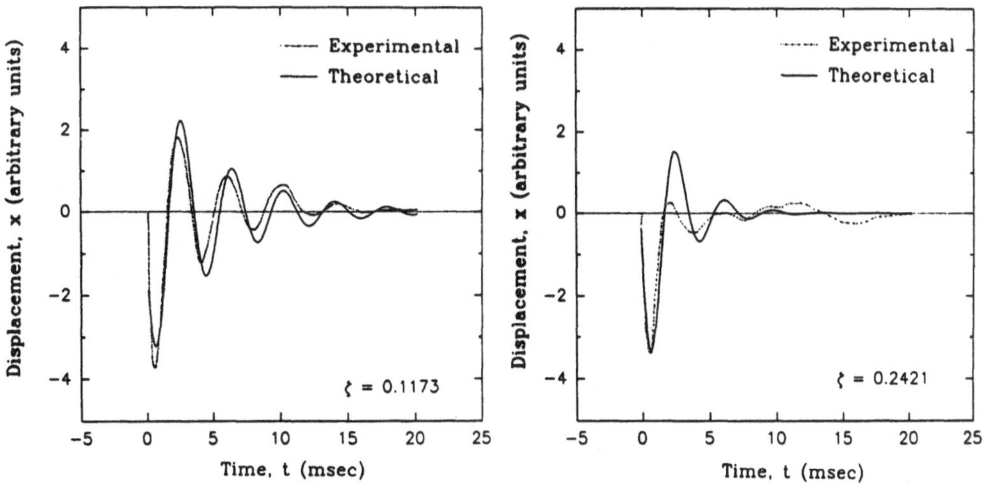

Fig. 5. Impulse test results.

the model. These results confirmed that our simple models were adequate for predicting the stiffness and damping of the magnetic bearing, and that the performance of the bearing was suitable for high speed rotation. This was confirmed by the next series of tests.

Rotational Tests

In order to verify that the magnetic bearing is suitable for high speed miniature turbomachinery, we proceeded to demonstrate the bearing stability at high rotational speeds. For these tests, the shaft was rotationally driven by the turbine located at one end. The shaft was positioned by the magnetic bearing at one end, and the pressurized bearing at the other end.

The shaft was stably and repeatedly run up to 4500 rev/s (270,000 rpm). The runout of the shaft for speeds greater than 3200 rev/s was constant and equal to about 0.3 μm (12 μ-inch) total indicated runout (TIR). This corresponds to about 0.6% of the radial gap in the bearings. Based on the stable and accurate positioning at high rotational speeds, we conclude that the bearing is suitable for high speed turbomachines. The integration of these bearings in a compressor has just gotten underway.

ACKNOWLEDGEMENTS

The present work to develop a miniature magnetic bearing was funded by NASA/Goddard Space Flight Center under contract NAS5-30854. Additional funding for the development of magnetic bearings for medium-sized cryogenic turbomachines was provided by DOE under contract DE-ACC01-87-ER80477. The authors would like to thank NASA and DOE for their support of this work.

REFERENCES

1. H. Sixsmith et al, Long-life, closed-cycle for space, in: "Proceedings of the Fourth International Cryocoolers Conference," Easton, MD (July 1987).

2. H. Sixsmith et al, Small turbo-brayton cryocoolers, in: "Advances in Cryogenic Engineering," St. Charles, IL (1987).

DEVELOPMENT OF EXTERNALLY PRESSURIZED THRUST BEARING FOR HIGH EXPANSION RATIO EXPANDER

N.Ino*, A.Machida*, K.Tsugawa*, Y.Arai*,
H.Hashimoto**, A.Yasuda***

* Mayekawa Mfg. Co. Ltd., MYCOM Advanced Technology Laboratory, Ibaraki, Japan
** Tokai University, Kanagawa, Japan
*** Super-GM, Osaka, Japan

ABSTRACT

The authors developed an expander for a helium liquefier of 100 L/h liquefaction capacity used for a superconducting generator. This paper focuses on the development of the shaft-bearing system, which uses a thrust bearing exerting a significant influence on the reliability of the turbine. An externally pressurized thrust bearing was used to support the high thrust load resulting from the high expansion ratio of the turbine. The characteristics of the bearing were estimated based on analytical solution of the theory of complex velocity potential. Good agreements were found between theoretical results and exprimental ones and the practicality of the design method was verified.

INTRODUCTION

THe final objective of the authors is to develop an expander for a liquefier having 100 L/h liquefaction capacity used for a superconducting generator. The rotary shaft system of the device is the most important part of the expander, having a major effect on stability and reliability at high expansion ratios and under operating mode changes. On such a system, the high thrust force, high revolution speed due to high expansion ratio, and the abrupt changes in thrust load due to changes in operation mode are concentrated primarily on the thrust bearing. The use of an externally pressurized thrust bearing (EPTB) as a thrust bearing and a tilting pad journal bearing (TPJB) as a radial bearing may be the best combination for coping with the operating conditions noted above. The authors successively developed a very stable and highly reliable expander using an annular collar thrust bearing with multi-feeding holes. The report given hereunder describes the details of development of an externally pressurized thrust bearing used for a high expansion ratio expander, including comparison of theoretical results and experimental ones on the bearing characteristics

BASIC EQUATIONS AND SOLUTION

Basic Equation for Lubrication of Hydrostatic Thrust Bearing

As is well known, the basic equation for lubrication of the hydrostatic thrust bearing is as given in the following Laplace's equation.

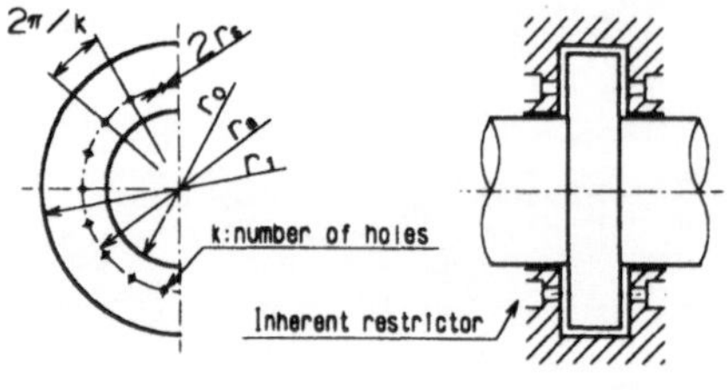

Fig.1. Annular collar thrust bearing with mutiple gas feeding holes

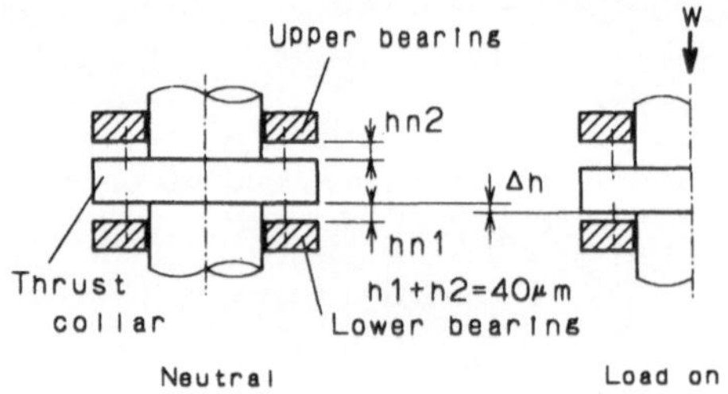

Fig.2. Relationship between upper and lower clearances of combined bearing

$$\frac{\partial^2 P}{\partial x^2} + \frac{\partial^2 P}{\partial y^2} = 0 \tag{1}$$

$$P = P_{comp}^{(1+n)/n} \tag{2}$$

where

x,y= Cartesian coordinate x,y, n= polytropic exponential, Pcomp= pressure distribution of thrust bearing, P= solution of equation(1)

Analytical Solution for Annular Collar Thrust Bearing with Multi-Feeding Holes.

Figure 1 shows an EPTB with multiple gas feeding holes and geometrical symbols to be used in the analysis. The inherent restrictors as illustrated in Fig.1 are employed in order to improve the stability against pneumatic hammer. From the theory of complex velocity potential [1,2,3], the solution of the equation (1) for the EPTB with multi-feeding holes is expressed in the equations (3) and (4), under assumption of an isothermal process of n=1.

$$\bar{P}_{comp} = P_{comp}/P_a = (\bar{K}_{1comp} \cdot \bar{A}_r + \bar{K}_{2comp})^{1/2} \tag{3}$$

$$P = P_{comp}^2 \tag{4}$$

In the equation (3) p_a is an environmental pressure and K_1,Ar,K_{2comp}, which are determined from the feeding pressure ratio Ps(=p_s/p_a) and bearing geometry, are described in Appendix 1.

There exists an optimum radius value r_a statisfying both the maximum load capacity and minimum flow condition and it is obtained from the following equation[4].

$$\bar{r}_a = r_a/r_1 = (r_0 \cdot r_1)^{1/2} / r_0 = (r_1/r_0)^{1/2} = (\bar{r}_1)^{1/2} \tag{5}$$

When the bearing geometry and supply gas pressure ratio Ps are determined, the restictor exit pressure ratio P_o(=P_o/P_a) is obtained from the following equations.

$$\bar{P}_0^2 = C_f \cdot (\Lambda s/k) \cdot \Phi_o \cdot \bar{A}_{k1} \cdot \bar{P}_s^2 + 1 \tag{6}$$

$$\Phi_0 = \{\frac{2\gamma}{\gamma-1}(P_o/P_s)^{2/\gamma}\{1-(P_o/P_s)^{(\gamma-1)/\gamma}\}\}^{1/2} \quad \textit{for unchoked flow} \tag{7}$$

$$\Phi_0 = \{\frac{2\gamma}{\gamma+1}(\frac{2}{\gamma+1})^{2/(\gamma-1)}\}^{1/2} \quad \textit{for choked flow} \tag{8}$$

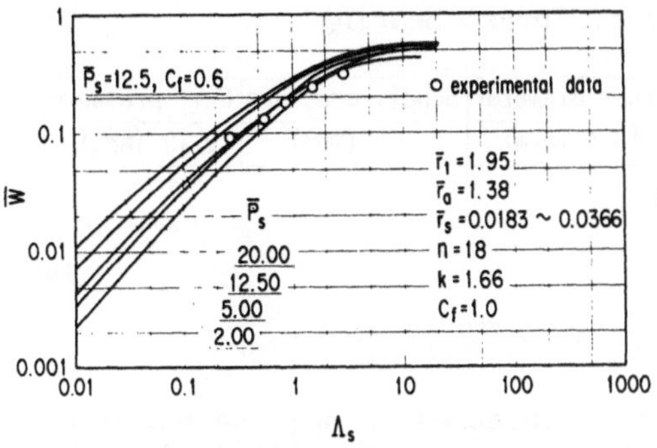

Fig.3. Dimensionless load vs. restrictor coefficient of single thrust bearing

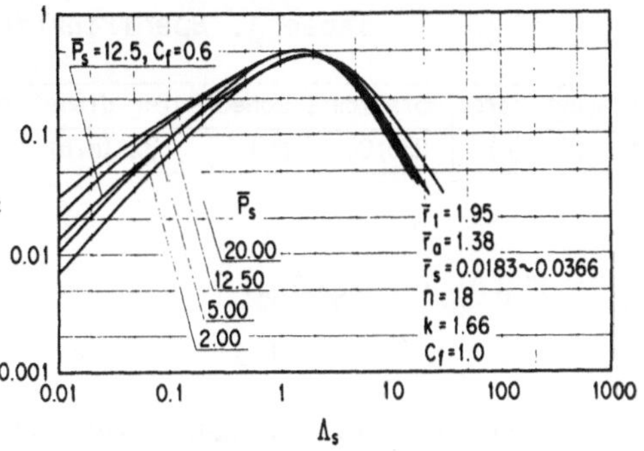

Fig.4. Dimensionless stiffnes vs. restrictor coefficient of single thrust bearing

where

C_f= the discharge coefficient of a restrictor, γ= specific heat ratio.

The variable Φ_0 changes in accordance with the restrictor nozzle choked or not. When P_0 is obtained, the pressure distribution Pcomp can be determined from the equation (3) and then the dimensionless load capacity W,dimensionless mass flow rate Qm, and dimensionless stiffness K of the single thrust bearing can also be determined from the following equations (9),(10) and (11).

$$\overline{W} = \frac{W}{\pi(r_1^2 - r_0^2)(P_s - P_a)} = \frac{2k}{\pi(\overline{r}_1^2 - 1)(\overline{P}_s - 1)} \int_1^{\overline{r}_1} \int_0^{\pi/k} (\overline{P}_{comp} - 1)\,\overline{r}\,d\overline{r}\,d\theta \tag{9}$$

$$\overline{Q}_m = Q_m / (\pi \cdot h^3 \cdot P_s^2 / 6 \cdot \mu \cdot R \cdot T_s) = C_f \cdot \Lambda s \cdot \Phi_o \tag{10}$$

$$\overline{K} = k / \frac{2}{3}\frac{\pi}{h}(r_1^2 - r_0^2)(P_s - P_a) = 3\frac{\partial \overline{W}}{\partial \Lambda s}\Lambda s \tag{11}$$

Where the restricter coefficient Λ_s, which is an important variable in the design of bearing, is defined as follows.

$$\Lambda_S = 12 \cdot \mu \cdot k \cdot r_s (R \cdot T_s)^{1/2} / (P_s \cdot h^2) \tag{12}$$

where

μ= gas viscosity, R= gas constant, Ts= inlet gas temperature, h= bearing film thickness, r_s= feeding hole radius

Combined Thrust Bearing

A single bearing is rarely used as a thrust bearing and a combined bearing as shown in Figs.1 and 2 is normally used. Provided that the collar is in a neutral position under no load condition, the film thicknesses can be regarded as h_{n1} and h_{n2}, the total film thickness as h_{ct} and the restrictor coefficient as Λ_{n1} and Λ_{n2}, respectively, as shown in Fig.2 in which the different characteristics were provided in the upper and lower bearings in order to discuss the problem generally. From the definition of total film thickness h_{ct},

$$h_{ct} = h_{n1} + h_{n2} \tag{13}$$

and from the force balance requirement between the upper and lower bearings, the following equation is obtained.

Table 1. Specifications of Thrust Bearing

Thrust force	Rotational speed	Outer diameter	Inner diameter	Supply pressure	Exit pressure
200 (N)	230,000 (rpm)	32 (m/m)	16.4 (m/m)	1.5 (Mpa)	0.12 (Mpa)

$$\overline{W}_1(\Lambda_{n1}) = \overline{W}_2(\Lambda_{n2}) \tag{14}$$

where the subscripts 1 and 2 refer to the lower and upper bearings.

Since h_{ct} is a given value, the film thicknesses h_{n1} and h_{n2} under a neutral condition can be found from equations (13) and (14). Consequently, the dimensionless load capacity W_c, flow rate Q_{mc} and stiffness K_c of the combined bearing are expressed by the following equations.

$$\overline{W}_c = (W_1 - W_2)/\pi\cdot(r_1^2-r_0^2)(P_s-P_a) = |\,\overline{W}_1(\Lambda_{s1}) - \overline{W}_2(\Lambda_{s2})| \tag{15}$$

$$\overline{Q}_{mc} = (Q_{m1}+Q_{m2})/(\pi\cdot h_{ct}^3\cdot P_s^2/6\cdot\mu\cdot R\cdot T_s)$$

$$= \overline{Q}_{m1}(\Lambda_{s1})\cdot(\overline{h}_{n1}-\varepsilon)^3 + \overline{Q}_{m2}(\Lambda_{s2})\cdot(\overline{h}_{n2}+\varepsilon)^3 \tag{16}$$

$$\overline{K}_c = (K_1+K_2)/(2\cdot\pi\cdot(r_1^2-r_0^2)(P_s-P_a)/3\cdot h_{ct})$$

$$= \overline{K}_1(\Lambda_{s1})/(\overline{h}_{n1}-\varepsilon) + \overline{K}_2(\Lambda_{s2})/(\overline{h}_{n2}+\varepsilon) \tag{17}$$

where, W_1, W_2, K_1, K_2, Q_{m1} and Q_{m2} are dimensionless characteristics of a single bearing and all other symbols used in the above equations are as defined in Appendix 2.

CALCULATION RESULTS AND DISCUSSION

Characteristics of Single Bearing and Combined Bearing

Characteristics of a single bearing are shown in Figs.3,4 and 5, in which practical range of Λ_s is considered as $0.05 < \Lambda_s < 5$. Fig.6 shows the relationship between the restrictor pressure ratio Po/Ps and Λ_s. It is found that the restrictor is in the choked condition when Ps>2 and Λ_s<0.7. Figures 7,8 and 9 show the characteristics of the combined bearing in the load on position shown in Fig.2. The solid line in the figure represents the combined chracteristics and the broken line represents the individual characteristics, respectively.

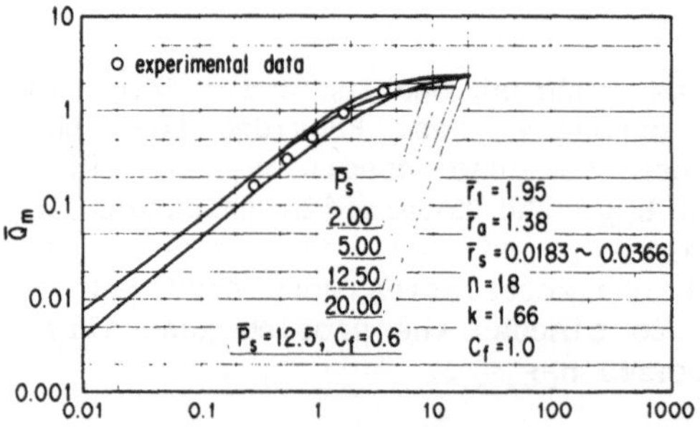

Fig.5. Dimensionless mass flow rate vs.restrictor coefficient of single thrust bearing

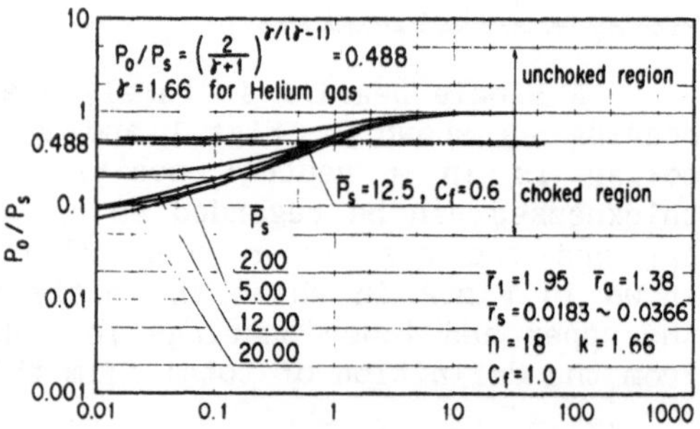

Fig.6. Restrictor pressure ratio vs.restrictor coefficient of single thrust bearing

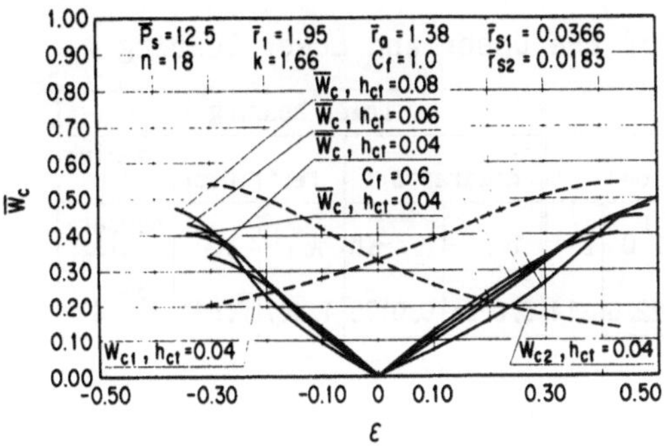

Fig.7. Dimensionless load vs. clearance ratio of combined thrust bearing

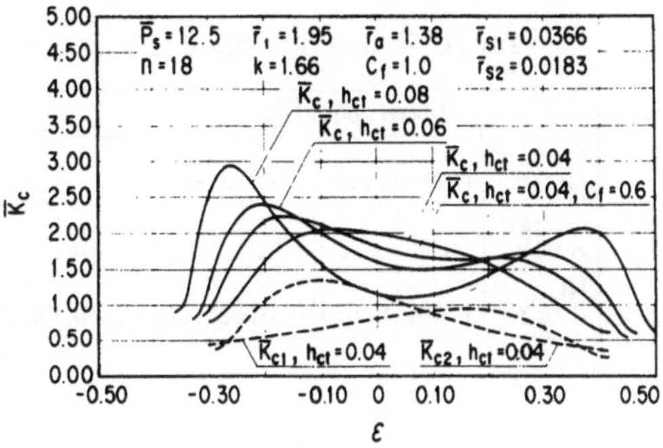

Fig.8. Dimensionless stiffness vs.clearance ratio of combined thrust bearing

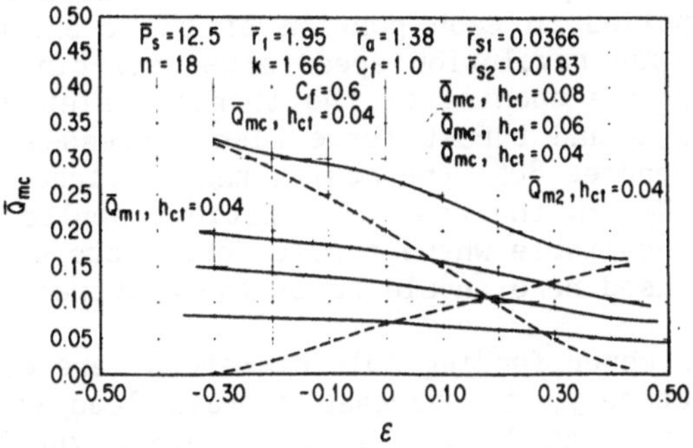

Fig.9. Dimensionless mass flow rate vs. clearance ratio of combined thrust bearing

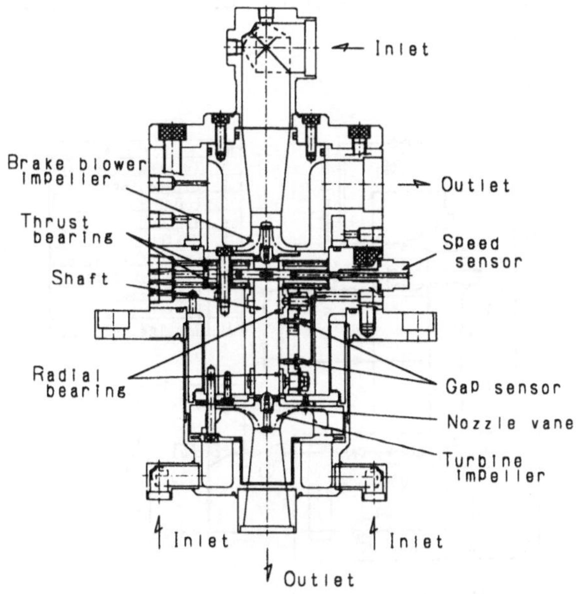

Fig.10. Helium gas expander

Table 2 Design Specifications of the Upper and Lower Bearing

Upper bearing								Lower bearing				
feeding hole	restrictor		clearance		common			clearance		restrictor		feeding hole
number × dia.	Λ_{n2}	Λ_{s2}	h_{n2}	$h_{n2}+\Delta h$	h_{ct}	hole P.C.D	ε	h_{n1}	$h_{n1}-\Delta h$	Λ_{n1}	Λ_{s1}	number × dia.
18× ϕ 0.3	1.467	0.820	0.0166	0.0222	0.04	ϕ 22.6	.14	0.0234	0.0178	1.467	2.534	18× ϕ 0.6

Design of Thrust Bearing

Table 1 shows the specifications of the thrust bearing. The supply pressure is the same as the high pressure of the He liquefier. The outer diameter ϕ32 mm of the collar was determined based on the rotational strength of the collar and the shaft diameter of ϕ16 mm based on the natural bending frequency. The thrust force was determined based on the presssure to be applied to each portion of the diameter of the turbine and blower wheels. The revolution speed after starting was maintained at a rated 230,000rpm. The opening of the turbine inlet valve increases as cooling down develops and thrust force also increases. When the turbine inlet temperature reaches 80K, the design thrust force acting downward is estimated to be 200N. In the case of practical application, however, an emergency stop is inevitable where thrust force acts upward. Accordingly, the thrust bearing used here should be of the combined type.

The lower and upper feeding hole diameters were designed at ϕ0.6 mm and ϕ0.3 mm respectively, to increase overall load capacity and reduce the quantity of gas supplied. Table 2 shows the combined bearing specifications when C_f=0.6

EXPERIMENTAL RESULTS AND OBSERVATIONS

Figure 10 shows the structure of the turbine expander used in this experiment while Figs.11 and 12 give the outline of the turbine expander test equipment and the configuration of the turbine shaft system, respectively.

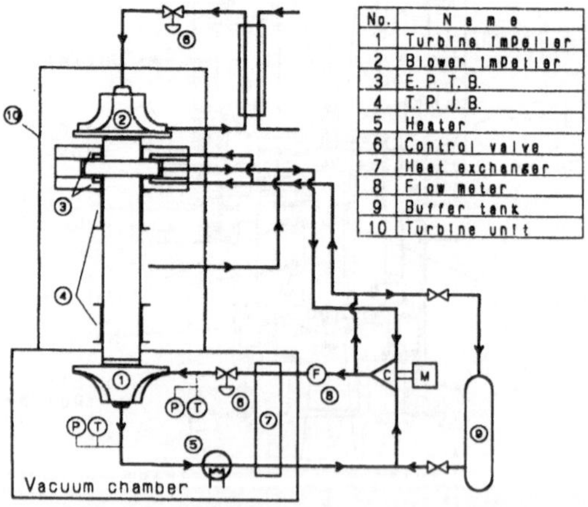
No.
Name
1 Turbine impeller
2 Blower impeller
3 E. P. T. B.
4 T. P. J. B.
5 Heater
6 Control valve
7 Heat exchanger
8 Flow meter
9 Buffer tank
10 Turbine unit
Vacuum chamber

Fig.11. Schematic diagram of testing equipment

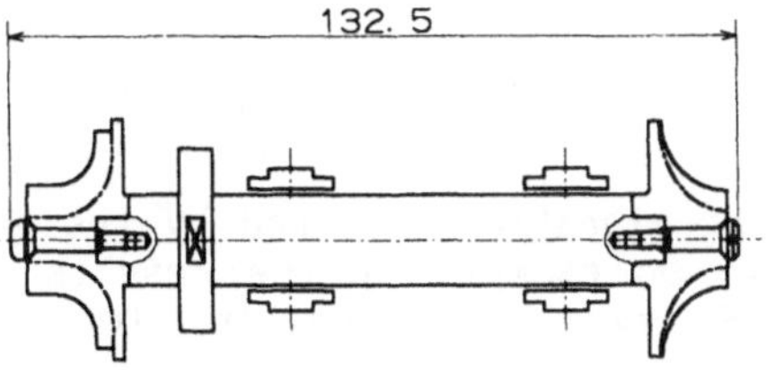

Fig.12. Schematic configuration of shaft bearing system

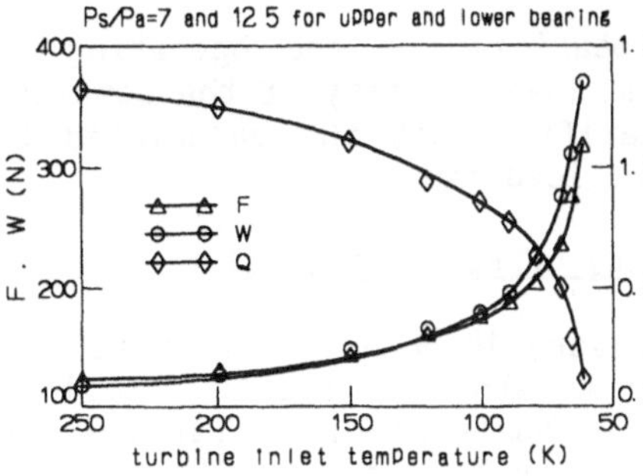

Fig.13. Comoarison of thrust force with load carring capacity

Thrust Bearing Characteristics

The measured results of load capacity and flow rate for the single thrust bearing using the experimental machine demonstrated in Fig.10 are shown as a plot in Figs.3 and 5 . Some papers have reported that the calculation results using the theory of complex velocity potential are reasonably consistent with experimental values when the bearing film thickness and the supply gas pressure ratio p_s(=Ps/Pa) are small[1,2]. On the contrary, when the supply gas pressure ratio and the bearing film thickness are large, restrictor loss is evident due to the inertia effect of fluid at the restrictor outlet[3,4]. As results of our experiment, the restictor loss has also been recognized as shown in Figs.3 and 5 . When counting backward the discharge coefficient of C_f from the results above, $C_f \approx 0.6$ can be worked out. It is reported that the loss can be reduced by improving the shape of the restrictor inlet[4].

It was observed from the characteristics shown in Figs.7,8 and 9 that the smaller value of $C_f(=0.6)$ gives the better performance than the larger value of $C_f(=1.0)$ when ε reaches a larger value.

Thrust Force and Bearing Film Thickness During Operation

The relationship between the thrust force, which is calculated from the measured results of the pressure applied on the shafting (turbine and blower) during operation, and the changes of turbine inlet temperature are shown in Fig.13 with the notation of Δ. On this occasion the relationship between the total measured flow rate Qmc of the upper and

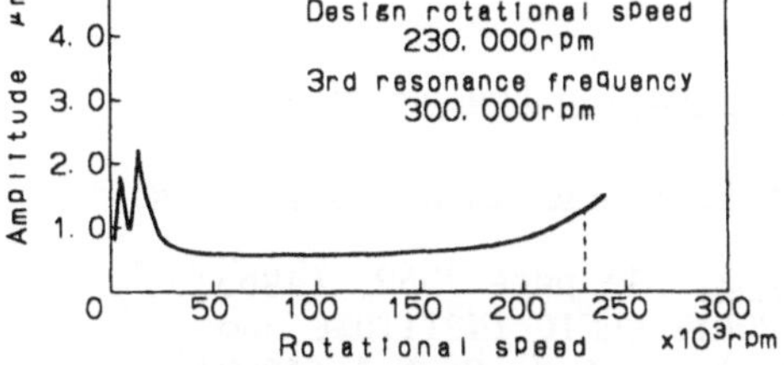

Fig.14. Starting characteristics of turbine shaft

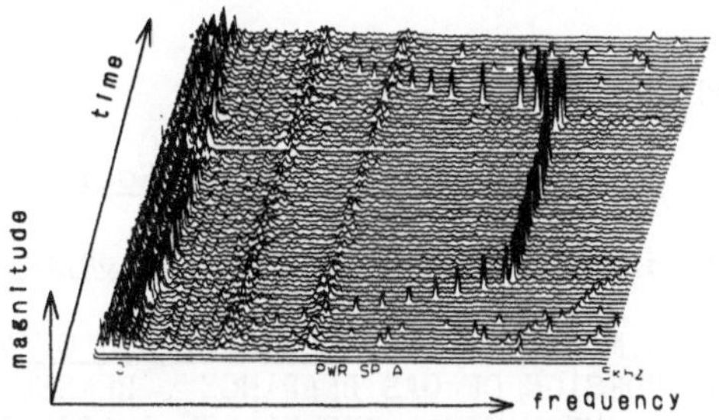

Fig.15. Results of frequency analysis of turbine shaft

lower bearings and the operating temperature is shown with ◇ and the measured load capacity of the combined bearing is demonstrated with ○. It was clarified that the calculated thrust force Δ agrees well with the measured load capacity ○.

<u>Stability of Rotating Shaft</u>

Figure 14 shows the results of shaft vibration measured by the gap sensors as shown in Fig.10. Fig.15 shows an example of the results of frequency analysis. From Figs.14 and 15, excellent stability for all operating conditions including starting, rated revolution speed, cooling down and stopping is observed. No flaws were found on the heat treated surface of the journal combined with the ceramic tilting pads after 200 start/stop cycles and then it was confirmed that the combination of shaft and ceramic bearing provides significant improvement in their contact damage.

CONCLUSIONS

The conclusions are summarized as follows:

a) The characteristics of the single and combined thrust bearings can be determined using the theory of complex velocity potential.

b) The experimental results of the single and c ombined thrust bearings agreed well with the calculated results and then the practicality of the calculations was confirmed.

c) The combination of TBJB and EPTB showed very stable operation for a wide range of operating conditions including starting, stopping, cooling down and rated operation.

d) The combination of shaft and bearing material is satisfactory and no damage on the journal surfaces was found after a number of repetitions of starting and stopping.

ACKNOWLEDGMENTS

This project was sponsored by the Agency of Industrial Science and Techonology of MITI, which commissioned NEDO to promote the work and was carried out by Super-GM and its member companies. The authors would like to express their appreciation to individuals of the above organizations for their assistance in the preparation of this paper.

REFERENCES

1. H.Yabe, <u>Journal of Japan Society of Lubrication Engineers</u>, 17: page 64, (1972).
2. H.Mori, H.Yabe, <u>Journal of Japan Society of Lubrication Engineers</u>, 8: page 11, (1963).
3. H.Mori, H.Yabe, <u>TRANSACTIONS OF THE JSME</u>, 31: page 1562, (1965).
4. "DESIGN OF GAS BEARINGS", RENSSELAER POLYTECHNIC INSTITUTE and MECHANICAL TECHNOLOGY INCORPORATED, 1: ch.5.3, Ch.5.1.(1967).

APPENDIXES

Appendix 1

$$\overline{K}_{1comp} = K_{1comp} / P_a^2 = (\overline{P}_0^2 - 1) / \overline{A}_{k1} \quad (1)$$

$$\overline{K}_{2comp} = K_{2comp} / P_a = 1.0 \quad (2)$$

$$\overline{A}_{k1} = \sum_{j=-\infty}^{\infty} \log \left\{ \frac{\sin^2 \overline{A}_{1s} + \sinh^2 \overline{B}_{0j}}{\sin^2 \overline{A}_{0s} + \sinh^2 \overline{B}_{0j}} \right\} \quad (3)$$

$$\overline{A}_{1s} = \frac{\pi \cdot \log \{ (\overline{r}_a - \overline{r}_s) \cdot \overline{r}_a \}}{2 \cdot \log (1 / \overline{r}_1)} \quad (4)$$

$$\overline{A}_{0s} = \frac{\pi \cdot \log \{ (\overline{r}_a - \overline{r}_s) / \overline{r}_a \}}{2 \cdot \log (1 / \overline{r}_1)} \quad (5)$$

$$\overline{B}_{0j} = \frac{\pi^2 \cdot j / k}{\log (1 / \overline{r}_1)} \quad (6)$$

$$\overline{A}_r = \sum_{j=-\infty}^{\infty} \log \left\{ \frac{\sin^2 \overline{A}_1 + \sinh^2 \overline{B}_j}{\sin^2 \overline{A}_0 + \sinh^2 \overline{B}_j} \right\} \quad (7)$$

$$\overline{A}_1 = \frac{\pi \cdot \log (\overline{r} \cdot \overline{r}_a)}{2 \cdot \log (1 / \overline{r}_1)} \quad (8)$$

$$\overline{A}_0 = \frac{\pi \cdot \log (\overline{r} / \overline{r}_a)}{2 \cdot \log (1 / \overline{r}_1)} \quad (9)$$

$$\overline{B}_0 = \frac{\pi \cdot (\theta + 2\pi j / k)}{2 \cdot \log (1 / \overline{r}_1)} \quad (10)$$

Appendix 2

$$\overline{h}_{n1} = h_{n1} / h_{ct} \quad (1)$$

$$\overline{h}_{n2} = h_{n2} / h_{ct} \quad (2)$$

$$\varepsilon = \Delta h / h_{ct} \quad (3)$$

$$\Lambda_{n1} = 12 \cdot \mu \cdot k_1 \cdot r_{s1} \cdot (R \cdot T_s)^{1/2} / P_s \cdot h_{n1}^2 \quad (4)$$

$$\Lambda_{n2} = 12 \cdot \mu \cdot k_2 \cdot r_{s2} \cdot (R \cdot T_s)^{1/2} / P_s \cdot h_{n2}^2 \quad (5)$$

$$\Lambda_{s1} = 12 \cdot \mu \cdot k_1 \cdot r_{s1} \cdot (R \cdot T_s)^{1/2} / P_s \cdot (h_{n1} - \Delta h)^2 = \Lambda n1 / \{1 - \varepsilon \cdot (1 - h_{n1} / h_{n2})\}^2 \quad (6)$$

$$\Lambda_{s2} = 12 \cdot \mu \cdot k_2 \cdot r_{s2} \cdot (R \cdot T_s)^{1/2} / P_s \cdot (h_{n2} + \Delta h)^2 = \Lambda n1 / \{1 - \varepsilon \cdot (1 + h_{n1} / h_{n1})\}^2 \quad (7)$$

DEVELOPMENT ON A LARGE HELIUM TURBO-EXPANDER WITH VARIABLE CAPACITY

T. Kato, Y. Kamiyauchi, E. Tada, T. Hiyama, K. Kawano,
M. Sugimoto, E. Kawagoe, H. Ishida, J. Yoshida, H. Tsuji,
* S. Sato, * Y. Nakayama, * I. Kawashima

Naka Fusion Research Establishment, Japan Atomic
Energy Research Institute, 801-1 Mukaiyama,
Naka-machi, Naka-gun, Ibaraki-ken, 311-01, Japan
* Kobe Steel Ltd., Chuo-ku, Kobe-shi, Hyogo-ken, 651, Japan

ABSTRACT

A large helium turbo-expander with variable-flow-capacity mechanism (an adjustable turbine) has been developed, which will effectively controls refrigeration power. A type of variable nozzle vane height is selected as the mechanism. Mechanical performance test of the mechanism was carried out under cryogenic temperature and fluid-dynamic performance was measured on connecting the adjustable turbine with an existing helium refrigerator. Turbine efficiency was measured by varying nozzle openings from 40% to 100% at the room temperature, 80 K and 40 K. Smooth operation of changing capacity and effective turbine power control without any degradation were verified.

INTRODUCTION

A thermonuclear fusion experimental reactor, for example, International Thermonuclear Fusion Experimental Reactor (ITER) requires a very large helium cryogenic system with refrigeration capacity of about 100 kW at 4 K region[1], therefore the development of larger key components of the cryogenic-system is indispensable. As one of these developments, Japan Atomic Energy Research Institute (JAERI) and Kobe Steel Ltd. have developed an adjustable turbine, which is available to directly control the flow rate and to provide effective control of refrigeration power.

The adjustable turbine corresponds to a 10-kW refrigerator and has the mechanism in which the nozzle opening can be changed by moving an upper side wall of the nozzle. Its performance test was done, then the mechanical and fluid-dynamic performances were determined by changing its capacity at the equivalent design condition with cold helium. In this paper, the effect of the adjustable turbine, its design and mechanism, and performance test results are described.

REQUIREMENT AND EFFECT OF AN ADJUSTABLE TURBINE

The operation scenario of the fusion experimental reactor requires changing the heat load of the cryogenic system over a large range, for example a plasma-burning operation mode and a standby mode imposes a heat load of about 100 kW and 30 kW, respectively[1]. Such a large heat load difference will be controlled by the unit operating configuration such as from 1 to 3 units of 30-kW size refrigerator. However, the practical refrigerator operation shall require delicate refrigerator power adjustment. For example, assuming that 20-kW operation which is less than the design refrigeration power by 10 kW, is required in the standby mode, the mass flow of the turbine will be firstly reduced, and the excessive refrigeration power will be used to liquefy into a liquid helium tank, or the heat load will be increased by an electric heater to adjust to the design value. Such operation methods cause waste of input energy in the sense of energy efficiency, because the loss of 1 W at 4 K is the same as 500 - 1000 W of input energy at ambient temperature for a conventional refrigerator. Therefore, the refrigerator, especially a large capacity one, should have the ability to control the power corresponding to the applied heat load without degrading efficiency. Such a refrigerator will be constructed by adopting the adjustable turbine.

In order to see how the adjustable turbine effects refrigerator control, let's compare it with the turbo-expander without variable-flow-capacity (the conventional turbine). Turbine power (L_t) is described by

$$L_t = \eta \cdot \dot{m} \cdot \Delta H_{ad}$$

Where, η is adiabatic efficiency, $\dot{m}$ is mass flow rate (g/s), and ΔH_{ad} is adiabatic head (J/g). To control the refrigeration power is just to control the turbine work, by changing the mass flow rate or the adiabatic head in the above formula. In conventional turbine, the mass flow rate and the adiabatic head are controlled by a valve which is generally located upstream of the turbine. The process of changing adiabatic head is as shown in Fig. 1. The pressure out of compressor P_i is reduced by closing the valve to be P'_i then helium is expanded to the turbine outlet pressure P_e with the efficiency of η. This process is traced as①-③-④ in Fig. 1 and the practical turbine head is indicated as $\Delta H'$. On the other hand, in the case of the adjustable turbine, since the turbine can control the variable nozzle opening like a valve, the expansion process can be from pressure P_i to pressure P_e resulting in the practical turbine head of ΔH traced as ①-② in Fig. 1. It can be recognized that ΔH is obviously larger than $\Delta H'$, if both turbine efficiency defined by $\eta = \Delta H/\Delta H_{ad}$ or $\eta = \Delta H'/\Delta H_{ad}$ are similar to each other. Remembering that the turbine work is described as a product of the mass flow and the turbine head, the adjustable turbine can reduce more flow rate than the conventional turbine at the same turbine work: therefore, the compressor input power can be saved and operation is more efficient.

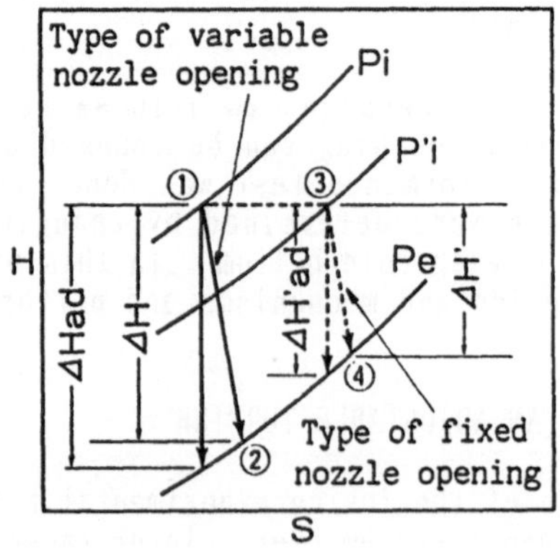

Fig. 1. The processes of reducing adiabatic head

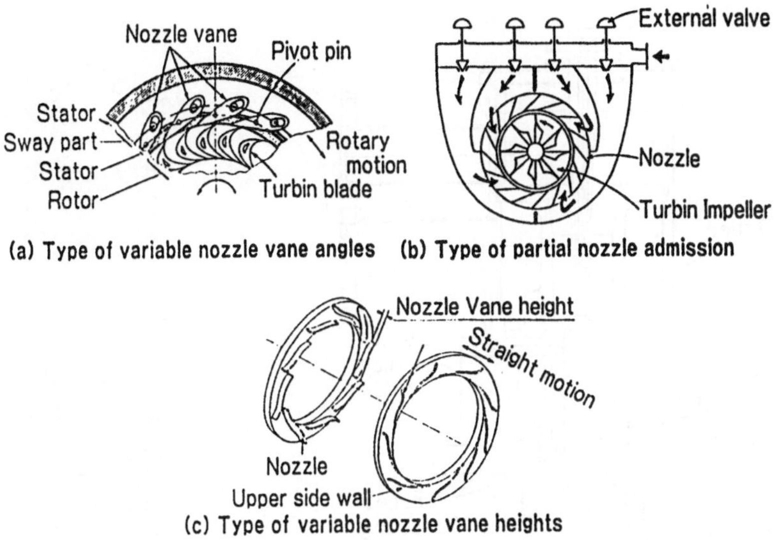

Fig. 2. Various systems of variable nozzle opening

DESIGN OF THE VARIABLE CAPACITY TURBO-EXPANDER

Structure and specifications

There is no previous example of applying the variable-flow-capacity mechanism to a helium turbine; however, the various systems shown in Fig. 2 are applied to radial in-flow turbines adopted to the turbo-charger of automobiles[2]. In Fig. 2, the type (a) of variable nozzle vane angles is a mechanism in which the nozzle vanes rotate simultaneously around pivot pins, where varying nozzle throat, and the flow-in angle of gas-entering the turbine blade can be changed. The type (b) of partial nozzle admission controls the flow quantity by partial admission due to the change of opening degree of the external valves. Type (c) of variable nozzle vane height adjusts the nozzle height by driving an upper side wall in the nozzle heights direction toward the fixed nozzle vane. The mechanism of type (c) is selected for the following reasons; In type (b), flowing to only a portion of the impeller circumference, high turbine efficiency shall not be expected and continuous capacity control shall not be obtained. In type (a), which is generally applied to the turbo-charger of the automobiles, the mechanism typically has great clearance, resulting in much leakage loss due to helium gas. While in the type (c), continuous control is available, compared with type (b), and leakage loss will be smaller than type (a) due to the simpler mechanism with less clearance. Fabrication of the helium turbine, which is relatively small, requires high accuracy, especially since the turbine is put into the low temperature helium with a large temperature difference; accordingly, the driving mechanism should be simpler.

The structure of the adjustable turbine is shown in Fig. 3. A pulse motor is adopted as a driving force for the mechanism. Rotary motion by the motor is turned to straight motion through the taper cam ring, which can move the upper side wall downward through the connecting drive shaft. On the contrary, when moving the upper side wall up, the motor rotation is reversed and the upper side wall is pushed up by the force of the upper driving spring. Through this mechanism, the nozzle openings can be changed from 0 to 130%, assuming that the rated design opening is 100%.

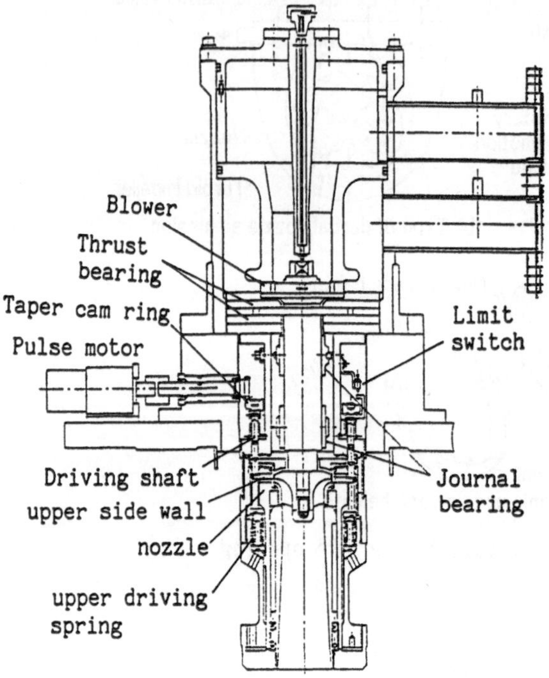

Fig. 3. Structure of a turbine with prototype variable nozzle opening mechanism

Fig. 4. Inner module of the turbo-expander with variable nozzle opening mechanism

The impeller with a shroud is adopted so as to prevent the expanding gas from passing through the tip clearance. Self-acting gas bearings are applied for journal and thrust bearings, which are effective for not only avoiding accident caused by impurities in the cryogenic helium gas but also attaining low friction loss. A tilting-pad mechanism is adopted for the journal bearings and a pump-in spiral groove type is used for the thrust bearing. Figure 4 is a photograph of the turbine. Design specifications at 100% nozzle opening are given in Table 1. The turbine will have the capacity for use in a 10-kW refrigerator.

Table 1. Design specification, at 100% nozzle opening

		Design
Type		Radial inward flow
Bearing	Journal	Tilting pad type gas bearing
	Thrust	Spiral grooved type gas bearing
Nozzle opening (%)		0 ~ 130
100% Nozzle opening		
Flow rate (kg/s)		0.356
Inlet temp. (K)		18
Inlet pressure (MPa)		0.63
Pressure ratio		5.8
Rotating speed (rps)		1121

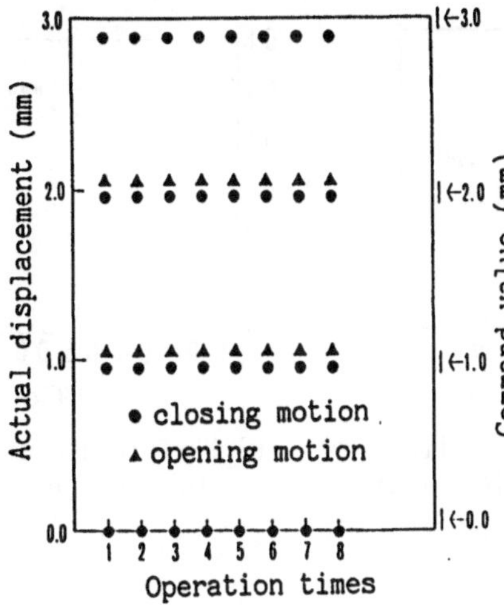

Fig. 5. Corresponding relation
between the command value and its actual displacement

PERFORMANCE TEST

Variable mechanism performance test

To confirm smooth operation of the mechanism under the cryogenic conditions, a mechanical performance test was carried out. Inserting a cryogenic fiberscope to the part of the nozzle side wall, the practical motion of the nozzle opening, corresponding with the opening command inputted to the computer, was directly displayed and observed on a TV monitor.

Several cyclic motions for opening and closing were performed at liquid nitrogen temperature and the relation between the actual nozzle opening and command value to the computer was observed as shown in Fig. 5. There is good agreement between them and smooth operation was confirmed. A small backlash, 85μm, was also observed; however, such a backlash would be easily corrected by the computer and the coincident operation between the actual opening and the opening command could be available in the system.

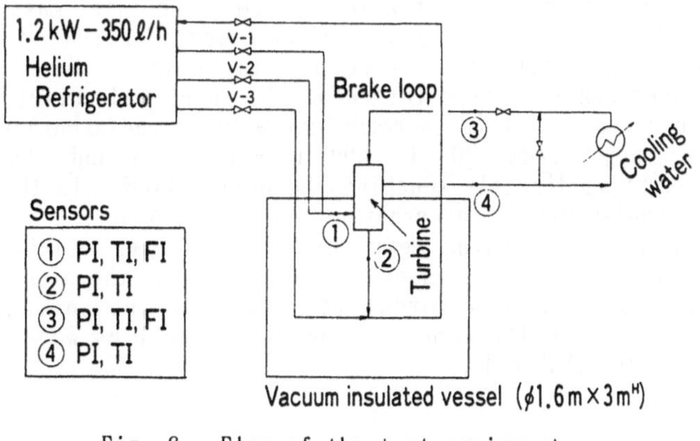

Fig. 6. Flow of the test equipment

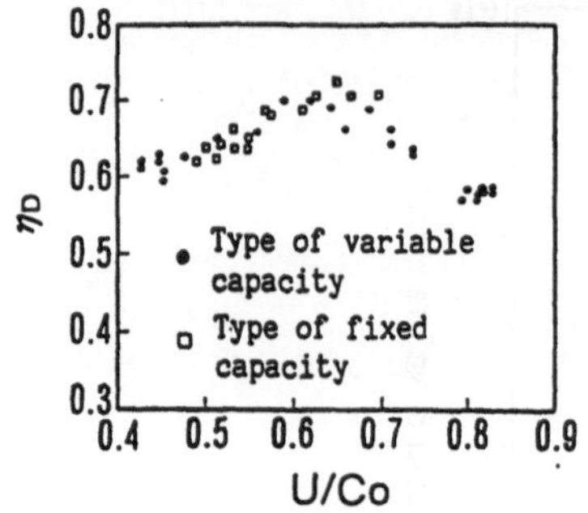

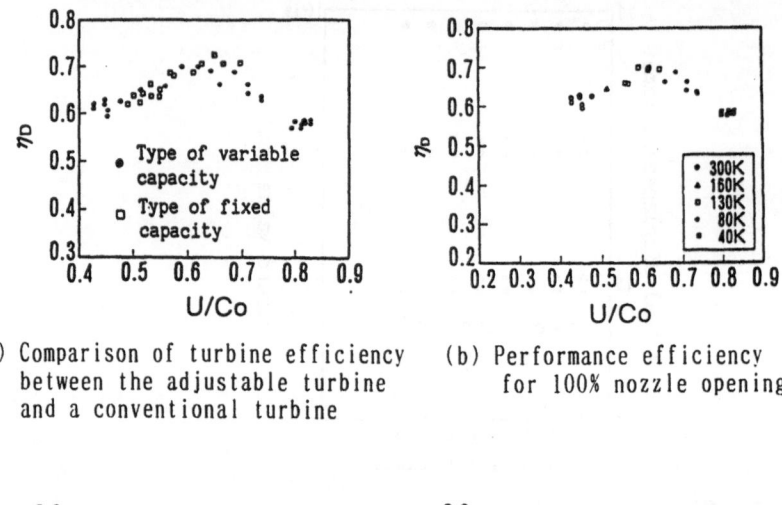

(a) Comparison of turbine efficiency between the adjustable turbine and a conventional turbine

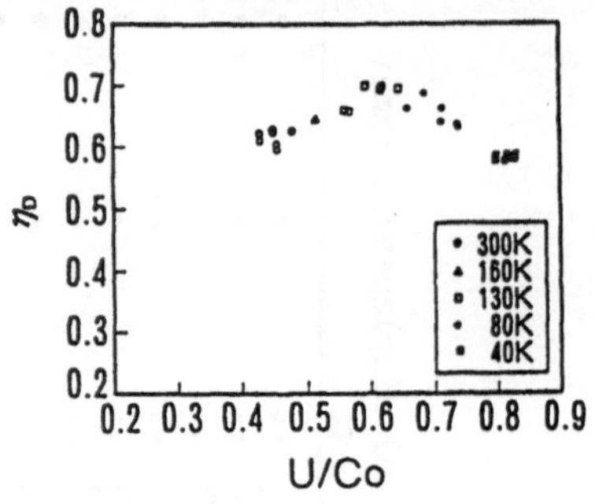

(b) Performance efficiency for 100% nozzle opening

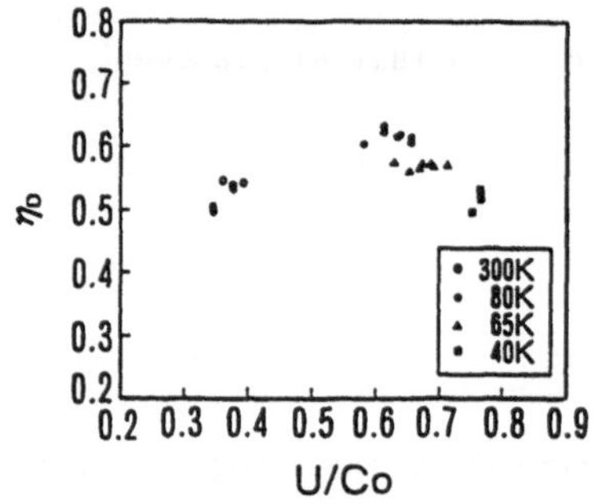

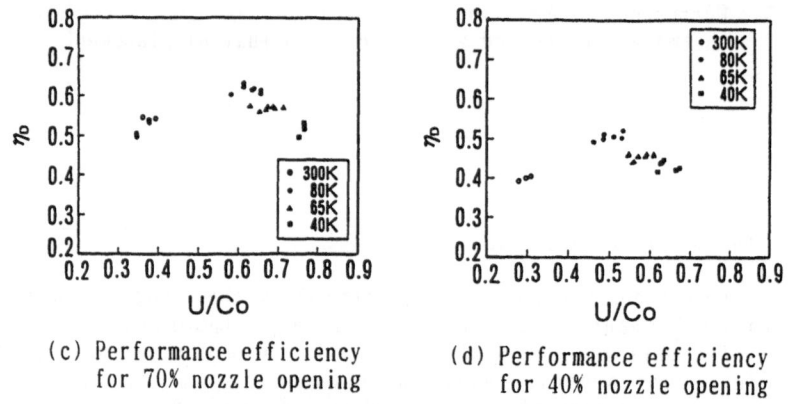

(c) Performance efficiency for 70% nozzle opening

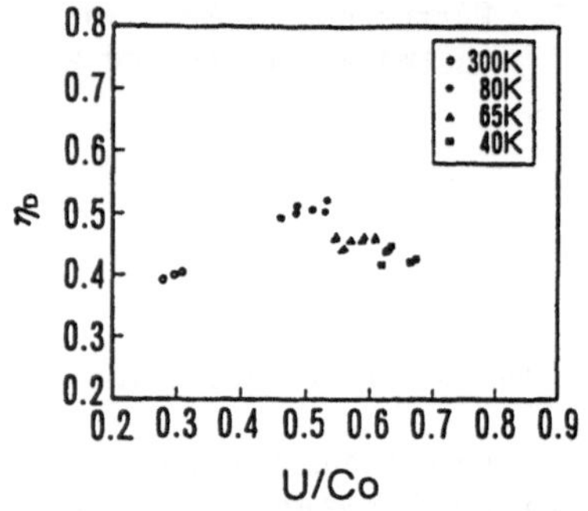

(d) Performance efficiency for 40% nozzle opening

Fig. 7. Characteristics of turbine efficiency

Fluid dynamics performance test

Test equipment. A cryostat of 1.6 m diameter and 3 m height was used as the test stand, with the turbo-expander mounted on its top flange. The JAERI's 1.2 kW/350-l/h refrigerator[3] supplied helium flow to the turbo-expander in various temperature ranges, and also supplied room temperature helium to the blower circuit and the thrust gas bearing system. The inlet mass flow rate and the pressure ratio through the turbo-expander were controlled by the valves installed in the distribution valve box as shown in Fig. 6. Delicate temperature measurements were performed at the inlet and outlet of the turbo-expander, where Pt-Co resistive thermometers with proved accuracy of less than 0.1 K in the range from 20 to 300 K were used and installed inside the piping so as to be directly touching the helium flow. In the brake blower circuit, thermocouples with an accuracy of around 0.1 K were used with the same installation method. The data, temperature, pressure, and flow rate, were automatically observed and recorded by pen recorders and the micro-computer data acquisition system which can immediately convert the voltage signals to physical quantities. In the computer, helium properties were calculated by using NBS helium data (NBS 631).

Test method and parameters. Measuring the design performance of a helium turbine generally requires an actual size helium refrigerator. In the test stand system mentioned in the previous section, however, the maximum inlet helium flow and pressure ratio are around 60 g/s and 2.0, respectively, which are much less than the design conditions by around one sixth in mass flow and one third in pressure ratio. Thus an equivalent performance test method is indispensable[4], in which the turbine inlet velocity triangle is set at the same level as the design specifications by arranging the inlet temperature, pressure, and pressure ratio. Adopting such a test method, the fluid dynamic performance test was carried out. As test parameters, the nozzle openings were set to three steps of 100%, 70% and 40%, respectively, the turbo-expander inlet temperature was controlled in the three ranges of room temperature, 80 K and 40 K, respectively, for each nozzle opening and the rotating speed was changed from 700 rps to 1000 rps in steps of 100 rps.

Test results. Equiping the adjustable turbine with a variable-flow-capacity mechanism increases the risk of leakage loss resulting in a low turbine efficiency. Checking such a degradation can be carried out by comparing the efficiency of the adjustable turbine with the efficiency of the conventional turbine at the design nozzle opening. Figure 7 (a) shows such a comparison where each efficiency is plotted as a function of blade-jet speed ratio, U/C_o. Both trends are similar indicating that there is no degradation on the adjustable turbine. The performance efficiency for several nozzle openings was measured as shown in Fig. 7 (b), (c) and (d), corresponding with nozzle opening of 100%, 70% and 40%, respectively. Reducing nozzle opening from 100%, causes an increasing flow loss due to a large steep expansion between the nozzle output and the impeller input. In practice, the smaller the nozzle opening, the lower the turbine efficiency. Efficiencies of 70%, 64% and 52%, are observed at nozzle opening of 100%, 70% and 40%, respectively.

The effect of reducing mass flow rate for the adjustable turbine was measured. Figure 8 (a), shows the mass flow characteristics as a function of the turbine power at several nozzle openings for 80 K operation. For example, turbine power of 500 W requires a mass flow rate of 27 g/s at 100% nozzle opening. However, at the same turbine power, the mass flow rate can be reduced to 21 g/s and 16 g/s at nozzle opening of 70%, and 40%, respectively. The turbine power is the same because the smaller nozzle opening gives a larger adiabatic head as plotted in Fig. 8 (b). Such relations explain the effect of the adjustable turbine saving the input energy to the refrigerator as mentioned before.

During for experimental period, total operation time reached around 46 hours, including 20 times of start/stop and nozzle-moving operation. Through the experiment, there was no mechanical trouble with the adjustable turbine.

CONCLUSION

JAERI and Kobe Steel Ltd. have successfully developed the adjustable turbine. Its performance test was carried out from the point of view of mechanics and fluid-dynamics. The turbine efficiency was measured in the large range of blade-jet speed ratio (U/C_o) from around 0.2 to 0.85. These results verified the mooth operation of the variable flow mechanism and effective turbine power control. Availability of such a turbine will advance the realization of the large and efficient cryogenic system required for the fusion experimental reactor.

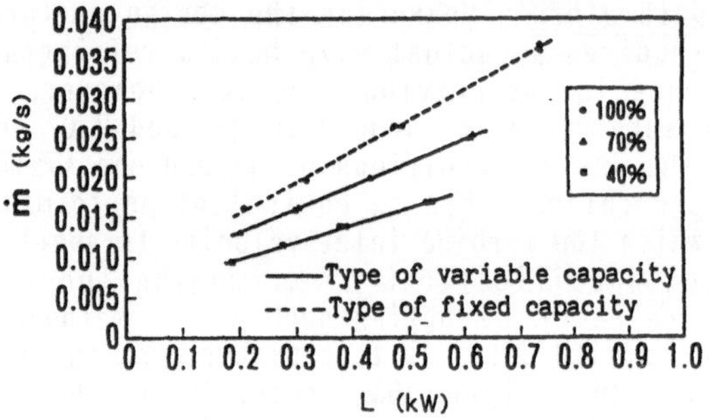

(a) Relationship between flow rate and turbine power

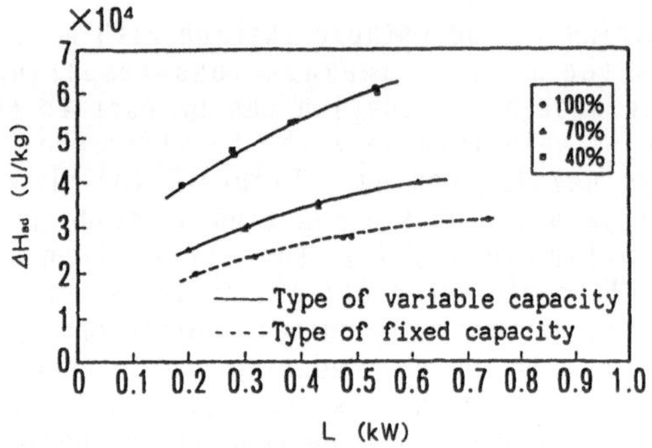

(b) Relationship between adiabatic head and turbine power

Fig. 8. Characteristics of turbine mass flow rate and adiabatic head as a function of turbine

ACKNOWLEDGMENT

The authors would like to thank Drs. M. Yoshikawa, T. Iijima and S. Shimamoto for their continuous encouragement on this work. The contribution to this work from the staffs of the superconducting magnet laboratory at JAERI and of the manufacture division at Kobe Steel Ltd. are greatly acknowledged.

REFERENCES

1. ITER CAD magnet group, "ITER Magnets," ITER Documentation series, No. 26, (1990), IAEA, Vienna.
2. O.E. Balje ; TURBO MACHINES, John Wiley & Sons, NY (1981), P.317
3. E. Tada et al., "350-L/h, 1.2 kW Helium Cryogenic System for Development of Fusion Technology," Proc. ICEC 9, P. 93 (1982)
4. T. Kato et al., "A Large Scale Turbo-expander Development and Its Performance Test Result," Advances in Cryogenic Engineering, Vol.35, P.1005, (1990)

DEVELOPMENT OF HIGH EXPANSION RATIO HELIUM TURBO EXPANDER

N.Ino*, A.Machida*, K.Ttsugawa*, Y.Arai*,
M.Matsuki**, H.Hashimoto+, A.Yasuda++

* Mayekawa Mfg. Co. Ltd., MYCOM Advanced Technology Laboratory, Ibaraki, Japan
** Nippon Institute of Technology, Saitama, Japan
\+ Tokai University, Kanagawa, Japan,
++ Super-GM, Osaka,Japan

ABSTRACT

The authors developed a high expansion ratio radial inflow turbine for a helium liquefier of 100 L/h capacity for use with a 70 MW superconductive generator. The following results were obtained from this development work: 1)Very stable and highly reliable operation was maintained through use of a externally pressurized thrust bearing and a tilting pad journal bearing. 2)It was found that in the case of a high expansion ratio turbine, a large nozzle loss is inevitable. 3)It was confirmed that the application of a conformal transformation method in modification of the nozzle setting angle is effective with a small cryogenic turbine having a large expansion ratio. 4)Approximately 70% efficiency was obtained, notwithstanding the high expansion ratio.

INTRODUCTION

The objective of the research is development of an expansion turbine of high reliability and high efficiency for a 100 L/h helium liquefier for a 70MW superconductive generator. As previously reported[1], the liquefying cycle is a Claude type with two parallel turbines. This report focuses only on the high temperature side expansion turbine used in the cycle. Numerous examples of application of low expansion ratio expanders, e.g., expansion ratios of 4∼8, have been reported. Reports on high expansion ratio helium gas expanders with expansion ratios of 10 or more are limited[2],however, and few reports make mention of the design method or performance of the turbine. The authors of this paper developed a high expansion ratio helium turbo expander and were able to clarify various points to consider in the design and other factors which affect performance. Details of these findings are given herewith.

SPECIFICATIONS AND DESIGN OF EXPERIMENTAL MACHINE

There is an established design method for the radial inflow turbine and the conventional radial turbine design method program was used to determine turbine geometry[3]. Table 1 gives turbine design specifications, Fig.1 shows the T-S diagram, Fig.2 gives the velocity diagram of the turbine, Fig.3 shows a full view of the impeller and Fig.4 shows the nozzle shape. Subscripts of pressures and temperatures in Table 1 are as

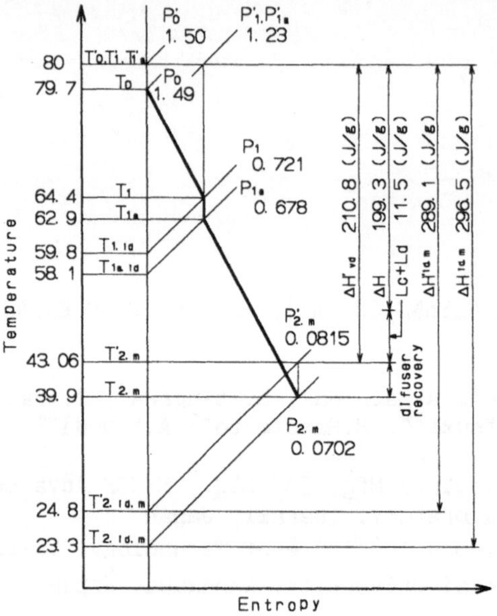

Fig.1. T-S diagram of turbo expander

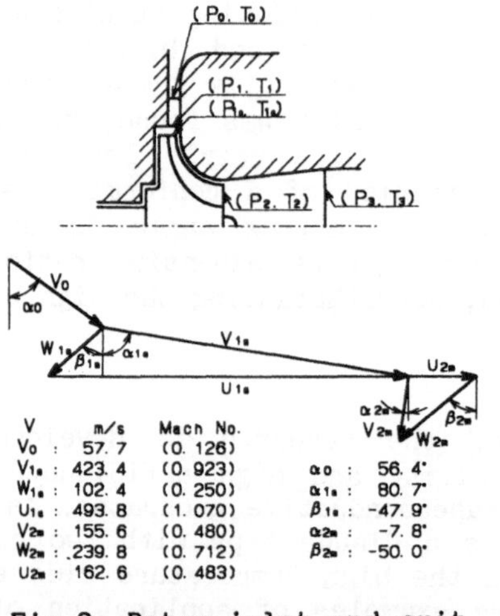

Fig.2. Rotor inlet and exit velocity diagram

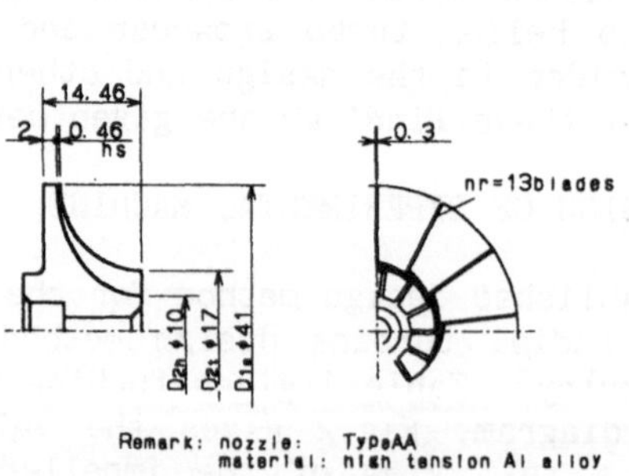

Fig.3. Schematic configuration of impeller

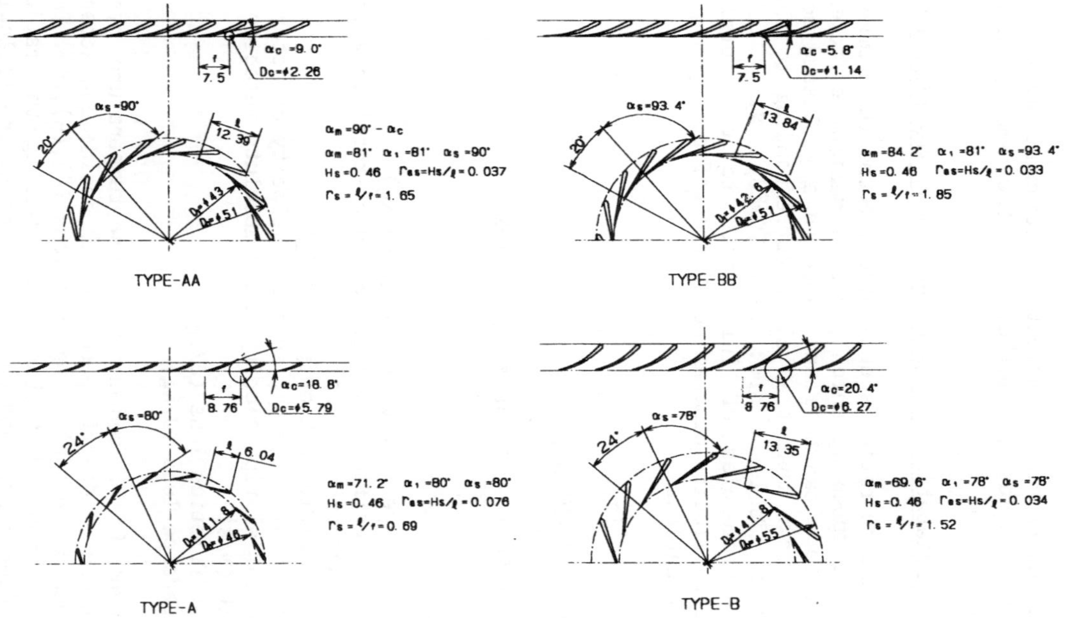

Fig.4. Conformal transformation of circular nozzle arrangement

Table 1 Design Specifications and Experimental Results

Item	Design	Result	Item	Design	Result	Item	Design	Result	Item	Design	Result
P'_o (MPa)	1.50	1.53	T'_o (K)	80.0	73.8	η_{vd}	0.729	0.695	e_s	0.238	0.339
P_{1a} (MPa)	0.677	0.467	T_{1a} (K)	62.9	55.0	Rr	0.630	0.409	e_r	0.691	0.282
P'_2 (MPa)	0.0815	0.127	T'_2 (K)	39.9	41.6	Mg(g/s)	20.5	24.0	r_a (mm)	0.082	0.05
Ns	0.296	0.284	ν	0.642	0.699	Nr(rpm)	230,000	228,000	r_r (mm)	0.05	0.03
n_s	18	×	n_r	13	×	γ (deg.)	6.0	×	r_d	1.93	×

defined in Fig.2. Superscript ' indicates a stagnant state. The following is an outline of the design method for the small, high expansion cryogenic turbine.

Specific Speed N_s, Adiabatic Efficiency η_{tt}, η_{vd}, Reaction Ratio R_r

When the expansion ratio is determined, round the number of maximum efficiency, specific speed N_s which produces maximum efficiency and the blade-jet speed ratio ν can be found based on a published source[4]. From this source it is known that the values corresponding to the operating conditions shown in Table 1 are: η_{tt}= 70~80%, N_s = 0.2~0.3, ν = 0.66~0.70. The efficiencies of inlet-total to exit-total adiabatic efficiency η_{tt}, and velocity diagram adiabatic efficiency η_{vd} are defined by the following equations.

$$\eta_{tt} = \Delta H \,/\, \Delta H'_{id,t} \qquad (1)$$

$$\eta_{vd} = \Delta H'_{vd} \,/\, \Delta H'_{id,t} \qquad (2)$$

$$\Delta H'_{vd} = \Delta H + L_d + L_c = U_{1a}\cdot V_{1a} - U_2\cdot V_{2m} \qquad (3)$$

where

- ΔH =turbine specific work output (J/Kg)
- $\Delta H'_{id,t}$=ideal specific work based on inlet-total to exit-total pressure ratio (J/Kg)
- $\Delta H'_{vd}$ =gas specific work output (J/Kg)
- U_{1a},V_{1a}=tangential velocity, absolute velocity of rotor inlet (m/s)
- U_2,V_{2m} =tangential tip velocity, mean absolute velocity of rotor exit (m/s)
- Ld =disc friction loss (J/Kg)
- Lc =clearance loss (J/Kg)

The efficiency used for a conventional low temperature cycle is η_{vd}. Specific speed N_s and blade-jet speed ν which decide characteristics of a turbine are defined by the following equations (4),(5),(6). The degree of reaction R_r is an important factor in determining the ratio of energy utilized by the rotor blade and can be defined by the following equation (7).

$$N_s = N\cdot Q_2^{1/2} \,/\, (\Delta H'_{id,t})^{3/4} \qquad (4)$$

$$\nu = U_{1a} \,/\, Co \qquad (5)$$

$$Co = (2\Delta H'_{id,s})^{1/2} \qquad (6)$$

$$R_r = 1 - \Delta H'_{1aid,s} \,/\, \Delta H'_{id,s} \qquad (7)$$

where

N =rotative angular velocity (rad/S)

Q_2 =volume flow rate at rotor exit (m^3/s)

C_o =ideal jet-speed based on inlet-total to exit-static pressure ratio (m/s)

$\Delta H'_{1aid,s}$=ideal specific work based on inlet-total to stator exit-static pressure ratio (J/Kg)

$\Delta H'_{id,s}$ =ideal specific work based on inlet-total to exit-static pressure ratio (J/Kg)

Modification of Nozzle Setting Angle α_s

The relationship between the theoretical flow angle α_1 and the setting angle α_s is important in the design of the nozzle. In the case of a circular nozzle, the mean gas flow angle α_m at the nozzle outlet is not identical with α_s but is $\alpha_s > \alpha_m$. Consequently, when the relationship is considered as $\alpha_s = \alpha_1$, the expected performance cannot be obtained. That is why it is necessary to modify the nozzle setting angle. Usually, a conformal transformation method is used to amend the setting angle[5]. As demonstrated in Fig.4, the setting angle α_s can be found by developing the circular nozzle arrangement into a linear nozzle arrangement.

The effectiveness of the conformal transformation method for the turbine nozzle using air as a medium under normal and high temperature conditions has been duly proven[6], but there is no literature available on the use of an expander with very low temperature helium gas. The values of α_1,α_s and α_m for the experimental nozzle are shown in Fig.4. A total of four types of nozzle A, B, AA and BB, were manufactured in order to compare performance.

Diffuser

The kinematic energy at the rotor outlet should be recovered using a diffuser. The best suited diffusing angle $\gamma(=\tan^{-1}(D_d/2 \cdot l_d))$ which minimizes the loss in pressure recovery is 5~6 degrees and the aspect ratio $r_{ad}(=l_d/D_d)$ is 1.4~3.3 . Since the diffuser outlet diameter D_d is identical to the outlet piping diameter, the diffuser length l_d can be determined from γ by setting D_d so that 10~20 m/s of piping flow velocity can be controlled.

Various Internal Losses

The losses to be considered in the calculation program are stator boundary layer loss L_s, rotor disc friction loss L_d, rotor shroud clearance loss L_c, rotor boundary layer loss L_r and rotor exit flow loss Le. Of the five, L_r and L_s are the largest and most significant. The loss coefficients, namely ε_s and ε_r, defined by the following equations (8) and (9) are used in determining the above two significant losses, described by equations (10) and (11).

$$\varepsilon_s=1-V^2{}_1 / V^2{}_{1id} \tag{8}$$

$$\varepsilon_r=1-W^2{}_{2m} / W^2{}_{2mid} \tag{9}$$

$$Ls=V^2{}_{1id}\cdot\varepsilon_s / 2 \tag{10}$$

$$Ld=V^2{}_{2mid}\cdot\varepsilon_r / 2 \tag{11}$$

where

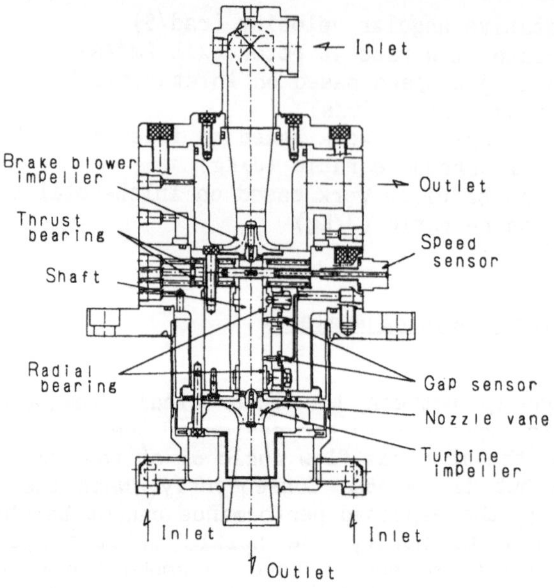

Fig.5. Helium gas expander

V_1, V_{1id} =absolute, ideal absolute velocity at stator exit (m/s)
W_{2m}, W_{2mid}=mean relative, ideal relative velocity at rotor exit (m/s)

Heat Losses due to Thermal Conductivity

Heat losses exist due to thermal conductivity from the axial high temperature side to the turbine low temperature side via the driven shaft, housing wall surface, bolts and spacers, etc. Calculation estimates of heat conductive amount produced values of 25W, 11W and 8W, respectively. Since turbine output≈4KW, the total heat loss is estimated at around 1% of the turbine output.

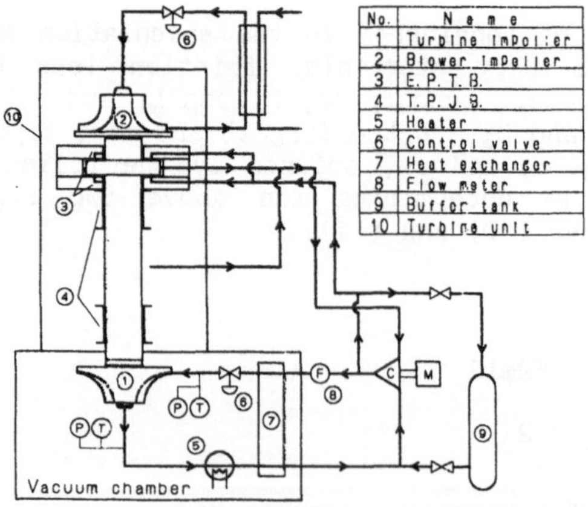
No. Name
1 Turbine impeller
2 Blower impeller
3 E.P.T.B.
4 T.P.J.B.
5 Heater
6 Control valve
7 Heat exchanger
8 Flow meter
9 Buffer tank
10 Turbine unit
Vacuum chamber

Fig.6. Schematic diagram of testing equipment

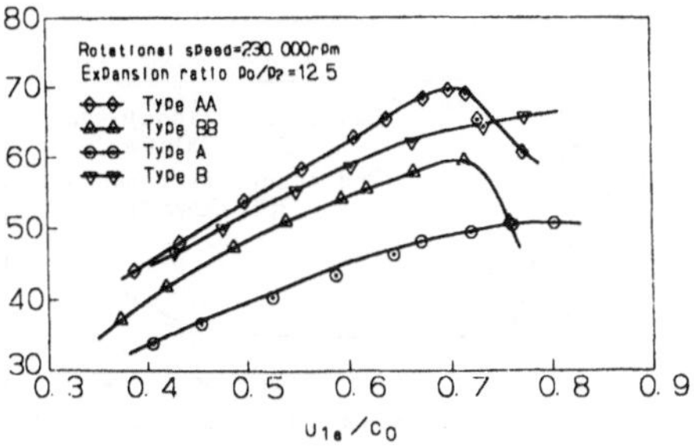

Fig.7. η_{vd} vs. u_{1a}/c_o with various nozzle tested

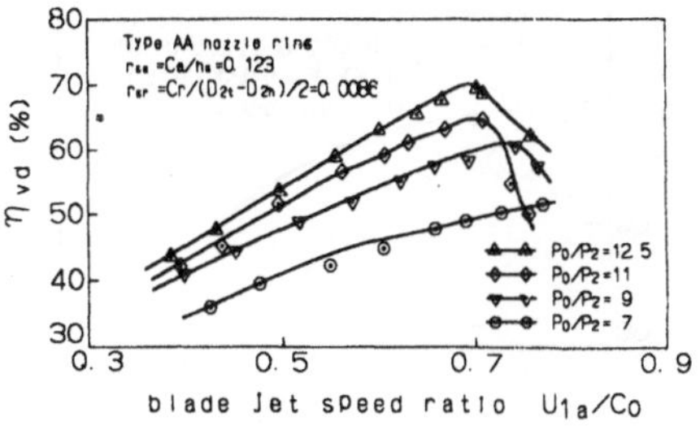

Fig.8. η_{vd} vs. u_{1a}/c_o with expansion ratio varied

Radial and Thrust Bearings

All radial and thrust bearings used in this turbine are gas bearings. The radial bearing is a Tilting Pad Journal Bearing (TPJB) because the vertical shaft does not need support for the weight of the shaft itself and it has good stability for the high revolution speed. On the other hand, the thrust bearing used is an Externally Pressurized Thrust Bearing (EPTB) most suited to supporting the high thrust load resulting from a high expansion ratio. The EPTB must also be a combined type composed of an upper and lower bearing for supporting reversible thrust forces.

Materials of Shaft and Rotor

In designing the shaft-bearing system, prevention of contact damage between the journal and bearing at start-up is very important. Shaft hardness and bearing material selection are therefore important considerations. The experimental machine used hardening heat treated Martensite stainless steel(JIS SUS 440C) for the shaft and a ceramic for the bearing. In consideration of the centrifugal force acting on the shaft collar and the natural bending frequency of the shaft, the collar diameter was set at Ø32mm, the shaft diameter at Ø16mm and the revolution speed at 230,000rpm.

EXPERIMENTAL EQUIPMENT AND METHOD

Figure 5 shows the structure of the helium gas expander used in the experiment and Fig.6 shows a schematic diagram of the testing equipment. Turbine revolution is controlled by the blower impeller 2 shown in Fig.6. The turbine casing body combined with the turbine shaft is connected to the vacuum chamber by a flange. Under normal operation, the opening of the turbine inlet valve 6 should be increased according to the drop in inlet temperature while maintaining constant turbine outlet pressure and revolution speed. After inlet temperature reaches a designated value, designed inlet temperature is maintained using a heater 5. Items measured during operation were the turbine outlet/inlet pressure and temperature, flow rate, revolution speed, and pressure and temperature of the nozzle excit. The vibration characteristics of the rotative shaft were probed using the gap sensor shown in Fig.5.

EXPERIMENTAL RESULTS AND OBSERVATION

Best Nozzle Setting Angle

Figure 7 shows the measurement results of efficiency η_{vd} versus blade-jet speed ratio ν for the four types of nozzle shown in Fig.4. From

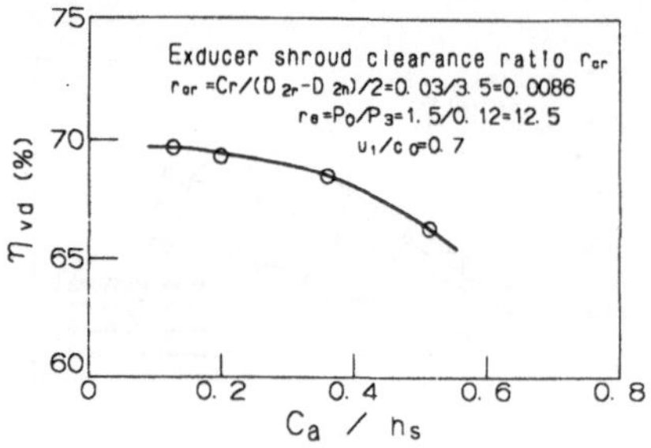

Fig.9. η_{vd} vs. axial clearance ratio c_a/h_s

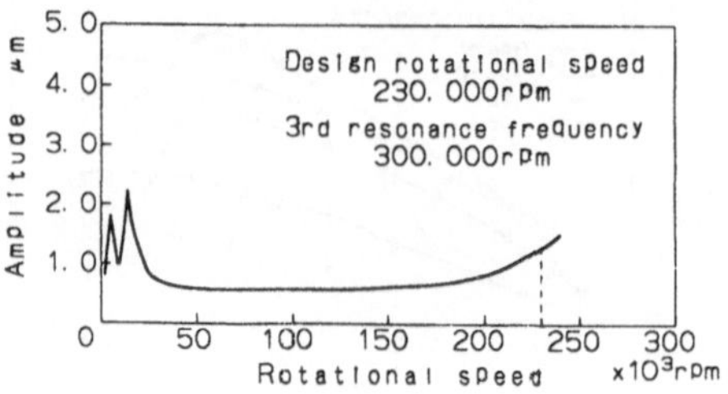

Fig.10. Starting characteristics of turbine shaft

Fig.7 we can see that the order of the efficiencies is AA>B>BB>A. Finally, it was determined that the type AA nozzle with a pressure surface side setting angle α_s adjusted so that the mean flow angle α_m derived from the conformal transformation method equals the theoretical flow angle α_1 shows the best efficiency. Figure 8 shows the relationship between η_{vd} and ν with the expansion ratio considered as a parameter when the type AA nozzle is used.

Maximum Efficiency and Hydrodynamic Losses

Table 1 gives a comparison of design values and measured values at the maximum efficiency point using the type AA nozzle. As the data shows, the efficiency η_{vd} is about 70% at an expansion ratio of 12.5 shown in Fig.8 and Table 1. The reason why the reaction ratio R_r is smaller than the design value may be due to the lower nozzle outlet pressure p_1, which was caused by stator loss coefficient ε_s larger than its design value in the stator passage. On the other hand, the rotor loss coefficient ε_r is smaller than the design value.

In the case of a small-scale turbine with a large expansion ratio, the hydrodynamic loss associated with the stator is unavoidable. Careful design and processing of the stator is therefore essential, taking into consideration the shape, surface smoothness and fluid loss.

Effect of Shroud Clearance

When studying the relationship between efficiency and shroud clearance, the influence of two factors, inlet axial clearance C_a and exit radial clearance C_r, must be taken into consideration. In the case of the very low temperature helium expander, the efficiency is influenced by changes in the axial clearance due to heat contraction of the shaft

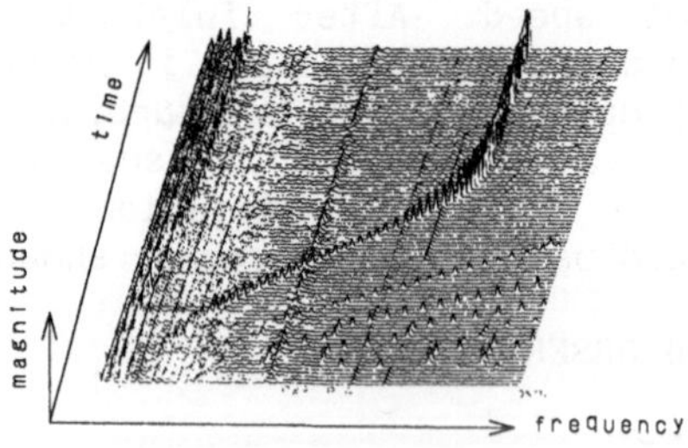

Fig.11. Results of frequency analysis of turbine shaft

and due to changes in thrust load from start-up to rated normal operation. Figure 9 shows the experimental results in this regard. Referring to the literature[7], it is reported that the influence of the axial clearance C_a is lower than that of the radial clearance C_r and that the efficiency drops linearly and gradually according to the increase of axial clearance Ca when the radial clearance C_r is constant. These experimental results show almost the same efficiency curve as that of the literature but the efficiency goes down significantly when the axial shroud clearance ratio C_a/h_s reaches the value more than 0.3. Consequently, it is essential to pay special attention to keep the axial shroud clearance ratio less than 0.3 with small-scale, large expansion ratio and low temperature turbines.

Bearings

Good stability and highly reliable operation were obtained through use of a combination of the TPJB and EPJB. Figure 10 shows the relationship between shaft vibration and revolution speed while Fig.11 gives the results of frequency analysis of the turbine shaft from start-up to rated revolution speed. From Fig.10, the first and second resonance points based on the stiffness of the EPTB and TPJB are evident just after start-up. The natural bending frequency of the shaft is 300,000 rpm and the revolution speed against the third resonance point is more than 30% higher than the rated revolution speed. Concerned about possible damage to the radial bearings when starting and stopping, the authors cycled through start/stop operation 200 times with no discernible affect on the bearings and the satisfactory combination of shaft and bearing materials was duly confirmed.

CONCLUSION

The results noted in the above report can be summarized as follows:

1) It was confirmed that the conformal transformation method is effective in detemining the nozzle setting angle of the helium gas expander for very low temperature usage.
2) With a high expansion ratio turbine, considerable fluid loss at the nozzle is inevitable. Consequently, special care must be taken in design to determine the shape of the nozzle as well as the final finish of the nozzle passage surface to minimize roughness.
3) It was proved that the axial shroud clearance has a significant effect on the efficiency of the small cryogenic turbine with large expansion ratio when the axial shroud clearance ratio reaches the value larger than 0.3 .
4) It was confirmed in a series of start/stop cycles that the combination of the EPTB and the TPJB exhibits very stable and highly reliable performance when used on a high expansion ratio, high revolution speed, very low temperature gas turbine
5) It was confirmed that the combination of hardened shaft and ceramic bearing provides significant improvement in reliability when starting and stopping.

ACKNOWLEDGEMENT

This project is sponsored by the Agency of Industrial Science and Technology of MITI,which commissioned NEDO to promote the work, and was carried out by Super-GM and its member companies. The authors would like to express their appreciation to individuals of the above organizations for their asssistance in the preparation of this paper.

REFFERENCES

1. N. Ino,et al,"Proc. Thirteenth Intl. Cryo. Engr. Conf.", Butterworth, Guilford, UK (1990), p.113.
2. J. C. Villard and F. J. Muller,"Advances of Cryogenic Engineering", Volume19, F-2, Plenum Press, New York (1973).
3. Carroll A. Todd and Sammel M. Futual, Jr., NASA TN D-5059, 3(1969)
4. Harold E. Rohlik, NASA TN D-4384, 2(1968).
5. Georg F. Wislicenus, "Fluid Mechanics of TURBOMACHINERY", McGraw Hill Book Co. Inc, New York and London, 1947.
6. Nagao Mizumachi, "A Study of Radial Gas-Turbines", Report of the Institute of Industrial Science, University of Tokyo, VOL.8, NO.1, 12(1958).
7. Samuel M. Futral Jr., Donald E. Holeski, NASA TN D-5513, (1970).

DEVELOPMENT OF A LARGE CENTRIFUGAL CRYOGENIC PUMP

T.Kato, H.Ishida, E.Tada, T.Hiyama, K.Kawano, M.Sugimoto,
E.Kawagoe, J.Yoshida, Y.Kamiyauchi, H.Tsuji

Japan Atomic Energy Research Institute
Naka, Ibaraki, Japan

N.Saji, H.Asakura, M.Kubota

Ishikawajima–Harima Heavy Industries Co., Ltd.
Koto, Tokyo, Japan

ABSTRACT

The world's largest centrifugal cryogenic helium pump has been developed, which can circulate the supercritical helium at 4 K and 0.9 MPa of the mass flow rate of 350 g/s with the pump head of more than 0.15 MPa at the design condition. The induction motor is adopted as the pump driving system and the pump impeller is made of aluminum precision casting. The bearing system adopts the self–acting gas bearing, which can be free from the shaft seal mechanism and maintenance for long term. Reducing the heat leak from the motor part to the impeller part, the motor system is cooled at 80–K region by liquid nitrogen. And the pump adiabatic efficiency has been attained to be 86% at the rated design condition.

Through the performance test at the room and the cryogenic temperature, the pump has verified a smooth and stable operation up to the maximum revolution of 32,000 rpm, satisfying the design specification. At the cryogenic temperature operation, the maximum pump head of more than 0.2 MPa and the maximum flow rate of 630 g/s were observed.

INTRODUCTION

The design of a fusion experimental reactor such as International Thermonuclear Fusion Experimental Reactor (ITER) requires the forced–flow cooled superconducting coil system cooled by supercritical helium (SHe) for both poloidal and toroidal coils, where the amount of the supplying SHe to the coils shall be from a few thousand to around ten thousand g/s[1)]. Therefore, the large SHe supplying system shall be required.

Two methods are taken into account for supplying SHe, one is the method supplying by the room temperature helium compressor in the refrigerator (R/T compressor method), the other is the method supplying by the cryogenic pump (cryogenic pump method), which is equipped in the liquid helium bath at the cold end of the refrigerator.

The R/T compressor method requires the large helium compressor in the refrigerator and the control of SHe flow shall be complicated due to the connection to the total heat balance of the refrigerator system. On the other hand, the cryogenic pump method will attain easier control due to less effect on the heat balance of the refrigerator and it is verified by the experience of operating the Demo Poroidal Coil (DPC) system[2)]. Though the additional heat load is applied to the refrigerator as a pumping work, the energy efficiency of the

Table 1.The Pump Design Specifications

Type	Centrifugal (Single Stage)
Fluid	Supercritical Helium (SHe)
Inlet Pressure	0.9 MPa
Inlet Temperature	4.0 K
Pump Head	0.15 MPa
Mass Flow Rate	350 g/s
Rotating Speed	26,000 rpm
Motor	Three-phase Induction Motor
Bearing	Foil Type Self-acting Gas Bearing

cryogenic pump method is better than that of the R/T compressor method if the adiabatic efficiency of the pump is better than 20%[3)]. Therefore, the cryogenic pump method will be adopted for the SHe supplying system of the fusion experimental reactor.

Japan Atomic Energy Research Institute (JAERI) and Ishikawajima-Harima Heavy Industries Co., Ltd. (IHI) have successfully developed the large centrifugal cryogenic helium pump, aiming to develop such a large SHe supplying system. The developed pump has a size of around one fifth for the pump required in the fusion experimental reactor since it would be used for the pump of Demo Poloidal Coil Facility (DPCF)[2)], where the SHe flow of around 350 g/s is required. Being carried out the performance test in DPCF at the cryogenic helium condition, the mechanical and the fluid-dynamic performance of the pump have been measured. The design and construction and the performance test results are described in the following section.

DESIGN AND FABRICATION

Basic Design

The design specifications of the pump are determined as listed in Table 1, corresponding with the requirements for the DPCF specifications[2)]. This pump adopts the centrifugal flow pump with a single stage. As a pump driving force, an inductive motor is adopted, having the 3.5-kVA power supply with high-frequency inverter to produce the maximum revolution of 33,000 rpm.

The basic design parameters such as impeller diameter and revolution are determined by applying the similarity method on a turbo-machine. There is generally strong correspondence between impeller efficiency and specific speed (Ns). Applying such a relation, the specific speed is determined to be 345 under the unit of (rpm, m, m^3/min), expecting to have high impeller efficiency of more than 80% in the range of 300 - 400. Head coefficient is also assumed to be around 0.4. Though the rated design mass flow rate is 350 g/s, the mass flow rate of 455 g/s, being larger than the design flow rate by 1.3 times, is selected as the flow rate at the point of the maximum impeller efficiency in order to maintain the high efficiency operation in the large flow region. Assuming that the flow of the supercritical helium is incompressible in the pump operating region, the design pump head is determined. From these parameters, the design revolution and the impeller diameter are determined to be around 26,000 rpm and 37 mm, respectively.

The cross-sectional view of the pump and the photograph of the pump rotor are shown in Fig. 1 and Fig. 2, respectively.

Impeller

The impeller, which is made by aluminum precision casting, has four blades and is equiped with a shroud cover to reduce leakage loss.

Rotor

As shown in Fig. 2, the center part of the rotor is used for the motor part, the both sides of the motor part are supported by the journal bearings, and at the both ends, the pump impeller and the thrust bearing are installed, respectively. The material of the rotor shaft

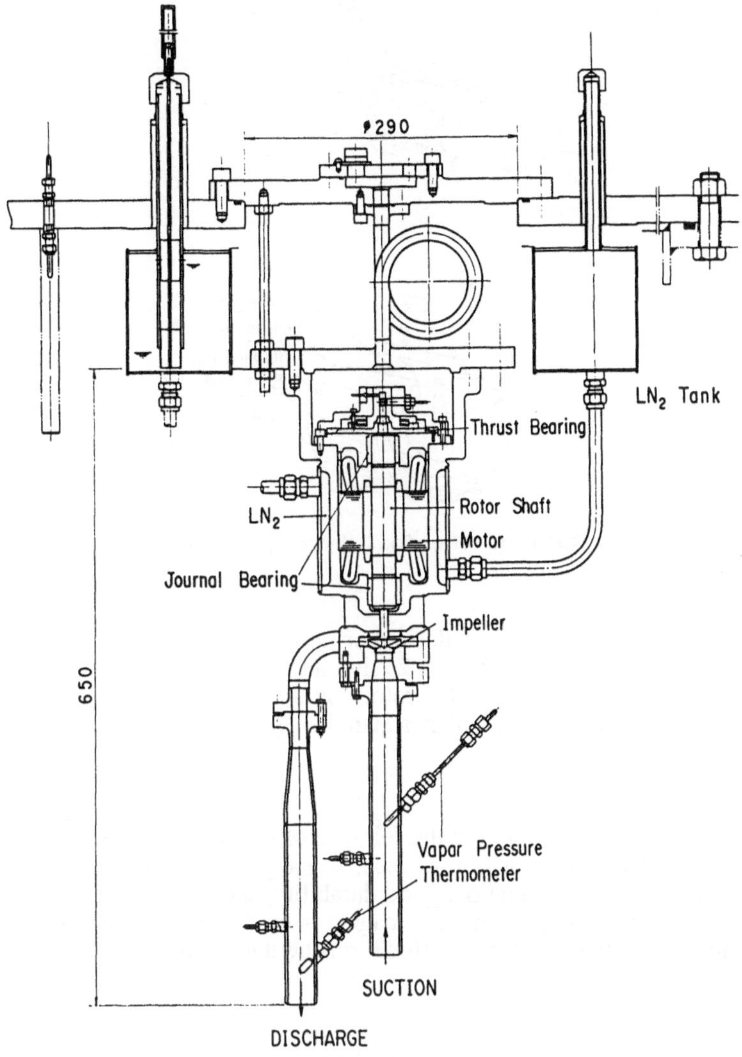

Fig. 1. The cross-sectional view of the pump

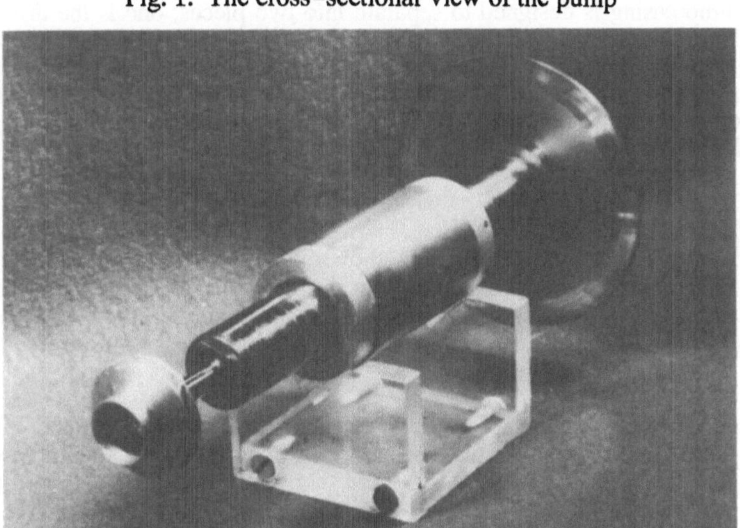

Fig. 2. The photograph of the pump rotor

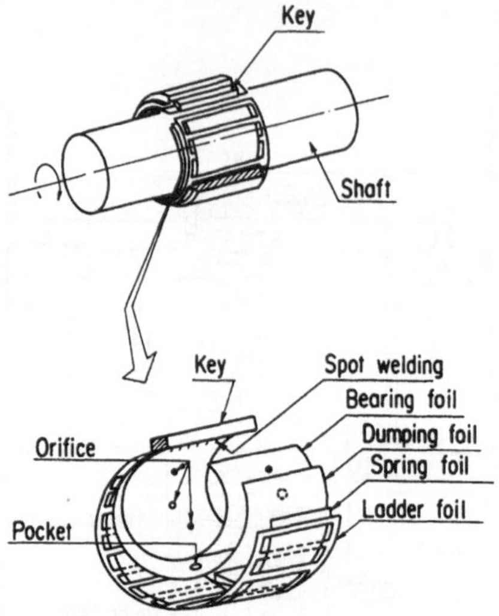

Fig.3. Detail of the foil type journal bearing

adopts austenitic stainless steel except the motor part which is martensitic stainless steel. The shaft diameter at the journal bearing is 25 mm and the shaft length is 265 mm, respectively. Analyzing the critical speed for the rotor, the third critical speed is calculated to be around 100,000 rpm, which is enough higher than the design revolution of 26,000 rpm.

Bearing

The bearings of the pump adopts the foil type of self–acting dynamic gas bearing for both the journal and the thrust bearings. A detail of the journal bearing is shown in Fig. 3, bringing the stable high speed operation, good durability, and easy adjustment. In the thrust bearing system, the ring–shaped permanent magnet is used to pull up the thrust disk of the 100–mm diameter, occurring a large frictional torque due to the rotor dead weight when initiating the rotation.

Pump Casing

The pump casing is designed to separate into two pieces, one is the driving system section, containing the motor, and the other is impeller section, where the impeller and the cryogenic communication pipings are located. The driving section is cooled by liquid nitrogen and both sections are connected with thin piping; accordingly the heat leak from the driving section to the impeller section is decreased to be less than 2 W, attaining the high adiabatic efficiency.

TEST FACILITY AND METHOD

The pump was tested at the DPCF as shown in Fig. 4, where the closed circuit loop was formed in the pump vacuum chamber, closing the valve (1) and (2) to separate from the coil system (the existing vacuum chamber) and controlling the pump load such as flow rate and pump head by the valve (3). In the initial condition, the pressure of the test circuit was fixed by controlling the valve (4).

When determining the pump adiabatic efficiency, delicate mesurments of temperature were required since the temperature difference through the pump should be very small due to the large pump flow rate. Vapor–pressure thermometers were used and installed at the inlet and the outlet piping.

In the vapor–pressure thermometer system, the volume ratio between the cryogenic sensor part and the room temperature pressure gauge part was designed to be around 200 at

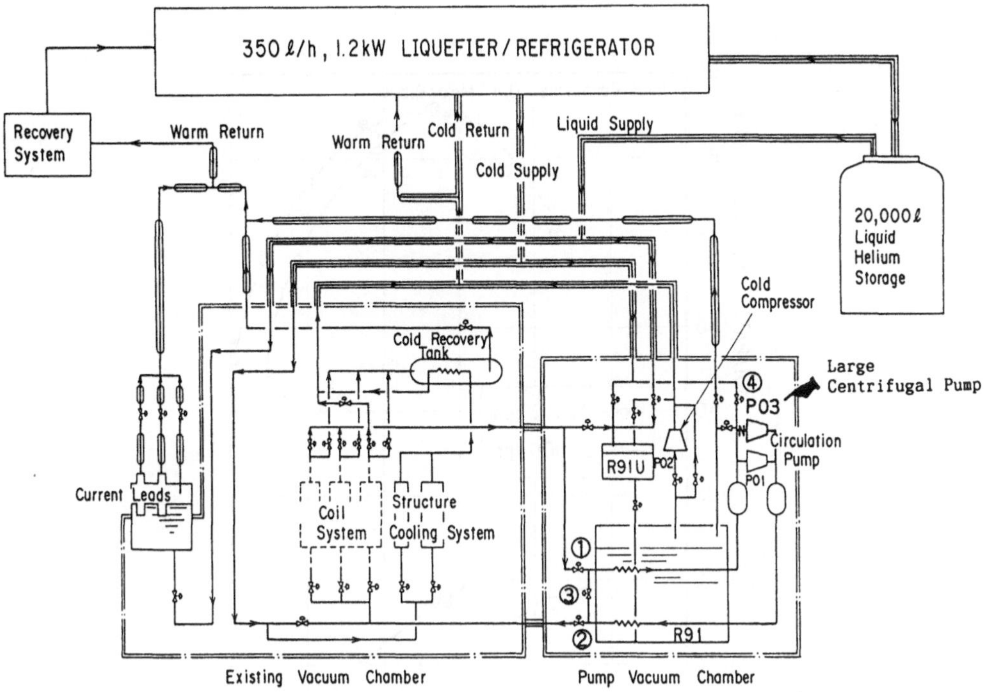

Fig. 4. The schematic flow diagram of the pump test circuit

the inlet temperature and 300 at the outlet temperature, respectively. The initial filling–up pressure was set to be 0.2533 MPa (2.5 atm), paying much attention of leak from the pressure gauges.

The germanium resistance thermometer located in the liquid helium bath was used for the calibration and the temperature measurements were performed after being agreed both the germanium and the vapor–pressure thermometer.

An orifice–type flow meter, having the accuracy of $\pm$ 5%[4], was used for mass flow measurement, where the helium density calculation was based on NBS 631. The performance test was conducted under the temperature of around 4.5 K and the pump inlet pressure of around 0.5 MPa.

PERFORMANCE TEST RESULTS

General Performance

The adiabatic efficiency and the pump head characteristics are shown in Fig. 5 as a function of mass flow rate. The inlet conditions were the pressure of 0.47 MPa and the temperature of 4.5 K at both the revolution of 27,000 rpm and 29,000 rpm, respectively.

The design specification was achieved at the revolution of 27,000 rpm which is larger than the design revolution of 26,000 rpm by 1,000 rpm. And efficiency of 86 % was achieved at the design mass flow rate of 350 g/s with the design pump head of 0.15 MPa.

When rotating the pump at 32,000 rpm, the maximum mass flow rate was measured to be 630 g/s with the pump head of around 0.1 MPa. The maximum pump head is restricted to be 0.2 MPa due to the thrust bearing design and the minimum controlled mass flow rate was determined to be around to be 100 g/s with the pump head of 0.2 MPa at 27,000 rpm. However, surging phenomenon was not observed in their flow ranges.

The pump heat load was also determined as a function of mass flow rate as shown in Fig. 6. The heat load at the design point was measured to be around 430 W.

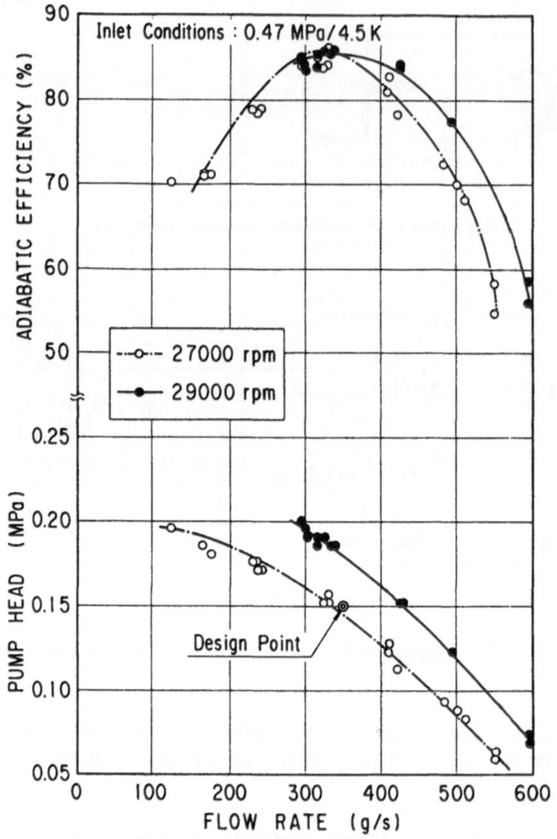

Fig. 5. The adiabatic efficiency and the pump head characteristics as a function of mass flow rate

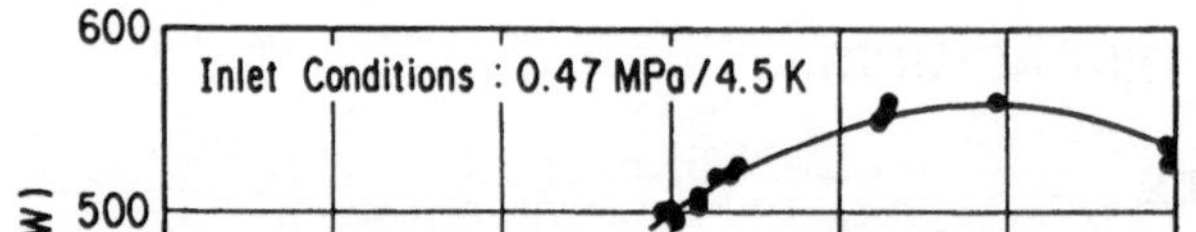

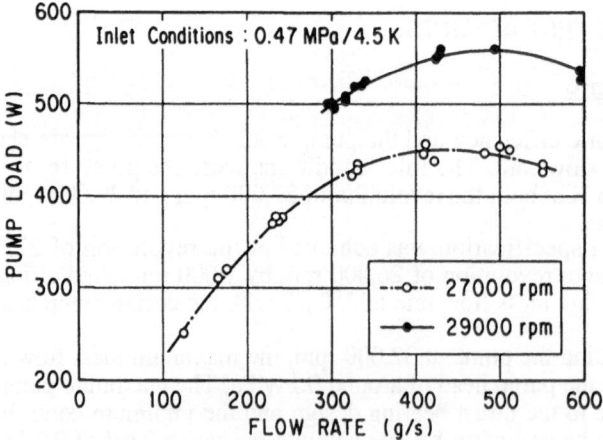

Fig. 6. The pump heat load characteristics as a function of mass flow rate

Through the performance test, the start/stop operation of more than 10 times and the total operation of around 50 hours were conducted. Any instability in the operation was not observed.

Adiabatic Efficiency

The adiabatic efficiency of the cryogenic pump is significant factor when the pump is adopted in the cryogenic system in the fusion experimental reactor as mentioned in the introduction.

The pump adiabatic efficiency is defined as follows:

$$\eta = \Delta H / \Delta H_{ad}$$

where the ΔH indicates the actual enthalpy rise by calculating from the measured inlet and outlet helium conditions and the ΔH_{ad} is the isentropic enthalpy rise by calculating from the inlet helium condition. Accordingly, the ΔH shall include heat leak though the pump motor part cooled by liquid nitrogen. Such a liquid nitrogen consumption, however, is not considered in the efficiency.

Through the experiments, the very high efficiency performance was observed, showing that the efficiency of more than 70 % was attained in the large region of the mass flow rate from 100 g/s to 500 g/s.The pump efficiency generally depends on the fabrication accuracy in the impeller shape and its surface roughness because the accurate fabrication and small surface roughness of the impeller produce the smooth distribution of flow velocity and pressure through the impeller.

In case of a water pump, the impeller diameter will be around 1 m if the pump is used in the water stream with the same Reynolds number as the cryogenic pump. In such a large diameter, the impeller can be fabricated with high accuracy and small surface roughness by using the conventional fabrication technique, resulting in the high efficiency of 90 % at the actual water pump.

On the other hand, in case of the cryogenic pump, the impeller diameter is much smaller than that of the water pump due to the cryogenic helium property. Our developed pump has only the impeller diameter of 37 mm which is around 1/30 of the water pump. Therefore, it is very difficult that the fabrication accuracy and the surface roughness should be relatively similar to those of the water pump.

To resolve such a issue, the aluminum precision casting method, in which the cast was fabricated within the tolerance of 2/100 mm, was used to fabricate the impeller. And surface roughness could be achieved to be less than 6.3 μm. Thus beautiful pump efficiency performance could be achieved.

CONCLUSION

JAERI and IHI have successfully developed the large centrifugal cryogenic pump, which will be available for the large supercritical helium supplying system in the fusion experimental reactor with scaling up. The pump performance test has given the beautiful pump characteristics as a function of the mass flow rate, appearing the maximum efficiency of 86% at the design condition. And the smooth controllability has also been verified.

ACKNOWLEDGMENT

The authors would like to thank Drs. M. Yoshikawa, T. Iijima, and S. Shimamoto for their continuous encouragement on this development. The contribution for the computer data acquisition system from Mr. M. Sato at Ishikawajima-Harima Heavy Industries Co., Ltd. is greatly acknowledged.

REFERENCES

1. ITER CDA magnet group. "ITER Magnets," ITER Documentation series, No. 26, ITER Vienna (1990).
2. T. Kato et al., "OPERATION PERFORMANCE OF DPCF IN THE TEST OF THE Nb–Ti DEMO POLOIDAL COILS (DPC–U1,U2)," Proc. MT–11, Tsukuba (1989) p. 1350.
3. T. Kato et al., "CRYOGENIC SYSTEM DESIGN AND ITS COMPO NENT DEVELOPMENT FOR FUSION EXPERIMENTAL REACTOR (FER)," Proc. of 16th Symposium on Fusion Technology, London, (1990).
4. M. Yoshiwa et al,. "CALIBRATION SYSTEM OF HEAD TYPE FLOW METER IN CRYOGENIC TEMPERATURE," Proc. ICEC9, Kobe (1982) p. 616.

ANALYSIS OF A LOW-TEMPERATURE MAGNETIC HELIUM PUMP

Coyne Prenger and Walter Stewart

Los Alamos National Laboratory
Los Alamos, New Mexico

ABSTRACT

In an effort to improve reliability of cryocoolers, concepts involving no moving parts are being investigated. One concept utilizes an Active Magnetic Regenerator, AMR, to produce refrigeration. However, circulation of the helium working fluid is required for operation of the device. Currently available helium pumps have moving parts and; therefore, result in poor reliability.

We propose a magnetically driven pump to provide the helium circulation for the AMR. The pump utilizes the magnetocaloric effect to produce an oscillatory helium flow and; has no moving parts. An analytical model has been developed to analyze the pump's performance in conjunction with an AMR operating between 7 and 20 K. At a frequency of 1 Hz a 0.5 liter pump can produce a 0.75 g/s flow rate at 20 K at an operating pressure of 5 atm.

At the liquid helium temperature a two-phase version of this pump would perform substantially better than the single-phase version. A design concept has been developed and will be presented along with the model results.

INTRODUCTION

Recent advances in the development of active magnetic regenerators (AMR) for use in low-temperature refrigerators have provided an opportunity for reducing the number of moving parts in cryocoolers.[1] Studies have shown[2] that a low-temperature magnetic stage coupled to a two-stage gas refrigerator provides a higher efficiency than an all gas system when spanning the 7 to 300 K temperature range. The potential exists for eliminating all moving parts from the low-temperature stage if the magnetic field can be cycled by charging and discharging the magnets and if a no-moving-part pump is available to circulate the helium working fluid. Operation of an AMR refrigerator[3] requires an oscillating flow of helium through the regenerator. A typical low-temperature stage would operate between 7 and 20 K and would require a helium flow rate of approximately 1 g/s at a frequency of 0.1 Hz. This would provide a cooling power of 1 W at 7 K. With these specifications a concept for a no-moving-part helium pump was developed.

PRINCIPLE OF OPERATION

Because of the need to eliminate moving parts from the cryocooler, a thermally driven pump was considered. As a natural complement to the AMR, the use of a magnetic system to provide the heat source was investigated. A schematic of the AMR refrigerator is shown in Fig. 1. The system consists of an AMR, operating between 7 and 20 K; two heat exchangers, one at 20 K and one at 7 K; a magnetically driven pump; and a surge tank.

The magnetic pump consists of an 0.25-L canister packed with uniform spheres of magnetic material (for this temperature range, either Gadolinium Gallium Garnet (GGG) or Disprosium Aluminum Garnet (DAG) is suitable) with a void fraction of 0.416 (for uniform packing). The canister is inserted into the bore of a superconducting magnet. The magnet is cycled to produce an oscillating magnetic field. The resulting magnetocaloric effect raises and lowers the temperature of the helium gas in contact with the magnetic material. Because the helium is in a closed system, the temperature change creates an oscillating pressure within the canister. By connecting a surge tank at the opposite end of the system, helium can be made to flow from the canister, through the heat exchangers and regenerator, and into the surge volume, upon magnetization, then from the surge volume, back through the system, and into the canister, upon demagnetization. This then accomplishes the required oscillatory helium flow. Excess heat produced by the pump is removed by the high-temperature heat exchanger and is rejected to the heat sink. No additional heat load is added to the AMR. In addition to analyzing the magnetic pump performance when the pump is located at the hot end of the AMR, as described above, the performance was also determined when locating the pump at the cold end of the AMR. In the latter case, the heat load from the pump must be absorbed by the AMR, and it is an additional heat load. A third possibility was also considered. It involved extending the temperature of the AMR to 4 K to permit a two-phase helium pump to be used at the cold end. In this case, the helium expansion occurs at constant tem-

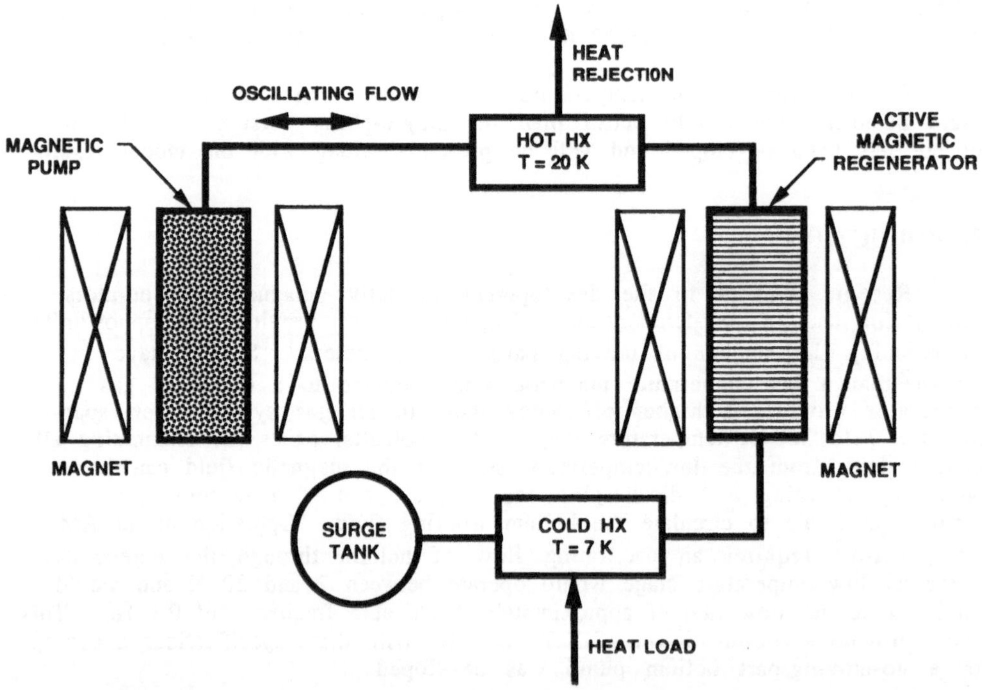

Fig. 1. Schematic of no-moving-part AMR refrigerator with magnetic pump at the hot end.

perature and no additional heat load is added to the AMR. The energy added by the pump is stored as latent heat. The two-phase system provides the highest pump performance with the lowest impact to the AMR. However, the AMR temperature range must be extended to 4 K to permit two-phase operation. Extension of the AMR's operating temperature range requires an increase in system capacity. An analytical model was developed to analyze pump performance for each configuration discussed.

ANALYTICAL MODEL

The analytical model assumes that the helium working fluid behaves as an ideal gas. The heat capacity for helium is included in the model as a function of temperature. The system volume was 0.5 L and includes the regenerator, two heat exchangers, the surge tank and the connecting tubing. The system volume was assumed to be twice the pump volume. In addition, the volume of the surge tank was assumed to be large compared with the other components. The magnetic field was assumed to be uniform throughout the pump canister, and the magnetization was assumed to be adiabatic. Thermal addenda for the helium was included, but thermal addenda for the canister was neglected--a reasonable assumption for operating temperatures below 20 K.

The model is a thermodynamic description of the process and does not include heat transfer or pressure drop effects. The temperature difference between the helium working fluid and the magnetic material is neglected and the flow resistance is assumed to be negligible. A finite heat transfer coefficient will degrade the pump performance from that predicted by the model, especially at a high frequency, when the helium residence time in the pump canister is reduced. Estimates of heat transfer coefficients and flow resistance and their effect on the analysis have not been addressed.

The model was used to calculate the helium mass expelled from the canister during each magnetic cycle. This can be converted into a mass flow rate (g/s) by multiplying by the cycle frequency in Hertz. Therefore, for a frequency of 0.1 Hz, the flow rate in g/s is one-tenth the mass change for the magnetic cycle. The pump performance results are presented in terms of mass change so they will be independent of operating frequency, which is determined by other factors such as eddy current heating and magnet charging efficiency.

RESULTS

Performance results are presented for the three system configurations discussed above. In all cases the volume of the pump canister was limited to 0.25 L because of constraints on magnet size.

Figure 2 shows results for the helium pump located on the high-temperature side of the regenerator. In this case the heat load from the pump is rejected to the 20-K heat sink, with no additional load on the AMR. Mass change is given as a function of magnet field strength, with maximum canister pressure as a parameter. The maximum canister pressure is a measure of the initial charge of helium in the system. The larger the initial charge, the higher the maximum pressure, and the larger the mass flow. As shown in Fig. 2, the mass change is less than that required for 1 g/s at a frequency of 0.1 Hz.

Relocating the pump to the low-temperature end of the system results in a significant increase in pump output, as shown in Fig. 3. For this case, flow rates near 1 g/s at a frequency of 0.1 Hz can be obtained. However, the pump load is added at the low-temperature end, resulting in additional load for the AMR. This additional load decreases the cooling power which offsets the improved pump performance.

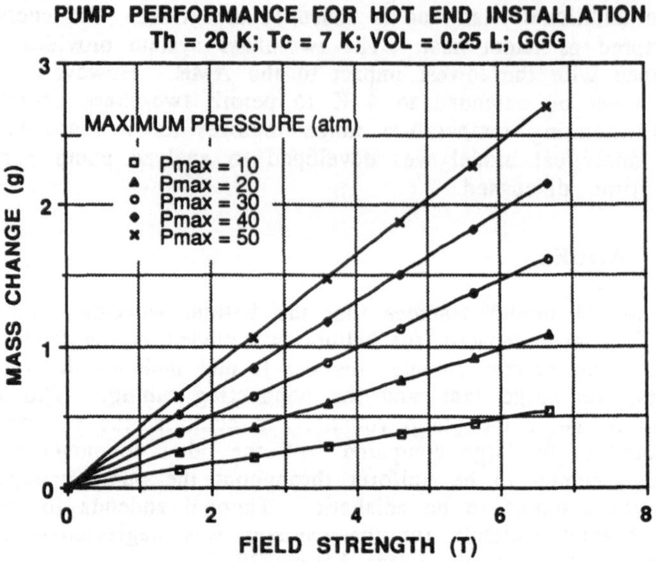

Fig. 2. Performance of magnetic pump installed at the hot end with maximum system pressure as a parameter.

By extending the AMR temperature down to 4 K and operating the magnetic pump in the two-phase region a significant pump output can be obtained with no additional load to the AMR. The helium expansion occurs isothermally by a phase change, and the pump energy is stored as latent heat within the canister. Figure 4 shows the pump performance for this case. A flow rate of 0.7 g/s can be obtained at a frequency of 0.1 Hz, with an operating pressure of approximately 1 atm. The improved pump performance is attained at the expense of extending the low-temperature of the AMR to 4 K.

CONCLUSIONS

- The no-moving-part helium pump is complementary to the AMR because of similarities in working material and magnet design.

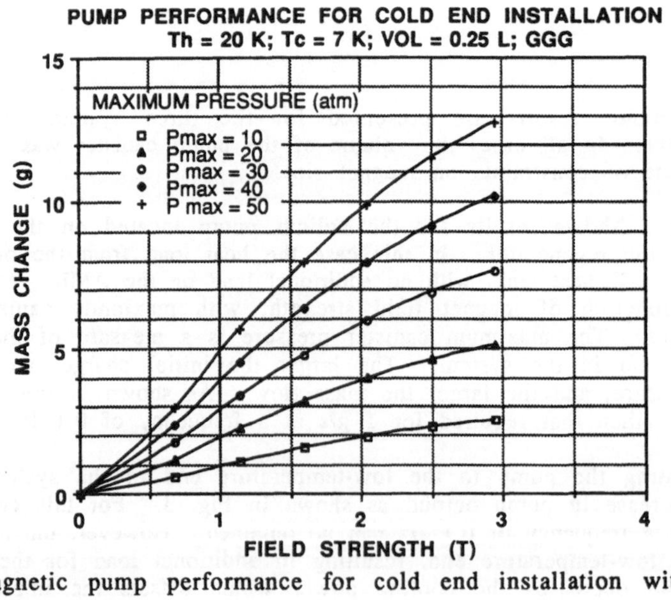

Fig. 3. Magnetic pump performance for cold end installation with pressure as a parameter.

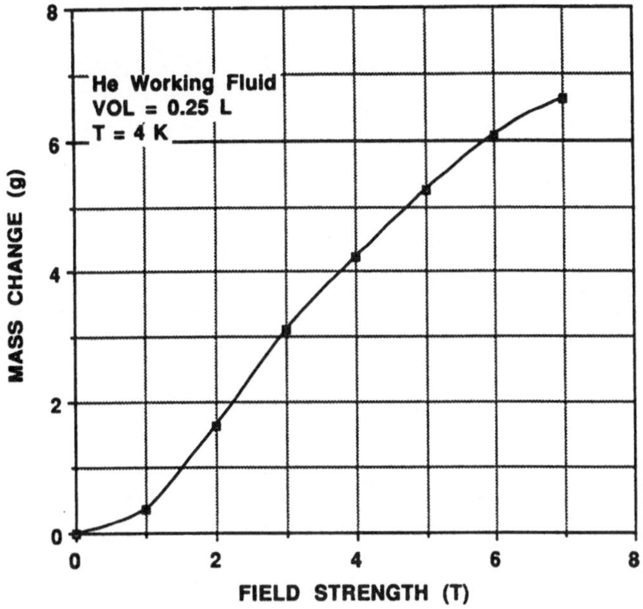

Fig. 4. Performance of a two-phase helium pump at 4 K.

- For the single-phase pump at 20 K, flow rates of 1 g/s at 0.1-Hz frequency can be obtained only with canister volumes approaching several liters if operating pressures are limited to 50 atm.
- Operation of a single-phase pump on the cold end of the AMR is not feasible because of the additional sensible heat load on the helium working fluid, which reduces cooling power.
- A two-phase helium pump operating at 4 K provides adequate helium flow with an 0.25-L canister volume and at low operating pressure. This requires extending the operating temperature of the AMR to 4 K.

ACKNOWLEDGEMENTS

This study was funded by the Air Force Space Technology Center through the David Taylor Research Center. We acknowledge the support of Geoffrey Green, the program manager.

REFERENCES

1. F.J. Cogswell and J. L. Smith, Jr, "Appropriate Thermodynamic Cycles for Static Magnetic Refrigeration," Fourth Interagency Meeting on Cryocoolers, Plymouth, MA (October, 21, 1991). (Restricted distribution)
2. J. A. Barclay and F. C. Prenger, "Operational Envelope for Magnetic Refrigerators," Advances in Cryogenic Engineering, **35B**, R. W. Fast, Ed. (Plenum Press, New York, 1990), pp.1097-1104.
3. C. P. Taussig et al., "Magnetic Refrigeration Based on Magnetically Active Regeneration," in Proc. of the Fourth International Cryocoolers Conf., Easton MD (1986), pp 79-88.

RECENT PROGRESS ON RARE EARTH MAGNETIC REGENERATOR MATERIALS

T. Hashimoto, M. Ogawa, A. Hayashi and M. Makino

Tokyo Institute of Technology, Department of Applied Physics
Oh-okayama, Meguro, Tokyo, Japan

R. Li and K. Aoki

Hiratsuka Lab., Sumitomo Heavy Industies, Ltd.
Yuuhigaoka, Hiratsuka, Kanagawa, Japan

ABSTRACT

Our group has reported several candidates for magnetic regenerator materials in the helium temperature range, such as, $(Er_{1-x}Dy_x)Ni_2$, $Er(Ni_{1-x}Co_x)_2$, ErNi and Er_3Ni. However, those Curie temperatures T_c's are more than ∿5.5 K and the specific heats C(T)'s of those systems become very small near 4.2 K. In the present investigation we have succeeded in developing a new low T_c material, $Er_{0.9}Yb_{0.1}Ni_2$ whose T_c is ∿4 K. Moreover, other new regenerator materials, such as $Er_{1-x}Yb_xNi$ and $Er_{1-x}Ho_xNi_2$ have also been investigated and usefulness of those compounds was verified. We have also studied the origin of the small C(T) of some Er compounds near 4.2 K and it has been made clear that the Schottky type specific heat reduces the absolute value of C(T) in low temperature range.

INTRODUCTION

To develop effective regenerator materials in low temperature range near 4.2 K, we have investigated some heavy rare earth magnetic compounds since 1985[1-16].

Presently, lead has been used as a regenerator material, which has several suitable physical properties for application as a regenerator. However, below ∿10 K the specific heat of the lead metal that originated from the thermal vibration of lattice decreases in proportion to T^3 and, as a result, the lead becomes ineffective for regenerator operation.

In the temperature range below ∿10 K, the most promising physical mechanism which gives a large specific heat is a magnetic phase transition from the disordered phase to the ordered one. On the basis of the phase transition theory it was concluded that the heavy rare earth compounds are hopeful for the regenerator material and several kinds of the magnetic systems have been investigated by our group[1-9,13-16].

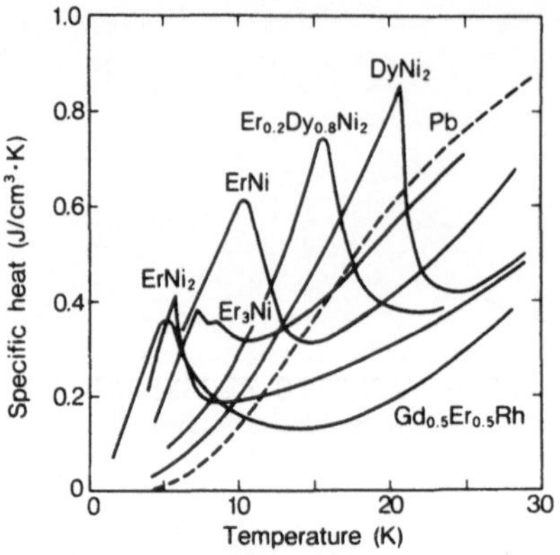

Fig.1 Specific heats of several kinds of the promising magnetic regenerator materials.

we have found several promising candidate magnetic systems applicable for the regenerator near and below ~10 K. Figure 1 shows typical examples of the specific heats of those systems, $Er_{1-x}Dy_xNi_2$, ErNi, Er_3Ni and $Gd_{0.5}Er_{0.5}Rh$[17], below 20 K. It is clearly shown that the specific heat of those systems is fairly larger than that of Pb below 10 K.

In the present paper we will show some experimental results of new magnetic materials which have the low Curie temperature T_c and also discuss the characteristics of the magnetic specific heats of Er compounds.

NEW LOW T_c MAGNETIC MATERIALS

In rare earth compounds the magnetic exchange interaction is the s-f interaction and, therefore, their Curie temperatures are mainly governed by the following two factors: $(g_J-1)^2J(J+1)$(called de Gennes factor) and the sinusoidal function F(x). Here, we noticed the first factor.

In Fig.2 are shown the rare earth element dependence of the factor $(g_J-1)^2J(J+1)$ and the experimental results on the rare earth element R dependence of T_c in the RNi_2 system. This figure clearly shows that the Curie temperature T_c depends on the factor $(g_J-1)^2J(J+1)$ in RNi_2 system. Therefore, in order to obtain low T_c material, we have used the $(Er_{1-x}Yb_x)Ni_2$ system.

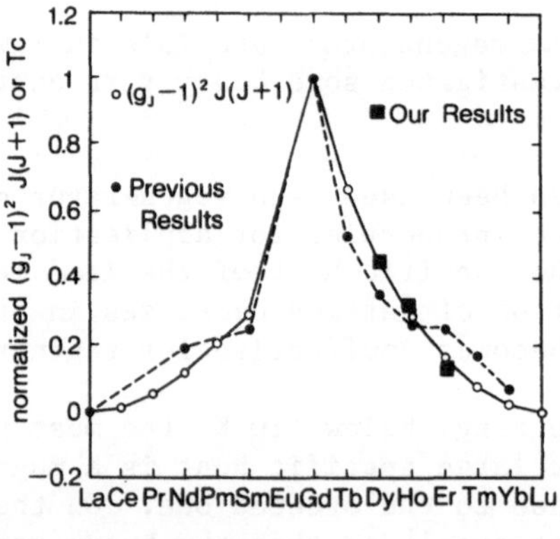

Fig.2 The rare earth (R) dependence of T_c of the RNi_2 system. Simultaneously, R dependence of the factor $(g_J-1)^2J(J+1)$ is shown. Those two kinds of R dependences almost coincide with each other.

Table 1. Curie Temperature T_c of the $(Er_{1-x}Yb_x)Ni_2$ System

concentration x	0	0.05	0.1
T_c from specific heat	5.68	4.55	4.07
T_c from magnetic measurement	6.20	5.26	5.10

The specific heats C(T) of $Er_{0.9}Yb_{0.1}Ni_2$, and that of $ErNi_2$ were measured. In $Er_{0.9}Yb_{0.1}Ni_2$, the peak near T_c in the temperature dependence curve of C(T) is fairly broad in comparison with that in $ErNi_2$. This broad peak is originated from the slightly inhomogeneous distribution of Yb ions in sample and, however the low T_c(T_c=4 K) can be actualized as shown in Table 1.

The substitution effect of Yb has been confirmed in another system. The ErNi has a larger magnetic specific heat in comparison with $ErNi_2$. Therefore, we have observed the specific heat of $Er_{0.9}Yb_{0.1}Ni$. In this case the T_c decreases from 10.5 K to 8 K by substitution of Yb(10%) for Er and, however, the absolute value of $C_J(T)$ does not decrease as shown in Fig.3.

Another experiment has been performed. Since a Holmium atom has the largest J-value which is an indispensable condition for selection of the regenerator material, the specific heat of the $(Er_{1-x}Ho_x)Ni_2$ system has been measured. The results are shown in Fig.4 and are very similar to those of the $(Er_{1-x}Dy_x)Ni_2$ system in concentration x dependence and in the large specific heat.

CONSIDERATION

We have shown several kinds of experimental results which verify usefulness of the low T_c magnetic material for the regenerator of a helium liquefier. A typical example is shown in Fig.5. This result is not our champion data and every technician can obtain it by exchanging the Pb particles for Er_3Ni powders in the 2nd stage regenerator of the common GM refrigerator.

The variation of the refrigeration capacity against the lowest refrigeration temperature shown in Fig.5 suggests two kinds of remarkable things. First, in this experiment only one kind of magnetic material Er_3Ni is used as the regenerator material and, therefore, it is presumed

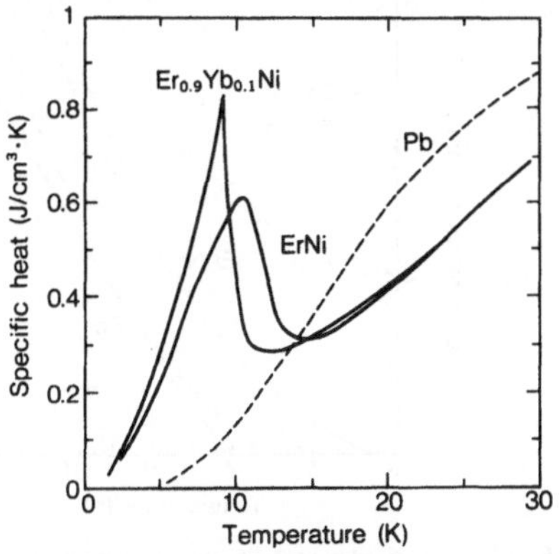

Fig.3 Temperature variations of the specific heats of $Er_{0.9}Yb_{0.1}Ni$ and ErNi. For comparison, the specific heat of Pb is also shown.

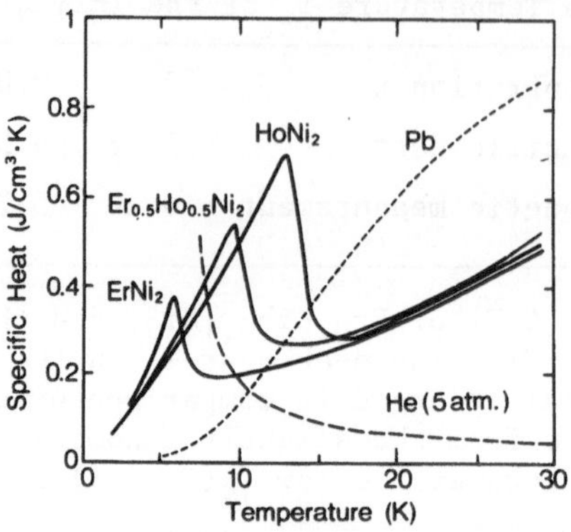

Fig.4 Temperature variations of the specific heats of the $Er_{1-x}Ho_xNi_2$ system. For comparison, the specific heats of Pb and He-gas (5 atoms) are also shown.

that the magnetic peak of Er_3Ni that appeared in the C(T) curve is very broad in comparison with a usual peak appearing in a typical magnetic phase transition temperature, T_c. Second, the lowest temperature is in a 3 K range, but the refrigeration capacity is very small near 4.2 K. This result could be explained by the fact that the heat capacity of Er_3Ni below 5 K is very small in comparison with that of He gas, and, therefore, the regenerator cannot operate effectively in this range.

So, first we will consider the origin of two kinds of thigs mentioned above and then consider how to cope with the small C(T) below ∿5 K. When the magnetic ion, such as Er, whose orbital moment is not zero is put in the anisotropic crystalline field, the energy states of spins are usually split. As a result, an anomaly (called the Schottky anomaly) is observed in the temperature dependence of C(T), which is a very broad peak of C(T) in comparison with that of the magnetic phase change. Therefore, the specific heats of those compounds are expressed by the superposition of those two kinds of C(T)'s, that is, the specific heat of the Schottky type $C_S(T)$ and that due to the magnetic phase change $C_J(T)$.

An example is shown in Fig.6. The crystal structure of $ErNi_2$ is a cubic Laves structure. Er ions are put in an anisotropic field and as the result, the spin levels in Er ions are split. Assuming that the spin states are split into two levels and the energy gap of those two is Δ, the specific heat $C_S(T)$ is expressed by

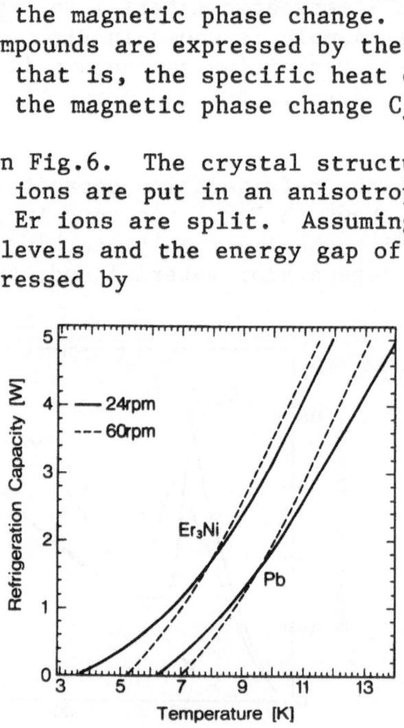

Fig.5 Variations of the refrigeration capacity against the lowest temperature of 2nd stage regenerator in G-M refrigerator in which Er_3Ni powders or Pb powders are filled.

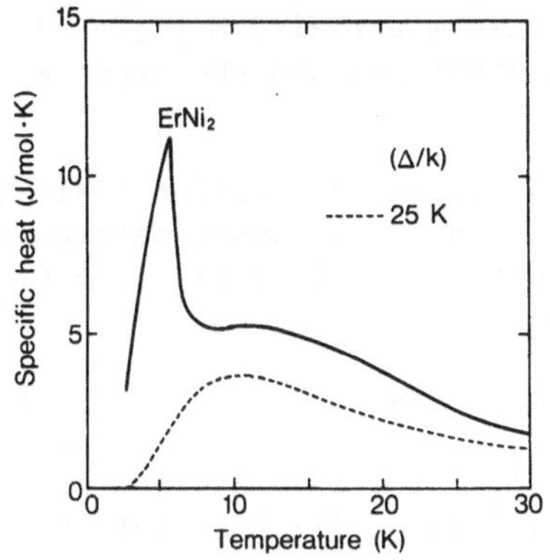

Fig.6 The magnetic specific heat of $ErNi_2$. The dotted line shows the Schottky anomaly, where we assume (Δ/k) is 25 K.

$$C_S(T) = R\left(\frac{\Delta}{kT}\right)^2 \frac{d_1}{d_0} \cdot \frac{\exp(-\Delta/kT)}{[1+(d_1/d_0)\exp(-\Delta/kT)]^2}$$

where R is the gas constant, k is the Boltzmann's constant, d_0 and d_1 are the degrees of degeneracy of the ground and the excited state.

In $ErNi_2$ d_0 and d_1 are 4-fold degeneracy and the (Δ/k) is 25 K. Using these values, we obtained $C_S(T)$ as the dotted line in Fig.6. This figure clearly shows that the total specific heat of $ErNi_2$ is expressed by the superposition of $C_S(T)$ and $C_J(T)$.

The specific heat of the Er_3Ni compound is also explained by the similar method. The large specific heat in the temperature range above T_c is originated from the Schottky anomaly, where Δ/k is more than 30 K. Since the resultant moment, which contributes to the entropy change at T_c, is not so large, the peak of $C_J(T)$ is thought to be very small.

On the basis of the above discussion, in order to get the large refrigeration power in the low temperature range, we have to select the large C(T) magnetic material which has no $C_S(T)$. However, in these materials the temperature width of $C_J(T)$ peak is very narrow below ∿10 K, and, therefore, the layer-structural regenerator[1] must be used and in this case the magnetic material whose $C_J(T)$ peak is very large near 4.2 K is necessary.

One of the promising cnadidates is $Er_{1-x}Yb_xNi$ system shown in Fig.2. The crystal structure of this compound is comparatively simple and, therefore, $C_S(T)$ is almost zero. As a result, large specific heat will be expected near 4.2 K.

CONCLUSIVE REMARKS

It has been verified that the magnetic material is very useful as the regenerator material below ∿10 K. Appling the magnetic material, such as Er_3Ni, to the regenerator in the GM refrigerator, we could reach to a 3 K range and, however, its refrigeration power at 4.2 K was very small (∿300 mW).

In order to obtain large refrigeration power, we have to use the magnetic material which has the large magnetic specific heat near 4 K.

Therefore, the new magnetic materials, $Er_{1-x}Yb_xNi_2$ and $Er_{1-x}Yb_xNi$, whose T_c's are ∿4 K have been developed and the experimental results are reported.

In some Er compounds whose crystalline field is anisotropic, the specific heat C(T) is expressed by the superposition of two kinds of C(T)'s, that is, the specific heat of the Schottky type $C_S(T)$ and that due to the magnetic phase change $C_J(T)$.

It has been made clear that $ErNi_2$ has a large $C_S(T)$ in the low temperature range and, as a result, its $C_J(T)$ becomes smaller than that expected from J = 15/2 (≡J-value of a free Er ion). While, in the $Er_{1-x}Yb_xNi$ system $C_S(T)$ = 0 and the large $C_J(T)$ have been observed near T_c.

Now, several promising magnetic materials have been developed and, hereafter, it is thought that the systematic investigation on the magnetic layer-structural regenerator is necessary.

REFERENCES

1. T. Hashimoto et al., "Recent progress in the materials for regenerator in the range from 4.2 K to 20 K", Proc. Intl. Cryog. Conf. (ICMC) p.667, Schenyang, China, (1988)
2. R. Li et al., "A new regenerator material $Er(Ni_{1-x}Co_x)_2$ with high specific heat in the range from 4.2 K to 20 K", Proc. Intl. Cryo. Engr. Conf., Butterworth, Guildford, UK (1988), p.423
3. T. Hashimoto, "Recent new application of the heavy rare earth magnetic material in the cryogenic field", Proc. of the 10th Intl. Workshop on Rare-Earth Magnets and Their Applications, The society of Non-Traditional Technology, Minatoku, Japan, p.317, (1988)
4. R. Li et al., "Specific heat of a new regenerator material $Er(Ni_{1-x}Co_x)_2$ at low temperatures", Proc. of the 10th Intl. Workshop on Rare-Earth Magnets and Their Applications, The society of Non-Traditional Technology, Minatoku, Japan, p.221, (1989)
5. T. Kuriyama et al., "Development of a two-stage GM refrigerator with rare earth matrix for cooling super conducting devices", Proc. of the 10th Intl. Workshop on Rare-Earth Magnets and Their Applications, The society of Non-Traditional Technology, Minatoku, Japan, p.335, (1989)
6. Y. Tokai et al., "New magnetic material R3T system used as heat regenerator for cryogenic application", Proc. of the 10th Intl. Workshop on Rare-Earth Magnets and Their Applications, The society of Non-Traditional Technology, Minatoku, Japan, (1989)
7. T. Hashimoto, "Application of the functional magnetic material for cryogenic engineering in Japan", Proc. of Intl. Conf. on CRYOGENICS and REFRIGERATION, p.273, (1989)
8. T. Kuriyama et al., "Development of 5 K GM refrigerator using rare earth compounds as a regenerator matrix", Proc. of Intl. Conf. on CRYOGENICS and REFRIGERATION, (1989)
9. M. Sahashi et al., "Regenerator material R3T system with extremely large heat capacities between 4 K and room temperature", Proc. of Intl. Conf. on CRYOGENICS and REFRIGERATION, (1989)
10. R. Li et al., "Measurement of ineffectiveness on regenerator packed with magnetic regenerator materials between 4 and 35 K", Advances in Cryogenic Engineering, Vol.35, New York, p.1183, (1990)
11. T. Kuriyama et al., "High efficient two-stage GM refrigerator with magnetic material in the liquid helium temperature region", Advances in Cryogenic Engineering, Vol.35, New York, p.1261 (1990)
12. M. Sahashi et al., "New magnetic material R3T system with extremely large heat capacities as heat regenerators", Advances in Cryogenic Engineering, Vol.35, New York, p. 1175 (1990)

13. R. Li et al., "Magnetic intermetallic compounds for cryogenic regenerator", Cryogenics 30 : 521 (1990)
14. T. Hashimoto et al., "Recent advance in magnetic regenerator material", Cryogenics 30 : 192 (1990)
15. T. Kuriyama et al., "Two-stage GM refrigerator with Er3Ni regenerator for helium liquefaction", Proc. of the Intl. Cryocooler Conf. to be published, (1990)
16. M. Ogawa et al., "Thermal conductivity of magnetic intermetallic compounds for cryogenic regenerator", Cryogenics 31 : 405 (1991)
17. K.H.J. Buschow et al., "Extremely large heat capacities between 4 and 10 K", Cryogenics 15 : 261 (1975)

MODELING THE ACTIVE MAGNETIC REGENERATOR

A. J. DeGregoria

Astronautics Corporation of America
Astronautics Technology Center
Madison Wisconsin

ABSTRACT

A time-dependent one-dimensional model of the Active Magnetic Regenerator (AMR) is described. The model assumes that the heat capacity of the pore fluid in the regenerator is negligible compared to the magnetic material. Measured magnetic material properties are used, including the effect of hysteresis. The variation of the fluid helium properties with temperature are included. Heat transfer between the fluid and bed is obtained from an emperical correlation, as is the pressure drop, axial conduction and axial dispersion.

Equations are presented and the numerical procedure used to solve them is discussed with emphasis on accuracy. The model has been applied to results from an AMR test device.

INTRODUCTION

A small (1 ton/day) magnetic hydrogen liquefier is under development.[1] It will operate from about 80 K, rejecting heat to liquid nitrogen, to 20 K, the liquefaction temperature of hydrogen. The principle it will operate on is that of the Active Magnetic Regenerator (AMR). The configuration currently envisioned is a system comprised of an upper and lower stage AMR bed. The AMR is also being developed for the lower stage of a demonstration 40 K, 50 W cooler, to cool proposed high T_c superconducting magnets. In addition to these ongoing projects, the AMR is being considered as a cooler for hydrogen slush production. The AMR would cool an auger producing slush at 13.8 K, while rejecting heat to a hydrogen bath at 20 K.

An AMR test refrigerator is in operation to develop the AMR subsystems for the above applications. Its operating range is from about 4 to 80 K. It has been run successfully between about 30 and 6 K, and between about 80 and 30 K.[2]

A model of the AMR has been developed to aid in the design the above refrigerators. It is in the process of being validated.

Advances in Cryogenic Engineering, Vol. 37, Part B
Edited by R.W. Fast, Plenum Press, New York, 1992

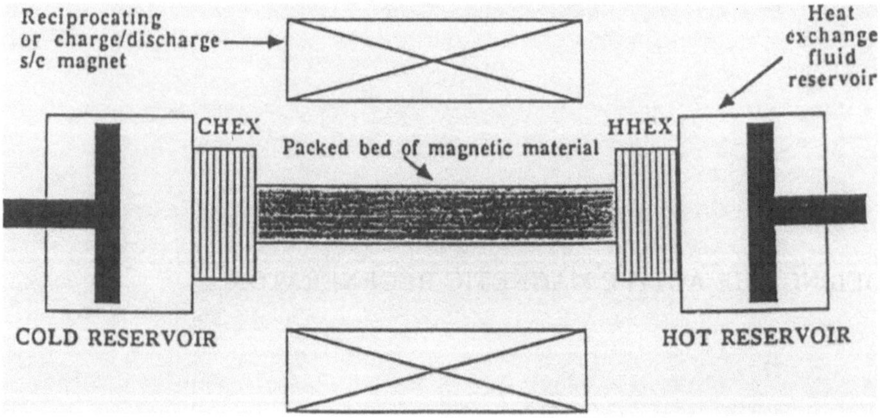

Figure 1: Schematic of the Active Magnetic Regenerator.

The reciprocating version of the Active Magnetic Regenerator (AMR) is shown in schematic in Fig. 1. The regenerator is shown in the form of a packed particle bed of magnetic material. The regenerator bed is sandwiched between the hot and cold heat exchangers. The heat transfer fluid is typically helium in the cryogenic regime, though hydrogen is a possibility in a hydrogen liquefier or slushifier. The pistons shuttle the fluid between the reservoirs.

A complete cycle of the AMR is as follows: With the fluid all in the cold reservoir, the magnetic material in the bed is adiabatically magnetized. The fluid is then passed from the cold to the hot reservoir. With the fluid now in the hot reservoir, the bed is adiabatically demagnetized. The fluid is then passed from the hot back to the cold reservoir.

Figure 2 illustrates the temperature profiles of the bed and fluid as a function of bed position. For simplicity, the bed heat capacity is assumed to be infinite since the temperature profiles do not change over the flow periods in that limit. The upper

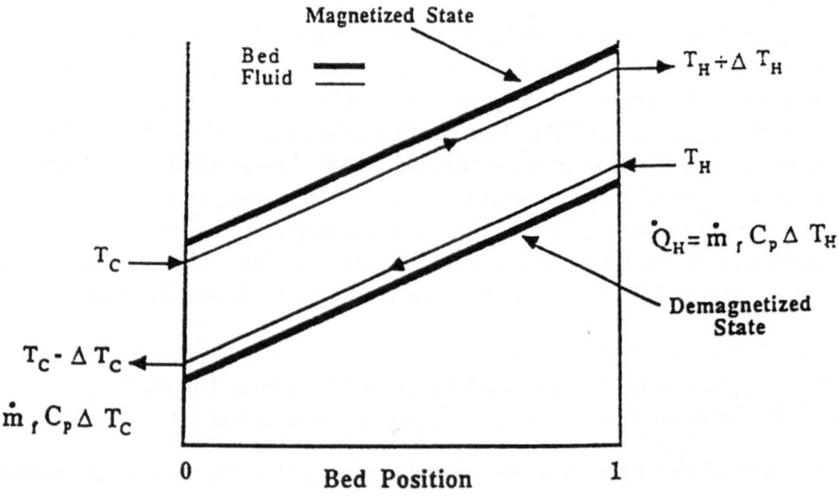

Figure 2: Temperature profiles for the bed and fluid in the AMR in the infinite bed thermal mass limit.

profiles correspond to the flow from the cold to the hot reservoir while the bed is magnetized. The fluid enters the bed at the temperature of the cold reservoir T_{cold} since it has just left the cold heat exchanger. Since the bed temperature rises going from the cold to the hot end, the fluid temperature rises as it exchanges heat with the bed and leaves the bed at a temperature greater than that of the hot reservoir T_{hot}. Passing through the hot heat exchanger, the fluid temperature drops to T_{hot} giving up an amount of heat

$$\dot{Q}_{hot} = \dot{m}_f c_p \Delta T_{hot} \tag{1}$$

After the flow is complete, the bed is adiabatically demagnetized, achieving the lower profile shown in Fig. 2. The flow proceeds then from the hot to the cold end. Upon entering the bed, the fluid temperature is T_{hot}. Exchanging heat with the bed, it drops to a value below T_{cold} at the cold end . Going through the cold heat exchanger, it absorbs heat equivalent to

$$\dot{Q}_{cold} = \dot{m}_f c_p \Delta T_{cold} \tag{2}$$

The AMR is unique among regenerative or recuperative magnetic or gas cycles in that an element of working magnetic material only goes through a simple non-regenerative Brayton cycle about a specific temperature determined by its location in the bed. The heat transfer fluid links these cycles together to span a large temperature range. The AMR has a significant advantage here in that any given element of magnetic material does not have to be active over the entire temperature range of the device.

In analyzing the AMR, we have taken the approach of first writing the general equations and attempting to solve them. We then implemented approximations as necessary to obtain a model which is useful as a design tool.

MODELING

A model which solves the complete physical problem in one-dimension for the AMR was developed by Barclay[3]. The computer code MRRAP integrates the full one-dimensional time-dependent equations using a finite difference scheme.

In order to conduct our design studies, it is necessary to make simplifications in these equations. If we assume that the pore fluid heat capacity is negligible, we have the following equations[4] relating the bed temperature t_b and fluid temperature t_f as a function of position x and time t:

$$hA(t_b - t_f) = \dot{m}_f c_p L \frac{\partial t_f}{\partial x} \tag{3}$$

$$hA(t_f - t_b) = M_b c_b \frac{\partial t_b}{\partial t} \tag{4}$$

where h is the heat transfer coefficient between the fluid and solids in the bed; A is the total cross-sectional area between the fluid and solids; L is the bed length; M_b and $\dot{m}_f$ are the bed mass and fluid flow rate; and c_b and c_p are the heat capacities of the bed and fluid.

Equations 3 and 4 can be cast into dimensionless form in terms of a reduced length Λ and reduced period Π, where

$$\Lambda = \frac{hA}{\dot{m}_f c_p} \tag{5}$$

$$\Pi = \frac{PhA}{M_b c_b}$$

P is the time period of the flow in either direction.

We have solved Equations 3 and 4 numerically. This version of the model is computationally intensive. We have therefore developed a model in which the bed mass is assumed to be infinite, corresponding to zero reduced period. This assumption eliminates the time dependence in the problem allowing us to obtain solutions very rapidly. The limit is useful since it represents the point of maximum refrigerator efficiency. It is also a physically realizable limit to a good approximation.

The equations in the infinite bed mass limit are

$$\frac{\partial t_{fh}}{\partial x} = \Lambda_h (t_{bh} - t_{fh}) \tag{6}$$

$$\frac{\partial t_{fc}}{\partial x} = \Lambda_c (t_{fc} - t_{bc}) \tag{7}$$

$$t_{bh} = t_{bc} + \Delta t_{ad} \tag{8}$$

$$h_h A P_h \frac{(t_{fh} - t_{bh})}{t_{bh}} = h_c A P_c \frac{(t_{fc} - t_{bc})}{t_{bc}} \tag{9}$$

The subscripts c and h refer to the cold blow (flow from the hot to cold reservoirs) and the hot blow (flow from the cold to the hot reservoirs), respectively; Δt_{ad} is the adiabatic temperature change of the magnetic material. Equation 9 is the algebraic equation which results from the two time dependent equations (see equation (4)) when we go to the infinite bed mass limit.

Equations 6 - 9 are solved numerically by a shooting method. We integrate from the cold end knowing that the hot fluid stream is at temperature t_{cold} and guessing at the temperature of the cold fluid stream. Integrating across the bed, we obtain a value for the cold fluid stream at the outlet which we want to be t_{hot}. We repeat the integrations adjusting our guess for the temperature of the cold fluid stream at the cold end until we obtain the matching at the hot end to some desired degree of accuracy.

Both versions of AMR models have been checked against each other in the appropriated limit. They have also been checked against known results for passive regenerators[4].

The heat transfer coefficient between fluid and bed is obtained from empirical correlations for packed particle beds.[5] Variations in properties with temperature of both helium[6] and the magnetic material are accounted for. Pumping losses, thermal conduction losses across the bed and axial dispersion are taken into account separately using well established empirical correlations from the literature[7,8,9].

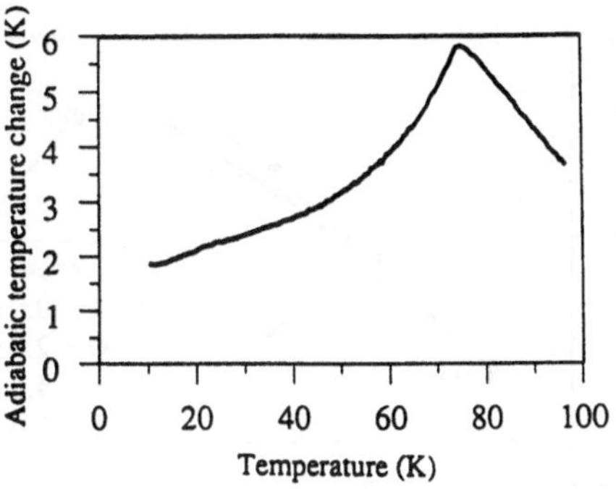

Figure 3: Adiabatic temperature change of $GdNi_2$ going from 0 to 7 T.

As an example, we show results for the material $GdNi_2$.[10] Figure 3 shows the measured adiabatic temperature increase for the material in going from 0 to 7 T. The performance of the AMR is sensitive to the thermodynamic properties of the magnetic material. We have found that the ideal magnetic material is one which has an adiabatic temperature change which is linearly proportional to the temperature:

$$\Delta t_{ad} = f(H)t \tag{10}$$

$GdNi_2$ should be a good material for an AMR below the Curie point of about 75 K.

Figure 4 shows computed temperature profiles over a complete AMR cycle. In this sample problem, the hot and cold fluid flow periods and the field change period are all 1.0 s. The helium gas flow is 0.53 g/s at an average pressure of 1.0 MPa. The bed length and cross section are 5 cm. and 2.84 cm^2, respectively. The particle size is 0.015 cm. The field change is 7 T.

Temperature profiles over the 4 parts of the AMR cycle are shown in the figure: Over the field increase (decrease), the lowest (highest) profile shows the bed temperature before the field increase (decrease), and the highest (lowest) profile shows the bed temperature after the field increase (decrease). Over the positive and negative

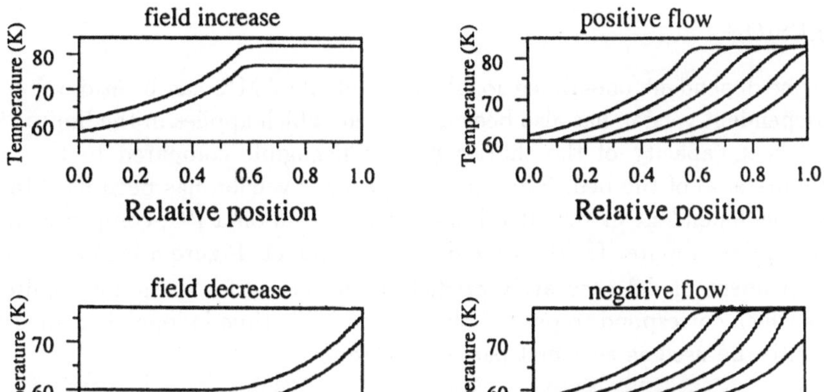

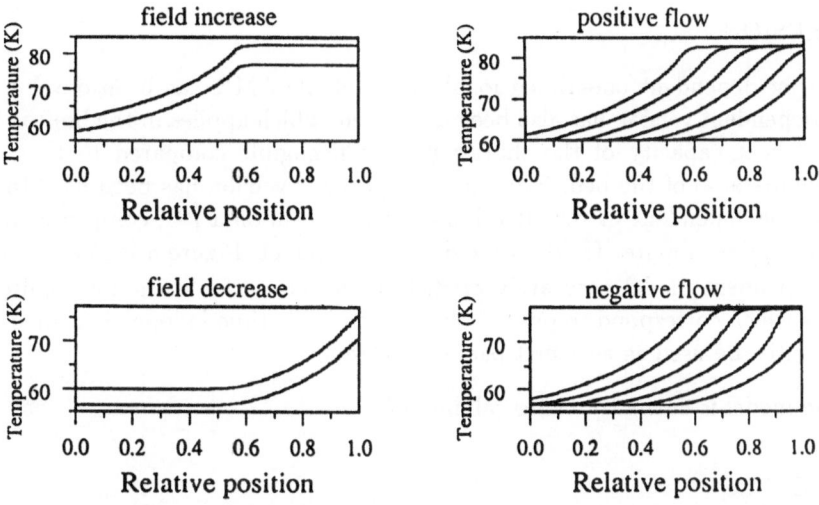

Figure 4: Bed temperature profiles over an AMR cycle.

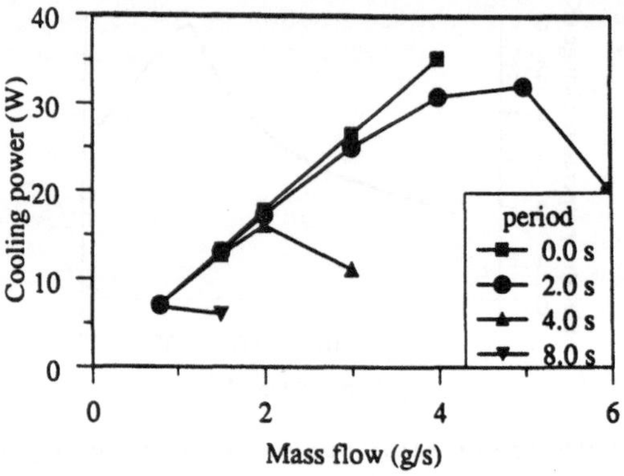

Figure 5: Load curves at 7 T for $GdNi_2$.

flow periods, profiles are shown at equal time intervals. The uppermost (lowermost) profile shows the bed at the start of the positive (negative) flow. Lower (upper) profiles in succession show successively later time intervals with the lowest (highest) profile representing the bed at the end of the positive (negative) flow period.

Figure 5 shows computed load curves for several values of the total period. The AMR parameters are the same as those described for Fig. 4, except for the variation in period and fluid flow. Note that the points represent the actual values. The connecting lines are there for continuity.

The results shown for zero period are from the time independent, zero reduced period, computations described above. The time dependent computations show consistency with this limiting result.

For non-zero period, the cooling power curves turns over at a particular values of mass flow. The reason for this is that the heat capacity of the fluid shuttling through the bed becomes greater than the heat capacity of the bed material. The bed can no longer "regenerate" the fluid. The zero period curve would also turn over at high enough mass flow, but for a different reason. At high enough mass flow, the heat transfer between the fluid and bed becomes poor.

CONCLUSION

A time-dependent one-dimensional model of the AMR has been developed. A time-independent version has also been developed, which applies in the limit in which the total heat capacity of the shuttle fluid is negligible compared to that of the magnetic material of the bed. The time-independent version has been used to verify the full time-dependent model. It takes minutes to run on a PC, compared to hours on a super-minicomputer for the time-dependent model. Figure 5 indicates that the time-independent model accurately predicts performance for mass flows quite close to values which correspond to peak cooling power. The time-independent model can, consequently, be used as as a first pass design tool.

The model is currently being validated.[2]

REFERENCES

1. D. J. Janda et al, Design of an active magnetic regenerative hydrogen liquefier, this conf. (1991)

2. A. J. DeGregoria et al, Test results of an active magnetic regenerative refrigerator, this conf. (1991)

3. J. A. Barclay, The theory of an active magnetic regenerative refrigerator, Proc. 2nd Conf. of Refrigeration for Cryogenic Sensor and Electronic Systems (NASA, Goddard Research Center), (1982).

4. F. W. Schmidt and A. J. Willmott, Thermal energy storage and regeneration, McGraw-Hill, 1981.

5. S. Whitaker, Forced convection heat transfer correlations for flow in pipes, past flat plates, single cylinders, single spheres, and for flow in packed beds and tube bundles, AIChE Journal Vol. 18, No. 2 (1972).

6. R. D. McCarty, Thermophysical properties of helium-4 from 2 to 1500 K with pressures to 1000 atmospheres, NBS technical note 631 (1972).

7. I. F. Macdonald et al, Flow through porous media - the ergun equation revisited, Ind. Eng. Chem. Fundam., 18, 199-208 (1970).

8. G. S. G. Beveridge and D. P. Haughey, Axial heat transfer in packed beds: Stagnant beds between 20 and 750C, Int. J. Heat Mass Transfer 14, 1093-1113 (1971).

9. S. Sarangi and H. S. Baral, Effects of axial conduction in the fluid on cryogenic regenerator performance, Cryog., 27 (1986).

10. C. B. Zimm et al, Materials for regenerative magnetic cooling spanning 20 K to 80 K, this conf. (1991).

TEST RESULTS OF AN ACTIVE MAGNETIC REGENERATIVE REFRIGERATOR

A.J. DeGregoria, L.J. Feuling, J.F. Laatsch, J.R. Rowe,
J.R. Trueblood, A. A. Wang

Astronautics Corporation of America
Astronautics Technology Center
Madison Wisconsin

ABSTRACT

The principle of the Active Magnetic Regenerator (AMR) is tested with an experimental refrigerator designed to operate within the temperature range of about 4 to 80 K. Applications, including helium and hydrogen liquefaction and hydrogen slush generation, are envisioned. The device uses a single moveable superconducting solenoidal magnet in persistent mode to alternately charge and discharge two in-line beds of magnetic material. Between magnet motions, a double-acting piston displacer moves heat transfer fluid in the form of helium gas through the beds, absorbing heat at the cold heat exchanger and rejecting heat at the hot heat exchanger.

A description of the refrigerator and performance results are presented. Comparisons to a detailed AMR model are shown.

INTRODUCTION

Magnetic refrigeration is a potentially compact, reliable and efficient technology for the low temperature cryogenic regime. Below about 20 K, paramagnetic materials can be used in conjunction with superconducting magnets to produce refrigeration in which the magnetic material follows an approximate Carnot cycle, spanning a temperature range greater than that of the refrigerator. Above 20 K, regeneration, recuperation or staging is necessary to achieve the typical temperature spans of interest. Below 20 K, these concepts can also be used in place of the Carnot cycle. Regenerative or recuperative cycles use regeneration or recuperation to permit the magnetic material to span a greater temperature than is otherwise possible. In staging, the magnetic material elements span temperature ranges less than the total temperature range of the refrigerator. By connecting the elements either in series or parallel, large temperature spans are possible.

The Active Magnetic Regenerator (AMR) is a refrigeration concept which uses either regeneration or parallel staging to achieve a large temperature span (depending on the relative heat capacity between heat transfer fluid and magnetic material). Referring to Fig. 1, an AMR cycle consists of the four following segments: bed magnetization, warming the magnetic material and the bed fluid by the magneto-caloric effect; cold to hot fluid flow through the bed, transferring heat to the hot heat exchanger; bed demagnetization, cooling the magnetic material and fluid; and hot to cold fluid flow through the bed, absorbing heat at the cold heat exchanger.

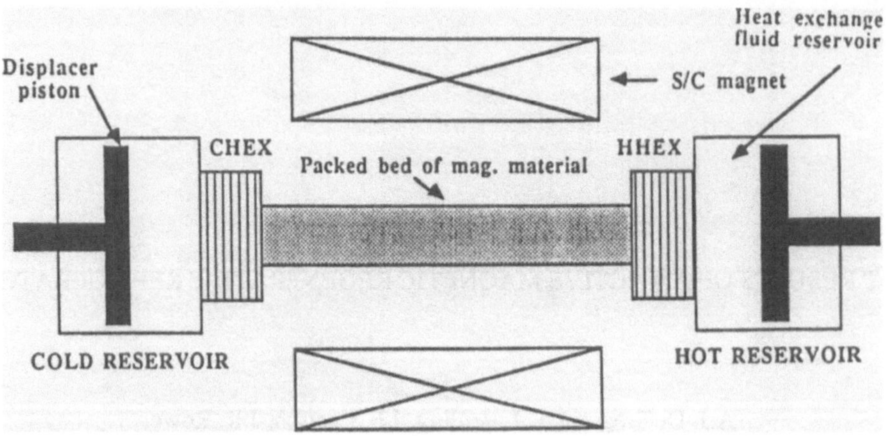

Fig. 1. Schematic of the AMR.

There are two limits in which the AMR can be shown to work well. In one limit, the heat capacity of the fluid in the bed is large compared to that of the magnetic material. The magnetic material follows a regenerative cycle in which the fluid in the bed acts as the regenerator. In the other limit, the fluid heat capacity is negligible compared to that of the magnetic material. The elements of magnetic material follow either non-regenerative Brayton or Carnot cycles connected in parallel by the heat transfer fluid. The former limit applies to devices which use either liquid helium or high pressure helium and is the limit in which the device was conceived.[1,2] The latter limit applies to the device described here.

There are two other experimental AMR devices designed to operate in the cryogenic regime.[3,4] These devices both use charge-discharge magnets and gadolinium gallium garnet (GGG) as the active magnetic material and are intended to span approximately a 15 K to 4 K temperature range. Both devices are designed to have fluid flow during part of the field change in an attempt to have the elements of magnetic material follow Carnot cycles.

Our device uses a reciprocating magnet in persistent mode designed to produce a uniform field over the bed. No fluid flow occurs during field change. The elements of magnetic material follow non-regenerative Brayton cycles. It has been shown that, in this case, in which the bed goes from a uniform high field to zero field, an 'ideal' magnetic material exists in which the adiabatic temperature change is linearly proportional to the absolute temperature.[5] Ferromagnetic materials below their Curie point temperature exhibit this approximate behavior. This gives us a wide range of materials to choose from and a wide range of temperatures over which this type of AMR can operate.

EXPERIMENTAL APPARATUS

Figure 2 illustrates the key components of the device. Two AMR beds are connected to a common cold heat exchanger (CHEX). The beds are connected to separate hot heat exchangers (HHEX) which are at opposite ends of an aluminum thermal bus. One end of the thermal bus is connected to the bottom of a Gifford McMahon (GM) refrigerator. A reciprocating magnet alternately magnetizes one bed while demagnetizing the other. When the magnet is stationary, a double-acting piston displacer forces gas through one of the HHEX through one of the beds, through the CHEX, through the other bed, through the other HHEX and back to the displacer. A heater is attached to the CHEX.

The NbTi solenoid magnet is contained in an annular stainless steel helium dewar. A soft iron flux-return moves with the magnet and contains any stray field which would otherwise produce eddy current heating in the dewar, ball bearing wheels which

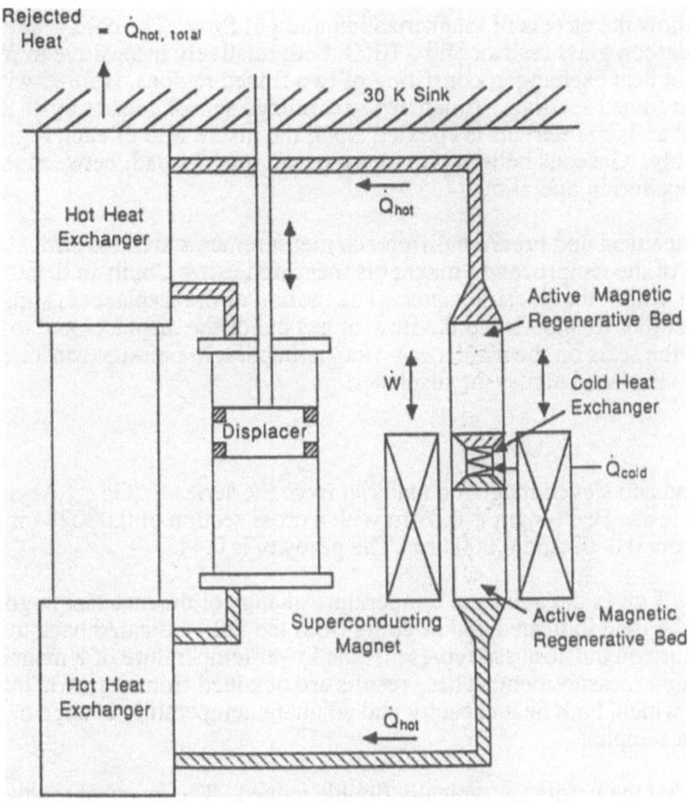

Fig. 2. Key components of the AMR experiment.

guide the magnet on tracks, etc. Separate linear actuators, at room temperature, drive the magnet and displacer. Two position control systems independently move the magnet and displacer piston. (For a more complete description of the magnet system, see [6].)

The displacer uses two circular spring-loaded ultra high density polyethylene seals mounted on the piston and sliding on a polished cylinder wall, to eliminate leakage.

The hot heat exchangers are made by cutting narrow closely spaced groves in a cylindrical aluminum bus in the axial direction then placing sleeves over them. By the appropriate plumbing, helium gas is forced through the groves to effect heat exchange.

The cold heat exchanger presents unique problems. Large low resistance conduction loops must be avoided to prevent eddy-current heating from the changing magnetic field. Also, a high heat capacity is necessary at operating temperatures below 10 K. To address both issues, a tube in shell heat exchanger was devised which uses fluid helium for heat capacity.

The GM refrigerator used in the tests is a Cryomech GB04. The apparatus is designed to accept the more powerful Cryomech GB37 also.

Accurate measurement of temperature, pressure difference and displacement are the primary requirements for performance testing of the AMR refrigerator. The device is instrumented with fifty temperature sensors including platinum resistors (PRT), carbon glass resistors (CGR), and other resistance thermometers (BRO) developed and calibrated in-house. In most cases, accuracy of +/- 0.2 K is adequate.

The most critical temperature measurements center on the cold heat exchanger, the hot heat exchanger, and the magnetic material beds. The heat exchanger temperatures are used as the cold and hot sink temperatures of the refrigeration cycle. Measurement of the temperature at different points along the regenerator beds is desirable to show the effects of magnetization and gas flow. The cold heat exchanger is fitted with a carbon glass resistor and a BRO, both relatively insensitive to magnetic fields. The hot heat exchanger, consisting of two finned regions, is fitted with PRTs and BROs on both finned sections. A narrow, streamlined sensor consisting of five BROs evenly spaced at 3 cm intervals is epoxied along the inside wall of each regenerator bed during assembly. Gaseous helium provides the heat transfer path between the material and the sensors during operation.

Displacement and pressure difference measurements are also critical. The displacement of the reciprocating magnet is measured using a built-in linear potentiometer within the linear actuator. The motion of the displacer is similarly measured. In order to understand the flow of gas out of the displacer and to check on blow-by past the seals on the displacer piston, a diaphragm pressure sensor (calibrated in liquid helium) is placed across the displacer.

RESULTS

Ground and sieved magnetic material from the series $Er_xGd_{(1-x)}Al_2$ is employed for the initial tests. Bed length is 0.09 m with a cross section of 0.000284 m^2. Particle sizes range from 0.0002 and 0.0004 m. The porosity is 0.44.

Figure 3 gives the adiabatic temperature change of the material in going from zero field to the field indicated and in going from the field indicated back to zero field. The temperature on the abscissa represents the lower temperature of a magnetization or demagnetization measurement. These results are obtained from our own 'heat capacity' apparatus, in which, both heat capacity and adiabatic temperature change data can be obtained for a sample.[7]

Note that the points are typically double valued. The lower of the two points shown for a given measurement represents demagnetization. A small amount of magnetic hysteretic heating is consequently indicated by the data. The AMR has been run with only field change (no flow) on this material. A steady rise in temperature consistent with the data shown in Fig. 3 is observed, confirming the hysteretic heating observed in the heat capacity apparatus on a much longer time scale.

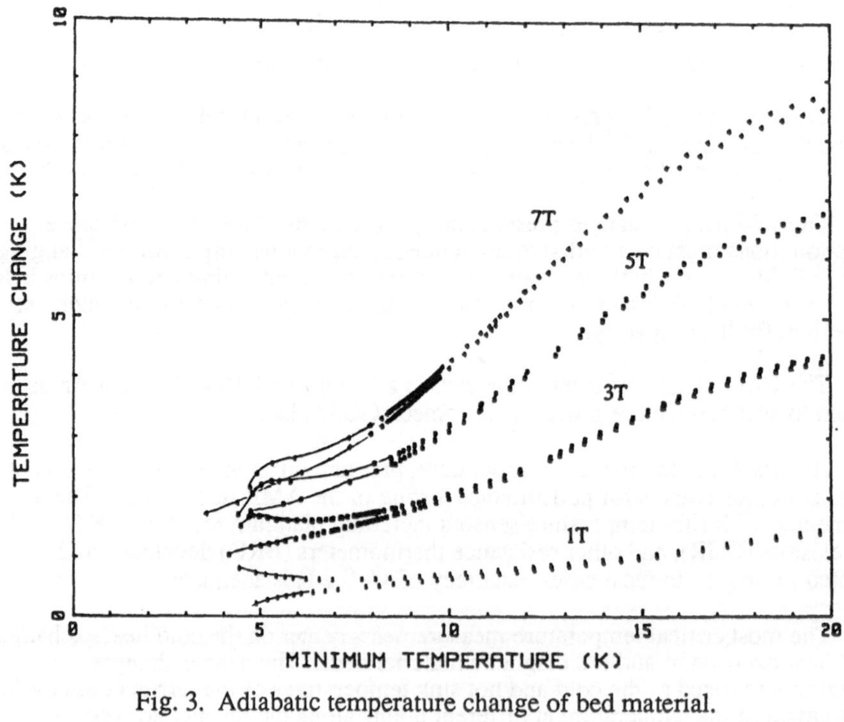

Fig. 3. Adiabatic temperature change of bed material.

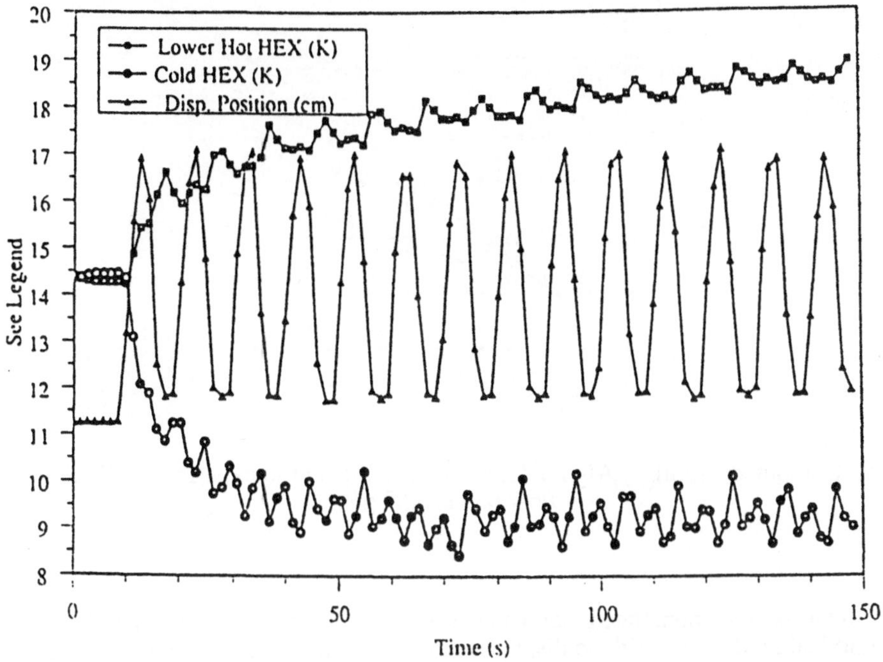

Fig. 4. HHEX and CHEX temperatures vs. time for a 3 T run. Displacer motion is superimposed.

Figure 3 indicates that the material has an adiabatic temperature change which is close to 'ideal' over the 10 K to 20 K temperature range. That is, the adiabatic temperature change is close to being proportional to the absolute temperature.[5]

AMR tests were run for discrete values of the magnetic field of 0.5, 1.0, 2.0, 3.0, 5.0 and 7.0 T. Figure 4 shows an experimental run at a field of 3.0 T, from startup. The period for the total AMR cycle is 10 s, here, and for all runs shown in this paper.

Figure 4 indicates the HHEX and CHEX temperatures were both at about 14.5 K, prior to starting the AMR cycle. After startup, the temperature difference is rapidly established. The CHEX achieves its bottom temperature of about 9 K after only about 7 cycles. The HHEX is continuing to rise throughout the 14 cycles show. The displacer motion, per blow, is 0.015 m/s for 3.3 s.

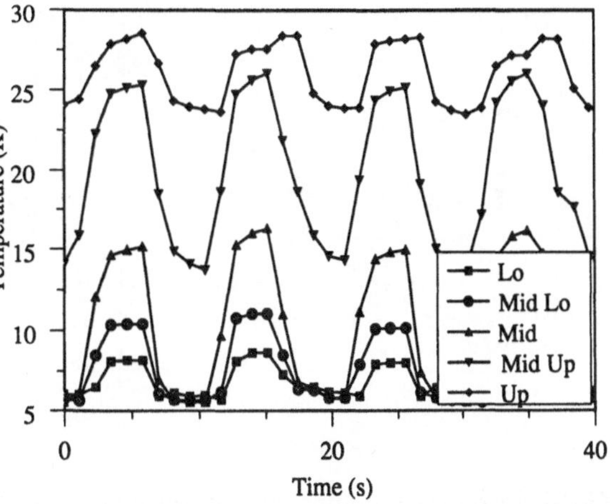

Fig. 5. Temperature vs. time in the bed at the five axial locations 0.0, 0.0125, 0.045, 0.0675 and 0.09 m from the cold end of the bed. The magnetic field is 5 T and the displacer motion is 0.018 m/s for 0.5 s.

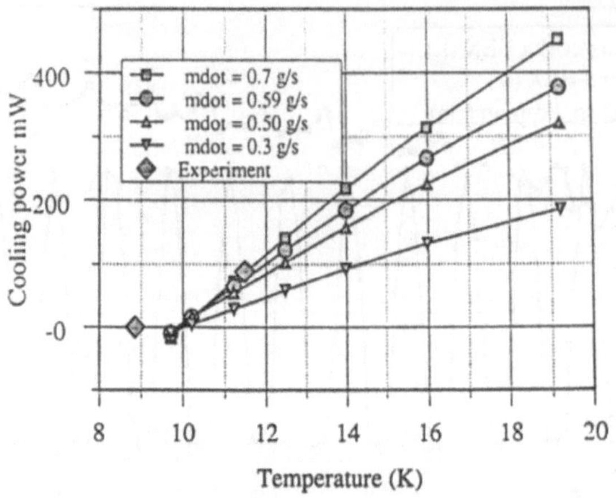

Fig. 6. Computed $Er_xGd_{(1-x)}Al_2$ 1 T load curves compared to experiment. Experimental flow rate is 0.59 g/s.

It has been demonstrated that the beds and the CHEX are thermally isolated from the liquid helium dewar. With no displacer motion or magnet motion, the temperatures in the bed and CHEX remain constant. With displacer motion and no magnet motion, the beds and CHEX achieve the temperature of the HHEX. The reverse cycle has also been run, in which, heat is pumped from HHEX to CHEX. The CHEX achieved a temperatures of 40 K before the cycle was shut down. (Since the CHEX is thermally isolated, its temperature continues to rise.)

The very rapid approach to steady state, with respect to the number of AMR cycles, indicates that the heat capacity of the total gas flowing through the bed, per blow, is comparable to the total heat capacity of the magnetic material in the bed.

Figure 5 shows a 5 T run with no load, after steady state. The five temperature sensors in the bed are displayed. Note that temperature fluctuations at points in the bed can exceed the adiabatic temperature change of the magnetic material. Heat transfer in the bed due to fluid flow can cause larger temperature fluctuations than those produced by the magnetic field change.

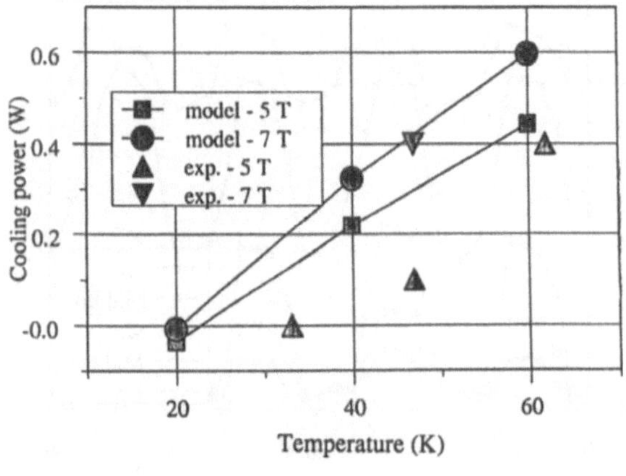

Fig. 7. Computed $GdNi_2$ load curves compared to experiment. Gas flow rate is 0.07 g/s.

COMPARISONS TO THE MODEL

A time dependent model of the AMR has been developed. Figure 6 shows computed load curves at a 1 T field for different gas flow rates through the beds (mdot). The cold and hot flow periods are 0.5 s. The field change periods are 4.5 s, yielding a total cycle period of 10 s. The average gas pressure is 0.1 MPa. The hot temperature is 19.25 K. Two experimental points are shown for comparison. Fluid flow is not well characterized in these test results. Based on the fill conditions, the gas flow for the experimental points is 0.59 g/s.

The model appears to under-predict the performance of the device by a small amount. It is interesting to note that, regardless of the gas flow, within reason, the no-load temperature predicted by the model is approximately 9.75 K. This is about 1 K above the experimental value.

The magnetic material $GdNi_2$ has also been employed in the refrigerator. Ground and sieved particles in the size range 0.0001 m to 0.0002 m are used. The cold and hot flow periods and the field change periods are all 2.5 s. The average gas pressure is 0.1 MPa. Figure 7 shows model load curves for this material compared to experiment for two fields. The hot temperature of the refrigerator is 77 K.

Figure 7 indicates that the material has a no load temperature span from about 77 K to about 33 K, at 5 T. The model predicts a no load cold temperature about 10 K below that observed. Additional tests are ongoing to better characterize the experimental load curves. The model appears to fit the 7 T data better than the 5 T data. Since only one experimental point exists at 7 T, this may be fortuitous.

Some possible reasons for the discrepancies between model and experiment are the following: The model assumes spherical particles in the bed with a uniform diameter. Ground and sieved material is actually used. Correlations for heat transfer coefficient, bed conduction and axial dispersion can, consequently, be in error in either direction. Also, the magnetic material heat capacity and adiabatic temperature change measurements are accurate to within about 10 %. Errors of this amount can produce comparable errors in total no load temperature span, for example.

CONCLUSION

The AMR test apparatus has been successfully assembled and tested. The materials used, $Er_xGd_{(1-x)}Al_2$ and $GdNi_2$, were predicted to work well as refrigerants in the approximate temperature ranges of 20 K to 10 K, and 80 K to 40 K, respectively. This is verified.

Results using $Er_xGd_{(1-x)}Al_2$ indicate the detailed model of the AMR is accurate to within about 10 % in predicting the no load temperature span of the refrigerator. Initial results using $GdNi_2$ indicate the model predicts the no load temperature span within 25 %. More testing with $GdNi_2$ is in progress.

ACKNOWLEDGEMENTS

We would like to thank Peter Claybaker for designing the apparatus which worked the first time. We would also like to thank Tom Stankey, Joe Johnson, Rick Lubasz, Joe Hertel, Ron Syphard, Randolph Pax, Steve Kral, Coyne Prenger, and John Barclay for technical and administrative assistance.

Partial funding was provided by U.S. Department of Energy contract #DE-AC02-90CE40895.

REFERENCES

1. J. A. Barclay, The theory of an active magnetic regenerative refrigerator, in: "Proc. of the 2nd Biennial Conf. on Refrigeration for Cryo. Sensors and Electronic Systems", NASA Goddard Space Flight Center, Greenbelt, MD (1982).
2. J. A. Barclay and W. A. Steyert, Active Magnetic Regenerator, US Patent 4,332,135, Jun. 1, 1982.
3. F. J. Cogswell, J. L. Smith, Jr., Y. Iwasa, Regenerative magnetic refrigeration over the temperature range of 4.2 K to 15 K, in: "Proc. Fifth Intl. Cryocooler Conf.", Monterey, California, (1988).
4. P. Seyfert, Research on magnetic refrigeration at CEA Grenoble, in: "Advances in Cryogenic Engineering", Vol. 35, Plenum Press, New York (1990).
5. C. R. Cross et al, Optimal temperature entropy curves for magnetic refrigeration, in: "Advances in Cryogenic Engineering", Vol. 33, Plenum Press, New York (1988), p. 767.
6. J. R. Trueblood et al, A vertically reciprocating NbTi solenoid used in a regenerative magnetic refrigerator, IEEE Trans. on Magnetics, Vol. 27, no. 2, p. 2384 (1991).
7. C. B. Zimm, P. L. Kral and J. A. Barclay, The magnetocaloric effect in erbium, in: "Proc. of the Fifth Intl. Cryocoolers Conf.", Monterey, California, (1988).

MATERIALS FOR REGENERATIVE MAGNETIC COOLING SPANNING 20K TO 80K

C.B. Zimm, E.M. Ludeman, M.C. Severson, T.A. Henning

Astronautics Technology Center
5800 Cottage Grove Rd., Madison, Wisconsin

ABSTRACT

Medium to high power cryocoolers based on the active magnetic regenerative concept (AMR) are under development for hydrogen liquefaction. Refrigeration performance is dependent on the temperature dependence of the adiabatic temperature change upon application of a magnetic field (ΔT_s). We report on measurements of ΔT_s and the field dependent heat capacity, C_B, of the ferromagnetic materials GdPd, GdNi, $GdNi_2$, and $(Er_{0.86}Gd_{0.14})Al_2$. The binary compounds have sharp ordering transitions encompassing all of the available magnetic entropy, whereas the ternary with two magnetic rare earths shows a broadened transition. $GdNi_2$ and GdPd have properties well suited for use in AMR stages operating from 80 K to 40 K and 40 K to 20 K, respectively.

INTRODUCTION

Interest in development of compact and efficient small scale (1000 kg/day) hydrogen liquefaction plants has led to the design of a 20 K to 80 K Active Magnetic Regenerative (AMR) magnetic liquefier[1]. Our chosen refrigeration cycle consists of two adiabatic legs in which the magnetic field is (alternately) increased or decreased, connected by two constant field legs during which a heat transfer medium (He gas) transfers heat to or from the magnetic regenerator. For the case that the heat capacity of the heat transfer fluid is constant and small compared to the bed material, it has been shown[2] that optimum refrigeration performance requires a magnetic regenerator bed with the property that the adiabatic temperature change with change in magnetic field, ΔT_s, is proportional to the absolute temperature, T.

We have chosen to use one material per stage of the AMR to simplify fabrication and to ensure rapid cooldown. The ΔT_s increasing with temperature is obtained by choosing a ferromagnetic material with a sharp Curie point (the ordering temperature) equal to the maximum operating temperature of the stage. Rare earth intermetallic compounds are good choices because their large values of the angular momentum J tends to produce a large product of ΔT_s and C_B, making a high refrigeration power density possible. Magnetic hysteresis is a potential problem if the material is operated well below its Curie point in cases where magnetic atoms in the structure have orbital angular momentum. We have thus chosen the three ferromagnetic gadolinium compounds, GdPd, GdNi, and $GdNi_2$ for study. Gadolinium is an S-state ion and it has been shown[3] that nickel is non-magnetic in GdNi and $GdNi_2$. We also compare these materials with $Er_{0.86}Gd_{0.14}Al_2$, a pseudo-binary mixture between aluminide compounds of the two magnetic rare earths gadolinium and erbium. We report measurements on

Advances in Cryogenic Engineering, Vol. 37, Part B
Edited by R.W. Fast, Plenum Press, New York, 1992

these four materials of the heat capacity at constant magnetic field, C_B, and the adiabatic temperature change upon change in field; these properties characterize the behavior of the magnetic material in the isofield and adiabatic legs of the AMR cycle.

EXPERIMENTAL

GdNi and GdPd are congruently melting compounds with the orthorhombic CrB structure. Our specimens were prepared by arc-melting the component elements in a argon atmosphere over a water-cooled copper hearth. Phase purity was verified by metallography. $GdNi_2$ is a Laves phase compound with the cubic $MgCu_2$ structure. Although $GdNi_2$ is reported to be incongruently melting[4], we obtained specimens of satisfactory phase purity by arc-melting. Attempts to form a specimen by sintering were not successful. $ErAl_2$ and $GdAl_2$ are congruently-melting $MgCu_2$ Laves phase compounds that exhibit solid-solid solubility. Arc-melted specimens of $Er_{0.86}Gd_{0.14}Al_2$ were prepared by first melting Gd and Er together, then reacting with Al. These specimens were prone to cracking during solidification. Standard powder metallurgy techniques and pressureless sintering were used to obtain a specimen for measurements.

The C_B and ΔT_s measurements were done on 10 to 20 g samples suspended on fine linen threads in vacuum. The specimens with aspect ratios of about 5 to 1 were aligned with the magnetic field to minimize demagnetization effects, which were not corected for. The measurements were done by applying a known heat pulse or magnetic field change to the sample and observing the resulting temperature change. Corrections were made for heat losses to the surroundings and for thermal addenda. Corrections for the magnetoresistance of the carbon-glass thermometer used were negligible. More details on the apparatus and method for doing the measurements have been reported previously[5].

RESULTS

Figures 1 and 2 show the measured C_B and ΔT_s of GdPd. Figures 3 and 4 show the corresponding measurements on GdNi. Limited heat capacity data on these

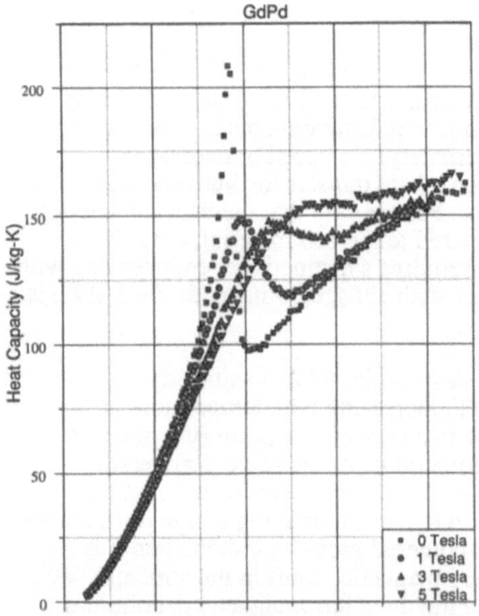

Fig. 1. The heat capacity of GdPd measured at various applied magnetic fields.

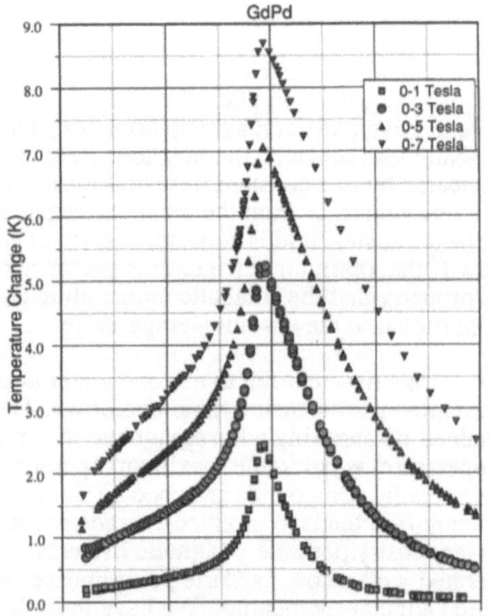

Fig. 2. ΔT_s for GdPd. Data is shown for both increase and decrease of magnetic field.

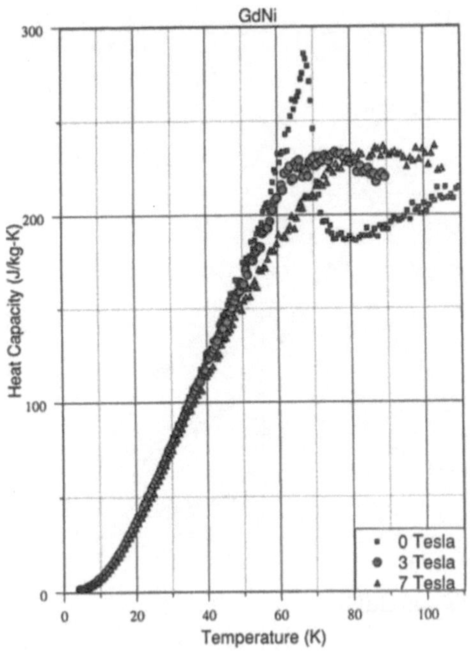

Fig. 3. Measured heat capacity of GdNi.

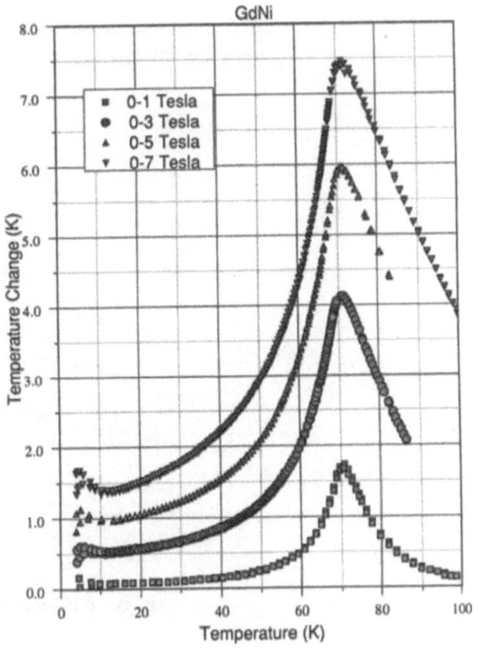

Fig. 4. Measured ΔT_s of GdNi.

compounds from a previous series of measurements has been reported[6,7]. Figures 5 and 6 show C_B and ΔT_s for $GdNi_2$. The data for GdPd, GdNi and $GdNi_2$ each show a sharp transition, at 38 K, 70 K, and 75 K, respectively. These are marked by a sharp lambda-type peaks in the zero field heat capacity and by peaks in the ΔT_s curves. The ferromagnetic character of the transition is verified by the action of applied magnetic fields in broadening the heat capacity peaks and moving them to higher temperature. The adiabatic temperature changes shows the desired monotonic increase in temperature below the Curie point, T_c, although the dependence is not strictly linear.

Figures 7 and 8 show the contrasting behavior of $Er_{0.86}Gd_{0.14}Al_2$. Note the zero-field heat capacity anomaly stretches from about 15 K to over 30 K, and the adiabatic

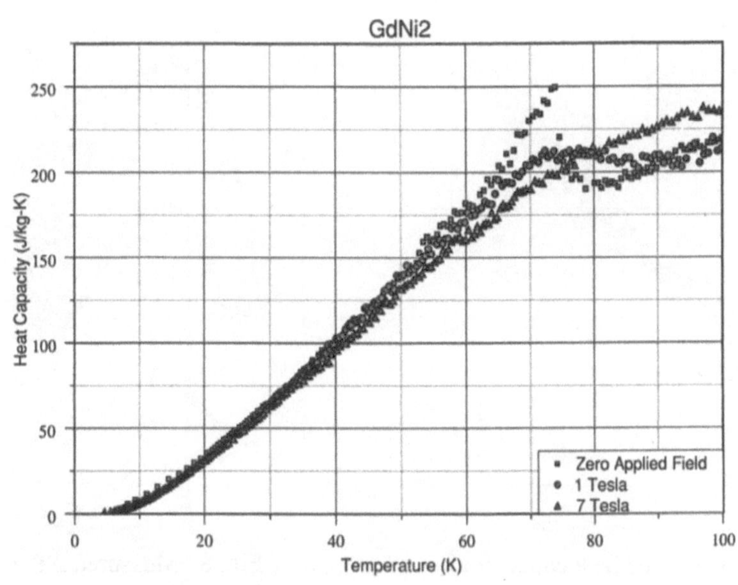

Fig. 5. Measured heat capacity of $GdNi_2$.

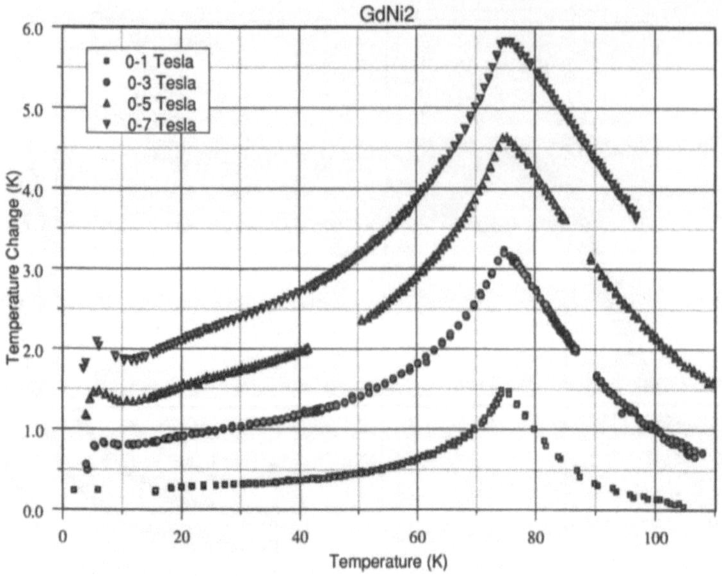

Fig. 6. Measured ΔT_s of GdNi2.

temperature change versus temperature curve is concave downward from 15 K to 35 K. Although the Curie point of this material as observed in the magnetic susceptibility[8] is about 40 K, the temperature dependence of ΔT_s precludes its use in an AMR much above 20 K.

DISCUSSION

The heat capacity of magnetic metals can be considered as the sum of three terms: C_l, the lattice heat capacity, C_e, the heat capacity due to conduction electrons, and C_m,

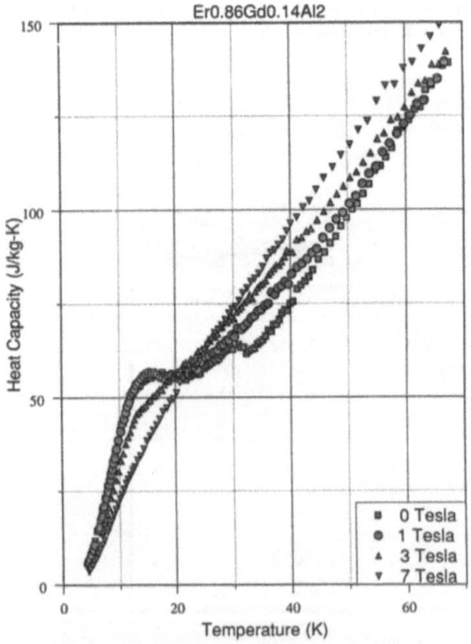

Fig. 7. Measured heat capacity of $Er_{0.86}Gd_{0.14}Al_2$.

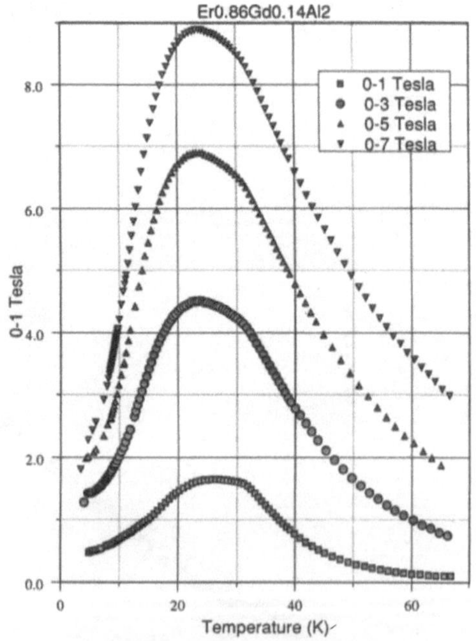

Fig. 8. Measured ΔT_s of $Er_{0.86}Gd_{0.14}Al_2$.

Table 1. Parameters derived from fitting heat capacity data

Compound	γ (J/kg-K^2)	θ_D (K)	2J+1
GdPd	0.05 +0.02	185 +5	8.03 +0.3
GdNi	0.075	205	7.73
$GdNi_2$	0.07	245	7.98
$Er_{0.86}Gd_{0.14}Al_2$	0.05	330	10.165

the magnetic heat capacity. Only the last term is field dependent and is involved in the magnetocaloric effects of interest here. The C_l is conventionally taken to be of the Debye form which goes as AT^3 at low temperature and saturates at a constant value above the Debye temperature; C_e is normally proportional to γT. Plots of C/T versus T^2 at low temperature should be linear with γ as an intercept and A as the slope. These plots work well for non-magnetic metals but for the materials studied here, only γ could be extracted because of curvature in the plots due to the onset of magnetic heat capacity. However, above the Curie point in zero field, the magnetic heat capacity becomes small, allowing fitting of the heat capacity to the full Debye form[9] plus a linear term γT determined from the low temperature C/T plot. The resulting values of γ and the Debye temperature are given in table 1. The raw heat capacity data were corrected for the presence of 5 at.% of rare-earth oxides before fitting; the 5% figure was estimated from metallography.

The fits to the lattice and electronic heat capacities may now be subtracted from the total heat capacity in order to extract the magnetic heat capacity. C_m may be checked using a relation for the magnetic entropy difference S_m between the completely ordered and disordered states of a quantized magnetic moment. The relation is S_m=Rln(2J+1), where J is the quantum number for the total angular momentum. The Gd-based ferromagnets studied here attain a fully ordered state at low temperature as seen by magnetization measurements[6,7,10]; assuming full disorder is attained a few degrees above the Curie point, integrating C/T from 0 to above the Curie point should give S_m and hence 2J+1. The values of 2J+1 obtained are given in table 1, and are close to the theoretical value of 8 for Gd. For erbium, the maximum theoretical value of 16 was not obtained, indicating the persistence of crystal field effects well above the Curie point; in $ErAl_2$, the lower-lying crystal field states[11] have a multiplicity of 10, which is close to the value of 2J+1 obtained.

The value of J obtained from S_m depended very critically on the choice of the Debye temperature; if it was chosen such that the dC_m/dT was zero well above T_c, consistently reasonable values of J were obtained, whereas if the mean value of C_m was made zero above T_c with non-zero dC_m/dT allowed, erratic results were obtained. Part of the difficulty in fitting occurs because the T_c's are high enough that the region where dC_l/dT is large is masked, so the Debye temperature fitting is not very sensitive. Another potential problem is that the Debye form is only approximate, especially for compounds of dissimilar atoms. For the case of non-conducting (hence C_e=0) magnetic compounds with low ordering temperatures, successful fitting has been done using a sum of two Debye functions, one ascribed to each type of atom[12]. Another method used to estimate the lattice heat capacity is to use measured values for an isostructural compound[12], such as LaNi for GdNi. With the exception of the dialuminides, no data in the temperature region of interest is available; a previous author[13] found for the dialuminides that this procedure did not work well.

Plotting the heat capacities of the isostructural compounds GdNi and GdPd in the reduced coordinates C_molar/R and T/T_c produces nearly identical curves (fig 9),

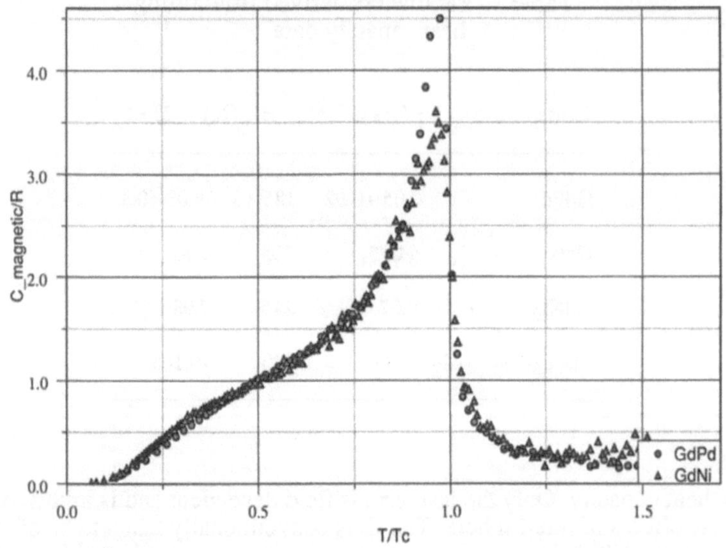

Fig. 9. The magnetic portion of the zero field heat capacities of GdPd and GdNi, plotted in reduced units. Tc is the Curie temperature, C_magnetic is per mole, and R is the gas constant.

showing that the ordering phenomena are similar despite the large difference in interaction strengths. Comparing the reduced heat capacities of GdNi, $GdNi_2$ and $Er_{0.86}Gd_{0.14}Al_2$ (fig 10) shows that orderings are of diverse character. Differing temperature dependences of the orderings of GdNi and $GdNi_2$ are not surprising considering the structures differ in symmetry, numbers of next-nearest neighbors, etc. $GdNi_2$ and $Er_{0.86}Gd_{0.14}Al_2$ have the same structure, but the latter contains two different magnetic rare earth atoms. If the Er and Gd atoms are randomly distributed on the rare-earth sites, the localized nature of the 4f rare-earth electrons would produce a variation in local magnetic environments and hence a broadened magnetic transition, as observed.

Comparing ΔT_s among the materials is more difficult because the presence of the magnetic field introduces another energy, mH, where m is the atomic magnetic moment.

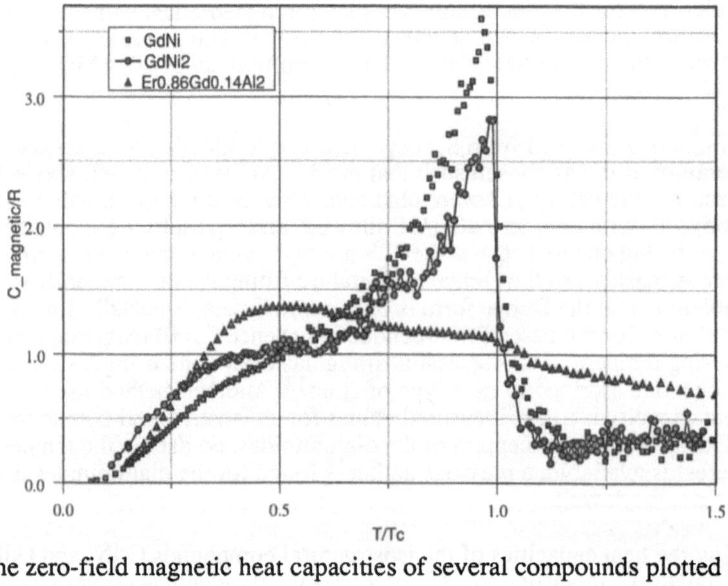

Fig. 10. The zero-field magnetic heat capacities of several compounds plotted in reduced units.

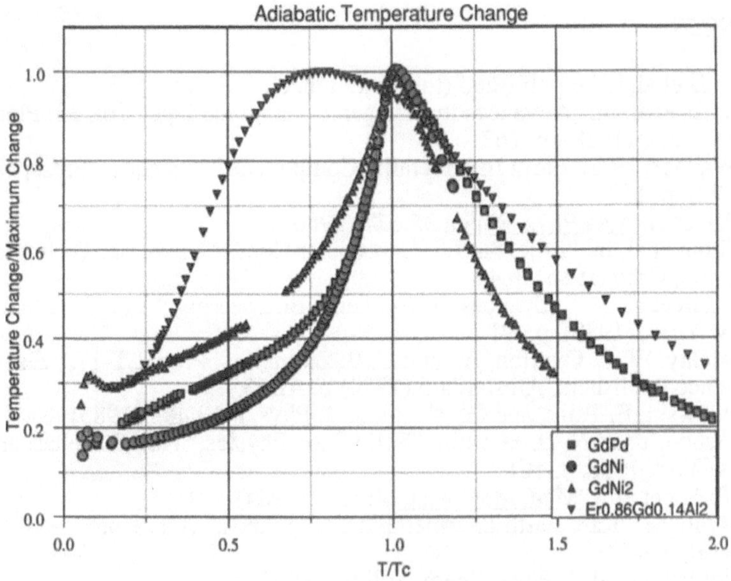

Fig. 11. ΔT_s for a 0-5 Tesla field change, plotted in reduced units.

Correcting for lattice effects is also no longer a simple matter of subtracting a lattice heat capacity. Plotting ΔT_s divided by its maximum value against T/T_c (figure 11) does show similar trends to that of figure 10. ΔT_s of GdNi and GdPd are similar; $GdNi_2$ is slightly broader and $Er_{0.86}Gd_{0.14}Al_2$ shows a wide peak.

From the standpoint of 20 K to 80 K AMR application, $GdNi_2$ is an excellent material to use in an upper stage. ΔT_s is approximately linearly proportional to temperature from 40 K to 80 K, and its maximal value of 0.9 K per Tesla allows use of of reasonable magnetic fields of 5 to 7 Tesla. GdNi has a larger heat capacity and ΔT_s than $GdNi_2$, but the rapid temperature dependence of ΔT_s below T_c makes it less suitable for use over a broad range of temperature. GdPd would be suitable in a lower stage spanning the remaining 40 K to 20 K region. In this temperature range, C_p of helium decreases with increasing temperature, suggesting that a ΔT_s increasing faster than linearly with T may be advantageous. $Er_{0.86}Gd_{0.14}Al_2$ is only suitable for use below 20 K because of the broadening of its magnetic transition. It has worked well in a 7 K to 20 K AMR device[14].

CONCLUSIONS

Several ferromagnetic materials have been studied for their utility as magnetic refrigerants in the 20 K to 80 K range. The binary Gd compounds GdNi, $GdNi_2$ and GdPd have been found to have single, sharp ferromagnetic ordering transitions encompassing all the available magnetic entropy. The mixed rare-earth pseudo-binary compound $Er_{0.86}Gd_{0.14}Al_2$ has been found to have a much broader magnetic transition which we ascribe to the disordered rare-earth sublattice and the localized nature of the 4f electrons.

$GdNi_2$ and GdPd have properties well suited for use in AMR stages operating from 80 K to 40 K and 40 K to 20 K, respectively. Further work is being done to find lower cost alternatives to GdPd.

ACKNOWLEDGEMENTS

Some of the materials covered in this study, as well as the methods of measurement, were suggested by Dr. J. A. Barclay, whose advice and encouragement are gratefully acknowledged. This work was partly funded by the U. S. Department of Energy, contract #DE-AC02-90CE40895.

REFERENCES

1. D. J. Janda et al, to be published (this conference).
2. C. R. Cross et al, in: "Advances in Cryogenic Engineering," Vol. 33, Plenum Press, New York, (1988), p. 767.
3. W. E. Wallace, "Rare Earth Intermetallic Compounds," Academic Press, New York (1973), p. 113
4. Y. Y. Pan et al, Acta Phys. Sinica 35:677 (1986)
5. C. B. Zimm et al, in: "Proc. 5th Int. Cryocooler Conf., "Universal Technology Corp., Dayton, Ohio (1988), p.49
6. C. B. Zimm et al, in: "Advances in Cryogenic Engineering," Vol. 33, Plenum Press, New York, (1988), p.791
7. J. A. Barclay, W. C. Overton, Jr., and C. B. Zimm, in: "Proc. LT-17," Elsevier Science Publishers, Amsterdam (1984) p. AL13
8. A. Chelkowski, E. Talik, and G. Wnetrzak, J. Phys. F 13:483 (1983)
9. N. W. Ashcroft and N. D. Mermin, "Solid State Physics," Holt, Rinehart and Winston, New York 1976, p. 461
10. M. R. Ibarra et al, J. Mag. Mag. Mat., 46:167 (1984)
11. W. E. Wallace, "Rare Earth Intermetallic Compounds," Academic Press, New York (1973), p. 43
12. J. A. Hofmann et al, J. Phys. Chem. Solids 1:45 (1956)
13. W. R. Johanson et al, J. Appl. Phys., 64:5892 (1988)
14. A. J. DeGregoria et al, to be published (this conference)

DESIGN OF AN ACTIVE MAGNETIC REGENERATIVE HYDROGEN LIQUEFIER

Dennis Janda, Tony DeGregoria, Joseph Johnson, Stephen Kral

Astronautics Technology Center
Astronautics Corporation of America
Madison, Wisconsin

Glenn Kinard
Air Products and Chemicals, Inc.
Allentown, Pennsylvania

ABSTRACT

Design of a two-stage, 1/10 ton/day demonstration Active Magnetic Regenerator (AMR) Hydrogen Liquefier operating between 77 K and 20 K is described. The device uses particle beds of rare earth intermetallic materials as the regenerators in each stage with helium gas as the heat transfer fluid. A detailed design of the device is described in this paper, with relevant physical and material properties considered. The magnetic field profiles and resulting forces on the particle beds and between magnets during operation are reviewed. The device is compact, efficient, and operates at slow speed for long life. It is the first phase of a six year development effort to produce an economical one ton/day magnetic hydrogen liquefier.

INTRODUCTION

During the energy crisis of the 1970's, magnetic refrigeration was identified as a technology that could conceivably provide highly efficient refrigeration above 1 K, perhaps to and above room temperature. The interest in MRs stemmed from the reversible nature and inherently high energy density associated with the magnetocaloric effect. Analyses have shown that these properties lead to devices that are potentially more efficient, compact, and reliable than conventional gas cycle refrigerators. For this reason, MR technology has been steadily pursued since the middle 1970's and has reached the engineering prototype stage. This paper outlines the design of a large scale Magnetic Hydrogen Liquefier, under joint development by Astronautics Corporation of America and the Department of Energy Office of Industrial Programs.

As a result of the Magnetic Heat Pump Workshop sponsored by the DOE Office of Industrial Programs and related marketing studies,[1] the liquefaction of hydrogen with liquid nitrogen precooling was selected as a good industrial market sector for initial commercialization of MR. Preliminary studies indicate that an MR liquefier system offers reduced capital and operating costs over conventional liquefiers. The initial development phase focuses on the design of a subscale liquefier. Once fabricated and demonstrated, a full-scale 1 ton/day liquefier will be developed.

SUBSCALE SYSTEM DESCRIPTION

A subscale (approximately 1/10 ton per day) magnetic liquefier has been specified to demonstrate the feasibility of magnetic refrigerators for liquefaction. The refrigeration is based on the Active Magnetic Regenerator (AMR) cycle already demonstrated at Astronautics Corporation.[2] Liquefier production is approximately 50 l/hr of liquid para-hydrogen at 20 K from precooled 77 K gas feedstock. A liquid nitrogen bath provides the precooling. The gaseous hydrogen feedstock is supplied at 4 MPa (40 atm) and approximately 41.5% para concentration. Helium pressure within the magnetic refrigerator is 2 MPa (20 atm). The subscale refrigerator incorporates two stages to span between 77 K and 20 K. A 2 sec. period for the AMR cycle operates on rare earth intermetallic packed particle beds. Initial performance modeling has been based on $GdNi_2$ and GdPd first and second stage beds, respectively. Superconducting NbTi solenoid magnets cooled by liquid helium provide a 7 T field for the first stage beds and a 5 T field for the second stage.

SYSTEM CONFIGURATION

AMR operation is characterized by bidirectional flow of a heat exchange gas through active magnetic regenerative beds. A discussion of the theory and operating principles of AMR devices is reviewed in "Initial Test Results of an Active Magnetic Regenerative Refrigerator."[2] The bidirectional nature of the helium flow through the AMR beds necessitates a valving and control arrangement to coordinate the "blow" and dwell periods of the AMR cycle. The flow schematic is relatively simple, as illustrated in Fig. 1. A helium circulation pump provides a continuous flow of 2 MPa (20 atm) gas through the refrigerator with 0.14 MPa (20 psi) boost. This accounts for the pressure drops throughout The

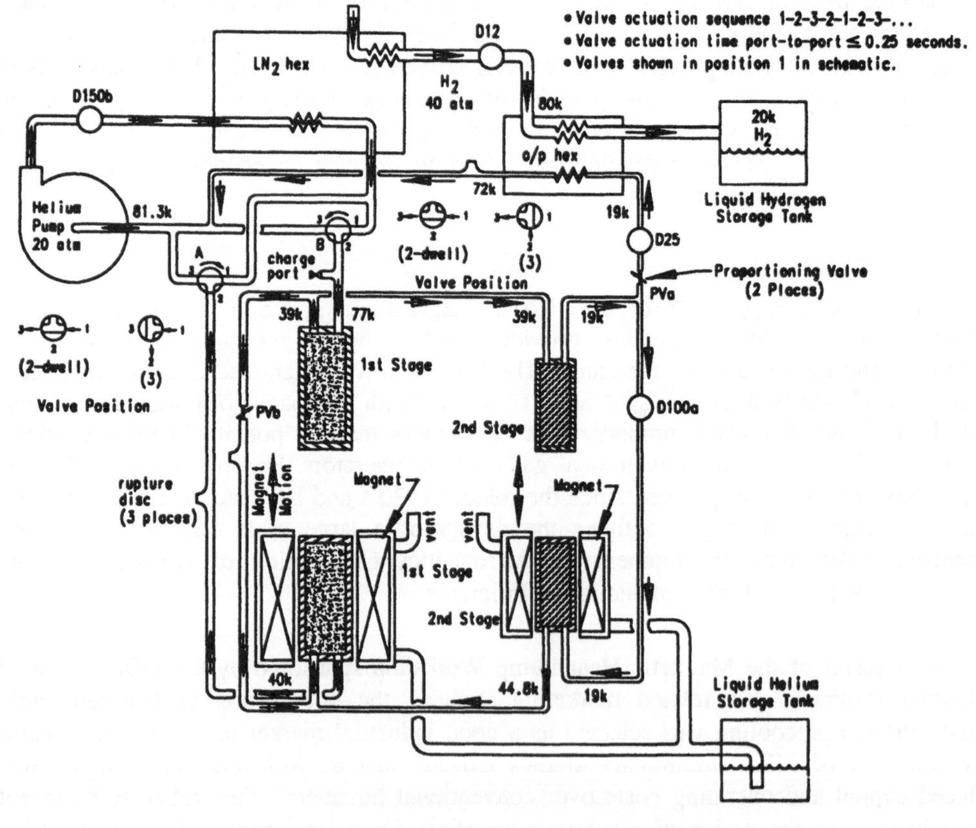

Fig. 1 Hydrogen liquefier flow schematic.

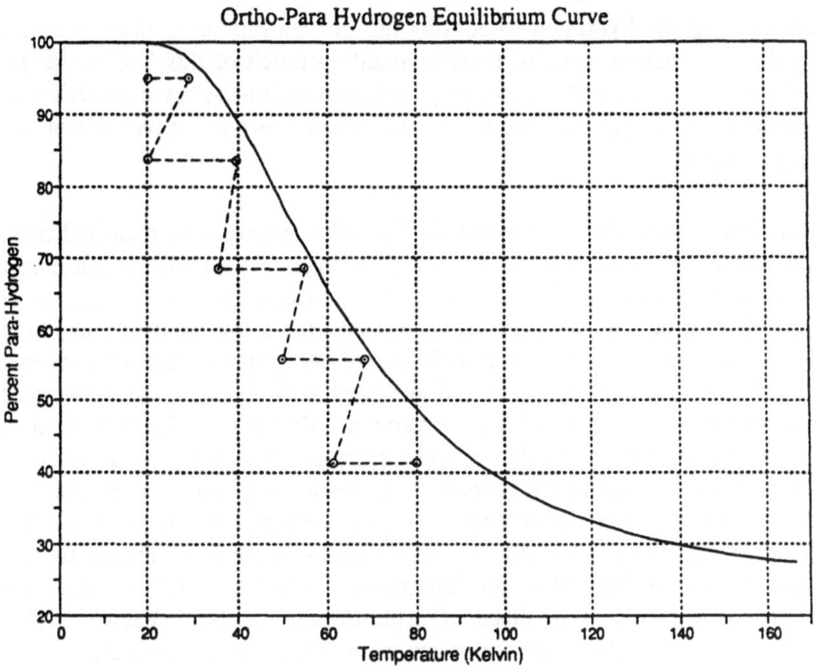

Fig. 2 Ortho-to-para conversion in hydrogen liquefier.

piping, valves, heat exchangers, and beds. During dwell periods (0.5 sec.) the valves are activated such that all helium flow is diverted directly to the pump inlet (position 2 in the schematic), thereby bypassing the refrigerator.

As depicted in the figure, flow passes through the demagnetized first stage bed and splits. A portion of the flow passes on to the second stage demagnetized bed while the remainder removes heat from the magnetized first stage bed and returns to the pump. Flow leaving the second stage demagnetized bed is split between the Ortho-Para Heat Exchanger (O/Phex) and the magnetized second stage bed. These flows also return to the helium pump inlet, and last 0.5 seconds. The flow is then diverted for a 0.5 sec. dwell period while the magnets shuttle to the previously demagnetized beds. Flow through the refrigerator is resumed in the opposite direction for 0.5 sec. and then diverted while the magnets shuttle back to the originally magnetized beds. This 2 sec. cycle is continuously repeated during liquefier operation. This flow arrangement results in unbalanced mass flows through the beds. A larger mass flow passes through the demagnetized beds than through the magnetized beds because some flow is diverted to the O/Phex without passing through the magnetized beds. This flow imbalance would normally result in warming of passive beds. The varying heat capacity of active magnetic beds, however, mitigates this reaction to the flow imbalance.

The flow diverted to the O/Phex not only cools the hydrogen gas to 20 K, but also removes the heat of conversion from ortho-to-para hydrogen. Ortho-hydrogen and para-hydrogen refer to the two energy states of the diatomic hydrogen molecule. These states are distinguished by the parallel (ortho), and anti-parallel (para) orientation of their nuclear spins. The relative equilibrium concentrations of these states varies with temperature, as depicted in Fig. 2. Conversion of ortho-hydrogen to para-hydrogen is a naturally occurring exothermic reaction. Room temperature hydrogen gas liquefied without ortho-to-para conversion will boil off at an initial rate of approximately 20% per day[3] as it naturally converts to 100% para-hydrogen in reaching its 20 K equilibrium state. Catalysts are incorporated to bring the hydrogen to equilibrium concentrations at intermediate temperatures during the liquefaction process, thereby avoiding much of this loss at the expense of an increased thermal load on the liquefier. Optimum ortho-to-para conversion is accomplished in incremental stages to remove the associated heats of conversion at the highest possible temperatures.

It is not always necessary to convert all the ortho-hydrogen to para-hydrogen. If the storage time before the liquid hydrogen is used is short, it then makes sense to limit the degree of conversion during the liquefaction process. Regardless of the degree of conversion to para-hydrogen during the liquefaction process, the ortho concentration in the liquid will eventually approach zero over time with loss of liquid due to heat of conversion causing boil-off. If the liquid is to be stored for long periods then the conversion before storage should be complete to 100% para-hydrogen. This avoids the boil-off losses associated with the heat of conversion that would otherwise take place in the liquid. In short, an optimal ortho-para concentration proportion exists for a given storage time.[4] Hydrogen to be used immediately after liquefaction, however, requires no conversion. To demonstrate a complete liquefaction system, the Astronautics liquefier simulates the production of 100% para-hydrogen by using an integrated Ortho-Para Helium Heat Exchanger (O/Phex) with electric heaters in place of ortho-to-para catalytic reactors. The heat of conversion is simulated by the heaters in place of the more costly catalysts.

Also, nearly all gas cycle hydrogen liquefiers use LN_2 for precooling of the H_2 process stream and to extract as much of the heat of conversion as possible before expansion. The magnetic liquefier incorporates this same feature. The ortho-to-para conversion below 77 K for the liquefier takes place in four steps in the O/Phex with the first occurring after the H_2 has been cooled to about 61 K and the last after the H_2 has been cooled to roughly 20 K. After this last conversion the hydrogen is approximately 95% para-hydrogen in content. The actual conversion steps are illustrated in Fig. 2.

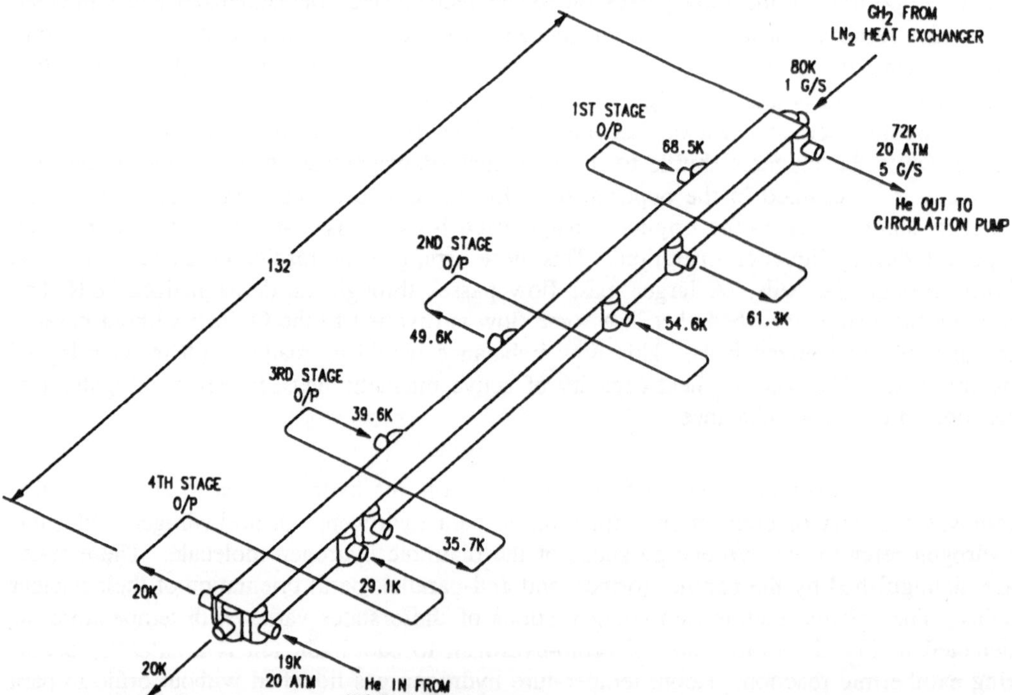

Fig. 3 Ortho-to-para heat exchanger.

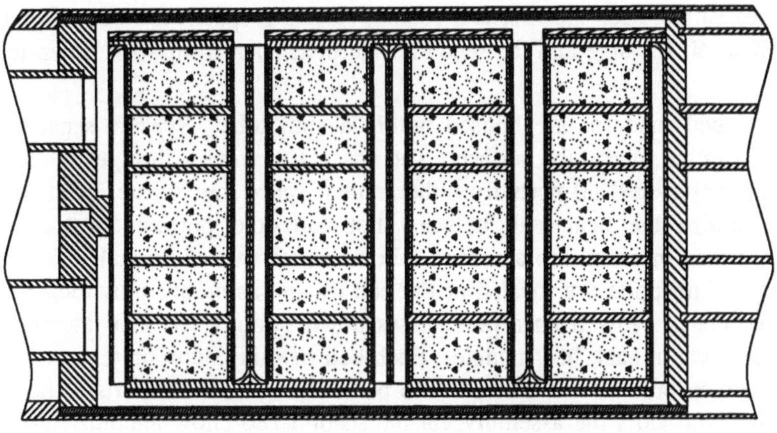

Fig. 4 Typical regenerator bed assembly.

The LN_2 heat exchanger precools the hydrogen process stream prior to entering the O/Phex heat exchanger. The LN_2 heat exchanger also precools the circulating helium gas after it exits the helium circulation pump to remove the heat of compression. This pump generates a 0.14 MPa (20 psi) boost to circulate the helium through the two magnetic refrigerator stages and the O/Phex. An LN_2 reservoir is coupled to the precooler heat exchanger to continuously replenish the liquid that boils off within the precooler during the heat exchange with the helium and hydrogen streams. Boiling heat exchange takes place in the bottom zone of the vertically oriented assembly at a rate of approximately 168 kg/hr (370 lbs/hr). Cold nitrogen boil-off vapor provides initial cooling to the hydrogen and helium flows in the top zone of the heat exchanger prior to being vented.

The O/Phex heat exchanger removes the heat of conversion in four stages while incrementally cooling this stream before and after each stage. Cooling is provided by the 5.0 g/s counter-flow of helium from the second stage of the magnetic refrigerator. The configuration of the heat exchanger is illustrated in Fig. 3. The hydrogen process stream is diverted from the heat exchanger for each separate ortho-to-para conversion. Reentrance of the heated hydrogen gas occurs at a location of roughly equivalent temperature in the heat exchanger. A relatively linear temperature gradient from 19 K to 80 K spans the 3.4 m (132 in.) length of this assembly. The hydrogen enters the heat exchanger with roughly a 41.5% para concentration from the LN_2 heat exchanger. After the first conversion stage the para concentration is boosted to 56%. After the second, third, and fourth stages it is boosted to 69%, 84%, and finally 95% para hydrogen.

MAGNETIC STAGES

The cooling of the H_2 process stream below 77 K in the O/Phex is provided by the magnetic stages of the liquefier. Helium cooled in the magnetic beds circulates through the O/Phex to extract heat from the H_2. The helium is cooled in two stages.

Each of the two stages is comprised of a pair of bed assemblies. The assemblies are composed of a stack of four individual bed modules housed within three concentric thin wall tubes. The configuration of one bed assembly is pictured in Fig. 4. All structural components in the assembly are fabricated from G-10 composite material. All joints are made with cryogenic grade epoxy resins. The first stage bed material is currently specified as 0.15 mm dia spherical $GdNi_2$ particles and the second stage is GdPd. These materials have been selected based on their adiabatic temperature change over the temperature ranges suitable for the first and second stages, respectively. The first stage beds span the 77 K to 40 K range and the second stage beds span from 40 K down to 20 K.

Individual bed modules are housed within a split tube. Spacers separate the outer module wall from the inner tube surface. The spacers provide a "bridge" between the two cylindrical surfaces to seal all ports from the annular space they create. This space is evacuated to minimize the heat leak between hot and cold portions of the beds. This subassembly is then inserted into a second tube to create two additional 180° annular flow spaces. Two helium ports are located at the end of this assembly providing for flow into and out of the beds. Flow from a port passes down through the 180° annular space and enters the side ports in each bed module. Each bed module incorporates a top and bottom flow header to distribute flow over the entire bed surface from the ports. The flow passes through the bed and exits the assembly via the second 180° flow annulus. Pressure drops are kept to a minimum by incorporating large cross-sectional flow areas where practical.

The bed structure must accommodate large magnetic forces and provide for helium flow into and out of the bed modules. The peak magnetic force on each of the first stage bed assemblies is approximately 285,000 N (64,000 lbs.) The actual net drive loads are significantly less due to the close proximity of the two bed assemblies to one another in each stage. In general, each bed assembly is attracted toward the magnet, regardless of its position between them. Therefore, the attractive force between the magnet and one bed is either partially or completely counteracted by the other bed. The bed structure nevertheless must support the full magnetic loads.

The magnetic loads result from the pair of superconducting solenoid magnets that generate the magnetic fields necessary for liquefier operation. The first stage magnet generates a uniform 7 T field throughout its bore length to alternately magnetize each of the two first stage beds. The second stage magnet similarly generates a 5 T field. Both magnets are of solenoid configuration, using NbTi wire. A large magnetic iron flux return is provided to contain the stray fields thereby minimizing the effects of eddy current heating in adjacent, stationary electrically conducting components. A relatively conservative current carrying capacity has been designed into the windings to limit the chances for quenches as the magnets reciprocate between beds.

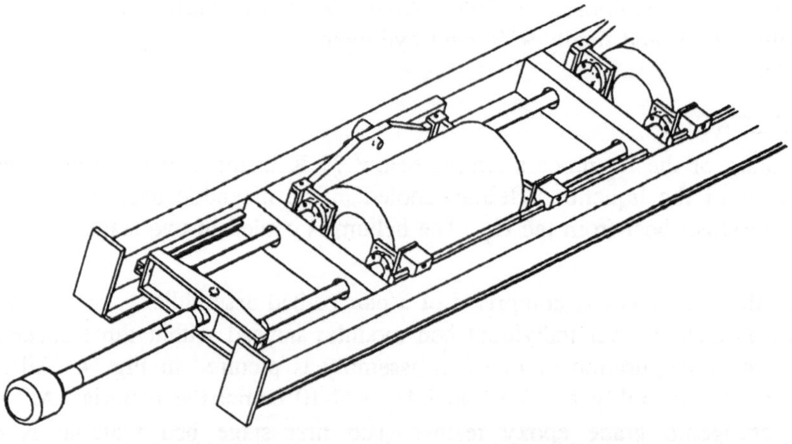

Fig. 5 Hydrogen liquefier magnetic stage.

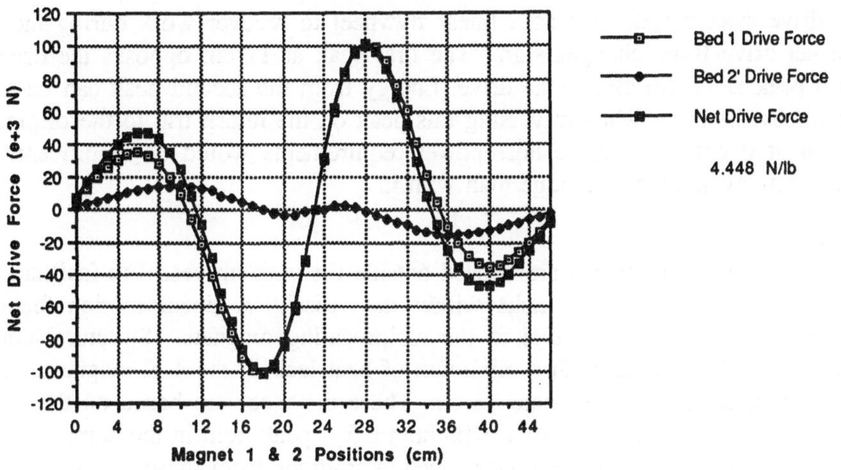

Fig. 6 Hydrogen liquefier net drive loads.

The magnets are wound on thin wall stainless tubes with iron flanges. After winding, the coils are encapsulated by an iron jacket that completes the flux return path between these two flanges. This assembly is mounted within a liquid helium dewar that is in turn thermally, but not structurally isolated from both the drive and linear guide system. The magnet windings, persistent mode switch, and joints remain immersed in liquid helium during operation. The helium level in the magnet dewar is constantly replenished via a flexible transfer line connected to a stationary helium reservoir. The heat leak into each magnet dewar from the drive and linear guides is estimated to be 1 W. Heat leak from the leads does not enter the magnet dewar during operation. These leads are demountable and remain disconnected while the magnets reciprocate. The configuration of a magnetic stage is illustrated in Fig. 5.

The magnets experience a changing flux during persistent mode operation due to the changing material permeability in each bore while shuttling between beds. This can result in AC losses in the windings and persistent mode decay. This may be largely offset by adding magnetic material between the bed pairs in each stage with careful attention to avoid significant eddy current heating. Magnet recharging is initiated when the fields decay below a threshold value.

The portion of the magnetic load imposed on the beds by the magnets that is not mutually cancelled must be counteracted by the reciprocating drive. Estimates of loads imposed on the drive system during operation of the liquefier are plotted in Fig. 6. The net peak drive load is approximately 100,000 N (22,500 lbs). The net drive force is zero at three distinct locations between the pair of bed assemblies in each stage. During the dwell periods when the magnets are parked at the centers of either of the two bed assemblies comprising each stage (ie. at the travel extremes) a net magnetic force must be sustained by the drive of approximately 8,000 N (1,800 lbs). This force tends to pull the magnets off the beds and assists the drive in accelerating the magnet at the start of motion. This force also assists the drive in decelerating the magnet at the end of each travel segment. It is during the peak loads at mid-travel that the drive must generate the most force and work. The time for one complete cycle of motion is 2 sec. This low frequency limits the inertial loads imposed on the drive as well as the accumulation of cycles.

The drive system functions as a linear flywheel to recover work during the periods when the net drive force changes sign. The first peak at 17 cm opposes the drive while the second peak at 28 cm assists the drive. Energy from the second peak can be stored to be expended by the drive when traversing this peak on the return trip to the original starting location at 0 cm. The drive horsepower requirements would be significantly larger without this energy storage technique than with it.

All major drive components are housed outside the vacuum vessel to facilitate servicing and minimize contamination risks within the vessel. A linear motion feedthrough transfers drive loads into the vessel to the reciprocating magnets. Structural composite tubes transfer the drive loads with a minimum of heat leak to the 4 K magnets. The first and second stage magnets reciprocate 46 cm. These distances are based on the minimum allowable bed separation resulting in a maximum 0.5 T peak field in the demagnetized bed with the magnet centered over the adjacent bed. Variations from these distances for either stage results in an increase in the demagnetized peak field or significant increases in net drive forces. The increase in net drive force is due to the decrease in magnetic "coupling" between beds as their separation distance increases.

CONCLUSIONS

The subscale hydrogen liquefier design described in this paper is based on a single-stage AMR proof-of-principle refrigerator currently under test at the Astronautics Technology Center. This device has performed beyond expectations and has clearly demonstrated viability of the concept. Improvements incorporated in the Hydrogen Liquefier design have been derived from laboratory test experience with this device. Magnetic materials testing of regenerator beds has confirmed the suitability of the chosen bed materials. The next phase of this effort will be to fabricate and test the liquefier in conjunction with Air Products and Chemicals Inc.

ACKNOWLEDGEMENTS

Astronautics engineering staff contributing to the design of the Hydrogen Liquefier at the Technology Center include: Alex Jastrab, Jerry Kaliszewski, Rick Lubasz, and Kurt Eckroth. Their contributions are greatly appreciated. Major funding for this effort has been supplied by the Department of Energy and Astronautics Corporation. The vision of both organizations has made it possible for this development to proceed.

REFERENCES

1. J. A. Waynert, A. J. DeGregoria, R. W. Foster, J. A. Barclay, "Evaluation of Industrial Magnetic Heat Pump/Refrigerator Concepts that Utilize Superconducting Magnets," Astronautics Corporation of America (ATC), Madison, Wisconsin, Final Report, Prepared for Argonne National Laboratory, Contract Number: 90232402 (1989).
2. A. J. DeGregoria et al, "Initial Test Results of an Active Magnetic Regenerative Refrigerator," Proc. 4th Interagency Meeting on Cryocoolers (1990) p. 277.
3. R. B. Scott, "Cryogenic Engineering," D. Van Nostrand, New York (1959), p. 290.
4. S. A. Sherif, et al, "Analysis and Optimization of Hydrogen Liquefaction and Storage Systems," University of Miami, Coral Gables, Florida (1990).

THE AXAF/XRS ADR: ENGINEERING MODEL

Aristides, T. Serlemitsos/GSFC; Marcelino SanSebastian, Evan S. Kunes/STX Corp.

Code 713, NASA/Goddard Space Flight Center
Greenbelt, MD 20771

ABSTRACT

A spaceworthy Adiabatic Demagnetization Refrigerator is under development at Goddard Space Flight Center as part of the X-Ray Spectrometer (XRS), an instrument on the Advanced X-ray Astrophysics Facility (AXAF). XRS will employ an array of 32 microcalorimeters capable of detecting X-rays in the energy range of 0.3 - 10 keV. In order to achieve a desired resolution of 12 eV, these detectors must be operated at a temperature of 0.065 - 0.100 K. An ADR must be used to cool these detectors in space.

A breadboard model was designed and built less than two years ago, and provided excellent results. We are presently at the development stage of the engineering model. Several changes have been made to the original design in order to improve the efficiency of the ADR, to reduce its weight, and to strengthen the salt pill suspension system so that the ADR can survive launch loads and have low sensitivity to microphonic inputs. We shall report on the results of these changes; what worked and what did not.

INTRODUCTION

The XRS[1] has been selected to be one of the experiments flown on AXAF[2]. An array of 32 microcalorimeters will be used to detect X-rays in the energy range of 0.3 - 10 keV. In order to achieve a desired resolution of 12 eV, the detectors must be operated at 65 mK. The only way this temperature can be achieved in space at the present time is through the use of an ADR. An ADR development started at GSFC in the early 80's[3]. In the past three years, when XRS became a viable experiment for the AXAF, the ADR development effort has been intensified. Various breadboard model ADR's have been built and tested extensively[4,5]. The experience acquired from these test lead to the design of the engineering model. Due to weight limitations the ADR magnet can produce a maximum field of only 1.9 T, which limits the ADR cooling power considerably. This along with the requirement that the experiment have a long lifetime, dictated that the ADR delivers the maximum possible efficiency. In the engineering model design a lot of attention has been paid to detail, in order to achieve the desired efficiency. However, efficiency must be achieved at no expense to the space worthiness of the ADR.

In this paper we are going to present a brief account of the recent history of the GSFC ADR development. We feel it is important to future ADR development efforts to point out various technics used along with what worked and what did not.

The main components of an ADR are the salt pill and the heat switch. The salt pill consists of the paramagnetic salt crystal which provides the magnetic cooling, and the thermal bus which provides thermal contact to the heat switch during magnetization, and to the detector cold stage during 'low' temperature operation. The heat switch is used to remove the heat generated in the salt during magnetization and transfer it to the dewar cryogen. Once the salt is magnetized and its temperature drops near the bath temperature, the heat switch is 'turned off' thermally isolating the salt pill. For an ADR used to cool very sensitive detectors in space, a very sturdy suspension is also necessary to protect it from damage during lift-off and to minimize microphonic input to the detectors.

As mentioned above, the development of the GSFC ADR started in the early 80's. However, it was not until XRS became a viable AXAF experiment, that the ADR development began in earnest. The first XRS ADR was built in 1988 to demonstrate the feasibility of cooling the detectors in space. The salt pill was constructed by sandwiching 3-mm thick salt crystal slabs and ribbons made with 150 μm copper wires. Both the crystal slabs and the wire ribbons were thoroughly coated with lacquer. The reason for the coating is to prevent water loss by the crystals, and to protect the copper wires against corrosion. A cylindrical shape was formed and enclosed in a fiberglass shell. Apiezon J-oil was used to fill the gaps and to provide better thermal contact between the salt and the thermal bus. Both ends of the salt pill were sealed using fiberglass disks and Stycast[6] 2850-FT epoxy. The wire bundle coming out of the salt pill was cold-welded by electroforming to a copper plate on which the modular heat switch was bolted. A gas gap heat switch is used for the XRS ADR. Figure 1 shows a general schematic diagram of the type of heat switch employed in the ADR. During the development phase, modularity was used whenever possible, to make it easier to experiment with various designs. The heat switch employed in the first XRS ADR differed from that shown in Figure 1 in that Stycast was used to attach the Vespel shell to the two endplates. In the breadboard models the Vespel shell was bolted to the end plates using indium O-rings as vacuum seals. Bellville washers were used to maintain pressure on the indium O-rings and avoiding vacuum leaks after cooldown.

The first XRS ADR served its purpose in that it showed that an ADR could be built which could reach temperatures below 0.1 K and have a reasonable hold time. Yet it became immediately obvious that considerable improvements were needed to achieve the goals of the XRS ADR program. The method used to construct the salt pill resulted in the salt occupying only ~80 % of the available volume reducing the cooling power of the ADR. The thermal

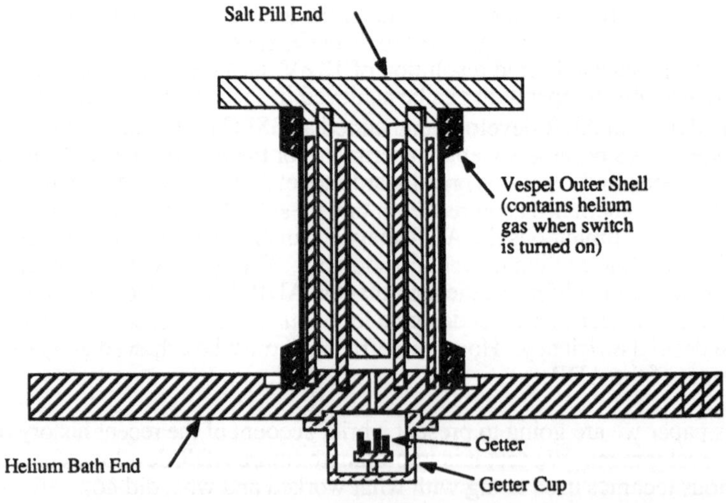

Figure 1. Schematic Diagram of the Breadboard Model Heat Switch

contact between the salt and the thermal bus was not very good. This results in a long time constant and not so fine temperature controlling. Also, the sealing of the salt pill was questionable. The salt used, ferric ammonium sulfate, undergoes decomposition above ~39°C. Stycast is quite viscous at room temperature, so filling the space between wires in a bundle of 4000 - 5000 wires is not easy. Less viscous epoxies may provide a better seal, but their behavior at low temperatures must first be studied. Finally, when electroforming is used to cold-weld the thermal bus wires to the heat switch interface plate, care must be taken to clean the wire bundle between this plate and the salt pill very thoroughly. Solution from the electroforming bath will cause corrosion to the copper wires if not cleaned completely.

The results obtained from that first XRS ADR pointed out to the need to make several improvements. The best way to achieve a good thermal contact between the thermal bus wires and the salt, is to grow the crystal around the wires. However, we were faced with the fact that ferric ammonium sulfate is extremely corrosive to copper. When the 150-μm copper wires are placed in the 37°C salt solution they disintegrate in a few hours. We made several wire strips, and gold-plated each with different film thicknesses, ranging from 7 μm to 38 μm, and kept them in warm salt solution for a week. We found that even the 25-μm gold film did not provide 100% protection. Only the 38-μm film did that. We proceeded to construct a cylindrical wire harness, 62 mm in diameter and 15 cm long, made up of a total of 6080 wires. On the heat switch end the harness wires were formed into a cylindrical bundle ~1.5 cm in diameter and passed through a central hole in an 7-cm diameter stainless steel plate. This plate had eight 8-32 threaded studs attached to it protruding ~2 cm toward the heat switch side. The wires were folded and evenly spread around the plate. Silver epoxy was liberally applied to the wires. A 3-mm copper heat switch interface plate, with a hole pattern matching the studs, was used to sandwich the silver epoxy and wires between it and the stainless steel plate. Even pressure was applied, until the epoxy had cured, with a 12-mm thick stainless steel plate using 8-32 nuts. 90% of the wires were cut at the edge of the stainless steel and copper plates. The remaining 10% were formed into four 150-wire bundles and wrapped around four copper rods which had been welded on the copper plate. These 15-cm rods serve as the thermal bus to the calorimeter cold stage. The wire bundles were cold-welded on the calorimeter thermal rods. The whole assembly was plated with 7 μm gold film. The wire harness, from the heat switch interface plate on, was then further gold-plated to a total film thickness of 38 μm for protection against corrosion. The salt pill thermal bus was then inserted into a fiberglass shell and sealed at the bottom with stycast. Salt crystal was grown around the thermal bus to the desired length and the top was sealed with a fiberglass end cap and stycast.

The thermal performance of this salt pill was excellent, and in two years it has shown no deterioration. One minor problem was that with this design, the two copper plates used to connect thermally the salt pill to the heat switch generate considerable eddy current heating to reduce the cooling power by 5-7 %. This, however can be minimized by slotting the two interface plates. Also, the indium used to provide better thermal contact between these two plates, traps some magnetic flux further reducing the cooling power of the salt pill. Again this effect can be minimized by using several narrow indium strips placed radially between the two plates. The biggest disadvantage of this salt pill was its elevated weight and cost. The extra gold-plating added ~650 g to the weight of the salt pill. The extra weight requires a much heftier suspension system which, in turn, adds to the overall weight of the ADR and also results in slightly higher parasitic heat load into the salt pill due to the heavier suspension cords that must be used. Of these two the increased weight is a serious problem, since XRS is already under pressure to reduce its overall weight.

In an effort to reduce primarily the weight , but also the cost of the ADR, we experimented using different wire for the thermal bus harness. Our tests showed that triple-FORMVAR insulated copper wire was not attacked by the salt solution, even when kept in a warm solution for a whole month. We, therefore, constructed a salt pill similar to the above, but using FORMVAR insulated wire. The results were disappointing. Our tests indicated poor thermal contact between the thermal bus and the salt. In order to assure ourselves that we did not damage the salt by accident, we melted the salt away and grew a new salt crystal using brand new salt solution. The results were the same. We, therefore, concluded that FORMVAR-insulated wire does not make good thermal contact with the salt, for unknown reasons. We also attempted to build a small salt pill with a thermal bus harness made out of aluminum wire, since our tests had shown that aluminum was not attacked by the salt solution. Our tests of that salt pill showed considerable trapping of magnetic field lines. Under no circumstances were we able to lower the temperature o that ADR below ~ 140 K. Unfortunately one of the two end cap of this salt pill was also aluminum. Therefore, we do not know whether the flux trapping that caused these difficulties was primarily in the wires, in the end cap, or in both. If time permits, we are planning to study this further in the future.

The experiences presented above, in relation to wire harness thermal buses, lead us to the decision to look for alternative ways to construct the salt pill.The engineering model ADR is presently under construction, so we have no data on it. A schematic diagram of the whole ADR is shown in Figure 2. We now present a brief description of the basic features of this design. The following considerations were very instrumental in arriving at this design. Copper surfaces, be it cylindrical or flat, could be protected against corrosion with no more that a 3.5 μm film of gold plating which adds much less weight and cost to the salt pill.

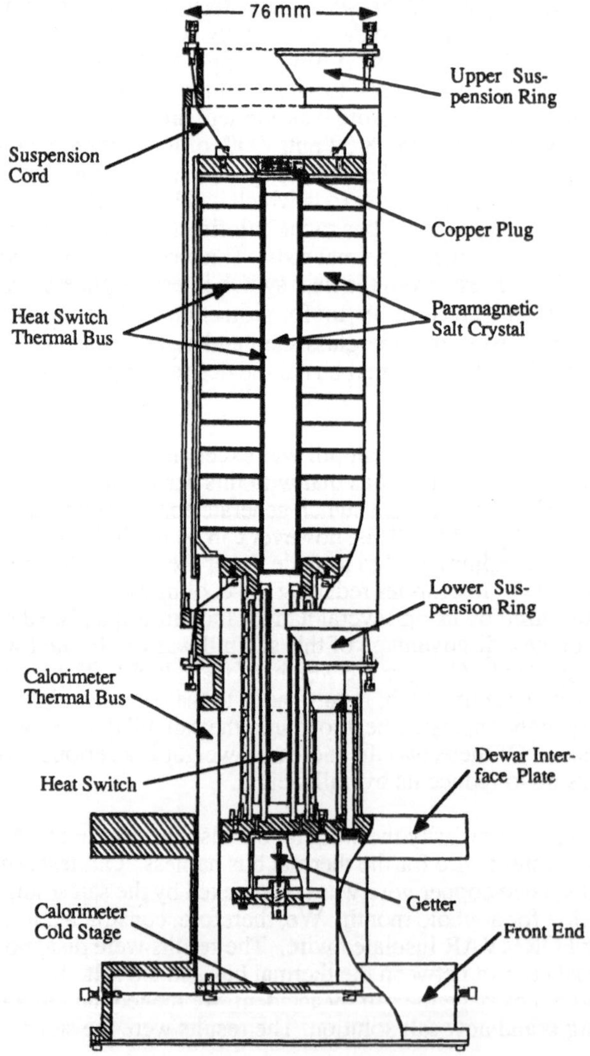

Figure 2. Engineering Model ADR

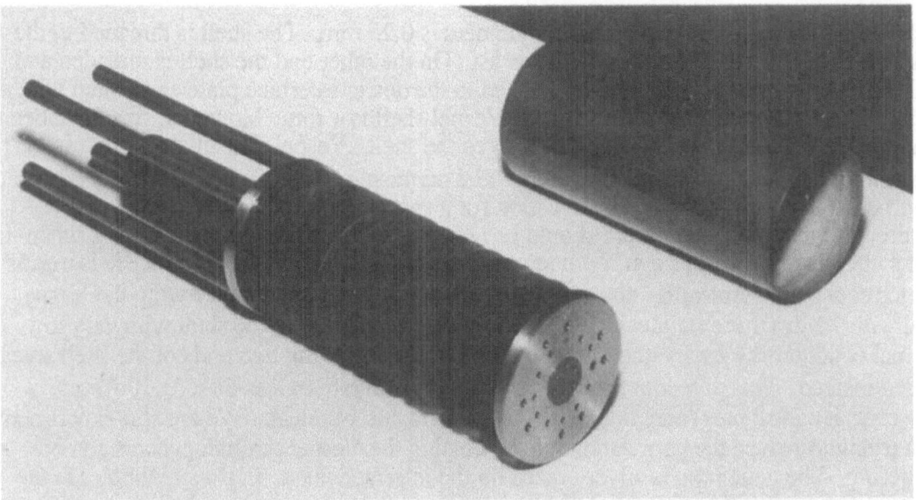

Figure 3. Engineering model Thermal Bus and Salt Pill Shell

Also, since ferric ammonium sulfate is a very good thermal conductor, especially when made up of larger size crystals, this thermal bus design would be as effective, if not more so, than the wire harness design in removing the heat from the salt during the magnetization phase. The copper wafers shown in Figure 2 are 0,25 mm thick and are evenly spaced every 9 mm. They are attached to the central cylindrical copper tube using pure indium solder. The tube is slit along two diametrically opposite lines. After the wafers are soldered to the tube, they are separated into two semicircles with a wire Electric Discharge Machine (EDM) so that the indium attachment will not crack. This splitting accomplishes two things. It eliminates large circular paths for electrons, thereby reducing significantly eddy current dissipation, and it interrupts the indium rings so as to eliminate magnetic flux trapping. The salt pill end caps are made out of stainless steel to minimize eddy currents. Four copper rods 9 mm in diameter and 16 cm long are press-fitted into four holes in the lower end cap and make up the thermal bus to the calorimeter cold stage. On the salt end of the end caps these rods are slotted and 1 cm wide, 0.5 mm thick copper strips are soldered into the slots. These strips run the length of the salt and are soldered to each copper wafer of the heat switch thermal bus. Figure 3 shows a photograph of the semiassembled thermal bus.

Also shown in Figure 3 is the fiberglass shell that encloses the salt pill. This shell is coated with ~2000 Å of copper and 2000 Å of stainless steel. The coatings serve two purposes. They render the fiberglass impermeable to water,to protect the salt from dehydration, and they provide a path for the heat in the shell to be transferred quickly to the salt pill rather than slowly leaching into it. The lower end cap is threaded into the fiberglass shell and sealed with stycast. Solution is then added and the crystal is grown in layers. When the crystal has filled the space within the shell, the top end cap is threaded into the shell, and is sealed with epoxy.

The breadboard model of the ADR was a modular design so it would facilitate the development effort. Each component of the ADR could be redesigned completely, without affecting the other parts. In the engineering model, some modularity was sacrificed in order to solve some of the problems observed in the previous models. As mentioned earlier the two copper plates interfacing the salt pill to the heat switch generated considerable eddy current heating. Since the eddy current dissipation is proportional to the fourth power of the radius, the eddy current losses can be minimized by reducing the radius of the high electrical conductivity metals used in the space where there is a high rate of change of the magnetic field. An enlargement of the heat switch portion from Figure 2 is shown in Figure 4. The two cylinders of the heat switch nearest the salt pill are welded directly on the copper tube at the center of the thermal bus. This provides excellent thermal link between the thermal bus and the heat switch, minimizing,at the same time, cooling power losses due to eddy currents and magnetic flux trapping. The heat switch cylinder - thermal bus tube combination is threaded into the stainless steel end cap and sealed with stycast. A Vespel[7] shell is used both to contain the helium gas and to keep the interlaced heat switch cylinders from touching each other. The nominal gap between cylinders is 0,25 mm. The shell is threaded to the stainless steel end cap and sealed with stycast. On the other end the shell is threaded and sealed to a brass adapter ring which is bolted to the dewar interface plate and sealed with an indium O-ring. The wall thickness of the Vespel shell is 1 mm. However, ribs have been machined at regular intervals to add rigidity to the shell. Vespel, like all polymers, is permeable to helium at room temperature. The permeation rate of ~ $5X10^{-6}$ standard cubic centimeters per second (sccs), is excessive for a space bound ADR. We have set a requirement that the gas pressure should be reduced by no more than a factor of 2 when the heat switch is under vacuum at room temperature for a period of six months. This means we had to reduce the permeation rate to ~ 10^{-7} sccs. To accomplish this we wrap the Vespel shell with 12-μm thick stainless steel foil. The foil is adhered to the shell with very low thermal conductivity epoxy and has an overlap of 6 mm. at the two ends of the shell stycast is used instead. This procedure reduces the permeation rate to less than $5X10^{-8}$ sccs. However, it contributes more than 3 μW to the parasitic heat leak. We are still experimenting with trying to reduce the permeation rate by coating the Vespel through evaporation or sputtering. The getter cup is silver brazed on the interface plate. Eight small holes at the bottom of the cup, offset from eight holes in the interface plate, permit the helium gas to enter the heat switch when the getter is heated. The getter, after baking, is mounted on the lid of the cup which is bolted on the cup and sealed with indium. The getter cup assembly operation is performed in a glove box containing helium gas.

As mentioned earlier, an important component of a space bound ADR is the suspension system. The system used to suspend the salt pill is shown in Figure 2. Two titanium rings on either side of the salt pill are kept apart by a titanium cylindrical shell 7.5 mm in diameter with a wall thickness of 0.75 mm. Each ring is 1.5 cm from the respective end of the salt pill. Loops made with 445 N test Kevlar[8] cord are hooked on notched bolts screwed into the stainless steel end caps, pass through holes at lower surface of the suspension rings and then attached to tensioning bolts. A stack of six Bellville washers, arranged back-to-back for maximum travel, is used between the tensioning nuts and the ring surface, so tension can be maintained at low temperatures. There are eight suspension loops at each end of the salt pill, each at a 45° angle with the end cap surface. Each loop is tensioned to ~200 N, Which is the force needed to flatten the Bellville washers. The tensioning is done gradually so as not to damage the heat switch. For this ADR the suspension system serves an additional purpose. In order to increase the hold time and, therefore, the efficiency of the ADR we extended the length of the heat switch. The length of the heat switch shell has been doubled to 9 mm reducing the parasitic heat leak from 11 to 5.5 μW. This, of course, has made it much more difficult to keep the heat switch cylinders from shorting especially when the ADR is cooled to

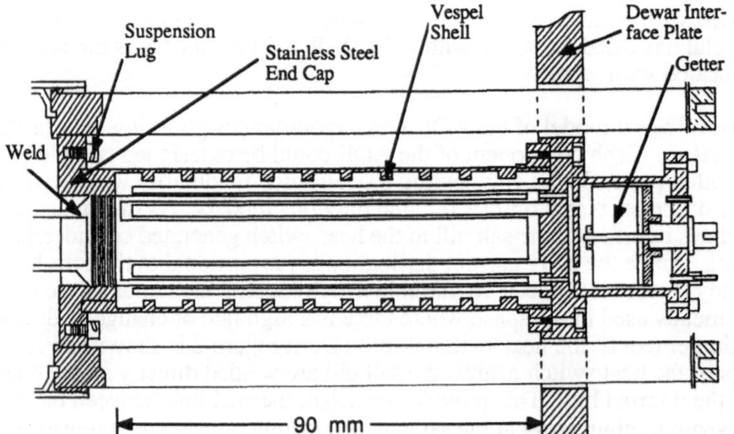

Figure 4. Gas Gap Heat Switch

LHe temperatures. During the suspension procedure an ohmmeter is used to indicate whether the two ends of the heat switch are shorted , or near shorted. Experience has shown that, once the suspension is done correctly and the suspension cords, including the ones at the calorimeter cold stage, are fully tensioned, there is no shorting, even after several thermal cycles of the dewar.

Finally, we must mention that for space experiments single point failures must be avoided. A redundant heat switch has therefore been baselined into the ADR design. At present, we are considering using another gas gap heat switch for redundancy. This, however, will almost double the parasitic heat leak. We are planning to experiment with a mechanical heat switch. Since, however, a mechanical heat switch, when it fails, it usually fails closed, no decision will be made to use such a switch unless we can prove total reliability of the design. As shown in Figures 2 and 3 a copper plug will be used to connect the salt pill thermal bus to the redundant heat switch. This plug is welded to the thermal bus tube and is screwed into the stainless steel end cap. In fact, this end cap, is screwed simultaneously to this plug and the fiberglass salt pill sleeve when the growth of the crystal is completed.

CONCLUSION

As mentioned earlier the engineering model ADR is currently under construction. However, a half-sized salt pill with full-sized heat switch has been built and tested with very good results. We are planning to continue trying to improve the XRS ADR performance. However, we feel that we already have a sound design. Both the wire harness thermal bus, and the engineering model designs produced ADR's that surpassed the original requirements. Thorough testing of both designs will continue until we are sure that the flight model will be the most efficient and reliable ADR we can use on the XRS.

REFERENCES

1. S.S. Holt,"X-Ray Spectroscopy of AGN with the AXAF 'Microcalorimeter'", *Astrophysical Letters and Communications* ***26*** 61, (1987).

2. M.C. Weisskopf, "The Advanced X-Ray Astrophysics Facility: An Overview", *Astrophysical Letters and Communications* ***26*** 1, (1987).

3. S.H. Castles, "Refrigeration for Cryogenic Sensors", *NASA Conference Publication 2287* ***389*** (1983).

4. A.T. Serlemitsos, et al., "Adiabatic Demagnetization Refrigerator for Space Use", *Advances in Cryogenic Engineering* **35** 1431, (1989).

5. A. T. Serlemitsos, et al., "A Spaceworthy ADR: Recent Developments", *Proc. SPIE* **xxx,** xxx (1990).

6. Stycast 2850-FT; Emerson and Cummings Inc., Canton MA 02021.

7. Vespel; a Dupont polyimide.

8. Kevlar, a Dupont Aramid.

MAGNETIC SHIELDING FOR A SPACEBORNE ADIABATIC DEMAGNETIZATION REFRIGERATOR (ADR)

Brent A. Warner, Peter J. Shirron, Stephen H. Castles, and Aristides T. Serlemitsos

Cryogenic Systems Development Section
NASA Goddard Space Flight Center
Greenbelt, Maryland

ABSTRACT

The Goddard Space Flight Center has studied magnetic shielding for an adiabatic demagnetization refrigerator. Four types of shielding were studied: active coils, passive ferromagnetic shells, passive superconducting coils, and passive superconducting shells. The passive superconducting shells failed by allowing flux penetration. The other three methods were successful, singly or together.

Experimental studies of passive ferromagnetic shielding are compared with calculations made using the Poisson Group of programs, distributed by the Los Alamos Accelerator Code Group of the Los Alamos National Laboratory. Agreement between calculation and experiment is good. The ferromagnetic material is a silicon iron alloy.

INTRODUCTION

The Goddard Space Flight Center (GSFC) is building an Adiabatic Demagnetization Refrigerator (ADR) for the X-Ray Spectrometer (XRS) for the Advanced X-ray Astrophysics Facility (AXAF). AXAF Electro-Magnetic Interference (EMI) specifications require the magnetic field at the edge of the experiment space to be below 0.08 millitesla. The magnet central field is 1.9 tesla. The edge of the experiment space is at a radius of approximately 60 centimeters from the magnet axis.

The four types of shielding we have studied are: active coils, passive ferromagnetic shells, passive superconducting coils, and passive superconducting shells. The passive superconducting shells failed. The other types of shielding show promise, singly or together. This paper deals mainly with passive ferromagnetic shielding, comparing measurements with calculations made using the Poisson Group programs.

ACTIVE SHIELDING

In active shielding, shield coils cancel the fringing field of the main coil. In the magnets we used, the shield coils were wound in series with the main coils. Active shielding is relatively light weight, compared to ferromagnetic shielding.. Unfortunately, active shields slightly reduce the magnet's central field. Also, active shielding may require a complicated design to reduce the fringing fields as desired. For example, an actively shielded magnet procured by GSFC has six shield coil sections.

That magnet is one of two actively shielded magnets built for GSFC by Cryomagnetics, Inc. The magnets come close to meeting AXAF EMI requirements. When the central field is 1.9 tesla, the maximum field at a radius of 60 cm is 0.1 millitesla.

Specifications for the active shielding were developed by GSFC with the help of a magnet calculation program written at the University of Southern California.[1]

PASSIVE SUPERCONDUCTING SHELLS

Passive superconducting shielding uses the Meissner effect: magnetic flux does not penetrate a superconductor. Thus, a superconducting shell could be a perfect shield. However, magnetic flux penetrates when the field strength reaches a critical value. For a type I superconductor, such as pure lead, the superconductor transitions to the normal, or resistive, state, as the field penetrates. When the magnetic field is removed, the material returns to the superconducting state--but with the penetrated flux "frozen" in place. Thus a failed superconducting shield becomes a field source. The flux remains frozen until the superconductor is heated above the superconducting-normal transition temperature. Heating the shell would not be acceptable in the AXAF spacecraft, since it would vaporize all of the liquid helium. However, superconducting shells are useful in some cases.[2]

We abandoned superconducting shells because of two problems. First, flux penetrated our test shields and was frozen in. Second, for fields too low to penetrate, the shell increased the strength of the fringing field on axis at the end of the magnet. This increase risked interfering with the XRS x-ray sensors, which will be placed on the magnet axis, a few centimeters outside the magnet. Solving the flux penetration problem would only worsen the on axis fringing field problem.

We tested two slightly different passive superconducting lead shields. Both were tested on an actively shielded magnets. Both shields consisted of lead foil tape on an aluminum support. The tape was 3M Co. type 422, composition 95.5% Pb, balance Sb and Sn.[3] Seams were made with conventional lead-tin solder. In both shields, the lead tape was placed on a cylinder around the magnet. In the original shield, the ends were partly closed by endcaps with holes in the centers. The second shield had no endcaps.

We do not know why the field penetrated the shields. The unshielded field at the shield location was below the critical field of lead at the operating temperature. Impurities alter critical fields[4]. Perhaps impurities lowered the critical field of our lead foil. Muething et al[2], found that the supercurrent peaked at the ends of their superconducting shields. Perhaps the current at the ends of our shields exceeded the critical current.

The other problem , which we measured only with the original shield, is that leakage flux was concentrated in the endcap holes. We detected this problem by measuring the on axis field at the outside end of the magnet mandrel. This fringing field decreased when flux penetrated the shield. In fact, removing the shield reduced the on axis fringing field.

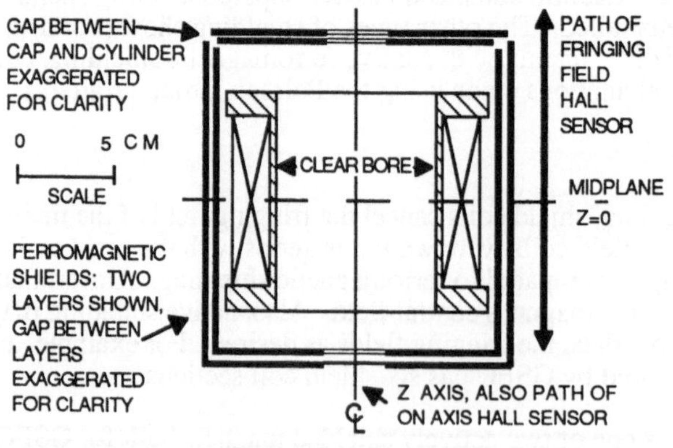

Fig. 1. Cross section of magnet with silicon iron shield.

FERROMAGNETIC SHIELDING

Ferromagnetic shielding uses shells of a ferromagnetic material, such as iron, to contain the magnetic flux. Ferromagnetic shielding is simple to design and build. However, ferromagnetic materials are dense, and ferromagnetic shielding is thus heavy compared to passive shield coils. Despite the weight disadvantage, ferromagnetic shielding of ADR's has been tested by P. Timbie et al at the University of California at Berkeley and by Alabama Cryogenic Engineering (ACE). Timbie et a tested a vanadium permendur shield, which they modelled using Poisson.[5] ACE tested a soft iron shield, which they modelled using their own software.[6]

Apparatus for Ferromagnetic Shield Tests

To test ferromagnetic shielding, we surrounded a superconducting magnet with a four-layer silicon iron shield. See Fig. 1. We then compared the resulting magnetic fields with calculations made by the Poisson Group programs.

The magnet is a single superconducting coil built by American Magnetics, Inc., in 1979. In the computer model, the physical dimensions of the coil are those supplied by AMI. The number of turns of wire was chosen to set the calculated central field equal to the measured central field. This value for the number of turns in within 4% of the value supplied by AMI.

The ferromagnetic shields are made of silicon iron, specifically Allegheny Ludlum Relay Steel #5. The composition is 2.25% Si, 0.40% Al, balance Fe. Silicon iron is readily available and easy to form. Silicon Iron has moderately high saturation, 1.79 tesla, with a permeability that is roughly constant down to liquid helium temperature.[7,8] The importance of temperature variation on shielding performance has been shown.[9,10] That similar alloys can vary greatly in response to cooling has also been shown.[11] Even samples which are similar enough in composition to be classed as the same alloy can vary in their responses to cooling.[12]

Our shield consists of four nesting cylinders, used separately or together. The smallest cylinder has an outer diameter of 15.2 cm. The largest has an outer diameter of 16.4 cm. The cylinders were made by rolling and welding sheet silicon iron, .157 cm thick. The cylinders have endcaps at both ends, of the same thickness. The center of each endcap is pierced by a 3.8 cm diameter hole. The bottom endcaps are welded to the cylinders, while the top endcaps are bolted to the magnet support rods. Thus the top endcaps are not as tightly joined to the cylinders as the bottom endcaps are. In one test we detected flux escaping from the joint at the top of the shield. After preliminary shielding tests, the shields were heat treated to match the samples measured by Ackermann et al (two hours at 843 C, furnace cool).

The magnetic field was measured with cryogenic Hall effect sensors. A Lake Shore Cryotronics model LGHA-321 sensor measured the field on the magnet axis. An F.W. Bell model BHT-921 sensor measure the fringing field outside the shield, along a line parallel to the magnet axis but 10 cm from the axis. Both sensors measured the field component parallel to the magnet axis, which was chosen as the z-axis.

Shielding Test Results

Effect of shield thickness. As shielding thickness increased, the fringing field decreased, and the central field increased. Figures 2 and 3 compare the central and fringing fields for the magnet with no shielding, with two layers of shielding, and with four layers of shielding. The magnet current was 5 amperes, producing an unshielded central field of .48 Tesla and an unshielded flux of 3 milliwebers.. The z axis is set along the magnet axis, with z=0 at the magnet center. A two layer shield decreased the fringing field at z=0 to 50% of its original value. A four layer shield decreased it to 10% of original. The central field at 5 amperes increased from .48 T with no shielding to .62 T with four layers of shielding.

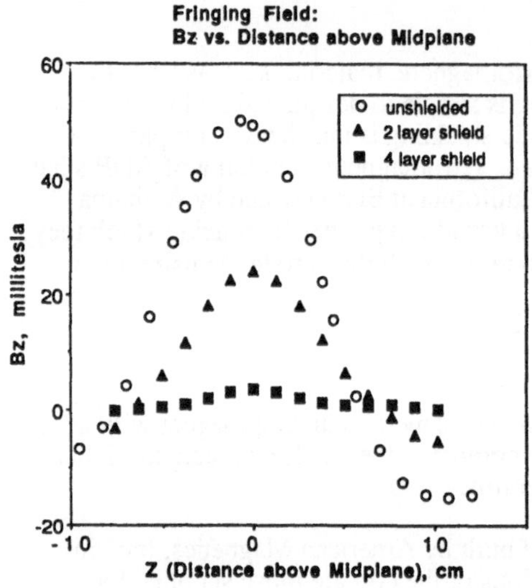

Figure 2. Measured fringing field profile comparing shielded and unshielded.

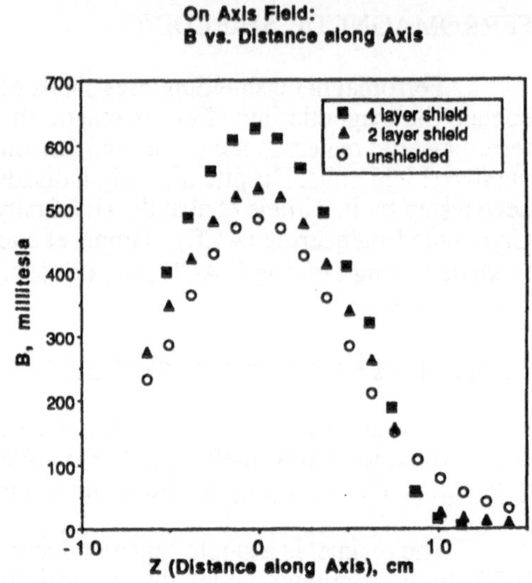

Figure 3. Measured on axis field profile with and without shielding.

Shielding improved the homogeneity of the field. For this apparatus, the increase was small, but shielding can improve homogeneity by an order of magnitude.[13]

Saturation and calculation. Figure 4 shows how the fringing field at z=0 varies with current. Flux penetrates the two layer shield at about 2 amperes and the four layer shield at about 4 amperes. The shield thickness was chosen with the help of a simple saturation approximation. The approximation greatly underestimated the required thickness, but the shield met its goals, thanks to a large safety factor. The approximation assumes that all the flux from the magnet connects through the shield, with no flux leakage--assuming, in other words, that the magnet and shield form a perfect magnetic circuit. To approximate the required thickness, set the flux from the magnet equal to the cross sectional area of the shield times the saturation induction of the shield material, then solve for the shield thickness.

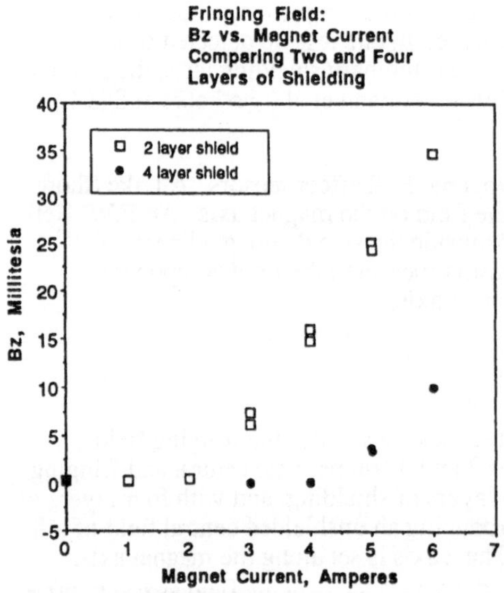

Figure 4. Fringing field as a function of current for 2 and 4 layer shields

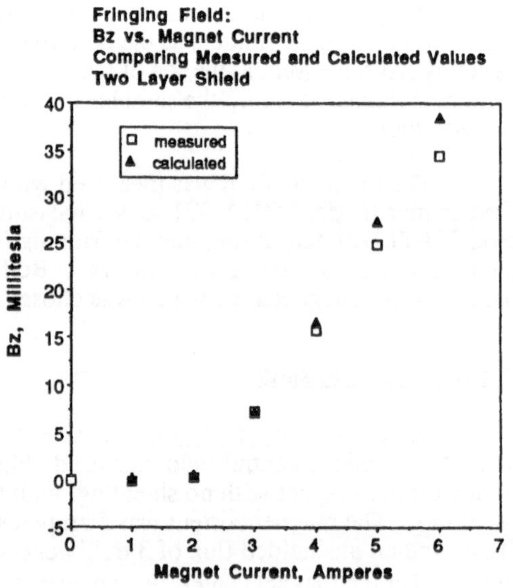

Figure 5. Fringing field as a function of current, comparing measured and calculated values

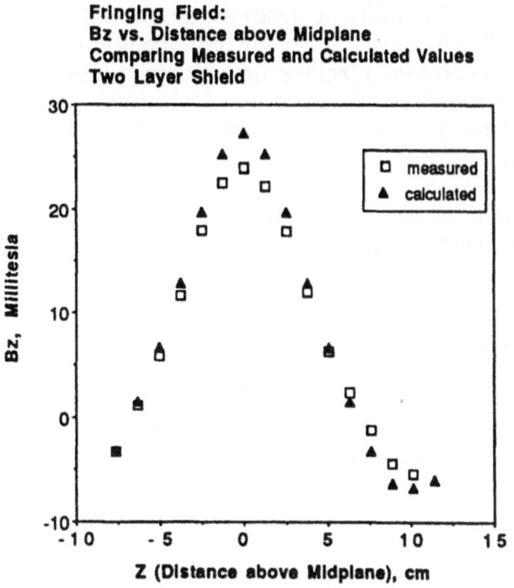

Figure 6. Fringing field profile, with shielding, comparing measured and calculated values.

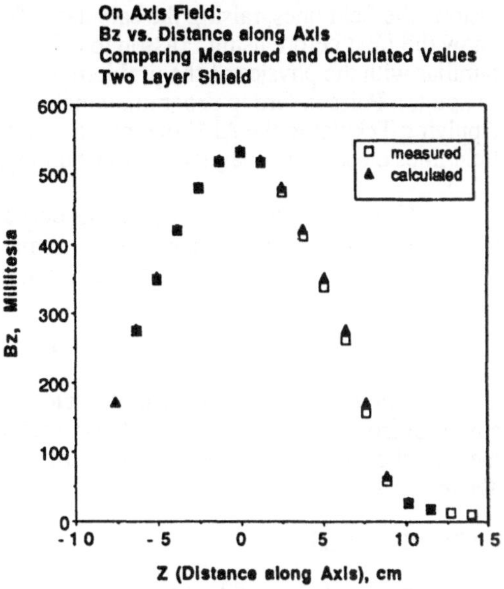

Figure 7. On axis field profile, with shielding, comparing measured and calculated values

This approximation underestimates the shield thickness needed for our shield shape. For example, the two inner silicon iron shields have an average diameter of 15.24 cm and a combined wall thickness of .314 cm. At saturation induction of 1.79 tesla,[7] total flux through the shields would be 2.69 milliweber. Total flux produced by the magnet at 4 amperes is approximately 2.7 milliweber. (This magnet flux was approximated from the measured on-axis field and the calculated off-axis field.) This shielding approximation implies that the shield saturates at a magnet current of 4 amperes. However, flux penetrates for currents above 2 amperes. Similarly, the approximation suggests that the four layer shield saturates at 8 A, while flux penetrates above 4 A. Thus, while this approximation may be useful for quick calculations, it should be treated as only a rough approximation.

Poisson. The Poisson Group programs are a series of magnetic and electric field calculation programs available from the Los Alamos Accelerator Code Group of the Los Alamos National Laboratory. We used the program Poisson, but not Pandira, the group's other magnetic field calculation program. Figure 5 compares calculated and measured fringing fields outside a two layer shield for various magnet currents. Clearly, Poisson is much more accurate than the simple saturation approximation. Figures 6 and 7 compare measured and calculated field values for the two layer shield for a magnet current of 5 amperes. Agreement is close, but not exact. Figures 8 and 9 show field profiles at the same locations for the magnet without shielding. Although the agreement on axis is good, the measured fringing field profile differs from the calculated profile. The magnitude of the field maximum is shifted by 2 1/2 millitesla, and the position is shifted by about 1/2 cm.

There are a number of possible explanations for the differences between the measurements and the calculations. First, the computer model of the magnet differs from the real magnet. The computer model has perfect midplane symmetry. The real magnet must be approximately symmetric, because it meets central field homogeneity specifications. Figure 8, however, shows that the fringing field is slightly asymmetric. This observed asymmetry could easily be produced by winding errors too small to have a noticeable effect on the central field.[14] When the magnet was built, no one foresaw this shielding study, hence no one measured the winding errors.

Second, Poisson requires boundary conditions for the field. Poorly chosen boundaries distort the field. For example, Figures 8 and 9 were calculated with the boundaries set more than 50 cm from the magnet. (Magnet coil size is 6.6 cm outer radius, 9 cm length.) The calculated field at z=0 and r=10 cm is 53 millitesla, close to the measured value of 49 millitesla. Confining the calculated field within 10 cm x 10 cm boundaries distorts the field lines, raising the calculated field to 176 millitesla. Does the user need to know the field before using Poisson to calculate it? Not exactly. But the user needs to be familiar with the physics of the problem. The user can gain a feel for the problem by using Tekplot, a Poisson Group program, which plots a two dimensional picture of the field lines. Applying Tekplot to the AMI magnet shows that the 10 cm x 10 cm boundary confined the field lines close to the magnet in a most unphysical way.

Third, the silicon iron hysteresis curve used with Poisson involved some guesswork (and thus some error.) The curve of Ackermann et al[8] was used below their maximum value of 1.7 T. For high fields, beyond saturation, the curve approaches a straight line, whose slope is given by

$$\Delta B = \mu_0 \Delta H.$$

For high fields, and for intermediate fields near saturation, the Poisson's internal hysteresis curve for iron was used. The two hysteresis curves are close each other at the splice. We feel that the combined curve is a reasonable approximation, in the absence of actual data above 1.7 Tesla.

PASSIVE COILS

The final type of shielding which this paper discusses is passive superconducting coil shielding. One advantage of passive superconducting coils is their low weight. Another advantage is that the passive coils adjust their currents automatically, and that resulting currents tend to cancel the dipole moment of the applied field. Cancelling the dipole moment is an an advantage because the dipole moment is the most slowly varying component. The main disadvantage of passive coils is the complexity of their thermal switches. A smaller disadvantage is that passive shield coils slightly decrease the central field. In one test, passive coils lowered the central field by about 1 %. In this study, passive shield coils were used only in combination with ferromagnetic shields. To make a passive shield coil, single core NbTi superconducting wire was coiled around a ferromagnetic shield. The ends of the wire were welded together, forming a continuous superconducting loop.

Flux which leaks from the ferromagnetic shield induces a current in the passive coil. The induced current produces a magnetic field which opposes the original applied flux. The current continues to flow as long as the coil remains superconducting. If the applied flux is

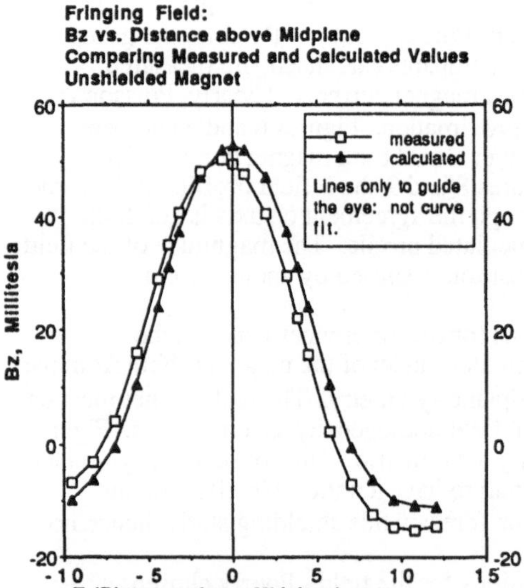

Figure 8. Fringing field profile for unshielded magnet, comparing measured and calculated values.

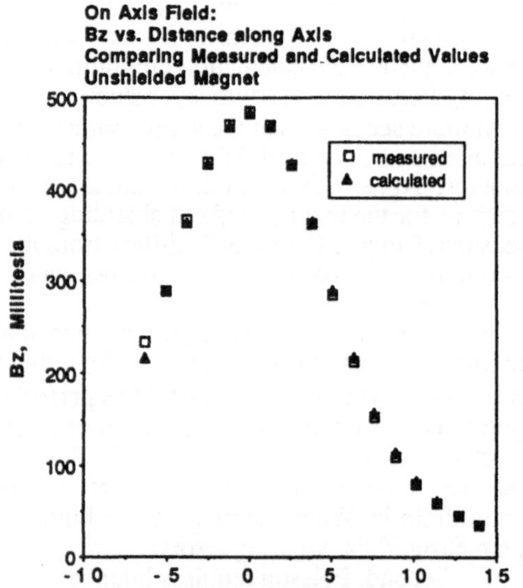

Figure 9. On axis field profile for unshielded magnet, comparing measured and calculated values

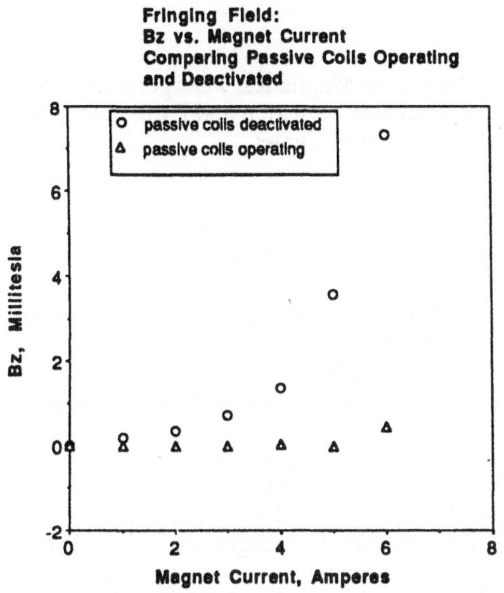

Figure 10. Fringing field as a function of magnet current, showing shielding effect of passive coils.

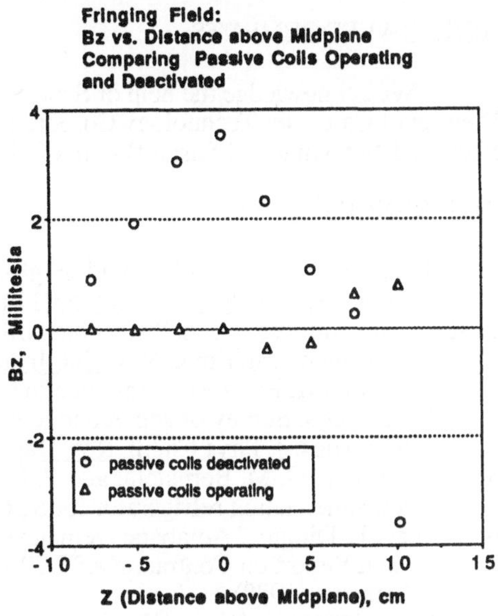

Figure 11. Fringing field profiles, showing shielding effect of passive coils.

high enough, the induced current exceeds the wire's critical current limit. The wire transitions to normal, and the current decays. When the applied field is dropped back to zero, the coils again oppose the change, "freezing" in some of the flux. This "frozen" flux can be eliminated by heating the wire above its superconducting-normal transition temperature. In our apparatus, the shield coil was immersed in the liquid helium bath that cooled the superconducting magnet. Instead of heating the entire wire, we heated only a section that was thermally isolated by a thermal switch. We used two successful thermal switch designs. One was an aluminum cylinder, sealed by indium O-rings and epoxy. The other, and more successful, was a block of plastic with a hole drilled into it. The wire and heater were inserted into the hole, which was then stuffed with Teflon tape.

Figures 10 and 11 show the effect of passive superconducting coils wound on a two layer silicon iron shield. The figure compares the fringing field with the coils superconducting and with the coils deactivated by heating the thermal switch. Clearly, the passive coils shield effectively. However, the shield coils lost superconductivity at fields lower than expected. It is possible that the weld joint had a lower critical current than the rest of the wire and thus triggered the loss of superconductivity.

CONCLUSIONS

We have identified three types of magnetic shielding useful at cryogenic temperatures: active coils, passive superconducting coils, and passive ferromagnetic shells. These methods can be used separately or together. For example, a study with Poisson suggests that a lightweight (2 kg) silicon iron shield could bring our present actively shielded magnet within AXAF EMI specifications. The magnet alone is 14 kg.

All three types of shielding were combined when an actively shielded magnet built by Cryomagnetics was placed inside a 6.3 kg ferromagnetic shield, made of Netic, an alloy developed by Magnetic Shield Corporation. Around the ferromagnetic shield were wound passive superconducting coils. When the central field of the magnet was 1.9 tesla, the field on the magnet midplane at 30 cm from the magnet center was less than 0.05 millitesla (0.5 Gauss). With the passive coils disabled, the axial field was 0.2 millitesla. With the ferromagnetic shield and passive coils removed, the field was greater than 0.4 millitesla. Thus, these three types of shielding can be used separately or together.

ACKNOWLEDGEMENTS

We acknowledge the help of Scott Smith of American Magnetics, Incorporated; Doug Dietrich of Carpenter Technology Corporation; Marcelino San Sebastian and Evan Kunes of STX; and Jim Ming and Susan Breon, of Goddard Space Flight Center.

REFERENCES

1. U. E. Israelsson and C. M. Gould, High-field magnet for low temperature low-field cryostats, Rev. Sci. Instrum. 55:1143 (1984).
2. K. A. Muething et al, Small solenoid with a superconducting shield for nuclear-magnetic-resonance near 1 mK, Rev. Sci. Instrum 53:485 (1982).
3. Sandy Holdorf, Private Communication, 3M Company, (1987).
4. B. W. Roberts, Survey of superconductive materials and critical evaluation of selected properties, J. Phys Chem. Ref. Data 50:581 (1976).
5. P. T. Timbie, G. M. Bernstein, and P. L. Richards, Development of an adiabatic demagnetization refrigerator for SIRTF, Cryogenics 30:271 (1990).
6. Michael L. Dingus, "Adiabatic Demagnetization Refrigerator for Use in Zero Gravity," Final Report on Contract NAS5-29418, Alabama Cryogenic Engineering, Huntsville, Alabama (1988).
7. F. W. Ackermann, W. A. Klawitter, and J. J. Drautman, Magnetic properties of commercial soft magnetic alloys at cryogenic temperatures, in: "Advances in Cryogenic Engineering," Vol. 16, Plenum Press, New York (1971), p. 46.
8. F. W. Ackermann and D. Dietrich, personal communication: Carpenter Technology Corporation internal memorandum used in preparing Ackermann et al.
9. R. F. Arentz and M. H. Johnson, Magnetic shielding in a cryogenic environment, Evaluation Engineering, 25:80 (1986).
10. D. L. Martin and R. L. Snowdon, Effect of temperature on magnetic shielding with Co-netic, Rev. Sci. Instrum. 46:523 (1975).
11. Y. Suzuki, E. Horikoshi, and K. Niwa, Magnetic properties and ferromagnetic shielding of Ni-Mo-Fe alloys at cryogenic temperatures, Fujitsu Scientific and Technical Journal 20:167 (1984).
12. I. B. Goldberg et al, Magnetic susceptibility of inconel alloys 718, 625, and 600 at cryogenic temperatures, in: "Advances in Cryogenic Engineering - Materials," Vol. 36, Plenum Press, New York (1990), p. 755.
13. Roger J. Hanson and Francis M. Pipkin, Mangetically shielded solenoid with field of high homogeneity, Rev. Sci. Instrum 36:179 (1965).
14. Scott Smith, Private Communication, American Magnetics, Inc. (1991).

PERFORMANCE OF A He II GAP HEAT SWITCH

Ali Kashani and Ben P. M. Helvensteijn

Sterling Federal Systems
Palo Alto, California

Randall A. Wilcox and Alan L. Spivak

Trans-Bay Electronics
Richmond, California

ABSTRACT

A 2 K He II gap heat switch has been developed as part of a 2-10 K adiabatic demagnetization refrigerator. Tests have been conducted on a prototype of the 2 K heat switch to characterize its performance. The prototype heat switch consists of a brass cylindrical load with an axial-blind-hole drilled through its center. A tube containing He II is placed inside the hole and maintains a 0.018 cm-wide gap with the brass load. In the on-mode of the heat switch the gap is filled with He II and in its off-mode the gap is emptied. This is accomplished with an activated carbon pump. With 0.1 mW applied to the load the on/off conductance ratio of the prototype heat switch is roughly 2500. For the switch to fully turn on, it takes about three minutes; however, its off-time is on the order of six minutes.

INTRODUCTION

Several future astrophysics missions planned by NASA, such as the Large Deployable Reflector (LDR), require a 2 K cryocooler. Work is on-going at NASA - Ames Research Center to develop an adiabatic demagnetization refrigerator (ADR) which would operate between 2 K and 10 K and would provide 0.1 W of cooling at 2 K.[1] The ADR would be employed as the last stage of a multistage cryocooler to achieve 2 K cooling. In the design of the ADR no moving components have been employed to increase reliability. The magnetic material selected for operating in the 2-10 K temperature range is Gadolinium Gallium Garnet (GGG). During part of the ADR cycle thermal contact has to be made between the GGG and the 2 K heat source. A He II gap heat switch has been developed to accomplish this task.

2 K He II GAP HEAT SWITCH

The concept of a He II gap heat switch has been previously tested.[2] The 2 K He II gap heat switch developed for the 2-10 K ADR is designed for placement inside the paramagnetic refrigerant. The heat switch, shown schematically in Fig. 1, is made up of several He II tubes which are placed in axial-blind-holes in the GGG, maintaining a small gap with the GGG. The He II tubes, which contain He II throughout the refrigeration cycle, are connected to the 2 K heat source. When the gaps are filled with He II, thermal contact between the GGG and the He II tubes is made. The GGG and the heat source are isolated from one another when the gaps are emptied.

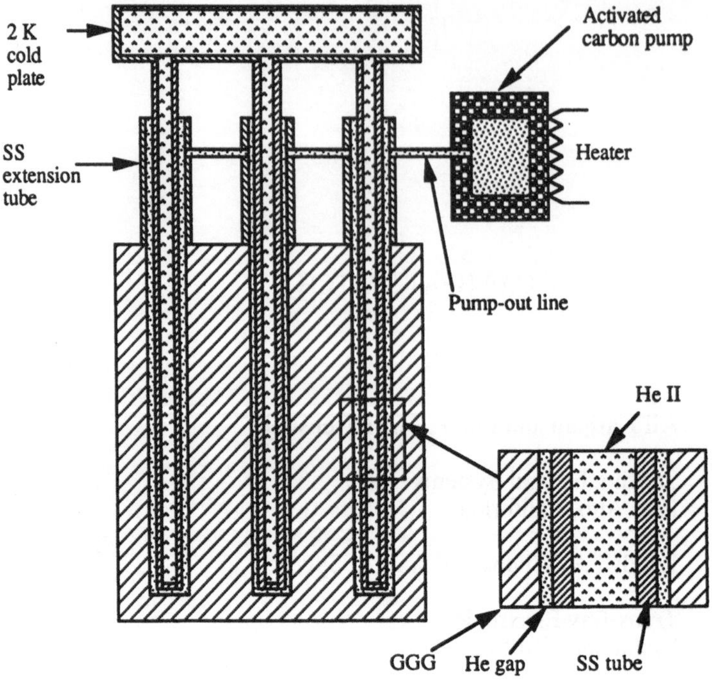

Fig. 1 Schematic of the 2 K He II heat switch as placed in the GGG.

The He II, which in the on-mode, provides the main heat transfer path between the heat source and the GGG is divided up into two volumes, i.e., the He II tubes and the He II gaps. This is to reduce the amount of helium gas that has to be condensed and evaporated during each refrigeration cycle. To reduce the thermal conductance of the heat switch in the off-mode, stainless steel extension tubes are used to seal the gaps. These tubes are bonded to both the GGG and the He II tubes, as illustrated in Fig. 1.

The supply and removal of helium to and from the gaps are accomplished by an activated carbon pump (ACP). The ACP adsorbs the helium gas when it is cooled and desorbs the gas when it is heated. To characterize the performance of the ACP, tests are being conducted on various activated carbon materials which are commercially available.[3]

2 K HEAT SWITCH PROTOTYPE

The performance of the 2 K He II heat switch is analyzed by testing its prototype. The prototype, shown schematically in Fig. 2, consists of a brass cylinder (load) with an axial-blind-hole drilled through its center. A stainless steel He II tube plugged at one end, is inserted into the hole. A gap is maintained between the He II tube and the brass load. The thermal link between the He II tube and the brass load is made by a stainless steel extension tube which slips over the He II tube as shown in Fig. 2. For the gap width to be uniform the inner diameter of the extension tube is selected to be the same size as the diameter of the blind-hole in the brass load. To seal the gap, one end of the SS tube is soldered to the brass load and its other end is soldered to the He II tube.

An activated carbon pump (ACP) is connected to the He II gap by use of a stainless steel pump-out tube. The amount of activated carbon glued to the inner surface of the ACP is 2 g. The activated carbon used in the ACP is courtesy of Barnebey-Cheney (type PE). A thermal link is made, between the ACP and the He II bath, by a brass rod to conduct the heat of adsorption.

Five carbon resistance thermometers monitor the temperature of the heat switch assembly at different locations. The thermometers T1 and T2 are placed at either ends of the

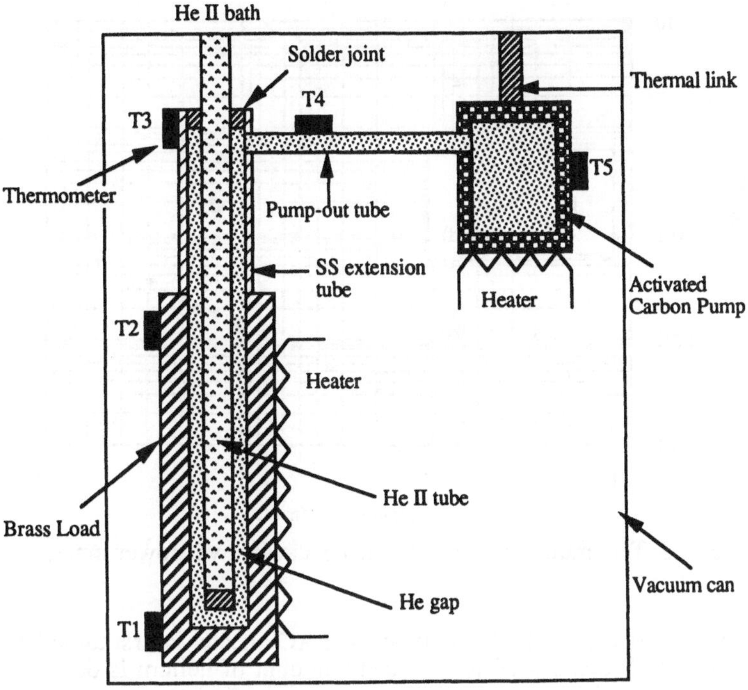

Fig. 2 Schematic of the He II heat switch prototype.

brass load. T3 is located where the He II tube and the SS extension tube are joined. Thermometers T4 and T5 are positioned on the pump-out tube and the ACP , respectively. Manganin-wire heaters are wound around the brass load and the ACP to control their temperatures. The heat switch assembly is put inside a vacuum can. The He II tube is open to the He II bath through a hole in the top flange of the vacuum can.

Results of the tests performed on a 2 K heat switch prototype with a gap size of 0.009 cm were presented in an earlier report.[4] The on and off conductances of that heat switch prototype were found to be close to the theoretical predictions and met the requirements of the ADR. However, the off-switching time of the heat switch was longer than twenty five minutes. To reduce its off-time the inner diameter of the pump-out tube was made larger and its length was shortened. This change had no significant impact on the off-time of the switch. It was concluded that the small gap size was the main reason for the slow off-time of the heat switch.

A second prototype heat switch was constructed having a gap size of 0.018 cm. In order to compare the off-time of the two heat switch prototypes the gap volume of the second prototype was made equal to that of the first prototype heat switch. The second prototype heat switch has a 4.13 cm-long brass load. The inner diameter of the blind-hole is 0.251 cm; whereas, the outer diameter of the He II tube is 0.216 cm. The SS extension tube is 2.0 cm long and has a wall thickness of 0.025 cm.

TEST RESULTS

At 2 K the off-conductance of the heat switch is tested having no helium gas present in the heat switch assembly. The off-conductance values obtained in the test are close to the conductance of stainless steel in this temperature range. This is an indication that there is no significant thermal contact between the brass load and the He II tube, except through the SS extension tube.

The heat switch assembly is then charged with helium gas to 5 atm and tested for its on-conductance at 2 K. The on-conductance of the heat switch is obtained for different values of

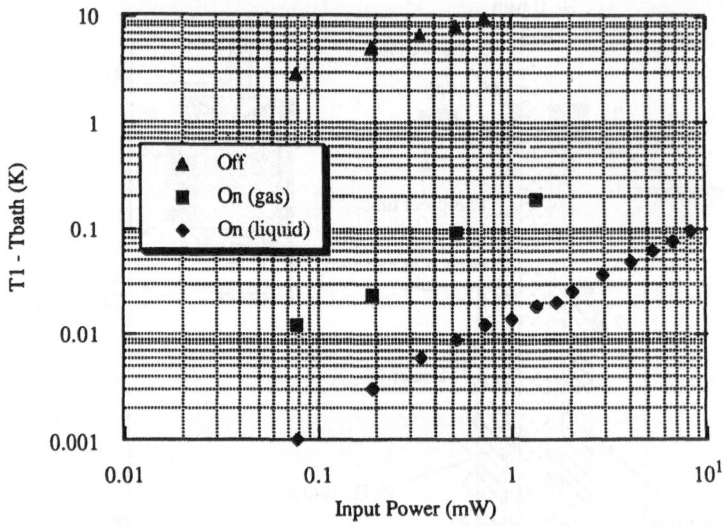

Fig. 3 Temerature drop across the heat switch vs. power input.

power applied to the brass load. In the on-mode, the ACP heater is first turned on. When the temperature of the ACP reaches 29 K, sufficient amount of helium is desorbed to initiate condensation in the gap. With the switch in the on-mode, the power applied to the load is increased and the change in the temperature of thermometer T1, above the bath temperature, is monitored. In Fig. 3, this temperature change is plotted as a function of the power applied to the load, for both the on-mode and the off-mode of the heat switch. The ratio of these two temperature changes, for any given power, represents the on/off conductance ratio of the heat switch at that power. This ratio is about 3000 at 0.07 mW and goes down to about 1000 at 0.6 mW.

The on-conductance of the heat switch when it is filled with helium gas is also plotted in Fig. 3. For this case the conduction across the gap takes place in the continuum regime. The on-conductance of the heat switch when filled with He II is about ten times higher than its on-conductance when it is filled with gaseous helium.

To test the on and off switching times of the heat switch, several transient tests were conducted while monitoring the traces of all five thermometers. In these tests a set power is applied to the brass load. When the load temperature reaches a desired value the ACP heater is turned on. As soon as the ACP starts to desorb the helium, the temperature of the load begins to decrease. Initially, the heat transfer across the gap is in the free-molecular regime. Once the ACP temperature reaches about 17 K, the load temperature decreases quite sharply. From this point on heat transfer across the gap is by the gas in the continuum regime. No further change in the temperature of the load is observed until liquid begins to condense in the gap. This occurs at an ACP temperature of about 29 K. At this point both thermometers on the load ,T1 and T2, indicate a decrease in the load temperature which is the result of liquid condensing in the gap. Furthermore, the temperatures of thermometers T3 and T4 start increasing. This is caused by the heat of condensation given off by the condensing helium gas.

After the gap is filled with liquid, the ACP heater is turned off and the helium is removed from the gap. This results in the evaporative cooling of the load and as a result its temperature is reduced to below the bath temperature. When the pump temperature reaches about 17 K the load temperature starts rising. This rise continues until the load temperature reaches the value it had reached just before the ACP heater was turned on. From the temperature traces of the load thermometers an on-time and an off-time can be estimated for the heat switch.

The temperature traces of all five thermometers are shown for two different load powers in Figs. 4 and 5. In Fig. 4, the load power is 0.08 mW; whereas, in Fig. 5 it is 3.0 mW. The time at which the different events in the tests take place are represented, in the bottom graph of each figure, by alphabetical letters. At time **a** power is applied to the load. At time **b,** large

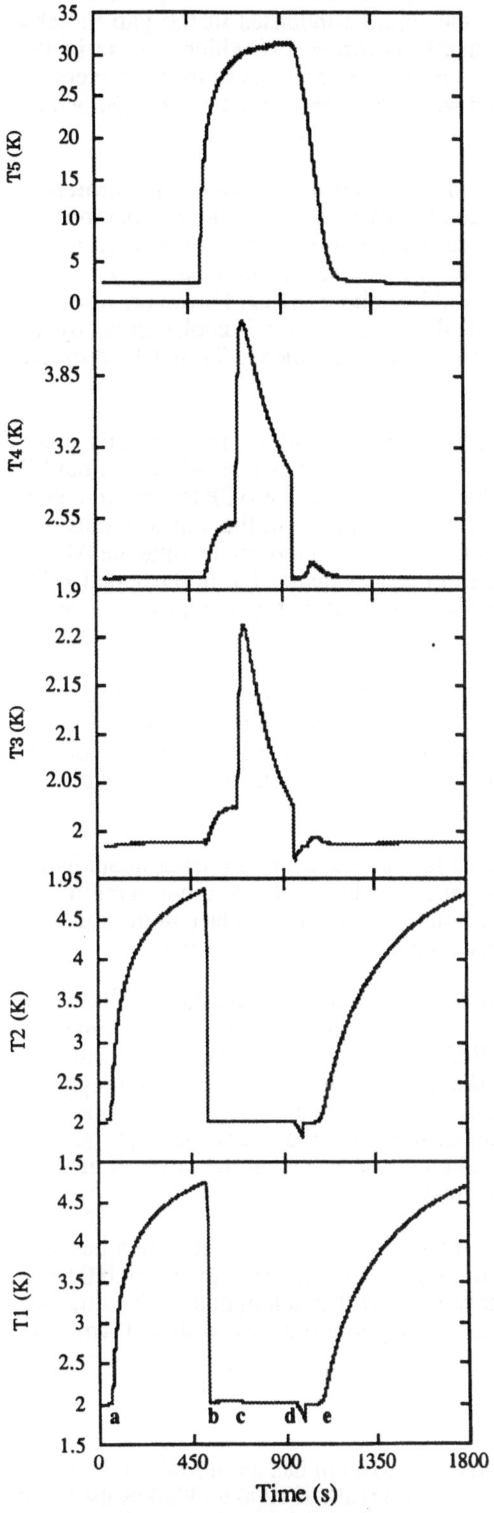

Fig. 4 Temperature traces of thermometers T1 - T5 when 0.08 mW is applied to the load.

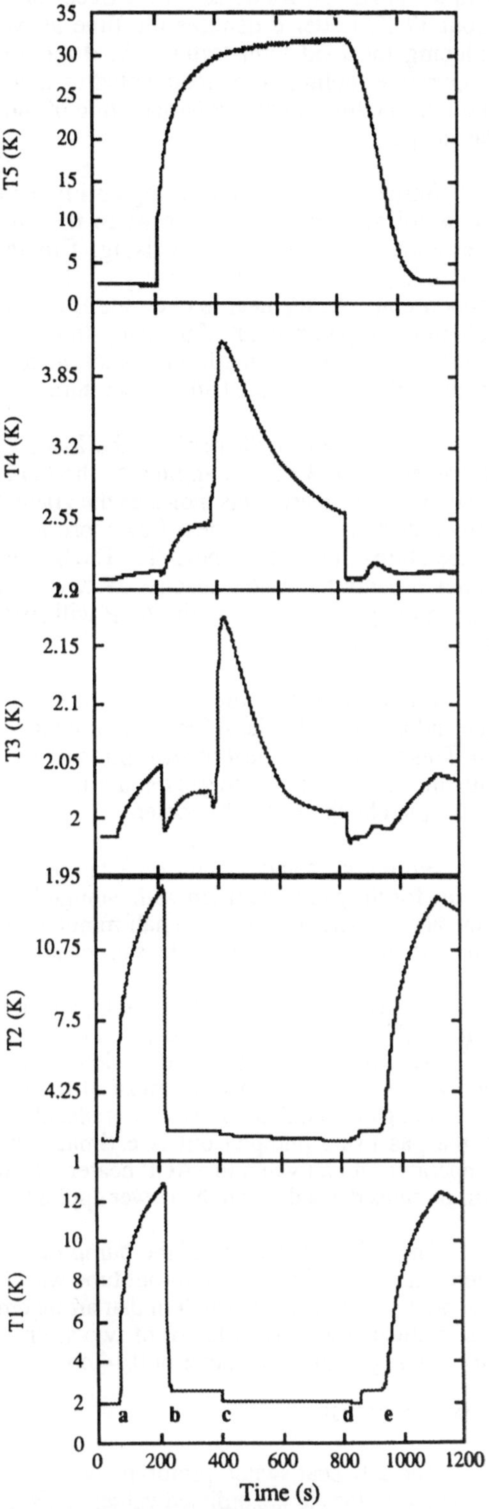

Fig. 5 Temperature traces of thermometers T1 - T5 when 3.0 mW is applied to the load.

temperature drops are noted on the traces of thermometers T1 and T2. At this point T5 is at about 17 K. Letter **c** denotes the time at which the liquid condenses in the gap thereby reducing the load temperature. At **d** the ACP heater is turned off which results in the evaporative cooling of the load, noted by a decrease in the temperatures of thermometers T1 and T2. The time when the temperature of the load starts to increase once again is shown by the letter **e**.

Some of the events occuring during the on and off cycles of the heat switch prototype are as follows. First, at time **c** when the liquid condenses in the gap, the temperature of thermometers T3 and T4 start rising. This is due to the latent heat of condensation as the helium begins to condense in the gap. If all the helium in the heat switch assembly were to condense at 2 K, the heat load on the He II tube would be about 0.18 J. However, when the helium gas is pumped out of the gap, starting at time **d**, the heat switch is cooled down by the evaporating liquid helium. In this event the temperatures of thermometers T1 and T2 decrease by about 0.2 K below the bath temperature.

Second, with 0.08 mW on the load, at time **b** both T1 and T2 reach steady state. However, when 3 mW is applied to the load, at time **b**, T1 reaches a steady value but T2 continues to decrease. This indicates the gradual filling of the gap as the ACP temperature goes above 29 K. At a higher heat flux more gas has to be condensed in the gap to reduce the temperature of the thermometer T2. The heat load on the heat switch from the time the ACP is turned on until the time it reaches 29 K is estimated to be less than 0.7 J. In the actual ADR system the gas desorbed by the ACP will first be cooled by a 10 K cryocooler to reduce the heat load on the ADR.

Finally, a short while after time **d**, when most of the helium has been pumped out of the gap and adsorbed by the ACP, a small temperature rise is noted by the thermometers T3 and T4. This is due to the heat of adsorption being released by the activated carbon and conducted into the heat switch by the gas remaining in the heat switch. In the ADR this heat will also be intercepted by the 10 K cryocooler.

From Figs. 4 and 5 an on-time can be estimated for the heat switch. It takes about half a minute for the gap to be filled with enough helium gas to conduct in the continuum regime. It then takes another two and one half minutes for the condensation to take place in the gap, i.e., time **b** to time **c**. This time can be shortened by heating up the ACP at a faster rate.

It takes more time for the prototype heat switch to fully turn off. After the ACP heater is turned off, it takes approximately two minutes for the ACP to go below 17 K. This time can be reduced by a double stage ACP, delineated in Ref. 1. By the time the ACP temperature is below 17 K it has pumped out most of the gas from the gap. However, there is enough gas left in the gap to conduct in the free-molecular regime. The time that it takes for the remainder of the gas to be pumped out is estimated based on how fast the brass load reaches the temperature it had when the ACP heater was turned on. This time is on the order of three to four minutes depending on the power applied to the load.

The effect of the size of the pump-out tube on the off-time of the heat switch has also been tested. Initially, the pump-out tube was 20 cm long with an inside diameter of 0.07 cm. The off-time of the heat switch in that arrangement was longer than ten minutes. This time has been reduced by nearly a factor of two after replacing the pump-out tube with a 10 cm-long tube, having an inner diameter of 0.14 cm.

CONCLUSIONS

The 2 K heat switch prototype with a gap size of 0.018 cm has an on/off conductance ratio that is close to the prdicted values. This ratio varies from about 1000 to 3000 as the input power is reduced from 0.6 mW to 0.07 mW. The on-time of the heat switch is on the order of three minutes. This time can be reduced by heating up the ACP at a faster rate. The off-time of the heat switch is on the order of six minutes. The ACP can be designed to pump out most of the gas present in less than one minute. However, the switch continues to conduct for an additional four minutes due to a small fraction of the gas remaining in the gap.

It is conceivable that the helium gas remaining in the gap gets adsorbed on the surface of the He II tube which is at 2 K. This would make it hard for the ACP to remove the remaining helium from the gap. A simple He II gap heat switch has been constructed which minimizes the 2 K cold surface. The off-time of this heat switch should give an indication as to how significant is the adsorption of the helium on the cold surface of the He II heat switch.

The off-time of the prototype heat switch has been shown to be a function of the size of the pump-out tube as well as the gap size; though, it appears that it is mostly dependent on the gap size. Increasing the width of the gap from 0.009 cm to 0.018 cm has reduced the off-time significantly.

The on/off conductance ratio of the 2 K heat switch is sufficient for the 2-10 K ADR. Its switching times need to be reduced further in order to meet the timing requirements of the ADR cycle.

REFERENCES

1. B. P. M. Helvensteijn and A. Kashani, Conceptual Design of a 0.1 W Magnetic Refrigerator for Operation Between 10 K and 2 K, in "Advances in Cryogenic Engineering," Vol. 35B Plenum Press, New York (1989), p. 1115.
2. J. P. Torre and G. Chanin, A Heat Switch for Liquid Helium Temperatures, Rev. Sci. Inst., 55:213 (1984).
3. B. P. M. Helvensteijn, A. Kashani and R. A. Wilcox, Activated Carbon Test Assembly, in: "Proc. Sixth Intl. Cryocooler Conf.," (1990) p. 103.
4. A. Kashani, B. P. M. Helvensteijn and R. A. Wilcox, Development of a 2 K He II Gap Heat Switch, in: "Proc. Sixth Intl. Cryocooler Conf.," (1990) p. 355.

A ^{3}He-GAP HEAT SWITCH FOR USE BELOW 2 K IN ZERO G

Pat R. Roach

Space Projects Division
NASA Ames Research Center
Moffett Field, California

Ben P.M. Helvensteijn

Sterling Federal Systems
Palo Alto, California

ABSTRACT

We have designed and tested a compact heat switch that has a simple design and a very large ON/OFF ratio. The design uses concentric cylinders of copper that can be fabricated with higher precision and with thinner web thickness than other designs. It is assembled with a technique that carefully controls the narrow gap between adjacent segments. These features allow a very large surface area for conduction to be fitted into a small volume. The conduction medium is liquid or gaseous ^{3}He which is put into or taken out of the switch by a small nearby charcoal pump in order to avoid an external mechanical pump and a long pump line.

Measurements of its performance down to 1 K show an ON/OFF conduction ratio of ~4000.

INTRODUCTION

We have been developing a ^{3}He-^{4}He dilution refrigerator for space applications[1] that will require the use of heat switches to couple and uncouple various chambers from the system heat sink and from each other. In order to maintain the advantages of no moving parts and of operation by charcoal pumps which the dilution refrigerator exhibits, we want to have a heat switch which uses gaseous or liquid helium as a thermal conduction medium that is pumped out by a charcoal pump when isolation between the two halves of the switch is needed.

For these tests liquid and gaseous ^{3}He was used as the conduction medium. We felt that the higher vapor pressure of ^{3}He at 1 K would make it faster to pump out of the switch than ^{4}He. In addition, we wanted to evaluate the suitability of ^{3}He gas in anticipation of needing a switch to operate at 0.4 K where ^{4}He has too low a vapor pressure to be used.

For our purposes it is necessary for the switch to be very compact and to have a large ON/OFF ratio. In addition, it is necessary for the OFF conduction to be very small. Because ^{3}He is not a very good thermal conductor, it is necessary to make the area across which the heat flows as large as possible and to make the distance the heat must flow through the helium as short as possible in order to achieve a good ON conduction. This suggests a design with a number of copper conductors of large surface area projecting from each end of the switch. These should then be assembled so that the surfaces from one end overlap those from the other end without touching them but with as small a gap as possible between them.

The compactness of the design is mainly a question of how thin the conductors can be made and how many of these plates can be squeezed into a given volume. For ease of machining and good mechanical stability of the resultant parts we felt that cylindrical geometry for the conductors was advantageous; it is extremely difficult to fabricate very thin, large plates of copper that will maintain their flatness and alignment to the degree necessary for this application.

Other designs of similar switches have either been able to relax the compactness requirement and use rather thick copper conductors[2,3], or they have been able to relax the low OFF conduction requirement and allow the conductors to touch in a few places[4]. This allows the gaps to be made smaller and allows the fabrication and assembly to be less critical.

HEAT SWITCH DESIGN

Figure 1 shows the design we have developed. It consists of a series of telescoping copper cylinders attached to each end of the switch in such a way that there is only a very narrow gap between adjacent cylinders when the two halves are assembled. The two halves are joined by a thin-wall stainless steel cylinder that serves to contain the helium but conducts very little heat between the two ends.

The overall length of the switch is 2.9 cm and the diameter of the housing is 1.5 cm. After assembly the gaps between adjacent cylinders are 0.01 cm and the total surface area across which heat flows is 26 cm^2. The wall thickness of the

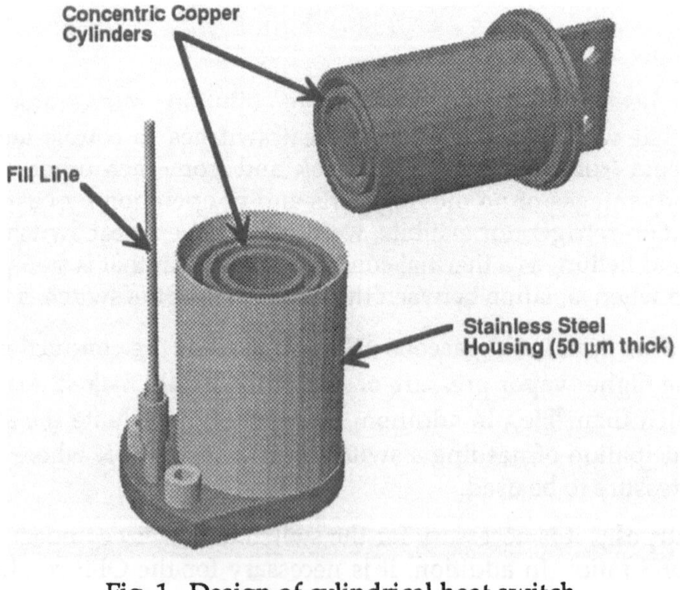

Fig. 1. Design of cylindrical heat switch.

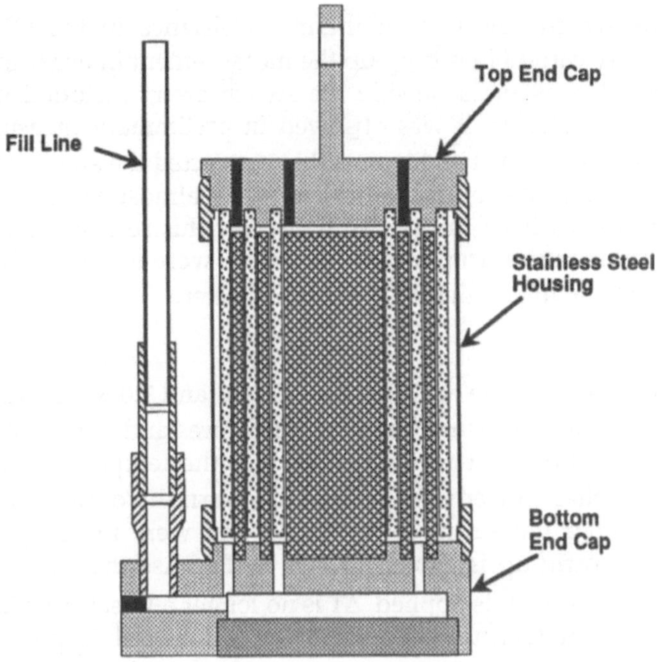

Fig. 2. Cross section of heat switch.

copper cylinders is 0.07 cm and their inner diameters are chosen to match standard reamer sizes with 1/16 inch increments. In this way the inner surfaces can be accurately reamed and the outer surfaces can be accurately turned on a lathe to achieve the precision needed to produce the close gap between cylinders that is desired when the cylinders are assembled.

The copper cylinders are silver soldered into shallow grooves in the copper end pieces. During this operation and also during the final assembly of the two halves of the switch the alignment of the parts is controlled by temporarily inserting a number of 0.01 cm diameter wires into the gaps between cylinders through small holes in the end caps; the holes in the top cap are later sealed with epoxy and the holes in the bottom cap become the access holes for helium to flow into and out of the gap spaces. These details are shown in a cross section view of the switch in Fig. 2.

When the helium is pumped out of the switch the only thermal path between the two ends is through the outer housing. Therefore, the housing was made from stainless steel and its wall thickness was made as thin as possible. We were able to make this only 50 μm thick by putting a thick-walled stainless steel tube of the correct inner diameter onto a tightly-fitting aluminum mandrel and then turning down the excess thickness on a lathe.

HEAT SWITCH TESTING

The switch was tested down to 1 K by mounting it on a ^{4}He pot that could be cooled to 0.9 K with no load on it. Pumping of the helium in the switch was provided by a large charcoal pump designed for other purposes. It was connected to the switch by a long pumping tube required in order to expedite the switch evaluation. These circumstances meant that it took quite a long time to pump the helium out of the switch and that no meaningful conclusion could be drawn about the cycling speed of the switch.

Because this switch has a small thermal resistance in the ON state, it is necessary to be very careful in setting up the measurement in order to be sure that extraneous thermal resistances outside the switch aren't included in the switch measurement. In particular, it was observed in preliminary measurements that the thermal resistances of the bolted joints that connected the switch to the ^{4}He pot on one end and to a heater on the other end were almost as big as the thermal resistance of the switch itself in the ON state. For this reason the thermometers measuring the temperature drop across the switch were mounted directly on the end caps of the switch, not on the ^{4}He pot or the heater.

RESULTS

Measurements were made at 77 K, 4.2 K, 2.0 K, and 1.0 K. Conduction of the switch with ^{3}He gas was measured at all temperatures and with ^{3}He liquid in the switch at 2.0 K and 1.0 K. It was observed that the temperature of the switch increased as the applied power went up due to a warm-up of the ^{4}He pot. Similar effects occur at all our measuring temperatures but were most apparent at 1 K. One effect of this warm-up is shown in Fig. 3. Because the temperature of the switch is changing as power is applied, ΔT is no longer a linear function of Q. It is easy to show that if the thermal conductivity of a material is proportional to T, $\kappa(T)=kT$ (as is nearly true in this case), then the expression for heat flow across a large temperature difference is:

$$\dot{Q} = \frac{kA}{2L}(T_2^2 - T_1^2) \qquad (1)$$

where T_2 and T_1 are the temperatures at the ends of the heat flow, A is the area of the heat flow path and L is the length of the path. Figure 4 shows the good agreement of the data with the form of Eq. 1. The slope of the straight line is the factor 2L/kA, so that the conductance of the switch, $\kappa(T)A/L$, is just 2T/(slope). The ON conductance of the switch, $\kappa_3 A_g/L_g$, is then $4.37 \times 10^{-2} \cdot T$ W/K, where κ_3 is the ^{3}He gas conductivity, A_g is the surface area and L_g is the length of the gas gap (A_g/L_g = 20.7 m for our switch).

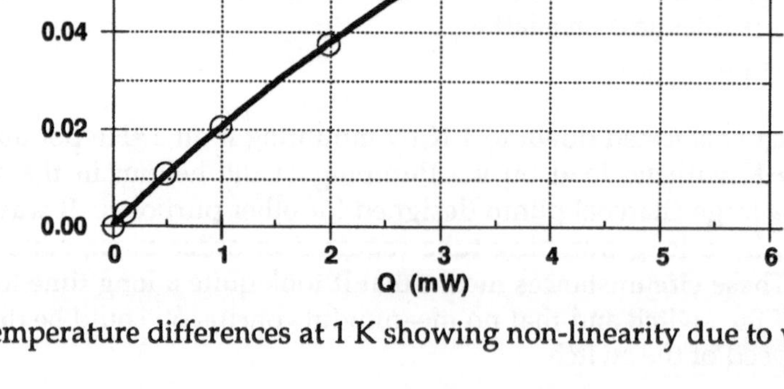

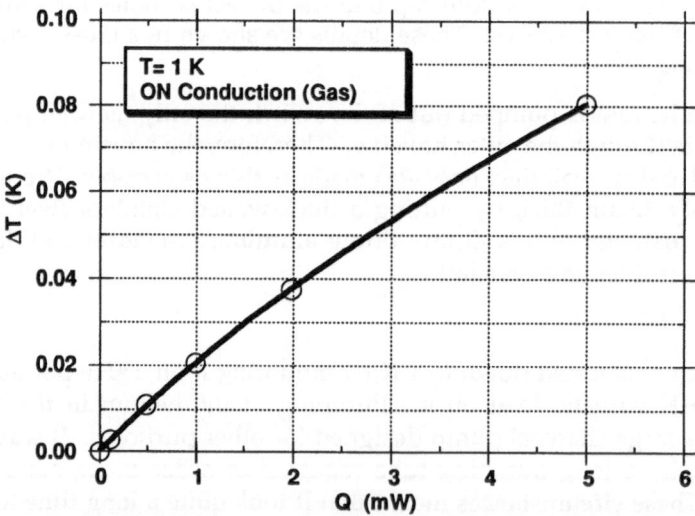

Fig. 3. Temperature differences at 1 K showing non-linearity due to warm-up.

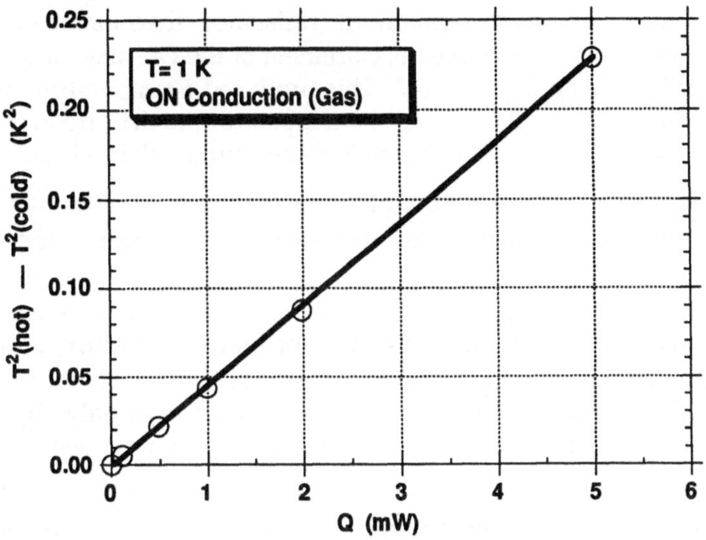

Fig. 4 Linearity of differences in T^2 for ON conduction.

The gas pressure in the switch was monitored by a thermocouple gauge at room temperature. It was coupled to the low temperature region by a 0.3 cm dia. tube so that thermomolecular pressure corrections would need to be made in interpreting its measurements at the lowest pressures. For the measurements reported here, the pressure in the switch was between 60 and 260 Pa. At these pressures the conduction was essentially pressure independent. A test at 1 K showed a large increase in the conductance in going from below 0.1 Pa to 3 Pa pressure in the switch, after which the conductance increased very little with higher pressures.

It took many hours to pump the gas out of the heat switch at 1 K in the configuration we had. The pumping line between the charcoal pump and the heat switch was very long in order to expedite the test on our dilution refrigerator system. Calculations suggest that this pumpout time should not be nearly so long even with the excessively long pump line. Part of the problem might be the slow desorption of a monolayer of helium from the switch surfaces.

In order to measure the OFF conduction, the switch was pumped overnight at 4.2 K. This consistently gave the same, very low conduction value. Very low powers were applied in this case and the temperatures of the cold end of the switch and the ^{4}He pot changed very little during the measurement. The OFF conductance of the switch, $\kappa_s A_h/L_h$, is $1.20 \times 10^{-5} \cdot T$ W/K, where κ_s is the conductivity, A_h is the cross-sectional area and L_h is the length of the stainless steel housing ($A_h/L_h = 1.6 \times 10^{-4}$ m for our switch). At 1K the ON/OFF conductance ratio is 3640 for ^{3}He gas.

At 2.0 K the pumping speed of the charcoal pump on the ^{4}He pot could be controlled somewhat by heating the charcoal and the base temperature for the measurement didn't vary as much as in the previous case. At 4.2 K the ^{4}He pot was not pumped but it did contain liquid helium that was in contact with the 4.2 K bath by a heat-pipe refluxing mechanism. Because of this effect the pot only warmed from 4.26 to 4.49 K when 20 mW was applied to the heat switch.

At 77 K our calibrated germanium resistance thermometers are rather insensitive so we obtained only one measurement of the ON state at the maximum power we could safely apply, 100 mW. This produced a temperature difference of 0.15 K across the switch while causing the temperature to drift up at a rate of 0.06 K/min. We don't feel that this data point is more accurate than ±10%.

By applying pressures greater than 1.2 kPa at 1 K and greater than 20 kPa at 2 K, liquid ^{3}He could be condensed into the switch. The effect of the condensing helium could clearly be seen as a large load on the ^{4}He pot. We could be sure when the switch was full by noting when the ^{3}He pressure stayed constant at a value much greater than the vapor pressure for that temperature, indicating that both the switch and the filling line beyond the ^{4}He pot were filled with liquid. Interestingly, the conduction of the switch with liquid was only slightly greater than that with gas. The time needed to remove the liquid was much longer, however.

Figure 5 summarizes the measurements at all the temperatures for which we have data. Between 1 and 2 K the ratio of ON and OFF conductions are ~3600 for the gas and ~4300 for the liquid. The OFF conductions are just what we would expect for typical conductivities of our thin-wall stainless steel housing. At higher temperatures the ON/OFF ratio is not as good because the ^{3}He gas conductivity does not go up as fast with temperature as the stainless steel conductivity.

ANALYSIS

The thermal conductivity of both ^{3}He liquid and gas as derived from our measurements are shown in Fig. 6. For comparison are shown previous

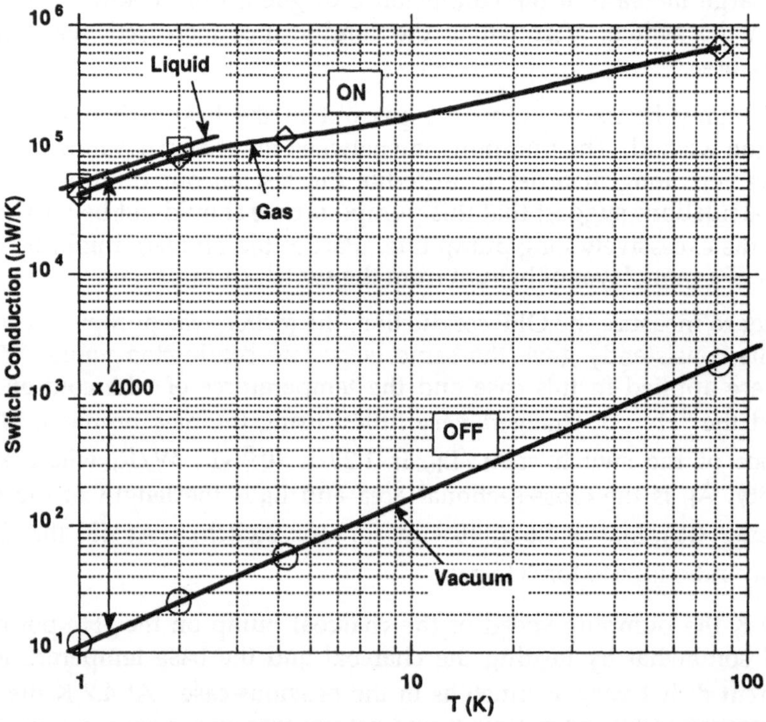

Fig. 5. Conduction of heat switch in OFF mode and in ON mode with both gaseous and liquid ^{3}He.

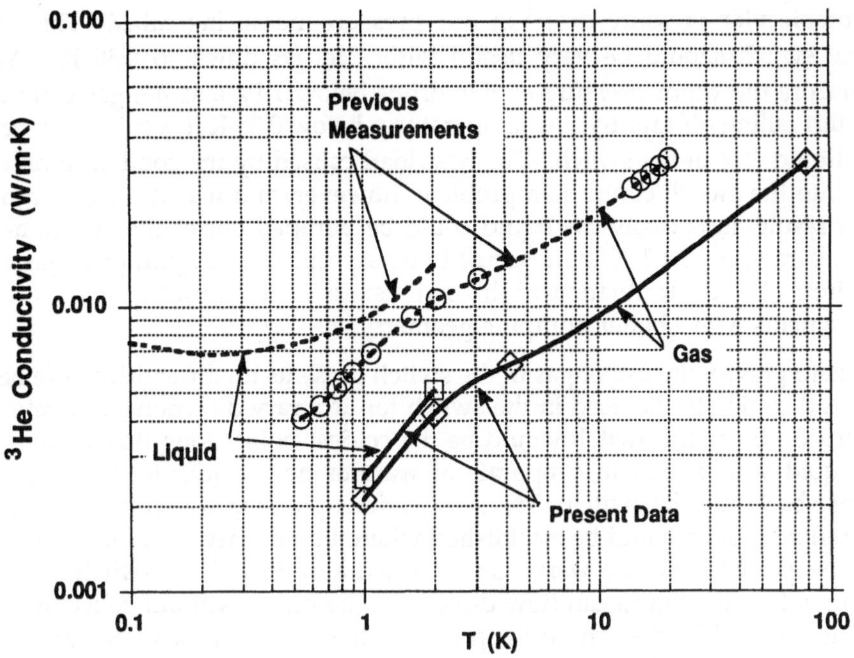

Fig. 6. Thermal conductivity of ^{3}He derived from present experiment.

measurements for liquid[5] and for gas[6]. The points for gas conductivity from 14 to 20 K are derived from viscosity measurements. The relationship[7] $\kappa = \varepsilon\ \eta\ C_V$ relates the conductivity, κ, to the viscosity, η, where $\varepsilon = 2.50$ for helium and $C_V = 3R/2M = 4160$ J/kg·K is the constant volume heat capacity where R is the molar gas constant and M is the molecular weight of ^{3}He.

Clearly, the conductivity results from our present data are about a factor of 3 too low. Since this discrepancy appears rather temperature independent over a wide temperature range, it is most likely due to an unexpected thermal resistance which has a temperature dependence that is similar to that of the helium. Many impure metals have conductivities that behave like this. In particular, if one estimates the effect of the silver solder at the end of the copper cylinders by assuming it has 1/100th the conductivity of copper and is 0.013 cm thick (much thicker than we would expect), it has enough thermal resistance to explain the discrepancy.

CONCLUSIONS

We have demonstrated the performance of a compact heat switch down to 1 K with both liquid and gaseous ^{3}He. Although it is quite compact, it has a very large surface area for conduction across a thin layer of helium, it has a very low conduction when OFF and the design prevents accidental touches between the conductors from opposite ends of the switch. The ratio of ON conduction to OFF conduction is ~4000 at 1–2 K, and we believe it would be 3 times higher if the silver solder joints at the ends of the copper cylinders were improved.

The use of this switch with liquid ^{3}He seems to have little advantage over gaseous ^{3}He at temperatures of 1 K and above. For lower temperatures we observed that the conduction is quite good with a gas pressure of 3 Pa (measured at room temperature, – it would be only 1.3 Pa in the switch because of the thermomolecular pressure drop[8] in our pressure-measuring tube). This means that the switch should be very useful with ^3He gas down to 0.35 K. At this temperature the vapor pressure of ^3He drops below 1 Pa and the gas conduction must start falling off rapidly. At temperatures below 0.35 K it would be necessary to use liquid ^3He in the switch. The heat load caused by the condensation of the liquid into the switch could be a problem, however, and it is not clear from our present results how easily the liquid could be pumped out of the switch at very low temperatures. At 1 K there seems to be a problem even pumping gas out of the switch. The determination of the seriousness of this problem awaits further measurements with a better pumping geometry.

In zero gravity the use of gas in the switch presents no difficulties. We believe that liquid could also be used in the switch for arbitrary directions of acceleration and zero G. Since the switch would be the coldest point along its pumping line, stray liquid would tend to evaporate at warmer points outside the switch and condense back into the switch. The most serious problem we would anticipate is that a refluxing flow could be established where liquid arrived at the hot charcoal pump, evaporated and the warm gas returned and heated the switch. This is a problem only if the liquid can flow directly to the pump without collecting in the pumping line and blocking it. If the pumping line is spirally wrapped around the switch then for any direction of force on the liquid stray liquid would have to collect somewhere in the line and the refluxing flow of gas would be blocked. In zero G the potential for this problem disappears.

The liquid could be pumped out of the switch without difficulty as long as the switch was at a high enough temperature for the vapor pressure to be able to push plugs of liquid toward the pump if such plugs were to fill the line.

ACKNOWLEDGEMENTS

We would like to thank John Paterson for expert machining of the small parts of the switch and Harry Dill for proving that a very thin-wall stainless steel tube could be made on the lathe by making the one we used.

REFERENCES

1. Pat R. Roach and Ben Helvensteijn, "Advances in Cryogenic Engineering", vol. 35, Plenum Press, New York, (1990), p. 1045.

2. C.K. Chan, *Jet Propulsion Laboratory Publication 87-7.*

3. D.J. Frank and T.C. Nast, "Advances in Cryogenic Engineering", vol. 31, Plenum Press, New York, (1986), p. 933.

4. J.P. Torre and G. Chanin, *Rev. of Sci. Instrum.*, **55**, 213 (1984).

5. D.M. Lee and H.A Fairbank, *Phys. Rev.*, **116**, 1359 (1959) and A.C. Anderson et al., *Phys. Rev. Lett.*, **6**, 443 (1961). (as summarized in J. Wilks, "The Properties of Liquid and Solid Helium", Clarendon Press, Oxford(1967) p. 432.)

6. Below 10 K: K. Fokkens et al., *Physica*, **30**, 2153 (1964). Above 10 K (from viscosity): E.W. Becker and R. Misenta, *Z. Physik*, **140**, 535 (1955).

7. S. Dushman, "Scientific Foundations of Vacuum Technique", John Wiley & Son, New York, (1962) chap. 1.

8. T. R. Roberts and S. G. Sydoriak, *Phys. Rev.*, **102**, 305 (1956).

TEST RESULTS OF AN ORIFICE PULSE TUBE REFRIGERATOR

Stephen F. Kral, Dallas Hill, Jon Restivo and Joseph Johnson

Astronautics Corporation of America
Astronautics Technology Center
Madison, Wisconsin

Peter Curwen, Warren Waldron and Howard Jones

Mechanical Technology, Inc.
Latham, New York

ABSTRACT

An orifice pulse tube refrigerator has been designed, fabricated and tested. An hydraulically actuated diaphragm compressor, which allows direct determination of compressor PV work, dynamic pressure amplitudes and phase shifts, is used to drive the refrigerator. Test results are presented and compared to analytic predictions and design specifications. Sources of potential performance improvement are identified and future testing goals are discussed.

INTRODUCTION

The orifice pulse tube refrigerator (OPT), shown in Figure 1, is the simplest mechanical gas-cycle refrigerator suited to practical applications at cryogenic temperatures. The essential components of the refrigerator are a room temperature compressor generating an oscillating pressure wave in the refrigerator's working fluid; a transfer line ; an aftercooler; a regenerator; a cold heat exchanger; the pulse tube; a hot heat exchanger; an orifice; and a ballast volume. The only moving parts in the OPT are in the compressor, at room temperature. There are no moving parts, no seals of any kind and no high tolerance components within the remainder of the device. With an appropriately designed compressor, the working fluid volume can be hermetically sealed, so that there is no possibility of contamination. The OPT has excited particular interest in those developing space qualified cryocoolers because its mechanical simplicity suggests high reliability, long operating life and low cost.

The OPT discussed here is part of our OPT program aimed at developing both single stage and multistage OPTs. The use of multistage OPT refrigerators to accept the heat rejected by magnetic refrigerators operating at low temperatures is a goal of this work, as well as the production of space qualified OPT cryocoolers. Our program includes experimental and prototype testing and the development of numerical and analytic models.

Analyses of the mechanisms underlying the refrigeration produced by the OPT have been published.[1,2,3] A number of unresolved issues still exist, including the influence of heat transfer to the wall of the pulse tube on the performance of the device and the flow dynamics of the working fluid in the pulse tube. The mechanical simplicity of the OPT belies the complexity of the fluid flow and heat transfer processes that occur during its operation. Of equal importance to

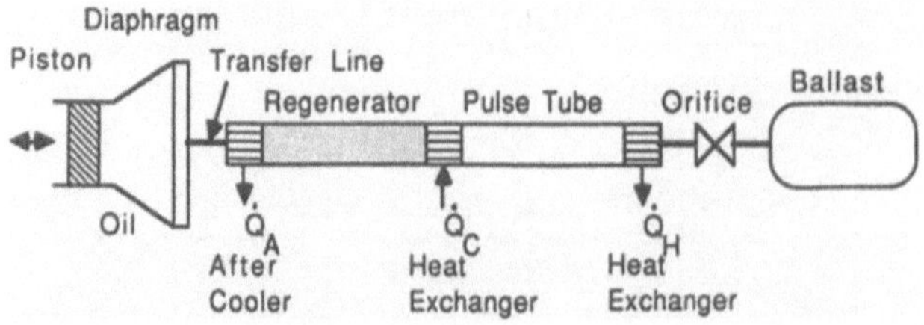

Fig. 1. Schematic diagram of pulse tube refrigerator.

understanding the principles of operation is the development of predictive and descriptive analytic and numerical models of the device so that efficient and accurate OPT design can occur.

Table 1 gives the range of OPT test conditions. The dimensions of the OPT are given in Table 2. All components except the heat exchangers are fabricated of 316 stainless steel (SS); the heat exchangers are OFHC copper. The screens in the heat exchangers facilitate heat exchange and act as flow straighteners.

Stainless steel was chosen for the pulse tube and regenerator body because of availability, ease of fabrication and cost. We estimate that the conduction loses through the regenerator and the pulse tube walls may be as large as 1.2 W. These losses can be reduced by 50% without the addition of working fluid contaminants by using titanium in place of stainless steel.

The performance of the OPT is dependent critically on the design of the regenerator. Tests with regenerators fabricated of 220 mesh SS screen and 0.3 mm Pb spheres were less satisfactory than those with the 400 mesh regenerator reported here. While the 400 mesh SS screen regenerator gave rise to a significant performance improvement, it has not been optimized. Performance improvement is expected with optimization because regenerator ineffectiveness is the largest single source of parasitic thermal load.

Table 1. Operating Conditions

Cooling Power at 80 K (W)	0 - 0.3
Operating Frequency (Hz)	10 - 20
Mean Pressure (MPa)	1.35 - 1.93
Dynamic Pressure Amplitude (Mpa)	0.19 - 0.26

Table 2. System Description

Component	Description
Compressor	MTI prototype hydraulically actuated10 cm^3 displacement, 0-2.1 MPa, 0-30 Hz
Transfer line	0.413 cm ID, 12.7 cm length
After cooler	1.27 cm ID, 1.27 cm length, water cooled copper
Regenerator	1.27 cm OD, 1.17 cm ID, 10.3 cm length, 316 stainless steel, filled with 1075-400 mesh stainless steel screen disks
Cold Heat Exchanger	0.826 cm ID, 1.52 cm length, filled with 40-100 mesh copper screen disks, resistance wire wound heater for cooling load
Pulse Tube	0.794 cm OD, 0.743 ID, 5.08 cm length 316 stainless steel
Hot Heat Exchanger	0.826 cm ID, 0.762 length, filled with pressed copper ribbon, water cooled
Orifice	Whitey micro-metering valve SS-22RS4 with vernier handle
Ballast	5.08 cm ID, 21.1 cm length

EXPERIMENTAL PROCEDURE

Tests of the device near to the design conditions were made possible by the use of a prototype, hydraulically actuated diaphragm compressor. This compressor operated with a fixed swept volume of 10 cm^3. In the prototype, a variable speed permanent magnet DC motor drives a piston which oscillates in the oil bath that hydraulically actuates the diaphragm. Operation to about 20 Hz was easily achieved. With this compressor, the pressure ratio in the OPT was somewhat smaller than our design specification. Nevertheless, because of the extreme flexibility and reliability of the compressor, performance over a very wide range of operating conditions was measured.

Temperatures are measured with thin film platinum resistance sensors, calibrated in our laboratories. A piezoelectric quartz pressure sensor measures the dynamic pressure in the compressor. Diaphragm displacement is measured with a noncontacting capacitance displacement probe. The dynamic pressure sensor and the diaphragm displacement sensor permitted direct measure of the compressor PV work. Pressure sensors at the pulse tube hot heat exchanger and in the ballast gave direct measure of pressure drop across the regenerator, pressure variations and hence mass flow into the ballast, and phase shifts across all important system components.

During performance testing the regenerator, cold and hot heat exchangers, pulse tube, orifice and ballast are placed in an evacuated vessel at a pressure of 10^{-6} torr or less. The assembly is wrapped in several layers of superinsulation, and the cold heat exchanger and adjacent tubing are separately wrapped with superinsulation. Temperature and pressure data are recorded and displayed on a computer controlled data acquisition system. The instantaneous diaphragm displacement and compressor head pressure are recorded and stored on a digital oscilloscope. These data are transferred to a computer to calculate the compressor's PV work. Tests consist of varying one of the system parameters: compressor frequency, average pulse tube pressure, thermal load on the cold heat exchanger, or orifice size, and measuring the remaining parameters when quasi-steady state conditions are established. The system is easily reconfigured, and measurements for variations on the component sizes reported here have been completed.

RESULTS

With pressure sensors in the pulse tube at the hot heat exchanger (just before the orifice), and in the ballast, pressure amplitude and phase measurements have been made for OPT operation during start-up, cool-down and at equilibrium. (The pressure sensor at the hot heat exchanger is removed when thermal performance measurements are made.) Figure 2 shows the amplitude and phase relationships of the dynamic pressures in the compressor, in the pulse tube, and in the ballast volume. The phase difference between the dynamic pressure in the pulse tube and that in the compressor is about 30°. The dynamic pressure in the ballast lags that of the pulse tube by about 90°. The mass flow rate through the orifice leads the dynamic pressure in the ballast by 90° and hence it is in phase with the dynamic pressure in the pulse tube, as expected. The amplitude of the mass flow rate through the orifice can be estimated from the

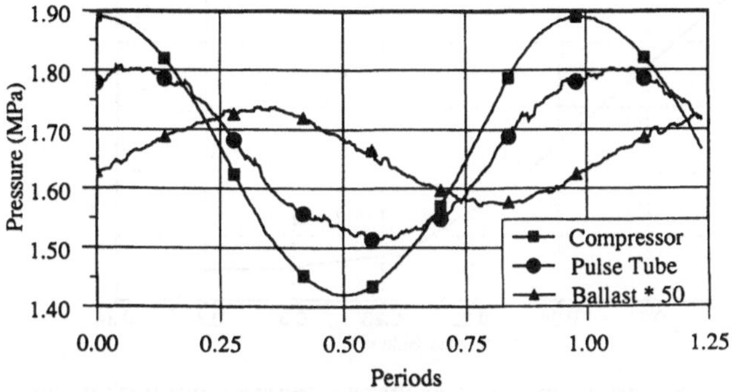

Fig. 2. Pressure distributions in pulse tube refrigerator.

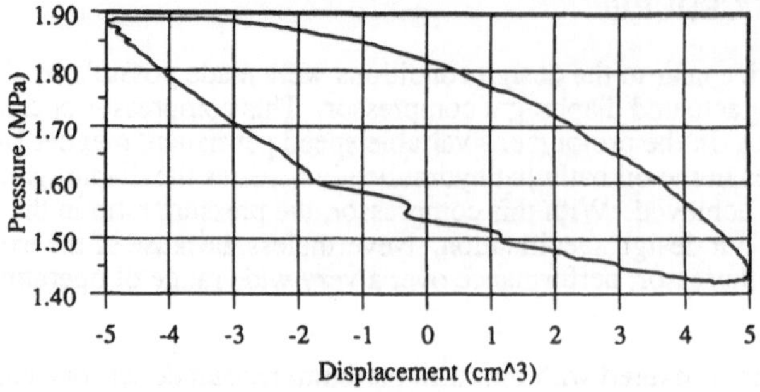

Fig. 3. PV work diagram of compressor operating at 14 Hz with 0.0 Watts cooling load.

amplitude of the dynamic pressure in the ballast, the volume of the ballast and the ideal gas law. For the case described here, the mass flow rate amplitude is about 18.4 mg/sec.

Figure 3 shows the PV diagram for the working fluid in the compressor measured when the cold heat exchanger is at its minimum temperature, with no external load, for operation at 14 Hz and a mean pressure of 1.65 MPa. (The peculiar structure in the lower branch of the PV diagram is thought to result from air in the compressor oil and is of no consequence. The load curve corresponding to this PV diagram is exhibited in Figure 7.) PV diagrams for the same frequency and mean pressure have been obtained for external thermal loads imposed on the cold heat exchanger of 1 W, 3 W and 5 W. For these external loads, ranging from 0 W to 5 W, the compressor PV work remains relatively constant at about 1.86 J. In general, we observe that PV work is not affected by variations in the external load or by changes in the mean pressure in the OPT. This suggests that parasitic thermal loads dominate the performance of this device. On the other hand, PV work increases with increasing frequency, as shown in Figure 4. This is due to a

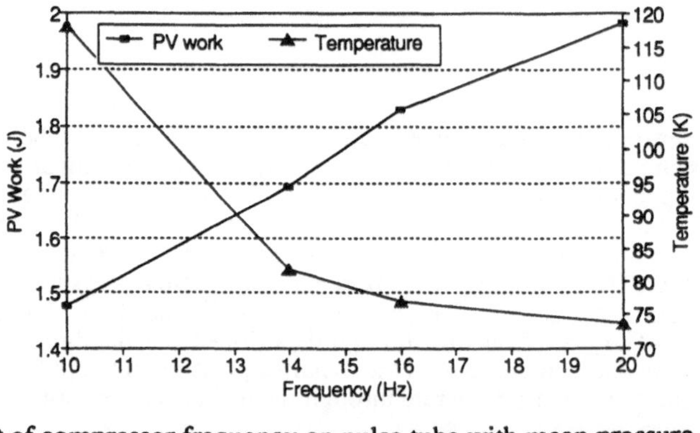

Fig. 4. Effect of compressor frequency on pulse tube with mean pressure of 1.65 MPa.

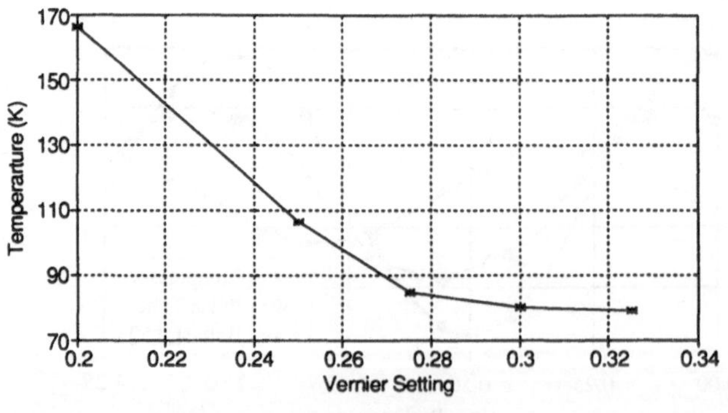

Fig. 5. Effect of orifice opening on cold heat exchanger temperature with mean pressure of 1.65 MPa and the compressor frequency of 14 Hz.

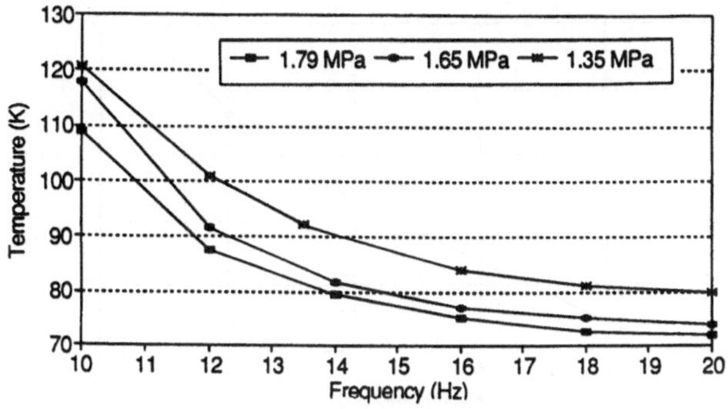

Fig. 6. Effect of frequency on cold heat exchanger temperature at three different mean pressures.

combination of effects, including the dependence of regenerator ineffectiveness on frequency, frequency dependent compressor hysteresis and, possibly, fluid mixing in the pulse tube.

The dynamic pressures shown in Figure 2 were taken for an orifice optimized for minimum operating temperature. Figure 5 illustrates the experimental optimization procedure. The equilibrium, no-load cold heat exchanger temperature is varied by varying the size of the orifice. A minimum temperature is determined. Once the minimum operating temperature is reached, the temperature is insensitive to small increases in the size of the orifice. Under external thermal load, small decreases in the cold heat exchanger temperature can be obtained with adjustments to the orifice size; the load results reported here are for orifice sizes fixed for minimum no-load temperature.

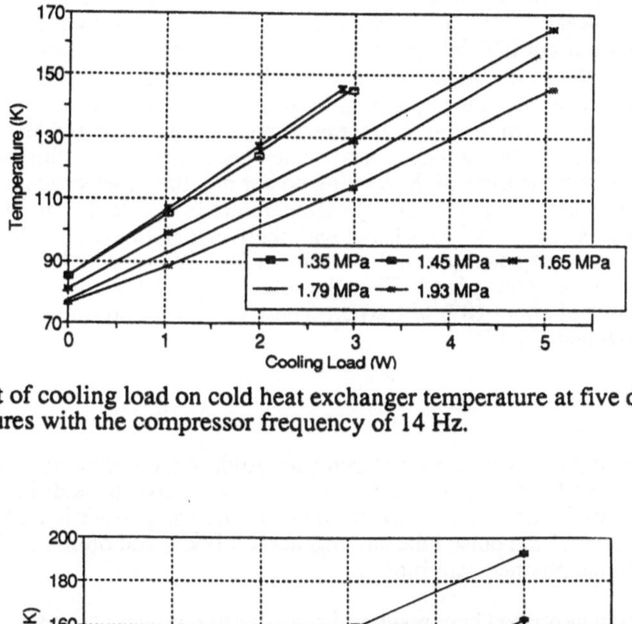

Fig. 7. Effect of cooling load on cold heat exchanger temperature at five different mean pressures with the compressor frequency of 14 Hz.

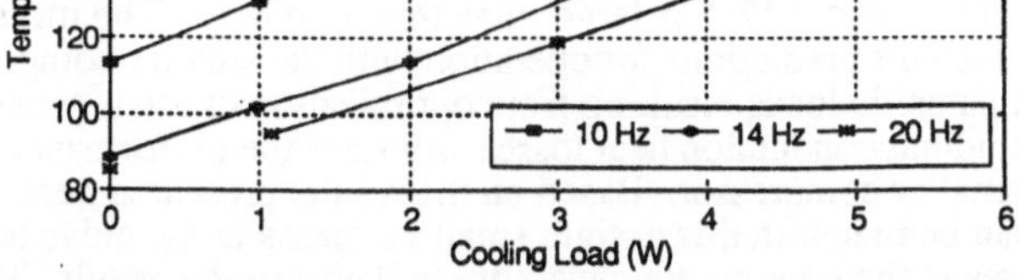

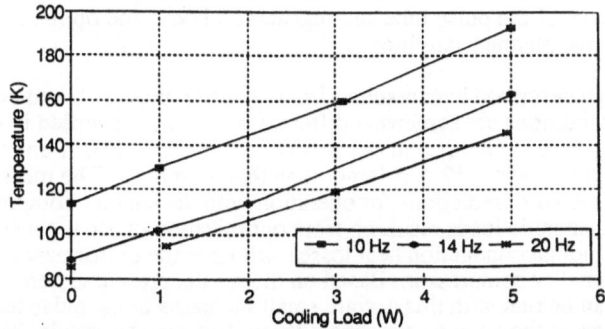

Fig. 8. Effect of cooling load on cold heat exchange temperature at 3 different compressor frequencies with a mean pressure of 1.65 MPa.

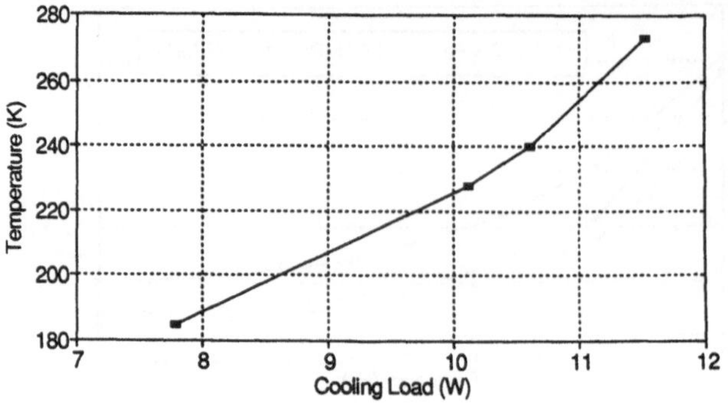

Fig. 9. Effect of large cooling load on cold heat exchanger temperature with a mean pressure of 1.65 MPa and compressor frequency of 14 Hz.

The minimum temperatures achieved with no external load for several operating pressures and frequencies is shown in Figure 6. Increasing the frequency at a fixed pressure or increasing the pressure at a fixed frequency decreases the minimum temperature. Figures 7 and 8 show the equilibrium temperatures as a function of external load. In Figure 7, with the frequency held fixed at 14 Hz, the load curves measured at various pressures are shown. In Figure 8, with the mean pressure fixed at 1.65 MPa, the load curves at various frequencies are shown. Finally, in Figure 9, the load curve for large external loads is shown. Data in Figure 9 were taken at 14 Hz and a mean pressure of 1.65 MPa.

These data are sufficient for a comparison of the performance of this device with published analytic models.[1,3] Using Radebaugh's adiabatic model, a refrigerator cooling power estimate can be made from pulse tube pressure amplitude, mass flow rate through the orifice, mean pressure and hot heat exchanger temperature. For our operating conditions, the cooling power is predicted to be about 0.93 W. Baks discusses corrections to the adiabatic model resulting from heat transfer to the wall. For the case where the hot heat exchanger is at 293 K, the cold heat exchanger is at 80 K, the operating frequency is 14 Hz and the mean pressure and pressure amplitude are as shown in Figure 2, these corrections are about 0.4 W. We have made measurements of the heat rejected at the hot heat exchanger under the conditions assumed in these calculations by direct measure of the cooling power of a thermoelectric cooler attached to the hot heat exchanger when the heat exchanger temperature is maintained at 293 K. The measured heat rejected is about 1 W, consistent with these calculations. The PV power measured under conditions where the rejected heat is 1 W is about 26.6 W. In the adiabatic model the rejected heat is equal to the cooling load. Thus the specific power for this cooler is 26.6 W/W, suggesting an efficiency of about 10% of Carnot. These results are preliminary, and additional measurements are now being made.

CONCLUSION

We are continuing performance measurements. Additional measurements of the heat rejected at the hot heat exchanger under load are being made to provide additional information about the parasitic thermal load. Measurements to determine compressor hysteresis are planned. Variations in the material of the pulse tube and regenerator body, and optimization of the regenerator are additional planned activities.

The OPT design described here resulted from a design process based on models we have developed. The test conditions are somewhat different from those specified in the design, principally in the compressor. Our design target was 1 W cooling at 80 K and the compressor swept volume specified was about 12.5 % larger than that used here. The measured performance is consistent with our model's predictions for operation with the smaller compressor. Small reductions in the total parasitic loads resulting from optimization of the regenerator and minimization of longitudinal conduction heat losses will raise the performance toward target values even with the smaller compressor. Based on the results presented here, we expect that the target design values can be met with this design; small increases in the pulse tube volume with accompanying increases in the compressor capability will ensure this result. These data are of also interest in our efforts to improve our design and performance predicting models based on understanding the OPT's operating principles.

ACKNOWLEDGEMENT

Mechanical Technology, Inc. is acknowledged gratefully for the loan of its prototype compressor. Gordon Haupt, Mark Powers and Charles Cross are all recognized for their contributions.

REFERENCES

1. R. Radebaugh, A review of pulse tube refrigeration, in: "Advances in Cryogenic Engineering," Vol 35, Part B, Plenum Press, New York, p. 1191.
2. G. M. Harpole, and C. K. Chan, Pulse tube cooler modeling, in : "Proc. Sixth Intl. Cryocoolers Conf.," Vol 1., G. Green and M. Knox, eds. , David Taylor Research Center, Bethesda, MD (1990), p. 91.
3. M. J. A. Baks et al., Experimental verification of an analytical model for orifice pulse tube refrigeration, Cryogenics 30:947 (1990).

MEASUREMENTS OF INSTANTANEOUS GAS VELOCITY AND TEMPERATURE IN A PULSE TUBE REFRIGERATOR

Marc David,* Jean-Claude Marechal and Pierre Encrenaz†

Laboratoire de radioastronomie, Ecole Normale Superieure, Paris
France

*also Cryophysics, Jouy-en-Josas, France

†also Observatoire de Meudon, Meudon Cedex, France

ABSTRACT

We report on measurements of instantaneous velocity in an orifice pulse tube (O.P.T.), a hybrid pulse tube (H.P.T.) and a double inlet pulse tube (D.I.P.T.). The available power at the cold end of the pulse tube is determined by at least three effects :

1) phase shift between volume flow and pressure fluctuations caused by the orifice (orifice effect),
2) surface heat pumping,
3) absorption of acoustic waves by the tube wall.

The objective of this project is to perform local measurements of velocity and temperature rather than the conventional collection of global data (pressure, working frequency and the average mass flow rate). From these local measurements we will determine the radial temperature and the velocity gradients. The results should permit us to obtain information on the orifice effect. In addition we will estimate the dimension of the thermal boundary layer and we will calculate the heat amount pumped by the surface heat pumping mechanism. Finally, we will be able to derive the amount of heat pumped by the absorption of acoustic waves from the measurements of instantaneous velocity. The understanding of the transient phenomena should give us the optimal working parameters of a pulse tube (O.P.T., H.P.T. or D.I.P.T.).

INTRODUCTION

The pulse tube refrigerator offers the potential of being developed into a small and reliable cryocooler[1,2]. The most useful types are the O.P.T.[3], the D.I.P.T.[4] and the H.P.T.[5], shown in Figure 1.

Advances in Cryogenic Engineering, Vol. 37, Part B
Edited by R.W. Fast, Plenum Press, New York, 1992

To date, only Storch and Radebaugh[6] have presented an analytical model of the O.P.T.. This model was later modified by Baks et al.[7]. However, their theories were not fully confirmed by their experiments. A better understanding of the different types of pulse tubes requires measurements of velocity and temperature profiles at different locations in the tube. The difficulty here lies in obtaining the velocity gradient measurement independent of temperature and with minimal disturbance.

First, this paper explains the theoretical objectives of such measurements and it describes an experimental method of precisely determining the radial temperature and velocity gradients.

PURPOSES OF THE MEASUREMENTS

Storch and Radebaugh had measured a cooling power a factor three to five times lower than what their theory had predicted. Baks et al had modified this theory by taking into account the heat transfer between the gas and the wall. Their experimental results, in which the negative influence of the regenerator losses was eliminated, were close to their predictions for small amplitudes of gas displacements. Except for the fact that both Baks and Storch had neglected refrigerator losses, their theories have been not fully satisfactory because they had considered the whole pulse tube as a thermodynamic system.

The purpose of our work is to devise an analytical model with a Lagrangian system. Therefore, we will try to explain the pulse tube by following different samples of gas during a cycle. This work is complicated mainly because of two reasons :

1) The system is always in a hydrodynamic transient state. Because of the flow straightener located at the bottom of the pulse tube, we have assumed that the velocity profile is flat at the tube entrance. Based on the theory of Landau and Liftchitz[8], we assume it will have a Poiseuille profile at a distance l from the tube entrance such that $l = aR_e$ where a is the tube radius and R_e the Reynolds number defined by $R_e = \rho Ua / v$ where ρ is the fluid density, U the fluid velocity and ν the kinematic viscosity. Obviously, the l magnitude is just an order of magnitude, and the boundary layer variation versus the distance covered by the working fluid is approximately known[8]. In our case, the average Reynolds number is 1000 and the tube radius is 19 mm. Consequently l is larger than the 200 mm tube length.

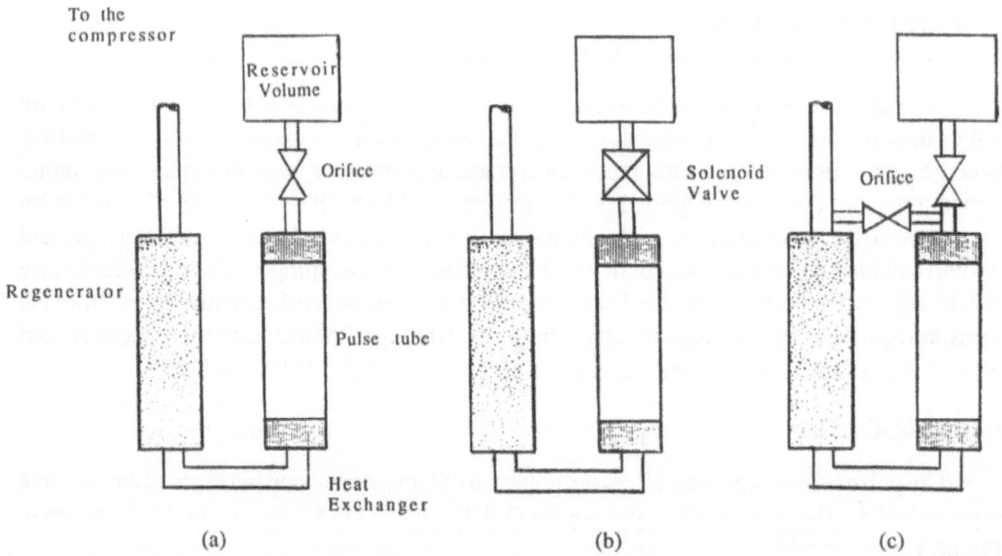

Fig. 1. Schematic of different types of pulse tube refrigerator: (a) orifice pulse tube, (b) hybrid pulse tube, (c) double inlet pulse tube.

It should be noted that this estimate is calculated for a constant velocity whereas the gas accelerations in the pulse tube are large.

2) The lag between the two gas velocities at the two ends of the tube which is induced by the orifice.

Local measurements of the velocity should permit us to devise an analytical model more pertinent than the others and to predict the velocity lags for the three types of pulse tube refrigerators. The determination of the velocity gradient will allow us to calculate the amount of heat Q_1 due to the viscosity ; likewise, the determination of the temperature gradient will allow the amount of heat Q_2, due to the thermal diffusivity, to be estimated. These two effects are currently neglected in the theories. Both measurements will be useful to estimate the term of cooling power Q_3 which combines the heat pumping mechanism of Wheatley with an additional heat transfer to the wall corresponding to the absorption of acoustic waves on normal reflection at a hard wall described by Baks.

$$Q_1 \approx \nu(\partial V_z / \partial r)^2 \tag{1}$$

$$Q_2 \approx (k / r)\partial(r\partial t / \partial r) / \partial r \tag{2}$$

$$Q_3 \approx (\tilde{P}^2 \, 2\pi f \, V_t / \gamma P_0 \, a)(\alpha / 2f)^{1/2}[G / 2 + (\gamma - 1) / 2]^{1/2} \tag{3}$$

where V_z is the gas velocity, k is the thermal conductivity of the gas, $\tilde{P}$ the pressure fluctuation, f the working frequency, V_t the tube volume, γ the specific heat ratio, P_0 the average component of the pressure, α is the thermal diffusivity of the gas and G a dimensionless measure for the temperature gradient along the wall.

$$G = 1 - (\nabla T_w \, \Delta x / \Delta T) \tag{4}$$

where ∇T_w is the temperature gradient along the pulse tube wall, Δx the amplitude of the displacement of the gas in the tube and ΔT the amplitude of temperature rise due to adiabatic compression of the gas.

METHOD OF MEASUREMENT

The two systems currently able to perform local measurements of velocity are laser anemometry and hot-wire anemometry. We have chosen the second system because of its low cost and its easy maneuverability. Also, using hot wire anemometer, gas temperature can be measured. The hot-wire anemometer parameters are as follows:

1) The velocity measurement must be independent of temperature. This is the main difficulty. The hot wire, whose temperature is always higher than the gas temperature, is cooled by both forced convection and temperature variations in the gas. A change in ambient temperature of 5 K for hot-wires in helium flows induces errors larger than or equal to 5 % in the velocity measurements. The error would be even greater in a pulse tube in which a 50 K temperature variation for a 1.5 pressure ratio can be expected.
The electrical schematic of the temperature compensated anemometer is shown in Figure 2. The velocity probe and the temperature compensating probe are identical. The temperature compensating circuit is calculated from the following equation for the velocity sensor (King's law) :

$$I^2 R_H / (R_H - R_C) = A + BU^n \tag{5}$$

where I is the current going through the probe, R_H the velocity probe compensating resistance, R_C the cold resistance of the velocity probe and U the fluid velocity. A, B and

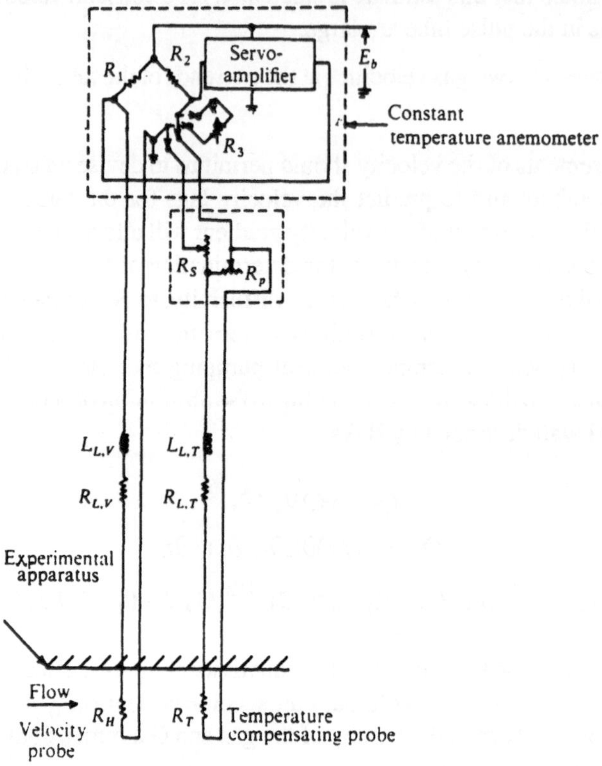

Fig. 2. Schematic of circuit utilized to compensate for ambient temperature variations. Equivalent resistance $R_3 = (R_S + R_P)\, R_T / (R_P + R_T)$

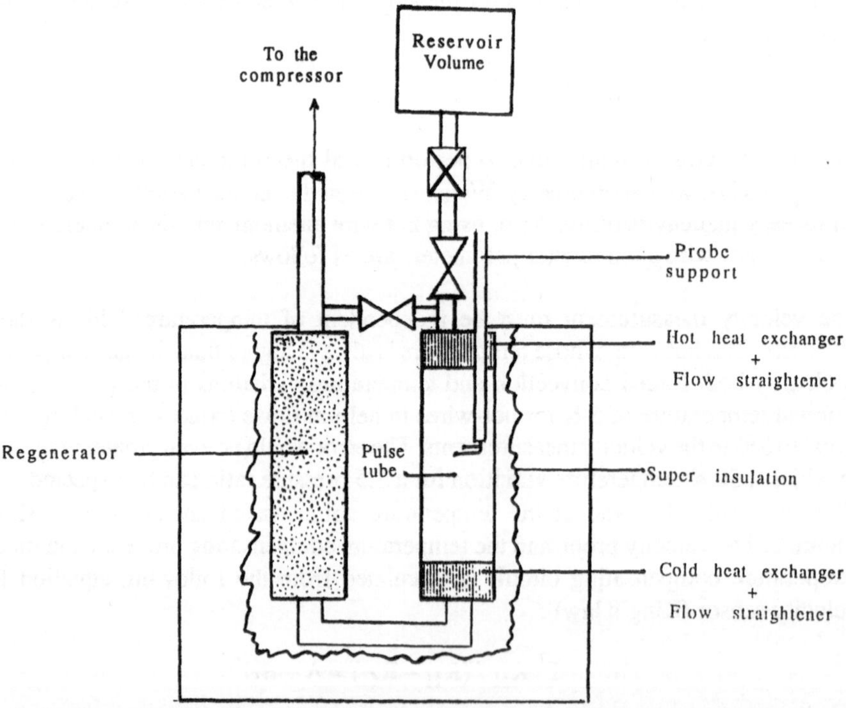

Fig. 3. Schematic of the apparatus designed to measure the velocity and temperature of the gas in pulse tube refrigerator.

n are nearly constant ; $R_C = R_H r_0(T)$ where r_0 is the overheat ratio. Because of constant temperature operation, the anemometer servo-amplifier maintains the probe temperature and hence its resistance R_H constant. Using King's law, Drubka et al[9] had demonstrated that the most efficient mode to keep the output voltage constant, i.e. independent of ambient temperature, was through maintaining a constant overheat ratio. Based on the bridge balance condition of the anemometer circuit in Figure 2, we can write

$$R_1 / R_2 = R_H / R_3 = M \tag{6}$$

where M is the bridge ratio and R_3 is the equivalent resistance of the temperature compensating circuit. The constant overheat ratio condition specifies the compensating circuit resistance with temperature necessary to maintain a proper operation. This can be written as:

$$dR_3 / dT = \left[R_{co}\, \alpha_c\, r_0(T)\right] / M \tag{7}$$

If we assume that the temperature dependence of R_C is such that :

$$R_c = R_{co}[1 + \alpha_c(T - T_0)] \tag{8}$$

where R_{C0} is the cold resistance of the velocity probe when T is equal to the probe reference temperature T_0. Hence, we can calculate the values of the series and parallel resistances. Finally, it is necessary to check that the current through the temperature compensating probe is low enough to assure that the probe resistance is independent of the fluid velocity.

The originality of this method lies in its ability to determine the parameters of the temperature compensating circuit without the knowledge of the temperature calibration. Using this method, these parameters may be determined knowing only the properties of the anemometer bridge and the temperature probes (i.e. the resistances and the temperature dependence coefficient).

2) The volume of the probe support must be low enough to minimize disturbance of the flow.

3) The hot-wire anemometer must be able to measure quickly variations in temperature and velocity up to a 8 Hz working frequency. In other words, the thermal penetration depth D_t must be higher than the wire diameter, and the thermal response time Á must be lower than 1 ms. The thermal penetration depth in a thermometer subjected to an oscillating flow is obtained from the relation :

$$D_t = [k / \pi \rho C_p f]^{1/2} \tag{9}$$

where f is the working frequency. The thermometer time response τ is calculated from

$$\tau = \rho V C_p / hA \tag{10}$$

where V and A are the volume and surface area of the thermometer and h is the heat transfer coefficient between the thermometer surface and the fluid.

TEST PROCEDURES

The scheme of the apparatus designed to measure the velocity and temperature of the gas in the pulse tube is shown in Figure 3. The tube has a 19 mm diameter and a 200 mm length. The probe support has a 6 mm outside diameter and it is located off-center in the pulse tube so that the gas velocity and temperature can be measured at different distances from the wall by simply rotating the probe support. The wires between the probe support and the probes have a 2 mm diameter and a 30 mm length.

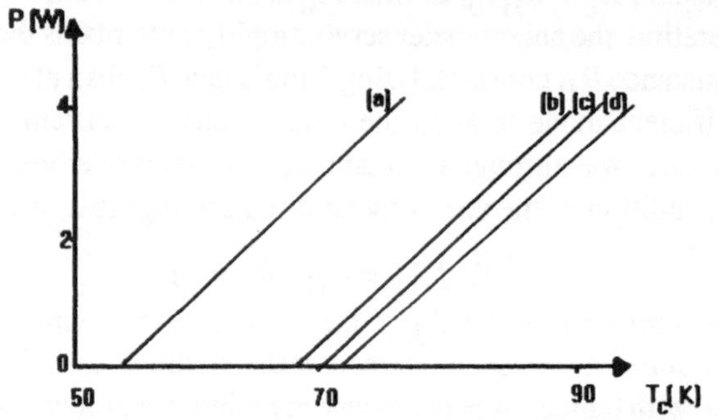

Fig. 4. The pulse tube performances (cold power versus temperature) versus the probe depth, (a) without the probe support, (b) the probe support is entirely inside the tube.

To estimate the perturbations due to the probe support, limit temperatures and cold powers were measured at different depths of the probe in the tube. Figure 4 shows that the probe support modifies the pulse tube performances, but the results versus the probe depths change to only a slight extent. When the anemometer was operating and the hot-wire heated, the results were the same as before. For the velocity measurements, we will assume that the velocity profile will not be disturbed when the gas flows from the cold end to the orifice. The system was made to compare the three types of pulse tube refrigerators and, eventually, the basic pulse tube[10], i.e. the pulse tube with all the orifices closed. The heat exchangers and connection tubes were designed to reduce void volumes.

The velocity measurements will be performed according to the previous process, and the gas temperature will be measured by supplying the temperature compensating probe with a weak 1 mA current in order to avoid self-heating. The probes are tungsten wires with a 5 μm diameter an a 3 mm length. The calibration curve for the tungsten resistance thermometers is linear over the range of temperature from 80 K to 400 K. The thermal penetration depth of the tungsten wires in the worst case, i.e. at the highest temperature and at the highest frequency, is more than 1 mm. The theoretical time response is always lower than 1 ms. It was calculated by using equation (5) and the Nusselt number based on the relation given by Collis and Williams[11] for hot-wires at low Reynolds number (from 0.01 to 140).

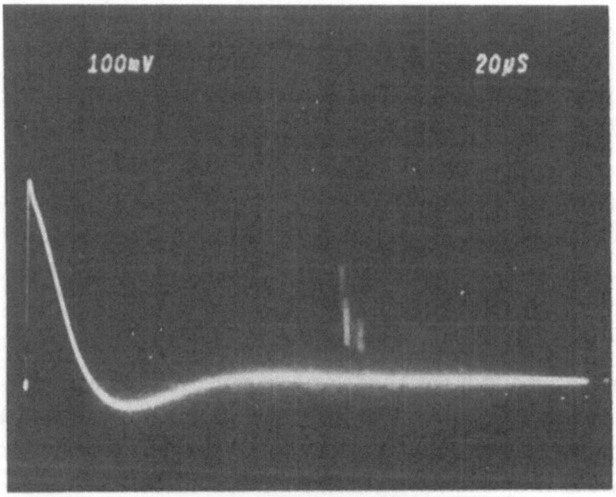

Fig. 5. Time response test of the anemometer.

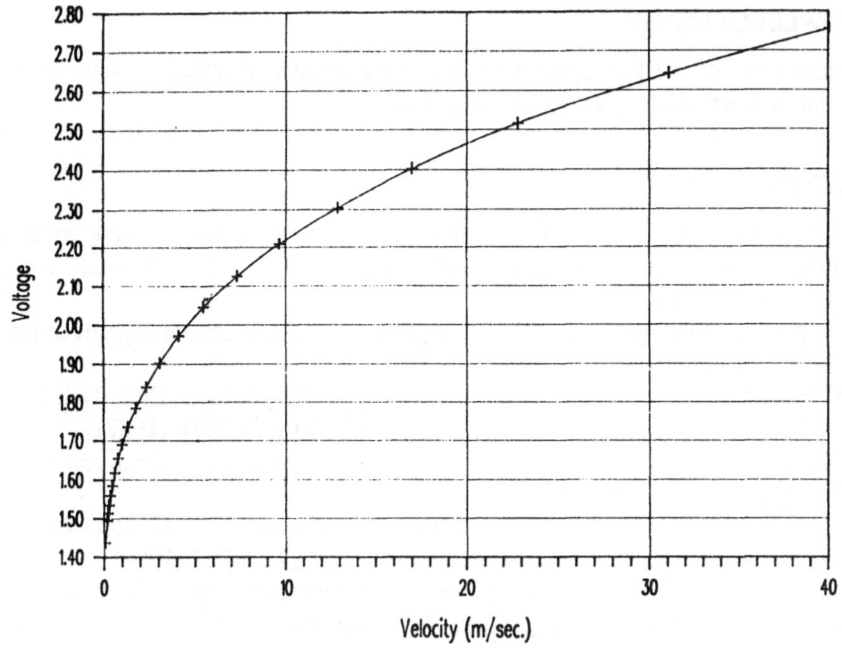

Fig. 6. Calibration curve of the hot-wire anemometer for steady-state flow.

For greater Reynolds number, heat exchanges between the wire and the gas are better, then h increases and τ decreases. The response of the probe was checked by applying a step function change in the Wheastone bridge. The higher input voltage caused the probe to self-heat. The response time, as shown in Figure 5, was lower than 100 µs. In steady state flow, the anemometer has been calibrated by a firm with a laser anemometer. The curve calibration of the probe velocity at constant temperature in air is shown in Figure 6. Durability of the probes was tested by subjected to both oscillating mass flow and oscillating pressure in the test apparatus shown in Figure 3.

To check the temperature compensating, we have compared the output voltages at room temperature and at liquid nitrogen temperature for a zero velocity. The difference between the two output voltages was lower than 10 mV which induced a velocity error lower than $3\ 10^{-2}$ m/s. At greater velocities, the error due to the temperature compensating is difficult to evaluate because many gas parameters change with temperature (density, viscosity and, then the coefficient of heat transfer between gas and probe). However, according to DISA, for a 5 µm tungsten wire, the velocity is reduced by 0.025 %/°C for a temperature rise. Before performing measurements inside the pulse tube, we will check the velocity calibration in an oscillating helium flow.

CONCLUSION

A best understanding of the different types of pulse tubes requires the measurements of the radial temperature and of the velocity gradient of the gas. A hot-wire anemometer with temperature compensating has been constructed and tested. The velocity probe and the temperature compensating probe were two 5 µm tungsten wires which had a short response time, a small thermal penetration depth and exhibited a high durability. The 2 mm diameter wires-support did not perturbate the gas flow. A test apparatus to perform measurements inside different types of pulse tubes had been constructed and successfully tested.

ACKNOWLEDGEMENTS

Some parts of the test apparatus were constructed and tested by Guy Jouve and Christophe Bouvier and they are greatly appreciated.

REFERENCES

1. R. Radebaugh et al., A comparison of three pulse tube refrigerators : new methods for reaching 60 K, "Advances in Cryogenic Engineering," Vol. 31 Plenum Press, New York (1986) p.779
2. J. Liang et al., Development of a single-stage pulse tube refrigerator capable of reaching 49 K, Cryogenics 30:.49 (1990)
3. R. Radebaugh, Pulse tube refrigeration. A new type of cryocooler, Proc. 18th Int. Conf. on low temp. phys. (Kyoto 1987), Jpn. J. Appl. Phys. 26:2016 (1987)
4. S. Zhu et al., A single stage double inlet pulse tube refrigerator capable of reaching 42 K, Cryogenics 30:257 (1990)
5. M. David and J. C. Maréchal, How to achieve the efficiency of a Gifford-Mac Mahon cryocooler with a pulse tube refrigerator, Cryogenics 30:262 (1990)
6. P. Storch and R. Radebaugh, Development and experimental test of an analytical model of the orifice pulse tube refrigerator, "Advances in Cryogenic Engineering," Vol. 33 Plenum Press, New York (1988) p.85
7. Baks et al., Experimental verification of an analytical model for orifice pulse tube refrigeration, Cryogenics 30:947 (1990)
8. L. Landau and E. Lifchitz in “Mécanique des fluides”, Editions Mir, Moscow (1971)
9. Drubka et al., in “DISA information”, N° 22, (Dec. 1977)
10. W.E. Gifford and R.C. Longsworth, Pulse tube refrigeration ASME paper n°63-WA-290 (Nov. 1963)
11. D.C. Collis and M.J. Williams, Two-Dimensional convection from heated wires at low Reynolds Numbers, J. Fluid Mech. (1959) p.357

MEASUREMENT OF THE PERFORMANCE OF A SPIRAL WOUND POLYIMIDE REGENERATOR IN A PULSE TUBE REFRIGERATOR*

Wayne Rawlins and Klaus D. Timmerhaus

Department of Chemical Engineering
University of Colorado
Boulder, Colorado

Ray Radebaugh and D.E. Daney

Chemical Engineering Division
National Institute of Standards and Technology
Boulder, Colorado

ABSTRACT

A regenerator for use in a pulse tube refrigerator has been constructed from a polyimide (polypyromellitimide or PPMI) whose small ratio of thermal conductivity to heat capacity make it a good candidate for a regenerator material in cryocoolers. The regenerator was fabricated using 25 μm thick photoresist strips bonded to a 50 μm thick sheet of PPMI. This composite sheet was wound in jelly-roll fashion around a mandrel and inserted into the regenerator housing. The photoresist strips, formed using a photolithographic technique, provided a 25 μm spacing for the axial flow of gas between each layer of PPMI. Ineffectiveness results are presented for this material under actual operating conditions in a pulse tube refrigerator and compared with a numerical model. The numerical model indicated that a polyimide regenerator would perform much better than one constructed of stainless steel screen, but the experimental results showed the opposite behavior. Measured values for the ineffectiveness were 0.003 for the stainless steel screen and 0.017 for the polyimide.

INTRODUCTION

Regenerator theory indicates that the best geometrical configuration for a regenerator is a parallel-plate arrangement.[1] Figure 1 from Radebaugh and Louie shows the relationship

---

*Research sponsored by NASA/Ames Research Center, Contribution of NIST, not subject to copyright.

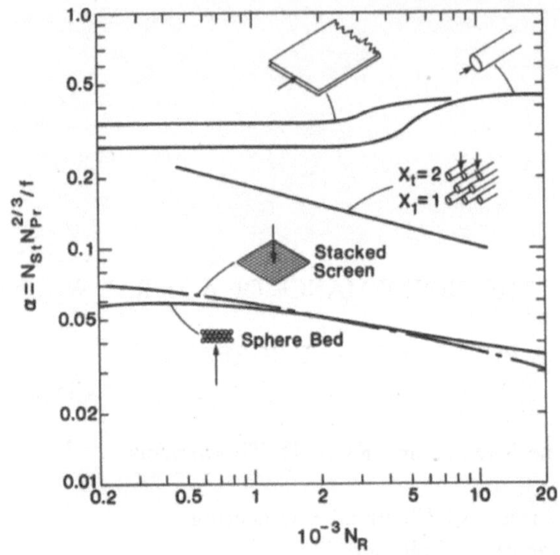

Fig. 1. Ratio of heat transfer and friction factor curves as a function of Reynolds number for several configurations.[1]

between α and the Reynolds number, N_R, where α is defined as

$$\alpha = N_{St} N_{Pr}^{2/3} / f , \tag{1}$$

N_{St} and N_{Pr} are the Stanton and Prandtl numbers respectively, and f is the friction factor. A high value of α is desirable because this provides a high rate of heat transfer for a specified pressure drop. In spite of this general theoretical information regenerators are still usually built with packed-screen or sphere geometries. The reasons are twofold. First is the problem of constructing a parallel-flow geometry with adequate heat transfer area while keeping the uniform spacing and thickness of intended channels to avoid concentration of flow in one region. Second, the materials commonly used for constructing cryogenic regenerators are metals, and the high thermal conductivity of a metal in a parallel-plate configuration would result in an excessive conduction loss at the cold end of the refrigerator.

To overcome these problems a method was devised for fabricating a regenerator with a parallel-plate geometry and 25 µm thick flow channels from a polyimide (polypyromellitimide or PPMI). Polyimide exhibits a thermal conductivity which is more than an order of magnitude lower than that of stainless steel at 80 K, and has a relatively high heat capacity. This low ratio of thermal conductivity to heat capacity makes it a good choice for a regenerator material. To study the feasibility of such a regenerator in a pulse tube refrigerator the performance characteristics of a stacked stainless steel screen and polyimide parallel-plate regenerator were compared using a computer program (REGEN2) developed by NIST personnel to aid in the design of regenerators. The results are given in Table 1, where ΔP/P is the ratio of the average pressure drop through the regenerator to the average regenerator pressure; NTU is the number of transfer units in a the half cycle; λ is the ineffectiveness for a half cycle and is defined as one minus the effectiveness; the thermal depth is the thermal penetration depth into the matrix divided half the thickness of the storage matrix; and $\dot{Q}_c$ is the conduction loss.

Table 1 indicates that the parallel-plate geometry has a much lower porosity accompanied by a lower void volume. It will thus exhibit higher pressure ratios than packed

Table 1. Results of REGEN2 comparing regenerators constructed from stainless steel screen and polyimide

	STAINLESS STEEL 7.9 OPENINGS/mm	POLYIMIDE PARALLEL PLATE
DIAMETER (mm)	15.9	28.6
LENGTH (mm)	100	45
VOID VOLUME (mm^3)	12,340	7582
$\dot{m}$ (g/s)	0.6613	0.6613
$\Delta P/P$	0.0048	0.0039
NTU	632.7	2848.3
λ	0.00309	0.000617
$\dot{Q}_c$ (W)	0.2989	0.4851
THERMAL DEPTH	15.54	5.11
POROSITY	0.67	0.33
NET REFRIGERATION (W)	0.165	0.8876

screens, which will result in higher refrigeration capacities for equivalent systems. The parallel-plate regenerator is also much more compact for equivalent heat storage capacity. Table 1 also shows that the calculated NTU is much higher and the ineffectiveness is five times lower for the polyimide parallel-plate regenerator than for the stainless steel regenerator. Therefore if such a regenerator was properly constructed it should result in considerable improvement in the performance of a pulse tube refrigerator.

CONSTRUCTION

The regenerator was wound, jelly-roll fashion, from 51 μm thick polyimide film having 25.4 μm thick photoresist ribs (see Fig. 2) aligned parallel to the flow direction. The photoresist ribs were formed in a photolithographic process which assured excellent uniformity in rib dimensions. The rib spacing was 0.40 mm and the rib width was 0.09 mm. The width of the film (length of flow passage) was 45 mm. The total length of film before being wound into a coil was 6042 mm. One end of the polyimide film was glued to a 6.4 mm diameter mandrel and wound around it with hand tension to a diameter slightly

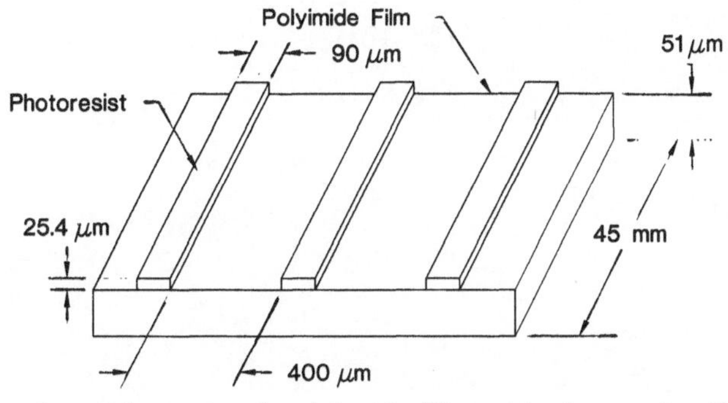

Fig. 2. Schematic of polyimide film with photoresist ribs.

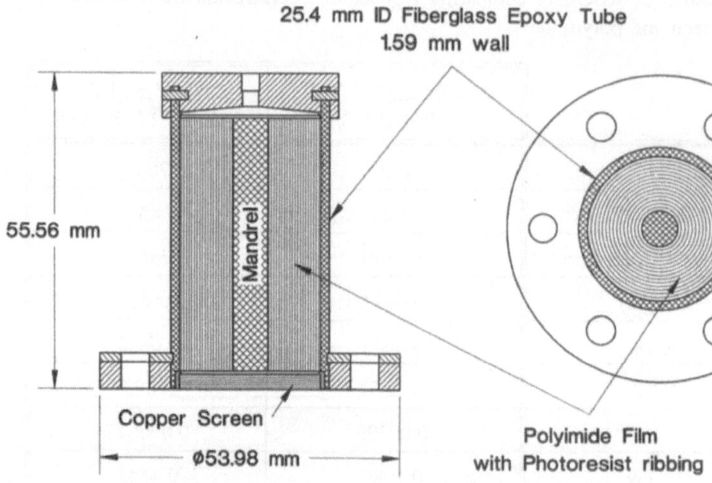

Fig. 3. Schematic of polyimide regenerator.

larger than the 25.4 mm inner diameter of a fiberglass epoxy tube, into which it was to be inserted. To ensure a tight fit of the polyimide winding in the regenerator housing the wound-up film was shrink-fit into the fiberglass epoxy tube by cooling it in LN_2. This ensured that there would be no channeling of the gas flow through the regenerator at the tube walls. A winding gap of 2.5 μm is typical for this method of assembly, which resulted in a flow passage of 27.9 μm, a total width of 4701 mm, and a length of 45 mm; see Fig. 3 and 4 for a schematic and photo of the assembly. Because the flow rate is proportional to the cube of the passage thickness, the gas flow which passed over the ribs rather than between them was only 0.2% of the total flow.

Before the regenerator was tested with oscillating flow, the pressure drop was measured under conditions of steady flow at both ambient and liquid nitrogen temperatures

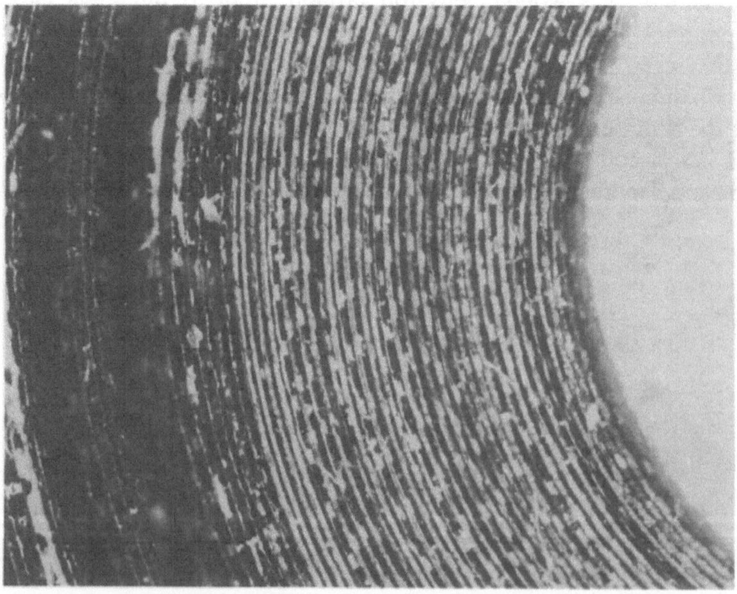

Fig. 4. Photo showing the rolled polyimide film at the end of that regenerator.

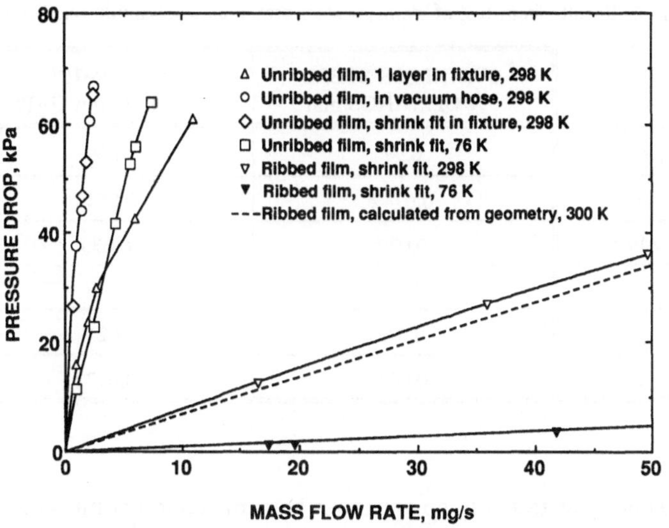

Fig. 5. Pressure drop vs mass flow rate for the polyimide regenerator in steady flow.

(Fig. 5). In addition, to study assembly techniques and winding gaps, a regenerator was wound from the film without the photoresist ribbing. Measurements of the pressure drop for this assembly are also given in Fig. 5. For the regenerator with the ribbed polyimide film the agreement between measured and calculated pressure drops is good, indicating uniform winding.

DISCUSSION

Regenerator performance is generally specified in terms of regenerator effectiveness, ε. The latter is defined as the dimensionless ratio of the actual heat transfer rate in the regenerator to the maximum possible heat transfer rate,[2,3] that is,

$$\varepsilon = \frac{\dot{Q}_{act}}{\dot{Q}_{max}} = \frac{actual\ heat\ transfer\ rate}{maximum\ possible\ heat\ transfer\ rate}. \tag{2}$$

To determine the effectiveness of a regenerator it is more convenient to define a regenerator ineffectiveness term λ given by[3,4]

$$\lambda = 1-\varepsilon = \frac{\dot{Q}_{max}-\dot{Q}_{act}}{\dot{Q}_{max}} \tag{3}$$

which reduces to

$$\lambda = \frac{\dot{Q}_{reg}}{\dot{Q}_{max}}, \tag{4}$$

where $\dot{Q}_{reg}$ represents the heat load on the regenerator. An energy balance for any differential section of the regenerator shows that the enthalpy flow is constant throughout the entire length of the regenerator as long as there are no radiative or convective losses to the external environment. The numerator of Eq. (4) can be calculated if the heat load at the cold end of the regenerator can be measured. The maximum possible heat transfer rate can be calculated from the known boundary conditions.

Table 2. Results from test of stainless steel screen and polyimide regenerators

	STAINLESS STEEL 7.9 OPENINGS/mm	POLYIMIDE WITH 25 μm GAPS
f (Hz)	6.97	6.69
$\dot{m}$ (g/s)	0.6742	0.8645
$\Delta P/P$	0.029	0.033
P_{max} / P_{min}	1.13	1.15
$\dot{Q}_{reg}$ (W)	1.15	8.21
λ	0.003	0.0172

The regenerator was tested in an apparatus constructed to permit these measurements to be made dynamically in an operating pulse tube refrigerator.[4] The enthalpy flow in the regenerator can be measured directly. This is achieved with the aid of laminar-flow heat exchangers designated as isothermalizers, which are located between the cold ends of the regenerator and the pulse tube. The isothermalizers are immersed in liquid nitrogen baths and the enthalpy flow is monitored by measuring the increase in the boil-off rate from the liquid nitrogen bath. The laminar-flow elements also permit the mass flow to be inferred from a measurement of the pressure drop across the units. A water-cooled heat exchanger at the warm end of the regenerator and the isothermalizers at the cold end of the regenerator control the entrance conditions at both ends of the regenerator. Absolute pressure is also monitored at the warm end of the regenerator and at the cold end of the pulse tube.

The test apparatus, which used helium as the working fluid, was operated at a mean pressure of approximately 2.2 MPa. The warm end of the regenerator was cooled with water to 290 K and the cold end of the regenerator was cooled to liquid nitrogen temperatures. The compressor supplying the oscillating pressure in the refrigerator had a swept volume of 25 cm^3 and was run at approximately 7.0 Hz. The orifice was an adjustable metering valve and was opened three turns.

RESULTS

Results for the polyimide parallel-plate regenerator are compared to those of an equivalently designed stainless steel screen regenerator in Table 2, using the design parameters in Table 1. Mass flow rate and absolute pressure at the cold end of the pulse tube with the polyimide regenerator are shown in Fig. 6. The pressure ratio in the pulse tube, which is the maximum pressure divided by the minimum pressure, is given as P_{max}/P_{min}. Table 2 shows that the performance of the polyimide regenerator did not achieve expectations. For both regenerators the pressure drops were greater than expected; otherwise the performance of the stainless steel screen regenerator was close to model predictions. The polyimide parallel-plate regenerator exhibited an ineffectiveness which was more than an order of magnitude greater than that predicted, and five times greater than that of the stainless steel regenerator. Even when the greater conduction loss was incorporated in the model predictions this could not account for the loss in performance. If channeling were occurring in the regenerator it could account for the reduction in performance; however, this assumption is not supported by the measured pressure drop under steady flow conditions. The pressure drop measured in the system during operation was higher than that predicted, which again indicates that channeling was not a factor. Another problem that needs further

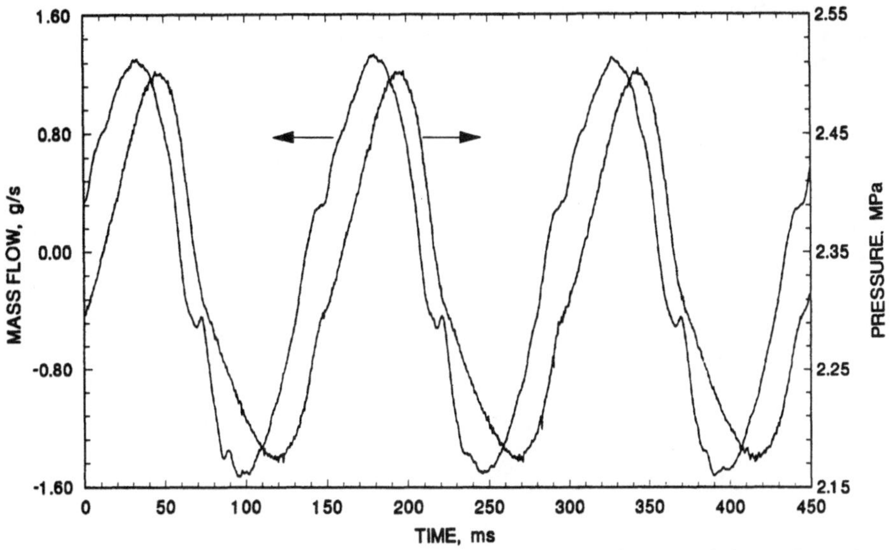

Fig. 6. Mass flow rate and absolute pressure at the cold end of the pulse tube.

investigation is the apparent shedding of particles by the polyimide regenerator. There are sensors (mentioned in previous work[4,5]) to measure temperature and mass flow at the cold end of the regenerator. These sensors are 1.5 mm long tungsten wires which have a diameter of 4 μm and all were quickly broken at start up, indicating particles may have been swept out of the polyimide assembly during start up.

CONCLUSION

A polyimide parallel-plate regenerator was constructed based on numerical predictions which promised superior performance compared to that of commonly used regenerator configurations and materials. Construction costs of the regenerator were relatively inexpensive, and fabrication was simple and did not require elaborate or expensive construction operations. Initial steady-flow tests indicated that the regenerator was constructed to design specification. Testing of the unit, however, led to very disappointing results. There is presently no plausible explanation for these results.

REFERENCES

1. R. Radebaugh and B. Louie, in: "Proceedings of the Third Cryocooler Conference", NBS Special Publication 698, U.S. Government Printing Office, Washington D.C. (1985), p.177.
2. W. M. Kays and A.L. London, in: "Compact Heat Exchangers", McGraw-Hill, New York (1984).
3. R. Radebaugh, D. Linenberger, and R.D. Voth, in: NBS Special Publication 607 (1981), p. 70.
4. W. Rawlins and R. Radebaugh, in: "Advances in Cryogenic Engineering," Vol. 35, Plenum Press, New York (1990), p. 1213.
5. W. Rawlins, R. Radebaugh, and K.D. Timmerhaus, to appear in: "Applications of Cryogenic Technology, Proceedings of Cryo'90," Plenum Press.

MINIATURIZATION OF A THERMOACOUSTIC PRESSURE WAVE GENERATOR

K.M. Godshalk

Superconductive Systems Group, Tektronix, Inc.
Beaverton, Oregon

ABSTRACT

The combination of a thermoacoustic pressure wave generator, or driver, with a pulse tube refrigerator shows great promise as a method to develop a long lifetime, no moving parts refrigerator. No moving parts refrigerators have been constructed[1,2] by combining thermoacoustic refrigerators and drivers. We will explore opportunities and methods for miniaturizing the thermoacoustic pressure wave generator. Power output, temperature gradient, and acoustic losses are several of the variables investigated which are affected by miniaturization. The results of this analysis will be used in designing and constructing a miniature thermoacoustic pressure wave generator.

INTRODUCTION

A thermoacoustic driver is a device that converts heat into acoustic power. Thermoacoustic drivers have the potential for providing a highly reliable, efficient means of producing pressure oscillations with no moving parts. One of the first thermoacoustic devices was the Sondhaus tube, observed by Sondhaus[3] in the 1850's. Thermoacoustic effects in the form of Taconis oscillations have long been familiar to those who use liquid helium. Current interest in thermoacoustic effects revived with the pulse tube refrigerator of Gifford and Longsworth[4], and thermoacoustic devices built by Wheatley[5] and Swift[1]. Rott[6] published a series of theoretical papers describing thermoacoustic effects. These papers have been the basis for pioneering theoretical and experimental[1,2,5] work done at Los Alamos National Laboratory (LANL).

A refrigerator requiring no moving parts may be constructed by combining a thermoacoustic driver with a thermoacoustic refrigerator or an orifice pulse tube refrigerator; Swift and Wheatley[1,2] have constructed combination thermoacoustic drivers and refrigerators. Since the thermoacoustic driver is a resonant device, its overall size is governed by the resonant frequency, and its practical use in some cases may be limited by the overall size.

The present work addresses the design rules for miniaturization of a thermoacoustic pressure wave generator. Devices resonant at ~500 Hz have been demonstrated[1,2]; our goal is to investigate the effects of increasing the frequency to ~2 kHz, and the resulting

Advances in Cryogenic Engineering, Vol. 37, Part B
Edited by R.W. Fast, Plenum Press, New York, 1992

design implications. This will involve examining methods of scaling thermoacoustic devices in order to gain insight into the relation between various design parameters, examining the power output possible, and considering the feasibility of manufacturing the components necessary for miniature thermoacoustic devices.

TYPICAL DEVICE

A schematic of a thermoacoustic driver designed to resonate at a frequency f = 540 Hz at a mean pressure P_m of 0.5 MPa of ^4He is shown in Fig. 1. The device consists of a tube of radius R whose overall length L is one quarter wavelength, and connects to a bulb at one end to form a quarter wavelength resonator. Inside the tube are two heat exchangers, and a thermoacoustic stack which begins a distance x_s from the closed end and extends for a length Δx. The interested reader should consult Swift[1] for a detailed description of the operation of a thermoacoustic driver. The thermoacoustic stack consists of parallel stainless steel plates of thickness 2ℓ spaced a distance $2y_o$ apart. The stack produces acoustic power by shifting the phase of the temperature and pressure oscillations; the spacing of the plates is three to four times the thermal penetration depth $\delta_k = (2\kappa/\omega)^{1/2}$, where κ is the thermal diffusivity and ω is the angular frequency. A temperature gradient is maintained across the thermoacoustic stack by heating the closed end of the device to a temperature T_h, while keeping the other end of the stack near room temperature (T_c) with a cooling water loop. The heat exchangers consist of closely spaced parallel copper plates. The device produces work W in the form of acoustic pressure oscillations, and shuttles a heat flux Q from the hot heat exchanger to the cold heat exchanger.

The design parameters given in the caption of Fig. 1 were arrived at using the program WAVEQ developed at LANL. This program operates in two modes: 1) given the geometry, mean pressure P_m and pressure amplitude P_a, heat flux Q, and hot temperature T_h, the program will find the cold end (ambient) temperature T_c and the acoustic work W produced, and 2) given the geometry, mean pressure and pressure amplitude, the target cold end temperature T_c, and target work W produced, the program will find the hot end temperature T_h and heat flux Q required to meet the desired target values of work and cold temperature. The second operation mode is used in this work.

Since the theory assumes laminar flow, additions were made to the program to calculate the boundary layer Reynolds number Re at several positions along the thermoacoustic stack; Re is given by[1,7] $u\delta_v/\nu$, where u is the gas velocity, δ_v is the viscous penetration depth, and ν is the gas viscosity. The transition to turbulence in oscillating gas flow (for frequencies less than 130 Hz) has been found to occur[7] at Re ≈ 500. The acoustic losses of the resonator due to viscous and compressive effects[1] were

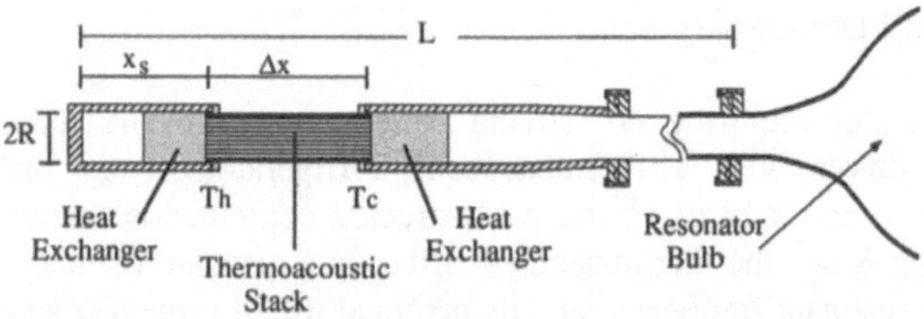

Figure 1. Schematic of a thermoacoustic driver. The overall length L is one quarter wavelength, and the resonator bulb is large enough so that the pressure oscillations in the bulb are negligible. The function of the basic parts, the thermoacoustic stack and the heat exchangers, is explained in the text. For a 540 Hz resonator with He gas, R = 1.9 cm, x_S= 6.0 cm, Δx = 8.0 cm, and L = 50 cm. The plates in the thermoacoustic stack are a distance 2yo = 0.078 cm apart and are of thickness 0.012 cm.

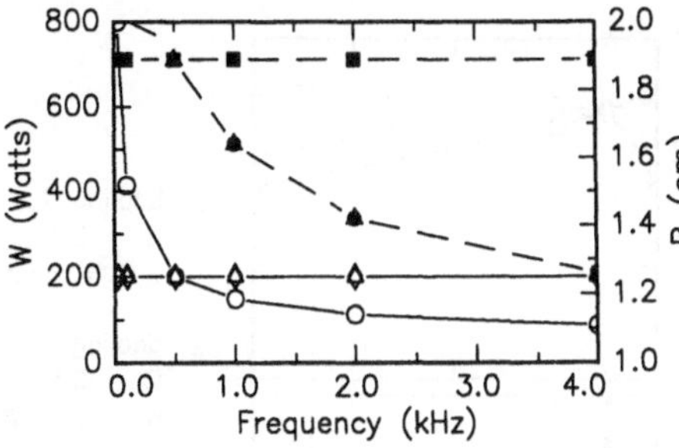

Fig. 2. The work W (open symbols) and the radius R (closed symbols) of the thermoacoustic device as a function of frequency for the three scaling methods discussed in the text. The circles are for scaling method 1, the triangles are for scaling 2, and the squares and diamonds are for scaling 3.

also calculated in an addition to the program, and will be referred to as W_l. As pointed out by Swift[1], the losses will be roughly proportional to the surface area of the resonator $2\pi RL$ times the thermal penetration depth δ_k.

SCALING METHODS

The first step taken in investigating the miniaturization of a thermoacoustic driver was to look at methods for scaling the work output and geometrical parameters. A lower work output is usually acceptable for smaller devices, so scaling the work produced to the cross section was investigated by keeping the power density W/R^2 constant. We also investigated keeping the work produced constant for all frequencies in order to gauge what effect this had on other parameters, such as T_h and W_l. The cold end temperature T_c = 300 K, mean pressure P_m = 0.5 MPa and pressure amplitude P_a = 0.1 MPa were kept constant for all cases. The position x_s and length Δx of the stack were scaled as 1/f, while the spacing $2y_o$ of the parallel plates in the stack was scaled as $(1/f)^{1/2}$. A convenient parameter in describing the thermoacoustic stack is $\Pi = \pi R^2/(y_o+\ell)$, which is the perimeter of the stack, i.e. $\Pi\Delta x$ = surface area. The work produced[1] by the stack is proportional to the surface area times δ_k, or to $\Pi\Delta x\delta_k$, so Π is convenient to use as a scaling parameter.

The frequencies investigated were 25, 100, 500, 1000, 2000, and 4000 Hz; the driver parameters were scaled to the values for the 500 Hz resonator described above. Three scaling methods were chosen: 1) Π and W/R^2 constant, 2) Π and W constant, and 3) R and W constant. Scaling method 3 results in Π increasing, since y_o is decreasing while R remains constant. The work produced and the device radius resulting from each of the three scaling methods are shown in Fig. 2. WAVEQ is used to find the resulting hot temperature T_h, the heat flux Q, Re, and resonator losses W_l for each method.

The results of the three scaling methods are shown in Fig. 3. Fig 3(a) shows a measure of the efficiency, W/Q, and Fig. 3(b) shows the percentage losses W_l/W. For all cases the fractional losses W_l/W decrease as the frequency increases; the losses for 25 and 100 Hz are off the scale. The decrease in W_l/W occurs because the dissipative surface area of the resonator is decreasing faster than the productive area of the thermoacoustic stack for the scaling methods used. For scaling methods 1 and 2, Π is constant, so the productive surface area decreases as Δx, or 1/f, while the dissipative surface decreases faster than 1/f since both R and L are decreasing. For scaling method 3, Π is increasing, so the productive area decreases at a rate less than 1/f, while the dissipative area is decreasing as 1/f since R is constant. In Fig. 3(a) we see that the efficiency decreases

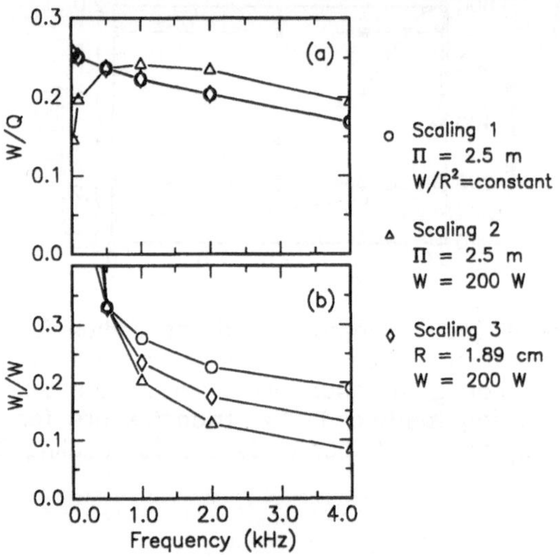

Fig. 3. (a) The ratio of the work W to the heat Q, and (b) the fractional acoustic losses W_l as a function of frequency for the three scaling methods discussed in the text.

slowly with increasing frequency, except for case 2 which shows a broad peak in the efficiency at approximately 1000 Hz.

The hot temperature T_h necessary for the device operation as the frequency changed for each scaling method was also investigated. For scaling method 1 with Π constant and both *W* and R decreasing with frequency, T_h remains roughly constant at 800 K. For scaling method 2 with Π and *W* constant, which results in W/R^2 increasing with frequency, T_h increases linearly as the frequency increases, from 800 K at 500 Hz, to 1600 K at 4000 Hz. For scaling method 3 with R and *W* constant, which results in Π increasing with frequency, T_h remains roughly constant at 800 K.

The boundary layer Reynolds number was found to depend on the frequency and to be approximately independent of the scaling method used. For 25 Hz we found Re ~ 1500, for 500 Hz Re ~ 350, and for 2000 Hz Re ~ 120. High Reynolds numbers were found at low frequencies primarily due to the fact that Re is proportional to the viscous penetration depth δ_ν, which increases as the frequency decreases, while *u* and ν remain roughly constant. Turbulence does not appear to be an issue in constructing miniature thermoacoustic drivers, since the critical Reynolds number for oscillating flow has been found[7] to be ~500, although the tests for critical Reynolds number have yet to be extended to high frequency oscillating flow.

DISCUSSION

The above scaling methods imply certain design requirements for the thermoacoustic stack and the heat exchangers. The plate spacing $2y_0$ of the thermoacoustic stack for the highest frequency is 0.276 mm; this spacing does not present much difficulty, especially since other parameters for the stack geometry, such as the radius and length, are also decreasing. One unknown factor is the ability of miniature heat exchangers to conduct large amounts of heat. A crude comparison of the required performance of the heat exchangers is to calculate the heat conducted per unit area of the heat exchanger for the different frequencies. An order of magnitude estimate of the surface area of a heat exchanger is the stack perimeter Π times the length of the heat exchanger, which we estimate as one quarter the length of the thermoacoustic stack. The heat conducted per unit area for the 500 Hz case would then be 17 kW/m^2; it is 83 kW/m^2 for 4000 Hz and scaling 1 above. These numbers are high, especially for the 4000 Hz case. A more reasonable number might be 10 kW/m^2, based on a convective heat transfer coefficient[8] of 200 W/m^2K and a temperature difference of 50 K. This implies a heat flux Q of ~60 W and work W of ~12 W for the 4000 Hz case, assuming 15% efficiency, as opposed to Q ~ 520 W for 4000 Hz and scaling method 1 above.

In light of the above comments regarding small heat exchangers, a fourth scaling method was investigated in which the driver radius R was scaled as 1/f, and the acoustic work output per unit cross sectional area, W/R^2, is kept constant. It was found that the hot temperature T_h remained roughly constant, while the acoustic losses increased dramatically, up to 60% of the work output for the 2000 Hz case. This is due to the fact that with the rapidly decreasing cross section, the productive surface of the thermoacoustic stack, equal to $\Pi\Delta x \propto R^2\Delta x \propto 1/f^3$, is decreasing faster than the dissipative surface of the resonator, which is equal to $RL \propto 1/f^2$. It is possible to decrease resonator losses[1] by increasing the mean pressure P_m in order to decrease δ_k and δ_v, and increasing Π at the same time by decreasing y_o, while keeping R constant. Calculations with WAVEQ show that this method is successful in decreasing the losses, but that the necessary hot temperature T_h increases as P_m increases (P_a and W were kept constant).

Several groups have investigated the properties of miniature heat exchangers. Hoopman[9] has constructed heat exchangers with channel diameters ranging from 0.01 mm to 1.0 mm; for 0.36 mm diameter channels they report a heat transfer to water of 1,620 kW/m^2 across a temperature difference of 50 K. Bier[10] and coworkers have also constructed microchannel heat exchangers, and for 0.1 mm x 0.078 mm channels they report a heat transfer to water of 1,350 kW/m^2 across a temperature difference of 60 K. Assuming that the heat transfer coefficient to a gas is typically[8] 50 times less than that to water, implies that the heat transfer to a gas would be 32 kW/m^2 and 27 kW/m^2, respectively, for the above two devices. Little[11] has constructed successful miniature counterflow heat exchangers for use with microminiature Joule-Thomson refrigerators which have a cooling capacity of 25-500 mW. The applicability of the miniature heat exchangers described above to oscillating gas flow, such as found in a thermoacoustic device, has yet to investigated experimentally, but is very promising.

Some further reduction in size of a thermoacoustic pressure wave generator may be gained by decreasing the average temperature of the device. For example, a device resonant at 2000 Hz at an average temperature of 400 K will have a quarter wavelength of 15 cm, while at a mean temperature of 200 K a quarter wavelength is 10 cm. This does not provide a dramatic decrease in length, but might be useful especially if a method of cooling to liquid nitrogen temperature is available.

In order to test the feasibility of making a small thermoacoustic device, and to investigate potential problems with operation at higher frequencies, a 6.0 cm long thermoacoustic driver was constructed. The device uses air at atmospheric pressure and produces audible acoustic energy when the cold end is cooled to liquid nitrogen temperature and the warm end is held at room temperature. The device is roughly one fourth the size of a similar device made by Wheatley[5], which resonated at 200 Hz. The present device is 6.0 cm long, the radius R = 0.75 cm, x_s = 2.5 cm, and Δx = 0.5 cm. The plates of the thermoacoustic stack are constructed of paper with a layer of 0.00254 cm (1 mil) thick polyimide tape, and the total thickness of the plates is 2ℓ = 0.01 cm.

Cardboard spacers 0.02 cm thick are used to separate the plates and GE 7031 varnish is used to glue the thermoacoustic stack together. The cold and hot heat exchangers are copper screen with a 0.01 cm wire diameter and a center to center wire spacing of 0.025 cm. The heat exchanger screens are soldered to Cu tubing, and the tubing sections with the heat exchangers are epoxied to either side of the thermoacoustic stack with Stycast 2850/FT epoxy. The device resonated at a frequency of 800 Hz, with a pressure amplitude of 0.008 psi, or 54 Pa, measured at a distance of 1 cm. The pressure was measured with Tektronix pressure probe[12] having a sensitivity of 5 V/psi. The construction of the thermoacoustic device was simple, although it was important to have the heat exchanger screens as close as possible to the thermoacoustic stack.

CONCLUSION

Several methods of scaling the work produced and the cross section of a thermoacoustic pressure wave generator to the resonant frequency were examined; the results provide insight into the relation between the work and cross section and other design parameters, such as the heat flux Q, T_h, and the acoustic losses. The chief design barrier to constructing miniature thermoacoustic devices is the design and construction of miniature heat exchangers which can transfer large amounts of heat. The total work produced by a miniature thermoacoustic device will be limited by the heat flux possible through the heat exchanger. Progress in the area of miniature heat exchangers is being made by several groups for other applications, and it is possible that similar design and construction methods could be used for the heat exchangers in a miniature thermoacoustic pressure wave generator.

A thermoacoustic device operating at 800 Hz was constructed using simple techniques, proving that there is no fundamental limit to thermoacoustic devices operating up to 800 Hz. As a result of these investigations into scaling methods, a successful thermoacoustic driver using helium as the gas and operating at 2000 Hz should run with an acoustic work of 38 W, and a heat flux Q of 250 W; this implies a heat transfer per unit area of 20 kW/m^2 through the heat exchanger. The design of such a heat exchanger requires further investigation. The results of the scaling methods discussed, and the resulting design implications will be used in designing and constructing an operating miniature thermoacoustic pressure wave generator.

REFERENCES

1. G.W. Swift, Thermoacoustic Engines, J. Acoust. Soc. Am. 84:1145 (1988).

2. J. C. Wheatley, G. W. Swift, and A. Migliori, The natural heat engine, in: "Los Alamos Science," Number 14 (1986).

3. C. Sondhaus, Ueber die Schallschwingungen der Luft in erhitzen Glasrohren und in gedeckten Pfeifen von ungleicher Weite, Ann. Phys. (Leipzig) 79:1 (1850).

4. W. E. Gifford and R. C. Longsworth, Surface heat pumping, in: "Advances in Cryogenic Engineering," Vol. 11, Plenum Press, New York (1966), p. 171.

5. J. Wheatley, T. Hofler, G. W. Swift, and A. Migliori, Understanding some simple phenomena in thermoacoustics with applications to acoustical heat engines, Am. J. Phys. 53:147 (1985).

6. N. Rott, Thermoacoustics, Adv. Appl. Mech. 20:135 (1980); N. Rott and G. Zouzoulas, Thermally driven oscillations, part IV: tubes with variable cross section, Z. Agnew. Math. Phys. 27:197 (1976).

7. P. Merkli and H. Thomann, Transition to turbulence in oscillating pipe flow, J. Fluid Mech. 68:567 (1975).

8. J. R. Welty, C. E. Wicks, and R. E. Wilson, "Fundamentals of Momentum, Heat, and Mass Transfer," John Wiley & Sons, New York (1976), p. 232.

9. T. L. Hoopman, Microchanneled structures, in: "Microstructures, Sensors, and Actuators", DSC-Vol. 19, D. Cho et.al. eds., American Society of Mechanical Engineers, New York (1990), pg. 171.

10. W. Bier, et. al., Manufacturing and testing of compact micro heat exchangers with high volumetric heat transfer coefficients," in: "Microstructures, Sensors, and Actuators", DSC-Vol. 19, D. Cho et.al. eds., American Society of Mechanical Engineers, New York (1990), pg. 189.

11. W. A. Little, Microminiature Refrigeration, Rev. Sci. Instru. 55:661 (1984).

12. Tektronix pressure probe P6610-1P, 1.0 psi maximum pressure, 40 kHz bandwidth, 5V/psi sensitivity.

ADVANCEMENTS IN CLOG RESISTANT AND DEMAND FLOW JOULE-THOMSON CRYOSTATS

J. W. Prentice, G. Walker, and S. G. Zylstra

General Pneumatics Corporation
Western Research Center
Scottsdale, Arizona

ABSTRACT

A cryostat composed of a special anti-clogging Joule-Thomson (J-T) expansion nozzle combined with demand flow regulation, derived from differential thermal expansion, was developed in the recent past. This new type of cryostat has proved to be remarkably rugged and highly resistant to blockage by contaminants in the gas flow. In addition, the cryostat has displayed surprisingly stable temperature characteristics.

The new cryostat design has now been applied to a variety of systems with different fluids and special requirements over a range of refrigeration capacities. This paper discusses some of the interesting situations and problems encountered in devising new variations of the basic design. Recent applications include use of liquid cryogen working fluid, modular designs, and variable capacity systems.

INTRODUCTION

Development of the General Pneumatics (GP) patented cryostat design has progressed at the Western Research Center since reported in a previous paper "Low Capacity Linde-Hampson Nitrogen Liquefier"[1] presented at the 1989 Cryogenic Engineering Conference. Following fabrication and testing of a cryostat for NASA (Fig. 1) capable of producing up to 37 watts of refrigeration at 85 K when supplied with nitrogen at 13.8 MPa, attention was turned toward producing a 1/4 watt Common Module size cryostat (Fig. 2).

Additional low capacity cryostats have been fabricated for use with gaseous nitrogen and argon, mixed gases, and liquid methane. Proposed uses have included deriving refrigeration while venting liquid hydrogen, to replacing a two-gas cooldown system with a single demand flow cryostat.

Advances in Cryogenic Engineering, Vol. 37, Part B
Edited by R.W. Fast, Plenum Press, New York, 1992

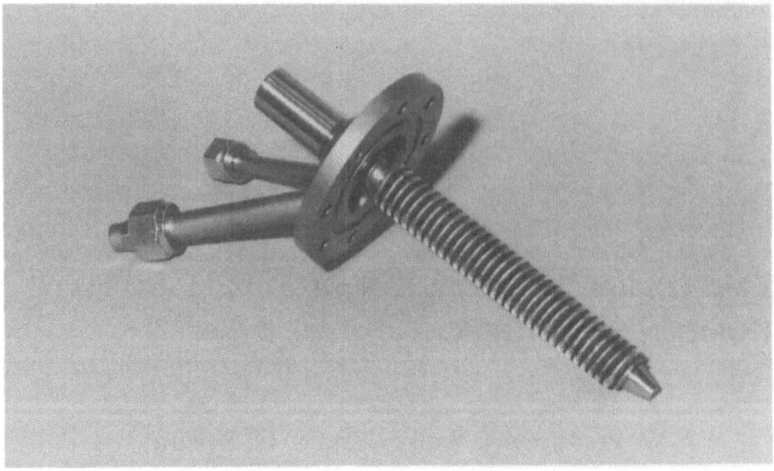

Fig. 1. Anti-Clogging Demand-Flow J-T Cryostat

DESIGN FEATURES

The GP cryostat designs are unique in that they feature a clog resistent J-T isenthalpic expansion nozzle. Additionally, the flow (hence, refrigeration capacity) is controlled by optional combinations of a demand flow feature and a manual micrometer-like adjuster. As experience has been gained, these and other features along with the related fabrication techniques have been refined.

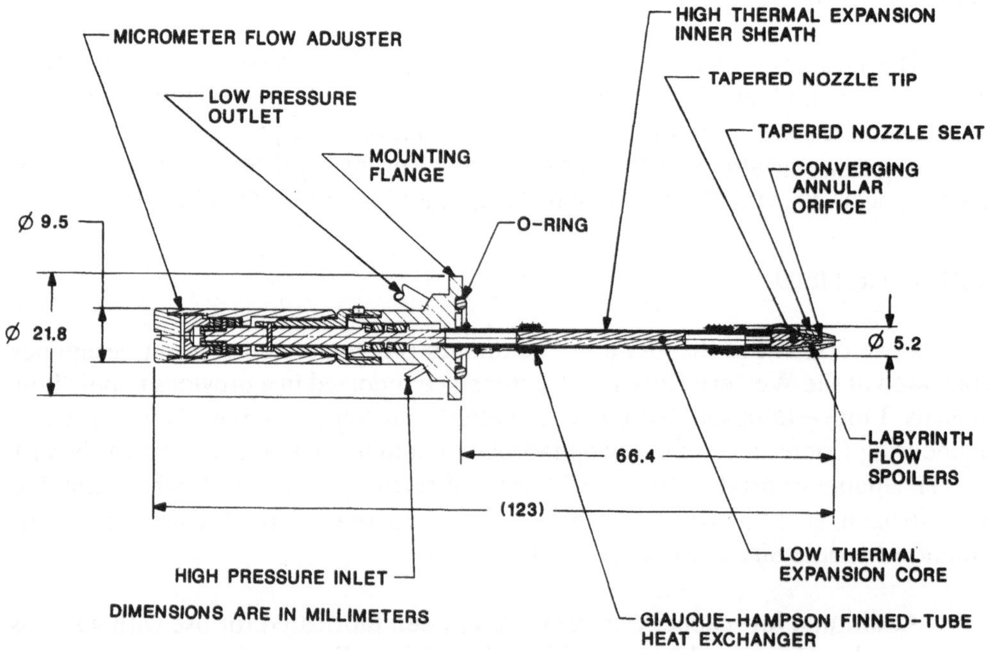

Fig. 2. Common-Module Type, 1/4 to 2 Watt Nominal, Adjustable, Self-Regulating Anti-Clogging J-T Cryostat

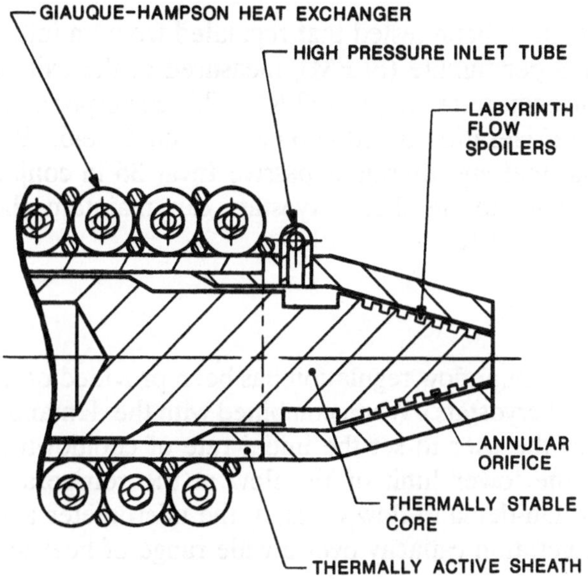

Fig. 3. Anti-Clogging Self-Regulating J-T Nozzle

Nozzle

The nozzle of the cryostat (Fig. 3) converges to an annular orifice formed by a tapered core. Circumferential grooves machined into the tapered core provide turbulence which allows wider flow passages and helps to break up and clear contaminants. This action combined with the large annular circumference relative to the flow area provides the patented clog resistent characteristic of the nozzle.

The included angle of the converging nozzle geometry influences adjustments to the mass flow rate. A 30 degree included angle increases/decreases the flow area more for each unit of linear movement of the core than does say a 15 degree angle. No appreciable difference in refrigeration capacity has been noted when comparing the 15 and 30 degree angles. Nozzle angles are therefore chosen based on flow control requirements.

During testing of a non-demand flow GP cryostat with a 30 degree nozzle, it was noted that the mass flow of the liquid methane working fluid increased slightly as the temperature at the nozzle decreased. The cause of the variation has been identified as the decrease in the kinematic viscosity with temperature. As the temperature decreased the dynamic viscosity increased, but the density increased more dramatically. Thus, the mass flow increased.[2]

Demand Flow Regulation

Demand flow regulation is the ability of a cryostat to achieve and maintain its refrigeration temperature with minimal flow while subjected to varying heat loads. The optional GP demand flow feature is achieved by differential thermal contraction of the materials supporting the nozzle and its core.

GP cryostats have been tested that regulated from an initial cooldown flow of 18 standard liters per minute (SLPM), measured at the exit, to 3 SLPM after achieving liquid nitrogen temperature (77 K). This corresponds to a change from 5.2 to 0.9 watts of refrigeration based upon a 7% liquid yield. By using thermally active 304 stainless steel and thermally passive Invar 36 in conjunction with a 15 degree included angle nozzle these cryostats can maintain their heat loaded temperature well within 1 K.

Manual Flow Regulation

Micrometer manual flow regulation has been provided on demand flow and non-demand flow GP cryostats. When combined with the demand flow feature, the micrometer enables the user to set the initial rate of cooldown flow and then, if needed, to adjust the lower limit of the flow at the refrigeration temperature. When used with a non-demand flow cryostat, the micrometer allows for accurate adjustment of refrigeration capacity over a wide range of heat loads.

Several micrometer configurations have been designed using differential threads to provide as much as 3024 equivalent threads per inch (108/112 pitch). One design using brass against stainless steel threads was found to bind at liquid methane temperatures near 111 K. Differential contraction in the longitudinal direction was suspected as the cause. The most recent micrometer design avoids this problem by using Inconel 718 against 316L stainless and by using a more tolerant 70/72 pitch differential thread (2520 equivalent threads per inch). Also, the thread arrangement allows the slightly more thermally active stainless threads (16×10^{-6} m/m·K) to contract away from the Inconel threads (13×10^{-6} m/m·K).

Heat Exchangers

GP cryostats employ helically wound finned tube heat exchangers constructed of 70/30 copper/nickel alloy. The heat exchanger is formed by wrapping the finned tubing around an inner sheath tube. A polyester thread is co-wound above and below the finned tubing to provide spacing and to direct the flow close to the finned tube. The number of wraps determines the overall length of the heat exchanger.

The internal diameter of the heat exchanger tube carries the high pressure flow to the nozzle. The configuration of the external finned low pressure side primarily determines the effectiveness of the contra-flow heat exchanger. Effectiveness can be calculated by the equation $\epsilon = T_{c2} - T_{c1}/T_{h1} - T_{c1}$.[3]

As the effectiveness approaches 100%, additional wraps of the finned tube have less influence due to the reduced temperature difference between the high and low pressure sides of the heat exchanger near the high temperature end.

The pressure differential (ΔP) across the length of the return flow side affects the temperature at the nozzle. Increasing the ΔP increases the temperature. Considerations that influence the ΔP are the return flow area, the size and existence of a co-winding, the flow rate, the density of the working fluid, and the overall length of the heat exchanger.

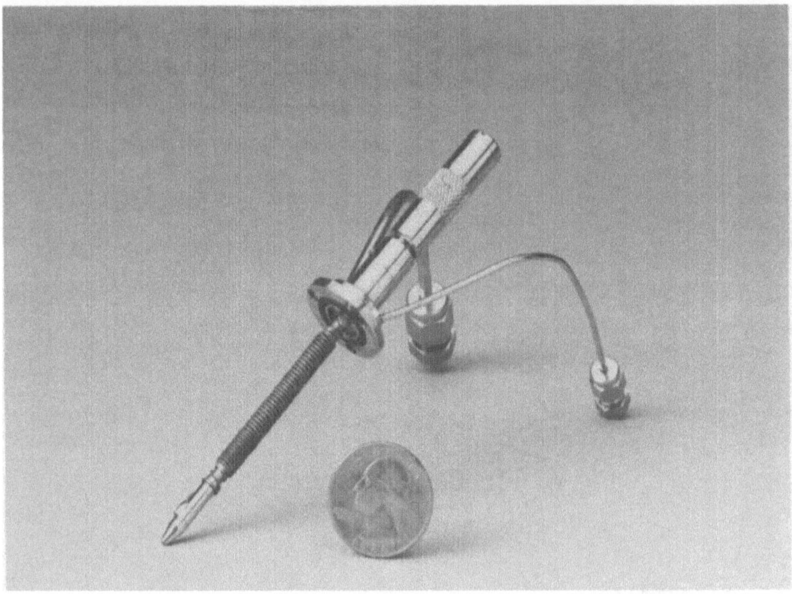

Fig. 4. Common Module Type J-T Cryostat

A GP cryostat (Fig. 4) was tested with several lengths of heat exchanger and two sizes of co-winding over a range of flow rates using nitrogen supplied at 13.6 MPa. For nominal flow rates of up to 20 SLPM, heat exchanger lengths of 35 or more wraps provided 97 to 99% effectiveness.

One cryostat with 39 wraps produced refrigeration at a maximum flow of 22 SLPM. With the same cryostat, 43 wraps did not improve the heat exchanger effectiveness, but did extend the maximum flow rate to 24.5 SLPM. Another configuration with 29 wraps was found to have an effectiveness of 92.3 to 98.5% depending upon the flow rate. Increasing the diameter of the co-winding from .127 mm to .254 mm increased the effectiveness by an average 2%.

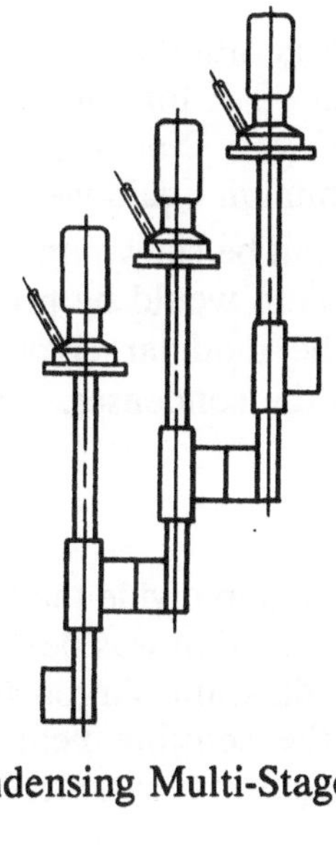

Fig. 5. Modular Pre-Condensing Multi-Stage J-T Cryostats Concept

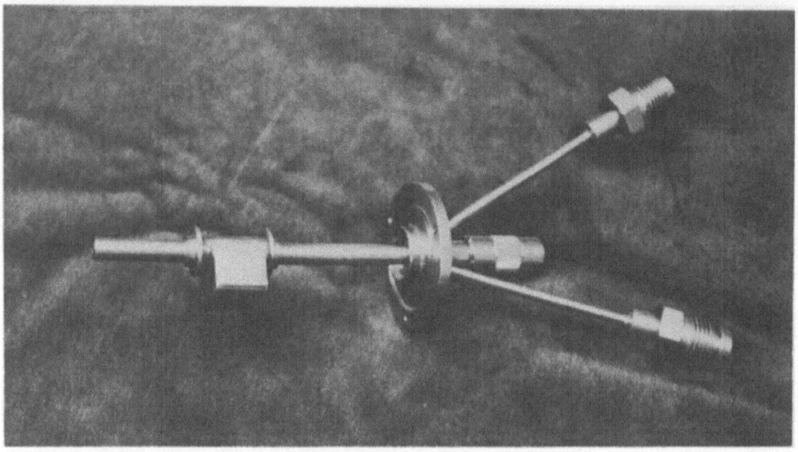

Fig. 6. Modular J-T Cryostat

MODULAR CRYOSTAT

The concept of a modular cryostat was proposed by Jack A. Jones of the Jet Propulsion Laboratory at a meeting in January 1990. It was suggested that a cryostat could be designed with a precooling boss such that a series of like cryostats could be linked together, each one precooling the next (Fig. 5). In this configuration cryogenic temperatures could be reached effectively in low pressure systems such as multi-stage sorption systems.

The first prototype modular cryostat was produced under a contract with Aerojet ElectroSystems (Fig. 6) for use in a sorption refrigerator.[4] The design objectives using argon as the working fluid and the initial test results are enumerated in the following paragraphs.

Refrigeration Capacity

It was desired that the refrigeration capacity at the cold end of the cryostat would be approximately 1 watt at 90 K for each 0.01 grams per second of argon flow over the range of .01 to .07 g/s. For a latent heat of vaporization of 162 Joules/gram , this requires a minimum liquid yield of 61.7 %. Argon enters the cryostat at 4 MPa and 300 K. It can be shown on a Temperature vs. Enthropy (T-S) diagram that only 4% liquid yield would be produced by a J-T cryostat under these conditions.[3] Therefore, the modular cryostat was designed to include a precooling section referred to as the condenser.

Condenser

The condenser was positioned to divide the finned-tube heat exchanger into an upper and lower section. The location was determined by an enthalpy balance analysis and confirmed by the T-S diagram. Argon gas enters the condenser at near 165 K having been cooled by the sensible heat extraction of the upper heat exchanger. Argon leaves the condenser as saturated liquid near 130 K. Additional heat is removed by the lower heat exchanger before the liquid is subjected to isenthalpic expansion through the nozzle.

Heat Exchanger

The working fluid passing through the clog resistant nozzle expands to the pressure of the coldwell producing a mixture of liquid and vapor. The pressure is dependant upon the outlet tube pressure plus the return flow pressure drop of the heat exchanger.

The heat exchanger was designed to minimize the return flow pressure drop to 6.8 KPa (1 psi), thus insuring a refrigeration temperature of 90 K or less. During testing the pressure was measured to be slightly more than 6.8 KPa for the maximum flow rate of .07 grams/second.

It was determined that the effectiveness of the upper heat exchanger could be as low as 80%. Any difference between the actual and 100% could be made up by extracting more heat through the condenser. Furthermore, any inefficiencies in the lower heat exchanger would pass on refrigeration to the upper heat exchanger and enhance its efficiency.

Test Configuration

The test equipment was configured to measure the resultant refrigeration capacity at the anticipated 90 K temperature cold end. An Omega Pt100 RTD inserted into a small copper block fitted around the coldwell provided temperature measurements. Correction to these measurements was needed to compensate for inadequate thermal anchoring.[5] The heat load required to vaporize the liquid argon (refrigeration capacity) was determined from multiplication of the voltage and current applied to a carbon resistor inserted into the copper block.

The rate of heat extraction from the condenser was also determined. A copper bar attached to the condenser by a #6-32 UNC screw provided a heat transfer conduit to a liquid nitrogen heat sink. The sink was fabricated by silver soldering a 3.18 mm (.125 in.) diameter stainless steel tube coiled around a 19 mm (.75 in.) diameter copper rod. A Minco heater button was attached to the copper bar to balance the heat load of the liquid nitrogen flow through the tube. An RTD was potted into the #6 screw to monitor the temperature at the condenser. In this way the heat extracted could be calculated as the heat capacity of the liquid nitrogen flow minus the heat load of the heater button required to maintain the desired temperature at the condenser. An inverted bell jar was adapted to provide vacuum insulation around the cryostat and the condenser heat sink.

Test Results

Test results (Table 1) indicated that both the modular cryostat and the test equipment functioned well. With argon supplied from 300 K at 4 MPa the cryostat was able to produce additional refrigeration at 90 K in excess of the amount provided by the condenser at 130 K.

Table 1. Test Results

Argon Flow (gram/sec)	Total (90 K) Refrigeration (Watt)	Condenser (130 K) Heat Extracted (Watt)	Refrigeration Gained (90 K) (Watt)
.013	0.74	0.5	.24
.019	1.96	1.5	.46
.030	2.60	2.1	.50

OTHER APPLICATIONS

Mixed Gases

A non-demand flow, Common Module size GP cryostat (Fig. 4) was used by JPL to test the effects on temperatures and cooling capacities when mixtures of fluids are used in J-T coolers.[6] The mixed fluids studied included varied combinations of neon, nitrogen, various hydro-carbons, argon, oxygen, carbon monoxide, carbon dioxide, and hydrogen sulfide. The manual adjustment feature proved to be an invaluable asset in conducting these laboratory tests.

Spacecraft Open-Cycle J-T Cooler

A demand flow GP cryostat was employed in another JPL program to design and test a J-T cryogenic cooler with extremely precise thermal stability for the Huygen Titan Probe. By using active feed-back temperature control of the cold head in combination with the self-regulating action of the J-T cryostat, a temperature stability of < 0.1 mK/min was achieved in laboratory experiments. This is over three orders of magnitude greater stability than required for typical J-T cooler applications.[7]

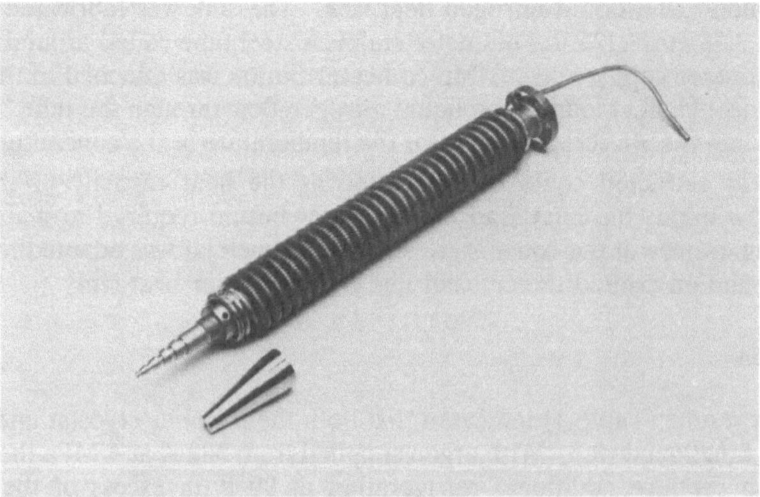

Fig. 7. Helium J-T Cryostat

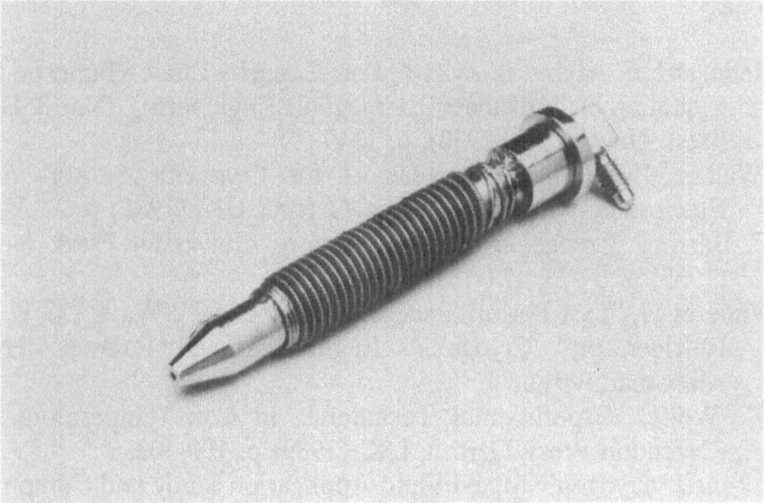

Fig. 8. Production Type J-T Cryostat

Helium J-T Cryostat

In an SDIO program managed by the U.S. Air Force Phillips Laboratory, a GP J-T cryostat (Fig. 7) is being used as the final stage of a helium liquefier. A four-stage Stirling cryocooler precools the gaseous helium down to 10 K before expansion through the J-T from 0.8 MPa to produce 1/3 watt of refrigeration at 4.2 K.[8] Since this is a closed system there is no need for the self-regulating feature but the anti-clogging design is deemed to be a critical attribute.

Production Design

In most applications the ultimate design of a J-T cryostat must have a very low thermal mass and minimal parasitic losses to provide efficient performance. This necessitates removal of the manual adjustment feature which is otherwise so useful in experimental work. A production version of the cryostat design, for a missile IR detector application, has been prototyped (Fig. 8) and is undergoing testing.

The specification requirements dictate a self-regulating design and the configuration as shown. Even with this short package it is anticipated that the required flow regulation range will be achieved by using Invar 36 and 316 CRES as the differential thermal expansion elements in the flow control mechanism.

SUMMARY

As experience is gained the clog resistant nozzle continues to prove its capability in varied applications. The demand flow and micrometer adjustment features are of additional benefit wherever precise temperature stability and flow control are required. The most recent development, the modular cryostat, will enable cryogenic temperatures to be reached effectively in low pressure, long life applications such as multi-stage sorption systems.

REFERENCES

1. K. Hedegard, E. Atkins, S. Zylstra, Low Capacity Linde-Hampson Nitrogen Liquefier, in: "Advances in Cryogenic Engineering", Vol. 35B, Plenum Press, New York, (1990), p. 1349.
2. G. Walker, "Miniature Refrigerators for Cryogenic Sensors and Cold Electronics", Clarendon Press, Oxford, UK, (1989) p. 23-35.
3. R. F. Barron, "Cryogenic Systems", Oxford University Press, New York, (1985) p. 69-73, 121-131.
4. L. Wade et al, "Test Performance of an Effecient 2 Watt, 137 K Sorption Refrigerator", Cryogenic Engineering Conference, Huntsville, Alabama (1991).
5. G. K. White, "Experimental Techniques in Low-Temperature Physics", Clarendon Press, Oxford, UK, (1989) p. 100-102.
6. J. A. Jones, "Cryogenic Mixed Fluid Application Study and Computer Code Development", JPL F-7977, Jet Propulsion Laboratory, Pasadena, California (1990).
7. S. Bard, J. J. Wu and C. Trimble, "Joule-Thomson Cryogenic Cooler with Extremely High Thermal Stability", AIAA 91-1427, American Institute of Aeronautics and Astronautics, Washington, D.C. (1991).
8. W. R. Ellison and G. Walker, "Low Capacity Reliquefier for Storage of Cryogenic Fluids", AFAL-TR-88-066, General Pneumatics Corporation, Scottsdale, Arizona (1988).

APPLICATION OF VANADIUM HYDRIDE COMPRESSORS FOR JOULE-THOMSON CRYOCOOLERS

R.C. Bowman, Jr.*, B.D. Freeman* and J.R. Phillips**

*Aerojet Electronic Systems Division, Azusa California

**Department of Engineering, Harvey Mudd College Claremont, California

ABSTRACT

The Joule-Thomson expansion of hydrogen gas offers efficient and reliable cryocoolers to produce temperatures between 10 and 50K. A critical component to the development of these devices is the metal hydride storage bed that provides a nonmechanical method to compress hydrogen gas via the reversible absorption by appropriate metals or alloys. A thermodynamic model has been used to calculate the impact of operational parameters such as input/output pressure ratios and bed temperature on energy balance and system efficiency. Detailed comparisons are reported for a compressor which utilizes vanadium metal as the sorbent for either hydrogen or deuterium where the unusually large isotope differences between the phase diagrams and thermal properties for VH_x and VD_x have been considered. The sensitivity of heat input requirements to the uncertainties in primary variables are described.

INTRODUCTION

Some proposed space surveillance missions will require long-term cooling of infrared focal planes to temperatures below 50K. Furthermore, the cryorefrigerators must be power efficient and functionally reliable throughout operational lifetimes. Sorption cryocoolers that utilize a closed-cycle circulation of hydrogen through a Joule-Thomson (J-T) expansion valve should permit development of long-life cryocoolers for temperatures between 50K and 10K. In fact, Philips Research Laboratories[1] demonstrated a hydrogen sorption cryorefrigerator nearly twenty years ago. A stable temperature of 26K with 1.0 watt (w) of cooling was reported[1] when the backside pressure of the J-T valve was maintained at 4 atm by hydrogen absorption in a $LaNi_5$ filled compressor bed. At least three other organizations[2-4] have

Advances in Cryogenic Engineering, Vol. 37, Part B
Edited by R.W. Fast, Plenum Press, New York, 1992

successfully built cryogenic refrigerators that used metal hydride compressors. All of these systems incorporated $LaNi_5$ or its substituted alloys as the hydrogen sorbent. However, further improvements in power requirements and system mass are needed before the hydrogen sorption cryocoolers can satisfy the criteria for space-based missions. The present paper addresses general thermal performance behavior of metal hydride compressors as related to J-T refrigerators and also evaluates the potential of vanadium as the hydrogen sorbent.

IDEAL VH_x SORPTION COMPRESSOR

Sorption refrigerators differ from most conventional ones in that the refrigerant is compressed by a thermochemical process rather than mechanically. The sorption compressor absorbs gas at a low pressure. The absorbed gas is then pressurized by heating the compressor. High pressure gas is desorbed and goes through a counter flow heat exchanger to the J-T expander. Cooling the depleted compressor returns it to the initial state. To provide refrigeration in the neighborhood of 20K, a system consisting of hydrogen as the refrigerant gas with vanadium as the sorbent material shows considerable promise. The sorption cycle for an idealized VH_x compressor is shown in Fig. 1. Two isotherms are shown, one at 292K and one at 394K. These correspond to plateau pressures of 0.3 MPa and 12 MPa respectively[5,6].

An energy balance was made for an ideal VH_x compressor in order to see what factors were important, particularly with respect to required heat input, and to what extent property uncertainties affected results. The compressor considered was assumed to consist of 100cc of solid sorbent and 100cc of void space (i.e., 50% voids). This corresponds to 611 grams of vanadium metal (12g-mol) and 12g of hydrogen absorbed and desorbed over the cycle. State points selected were desorption at 12 MPa and 394K; absorption at 0.3 MPa and 292K. If the following cycle is assumed: absorption 20 min., heating 10 min., desorption 20 min., and cooling 10 min., the compressor corresponds roughly to 1.6 watts of cooling at 25K.

For the overall cycle and for each step, the following equation should apply:

$$\Delta U = Q - W \tag{1}$$

where ΔU = change in internal energy, Q = heat in or out, W = work in or out.

For the complete cycle around a-b-c-d in Figure 1, we have $\Delta U=0$ and $Q=W$. That is, there is a net input of heat required to account for the net flow work output resulting from the absorption and desorption steps.

Results of this analysis are summarized in Table 1. These results are based on the thermodynamic data for the VH_x sorption developed by Flanagan[5,6] and on heat capacity relationships for VH and VH_2 developed as part of this effort. As can be seen, $\Delta U=0.6$

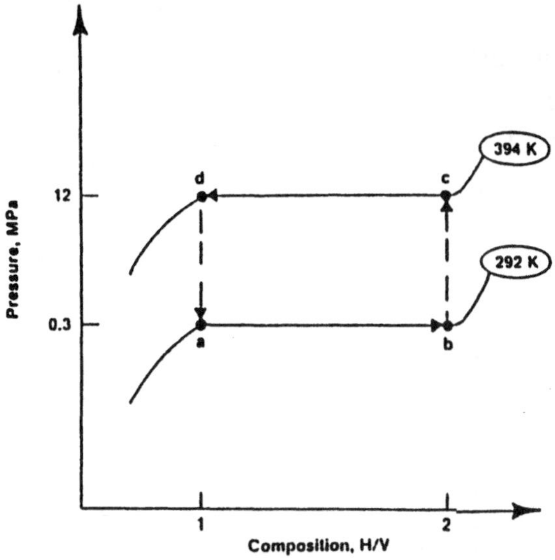

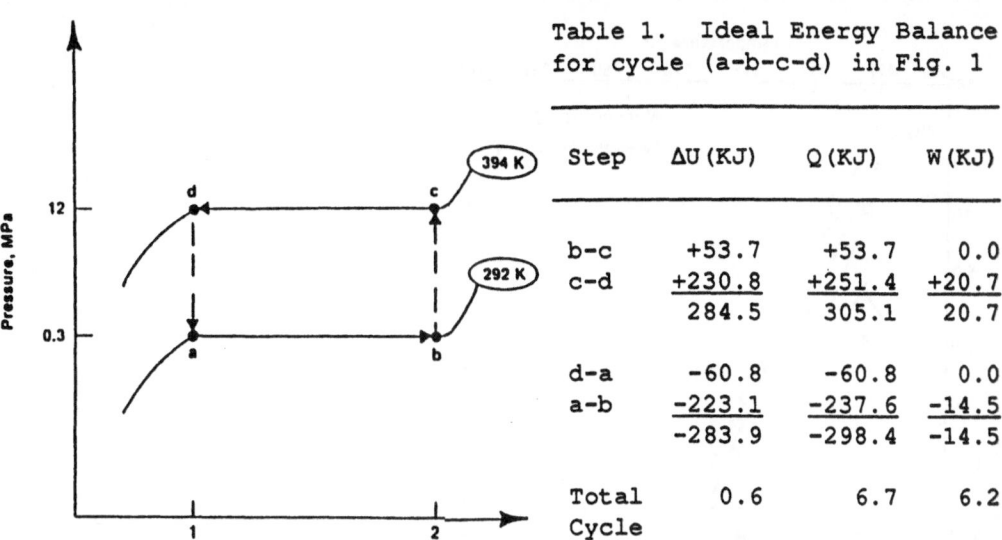

Fig. 1. Ideal VH_x Compressor Cycle (a-b-c-d) between two isotherms.

Table 1. Ideal Energy Balance for cycle (a-b-c-d) in Fig. 1

Step	ΔU (KJ)	Q (KJ)	W (KJ)
b-c	+53.7	+53.7	0.0
c-d	+230.8	+251.4	+20.7
	284.5	305.1	20.7
d-a	-60.8	-60.8	0.0
a-b	-223.1	-237.6	-14.5
	-283.9	-298.4	-14.5
Total Cycle	0.6	6.7	6.2

KJ, which gives an overall heat balance closure within 1% for the total cycle, and Q-W=0.5 KJ. Thus, the conditions of eqn. (1) are almost exactly satisfied for the entire cycle in Figure 1.

The steps presented in Table 1 correspond to those shown in Fig. 1. A net heat input for the cycle of 6.7KJ is required. The primary heat input processes are (1) desorption of hydrogen during constant volume heating which corresponds to 4.9% of the total heat requirement, (2) heating of VH_x metal at constant volume during step b-c corresponding to 12.6%, and (3) constant pressure desorption of hydrogen corresponding to 82.5% of the required heat. In this idealized analysis, heating of the vessel containing the powder bed was not considered. By far the most important variable is the heat of desorption of hydrogen from vanadium for this system, which is approximately 20 KJ/gH. Uncertainties associated with this value are discussed in the next section.

THERMOPHYSICAL PROPERTIES OF VANADIUM HYDRIDE AND DEUTERIDE

In order to evaluate the performance potential of vanadium hydride (or deuteride) sorption beds in Joule-Thomson cryorefrigerators, various thermophysical properties of these materials must be specified. The VH_x and VD_x systems[5] exhibit remarkably large isotope effects in the structures, composition ranges, and transition temperatures of the different phases. The broad two-phase regions[5] between x=0.9 and x=1.9 indicate these systems should be attractive candidates for hydrogen compressor beds. Furthermore, Fig. 2 shows the plateau pressures are strongly temperature dependent when the results due to Flanagan, et al[5,6] are plotted. It is clear from Fig. 2 that the VH_x and VD_x plateaus generate greater compression ratios than the plateaus[7] of $LaNi_5H_x$, which is often used for hydrogen storage applications. Since heating $VH(D)_x$ to lower temperatures than

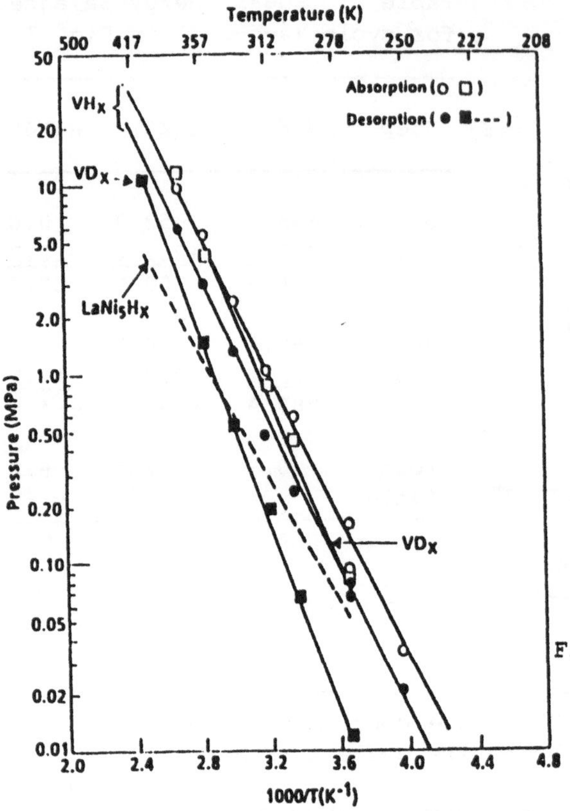

Fig. 2. Absorption and desorption pressures for the second plateau regions of VH_x and VD_x.

$LaNi_5H_x$ will produce an equivalent hydrogen inlet pressure to the JT-expander, vanadium should be a more efficient sorbent material than $LaNi_5$. In addition, the $VH_{0.9}$-$VH_{1.9}$ plateau gives a 40% larger hydrogen storage capacity per gram of sorbent. Since VH_x or VD_x are binary hydrides (deuterides), the disproportionation reactions that have plagued $LaNi_5H_x$ during extended cycling[8] will not cause problems.

However, the moderate level of hysteresis between the absorption/desorption isotherms of VH_x and VD_x does decrease the practical compression ratio that can be achieved. Fig. 2 shows that the VD_x plateaus possess larger hysteresis ratios than VH_x.

Determination of the power inputs required to operate the sorption compressor involves accurate heats of reactions (ΔH) as well as heat capacities (Cp) for the hydride phases on the either side of the plateau. Luo, et al.[5] have used calorimetry to measure $\Delta H = 19.1 \pm 0.3$ kJ/g-atom H and 23.0 ± 0.4 kJ/g-atom D for the second hydride and deuteride plateaus, respectively. However, reliable experimental values of Cp for the appropriate VH_x and VD_x phases are not available over the temperature range 250K to 420K for operation of a sorption compressor. The Dulong-Petit rule that each atom contributes $3k_B$ (where k_B is the Boltzmann's constant) to the specific heat is too crude for metal hydrides in this temperature interval. Consequently, we used a well-established numerical approach[9,10] to estimate the Cp values for β'-$VH_{0.90}$, γ-$VH_{1.92}$, α'-$VD_{0.90}$, and ϵ-$VD_{1.92}$ from measured physical properties. These parameters are described more completely elsewhere[11].

THERMAL MODEL OF HYDRIDE COMPRESSOR BEHAVIOR

Transient thermal modeling of compressor elements was achieved using the thermal package SINDA '85/FLUINT version 2.3 on a VAX mainframe. The model was run using the explicit forward differencing method. Time steps were determined by an automatic time step function in SINDA to maintain solution stability and speed processing time. Additionally, a maximum limit for ΔT of 1.0K was set to approximate more closely the exact solutions in numerical simulations. The compressor element was taken to be a cylinder assumed to be approximated by a 2-dimensional model in the radial and longitudinal directions.

The primary model used for absorption and desorption is a kinetic rate model. Flanagan et al[6] have determined that rates for these processes are proportional to the natural logarithm of the ratio of the prevailing pressure divided by the plateau equilibrium pressure. They also determined initial rate constants. Figure 3 shows several predicted loadings based on such a kinetic model using different values for the heats of hydrogen absorption. The graphs show the total moles absorbed during an absorption step with the cooling fluid flowing through an annulus on the outside of the powder bed.

As can be seen in Fig. 3 the rates of absorption start off rapidly before tapering off as time increases. This can be explained by realizing that as the outer nodes, which are cooled more readily, become fully loaded with hydrogen the overall rate of absorption decreases.

In Fig. 4 temperatures are shown for particular node locations as a function of their position for 10 minutes into the absorption step. The temperatures show a logarithmic progression, increasing as the nodes get closer to the center.

Fig. 5 shows the effect of varying the effective thermal conductivity of the VH_x bed on key bed temperatures. This study was done using a constant absorption rate model (different from the kinetic model in that all nodes are assumed to absorb gas at a constant and equal rate). It was estimated that the K_{eff} of a VH_x bed of 50% void volume would be about 3 W/mK. Thus, by increasing K_{eff}, therefore reducing bed temperatures, the time of the absorption step can be shortened. A common experimental practice has been to mix VH_x with a material such as copper to improve the heat transfer characteristics of the bed[6].

Heats of absorption and desorption of hydrogen from the VH_x bed are the single most important variables of the system when it comes to determining heat input required for the process. They also change the rate of absorption (or desorption) because heat removal and bed temperatures affect these rates. In Fig. 3 we have shown the effect of a parametric variation of ΔH of absorption on the time required to load the VH_x bed. As can be seen, a 10% variation in the heat of absorption does have a noticeable effect on the rate of absorption.

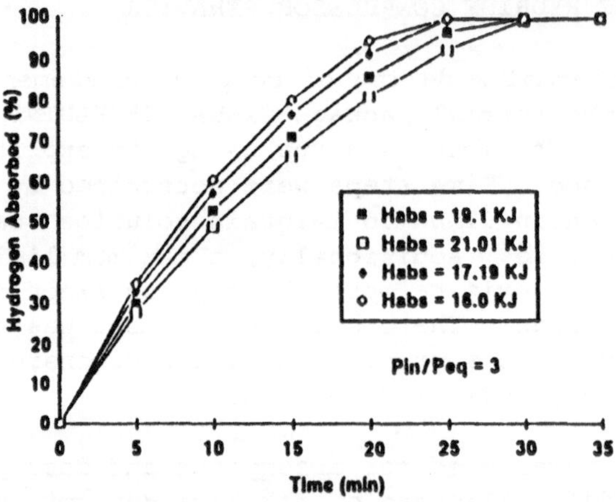

Fig. 3. Hydrogen loading on VH_x compressor bed.

PRELIMINARY OPTIMIZATION

Optimization and final design of a vanadium hydride based refrigeration system will result when the model described in the previous section is finished, the four compressor steps from Figure 1 are linked, and the model simulates continuous compressor operation. In addition, regeneration[12] must be considered in the development of an optimized system.

At this point, two preliminary optimization studies have been carried out assuming ideal conditions. The first was to compare the relative characteristics of deuterium and hydrogen as the refrigerant gas. The second was to determine the optimum high pressure for the system taking into account heating of the compressor vessel. Results of the latter study are described elsewhere.[11]

Calculations[11] predict that hydrogen produces significantly more refrigeration than does deuterium when compared on a unit mass basis. However, according to Flanagan[5,6] the heat required

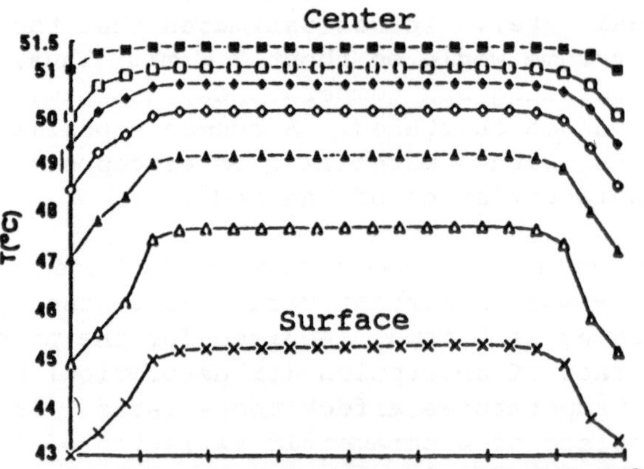

Fig. 4. Temperature profile after 10 minutes of H_2 absorption.

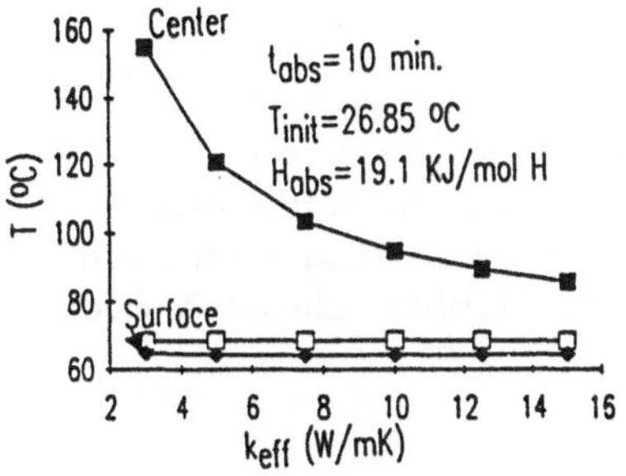

Fig. 5. Effect of bed thermal conductivity (K_{eff}) on key temperatures.

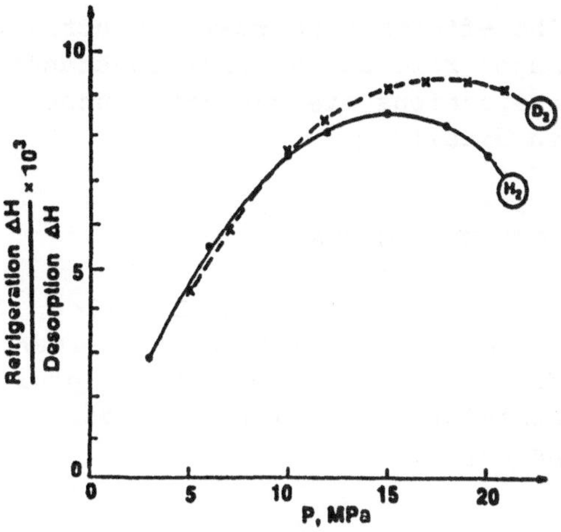

Fig. 6. Relative refrigeration produced by hydrogen and deuterium from VH_x and VD_x.

for desorption is also quite different for hydrogen and deuterium; (i.e., 20.5 KJ/g for hydrogen vs 12.0 KJ/g for deuterium). This is essentially due to the fact that the neutron associated with deuterium does not participate chemically in the desorption process.

The quantity of greatest importance in evaluating a refrigerator system is the refrigeration produced relative to the heat required to the process. Figure 6 shows refrigeration produced divided by heat of desorption for both hydrogen and deuterium as a function of inlet pressure. As can be seen, below about 12 MPa the curves are very close with hydrogen slightly favored. However, above 12 MPa deuterium is preferred.

Lower inlet pressures are favored because they will result in lower cost compressors of lower thermal mass, an important consideration during the compressor heating step. Thus the final system design is likely to call for inlet pressures below 12 MPa. In this region, hydrogen is slightly advantageous.

CONCLUSIONS

Vanadium has been shown to be a viable sorbent for hydride compressors in sorption cryorefrigerators. Although either normal hydrogen or deuterium could be handled in the Joule-Thomson closed-cycle, hydrogen is slightly preferred for applications using output pressures below 10 MPa and cold stage temperatures below 25K.

A transient thermal analysis model has been developed for more systematical assessments of hydride compressor performance. Preliminary results confirm the importance of the hydrogen absorption and desorption properties of the sorbent metal on the achievement of efficient operation of the compressor elements.

The effective thermal conductivity of the hydride bed plays a major role on the time constants of the system. More extensive evaluations are currently underway and will be reported elsewhere in detail.

ACKNOWLEDGMENTS

We wish to thank B. Axley, J. Burden, R. Caraveo, F. Cleveland, J. Jansen, N. Sherman, P. Sywulka, L. Wade and Y. Louie for their assistance and helpful comments regarding sorption behavior, thermodynamic modeling and the 'SINDA' thermal modeling package.

REFERENCES

1. H.H. van Mal and A. Mijnheer, "Proc. ICEC", Vol. 4, IPC Science & Technology Press, Guildford, UK, (1972) p 122.
2. J.A. Jones and P.M. Golben, Cryogenics 25: 212 (1985).
3. K. Karperos, "Proc. 4th Intern. Cryocooler Conf.", Easton, MD, September 25-26, 1986, p 1.
4. T. Kumano, B. Tada, Y. Tsuchida, Y. Kuraoka, T. Ishige, and H. Baba, Z. Physk. Chem. N.F. 164: 1509 (1989).
5. W. Luo, J.D. Clewley, and T.B. Flanagan, J. Chem. Phys. 93: 6710 (1990).
6. T.B. Flanagan, Private communication of unpublished data.
7. E.L. Huston and G.D. Sandrock, J. Less-Common Met. 74: 435 (1980).
8. J.-M. Park and J.-Y. Lee, Mat. Res. Bull. 22: 455 (1987).
9. H.E. Flotow and D.W. Osborne, J. Chem. Phys. 34: 1418 (1961).
10. M. Moss, et al., J. Chem. Phys. 84: 956 (1986).
11. B.D. Freeman, R.C. Bowman, Jr., and J.R. Phillips, Proc. of 1991 Space Cryogenics Workshop, Cleveland, OH, June 1991 (to be published).
12. L. Wade, P. Sywulka, M. Hattar, and J. Alvarez, "Advances in Cryogenic Engineering", Vol. 35b, Plenum, New York, (1990) p.1375.

THE CHARACTERIZATION OF SEVERAL HIGH SPECIFIC HEAT REGENERATOR MATERIALS IN A GIFFORD MCMAHON REFRIGERATOR

R. A. Ackermann

General Electric Research and Development Center
Schenectady, New York

ABSTRACT

Recent developments in the use of low-temperature, high specific heat materials in Gifford McMahon (GM) refrigerators has produced remarkable demonstrations of ultra-low temperatures. These materials, consisting of heavy rare earth and ceramics, exhibit a magnetic phase transition leading to a sharp rise in specific heat at the transition temperature. From recently published data, it is clear that these materials improve refrigerator performance, but the key material and operating parameters affecting the reported performance improvements are not clearly defined. Therefore, with this goal in mind, the General Electric Research and Development Center has manufactured a number of high specific heat materials and tested them in two GM refrigerators.

Spheres from 5 materials were fabricated and tested in both pneumatic and mechanical drive refrigerators. The materials consisted of 4 rare earth materials with transition temperatures ranging from 7.5 to 14 K, and one ceramic material with a sharp specific heat spike at 6.0 K. Spherical particles in the size range from 150 to 500 μm were produced by a spark erosion and gas atomizing manufacturing process. The test results identified several critical performance parameters important to the commercial use of these materials, and demonstrated a capacity increase of 150% at 10 K. Disappointingly, the ultra-low temperatures reported by other researchers were not achieved in these studies. We present the results of this testing, along with a discussion of the difficulties encountered in the manufacture and use of these materials.

INTRODUCTION

Recently, papers by Kuriyama, Hakamada, Nakagome et al.[1], have reported temperatures below 4.2 K with a GM refrigerator using rare earth regenerator materials. The startling results reported by these researchers led to the development of a program at GE which evaluates the potential of these materials to improve the performance of refrigerators used in magnetic resonant imaging systems. GM refrigerators are now used on all GE magnetic resonant imaging systems to reduce helium consumption by cooling two radiation shields surrounding the superconducting magnet. The following objectives were defined for the program.

1. Evaluate performance improvements at 10 K.
2. Evaluate manufacturability of rare earth materials.
3. Evaluate material endurance in a refrigerator.

Table 1. High Specific Heat Regenerator Materials

Material	Transition Temperature (K)	Specific Heat ($J/cm^3 - K$)	Thermal Diffusivity (cm^2/s)
Helium	@ 10 K 6 atm	0.48	0.0008
Lead	@ 10 K	500	3.13
ErNi	10.5	0.61	–
Er_3Ni	7.4	0.40	–
$ErNi_2$	6.2	0.42	0.0286
$(ErDy)_{1.0}Ni_2$	14	0.75	0.0267
GdRh	19	0.91	–
Nd	7.7	0.31	–
Ceramic	7.8	1.50	0.0001

To achieve these objectives, a test matrix was established which would provide data on the effects of the following parameters on capacity at 10 K: material transition temperature, manufacturing process, and packing volume and density.

Table 1 lists the materials that were tested. The important properties of specific heat and thermal diffusivity are also presented in Table 1 along with a comparison to lead and helium at 10 K. These materials were selected because of their known high specific heat spikes. The rhodium alloys were later dropped from the list because of their prohibitive cost, and the noedium was dropped because of its highly reactive nature that would have made it difficult to machine and handle. The other materials were considered viable candidates for commercial use and were included in the program.

MANUFACTURING AND TEST PROCEDURES

All of the rare earth materials tested were produced internally and powerized by either a spark erosion[2] or gas atomizing process. The powdered particles were sifted to a size range of 150 to 500 μm. The regenerator was loaded by replacing an equivalent volume of lead with the test material. Different amounts of material and packing densities were used to establish an optimum packing configuration. When more than one rare earth material was used to pack the regenerator, the placement and amount of material was determined by matching the material's transition temperature with the equivalent temperature location in the regenerator. A linear temperature profile between the first and second stage temperatures was used to establish the equivalent temperature locations in the regenerator.

The testing was performed in both a pneumatically driven GM refrigerator with a stationary second-stage regenerator and a mechanically driven refrigerator with a moving second-stage regenerator. The cyclic rate of the refrigerator, pressure ratio, and displacer stroke were varied to obtain optium operating conditions. For each test, a load map was generated to establish refrigerator performance over a wide range of operating conditions.

The two manufacturing procedures used, (the spark erosion[2] and gas atomizing), are schematically portrayed in Fig. 1. For each procedure, because of the brittle nature of these materials, true spherical particles were not achieved. The aggregate of material consisted of spherical particles and shatters as shown in Fig. 2. The atomizing produced a better

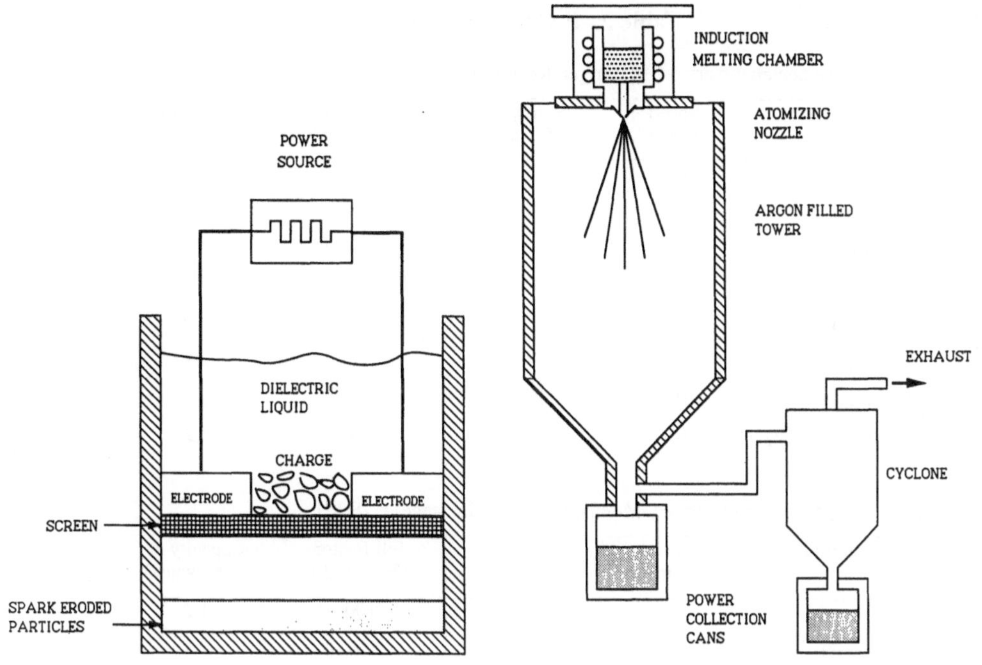

(a) (b)

Figure 1. (a) Metal power spark erosion cell, and (b) metal powder gas atomizer.

aggregate of spherical particles, but the yield of spheres in the size range required was much smaller than that for the spark erosion process. A typical yield of particles in the size range of 150 to 500 μm from the atomizer was only 10 to 15% of the processed material, while the yield from the spark erosion process was 60 to 70%.

The two manufacturing processes also had a significant effect on the performance of the refrigerator as shown in the following section.

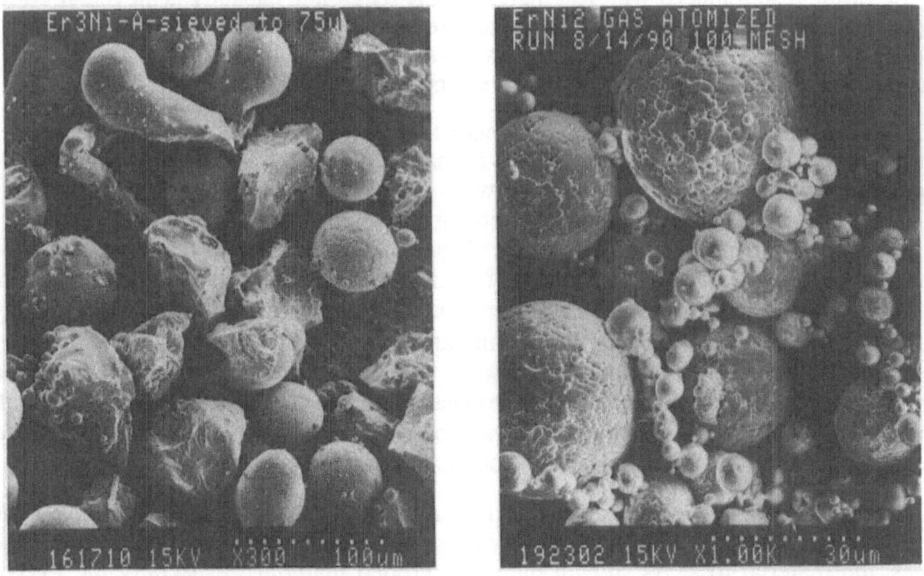

(a) (b)

Figure 2. (a) Spark eroded Er_3Ni, sieved to 125 μm, and (b) atomized $ErNi_2$, sieved to 125 μm.

Table 2. Performance Results

A. Two stage pneumatic drive GM refrigerator [T (1st stage) = 26 K]

Material	Speed (rpm)	Porosity (%)	Fill Factor (% by vol.)	Capacity at 10 K (watts)
Lead	144	38	100	2.0
ErNi	144	45	45	3.5
Er_3Ni	144	47	45	2.8
$ErNi_2$	144	45	30	2.8
(ErDy) Ni_2	144	41	33	3.9
Ceramic *	100	50	17	2.7
Ceramic+Er_3Ni	100	50/45	15/32	27
Er_3 Ni+(ErDy)Ni_2	144	48/41	16/38	4.2

B. Two stage mechanical drive GM refrigerator [T (1st stage) = 29 K]

Material	Speed (rpm)	Porosity (%)	Fill Factor (% by vol.)	Capacity at 12 K (watts)
Lead	72	38	100	1.2
Er_3Ni	72	47	46	23

* Material supplied by Ceramphysics, Inc

TEST RESULTS

With regards to our three objectives, the results obtained clearly showed that major improvements in capacity at 10 K can be achieved with these materials. A capacity increase from 2.0 to 4.2 W was achieved with a combination of Er_3Ni and $(ErDy)_{1.0}Ni_2$ material. In each case the maximum improvement occurred with a porosity of 40 to 45%, and with less that 50% of the lead replaced with test material. The maximum capacity achieved at 10 K, along with the optimum porosity and fill factor are given in Table 2. This table also shows that similar improvements in capacity, as compared to lead, were achieved with a mechanical drive refrigerator although it did not achieve the same low temperatures.

The test results also provided clear evidence that the single most important parameter affecting both performance and endurance was material manufacturability. Because of the brittle nature of the rare earth materials we were unable to obtain pure spherical particles from either the spark erosion or gas atomizing processes. This manifested itself in a variation in performance as shown in Fig. 3, and in the erosion of material from the regenerator during operation. The curves in Fig. 3 represent a back to back comparison of $(ErDy)_{1.0}Ni_2$ manufactured from the spark erosion and gas atomizing processes. An X-ray diffraction and electron micrographic examination of the two regenerator aggregates showed that the only difference existing between the two aggregates was the quality of the spherical particles. The atomizing process produced a much more spherical aggregate with fewer shatters than achieved from the spark erosion process. A similar result was also achieved for an Er_3Ni sample, prepared by both manufacturing processes.

As shown in Table 3, the manufacturing process also had a major impact on the durability of the material. 200-hour durability tests were conducted to measure performance changes and material loss. As shown, the spark eroded materials, especially the Er_3Ni, resulted in greater material loss. The weight loss resulted in a black powder that deposited itself throughout the cold head and eventually resulted in plugging the heat exchanger or the valve components. Furthermore, the loss of Er_3Ni from the mechanical drive refrigerator, with a moving regenerator, appeared to be more severe than for the pneumatic drive

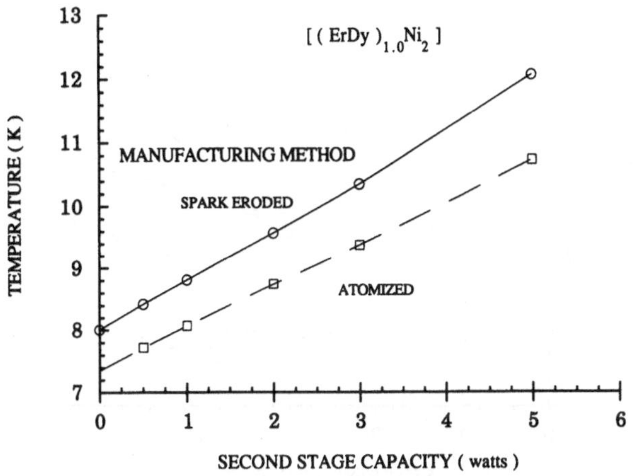

Figure 3. Comparison of regenerator material performance.

refrigerator with a stationary regenerator. In the case of the $(ErDy)_{1.0}Ni_2$ atomized material, which produced better spherical particles, the loss of material from the regenerator was not distinguishable in either refrigerator. This observation leads to the conclusion that the erosion is caused by the movement of the regenerator material, as a result of the inertia and flow forces acting upon the aggregate. Therefore, the quality of the spherical particles is a key manufacturing parameter in the use of these rare earths as regenerator materials.

In conjunction with our goal of evaluating the performance of these materials at 10 K, their low-temperature performance was also measured. Dissappointingly the ultra-low temperatures reported by other researchers were not reached; however, a temperature reduction of more than 2 K was achieved. The second-stage load curves for the materials giving the greatest capacity at 10 K, and the lowest temperatures are shown in Fig. 4. As shown, the lowest temperature achieved was 6.2 K with a combination of ceramic and Er_3Ni material. Speed and pressure were varied during these tests but had little effect on the results obtained. In each case, with the exception of the ceramic material, the highest capacity was obtained at the design speed of the machine. For both refrigerators, the speed varied by 30%, and the suction pressure from 65 to 100 psia, with a corresponding

Table 3. Regenerator Material Weight Loss During 200-Hour Endurance Run

Material	Manufacturing Process	Weight Loss (g)
1.0 Pneumatic Drive With Stationary Regenerator		
Lead		0.00
Er_3Ni	Spark Eroded	0.45
Er Ni_2	Atomized	0.05
$(ErDy)_{1.0}Ni_2$	Spark Eroded	0.20
$(ErDy)_{1.0}Ni_2$	Atomized	0.00
Ceramic		0.05
2.0 Mechanical Drive With Moving Regenerator		
Er_3Ni	Spark Eroded	4.45
$(ErDy)_{1.0}Ni_2$	Atomized	0.00

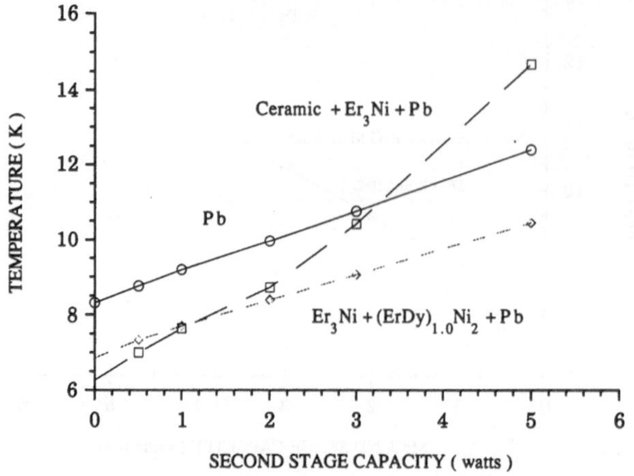

Figure 4. Comparison of regenerator material performance.

discharge pressure variation of 240 to 300 psia. For the ceramic material, the lowest temperature was achieved at a lower speed, which was attributed to the low thermal diffusivity of the ceramic.

The question that arises in achieving these low temperatures with a GM refrigerator is : "What is the available refrigeration produced and how does this compare to the inherent thermal losses in the system?" The model used to analyze this is shown in Fig. 5. The model assumes that gas leaves the regenerator and enters the expansion volume at the refrigeration temperature. The gas is then expanded adiabatically in the expansion volume and passed through the heat exchanger were it is warmed to provide the refrigeration effect. The gas leaves the heat exchanger at the refrigeration temperature and enters the regenerator to complete the cycle. The maximum available refrigeration was computed from the enthalpy change of the gas as it is heated from state 4 to state 5. The pressure change considered was from 6 to 20 bar and the working fluid was helium.

From the model, the maximum available refrigeration as function of the refrigeration temperature is shown in Fig. 6. Along with the maximum available refrigeration, the figure presents the conduction heat leak computed for the instrumentation leads, tubing, and second-stage displacer, assuming a first-stage temperature of 25 K. As seen, the available refrigeration below 4 K is very small and when heat leaks and a small regenerator inefficiency (I_e = 2%) are included, very little usable refrigeration exist below 6 K.

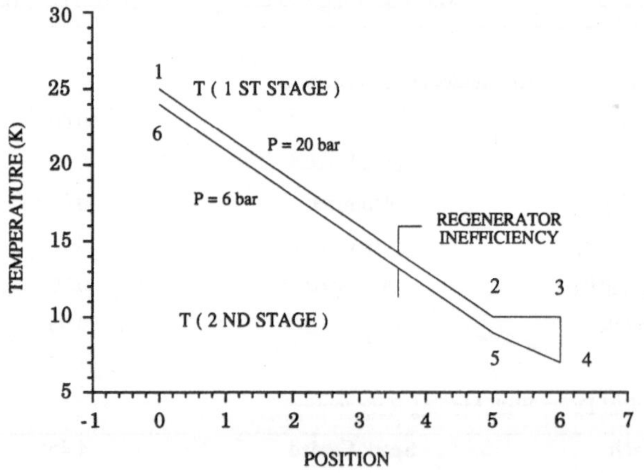

Figure 5. Second stage Gifford McMahon refrigerator model.

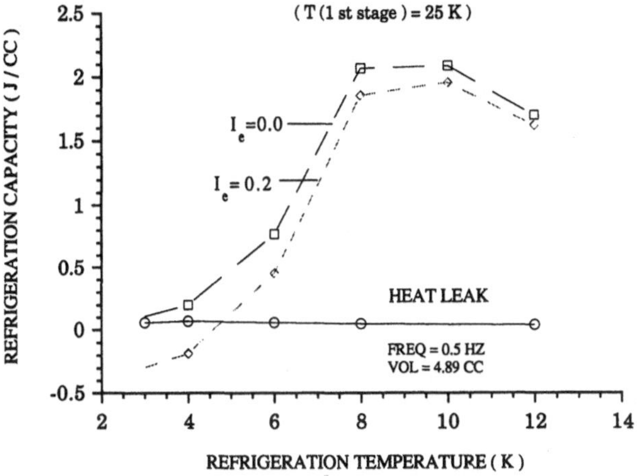

Figure 6. Maximum available refrigeration as a function of temperature.

Secondly, the analysis shows that a regenerator inefficiency of 5% produces a bottom temperature of 8 K, as actually achieved with lead, and a 150% improvement in regenerator inefficiency is required to achieve temperatures below 6 K. Therefore, considering the magnitude of the improvement required, it is not suprising that we could not achieve temperatures below 6 K by simply changing regenerator material. To achieve the ultra-low temperatures reported requires a thorough evaluation of the thermal losses in the refrigerator as well as improving the heat capacity of the regenerator.

CONCLUSIONS

The use of rare earth and ceramic materials with high specific heats at temperatures below 20 K provides the potential, for the first time, to achieve liquid helium temperatures with a GM machine. This is an important accomplishment that will assist the commercialization of low-temperature superconducting devices. For higher temperature applications, in the neighborhood of 10 K, the results will be equally important by improving efficiency and capacity at minimum cost. The following conclusions were drawn from this work regarding their use at 10 K.

1. High specific heat materials can effectively increase the refrigeration capacity at 10 K.

2. The manufacturability of the materials and the ability to produce high-quality spherical particles economically is the major challenge to their commercial use. It appears that both the performance and erosion of material from the regenerator is a function of this parameter.

3. The refrigeration capacity in a GM refrigerator below 6 K is very small and the use of GM refigerators at very low temperatures will require very sophisticated cryogenic system designs.

REFERENCES

1. T. Kuriyama, Hakamada, Nakagome et al., High Efficiency Two Stage GM Refrigerator with Magnetic Material in the Liquid Helium Temperature Region, in: "Advances in Cryogenic Engineering ," Vol. 35B, Plenum Press, New York, (1990), p. 1261.
2. J. L. Walter, "On the Preparation of Powder by Spark Erosion", General Electric Corporate R&D Center Technical Information Series, 88CRD202, (1988).

PRODUCTION OF SPHERICAL POWDERS OF RARE EARTH INTERMETALLIC COMPOUNDS FOR USE IN CRYOCOOLER REGENERATORS

Evan M. Ludeman and Carl B. Zimm

Astronautics Technology Center
Madison, Wisconsin

ABSTRACT

Magnetic rare earth intermetallic compounds have properties of great utility in the construction of cryocooler regenerators. The magnetocaloric effect observed in these materials is the basis for regenerative magnetic refrigeration, and large magnetic heat capacities have been used to enhance low temperature performance of Gifford-McMahon cryocoolers. Development of reliable rare earth intermetallic regenerators has been limited in large part by poor mechanical properties of these materials and their consequent breakdown in service.

We report on a new method for producing well formed spherical powders from the melt in the 0.3 - 0.75 mm size range via a low pressure, low velocity jet atomization technique. The method is adaptable to a wide range of materials and is limited chiefly by compatibility of the melt with the tungsten crucible assembly. Initial results of particulation experiments with $GdNi_2$ and Er_3Ni are reported.

INTRODUCTION

Recent developments in regenerative magnetic and very low temperature gas cycle regenerative cryocoolers have stimulated interest in the production of high quality regenerator beds containing rare earth intermetallic compounds.[1,2,3] Meeting simultaneous needs for high heat transfer area per unit volume, low axial thermal conductance, low pressure drop, low porosity and uniformly distributed flow, together with the desire to maintain simplicity and durability have led to the use of packed particle bed geometries.[4,5] While our experience with a packed bed of crushed and sieved granules of the rare earth intermetallics $Er_{0.86}Gd_{0.14}Al_2$ and $GdNi_2$ have shown excellent regenerator performance and minimal degradation over short operating times in an active magnetic regenerator refrigerator test device[6], experience to date with gas cycle cryocooler regenerators has shown that crushed granular materials are highly susceptible to breakdown in use and the abrasive dusts generated pose a very real hazard to the longevity of the cryocooler.

Spherical powders offer several important advantages over granular type materials. First, comminution in these generally brittle materials proceeds via a crack nucleation and propagation mechanism. As a result, crushed granular materials will have a high residual crack concentration and many physical asperities. An example is the crushed di-aluminide magnetic refrigerant shown in Figure 1. Qualitatively, such particles are somewhat difficult to classify by conventional sieving due to high aspect ratios, further, they are also difficult to pack in a stable arrangement in a regenerator bed. Small rotational disturbances as well as erosion of asperities will cause the packing factor to increase in use. Settling in the regenerator is to be avoided as experience

Figure 1. SEM photo of granular regenerator material $Er_{0.86}Gd_{0.14}Al_2$.

has shown that a loosely packed material allows for considerable movement of particulates. Severe attrition is the predictable result. Secondly, spherical particulates offer an improvement in packing behavior and stability due to their rotational symmetry, freedom from sharp edges and asperities, and smaller concentration of internal and surface cracks. Experience has also shown that classification via sieving is very easily accomplished with spherical particulates.

Rare earth intermetallics have been produced as spherical particulate by centrifugal atomization[2], two fluid (inert gas) atomization[7,8], free fall remelt[9] and molten salt processes.[10] The latter two processes both suffer from the requirement of having a well sized granular feed material, typically produced via comminution. This results in an unfortunate waste product in the form of excessively fine material which is difficult to reprocess due to a considerable potential for contamination. Throughput is typically small for these processes as well, and in the case of molten salt processes, severe oxidation can result from impurities in the salt. Centrifugal and gas atomization processes offer the advantage of forming particulate directly from the melt, however both of these techniques suffer difficulties as well. A substantial fraction of the material produced by the gas atomization process is in the form of flakes and broken or hollow spheres, necessitating a "sorting" process to separate the well formed particulate. Reported yields are generally below 30% and minimum batch sizes are in the 5 kg range which has inhibited further work due to cost considerations. Details such as process yield and material quality have not yet been reported for the rotating disk atomization technique as applied to the manufacture of magnetic earth alloys for regenerator use. Sahashi[2] has noted that the low melting temperature of Er_3Ni (845 ^oC) was an important factor in obtaining an acceptable yield. If this is indeed the case, the utility of this technique may be extremely limited in scope due to the refractory nature of the majority of rare earth intermetallic compounds.

METHODOLOGY AND RESULTS

We have chosen to explore an alternative technique for forming spherical powders directly from the melt. Figure 2 shows the apparatus schematically. This apparatus is designed to melt air sensitive alloys in a low contamination environment and, at the appropriate time, form a laminar jet of this material exiting an orifice in the bottom of the crucible. We rely on the Rayleigh instability[11] of the jet to form spherical droplets which cool and solidify during free fall to form spherical particulates. This approach offers the ability to work with convenient batch sizes for development purposes. The practical range for our device is approximately 10 to 250 grams of material, depending on density. Additionally, the particulate is formed at relatively low velocities (< 10 m/s) and the potential for particulate damage due to particle/particle and particle/wall collisions is reduced compared with other melt particulation techniques.

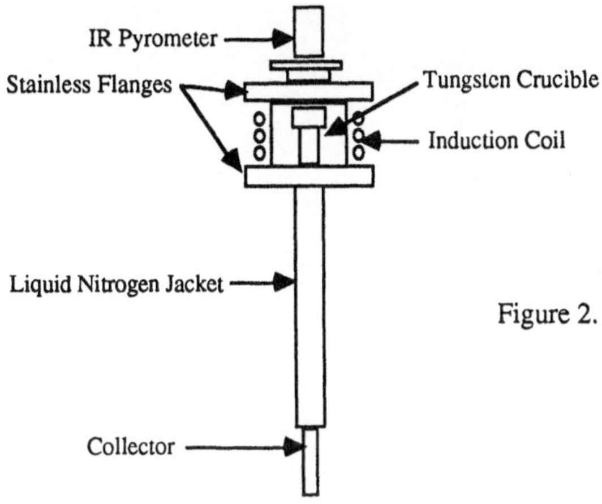

Figure 2. Particulation Apparatus Schematic

This technique is limited in scope chiefly by two factors; 1) reactivity of the crucible material and 2) potential for clogging of the orifice under certain conditions discussed later.

SEM photomicrographs of $GdNi_2$ produced by this method are shown in Figures 3 and 4. Figures 5 and 6 show similarly processed Er_3Ni. In both cases, the material has

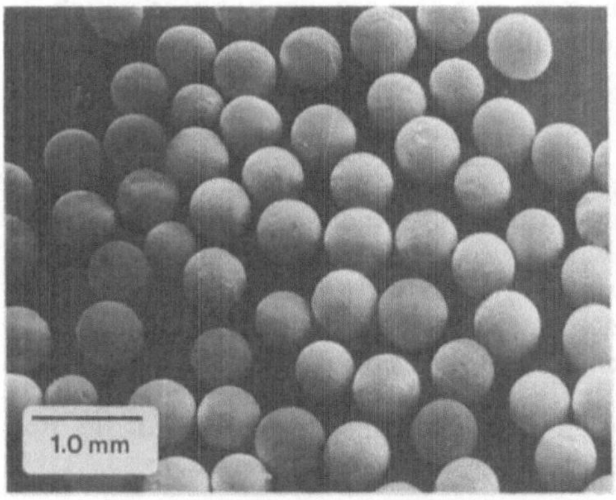

Figure 3. Spherical $GdNi_2$ formed by low velocity melt particulation.

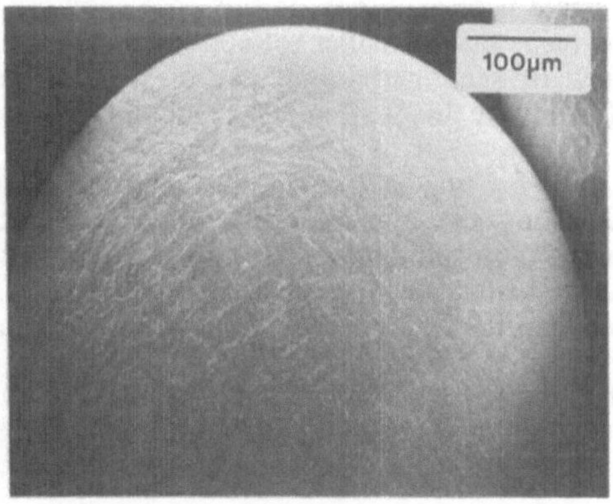

Figure 4. Detail of $GdNi_2$ particle.

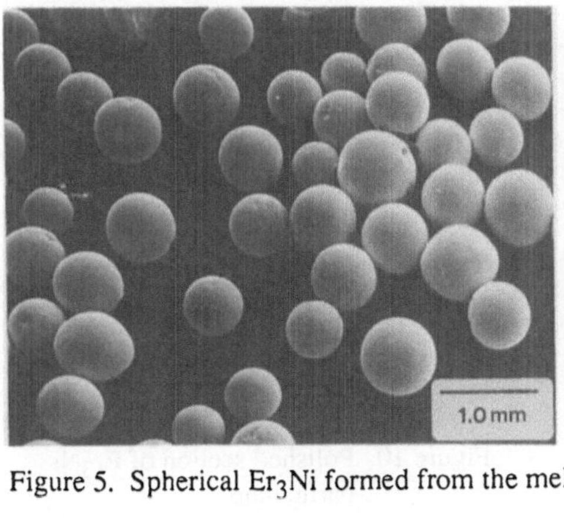

Figure 5. Spherical Er_3Ni formed from the melt.

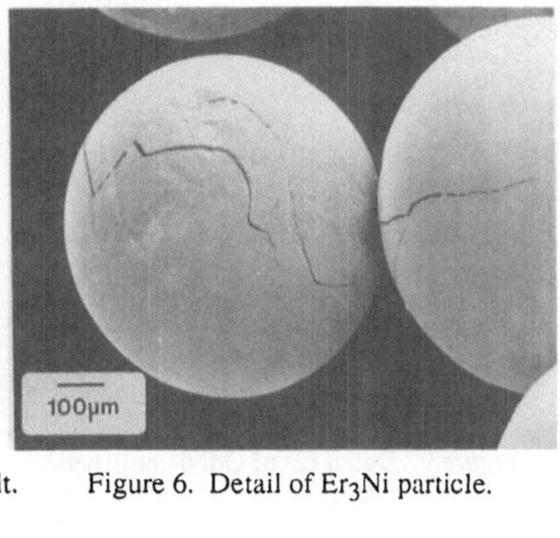

Figure 6. Detail of Er_3Ni particle.

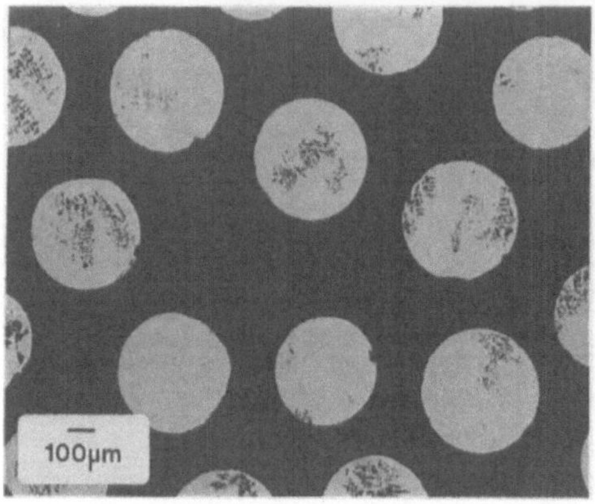

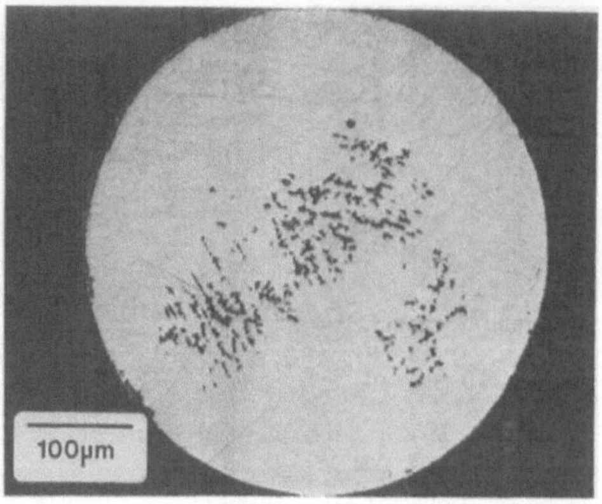

Figure 7. Polished section of $GdNi_2$ particulate. Figure 8. Detail (1) of $GdNi_2$ polished section.

been subjected to a considerable degree of handling and the surface damage and cracks visible are thought to be primarily due to particle/particle impacts. On the whole, $GdNi_2$ appears to be a far more durable material in this form. Figures 7 - 9 are optical photomicrographs showing polished sections of $GdNi_2$ particles at successively higher magnification. Although solidification voids are present to some extent, we note an absence of the extremely large internal voids seen by other researchers using gas atomization techniques. At the higher magnifications, the lamellar structure of the particle is clearly seen, in this case showing the presence of a secondary phase which on the basis of its smaller reflectivity is most likely GdNi. Figure 10 is a low power SEM view of sectioned Er_3Ni particles, again showing the absence of large internal voids afforded by this particulation technique.

By way of confirming, at least in a semi-quantitative manner, our observations regarding the apparent difference in durability of $GdNi_2$ and Er_3Ni, we conducted uniaxial crushing tests of individual particles, which we summarize in Figures 11 and 12. As indicated on the plots, some of the data were collected with the particle interleaved between 0.15 mm brass shims, while the balance were crushed on bare steel platens. No significant effect was attributed to the increased contact area provided by the brass shims and this suggests that failure is initiated internally rather than at the surface. What is unmistakable, however, is the dramatic difference in the load bearing ability of the two different materials. Clearly, Er_3Ni is a very weak and easily friable material.

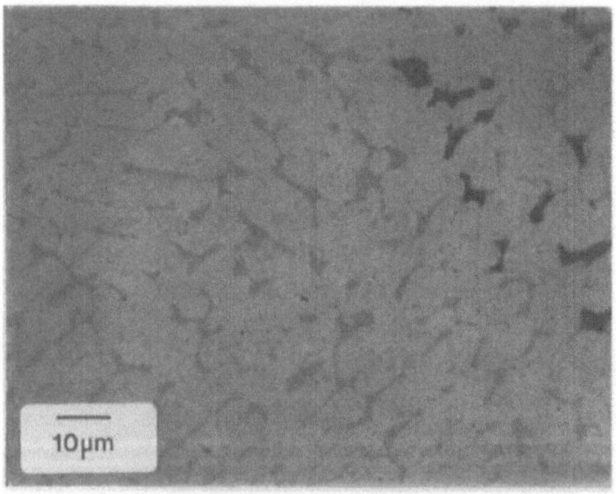

Figure 9. Detail (2) of $GdNi_2$ polished section.

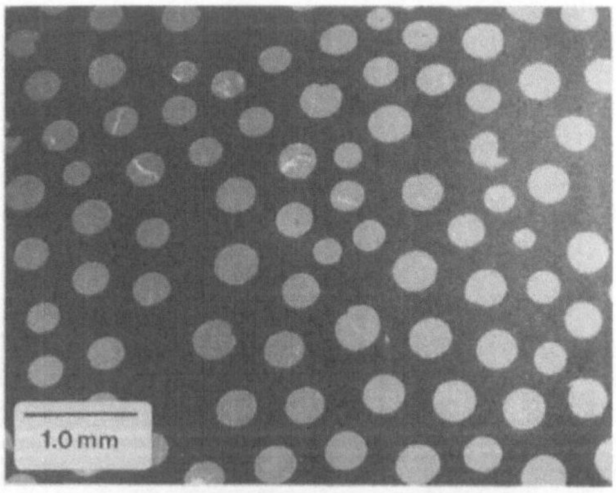

Figure 10. Polished section of Er_3Ni particulate.

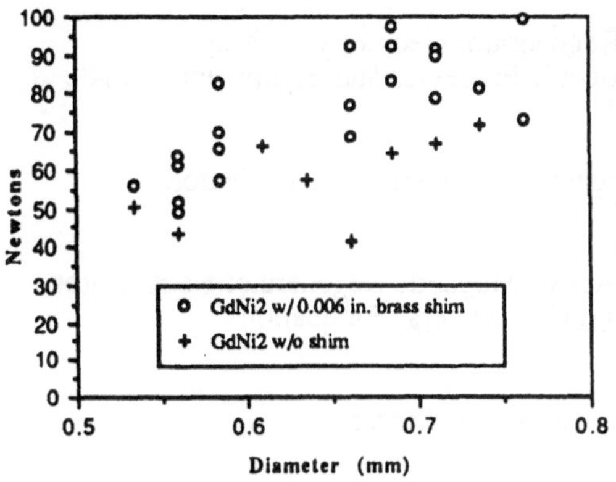

Figure 11. Crushing load vs. particle size for $GdNi_2$.

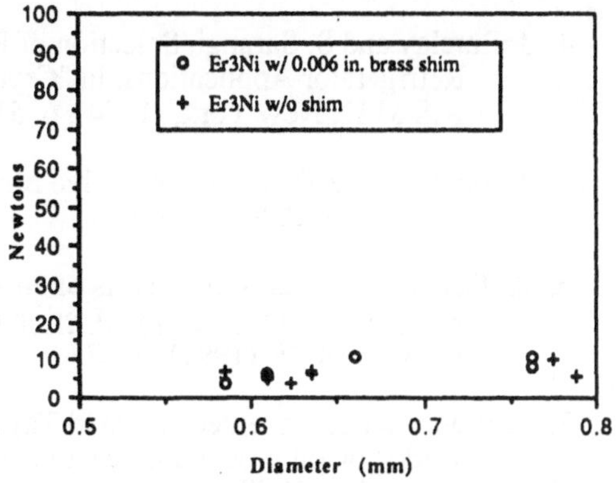

Figure 12. Crushing load vs. particle size for Er_3Ni.

One rather curious aspect of our work to date is that the expected Rayleigh correlation of mean particle size (MPS) to orifice size is not being followed in this system. We have observed an MPS/orifice size ratio nearly double that expected, which corresponds to a disturbance wavelength of *8 times* greater than predicted. At present we have no explanation to offer for this unexpected result, although experiments have shown it to be essentially independent of melt composition, viscosity and jet velocity. This of some concern in that the particle size range of interest for high power regenerative magnetic cryocoolers is typically 0.15mm. Thus far, we have not succeeded in producing material with an MPS below about 0.4mm due to nozzle clogging problems experienced with very small orifices.

CONCLUSIONS

A new melt particulation process based upon the natural instability of a laminar fluid stream has been demonstrated for the rare earth intermetallic compounds Er_3Ni and $GdNi_2$. Analysis of the resulting materials via metallography shows that the materials are free of large internal voids and defects although in the case of $GdNi_2$ some relatively minor solidification voids existed. Significant differences were found in the compressive load bearing capabilities of particulate formed from these two materials and the very low crushing strength of Er_3Ni corroborates the extensive amount of damage observed as a result of normal shipping and handling techniques.

ACKNOWLEDGMENTS

The authors wish to acknowledge the assistance of Kamin Mahoney in metallographic specimen preparation and mechanical testing.

REFERENCES

1. C. Zimm et al, Materials for Regenerative Magnetic Cooling Spanning 20K, to 80K, to be published in "Advances in Cryogenic Engineering", Vol. 37, Plenum Press, New York.

2. M. Sahashi et al, New Magnetic Material R_3T System with Extremely Large Heat Capacities Used as Heat Regenerators, in "Advances in Cryogenic Engineering", Vol. 35B, Plenum Press, New York, (1990), p. 1175.

3. D. Janda et al, Design of an Active Magnetic Regenerative Hydrogen Liquefier, to be published in "Advances in Cryogenic Engineering", Vol. 37, Plenum Press, New York.

4. J. Barclay and S. Sarangi, Selection of Regenerator Geometry for Magnetic Refrigerator Applications, in "Cryogenic Processes and Equipment - 1984", A.S.M.E., New York, (1984), p. 51.

5. J. Barclay, The Theory of an Active Magnetic Regenerative Refrigerator, NASA-CP -2287, (1983).

6. A. DeGregoria et al, Initial Tests of an Active Magnetic Regenerator Refrigerator, Proc. Fourth Interagency Meeting on Cryocoolers, Publication DTRC 91/003, (1991), p. 277.

7. F. Biancanello, Presented at David Taylor Research Center workshop on processing reactive metal powders, November, 1990.

8. I. Anderson, Presented at DTRC workshop, November 1990.

9. J. Perepezko, E. Ludeman and D. Thoma, unpublished.

10. R. Ardizone, Presented at DTRC workshop, November 1990.

11. J. Rayleigh, "Theory of Sound", 2nd edition, MacMillan and Co, Ltd., London, (1926).

ANALYSIS OF RARE EARTH COMPOUND REGENERATORS OPERATING AT 4 K

H. Seshake, T. Eda, K. Matsumoto, and T. Hashimoto

Department of Applied Physics
Tokyo Institute of Technology
Meguro, Tokyo, Japan

T. Kuriyama and H. Nakagome

Energy Science and Technology Laboratory
Toshiba Research and Development Center
Ukishima, Kawasaki, Japan

ABSTRACT

This paper describes the analysis of regenerators which consist of rare earth compounds. In a regenerative cycle refrigerator, refrigeration loss is larger than useful refrigeration capacity at liquid helium temperature, and the main loss is regenerator loss. The purpose of this paper is to establish a numerical calculation method which gives useful information for designing high efficiency regenerators. The model regenerator consisted of two regenerator compounds: Er_3Ni, which is hot-side material, and another material such as $ErNi_2$, $Er_{0.75}Dy_{0.25}Ni_2$, ErNi, or $Er_{0.9}Yb_{0.1}Ni$. The regenerator efficiency was calculated for various ratios of the two materials. Regenerator efficiencies for various combinations of rare earth compounds are systematically discussed. The maximum regenerator efficiency was obtained at 4 K with the condition that Er_3Ni was 45 % and $Er_{0.9}Yb_{0.1}Ni$ was 55 % of the regenerator volume. Moreover, it was shown that a refrigerator with this optimized regenerator could achieve a 52 % larger refrigeration capacity than one with an all-Er_3Ni regenerator.

INTRODUCTION

Regenerative refrigerators which use magnetic materials for regenerators have been studied in some papers[1-3]. Kuriyama et al[4] have succeeded in the liquefaction of helium with a two-stage GM refrigerator. Nagao et al[5] have succeeded in the generation of superfluid helium with a three-stage GM refrigerator. For the practical utilization of GM refrigerators, such as shield cooling of superconducting magnets, it is crucial to the improve the performance of refrigerators in the liquid helium temperature region.

Reduction of refrigeration losses is indispensable to improve refrigeration performance. There are various refrigeration losses such as the loss generated from regenerator inefficiency, thermal conductivity loss, shuttle heat transfer loss, and so on. The authors focused on the regenerator loss, because we have found that the regenerator loss plays the most important role in refrigeration losses[4]. In this paper, we analyze a numerical model of a regenerator so that we can compare various regenerator efficiencies, and then derive important information to improve regenerator efficiency with magnetic materials.

There are few studies of regenerators with more than two kinds of magnetic material. In this complex regenerator case, optimization is complicated because there are various combinations of regenerator materials and also matrix structure. The $Er_xDy_{1-x}Ni_2$ and $Er_xYb_{1-x}Ni$ intermetallic compound systems were selected for our regenerator material, which were shown to be promising materials by Li et al[6] and Hashimoto et al[7]. The authors compare the regenerator efficiencies for various combinations of these intermetallic compounds and various volume ratios.

NUMERICAL METHOD

First of all, the model regenerator is elucidated. To derive the governing equations for the regenerator, a constant convective heat transfer coefficient through the regenerator and negligible small void volume are assumed.

In liquid helium temperature region, the specific heats of both helium gas and regenerator matrix are highly dependent on temperature. To analyze the regenerator at this temperature, we cannot help taking these temperature-dependent specific heats into consideration. The governing equations are expressed as (1),(2) from the energy conservation law and equation of continuity[8]:

$$\frac{\partial H_g}{\partial x}=\frac{hA_wL}{\dot{m}}(T_s-T_g) \tag{1}$$

$$\frac{\partial H_s}{\partial t}=\frac{hA_wL\tau}{M_s}(T_g-T_s) \tag{2}$$

T_g	: temperature of gas	T_s	: temperature of regenerator matrix
H_g	: enthalpy of gas	H_s	: enthalpy of matrix
h	: heat transfer coefficient	A_w	: heat transfer area per unit length
M_s	: mass of regenerator matrix	L	: regenerator length
$\dot{m}$	: mass flow rate of the gas	τ	: half cycle period
x	: reduced length of the regenerator (x/L)	t	: reduced time (t/τ)

This pair of equations needs to be expressed in a suitable numerical scheme for computer calculation. The authors follow the same method[9] to express the partial differential equations in finite form. The regenerator location and process period were divided into 150 grid points. The values of H_g and H_s are tabulated in computer memory at every integer value of temperature as in the Reference 9. The regenerator efficiency is defined as enthalpy efficiency (η) as follows in,

$$\eta=\frac{\Delta H_{real}}{\Delta H_{ideal}} \tag{3}$$

where ΔH_{real} and ΔH_{ideal} are respectively the actual and ideal enthalpy changes of the gas.

In the region where the gas can be treated as an ideal gas, the enthalpy changes of high pressure flow and low-pressure flow are equal to each other. But in the region where the gas cannot be treated as an ideal gas, the enthalpy changes of high-pressure flow and low-pressure flow aren't the same any more. In the calculating condition discussed below, the enthalpy change of high-pressure flow is smaller than that of low pressure. When applying the energy conservation law, it is correct to put the smaller enthalpy change into the denominator. Therefore, in this paper, we selected the ideal enthalpy change of high-pressure flow for the denominator of Equation (3).

NUMERICAL MODEL

A numerical regenerator model is shown in Fig. 1. The gas pressures are 2 MPa for high pressure and 0.8 MPa for low pressure, and the temperature is fixed at 30 K for the high-temperature end and 4 K for the low-temperature end. We put Er_3Ni in high-temperature part of the regenerator and the other rare earth compound (RNi_x) in low-temperature part.

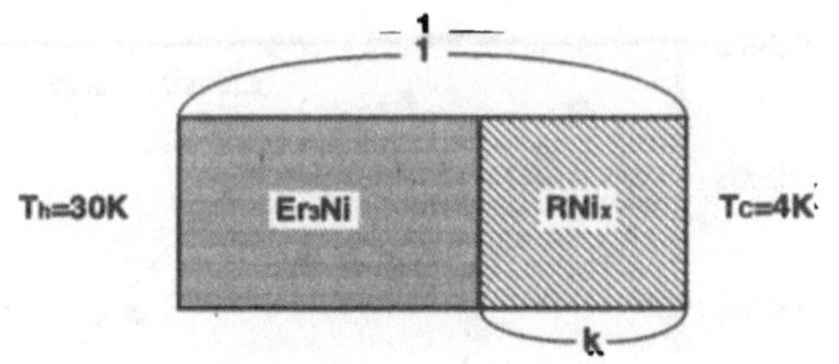

Fig. 1 The model regenerator. This regenerator has a hybrid structure with Er_3Ni and RNi_x. Er_3Ni is placed in the high-temperature part and RNi_x in the low-temperature part of the regenerator.

All matrices are assumed to be spheres of 0.25 mm diameter. The model regenerator is 32 mm in diameter and 85 mm in length, which is the same size as the 2nd-stage regenerator of our GM refrigerator. The mass flow rate $\dot{m}$ is 2 g/s, half cycle period τ is 0.5 s, and convective heat transfer coefficient h is 0.05 W/cm^2-K. In this calculation, the parameters are RNi_x and k, which is the ratio of RNi_x volume to the total regenerator volume.

RESULT AND DISCUSSION

The calculated efficiencies are discussed in the following three sections.

$Er_xDy_{1-x}Ni_2$

In this section the authors discuss the effect of the temperature-dependent specific heat regenerator efficiency by using $Er_xDy_{1-x}Ni_2$ for RNi_x whose specific heat changes systematically with x. The volumetric specific heats of $Er_xDy_{1-x}Ni_2$ are shown in Fig. 2. As shown in this figure, the temperature at which the specific heat reaches its peak increases with decreasing x. Moreover, the specific heat peak becomes higher with decreasing x.

The regenerator efficiencies are shown in Fig. 3. This figure shows that the efficiency of the regenerator increases with k when x is equal to 1 and 0.75. The maximum efficiency is obtained at k_{max} = 0.15 for x = 1 and k_{max} = 0.5 for x = 0.75, respectively (k_{max} is defined as the k value at which the maximum efficiency is obtained). The specific heat of $ErNi_2$ is larger than that of Er_3Ni below 7 K, and the specific heat of $Er_{0.75}Dy_{0.25}Ni_2$ is larger than that of Er_3Ni below 12 K. Therefore, $ErNi_2$ is useful in a narrower temperature region than $Er_{0.75}Dy_{0.25}Ni_2$, and this causes that the k_{max} of $ErNi_2$ to be smaller than that of $Er_{0.75}Dy_{0.25}Ni_2$. When x is 0.5 the efficiency decreases with k. This is because $Er_{0.5}Dy_{0.5}Ni_2$ has not as large a heat capacity as Er_3Ni at liquid helium temperature, as shown in Fig. 2. Hence, $Er_{0.5}Dy_{0.5}Ni_2$ cannot absorb enough heat from helium gas in this temperature region. Accordingly, the $Er_{0.5}Dy_{0.5}Ni_2$ regenerator is inferior to the all-Er_3Ni regenerator.

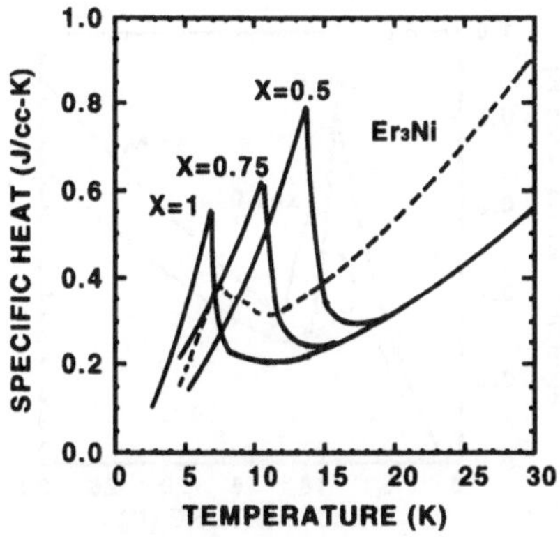

Fig. 2 The volumetric specific heats of Er_3Ni and $Er_xDy_{1-x}Ni_2$

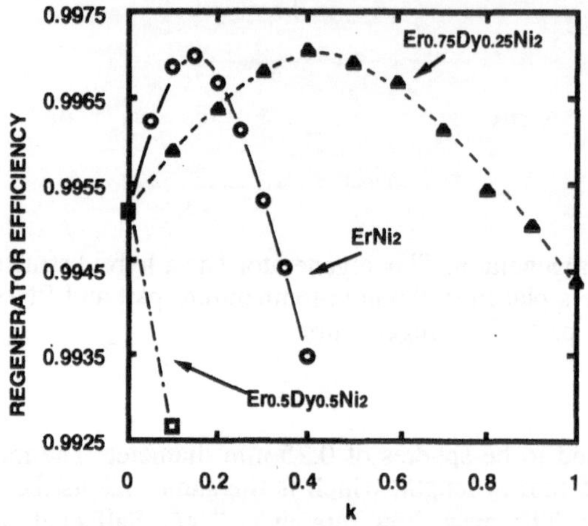

Fig. 3 Regenerator efficiencies of the hybrid structure regenerator with Er_3Ni and $Er_xDy_{1-x}Ni$ as functions of the ratio of $Er_xDy_{1-x}Ni_2$ volume to the total regenerator volume.

$Er_xYb_{1-x}Ni$

The volumetric specific heats of $Er_xYb_{1-x}Ni$ are shown in Fig. 4. This figure shows that the temperature at which the specific heat has its peak decreases with decreasing x. The maximum efficiencies are obtained at $k_{max} = 0.65$ for $x = 1$ and $k_{max} = 0.55$ for $x = 0.9$, respectively. The efficiency of the $Er_{0.9}Yb_{0.1}Ni$ regenerator at k_{max} is 0.55 has the largest value of all the rare earth compound regenerators discussed in this paper.

Figure 5 shows the temperature profiles of the matrix in the regenerator. In this figure, line A shows the matrix temperature profile right after the high-pressure gas flowed through the regenerator. Line B shows the matrix temperature profile of the matrix right after the low-pressure gas flowed through the regenerator. The specific heat of $Er_{0.9}Yb_{0.1}Ni$ is much larger than that of Er_3Ni at liquid helium temperature, as shown in Fig. 4. From the Fig. 5, it can be recognized that the variation of temperature near the cold outlet in the $Er_{0.9}Yb_{0.1}Ni$ regenerator is smaller than in the all-Er_3Ni regenerator. Hence, in the $Er_{0.9}Yb_{0.1}Ni$ regenerator, helium gas is discharged at a lower temperature from the cold outlet than in the all-Er_3Ni regenerator.

From the viewpoint of the regenerator matrix selection, the magnitude of the specific heat at liquid helium temperature has a large effect on the regenerator efficiency.

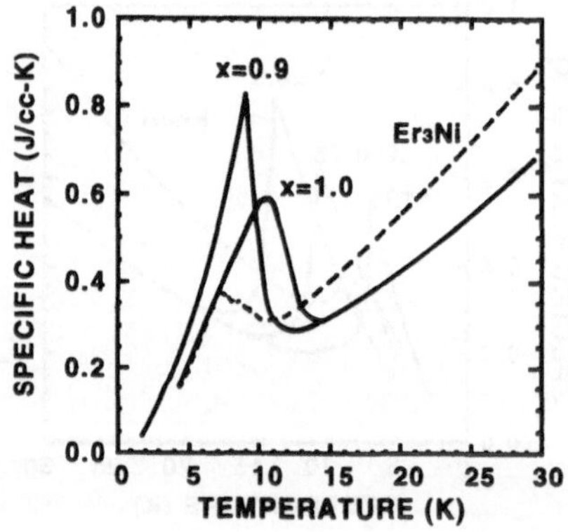

Fig. 4 The volumetric specific heats of Er_3Ni and $Er_xYb_{1-x}Ni$.

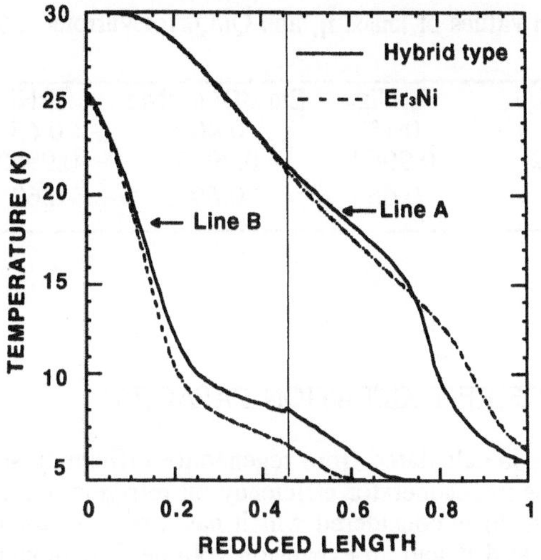

Fig. 5 Temperature profiles of the regenerator matrix at the end of each process. Hybrid type is the regenerator which consists of 45 % Er_3Ni and 55 % $Er_{0.9}Yb_{0.1}Ni$. The fine vertical line shows the boundary of matrix in the hybrid regenerator.

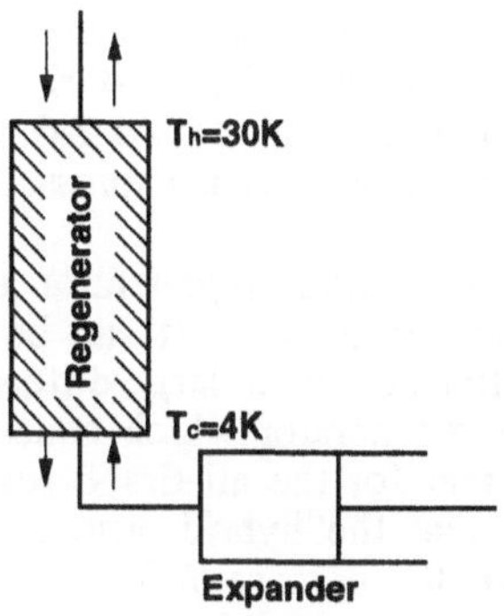

Fig. 6 The model refrigerator. This refrigerator consists of a regenerator with high temperature at 30 K, low temperature at 4 K, and an expander.

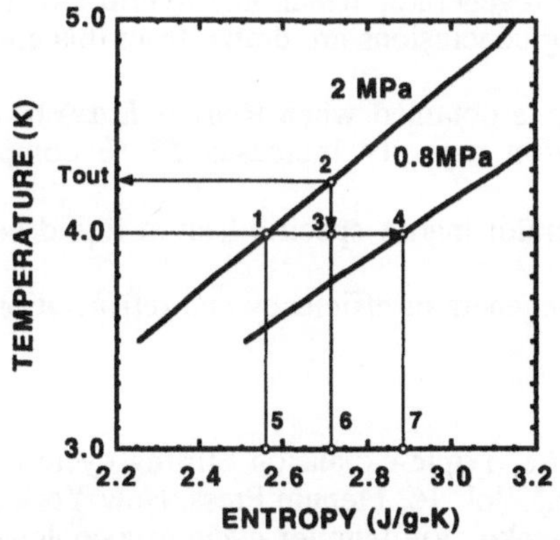

Fig. 7 Temperature - Entropy diagram of helium. This figure also shows the refrigeration cycle of the model refrigerator.

1 → 4 : ideal refrigeration cycle

2 → 3 → 4 : real refrigeration cycle

Table 1. Calculated values of k_{max}, η, and Q_r/Q_i for various regenerator materials

	Er_3Ni	$ErNi_2$	$Er_{0.75}Dy_{0.25}Ni_2$	ErNi	$Er_{0.9}Yb_{0.1}Ni$
k_{max}	0	0.15	0.40	0.65	0.55
η	0.9952	0.9967	0.9970	0.9962	0.9973
Q_r/Q_i	0.45	0.65	0.66	0.62	0.69

THE CALCULATION OF REFRIGERATION CAPACITY

Refrigeration capacity is calculated from regenerator efficiency so as to make clear the effect of small differences in regenerator efficiency on refrigeration performance. A model refrigerator (shown in Fig. 6) is considered which has a regenerator with high temperature at 30 K, low temperature at 4 K and an expansion volume. The model refrigeration cycle is explained with the help of Fig. 7, which is the entropy - temperature diagram of helium.

First of all, we consider the ideal refrigeration cycle as follows. With the ideal regenerator, the gas flows to the expansion volume at 4 K (point 1 in Fig. 7). Then, the gas expands isothermally (process 1 - 4). The area of rectangle 1-4-7-5 represents the ideal refrigeration capacity (Q_i). Next, we consider the real refrigeration cycle. With the inefficient regenerator, the gas flows to the expansion volume at a higher temperature (T_{out}) than 4 K (point 2). In our model cycle, the gas expands adiabatically to 4 K (process 2 - 3), then expands isothermally (process 3 - 4). The area of rectangle 3-4-7-6 represents the real refrigeration capacity (Q_r). The reduced refrigeration capacity is expressed as Q_r divided by Q_i.

Table 1 represents k_{max}, the maximum regenerator efficiencies, and the reduced refrigeration capacities for various regenerators. It can be recognized from table 1 that small differences in regenerator efficiency make large differences in refrigeration capacity. The refrigeration capacity for the regenerator which consists of 45 % Er_3Ni and 55 % $Er_{0.9}Yb_{0.1}Ni$ is 52 % larger than that for the all-Er_3Ni regenerator. From the calculated refrigeration capacity, it is clear that the hybrid structure regenerator with Er_3Ni and $Er_{0.9}Yb_{0.1}Ni$ is the most effective to increase refrigeration capacity at liquid helium temperature. Moreover, the calculating method discussed above is very useful in studying the effect of regenerator efficiency on refrigeration capacity.

CONCLUSIONS

The efficiency of the regenerator which had hybrid structure with Er_3Ni and RNi_x was calculated. The following conclusions are drawn from this analysis.

(1) The maximum efficiency is obtained when RNi_x is $Er_{0.9}Yb_{0.1}Ni$ and k is 0.55. In this condition, the refrigeration capacity increases 52 % compared with the all-Er_3Ni regenerator.
(2) The magnitude of regenerator matrix specific heat at liquid helium temperature affects regenerator efficiency.
(3) A method for analyzing regenerator efficiency and refrigeration capacity is established.

REFERENCES

1. A. Daniel and F. K. du Pre, Triple-expansion Stirling cycle refrigerator, in: "Advances in Cryogenic Engineering," Vol. 16, Plenum Press, New York (1971), p. 178.
2. Y. Matsubara and M. Kaneko, Vuilleumier cycle cryocooler operating below 8 K, in: "Proc. Third Cryocoolers Conf. ," National Bureau of Standards, Boulder, Colorado (1984), p. 234.
3. H. Nakashima, K. Ishibashi, and Y. Ishizaki, Development of 4 - 5 K cooling Stirling cycle refrigerator, in:"Proc. Fourth Intl. Cryocoolers Conf. ," David Tailor Naval Ship R&D Center, Annapolis, MD (1986), p. 263.

4. T. Kuriyama et al, Two-stage GM refrigerator with Er_3Ni regenerator for helium liquefaction, in: "Proc. Sixth Cryocooler Conf. ," to be published.
5. M. Nagao et al, Generation of superfluid helium by a Gifford-McMahon cycle cryocooler, in: "Proc. Sixth Cryocooler Conf. ," to be published.
6. R. Li et al, Magnetic material for cryogenic regenerator, in: "Proc. JSJS III," Okayama, Japan (1989), p. 84.
7. T. Hashimoto et al, Recent progress on the rare earth magnetic regenerator material, in: "Advance in Cryogenic Engineering," Vol. 37.
8. P. Rios and J. L. Smith, Jr. ,The effect of variable specific heat of the matrix on the performance of thermal regenerators, in: "Advances in Cryogenic Engineering," Vol. 13, Plenum Press, New York (1968), p. 566.
9. R. K. Sahoo and S. Sanragi, Effect of temperature-dependent specific heat of the working fluid on the performance of cryogenic regenerators, Cryogenics 25: 583 (1985).

DEVELOPMENT OF NEODYMIUM AND Er_3Ni REGENERATOR MATERIALS

Louis F. Aprigliano, Geoffrey Green, James Chafe,
Lisa O'Connor, Frank Biancanello*, and Steve Ridder*

David Taylor Research Center
Bethesda, Maryland

*National Institute of Standards and Technology
Gaithersburg, Maryland

ABSTRACT

Neodymium and Er_3Ni powders were produced and examined as possible candidates for use as regenerator matrices in the regenerative heat exchanger of the Gifford-McMahon cycle refrigerator. In the case of Er_3Ni, crushed powders were brittle and angular in shape, molten salt produced spheres were heavily oxidized, and gas atomized powder had a low yield (3 %) and a large fraction of hollow particles. Neodymium powder, which was produced by the rotating electrode method, was smooth and spherical in shape and had a high yield (15 %) of solid particles.

INTRODUCTION

The Gifford-McMahon (GM) refrigeration cycle can be used to meet the needs of many systems that require cryogenic cooling. The GM cycle makes use of the regenerative heat exchanger. A crucial component of the regenerative heat exchanger is the regenerative matrix. The function of the regenerative matrix is depicted in Fig. 1. A typical regenerative matrix consists of lead shot measuring approximately 0.23 mm (0.009 inch) in diameter.

The heat capacity of the regenerative matrix is an indicator of its performance. A plot of heat capacity versus temperature for lead is shown in Fig. 2[1]. In the case of lead, the drop-off in the heat capacity below 12 K is a significant factor that limits its effective cooling to 10 K. The heat capacities of the candidate regenerative matrices, neodymium[2] (Nd) and erbium-3-nickel[3] (Er_3Ni), are shown in Fig. 2. Both show a favorable heat capacity spike at approximately 6 K and both retain some heat capacity to 4 K. This feature makes them an attractive alternative to lead in the temperature range of 4 to 10 K.

Making powders of Nd and Er_3Ni is not a trivial process. Both are highly reactive with oxygen which makes them difficult to produce as quality powders. Furthermore, the Er_3Ni is extremely brittle and there

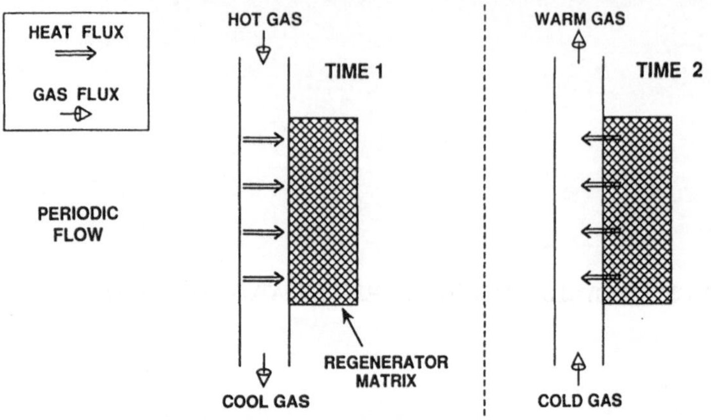

Figure 1. Schematic representation of the function of a regenerator matrix in a regenerative heat exchanger.

is some concern that the long term operation in the regenerative heat exchanger of a GM cycle can break the powder down into very fine particles. These fine particles can escape from the regenerative heat exchanger and adversely affect other GM cooler components (e.g. the displacer, seals, and valves).

The purpose of this study is to evaluate the quality of various sources of Er_3Ni and Nd powders. We are doing this because of the above mentioned features in the heat capacity of these two materials and because regenerative matrices made with powdered Er_3Ni have been reported[3,4,5] to improve the performance of the GM cycle. However, the work with Er_3Ni has provided very little information on how the powders were produced, on the process yields or on the quality. In addition to addressing these issues, we have studied the performance of these powders in a GM cooler. Those results will be reported[6] separately at this conference.

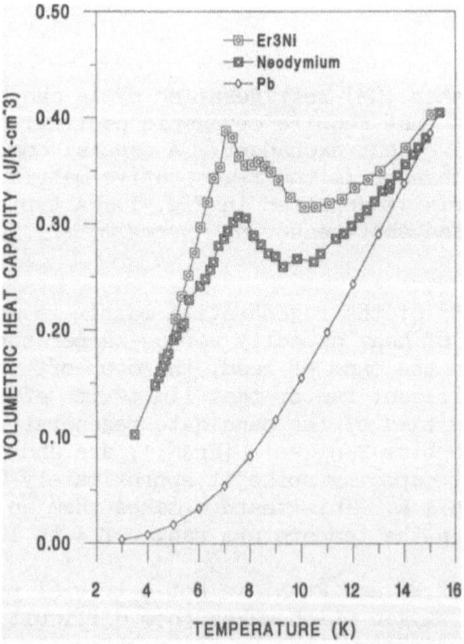

Figure 2. Heat capacity versus temperature for lead, neodymium, and Er_3Ni.

EXPERIMENTAL PROCEDURE

Er_3Ni

Commercial sources were used to obtain Er_3Ni in the form of crushed powder and molten salt bath produced spheres. In addition, the Gaithersburg branch of the National Institute of Standards and Technology (NIST) produced spherical powders of Er_3Ni using an inert gas atomization process. The crushed powder is made by breaking up a solid bar of Er_3Ni until powders that will pass a 0.25 mm (0.010 inch) sieve are produced. The molten salt bath involves melting a supply of Er_3Ni nuggets and allowing the molten particles to slowly settle and drift in a column of molten salt. Inert gas atomization involves evacuating a large cylindrical chamber -- 1.6 m (46 inch) inside diameter by 2.7 m (9 feet tall) -- and then backfilling to one atmosphere of argon. At the top is a melt chamber with an induction melting crucible supported on a stainless steel separating plate. A stopper rod is used to contain the liquid metal in the crucible. When the stopper rod is raised the molten metal flows through a ceramic delivery tube. The liquid forms a sheet at the nozzle tip which is sheared by the atomizing gas into first ligaments, then dumbbells and finally droplets. The droplets fall in the main chamber and solidify into spherical powders which are collected at the base of the cyclone separator.

Neodymium

A commercial source was used to obtain Nd in the form of spherical powders by the method of rotating plasma electrode (RPE). For this procedure, a bar of Nd, 6.35 cm (2.5 inches) in diameter and 25.4 cm (10 inches) long, is machined from a commercially supplied ingot of Nd. The bar is rotated along its long axis at high speed in a chamber filled with argon. A metal electrode is brought close to the end of the Nd bar and an electrical arc is struck. The combination of the heating and the rotational forces causes molten particles of Nd to be spun off and to solidify into spherical powders.

(An attempt has not been made to make Er_3Ni by the RPE method. Er_3Ni is very brittle and would be difficult to machine with the dimensional tolerances needed for the electrode in the RPE method.)

Examination

Optical and electron microscopy were utilized to examine the condition of the powders. The powders were embedded in epoxy and then in bakelite for polishing to a flat surface. This allowed the interior of the particles to be examined for voids and allowed for the use of the electron microprobe to study the composition and microstructure of the powders. The scanning electron microscope was used to study the surface topography of the powders and to measure their size.

RESULTS

Er_3Ni

Scanning electron and electron microprobe examination showed the crushed powder to be very angular in shape (Fig. 3) and the molten salt powder to be coated with a heavy oxide layer (Fig. 4). Neither features are considered desirable in a good regenerator matrix. The spray atomized powder was spherical and was free of a heavy oxide, but the insides of many of the spheres were hollow (Fig. 5). An x-ray diffraction analysis of this powder showed it to be predominately Er_3Ni with a minor amount of Er_2O_3.

Figure 3. Er_3Ni crushed powder; scanning electron micrograph.

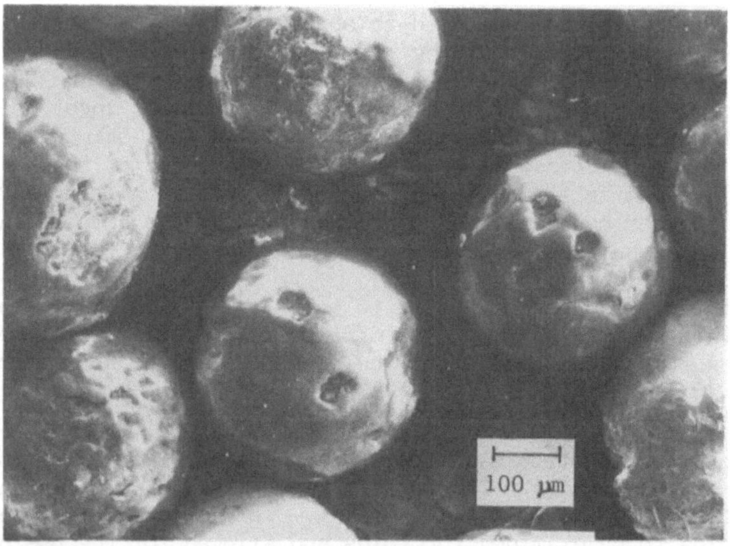

Figure 4. Molten salt bath produced Er_3Ni; scanning electron micrograph.

For the following reasons, hollow spheres are not preferred for use in a regenerator matrix. A powder size of 0.23 mm (0.009 inch) diameter is considered optimum for a regenerator matrix in a regenerative heat exchanger. During each cycle the regenerator matrix must alternately transfer heat to and from the helium that is compressed and expanded in the GM cooling cycle. If the powder particles in the regenerator matrix are too small or too large, the particle mass is not being used effectively. Likewise, if the particles are hollow, the mass needed to optimize the heat flow during each cycle is no longer present.

The inert gas atomized powder had a low yield (approximately 3 %) in the size range of 0.12 to 0.25 mm (0.005 to 0.010 inch). Most of the

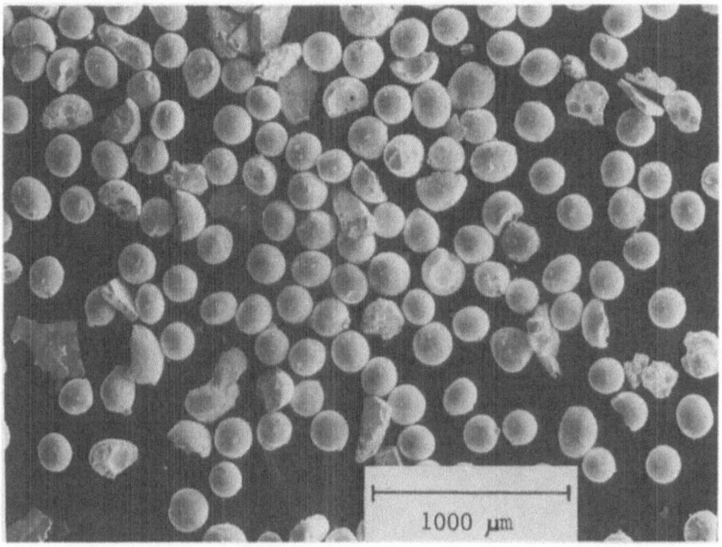

a. Scanning electron micrograph.

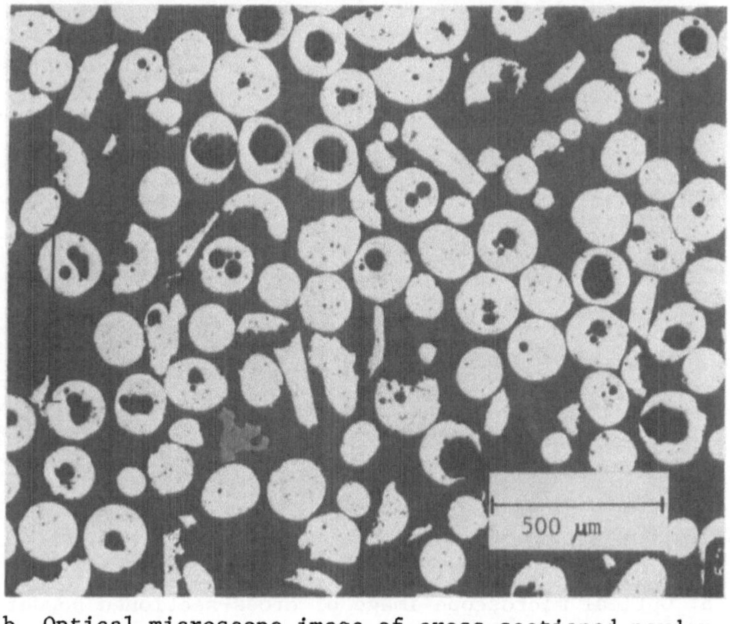

b. Optical microscope image of cross-sectioned powder.

Figure 5. Micrographs of gas atomized Er_3Ni.

particles are much smaller than this. As currently designed, the NIST gas atomizer favors the production of finely size particles. However, it can be modified to produce a higher yield of larger particles. Such a modification is being planned.

Neodymium

The Nd powder was visually bright and shiny and could be handled in room air for brief periods without violent or excessive oxidation. Scanning electron microscope examination showed the Nd powder to be very smooth in surface texture and to have a high proportion of spherical particles (Fig. 6). Electron microprobe examination showed that the

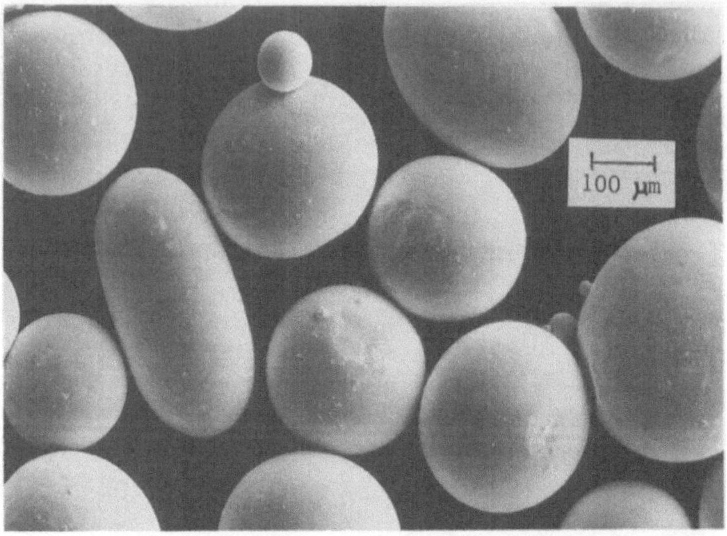

a. Scanning electron micrograph.

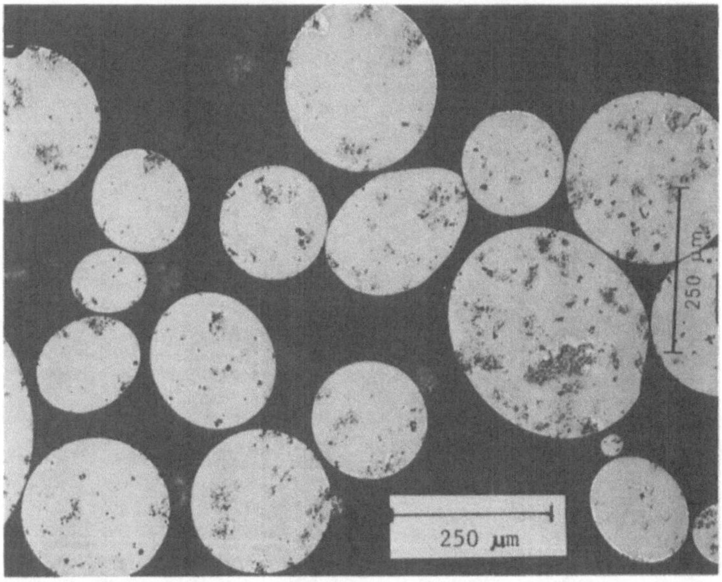

b. Optical microscope image of cross-sectioned powder.

Figure 6. Micrographs of neodymium powder produced by the rotating plasma electrode method.

particles did not have an excessively thick oxide layer and were solid throughout their cross-section. The yield (approximately 15 %) was acceptable in the size range 0.12 to 0.25 mm (0.005 to 0.010 inch). Since this yield was obtained on the first run, we believe that continuing efforts will further improve the yield.

CONCLUSION

1. Spherical Nd powder in the size range of 0.12 to 0.25 mm (0.005 to 0.010 inch) can be made with high yields (approximately 15 %) by the rotating plasma electrode method.

2. Due to the combined brittle and highly reactive nature of Er_3Ni, only the inert gas atomization process was able to produce spherical powder without excessive oxidation. However, more process experimentation and system modifications are needed if the yield of solid particles with sizes greater than 0.25 mm (0.005 inch) is to be increased.

REFERENCES

1. Corruccini, R. J. and J. J. Gniewek, Specific Heat and Enthalpies of Technical Solids at Low Temperatures", NBS Monograph 21, (1960).

2. Zimm, C. B., P. M. Ratzmann, J. A. Barclay, G. F. Green, and J. N. Chafe, "The Magnetocaloric Effect in Neodymium, Advances in Cryogenic Engineering, Materials, Vol 36A, Plenum Press, New York, 1990.

3. Sahashi, M., Y. Tokai, T. Kuriyama, H. Nakagome, R. Li, M. Ogawa, and T. Hashimoto, "New Magnetic Material R_3T System with Extremely Large Heat Capacities Used as a Heat Regenerator", Advances in Cryogenic Engineering, Vol 35B, Plenum Press, New York, 1989.

4. Kuriyama, T., R. Hakamada, H. Nakagome, Y. Tokai, and Y. Sahashi, "High Efficient Two-Stage GM Refrigerator with Magnetic Material in the Liquid Helium Temperature Region", Advances in Cryogenic Engineering, Vol. 35, Ed. R.W. Fast, Plenum Press, New York, 1989.

5. Nakagome, H., R. Hakamada, M. Takahashi, and T. Kuriyama, "Highly Efficient 4 K Refrigerator (GM Refrigerator with JT Circuit) Using Er_3Ni Regenerator", Proceedings of the International Cryocooler Conference, 1990.

6. Chafe, J. N., G. Green, and P. Gifford, "The Low Temperature Performance of a Three Stage Gifford-McMahon Cryocooler", Proceedings of this conference.

The Low Temperature Performance of a Three Stage Gifford-McMahon Cryocooler

J.N. Chafe and G. Green

David Taylor Research Center
Annapolis Md. 21401

P. Gifford

Cryomech Inc.
Syracuse NY. 13210

ABSTRACT

In an effort to produce compact, reliable and easily operated cryocoolers which can provide refrigeration below 10 K, an experimental investigation of low temperature regenerators was undertaken. As part of this program, a three stage G-M cycle refrigerator was designed and constructed. This apparatus was designed so that the whole length of its third stage regenerator operates at temperatures below 10 K. Instrumentation was included which provided temperature, pressure and volume data for each stage. Test results for Pb, Nd and Er_3Ni regenerators are presented. These test results include cool-down performance as well as refrigeration power at various temperatures.

INTRODUCTION

The United States Navy has an ongoing program to develop superconducting motors and generators for possible use as propulsion systems in ships and submarines. The work presented in this paper is part of this effort and is sponsored by The Office of Naval Technology.

A superconducting propulsion system which can withstand the demands of long sea voyages and combat conditions will require an extremely rugged and reliable cryogenic refrigeration system. The Gifford-McMahon (G-M) cycle refrigerator is a possible candidate for such a refrigeration system because of its relatively small size, simplicity, good reliability and ruggedness. Walker provides a description of the G-M cycle refrigerator[1]. Traditionally however, the lowest temperatures which G-M cycle machines could achieve have been limited to approximately 10 K. This poses a definite problem since the most widely used superconducting magnet systems operate at temperatures between 4 and 5 K. Thus in order for G-M cycle refrigerators to be a viable option for the Navy, they must be able to achieve temperatures below 5 K.

The primary goal of this research was to find ways which allowed the G-M cycle refrigerator to operate at lower temperatures. The most apparent factor which limits the

Advances in Cryogenic Engineering, Vol. 37, Part B
Edited by R.W. Fast, Plenum Press, New York, 1992

temperatures of the G-M cycle refrigerators is the low temperature regenerative heat exchanger. For this reason we concentrated our efforts on improving the performance of this heat exchanger (regenerator). We have been encouraged by the results of two research groups working on improving the low temperature performance of G-M cycle refrigerators[2, 3].

The focus of this paper is on the effect which matrix materials have on the temperature and cooling power performance of low temperature regenerators for G-M cycle refrigerators. This paper presents the test results of several matrix materials (i.e. lead, Er_3Ni and Neodymium). These tests compare the performance of regenerators which differ primarily in thermal properties (i.e. total heat capacity, thermal diffusivity, thermal conductivity, etc.). The overall geometry (volume, length and diameter), material geometry (particle size and shape) and operating conditions (pressures and temperatures) of these regenerators were kept as consistent as our apparatus and budget allowed.

A considerable effort was placed on the material development work which was done in conjunction with the regenerator testing and is being presented and published simultaneously at the 1991 Cryogenic Engineering Conference[4].

APPARATUS

For this research, a special apparatus was designed and built by Cryomech Inc. to test the low temperature regenerators in a G-M cycle refrigerator. This apparatus is essentially a three stage G-M refrigerator which was designed so that the third stage regenerator could be interchanged easily. Three separate third stage regenerators were constructed by identical methods, the only difference being that each used a different matrix material.

The apparatus was fully instrumented so that temperatures, gas pressures and displacer motion could be monitored for all three stages. In addition, a variable speed drive motor and a throttling valve plate were installed for more flexible control of the operating parameters.

Refrigerator

The first two stages of the three stage apparatus were based on a commercial G-M refrigerator design which produces cooling power at 10 K. It was anticipated that the full length of the third stage regenerator would operate below 10 K.

In this apparatus, all three regenerators were stationary, and each was paired with its own displacer. Figure 1 is a schematic of the apparatus. The displacers were operated pneumatically, and their motion was controlled by a rotary valve which also controlled the flow of gas through all three regenerators. The sizes and strokes for each stage of the apparatus is presented in table 1.

The use of a small D.C. gear-motor allowed the speed of rotation of the valve to be varied from about 100 RPM to less than 5 RPM. The valve was designed so that for each revolution, the refrigerator experienced two full cycles. Thus the cyclic speed of the apparatus could be varied from about 10 cycles/min. to 200 cycles/min. Although the rate of revolution of the valve could be changed, the timing between the motion of the displacers and the flow of gas in and out of the apparatus was fixed by the rotary valve.

After the initial operation of the apparatus, it became apparent that fine tuning the pneumatic drive system so that all three displacers moved in unison would be difficult. To aid in this fine tuning process, an additional valve plate, which placed a throttling valve in

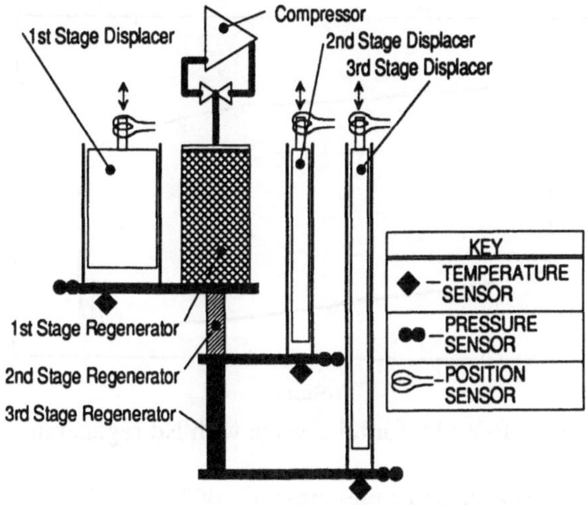

Fig. 1 Three stage G-M apparatus.

the flow path for each displacer control port, was installed. This valve plate gave a small amount of control and aided in balancing the motion of the displacers.

Instrumentation

Several instruments were added to the apparatus which are not normally used on commercial G-M refrigerators. The location of the instruments are schematically shown in fig. 1. The purpose of these instruments were to provide information helpful in understanding the processes which occur in the refrigerator during operation. The instrumentation consisted of silicon diode temperature transducers on each of the three stages, piezoresistive pressure transducers located at the cold regions of each stage and linear variable differential transformers (LVDTs) which measured the motion of the three displacers.

The pressure transducers and the LVDTs were used to display pressure vs. volume diagrams for each stage as the apparatus was being operated. Figure 2 is a typical P-V diagram for the third stage. These data were recorded with a Nd regenerator operating at a

Table 1 Critical dimensions of the apparatus

	First Stage	Second Stage	Third Stage
Regenerator Length	140 mm	76 mm	90 mm
Regenerator Diameter	57 mm	19.0 mm	19.0 mm
Regenerator Material	Phosphor-Bronze Screens	Lead Spheres (0.16 mm Dia.)	Pb (113.7 grams), Nd (69.4 grams), or Er_3Ni (75.9 grams)
Displacer Stroke	11.5 mm	16.5 mm	12.75 mm
Displacer Length	160 mm	254 mm	362 mm
Displacer Diameter	57 mm	19.0 mm	19.0 mm

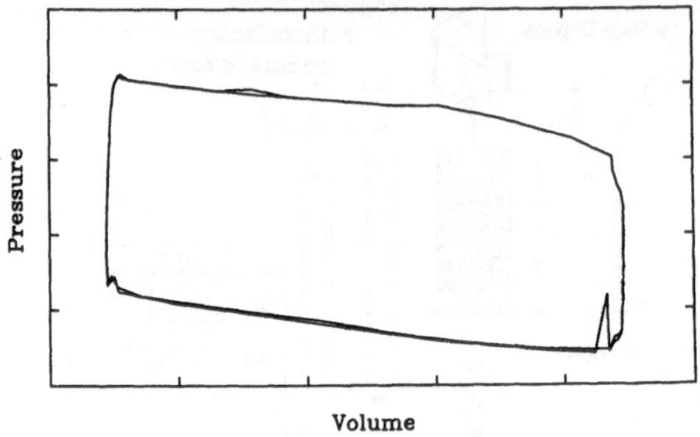

Fig. 2 P-V plot for third stage with Nd regenerator.

cyclic speed of 120 cycles/min. and a pressure ratio of 13.5:1. Diagrams like this provided a way to quickly assess if the apparatus was operating properly. The temperature transducers provided a quantitative method in assessing the performances of the various regenerators. They also gave some insight into how the temperature varied over a cycle. Figure 3 is a plot of temperature versus time for the cycles shown in fig.2. As can be seen, there is a significant variation in temperature during a cycle.

Low Temperature Regenerator

The interchangeable character of the third stage regenerator permitted the comparative testing of different regenerator materials specifically selected to operate at temperatures below those which conventional G-M regenerators operate. The material typically used in these regenerators is either lead or lead alloyed with small amounts of antimony. In addition to this regenerator material, we tested Neodymium (Nd) and Erbium-3-Nickel intermetallic compound (Er_3Ni). All three regenerator materials were in spherical powder form, each with a diameter of approximately 0.15 to 0.23 mm. The packing factor (i.e. the ratio of the actual volume of matrix material to the total volume available for matrix material) in the Pb regenerator was 66.9% , that in the Nd regenerator was 66.2% and that for the Er_3Ni was only 54.8%. The low packing factor in the Er_3Ni was attributable to hollow and irregularly shaped spheres and the use cotton fabric to prevent the leakage of material from the regenerator.

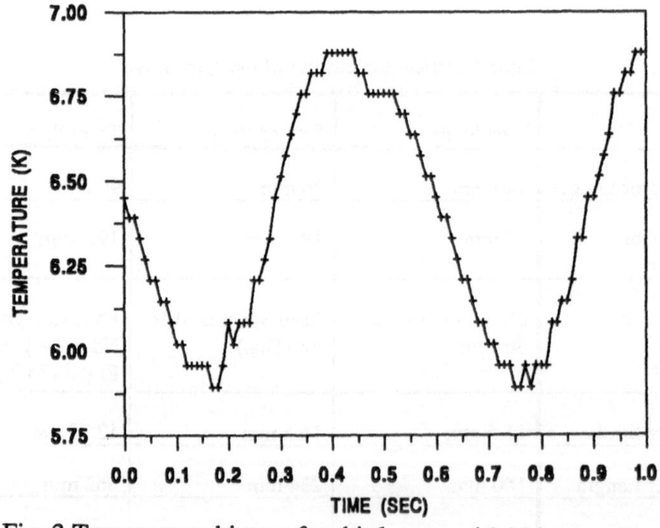

Fig. 3 Temperature history for third stage with Nd regenerator.

Figure 4 shows the volumetric heat capacities of the different low temperature regenerator materials as functions of temperature . This figure illustrates the fact that as the temperature falls below 10 K the heat capacity of lead diminishes rapidly. Both the Nd and the Er_3Ni materials have a greater heat capacity than lead below 10 K and they also have a relative peak in their heat capacities at about 8 K. The peak is somewhat larger for the Er_3Ni than for the Nd.

RESULTS

The three low temperature regenerators were evaluated under several circumstances. First, their cool-down characteristics were recorded. Second, the no-load temperature performance was measured as a function of cyclic speed, while keeping the pressure ratio constant. Third, the cooling power characteristics were measured while keeping both the speed and the pressure ratios constant.

Figure 5 shows a typical cool-down curve for the apparatus. As can be seen, the first stage cools down the quickest, followed next by the second stage and then the third stage. Complete cool-down occurred in about 90 minutes irrespective of which material was used in the third stage regenerator.

Tests were performed to measure the lowest achievable temperatures in this apparatus with the different regenerator matrix materials. These tests were performed using a constant pressure ratio of 13.5:1 while keeping the low pressure at 116.5 kPa. The results of these tests are shown in fig. 6. As can be seen, the Er_3Ni regenerator was able to produce the lowest temperatures, about 6.0 K, and did so at about 104 cycles/min. The Nd regenerator was able to achieve the next lowest temperatures, around 6.5 K, at speeds around 120 cycles/min. The coldest temperature the lead regenerator could produce was about 7.1 K at 120 cycles/min.

Tests were also performed to measure the cooling power of the three low temperature regenerators. Figure 7 presents the data. The apparatus was operating at a pressure ratio of 13.5:1 with the low pressure fixed at 116.5 kPa and the cyclic speed at 130 cycles/min. As was expected from the previous tests, for zero input power, the Er_3Ni regenerator produced

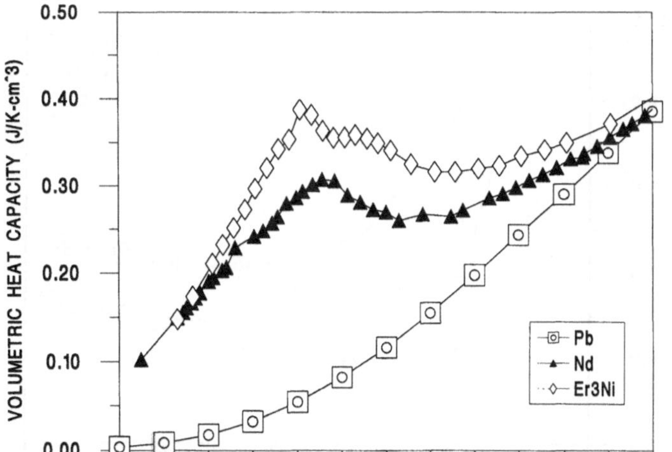

Fig. 4 Volumetric heat capacities of Pb, Nd, and Er_3Ni.[5, 6, 7]

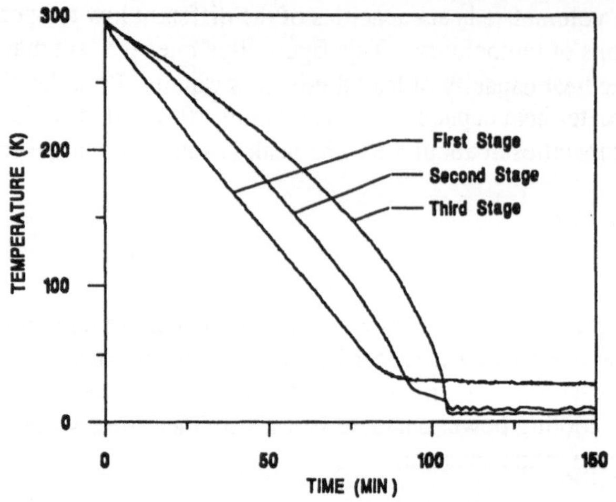

Fig 5 Cool-down history for three stage G-M apparatus with Pb regenerator.

the lowest temperatures, followed by Nd and then Pb. At 1 watt cooling power, both the Er_3Ni and the Nd regenerators operated at about 7 K while the Pb regenerator operated at about 8.2 K. At 2 watts cooling power, the Nd and Er_3Ni regenerators again operated at similar temperatures, slightly under 8.0 K, and the Pb regenerator operated at around 9.0 K.

SUMMARY

The low temperature regenerator tests showed that Nd and Er_3Ni perform better than Pb thermodynamically. We attribute this increase in performance to the higher volumetric heat capacities of these materials. These increases in performance in our apparatus were modest however, and the achieved temperatures and refrigeration powers were not sufficient to fulfill the Navy's needs. Other researchers have shown better performance (lower temperatures and higher cooling powers) using powder Er_3Ni matrix materials for their low temperature regenerator. The results of our tests indicate the improvement which these researchers achieved is due only in part to the material used in the low temperature regenerator. There must be other significant factors which are critical to improving the low

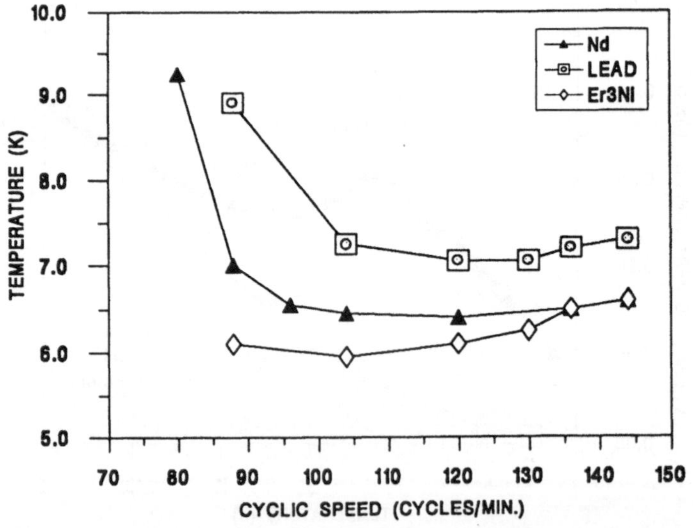

Fig. 6 No load performance of the three stage G-M apparatus.

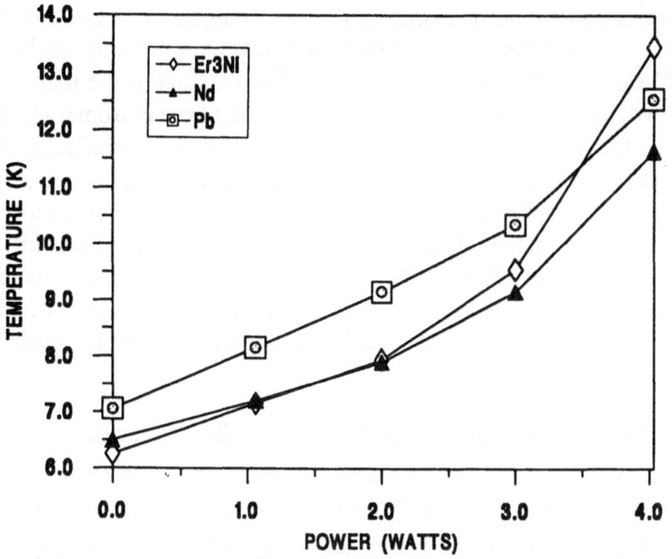

Fig. 7 Cooling power performance of the three stage G-M apparatus.

temperature performance and these factors need to be investigated in the future. The most likely of these factors include cyclic speed, displacer speed, external pressure drop, regenerator geometry and excess void volume.

The results of our tests did indicate that cyclic speed had a strong effect on regenerator performance. However it affected the higher temperature regenerators differently than the low temperature regenerators.The apparatus did not operate well at very low cyclic speeds because the first and second stages were designed to operate at relatively high speeds (144 cycles/min.). As the speed in our apparatus was lowered, the temperatures of the first and second stages climbed. This meant that the warm end of the third stage regenerator had to operate at non-optimal temperatures and it had to operate across increasing temperature differences. The ability to vary the cyclic speed, without seriously affecting the upper stage temperatures would be very helpful in evaluating the effect of cyclic speed on low temperature regenerators.

The speed at which the displacers moved may have reduced the performance of our apparatus. We were not able to measure the effect of displacer speed with our apparatus because it used a pneumatic drive system for displacer motion. While this type of drive system had many advantages, the ability to carefully control the speed was not one of them.

We also had indications that our apparatus may have had pressure drop problems in the gas lines between the compressor and the cold head, especially on the low pressure side. If this was the case, the pressure ratio that the refrigerator operated at was effectively diminished. This problem could be solved with larger diameter gas lines and a large accumulator on the low pressure side.

Other factors which need to be studied are the effect of the void volume in the expansion space and regenerator geometry . A large increase in heat capacity could be achieved simply by using a larger regenerator. Forming the matrix material into other configurations (e.g. perforated plates, ribbon gap, stacked screens, etc.) may also have significant benefit.

While the thermodynamic performance of the Er_3Ni regenerator was superior to the other regenerators tested, there are several reasons why this material may not be the most suitable for use in G-M cycle refrigerators. The primary disadvantage is the matrix rapidly abrades and breaks down into finer and finer sizes. This is due to the extremely brittle nature of this material and the fact that many of the particles are hollow and erratically shaped. When this breakdown occurs, the regenerator begins to "leak" the matrix material. As the matrix material leaves the regenerator, it flows to various parts of the refrigerator which causes premature wear in the seals and bearing surfaces. In addition, the material which remains in the partially full regenerator breaks down even more rapidly. Another major problem with this material is its high cost which is attributable to the costs of the raw material, the difficult manufacturing processes, and the low percentage of usable powder which results from these processes.

The regenerator constructed with Nd matrix material did not perform as well as the Er_3Ni regenerator but the differences were modest. The mechanical properties of the Nd are superior to Er_3Ni and somewhat similar to Pb in the fact that spheres are solid and resistant to abrasion. This produces a much more reliable regenerator. While the cost of the raw material is high, there does exist a commercial process for the relatively inexpensive manufacture of good quality spheres. This process yields a reasonably high percentage of usable powder. Nd does have a tendency to oxidize, especially when exposed to the high temperatures in the manufacturing process. However our experience shows that the material will not break down significantly if stored in an environment where the presence of oxygen is minimal.

REFERENCES

1. G. Walker, "Cryocoolers, Part I: Fundamentals" Plenum Publishing Corp., New York (1983). pps. 245-261.
2. T. Kuriyama et al., Two-Stage GM Refrigerator with Er_3Ni Regenerator for Helium Liquefaction, in: "Proceedings of the Sixth International Cryocoolers Conference,Vol. II", Report# DTRC-91/002, David Taylor Research Center, Bethesda Md (1991), pg. 3.
3 T. Inaguchi et al, Two -Stage Gifford-McMahon cycle Cryocooler Operating at about 2 K, in:, "Proceedings of the Sixth International Cryocoolers Congerence, Vol. II",Report #DTRC -91/002, David Taylor Research Center, Bethesda Md.(1991), pg. 25.
4. L. Aprigliano et al, Development of Neodymium and Er_3Ni Regenerator Materials, in: "Advances in Cryogenic Engineering", Vol. 37, Plenum Press, New York, (1992).
5. M. Sahashi et al, New Magnetic Material R_3T System with Extremely Large Heat Capacities Used as Heat Regenerators, in: "Advances in Cryogenic Engineering", Vol 35B, Plenum Press, New York, (1989), p.1175.
6. C. B. Zimm et al, The Magnetocaloric Effect in Neodymium, in: "Advances in Cryogenic Engineering (Materials)", Vol. 36A, Plenum Press., New York (1989), p. 763.
7. R.J. Corruccini and J. J. Gniewek, "Specific Heats and Enthalpies of Technical Solids at Low Temperatures", NBS Monograph 21, (1960), p. 9.

VIBRATION CHARACTERIZATION AND CONTROL OF MINIATURE STIRLING-CYCLE CRYOCOOLERS FOR SPACE APPLICATION

R.G. Ross, Jr., D.L. Johnson and V. Kotsubo

Jet Propulsion Laboratory
California Institute of Technology
Pasadena, California

ABSTRACT

A number of near-term precision space-science instruments have baselined the use of miniature long-life space Stirling-cycle cryocoolers. In support of these instruments, JPL is conducting an extensive cooler characterization test and analysis program focused at developing special sensitive performance measurement techniques and identifying means of improving cooler performance. This paper provides a summary overview of the vibration characteristics of split Stirling cryocoolers of the Oxford type and describes means being developed to achieve vibration levels consistent with the exacting requirements of sensitive infrared spectrometer instruments currently under development for NASA applications. A key emphasis of the paper is on exploring both active and passive means of reducing the residual upper harmonics of the drive frequency that remain with nulled back-to-back compressor and displacer units. Vibration supression results, measured with JPL's unique 6-degree-of-freedom force dynamometer, are presented for the 80K Stirling cooler manufactured by British Aerospace.

INTRODUCTION

The emerging line of second generation miniature Stirling-cycle cryocoolers, which are building on the successful Oxford University ISAMS cooler[1,2], are ideally suited to meeting the cryogenic cooling demands of a growing number of NASA space-science instruments. To satisfy the demanding application requirements, these emerging Stirling cryocoolers are successfully addressing a broad array of complex interface requirements that critically affect successful integration to the sensitive instrument detectors. Low vibration and EMI, and improved cooling performance at lower temperatures (55 to 60K) are particularly important parameters.

In January of 1990, JPL took delivery of one of the first long-life 80K Stirling cryocoolers manufactured by British Aerospace (BAe)[3], and began an extensive characterization activity designed to learn from and build upon the Oxford-heritage in a collaborative program with industry to meet the demands of NASA's near-term space-science instruments. Research has focused on all aspects of cryocooler performance including vibration[4], EMI[5], thermal performance[4,6], and reliability[4].

This paper focuses on cryocooler vibration characteristics and candidate strategies for vibration control in sensitive space-science instruments. Problems occur when cooler-generated vibration excites elastic deflections and resonances within the instrument structure and components that either adversely affect optical alignment, or generate spurious electrical signals. The latter are generated when electrical current-carrying or capacitively-coupled components undergo relative motions. Instruments such as the NASA Earth Observing System (Eos) Atmospheric Infrared Sounder (AIRS), a precision LWIR spectrometer, require focal-plane mechanical stability on the order of 1μm; similar requirements are proposed for instruments such as cryogenically cooled Ge-detector gamma-ray spectrometers.

VIBRATION CHARACTERISTICS OF STIRLING COOLERS

In characterizing cooler-generated vibration it is useful to speak in terms of the peak vibratory force imparted by the cooler into its supports when rigidly mounted. This force is the reaction force to moving masses within the cooler that undergo peak accelerations during various phases of the cooler's operational cycle. The accelerations can be from controlled motion such as the reciprocating sinusoidal motion of the Stirling compressor piston and displacer, or natural vibratory resonances of the cooler's elastic structural elements. Although no formally agreed upon requirements exist for acceptable vibratory force levels, a value on the order of 0.2 N (0.05 lbs) has gained acceptance as a reasonable design goal.

To help quantify and understand the force levels generated by present cooler designs, JPL has developed the 6-degree-of-freedom force dynamometer facility shown in Figure 1. This dynamometer has a frequency range from 10 to 500 Hz and a force sensitivity from 0.005 N (0.001 lb) to 445 N (100 lbs) full scale. During operation the forces and moments generated about each of the cooler's axes are available simultaneously for real-time quantitative analysis.

Figure 2 illustrates typical vibration-force spectra measured at JPL[4] for single and back-to-back 800 mW 80 K BAe Stirling compressors with low-distortion drive electronics. Notice that considerable vibratory force is present at upper harmonics of the 40 Hz drive frequency and that these upper harmonics do not readily cancel when two coolers are run phase-locked back-to-back for momentum cancellation.

Fig. 1. JPL's force dynamometer facility with BAe coolers undergoing vibration characterization

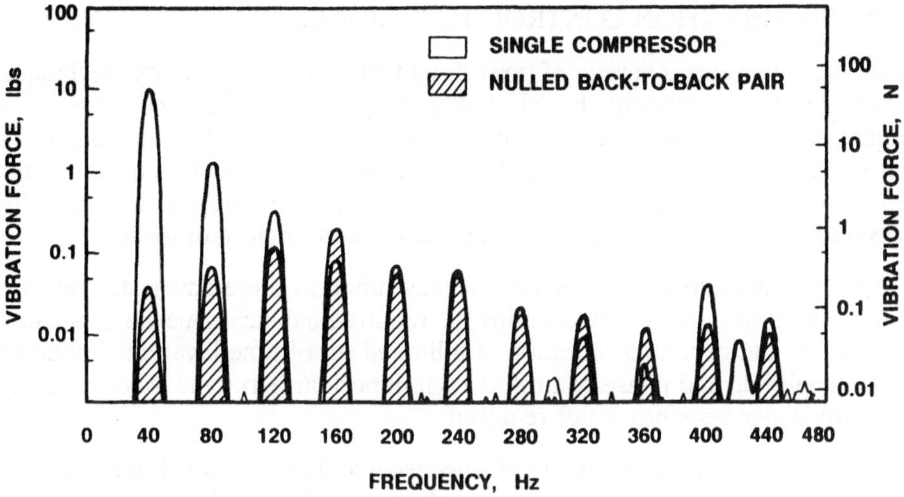

Fig. 2. Vibration force spectra for single and back-to-back BAe 80K compressors

Figure 3 illustrates similar data obtained for single and back-to-back BAe Stirling displacers that mate to the above compressors. Notice that the displacer vibration level is similar to that of the compressors despite its significantly (20X) smaller moving mass. It is hypothesized that this greater relative noise level in the displacer is the result of the displacer being predominantly driven by the gas pressure wave from the compressor, and only secondarily modulated by the displacer drive electronics for accurate phase and stroke control.

Although these vibration levels for back-to-back coolers are near the 0.05 lb goal, the level of vibration cancellation displayed is highly dependent on careful manual nulling with the visibility provided by the force dynamometer and spectrum analyzer. In a flight instrument, means must be found to actively provide this nulling function over the life of the instrument, and hopefully to additionally lower the remaining vibration through incorporation of advanced vibration control techniques.

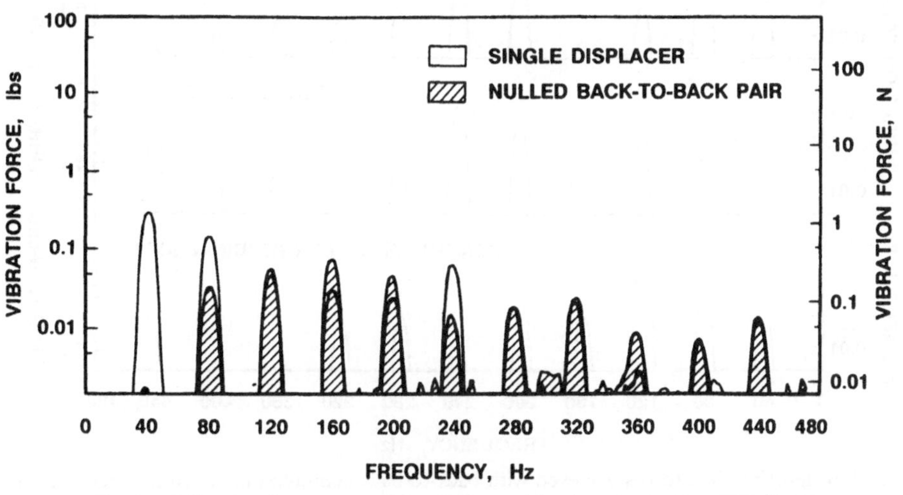

Fig. 3. Vibration force spectra for single and back-to-back BAe 80K displacers

ADVANCED VIBRATION CONTROL TECHNIQUES

In the last five years a variety of powerful adaptive vibration control techniques have been developed, generally based on digital signal processing techniques. These techniques[7] have been addressed to applications as widely varied as automotive noise control and control of vibration of rotating machinery. The general concept is to drive a servo-mechanism in such a manner as to cancel the objectionable vibration as measured by appropriate transducers at one or more important locations.

The back-to-back space cryocooler is ideally suited to these advanced control techniques because it inherently contains the required balancing mechanism (the second cooler) and generates vibration at a limited set of time-invariant frequencies; thus only amplitude and phase control at each important harmonic is required, and adaptive frequency response is not required.

As shown in Fig. 4, the feasibility of selectively nulling individual harmonics has been demonstrated[4] using a technique whereby the fundamental oscillator drive to one of the two back-to-back coolers is modified slightly with the addition of low-level phase-locked signals at multiples of the drive frequency. The amplitude and phase of each of these harmonics is manually adjusted to cancel the residual vibration between the two coolers at the selected frequencies. With the assumption that this nulling condition is stable, one first-order method of vibration control is to permanently incorporate this modified sine-wave into the cooler drive electronics; to deal with operational dependencies, a separate set of nulling parameters can be stored for each important cooler operating state. A recognized limitation of this approach is that, in the post-launch thermal-vacuum space environment or as the cooler ages, the required nulling parameters may change.

One approach to assuring long-term vibration cancellation is to incorporate some form of adaptive selection of the nulling parameters based on continuous or periodic feedback of real time vibration data. Such methods[7] and other digital control techniques under active development elsewhere within the cryocooler community[8] are showing considerable promise in early feasibility demonstrations. The possible

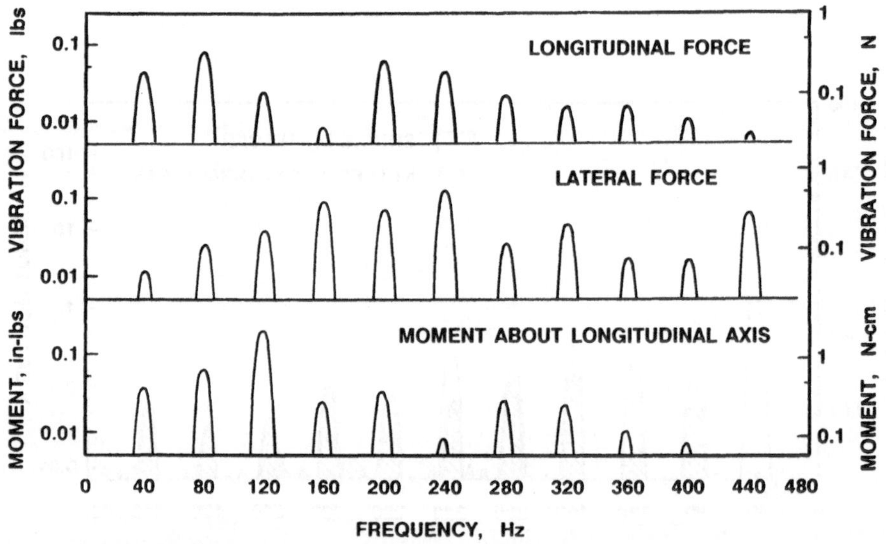

Fig. 4. Low-level residual forces achieved with back-to-back compressors nulled at three harmonics (40, 120, and 160 Hz)

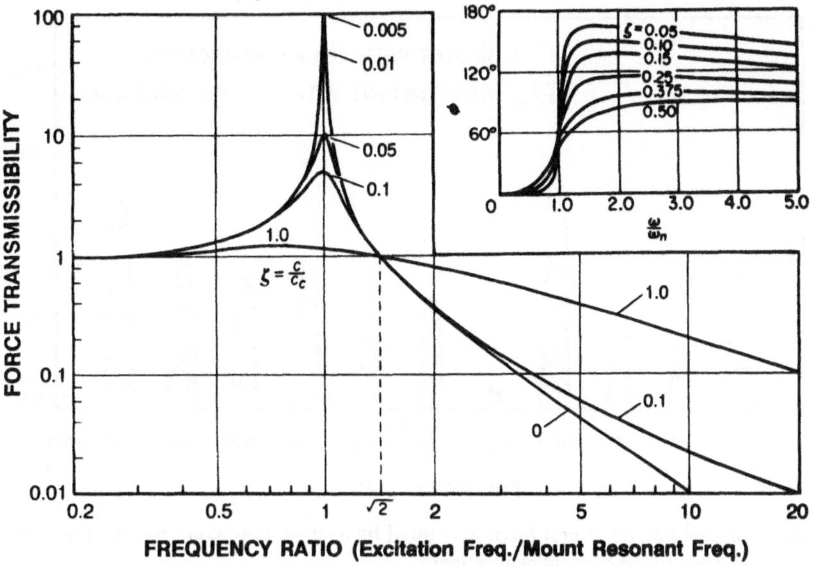

Fig. 5. Amplification of cooler-support interface forces as a function of frequency ratio and damping

limitation of these approaches is the extensive growth in electronics complexity associated with extensive digital signal processing. Selection of the appropriate feedback signal -- compressor piston position, vibration force, or cooler acceleration -- is important, as is making the adaptive control insensitive to disturbances originating from outside the cryocooler (such as from a nearby cooler).

VIBRATION CONTROL THROUGH PASSIVE TECHNIQUES

Although significant advances are being made in reducing cooler vibration through advanced control techniques, passive isolation and damping are also high-leverage and attractive means of minimizing the deleterious effect of cooler vibration. Because of the continuous--as opposed to transient--nature of cooler excitation, the vibration amplitude of instrument elements will increase without limit until the level of excitation energy per cycle matches the energy dissipation due to internal damping. Figure 5 displays this classical response relationship for a single-degree-of-freedom system as a function of the ratio of the drive frequency (cooler vibration harmonic) to the resonant frequency of the responding system (science instrument resonance). For damping ratios between 0.005 and 0.05, typical of aerospace structures, the cooler vibration force is seen to be amplified by factors from 10X to 100X. Unfortunately the level of damping is often lowest for cryogenic structures in vacuum environments; thus, without special consideration these 10X to 100X amplifications are likely to occur in space cryocooler applications.

To quantify the improvement possible with passive techniques, the force amplitude response of a BAe Stirling compressor was carefully measured in the lateral (normal to the piston axis) direction when mounted on a representative rigid aluminum structure with a first mode resonant frequency around 160 Hz. The cross-hatched curve in Fig. 6 displays the order of magnitude vibration reduction obtained when an simple viscoelastic passive damping device was attached to the exterior of the cooler compressor. This relatively high level of improvement, combined with the simplicity and broad-spectrum applicability of passive damping, suggests that incorporation of passive damping techniques should be carefully considered in space-science instruments.

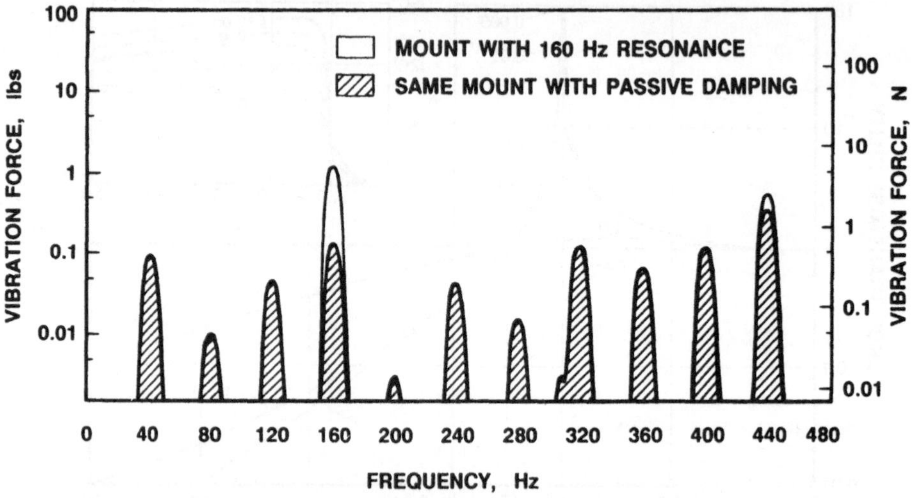

Fig. 6. Effect of damping on lateral force generated by a single BAe compressor mounted on a structure with 160 Hz resonant frequency

As an alternative, or complement, to added damping, Fig. 7 displays the vibration response reduction achieved by optimally positioning the structural resonant frequency midway between cooler drive harmonics, thus minimizing the cooler-structure cross-coupling. The problem is that the close spacing of the cooler vibration harmonics severely constrains the maximum uncoupling that can be achieved to about a 10% frequency offset; as can be noted from Fig. 5, this offset still corresponds to a force amplification factor of around 5x. An additional concern is the stability and predictability of the instrument structural resonant frequencies in the post-launch environment.

A third passive vibration suppression technique is classical vibration isolation achieved by suspending the cooler from the instrument structure with a very compliant mount. This involves operating on the right-hand side of Fig. 5, with the resonant frequency of the cooler/support system well below the cooler excitation frequency. This approach replaces transmitted force between cooler and instrument

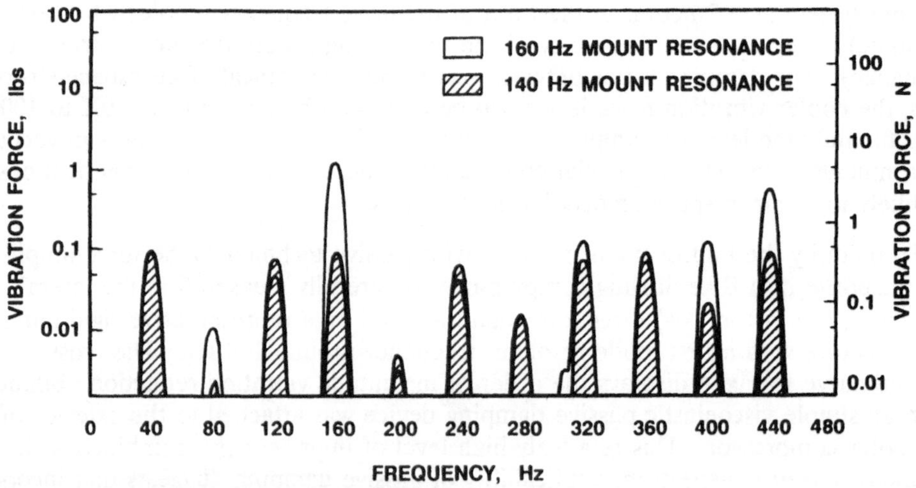

Fig. 7. Effect of offsetting structural resonance to 140 Hz so as not to couple with BAe compressor harmonic at 160 Hz

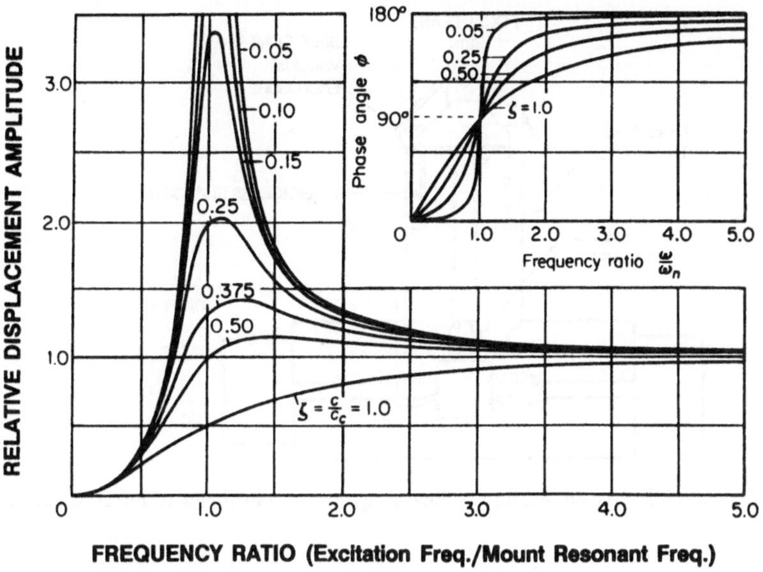

Fig. 8. Displacement amplitude response of soft-mounted cooler as a function of isolation stiffness and damping parameters

with near-constant relative movement between the two as described by Fig. 8. For low damping, the transmitted force with such a system reduces to the spring constant of the suspension system times this relative movement; the force thus falls off with increasing frequency ratio as shown in Fig. 5.

The challenge with passive vibration isolation is to achieve both vibration isolation and launch load survivability without requiring latches during launch. To prevent excessive coupling to launch vehicle resonances, many spacecraft (eg. the NASA Eos platforms) prohibit instrument resonances below 50 Hz; this greatly restricts low-frequency vibration isolation on such systems, or may necessitate complex latching systems. For the displacer, passive isolation is further complicated by the need to minimize motion of the cold finger tip with respect to the remainder of the detector cryostat assembly so as to minimize vibration conducted directly down the cold plumbing to the detector. This is discussed further in the next section.

COLD FINGER VIBRATION ISSUES

Because the cooler cold finger must be attached to the sensitive instrument detector by a high conductivity thermal link, transmission of vibration from the cold finger to the detector is also an important issue that must be carefully addressed. In a representative cryostat assembly, such as shown in Fig. 9, resonant frequencies are likely to occur near the strong harmonics of the cooler drive frequency because of the substantial mass of the thermal conductors in combination with the minimum-gage structural supports required to achieve high levels of thermal isolation. This likelihood of low-frequency resonances increases the probability of a highly resonant response to cold-finger vibration input.

For both vibration isolation as well as accommodation of differential expansion motions upon cooldown, flexible foil or wire thermal links -- shown in Fig. 9 -- are normally used to decouple the cold finger motion from the thermal link. However, even with this good isolation, minimum cold-finger vibration remains important.

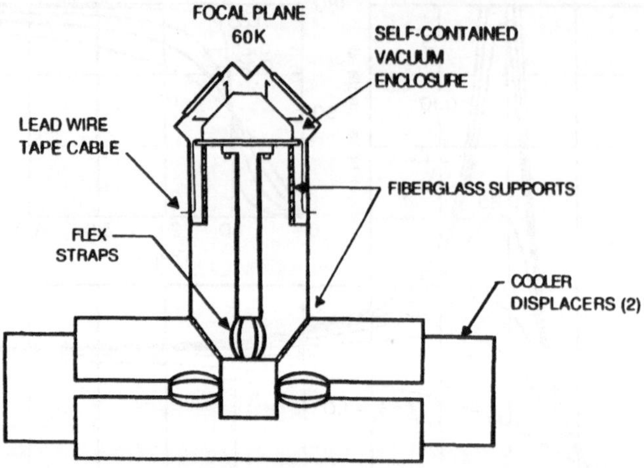

Fig. 9. Example cryostat assembly integrating cooler cold finger to infrared detector

To assess the expected level of cold-tip vibration with a rigid displacer mount, micro-accelerometers were attached to the tip of the BAe 80K cooler cold finger. Displacement spectra, computed from the measured acceleration spectra, are presented in Fig. 10 for both the lateral and longitudinal (finger-axis) directions. The larger 2.5 μm motion in the finger-axis direction at 40 Hz agrees well with the computed dilation of the cold finger in response to the 40 Hz fluctuating pressure of the Stirling cycle.

SUMMARY

Meeting the performance goals of near-term space-science instruments places demanding requirements on long-life space Stirling-cycle coolers. One of the most challenging requirements is achieving acceptably low levels of vibration of the instrument detector and sensitive optical and electronic elements. Advanced development efforts have begun to address these challenges with a broad spectrum of both passive and advanced active control techniques. Major reductions in cooler generated vibration are expected as these concepts mature into flight hardware for the emerging space-science instruments such as the Eos AIRS.

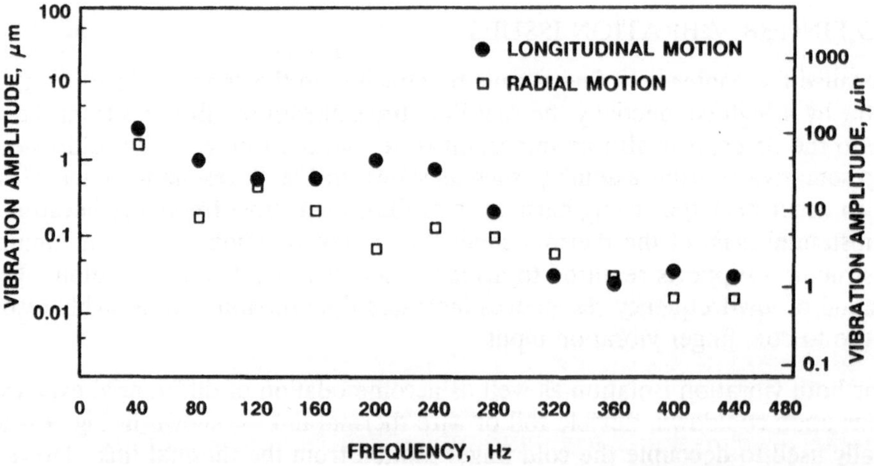

Fig. 10. Measured vibration amplitude at BAe cold finger tip with rigidly mounted displacer

ACKNOWLEDGEMENT

The work described in this paper was carried out by the Jet Propulsion Laboratory, California Institute of Technology, under a contract with the National Aeronautics and Space Administration. The financial support of JPL's Eos-AIRS instrument is graciously acknowledged.

REFERENCES

1. S.T. Werrett, et al., "Development of a Small Stirling Cycle Cooler for Space Applications", Adv. Cryo. Engin., Vol. 31 (1986), pp. 791-799.
2. G. Davey and A. Orlowska, "Miniature Stirling Cycle Cooler", Cryogenics, Vol. 27 (1987), pp. 148-151.
3. C.A. Lewis, "Long-life Stirling-cycle coolers for Applications in the 60-110K Range: Vibration Characterization of Thermal Switch Development", SAE Paper 891496, Society of Automotive Engineers, Warrendale, PA (1989).
4. R.G. Ross, Jr., D.L. Johnson and R.S. Sugimura, "Characterization of Miniature Stirling-cycle Cryocoolers for Space Application", Proceedings of the 6th International Cryocooler Conference, Plymouth MA, October 25-26, 1990, David Taylor Research Center Document DTRC-91/002, pp. 27-38.
5. D.L. Johnson and R.G. Ross, Jr., "Electromagnetic Compatibility Characterization of a BAe Stirling-cycle Cryocooler for Space Application", Proceedings of the 1991 Cryogenic Engineering Conference, Huntsville, AL, June 11-14, 1991.
6. V. Kotsubo, R.G. Ross, Jr. and D.L. Johnson, "Thermal Performance Characterization of Miniature Stirling-cycle Cryocoolers for Space Application", Proceedings of the 1991 Cryogenic Engineering Conference, Huntsville, AL, June 11-14, 1991.
7. L.A. Sievers and A.H. von Flotow,"Comparison and Extensions of Control Methods for Narrowband Disturbance Rejection", Active Noise and Vibration Control--1990, ASME Winter Annual Meeting Proceedings NCA-Vol. 8, American Society of Mechanical Engineers, New York, pp. 11-22.
8. J-N. Aubrun, et al., "A High Performance Force Cancellation Control System for Linear Drive Split Cycle Stirling Cryocoolers", Proceedings of the 1991 Cryogenic Engineering Conference, Huntsville, AL, June 11-14, 1991.

A HIGH-PERFORMANCE FORCE CANCELLATION CONTROL SYSTEM FOR LINEAR-DRIVE SPLIT-CYCLE STIRLING CRYOCOOLERS

J-N. Aubrun, R. R. Clappier, K. R. Lorell, T. C. Nast, and P. J. Reshatoff, Jr.

Lockheed Research & Development Division
Lockheed Missiles & Space Company, Inc.
Palo Alto, California

ABSTRACT

Miniaturized linear-drive Stirling-cycle cryocoolers designed with noncontacting parts are ideal for long-life cryogenic cooling onboard a wide range of spacecraft. These cryocoolers consume little power, have an almost indefinite operational life, and require no expensive ground handling equipment or procedures. A major problem in applying these cryocoolers to sensitive focal-plane instruments is the vibration induced by the reciprocating motion of internal components in both the compressor and displacer. Development of a Stirling cryocooler system optimized for minimal residual vibration has been a major goal at the Lockheed Research & Development Division.

This paper describes the analysis, development, and laboratory test of a unique hybrid analog/digital control system that can reduce residual forces to the millipound range. The controller utilizes digitally synthesized waveforms that have no harmonic distortion and whose phase and amplitude can be controlled with exceptional stability and precision. A proprietary Lockheed-developed algorithm is used to force the cooler reciprocating masses to follow this waveform virtually without error. Laboratory data from force/torque load cell instrumentation that verifies system performance down to millinewton residual force levels will be presented.

INTRODUCTION

Cryogenic cooling is required for long-lifetime spaceborne instruments such as infrared and gamma-ray spectrometers. Requirements for lifetimes of 5 to 10 years are becoming common. The NASA Earth Observing System (EOS) platform will utilize numerous instruments that require cooling in the 80-K range and one system as low as 30 K. NASA's present plan to provide this long-term cooling is based on cryocoolers utilizing technology developed by Oxford University and others. These split Stirling systems use linear motors to operate. If unbalanced, the motors produce large forces which in nearly all cases are excessive for spacecraft and sensors. The maximum allowable vibration from the compressor is in the range of 0.09 to 0.22 N. This requires attenuation of about 400 below the uncompensated force, which is on the order of 37 N.

Numerous passive and active methods have been used to reduce this vibration. In some cases, such methods can reduce the vibration at the drive frequency to nearly acceptable levels. Higher harmonics, however, generally become predominant and exceed allowable limits. This paper describes the development of a unique hybrid analog/digital controller which has demonstrated suppression of the induced force to well below the required levels. The design and testing were conducted as part of a joint development program of Lucas Aerospace and Lockheed Missiles & Space Company, Inc., for long-life space coolers.

SYSTEM DESCRIPTION

A split-cycle cryocooler consists of three main components: a compressor which provides a pressure pulse into the helium working fluid, an displacer/cold finger in which heat is exchanged between the device being cooled and the cooler working fluid, and an electronic control system which provides the drive signals to the compressor and the displacer. Figure 1 is a schematic of a refrigerator system showing all three components connected together. The control system receives signals from the position sensors in the compressor and displacer and returns drive signals to both units. The compressor and displacer operate using reciprocating motions of their internal components: the piston in the case of the compressor and the stack of gauze screens inside the cold finger in the case of the displacer.

The main purpose of the cooler is to provide adjustable cryogenic refrigeration. The heat pumped is a function of the compressor piston amplitude, the displacer piston amplitude, and the relative phase between compressor and displacer motion. Because of the nonlinear nature of the fluid dynamics involved, both compressor and displacer motions must be actively controlled. In addition, to provide vibration cancellation, the motion of one compressor/displacer system must be precisely compensated by another system. Inertial and gas forces generated by a single compressor can be as large as 70 N, and up to 13 N for the displacer.

VIBRATION DYNAMICS

Because of the dynamic laws of action and reaction, the forces responsible for the piston motion are also reacting on the body of the cooler, thus creating an important source of vibration. The most important component is the electrodynamic force F_e created by the linear motor driving the piston. A second force results from the compression of the gas, creating an effective spring that reacts on the cylinder top. Less important components are the spiral spring supporting the piston (usually 20 times softer than the "gas" spring), the motion of the mass of gas through compressor output, and the piston/cylinder gap or friction effects.

Preliminary studies have lead to a simplified model for the compressor dynamics. The basic dynamic equations of the system are:

For the compressor piston: $$M_p \ddot{x}_p = F_v \quad (1)$$

For the total cooler: $$M_c \ddot{x}_c = -F_v \quad (2)$$

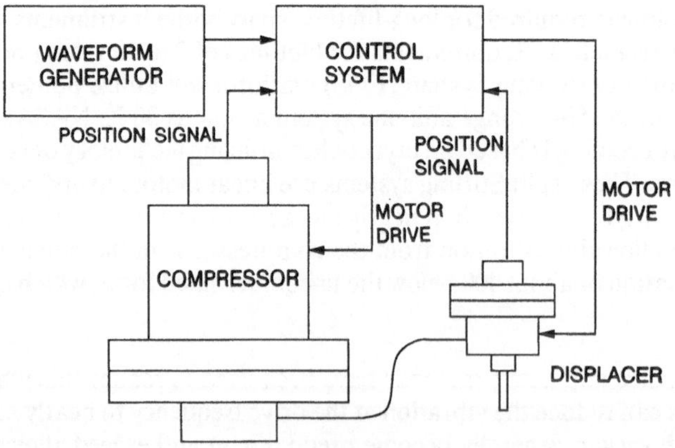

Fig. 1. Refrigerator system.

M_p may include the moving mass of the spiral springs and the gas. The force F_v that moves the piston is in general a function of the relative displacement $x = (x_p - x_c)$ and is the same force that moves the whole cooler, producing the undesired vibration. Since the mass M_c of the cooler is usually much larger than the piston mass M_p, the cooler displacement x_c is much smaller than the piston displacement x_p, and the inertial displacement x_p can be equated with the relative displacement x.

The piston dynamics can be approximated by a mass-spring system in series with a damper to reflect the gas dynamics. The flexural support of the piston is modeled by K_s, the gas compressibility is modeled by K_g, the gas viscous properties are modeled by D_g, and the gas dynamics are represented by the state variable x_1. With these assumptions, Eqs. (1) and (2) become:

$$M_p \ddot{x} = F_v \tag{3}$$

$$D_g \dot{x}_1 = K_g (x - x_1) \tag{4}$$

$$F_v = F_e - K_s x - K_g (x - x_1) \tag{5}$$

These equations show that the vibration is entirely determined by the motion of the piston given by Eq. (3). Thus, Eq. (3) offers a clue about how to reduce this vibration: if another mass is moved in exactly the opposite way, except perhaps for a scale factor, the total force on the system can be zeroed out.

PERFORMANCE REQUIREMENTS

The performance requirements of the system are dictated by both thermodynamic and vibration considerations. In the system described in this paper, the piston has a mass of 0.167 kg and a maximum stroke of ±3.5 mm. Assuming an operating frequency of 55 Hz, the inertial force required to accelerate the piston is approximately 70 N. Additional force is required to compress the gas and overcome viscous friction. The current design assumes an actuator capable of producing a maximum force of about 80 N.

Temperature is a strong function of the compressor stroke. Under normal operating conditions, the sensitivity of temperature to stroke is about 28.6 K/mm. If the temperature output of the cooler is required to be held to ±1 K, then the control system must control the stroke to ±35 μm, or about 0.5% of full stroke.

Temperature is also a function of the displacer stroke. Under normal operating conditions, the temperature sensitivity is about 36 K/mm. Thus, the displacer control system must be able to maintain the stroke amplitude to better than 28 μm, or about 1.8% of full stroke.

In addition, the temperature is also sensitive to the phase difference between compressor and displacer motion. Experiments have shown that optimal efficiency is obtained with a phase difference of 65°. If the phase varies by ±5°, the resulting change in temperature is 0.5 K. Thus, the displacer must be properly phased to within ±10° of the reference sine wave applied to the compressor.

The most stringent requirements placed on the control system come from the vibration cancellation requirements. Dynamical analysis of a system of two identical masses of mass M, moving in opposite sinusoidal motion, show that the residual root mean square (RMS) force F_r acting on the supporting structure is a function of the maximum stroke x, the driving frequency $\omega = 2\pi f$, the amplitude error dx, and the phase matching error between the two masses $d\phi$. The amplitude and phase of a matched pair must be controlled with enough accuracy to satisfy the following condition:

$$\left(\frac{dx}{x}\right)^2 + \left(d\phi\right)^2 < \left(\frac{Fr}{M\omega^2 x}\right)^2 \tag{6}$$

Table 1. Control System Accuracy Requirements

Component	Thermal Accuracy		Vibration Accuracy	
	Amplitude	Phase	Amplitude	Phase
Compressor	0.5%	7°	0.23%	0.13°
Displacer	1.8%	7°	1.4%	0.08°

If the temperature must be held to within 1 K, the compressor residual force to within 0.22 N, and the displacer residual force to within 0.027 N, then control must achieve the position accuracies shown in Table 1.

The control system performance is clearly driven by vibration cancellation. The required level of performance will be difficult to achieve using a conventional controller. In addition, the nonlinear nature of the system tends to produce harmonics which increase the vibration. The uncompensated contribution to vibration force of the nth harmonic is:

$$F_n = F_0 H_n n^2 \tag{7}$$

where F_0 is the maximum force applied to the piston, and H_n is the relative amplitude of the harmonic. Thus, the force produced by the first harmonic is 4 times greater than that of the fundamental if they both have the same amplitude in position. This places additional constraints on the control system.

ACTIVE FORCE CANCELLATION

A number of techniques have been used to minimize the forces generated during cryocooler operation. Tuned passive dampers on springs that resonate at the motor drive frequency can reduce the vibration by a factor of 20. Mounting two compressors or displacers back-to-back and driving them with appropriately similar waveforms can provide even better results. If the use of a second compressor or displacer is undesirable for systems reasons, the vibration can be reduced by an active balancer. This is a second linear motor/mass/spring system that produces the same force as a second compressor or displacer.

These techniques are either passive or open loop. That is, the current drives to the various motors (compressors, expanders, or balancers) are sine waves whose amplitude and phase have been adjusted for best results. The resulting motions of the masses, however, never follow their drive signals. The residual force is mostly due to the fact that the two opposing masses do not respond identically to a given drive signal because of the nonlinearities of the gas dynamics, magnetic materials, and electronic drivers. Some of the major components of the residual force are the harmonics that result from this nonlinear response.

In the approach developed by Lockheed, feedback techniques force the motion of the moving masses to follow identical waveforms. In each unit, a sensor measures the precise position of the mass, and the control system makes sure that this motion exactly follows the commanded waveform. As discussed above, the vibratory forces can be related almost entirely to the inertial motion of the opposing masses; thus, cancellation is actively enforced by controlling this motion. This technique eliminates the harmonic generation, since the control system can now overpower the nonlinear effects that would have otherwise produce differences in mass motions. If a pure sine wave is used for command, then harmonics are never even produced. But if other waveforms must be used, their harmonic content will also be canceled.

The key in this new approach is the accuracy with which the control system is able to enforce motion. As shown above, a classical servo loop must be very accurate to achieve the necessary motion control. This requires large bandwidths, which can pose difficult design problems. The solution developed by Lockheed is a combination of analog and digital adaptive servo loops that form a position motion control system which can meet these requirements.

Long-term changes in the systems dynamics and the electronic components can alter the amplitude or phase relationship between two back-to-back mass systems, and the original setting for commanding the moving mass motions may no longer be optimal. The ultimate closed-loop technique for vibration cancellation is to directly sense the vibration and adjust the motion of the compensating system accordingly. In this case, the residual vibration as measured by an accelerometer or a load cell is used to correct the motion of one of the masses to exactly null the measured vibration.

EXPERIMENTAL SETUP

All of the vibration experimental data were taken from cryocoolers working under normal refrigeration conditions. The compressor residual force data were obtained from a pair of compressors mounted back-to-back on an aluminum fixture suspended by springs. Three piezoelectric accelerometers were mounted around the perimeter of the compressor flange with 120° spacing. The two displacers were mounted on a similar platform, with only the two transfer tubes connecting the assemblies. Figure 2 shows the setup, including the instrumentation and power cables as well as the vacuum shrouds that allow cryogenic operation.

Some of the tests were repeated on a single-axis dynamometer using three load-cell force transducers. Figure 3 shows the back-to-back compressors installed on the dynamometer. The dynamometer could not accommodate the compressors and the displacers simultaneously, and was therefore less convenient to use. As a result, it was used only to confirm the data from the accelerometer test stand. For future work, we plan to build a six-axis dynamometer.

The force information was based on the analog average of the outputs of the charge amplifiers for each of the three accelerometers. This average was passed through a simple noise filter and analyzed by a Hewlett Packard 3562A Dynamic Signal Analyzer. This device measured the force amplitude at the motor fundamental frequency and the first six harmonics. It also formatted the data for digital plotting and floppy-disk storage.

Fig. 2. Experimental setup using accelerometers.

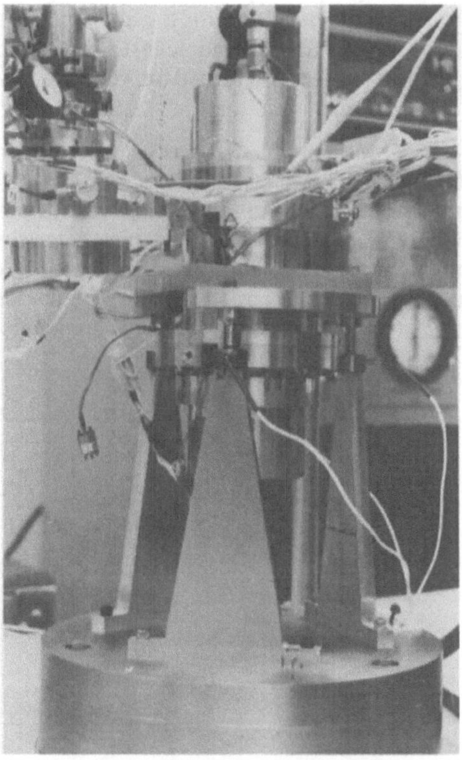

Fig. 3. Experimental setup using a dynamometer.

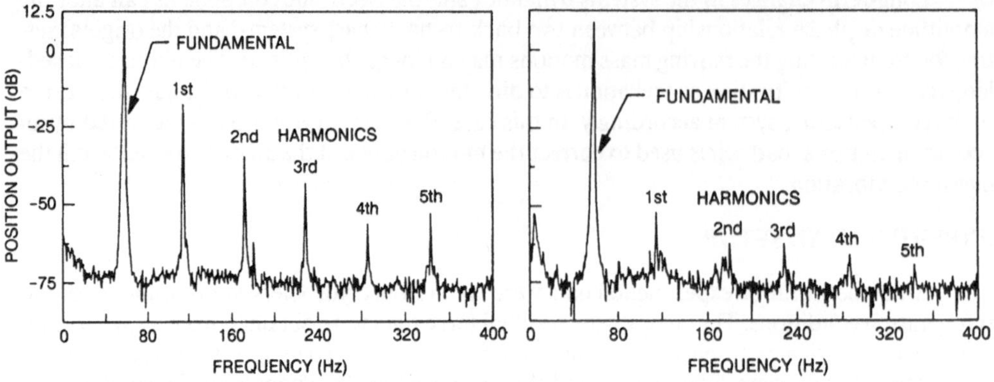

a. Analog control only. b. Analog/digital control.

Fig. 4. Piston position signal spectrum.

Motor axial force was measured with the accelerometers mounted directly to the aluminum fixture. Transverse force was measured by mounting the accelerometers on aluminum blocks that held the accelerometers normal to this first axis. Torque measurements about the common piston axis were made by rotating the accelerometer mounting blocks.

The dynamometer output was processed in the same way as the accelerometer output by the same electronics. The dynamometer, however, was used only to measure axial forces.

EXPERIMENTAL RESULTS

Laboratory data were gathered to demonstrate the effect of the hybrid analog/digital control system on the motor piston position and the degree to which residual vibration can be reduced by tight piston position control. Figure 4a shows the stroke amplitude spectrum of compressor piston position with the analog control system operating, but with the digital section turned off. The total harmonic distortion is -25 dB or 5.6%. Figure 4b shows the same situation with the digital section turned on. Here the total harmonic distortion is -59 dB or 0.11%. This is a 50 to 1 reduction in total harmonic distortion.

Vibration reduction was measured by first operating a single, uncompensated compressor. Figure 5 shows the results from a compressor running with a 4.0-mm stroke. The resulting axial force is 37.7 N. In Fig. 6, two compressors were operated back-to-back using only the analog control system to control the piston position, but with the second compressor stroke set for minimum vibration. The largest-amplitude frequency component has been reduced by 164 to 1, and the root sum square (RSS) of the fundamental and the first six harmonics has been reduced by 151 to 1. The largest frequency component is 0.23 N.

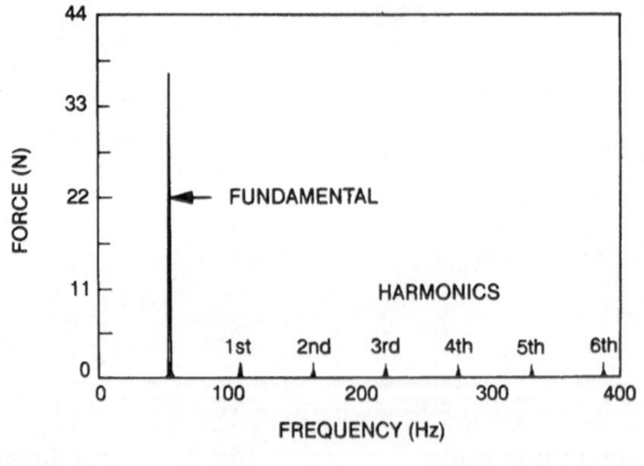

Fig. 5. Axial force, uncompensated compressor.

Table 2. Summary of Vibration Experimental Results (RSS of Fundamental and First Six Harmonics)

Measurement	Uncompensated Cryocooler (N)	Analog Controller		Hybrid Controller	
		Force (N) Torque (N-m)	Ratio	Force (N) Torque (N-m)	Ratio
Compressor(s) Axial	37.7	0.249	151	0.050	752
Compressor(s) Transverse	0.036	0.059	0.6	0.030	1.2
Compressor(s) Torque	0.0037	0.0039	0.9	0.0039	0.9
Displacer(s) Axial	2.51	0.014	185	0.005	503
Displacer(s) Transverse	0.008	0.003	3.0	0.002	3.7
Displacer(s) Torque	0.00125	0.00009	14	0.00006	19

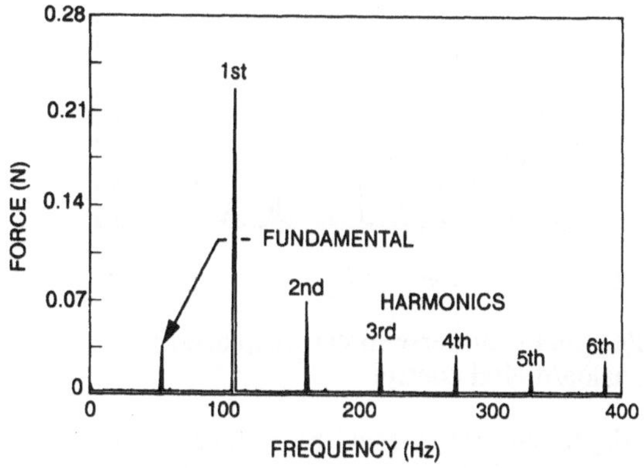

Fig. 6. Residual axial force using analog control only.

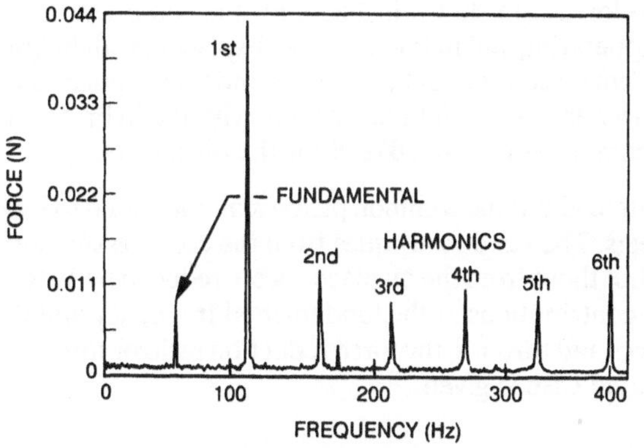

Fig. 7. Residual axial force using analog/digital control.

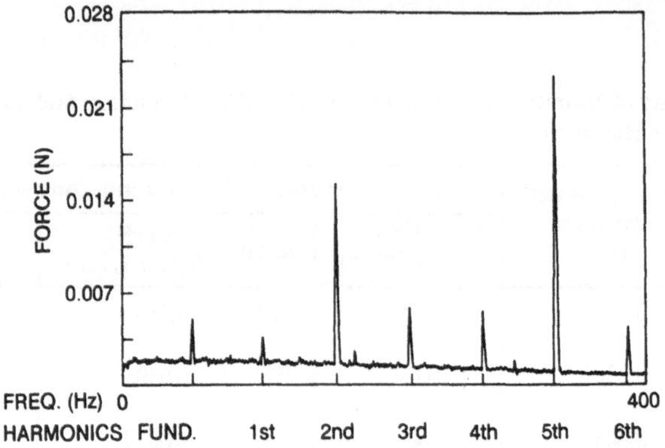

Fig. 8 Residual transverse forces; compressors with analog/digital control.

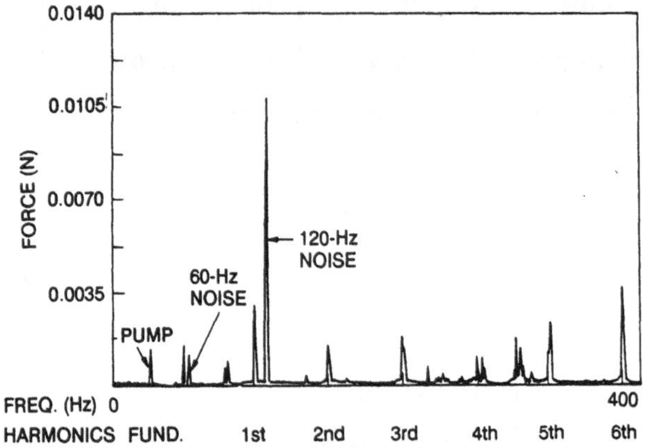

Fig. 9. Residual transverse forces; displacers with analog/digital control.

With the addition of a digital control system, the residual force is further reduced as shown in Fig. 7. Now the largest-amplitude frequency component has been reduced by 864 to 1, and the RSS of the fundamental and the first six harmonics has been reduced by 752 to 1. The largest frequency component is 0.044 N.

The residual transverse forces are shown in Fig. 8 for the compressors with full hybrid analog/digital control system operating and in Fig. 9 for the displacers with hybrid system operating. The hybrid system does not directly affect these forces and thus reduces them at most by a factor of 3.7. These forces, however, are small to begin with; with the hybrid system operating, they are only 0.03 N for the compressors and 0.002 N for the displacers.

The torques were measured about the common piston axis and found to be very low. Table 2 shows the results of these tests. The residual torques from the compressors were not reduced by the hybrid control system, but those from the displacers were reduced by 19 to 1. In each case, the results are the RSS of the contributions at the fundamental frequency and the first six harmonics. In addition to the forces and torques, the force reduction ratio or torque reduction ratio compared to the uncompensated case is given.

CONCLUSION

The experimental results demonstrate that the hybrid analog/digital control system meets the thermodynamic and vibration requirements with a good margin. The analog system almost meets them, but even if it were improved, there would be little or no margin. In an environment where consistant operation is required for long periods of time, the hybrid system is clearly superior.

COLD-TIP OFF-STATE CONDUCTION LOSS OF MINIATURE STIRLING CYCLE CRYOCOOLERS

V. Kotsubo, D.L. Johnson and R.G. Ross, Jr.

Jet Propulsion Laboratory
California Institute of Technology
Pasadena, California

ABSTRACT

For redundant miniature Stirling-cycle cryocoolers in space applications, the off-state heat conduction down the coldfinger of one cooler is a parasitic heat load on the other coolers. At JPL, a heat flow transducer specifically designed to measure this load has been developed, and measurements have been performed on the coldfinger of a British Aerospace 80 K Stirling cooler with the tip temperature ranging between 40 and 170 K. Measurements have also been made using a transient warmup technique, where the warmup rates of the coldtip under various applied heat loads are used to determine the static conduction load. There is a difference between the results of these two methods, and these differences are discussed with regard to the applicability of the transient warmup method to a non-operating coldfinger.

INTRODUCTION

An increasing number of astronomical and earth-observing space detectors have cryogenic cooling requirements, leading to a high demand for miniature space-qualified cryocoolers. These coolers must not only have high reliability and 5 - 10 year lifetimes, but must also be interfaceable to sensitive space instruments. The frontrunner in the technology to provide cooling in the 50 K - 150 K temperature range is a split-Stirling-cycle cryocooler based on a design originated at Oxford University[1,2]. This cooler uses tight tolerance, non-contacting clearance seals on the piston and displacer, linear motors, and spiral diaphragm flexure springs to maintain alignment of the piston and displacer within their clearance seals. The Jet Propulsion Laboratory (JPL) has a comprehensive test and analysis program underway to understand this cryocooler with regard to reliability and integration into space-instrument systems.[3] Among the issues being addressed are output vibration levels[4], EMI[5], thermal performance, life testing, and developing non-obtrusive diagnostic methods for detecting failure mechanisms.

One issue regarding systems integration is the off-state (non-operating) coldtip conduction heat load, where, in systems using redundant coolers without heat switches, the heat conducted down the coldfinger of a non-operating cryocooler is a parasitic load on the operating coolers. For various versions of the Oxford-heritage 80 K cooler, the reported values of this heat load range from 250 mW to 550 mW for coldtip temperatures near 55 K . Because this level of parasitic heat load is a substantial fraction of the available cooling power at these temperatures, accurate data of the heat load is essential for systems design.

A static conductance measurement technique has been developed by JPL to measure the coldfinger conduction load in a test setting that simulates the flight configuration of redundant coolers. In the test configuration, the coldfinger of the test cooler is enclosed in a vacuum housing together with the coldfinger of a second cooler that is used to cool the coldtip of the non-operating test cooler down to typical flight operating temperatures. The heat load of the test cooler is measured using an absolute heat flow transducer specifically designed for this application. This method has been applied to the BAe Oxford-heritage 80 K cooler, and the experimental details and results are presented in the first section of this paper.

A series of experiments have also been performed using the transient warmup technique[6] commonly used in the cooler industry. In this method, the cooler is first operated to cool its coldtip to its base temperature. The cooler is then turned off, and a known heat load is applied to the coldtip. By measuring the warmup rates with several different applied heat loads, the off-state conduction load can be determined. These measurements were made on the same BAe cooler used in the static conductance experiments in order to make a comparison of these two techniques. The second section of this paper presents these measurements, and discusses the differences in the results between this method and the static conduction method.

STATIC CONDUCTANCE MEASUREMENTS

The static conductance measurements involved the use of a second cryocooler to cool the coldfinger of the non-operating BAe cooler, with the two coldtips thermally linked to each other through the absolute heat flow transducer. The parasitic heat flow down the test coldfinger passes through the transducer enroute to the other cooler, providing a direct measurement under true steady-state conditions.

In the experimental arrangement, shown in Fig. 1, a 10 K Gifford-McMahon (GM) cryocooler, with its temperature regulated using a resistive heater and temperature controller, provided the cooling for the BAe coldtip. Both coldfingers were contained within a vacuum housing and linked through a thermal strap made of parallel strips of copper foil. To minimize the radiation heat leak, the BAe coldfinger was loosely wrapped with several layers of 0.009 mm doubly aluminized embossed Kapton. Because of the cryopumping of the GM cryocooler, the vacuum within the housing was in the 10^{-6} - 10^{-7} Torr range.

The heat flow transducer, shown in Fig. 2, is a thermal shunt made of a short section of german silver, a high thermal resistivity material. By measuring the temperature drop (ΔT) across the shunt, the heat flow Q_p can be determined from:

$$Q_p = \kappa \Delta T$$

where κ is the thermal conductance of the resistive element. Two copper end-pieces were silver soldered to the ends of the german silver, insuring isothermal

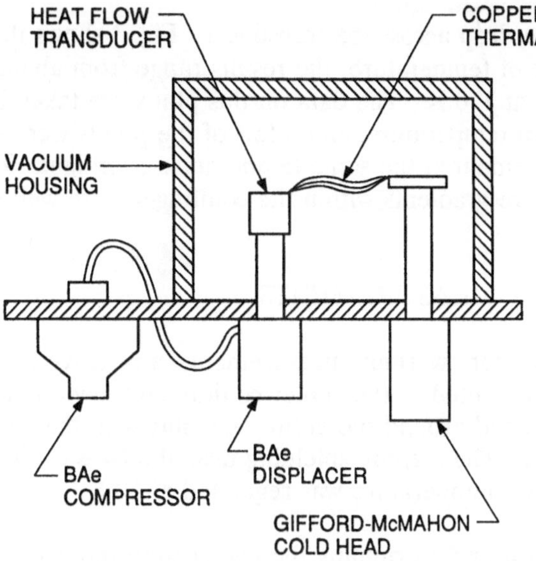

Fig. 1. Schematic of the static conduction test apparatus.

boundaries to create uniform heat flows across the resistive element. Diode temperature sensors (Lakeshore DT-470) were epoxied into each end cap, and an aluminum radiation shield enclosed the transducer. The radiation shield was heat sunk to the side of the transducer thermally anchored to the GM cryocooler so that the radiation heat load did not pass through the german silver shunt. The transducer was calibrated by thermally anchoring only one end to the GM cryocooler, applying measured heat loads with a resistive heater to the other end, and recording the resulting temperature differences.

Data were taken by regulating the GM cooler at a particular temperature, waiting several minutes for the BAe coldtip temperature to stabilize, and then re-

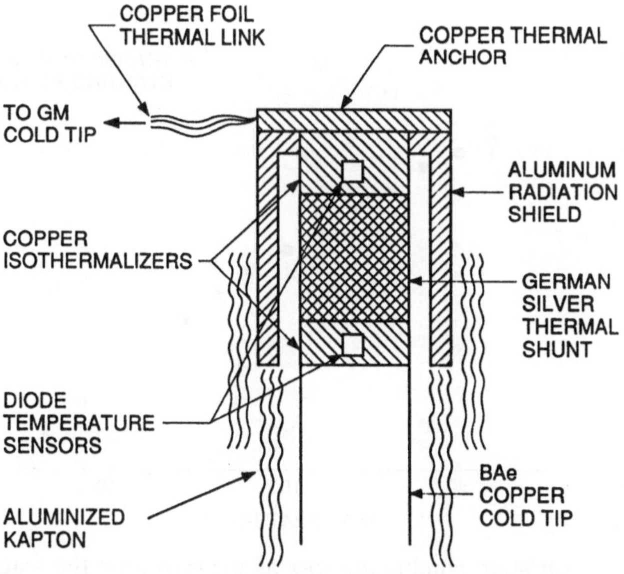

Fig. 2. Schematic of the absolute heat flow transducer as mounted on the BAe coldtip.

cording the temperature drop across the transducer. Fig. 3 shows the static conduction loads as a function of temperature; the results range from about 340 mW at 40 K to about 210 mW at 170 K. The data on this plot were taken both stepping up and stepping down in temperature, and a few of the points were taken after allowing the system to remain at the same temperature overnight. The repeatability indicates that temperature gradients within the coldfinger were well stabilized.

TRANSIENT WARMUP MEASUREMENTS

To perform the transient warmup measurements, a resistive heater, imbedded in a copper block, was mounted to the cooler coldtip, and several layers of aluminized Kapton were wrapped around the entire assembly with the edges sealed with aluminized Kapton tape. The Kapton shielding near the BAe coldtip was heat sunk to the GM coldtip, whose temperature was regulated at 50 K.

The data were taken by first running the cooler until the coldtip reached 50 K, stopping the cooler, then timing the warmup over set temperature intervals with a measured heat load applied to the coldtip as it warmed up to 95 K. The results, heat load vs. $1/\Delta t$, are shown in Fig. 4, where Δt is the time required to warm the tip over a 10 K temperature range centered about the temperature of interest. For example, for the 80 K curve, Δt corresponds to warming from 75 K to 85 K. Extrapolating to $1/\Delta t = 0$ gives the off-state conduction load, and the results are plotted in Fig. 3. The loads ranged from 390 mW at 60 K to 320 mW at 90 K, and are significantly higher than that using the static conductance technique.

To understand the difference between these two results, a simple model of the warmup technique can be made. The warmup time, Δt, for a given coldtip tempera-

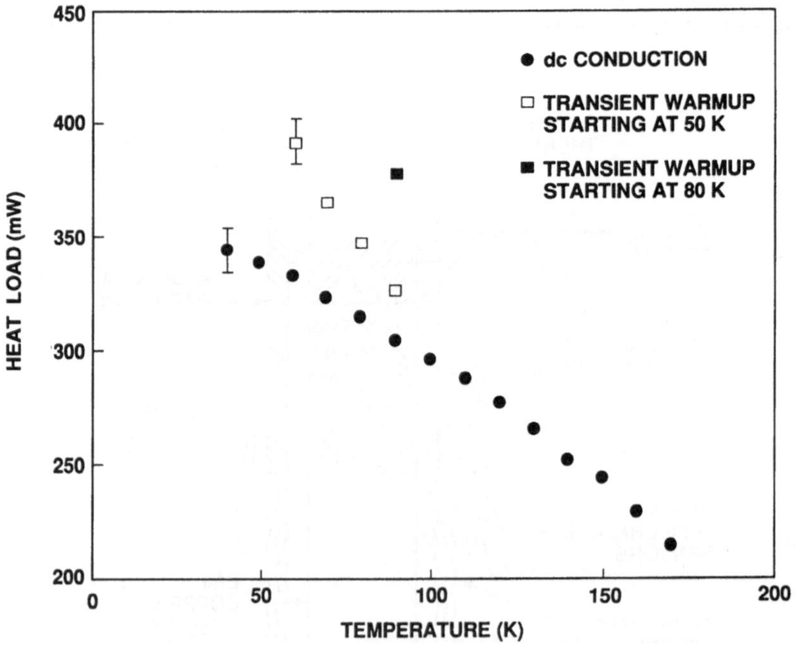

Fig. 3. Coldfinger off-state conduction loss as measured by the static conduction method and the transient warmup method.

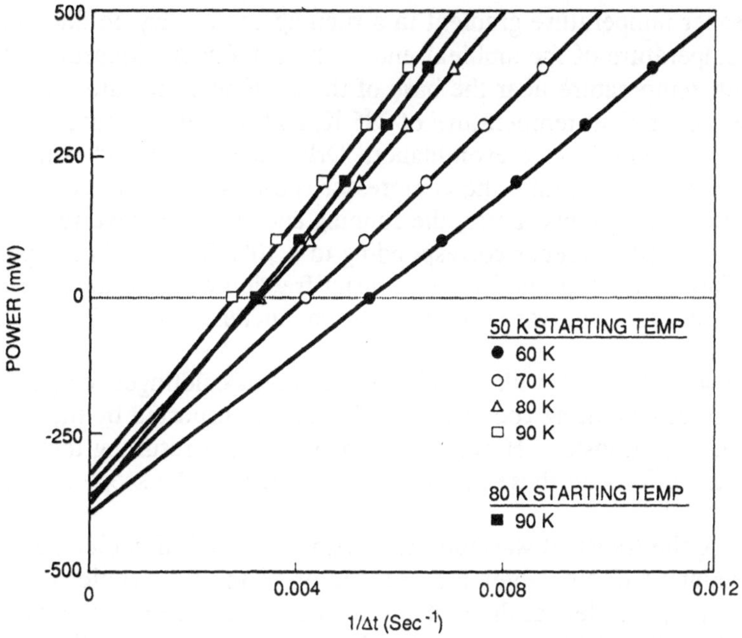

Fig. 4. Results of the transient warmup measurements.

ture rise ΔT is given by:

$$C\frac{\Delta T}{\Delta t} = Q_h + Q_p$$

where Q_h is the applied heat load, and Q_p is the parasitic heat load given by:

$$Q_p = K_{eff}\frac{dT}{dx}$$

Here K_{eff} and dT/dx are the thermal conductance and temperature gradient of the coldfinger near the coldtip. The heat capacity of the coldtip, C, is provided by the copper heater block and a copper mass soldered onto the end of the coldfinger. Because of the cold-finger MLI, the radiation heat load is assumed to be small. If dT/dx is constant within the 10 K timing intervals, independent of the applied heat load, then $1/\Delta t$ should be linear in Q_h. This linearity is apparent in Fig. 4. As a further consistency check, the slopes of the lines, which should be equal to $C\,\Delta T$, agree with estimates of the heat capacity of the copper masses and have the same temperature dependence as the specific heat of copper. These arguments suggest that dT/dx is being measured correctly with this method, and that dT/dx is higher in the coldfinger of a running cooler than in a non-operating cooler.

To investigate this, two different data sets were taken, timing the warmup between 85 K and 95 K, but with one set starting with the coldtip cooled to 50 K, and the other cooled to 80 K. These two sets of data are shown as the ■ and □ in Fig. 4, and the resulting measured heat loads are shown in Fig. 3. The run starting with the tip initially cooled to 80 K gave the higher load, while the run starting with the tip cooled to 50 K approached the static conductance values. This suggests that in warming up from 50 K to the 85 K - 95 K timing interval, the coldtip has relaxed into a temperature profile more similar to that of a non-operating cooler.

The steeper temperature gradient in a running cooler may be due in part to the higher temperature of the ambient end of the coldfinger. Measurements of the displacer body temperature near the base of the coldfinger indicate that it is 10 K - 15 K above the ambient temperature of 295 K, and the temperature of the warm end of the regenerator may be even higher. Orlowska and Davey[6] estimate that the average gas temperature within the compressor is 320 K, so if this is also the regenerator warm end temperature, the running cooler would have temperature gradients that are 10% steeper corresponding to a 10% increase in the parasitic heat load. This would account for a significant fraction of the difference between the results of the transient warmup and static conductance methods.

A running cooler is also likely to have a different coldfinger temperature profile than a non-operating cooler due to the heat transported by the moving gas and by shuttle heat transfer. However, an analysis of the resulting temperature profiles and the effects on these measurements is beyond the scope of this work.

In making the transient warmup measurements, the GM cooler was used to cool the aluminized Kapton shielding near the tip of the BAe cooler to 50 K. If the GM cooler coldtip was left at the ambient temperature of 295 K, the measured heat loads were about 30 mW higher than the above results. Other preliminary tests with differing radiation shielding arrangements also suggest an uncertainty in the amount of residual radiative heat leak, conservatively estimated on the order of 30 mW, in the experimental configurations used in these experiments.

SUMMARY

The integration of miniature split-Stirling-cycle cryocoolers into space-instrument systems requires accurate data on the off-state parasitic conduction load of the coldfinger, particularly for systems using redundant coolers without heat switches. This load has been measured for the British Aerospace version of the Oxford University 80 K cooler using a static conduction method with an absolute heat flow transducer developed at JPL. The values obtained ranged from 340 mW at 40 K to about 210 mW at 170 K.

Measurements were also performed using a transient warmup technique. This method was shown not to measure the off-state conduction load accurately, as the results were consistently higher than those from the JPL static conductance measurements. The disagreement is likely due to the steeper temperature gradient in the coldfinger of a running cooler as compared with a non-operating cooler, due in part to the higher warm end temperature of a running cooler.

ACKNOWLEDGEMENTS

This work was carried out by the Jet Propulsion Laboratory, California Institute of Technology, under a contract with the National Aeronautics and Space Administration. The support of JPL's Eos-AIRS instrument is gratefully acknowledged.

REFERENCES

1. S.T. Werrett et al., "Development of a Small Stirling-Cycle Cooler for Space Applications," Adv. Cryo. Engin., Vol. 31 (1986) p. 791.

2. G. Davey and A. Orlowska, "Miniature Stirling-Cycle Cooler," Cryogenics, Vol. 27 (1987) p. 148.
3. R.G. Ross, Jr., D.L. Johnson and R.S. Sugimura, "Characterization of Miniature Stirling-cycle Cryocoolers for Space Application", Proceedings of the 6th International Cryocooler Conference, Plymouth MA, October 25-26, 1990, David Taylor Research Center Document DTRC-91/002, p. 27-38.
4. R.G. Ross, Jr., D.L. Johnson and V. Kotsubo, "Vibration Characterization and Control of Miniature Stirling-Cycle Cryocoolers for Space Application," Proceedings of the 1991 Cryogenic Engineering Conference, Huntsville, AL, June 11-14, 1991.
5. D.L. Johnson and R.G. Ross Jr., "Electromagnetic Compatibility Characterization of a BAe Stirling-Cycle Cryocooler for Space Application," Proceedings of the 1991 Cryogenic Engineering Conference, Huntsville, AL, June 11-14, 1991.
6. A.H. Orlowska and G. Davey, "Measurement of losses in a Stirling-Cycle Cooler," Cryogenics, Vol. 27 (1987) p. 645.

ELECTROMAGNETIC COMPATIBILITY CHARACTERIZATION OF A BAe STIRLING-CYCLE CRYOCOOLER FOR SPACE APPLICATION

Dean L. Johnson and Ronald G. Ross, Jr.

Jet Propulsion Laboratory
California Institute of Technology
Pasadena, California

ABSTRACT

The intended use of Stirling-cycle cryocoolers to cool infrared and submillimeter imaging instruments on 5- to 10-year missions brings with it major challenges to cryocooler development. In particular, the voice-coil driven cryocoolers need to be electromagnetically compatible with the host instrument's detectors as well as with neighboring instruments; specifically the cryocoolers must not generate levels of interference that degrade performance or cause malfunction of the cooled imaging detectors, payload instruments, or host spacecraft.

To support the design and successful operation of NASA space instruments, the Jet Propulsion Laboratory (JPL) has an ongoing extensive cryocooler characterization, test and analysis program to identify cryocoolers capable of meeting the stringent requirements. The characterization activity focuses on sensitive performance measuring techniques for quantification of thermal performance, vibration, electromagnetic compatibility (EMC), and life-limiting reliability degradation mechanisms. This paper describes the EMC measurements of a British Aerospace (BAe) 80-K Stirling-cycle cooler. The measurements, performed in the JPL EMC test facility, include DC magnetic field characterization, radiated magnetic and electric field emissions, and conducted emissions on the internal lines between the cooler electronics and the cooler. The measurements conform to both the MIL-STD-461C specifications as well as to the specifications for the NASA Earth Observing System (Eos).

INTRODUCTION

The intended use of mechanical cryocoolers to provide continuous cooling to infrared and submillimeter imaging instruments for multi-year missions requires the suppression of the cooler's vibrational motion and electromagnetic emissions to levels that will not interfere with the instrument detector, or with neighboring instruments. The Atmospheric Infrared Sounder (AIRS) instrument, for example, which is scheduled for Platform A of the Earth Observing System (Eos), is a grating-array spectrometer using sensitive HgCdTe detector arrays cooled to 60K with multiple coolers. While the detector array's vibrational susceptibility level is quite challenging (on the order of 1 micron to prevent image blur), the detector and cooler's level of electromagnetic compatibility needed to prevent image degradation is less well understood.

As part of the JPL cryocooler characterization program[1-3] measurements of the electromagnetic signature of a British Aerospace (BAe) 80K cooler have been made to provide an early indication of the level of electromagnetic compatibility of the cooler with the host spacecraft and its payload instruments. This paper focuses on measurements of the cryocooler's DC magnetic field and the radiated AC magnetic and electric field emissions. Measurements were made with the bare cooler -- no mu-metal shielding was attempted to lower the magnetic field levels. The cooler's ground support electronics were placed outside of the measurement facilities and were not included in the radiated emissions measurements. The measurement results are compared to both military specifications (MIL-STD-461C)[4] and to the October 1990 General Instrument Interface Specifications for Eos[5]. The Eos specifications are concerned with the electromagnetic interference (EMI) that any instrument puts out, and how the EMI may affect the host observatory or neighboring instruments. Of perhaps greater concern is the effect the cooler EMI will have on the detector it is to cool, since the cooler's expander will be operating very near the detector. Thus measurements taken at short range (7 cm) are also included in this paper, and are discussed with respect to detector sensitivities.

CRYOCOOLER ELECTROMAGNETIC STRUCTURE

The cryocooler is a mechanically resonant system that operates much like a loudspeaker. Both the compressor and displacer units have spring-suspended drive assemblies driven via a moving coil in a permanent magnetic field. Mechanical motion is generated by applying an alternating current through the coil at a frequency chosen to be near the mechanical resonance of the compressor, this minimizes the required drive power. The displacer is pneumatically driven, with the linear motor used primarily to control the stroke amplitude and phase angle relative to the compressor stroke. A flight cooler is likely to be driven with a

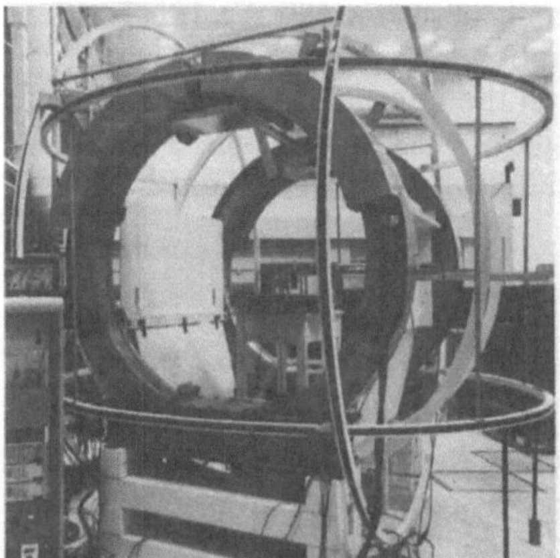

Fig. 1. DC magnetic field measurement facility.

Fig. 2. RF-shielded room for radiated emissions measurements.

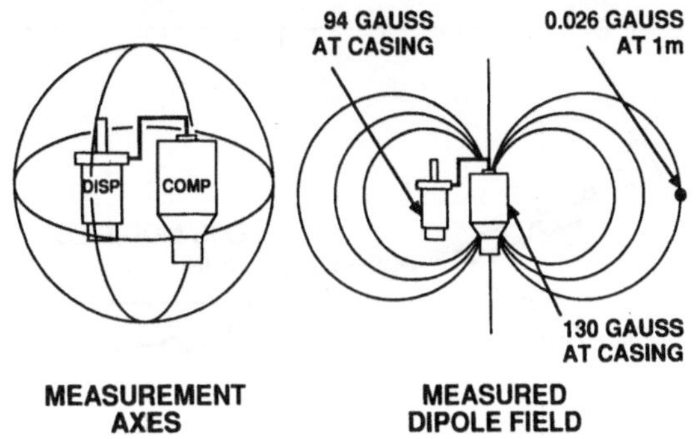

Fig. 3. The cryocooler DC magnetic field characterization.

switching power supply (having switching rates in the tens of kHz) using a synthesized sine wave to provide low distortion, low vibration operation. Electromagnetic position sensors (having excitation frequencies in the kHz range) are used to monitor the position of the linear drive assemblies.

ELECTROMAGNETIC COMPATIBILITY TEST FACILITY

All tests were made at the JPL EMC test laboratory. This laboratory is used for testing all JPL instruments for Deep Space missions. The magnetic field measurements were made in a DC magnetic measurement facility (Fig. 1). Utilizing three sets of Helmholtz coils, this facility is capable of suppressing all background magnetic fields to the 1 nanoTesla (nT) level over a volume of 1 m^3. The radiated emissions measurements were performed in a steel RF-shielded room (Fig. 2) with the facility electronics and cooler support electronics located in an adjacent room. The EMI data were obtained both with and without the cooler operating to measure cooler-contributed EMI relative to the ambient background levels.

ELECTROMAGNETIC COMPATIBILITY MEASUREMENTS

DC Magnetic Field

Both the compressor and displacer units use permanent magnets with iron pole pieces to provide the magnetic circuit for the drive coil. The resultant DC magnetic dipole field of each component has a $1/R^3$ dependence. Measurement of this field was made with the cooler placed on a rotating platform within the DC magnetic field measurement facility. Three-axis global mapping of the cooler's DC magnetic dipole field was performed by rotating the platform through 360° while 1) the cooler was lying on each of its three orthogonal surfaces, and 2) the cooler was repositioned through various angles relative to one surface. The measurements were repeated for both a nonoperating and operating cooler, with nearly identical results. The maximum DC field strength measured at 1 m from the cooler was 2.6 μT (0.026 gauss); this is half the allowable 5 μT magnetic field strength as specified for an Eos instrument. Additional field measurements were made at the compressor and displacer casings using Hall generators and yielded 13000 μT and 9400 μT maximums, respectively (Fig. 3).

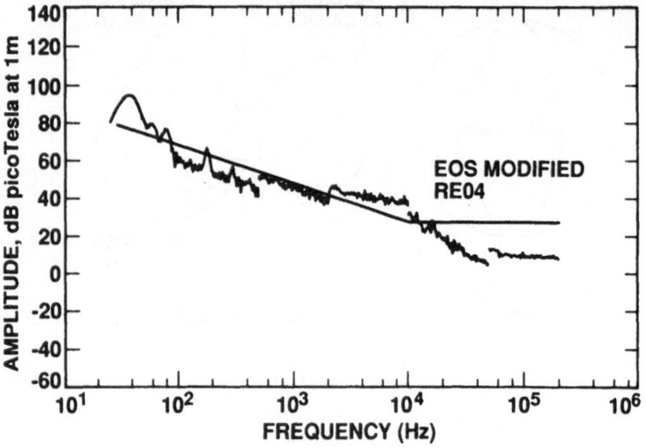

Fig. 4. The cryocooler radiated AC magnetic field emissions measured at 1 meter.

Radiated AC Magnetic Field Emissions

The cryocooler was placed in the steel RF shielded room for radiated emissions testing. Low distortion drive electronics were placed outside the shielded room and connecting cabling was fed through a bulkhead plate to the cooler. The cabling consisted of twisted pairs of leads and was sheathed in aluminum foil and grounded to the copper laminated table top to minimize any contributing radiation. The cooler was placed on the copper table top and also grounded to it. The cooler was operated at nominal compressor/displacer amplitudes (7.2 mm/2.6 mm, respectively) for the radiated magnetic field emission measurements.

Two sets of measurements were made: 1) at a 1-m distance corresponding to an Eos specification using a modified MIL-STD-461C RE04 test method, and 2) at

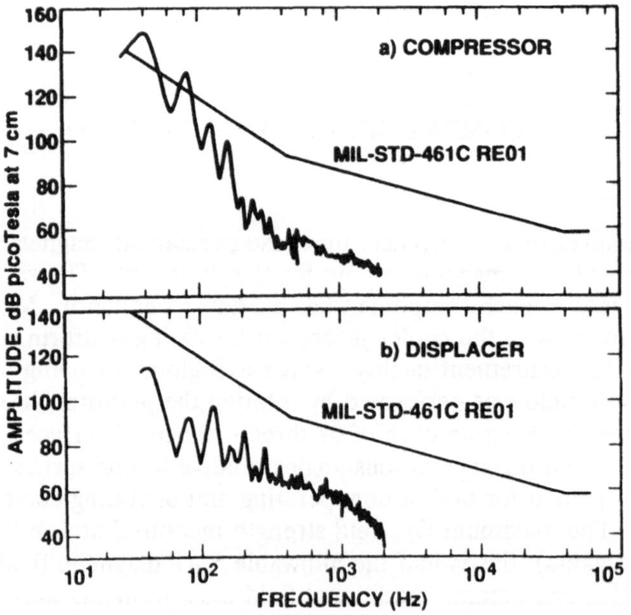

Fig. 5. The cryocooler radiated AC magnetic field emissions measured at 7 cm from a) the compressor, and b) the displacer.

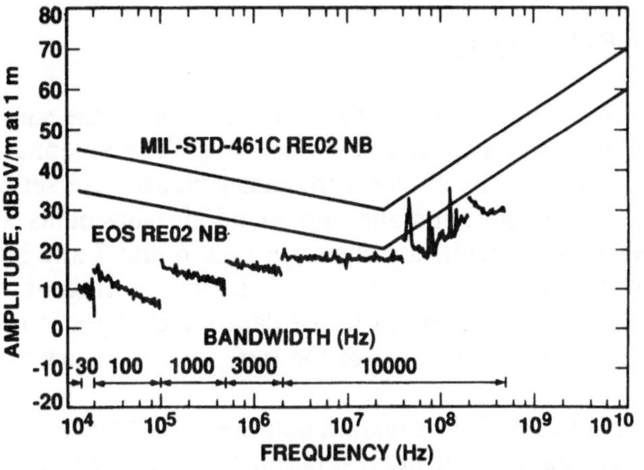

Fig. 6. The cryocooler narrowband radiated electric field emissions measurements.

a 7-cm distance corresponding to the MIL-STD-461C RE01 (there is no corresponding Eos requirement). Fig. 4 shows the radiated magnetic field emissions of the operating cooler as measured at 1 m. The data are plotted in dB above 1 pT and are compared to the preliminary accepted specification for equipment emissions as set for Eos[5]. The peaks at 40 and 80 Hz correspond to the first two harmonics of the cooler operating frequency. The other low frequency peaks seen are the odd harmonics of the 60 Hz line frequency, and are at the same level as measured in the pre-test room ambient measurements. In fact, at a distance of 1 m, only the compressor emission levels at 40 and 80 Hz are at levels measurable above the room ambient level; the emission levels for the displacer were not measurable above room ambient. The breaks in the measured data are due to changes in the amplifier gain and bandwidth settings.

The 7-cm measurements were made using the same loop detector, but at a distance of 7 cm from the casing of either the compressor or the displacer unit. Measurements of each component were made with the other component not running so as not to contaminate the electromagnetic environment during the tests. The results for both the compressor and the displacer are shown in Fig. 5. Radiated EMI discernable above ambient levels are limited to the frequency harmonics below 1000 Hz.

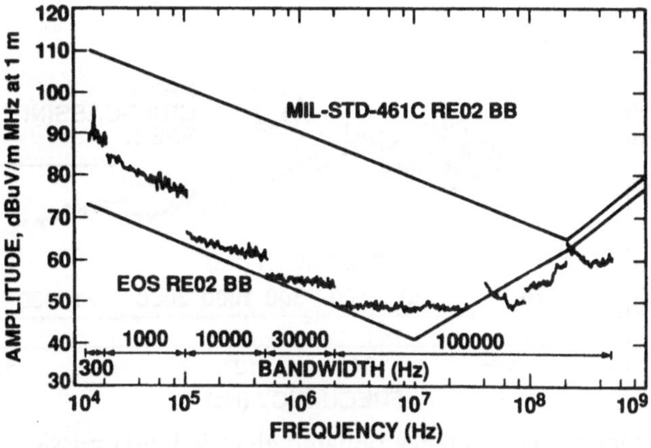

Fig. 7. The cryocooler broadband radiated electric field emissions measurements.

Radiated Electric Field Emissions

Radiated electric field emissions were measured using the same general procedure as used with the magnetic field emissions. The antennas used for measuring the emissions at the different frequency bands were set up at a 1-m distance (Fig. 2). The narrowband and broadband measurements were made from 14 kHz to 500 MHz. The results are shown in Figs. 6 and 7 and have been compared to the MIL-STD-461C RE02 curves and the more stringent, modified RE02 requirements for Eos. Discontinuities in the data are changes in the antennas, amplifiers, and bandwidths used to cover the different frequency bands.

Low frequency electric field emission data were also measured for sake of completeness. These data are of interest because of the low frequency operation of the cryocooler. Fig. 8 shows the results for the low frequency narrowband emissions of the cryocooler as measured at 1 m. No MIL-STD specification for this frequency range exists, and while not called out in the Eos guidelines, this frequency range is commonly called out by individual spacecraft, each with their own specifications. The NASA CRAF-Cassini mission specifications have been included in Fig. 8 for comparison. The 40, 80 and 120-Hz harmonics in the data are at levels above ambient background; the remainder of the data, including the odd harmonics of the 60-Hz line frequency, are all at the background levels. As in previous data, the breaks and discontinuities in the data represent amplifier, bandwidth, or antenna changes. The frequency range between 3 and 10 kHz represents a range where the best available antenna was not well matched to that frequency range.

Conducted Emissions

Conducted emissions measurements from 30 Hz to 50 kHz were made by placing a current probe around one line of the drive power cable. These measurements were made looking at different sets of ground support electronics for comparison purposes of the electronics only. Since these are not flight qualified electronics, the measurement results have not been included here. The purity of

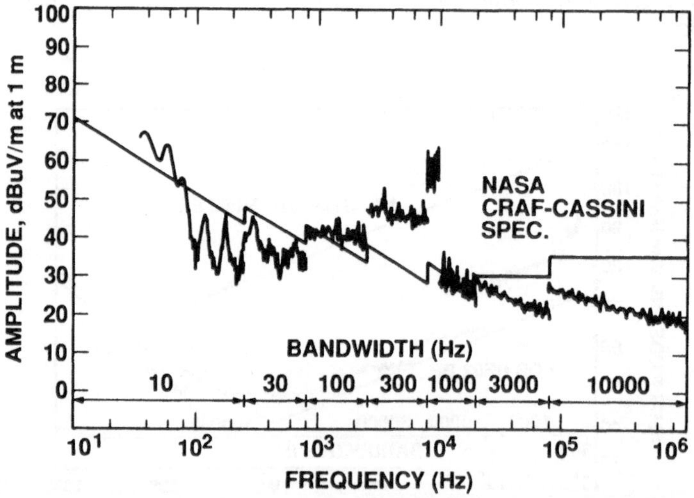

Fig. 8. The cryocooler low frequency radiated electric field emissions measurements (the NASA CRAF-Cassini specifications curve is shown for comparison).

the power supply drive current determined the conducted emission spectral purity in the data. What was representative in the data was the 3 ampere level of the 40-Hz harmonic, as this is the drive current required for the compressor using a 28 V power supply. The Eos requirements for the conducted emissions are for 3 amperes (at 120 V) for frequencies below 2 kHz.

DISCUSSION

The instrument interface requirements for Eos state that a science instrument on the platform shall not produce a magnetic field whose strength is in excess of 5 μT when measured a distance of 1 m from the perimeter of the instrument. The 5-μT value is an order of magnitude lower than the earth's magnetic field strength and two orders of magnitude lower than the magnetic field strength of the magnetic torquers used on the spacecraft platform. This value insures that the magnetic field does not interfere with the operation of other spacecraft instruments, nor produce significant magnetic torques on the overall spacecraft.

The 2.6-μT DC magnetic field strength measured for the cooler depicts a worst-case scenario because of the alignment of the magnetic dipole fields of the adjacent compressor and displacer. Even so, the measured value for the single cooler falls within the Eos requirement. The use of multiple coolers on the AIRS instrument will necessitate the revisiting of the DC magnetic field measurements when the cooler configuration on the instrument is established.

The radiated AC magnetic field emissions from the cooler as measured at 1 m were found to be above the ambient background levels for the 40- and 80-Hz harmonics only. These two harmonics are also seen to be above the Eos RE04 specification for the equipment radiated emissions. Similar measurements at a 7-cm distance when operating compressor and displacer individually showed measurable radiated emission levels below 1 kHz, with the radiated levels for the first few harmonics of the expander being 10 to 20 dB lower than that of the compressor. For comparison purposes, the 7-cm measurements have been plotted against the MIL-STD-461C RE01 curve in Fig. 8.

The results of the magnetic field measurements indicate that some level of shielding with mu-metal (perhaps 0.7 to 1.4 mm thick) will be required to lower the AC magnetic field levels to below the Eos specification. The thickness and geometry of the shielding will depend on the number of coolers required on the Eos instrument and how they are configured. A modest mass penalty may be imposed on the instrument by the requirement that the shield be made thick enough to facilitate the necessary shielding and have a fundamental vibration frequency high enough not to couple into the cooler vibration harmonics.

With the exception of a few narrowband electrical noise signals generated by the cooler support electronics in the MHz range, the radiated AC electric field emissions above 1 kHz were not measurable above ambient background. The radiated electric field emissions were well below MIL-STD curves for both narrowband (Fig. 7) and broadband (Fig. 8) cases and fall within specification for Eos in only the narrowband case. The Eos specification for broadband electric field emissions are presently more stringent than the present test equipment can measure for ambient background levels. The low frequency electric field emissions (Fig. 8), for which there is no present Eos specification, show the highest levels of emissions at the 40- and 80-Hz harmonics. If required, electrostatic shielding can be placed around instrument electronics to reduce the level of capacitive coupling of the electric field to the electronics.

These initial measurements of the cooler's radiated magnetic and electric field emissions provide the needed data to make estimates of the levels of induced voltages that may be coupled into the detector signal. These estimates can be made using near-field, common-mode coupling equations[6]. Assuming two operating coolers located 20 cm from a detector, the 40-Hz magnetic flux density and 40-Hz electric field intensity at the detector are 1 μT and 0.06 V/m, respectively. If, for example, these field levels are imposed on a 10-cm long unshielded twisted pair of leads having a separation of 1 mm, the induced voltages will be 10 nV for the magnetic field emissions and 0.001 nV for the electric field emissions. These order of magnitude values for the induced voltages are small compared to the 1-mV level of the detector signal, and do not raise any immediate concern for the initial instrument design. Future integrated detector/cooler tests will have to be conducted to determine final shielding requirements and insure electromagnetic compatibility.

The results of the conducted emissions measurements with the ground support electronics show Eos requirements can be met. The whole issue regarding powerline conducted emissions, voltage/current ripple and powerline voltage transients using the flight qualified electronics must still be considered. However, powerline transients due to the cooler operation should be of minimum concern because normal cooler operation uses soft startups whereby the cooler is powered up slowly until the piston stroke is increased to its operating amplitude. In addition, the drive power required to operate the cooler increases slowly from about 50% of full power to full power as the displacer cools from ambient temperature to 55 K.[1]

SUMMARY

Early EMC testing provides a sensitivity check on the generated EMI level and indicates whether mu-metal shielding or electrostatic shielding is required and sufficient to insure EMI levels are compatible with spacecraft instruments. The measured EMI of the BAe 80 K cooler meets specifications as set by the General Instrument Interface Specifications for Eos with the exception of the 40- and 80-Hz frequency harmonics of the radiated AC magnetic field emissions. Here mu-metal shielding will be required to reduce the emission levels to those specified for Eos. Even though the low frequency radiated magnetic field emissions are above the levels set by Eos, the estimated coupling of the cooler EMI to the detector indicate the EMI-induced voltages are orders of magnitude below the detector signal level and should not cause a degradation of the detector signal quality.

ACKNOWLEDGEMENT

The work described in this paper was carried out by the Jet Propulsion Laboratory, California Institute of Technology, under a contract with the National Aeronautics and Space Administration. Particular credit is due P. Narvaez and A. Whittlesey of JPL, who conducted the EMC measurements. The financial support of JPL's Eos-AIRS instrument is graciously acknowledged.

REFERENCES

1. R.G. Ross, Jr., D. L. Johnson and R. S. Sugimura, "Characterization of Miniature Stirling-cycle Cryocoolers for Space Aapplication", Proceedings of the 6th Intl. Cryocooler Conf., Plymouth, MA, DTRC-91/002, David Taylor Research Center (1991), pp. 27-38.
2. R.G. Ross, Jr., D.L. Johnson, and V. Kotsubo, "Vibration Characterization and Control of Miniature Stirling-cycle Cryocoolers for Space Application", presented at the 1991 Cryogenic Engineering Conference, Huntsville, AL, June 11-14, 1991.
3. V. Kotsubo, R.G. Ross, Jr. and D.L. Johnson, "Thermal Performance Characterization of Miniature Stirling-cycle Cryocoolers for Space Application", presented at the 1991 Cryogenic Engineering Conference, Huntsville, AL, June 11-14, 1991.
4. Electromagnetic Emission and Susceptibility Requirements for the Control of Electromagnetic Interference, MIL-STD-461C, Department of Defense, Washington, DC (1986).
5. General Instrument Interface Specification for the Eos Observatory, (UID101) CAGE No. 49671, GE Aerospace, Princeton, New Jersey, October 12, 1990.
6. D.R.J. White and M. Mardiguian, EMI Control Methodology and Procedures, 4th Ed., Interference Control Technologies, Don White Consultants Inc., Gainsville, Virginia (1985), pp. 6.6 - 6.12.

A THIRD ORDER COMPUTER MODEL FOR STIRLING REFRIGERATORS

S.W.K. Yuan and I.E.Spradley

Research and Development Division
Lockheed Missiles and Space Co.
Palo Alto, California

ABSTRACT

A third order split-Stirling refrigerator model has been developed under an Independent Research program at Lockheed's Research and Development Division. The model consists of more than 90 nodes, with most of them inside the regenerator of the displacer where the temperature gradient is large. Conservation of mass, momentum and energy are solved at each node until the solution converges. The model is programmed using Continuous System Simulation Language (CSSL), which lends itself to solving a large number of complex differential equations simultaneously. Some special features of the computer program include: Smith's complex Nusselt number for heat transfer in the compressor and expansion space; transport in clearance gaps of the compressor and displacer pistons; Kays and London's correlations between heat and mass transfer in the regenerator; Amar and Cannon's pressure drop in the regenerator screens; Gorring's thermal conductivity of heterogeneous materials for the regenerator matrix; laminar and turbulent transport in the transfer line; and entrance/exit effects in all major contractions and expansions. Typical outputs of the model include temperature, pressure and gas flowrate throughout the system. Other outputs are P-V work at the compression and expansion space, and heat balance over the entire refrigerator unit. This model has been validated against experimental results of an Oxford type Lucas Stirling refrigerator with excellent agreement being found between the two.

INTRODUCTION

Analyses of Stirling refrigerators range in degrees of complexity from the back-of-the-envelope type calculations to complicated nodal network analysis. For a rough estimation, the refrigeration capacity (Q_R) of a refrigerator can be expressed as

$$\dot{Q}_R = C\,\omega T_E\,P_{MEAN}\,V_E\,/\,2\pi \tag{1}$$

Advances in Cryogenic Engineering, Vol. 37, Part B
Edited by R.W. Fast, Plenum Press, New York, 1992

where constant C approximately equals to $1x10^{-4}$ K^{-1}. The above equation is valid for large systems with temperature ranges from 80 to 120 K[1], and is known as the zeroth order analysis.

The classical analysis of the ideal Stirling cycle has been given by Schmidt[2] (first order approach). He assumes an isothermal process with no pressure drops in the system. A steady state process is also assumed together with a perfect regenerator. The basic Schmidt analysis is not too useful for designing refrigerators because it lacks information about the irreversible losses.

A second order analysis consists of the basic Schmidt cycle plus decoupled loss terms[3-5]. These loss terms include static (conductive) heat loss, shuttle loss, regenerator ineffectiveness, pressure drop, and pumping loss. There are a number of second order computer models in the public domain. Lucas Aerospace Co has recently developed a model called CMOD[6]. Some of its results will be published in Reference 7 together with the validation of the present model.

The computer program discussed in this paper belongs to a third order analysis. This approach was first pioneered by Finkelstein[8] in the mid 70's. It requires breaking the machine into a number of nodes. Majority of the nodes reside within the regenerator where the temperature gradient is large. Equation of continuity, momentum and energy (for both gas and solid)

$$\frac{\partial \rho}{\partial t} = -\frac{\partial}{\partial x}\left(\frac{\dot{m}}{A_g}\right) \tag{2}$$

$$-\frac{\partial}{\partial t}\left(\frac{\dot{m}}{A_g}\right) = \frac{\partial P}{\partial x} + \frac{\partial}{\partial x}\frac{1}{\rho}\left(\frac{\dot{m}}{A_g}\right)^2 + \left(\frac{A}{LA_g}\right)\left(\frac{\dot{m}}{A_g}\right)\frac{|\dot{m}|}{A_g}\frac{f}{2\rho} \tag{3}$$

$$\frac{\partial}{\partial t}(\rho u) = \frac{h_t A}{L A_g}(T_m - T) - \frac{\partial}{\partial x}\left(\frac{\dot{m} h}{A_g}\right) + \frac{\partial}{\partial x}\left(k_g \frac{\partial T}{\partial x}\right) \tag{4}$$

$$\rho_m c_m \frac{\partial T_m}{\partial t} = \left(\frac{\varepsilon}{1-\varepsilon}\right)\frac{h_t A}{L A_g}(T - T_m) + \frac{\partial}{\partial x}\left(k_m \frac{\partial T_m}{\partial x}\right) \tag{5}$$

are solved at each node until the solution converges. This method also requires equations of state, empirical fomulaes for heat transfer and friction factor which will be discussed in the following sections.

THE MODEL

Figure 1 shows the schematic diagram of the Lucas Stirling refrigerator. It consists mainly of a displacer and a compressor connected by a transfer line. Two linear motors are present, one in the compressor and the other in the expander housing. A molecule of gas at the compression chamber can either flow into the transfer line (towards the displacer), or it can penetrate the clearance gap and enters the compressor housing. Similarly, a molecule of gas at the displacer plenum (where the transfer line meets the expander) can either enter the

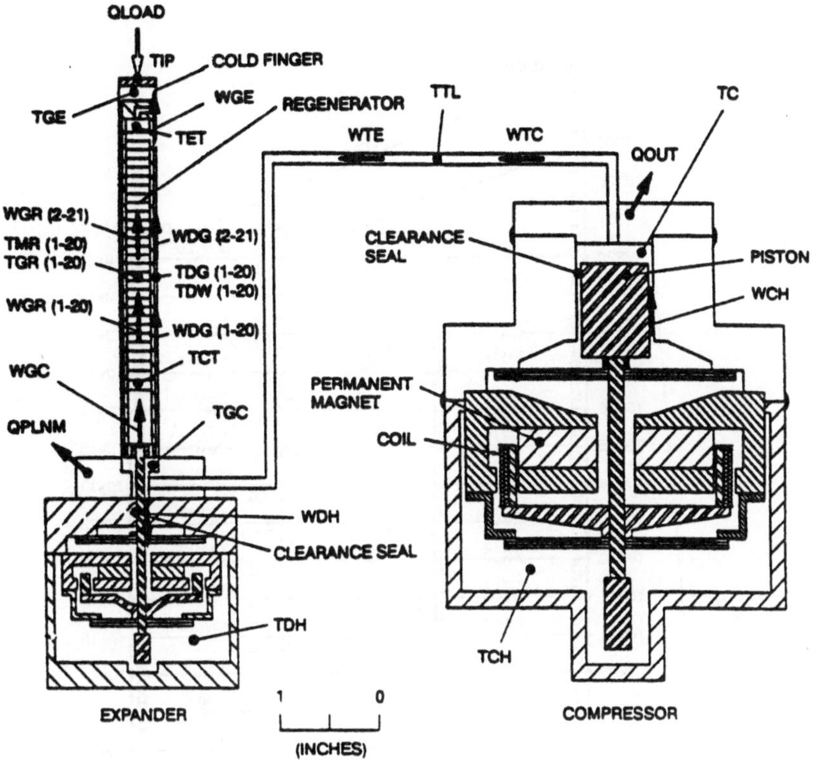

Figure 1 Lucas Stirling Cryocooler Schematic Diagram

displacer housing through the clearance gap, or it can flow into the expansion space via the regenerator or the displacer gap (between the regenerator and the coldfinger tube). Heat transfer takes place throughout the system by enthalpy transport. Heat transfer also occurs between the gas and regenerator matrix or the walls of the refrigerator. Figure 2 is a nodal diagram of the Lucas refrigerator. The majority of the nodes are located within the regenerator where the temperature gradient is large.

The Stirling Program of Refrigerator Model (SPRM)

The computer model developed under Lockheed's Independent Research program is written in Continuous System Simulation Language (CSSL). CSSL is an advanced language that can be used to solve large number of differential equations simultaneously. With four equation (Eq. 2 - 5) to be solved at each node and the number of nodes in the model, this language proves to be very useful.

Heat Transfer Coefficient in Compression and Expansion Space

Unlike most of the second order analysis, the present model does not assume isothermal or adiabatic compression. Heat transfer takes place between the gas and the wall according to the thermodynamic conditions of the system. Kornhauser and Smith[9] found that the heat transfer during compression and expansion is out of phase with the wall-bulk gas temperature difference, that the Newton's law of convection is inadequate in describing this phenomena. They express the heat transfer between the gas and the compressor (or expander wall) wall as a function of a complex Nusselt number.

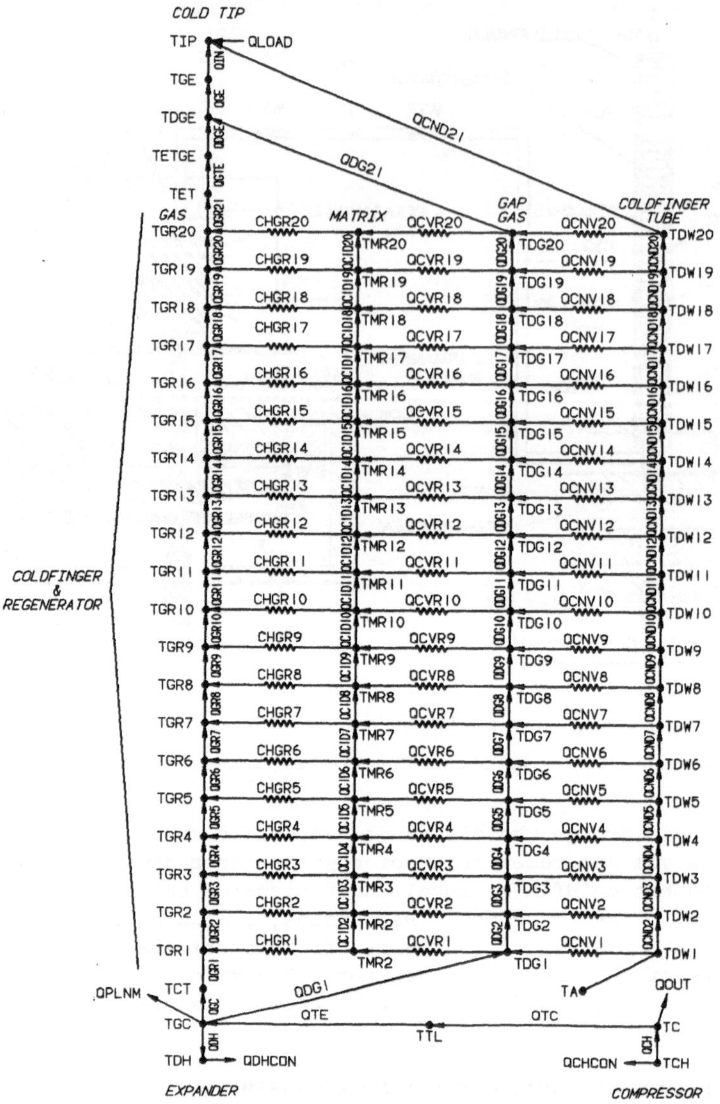

Figure 2 Nodal Diagram of Stirling Refrigerator Computer Model

$$\dot{Q} = \frac{k\,A}{D_h}\left[Nu_r(T - T_W) + \frac{Nu_i}{\omega}\frac{dT}{dt}\right] \tag{6}$$

With PeD_h/L_s larger than 100, the real and imaginary parts of Nusselt number are approximately equal and can be expressed by a power law

$$Nu_r = Nu_i = 0.98(Pe\,D_h / L_s)^{0.59} \tag{7}$$

For the Lucas Stirling refrigerator being modeled, 7 mm displacer and 10 mm compressor piston, $PeD_h/L_s >> 100$.

Heat Transfer and Pressure Drop in the Regenerator

The model makes use of Kays and London's experimental correlation[10] between heat and mass flow to calculate the heat transfer in the regenerator. Radebaugh[11] combined their results for various geometries in a single plot. For a given Reynold's number, the model calculates the Stanton number which in turn calculates the heat transfer coefficient.

As for the pressure drop, we elect to use the Amour and Cannon's expression[11] over that of Kays and London's. The main reason behind this is that Amour and Cannon's equation can correlate more type of screens (e.g. plain square, full twill, Fourdrinier, plain Dutch and twill Dutch). It also contains more characteristics of the screen (e.g. porosity, area to volume ratio, pore size etc.). According to Armour and Cannon, the friction factor across screens can be written as

$$f = \frac{\alpha}{N_{Re}} + \beta \tag{8}$$

where

$$f = \frac{\Delta P\, \varepsilon^2 D_p}{L \rho v^2}, \quad N_{Re} = \frac{\rho v}{\eta a^2 D_p}$$

Heat Conduction Along the Matrix of the Regenerator

The regenerator is randomly packed with a stack of screens where contact and even sintering of the matrix material is possible. Axial heat conduction along the regenerator matrix is thus very difficult to predict. Gorring and Churchill[12] have proposed a number of equations for thermal conductivity of heterogeneous material. For heat conduction through a square array of uniformly sized cylinders, they suggest the following empirical equation for the combined thermal conductivity,

$$k_x = k_g \frac{\chi - \varepsilon}{\chi + \varepsilon} \tag{9}$$

where

$$\chi = \frac{(1+\nu)}{(1-\nu)}, \quad \nu = \frac{k_m}{k_g}$$

for continuous media of the matrix material, the conductivity can be expressed as

$$k_x = k_m(1-\varepsilon) + k_g \varepsilon \tag{10}$$

the actual thermal conductivity should be somewhere between Equation (9) and (10).

Fluid Dynamic within the Refrigerator

For fluid flow through annular gaps of the refrigerator (e.g. compressor and displacer piston clearance gap), the following equation is used

$$\dot{m} = \frac{S^3 D \pi}{12} \frac{\rho \Delta P}{\eta L} \tag{11}$$

Frequency = 58 Hz
Radial Piston Clearance = 7.5 μm
Phase Angle = 80°
Compressor Housing = 320 K
Cold Tip Heat Load = 0.5 W

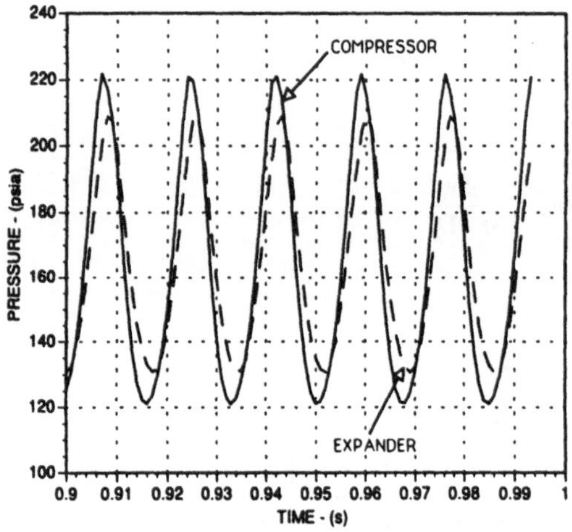

a. Compressor and Expander Piston Motion

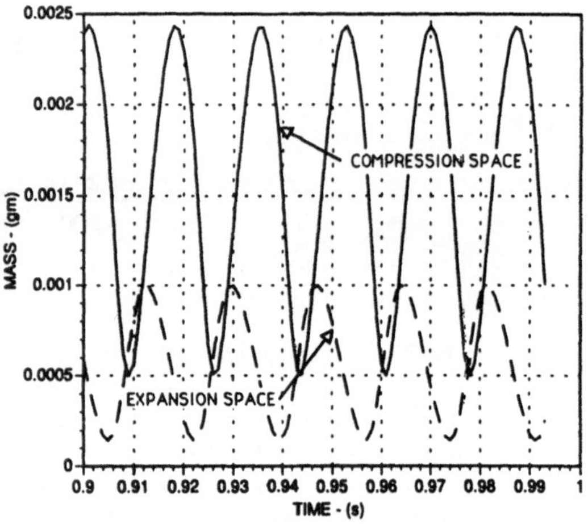

b. Temperature Response in Compressor and Expander

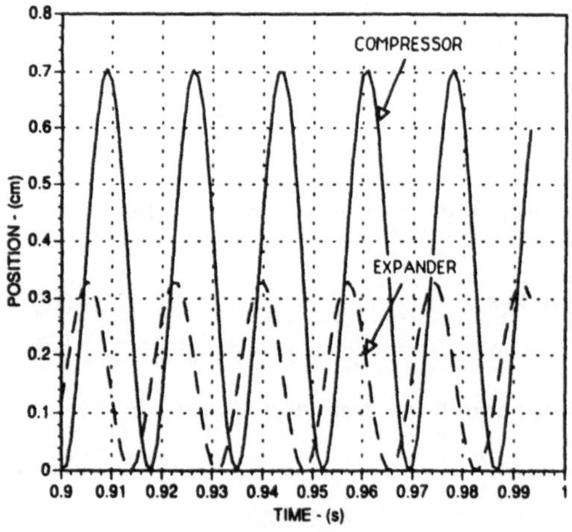

c. Pressure Response in Compressor and Expander

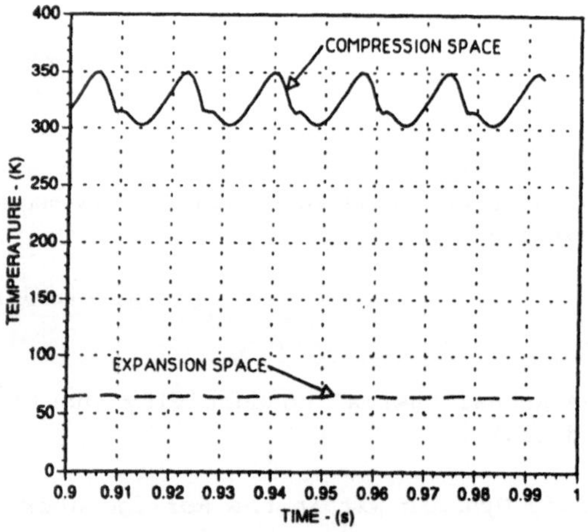

d. Mass Flowrates in Compressor and Expander

Figure 3 Typical Stirling Cryocooler Program Output

According to this equation, the mass flow in the gaps is proportional to the cube of the gap size.

Throughout the rest of the system, we use Hagen-Poiseuille (laminar) or Blassius (turbulent) equations for tube flow. This process is highly iterative, as the Reynold's number which determines the flow regime is a function of the friction factor or the velocity. This is one of the most time-consuming processes in the computation, and possibly the main reason for the excessive run-time of this model. Entrance effects are also considered wherever appropriate.

RESULTS AND DISCUSSION

The model can be run on a VAX 8600, IBM 6000 workstation or CRAY. The output from the computer program consists of the following groups of results:

1) Temperatures, pressures, mass flow, and heat flow as a function of time and position.
2) Energy balance over the entire system, including various loss terms.
3) Integrated quantities such as P-V work at the compressor and displacer.
4) Performance parameters, e.g. coefficient of performance, efficiency etc.

Figure 3a shows the pistons positions as a function of time, the maximum stroke for the compressor and the displacer are 7 and 3 mm respectively. The difference between the amplitude of the two piston motions is the phase angle. Typical outputs of the program e.g. temperature, pressure and flowrates are given in Figure 3b to 3d.

CONCLUSIONS

The third order nodal network computer model described in this paper realistically simulates the behavior and the characteristics of a split-cycle Stirling refrigerator. It computes the performance of the machine without any fudge factors or scaling. Although a number of third order analysis are available in the literature, unfortunately most of them are neither well documented nor have they been validated by experimental results. The present model has been validated extensively by the experimental results of a Lucas built Stirling refrigerator[7]. Thus, it allows one to perform parametric studies to optimize the efficiency of the Stirling refrigerator by selecting various structural designs (dimensions and materials) and operating conditions (frequencies, phase angle, strokes etc) with great degree of confidence.

NOMENCLATURE

A	total area for heat transfer
A_g	gas flow cross-sectional area
a	area to volume ratio for screen
C	constant for Equation (1)
c	specific heat capacity
D	diameter
u	internal energy
V	volume
v	velocity
x	distance

D_h	hydraulic diameter		*Greek*	
D_p	pore size of screens			
f	friction factor		α	constant = 8.61
k	thermal conductivity		β	constant = 0.52
h	enthalpy		ε	porosity
h_t	heat transfer coefficient		η	viscosity
L	length		ρ	density
L_s	stroke length		ω	angular frequency
m	mass flowrate			
N_{Re}	Reynold's number for screen flow		*Subscript*	
Nu	Nusselt number			
P	pressure		E	expansion space
Pe	Peclet number		g	gas
Q_R	refrigeration capacity		i	imaginary
S	gap size		MEAN	mean value
T	temperature		m	matrix
t	time		r	real
			w	wall or matrix

REFERENCES

1. G. Walker, Design guidelines for Stirling cryocoolers, *Cryogenics* 23:113 (1983).
2. G.Schmidt, Theorie der Lehmannschen Calorischen machine, *Z. Ver. dt. Ing.* 15:1 (1871).
3. P.A. Rios and J.L. Smith,Jr., An analytical and experimental evaluation of the pressure drop losses in the Stirling cycle, *Transactions of the American Society of Mechanical Engineers Journal of Engineering for Power*, April, 1982-8.
4. W. Martini, "Stirling engine design manual". NASA Report No. CR135382 (NTIS No. N78-23999).
5. M.Weiss, G.Walker, and R.Fauvel, Microcomputer simulation of Stirling cryocoolers, "Proceedings of the International Cryogenic Engineering Conference," Butterworth Scientific, Guildford, 1988.
6. Lucas Aerospace Co., private communications.
7. S.W.K.Yuan, I.E.Spradley and T.C.Nast, Computer simulation model for Lucas Stirling refrigerators, to be presented at the Space Cryogenics Workshop, Cleveland, Ohio, 1991.
8. T.Finkelstein, Computer analysis of Stirling Engines, in: "Advances in Cryogenic Engineering," Vol. 20, Plenum Press, New York, (1975), p.269.
9. A.A.Kornhauser and J.L.Smith, Jr., Application of a complex Nusselt number to heat transfer during compression and expansion,
10. W.M.Kays and A.L.London, "Compact heat exchangers", 2nd edition, McGraw-Hill, New York, 1964.
11. J.C.Armour and J.N.Cannon, Fluid flow through woven screens, *AIChE Journal*, 14:415 (1968).
12. R.L.Gorring and S.W.Churchill, Thermal conductivity of heterogeneous materials, *Chemical Engineering Progress*, 57:53 (1961).

STRUCTURAL AND THERMAL INTERFACE CHARACTERISTICS OF STIRLING CYCLE CRYOCOOLERS FOR SPACE APPLICATIONS

R. Boyle, E. James, P. Miller, V. Arillo, L. Sparr, S. Castles

National Aeronautics and Space Administration/Goddard Space Flight Center, Greenbelt, Maryland

ABSTRACT

Integration of a Stirling cycle cryocooler into a flight system will require careful attention to the thermal, structural, and electrical interfaces between the cryocooler, the instrument and the spacecraft. These issues are currently under investigation by National Aeronautics and Space Administration/Goddard Space Flight Center (NASA/GSFC) personnel in laboratory tests of representative longlife cryocoolers.

An 80K British Aerospace cryocooler has been instrumented as a testbed for vibration control systems characterization. Initial vibration data using a new six degree-of-freedom force dynamometer is presented in this report. Interface forces from the cryocoolers will be measured for different open-loop and closed-loop control systems. A 65K cryocooler manufactured by Stirling Technology Inc. will also be instrumented and undergo similar testing. Cryocooler characteristics will be categorized as either generic or design specific. Control systems methodologies to minimize vibration will be evaluated.

A thermal-vacuum test facility has been refurbished to characterize the performance of the cryocoolers in a simulated space thermal vacuum environment.

A flexible thermal strap which can be used to interface a cryocooler to a detector assembly has been designed and is currently undergoing manufacture. This flexible strap will be characterized thermally and mechanically.

INTRODUCTION

Many space flight instruments, particularly those operating in the infrared spectrum, require cooling at cryogenic temperatures. Past NASA experience has demonstrated the useful lifetime of these space flight instruments to be limited in most cases by the lifetime of the refrigeration system. Systems such as the Infrared Astronomical Satellite, the Cosmic Background Explorer, and the Broad-Band X-Ray Telescope used expendable stored cryogen systems, for which the maximum demonstrated on-orbit lifetime was only eleven months.

A reliable, closed-cycle mechanical cryocooler will offer spacecraft instrument designers extended useful cryogenic lifetimes of five years or more. However, instrument integration issues and concerns specific to mechanical cryocoolers (i.e. thermal, structural, and electrical interfaces) must be adequately understood and addressed; these issues and concerns are currently under investigation by GSFC cryocooler personnel.

Advances in Cryogenic Engineering, Vol. 37, Part B
Edited by R.W. Fast, Plenum Press, New York, 1992

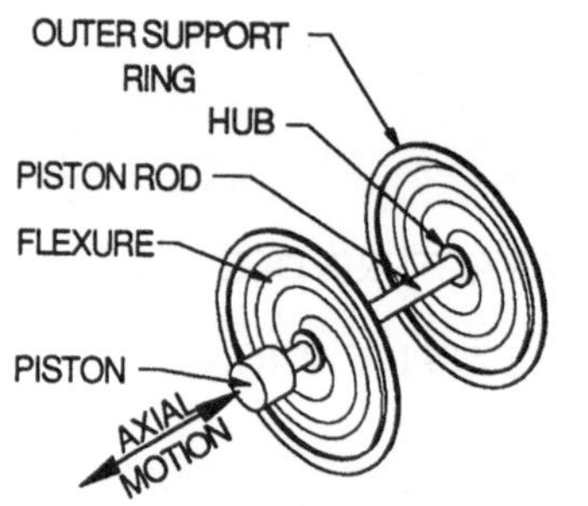

Figure 1 Schematic drawing of flexure spring.

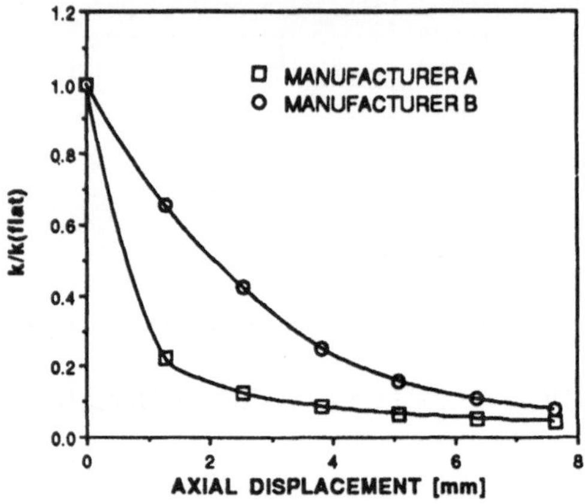

Figure 2 In-plane stiffness of a flexure spring

VIBRATION CHARACTERISTICS

An issue of critical concern to the designer of an optical instrument is the amount of vibration that is introduced by a cryocooler when integrated into a flight instrument assembly. Mechanical cryocoolers which are most likely to be used for flight instrument applications in the next decade are Stirling cycle machines, which typically have two or more pistons reciprocating at a frequency between 20 to 60Hz. These reciprocating masses could potentially produce a force on the instrument structure on the order of 50 Newtons.[1] Counterbalanced pistons moving in opposition, powered by simple drive electronics typically can reduce this force by an order of magnitude. General consensus in the cryocooler community is that flight instruments will require vibration levels perhaps another order of magnitude lower, or about 0.2N.

NASA/GSFC personnel are pursuing two approaches to understand, characterize and quantify the sources of cryocooler vibration. The first approach is analytical, in which the non-linear dynamics of the interior of a Stirling cryocooler will be modelled in sufficient detail to understand some of more dominant second-order causes and effects of vibration. An example of this problem is the flexure spring used in Oxford style machines, shown schematically in Figure 1. Groups of these flexures are used to support the moving components in Oxford machines, which rely on the in-plane stiffness of the flexure to prevent touch contact by maintaining tight tolerances on the radial clearances around the

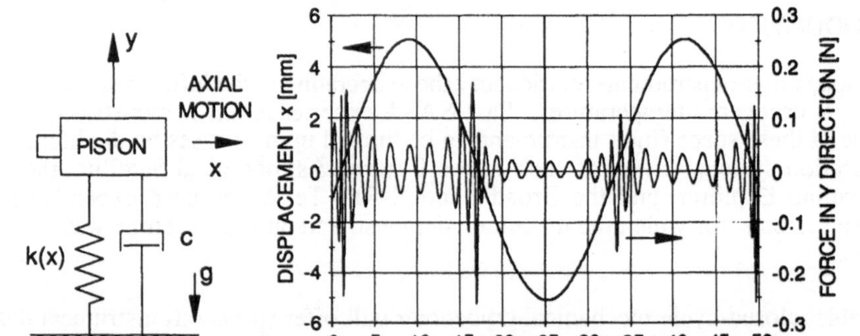

Figure 3 Simple spring-mass model of cryocooler piston oriented horizontally relative to gravity.

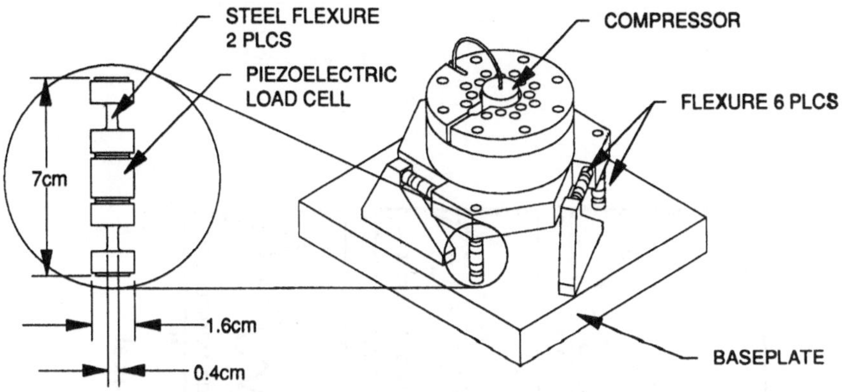

Figure 4 Six-degree-of-freedom force dynamometer for BAe 80K compressor. Displacer is mounted separately.

pistons. The flexures are initially very stiff in-plane, but when they are flexed out of the plane, the in-plane stiffness falls off very rapidly (Figure 2). When this cyclically varying stiffness is introduced into a simple dynamic model, as in Figure 3, we find that the mass will be set into motion, and will produce forces on the same order of magnitude as the instrument requirements. Note that damping significantly reduces the high-frequency response of the mass, but does not significantly reduce the magnitude of the maximum force in this model. We are adding additional detail to these simple models in small increments to assist in understanding details such as the role of squeeze film damping, gas spring effects, and asymmetric loading of the spring stacks.

The second part of this effort is to characterize existing Stirling cycle machines. An 80K British Aerospace cryocooler has been instrumented as a testbed for vibration control systems characterization. A six degree-of-freedom force dynamometer has been constructed for the BAe cryocooler (Figure 4), with the intent of using the dynamometer as a full-time test fixture. Initial data from the BAe cryocooler in a vertical orientation (Figure 5 & 6) shows similar characteristics to the machine characterized at JPL.[1] The same fixture has also been rotated 90° to collect vibration data on the cryocooler while it is horizontal. A comparison of the lateral force on the cryocooler when it is running in each orientation shows some increase in the peak values (Figure 7), and shows some of the characteristics seen in Figure 3. Completion of this characterization is expected during

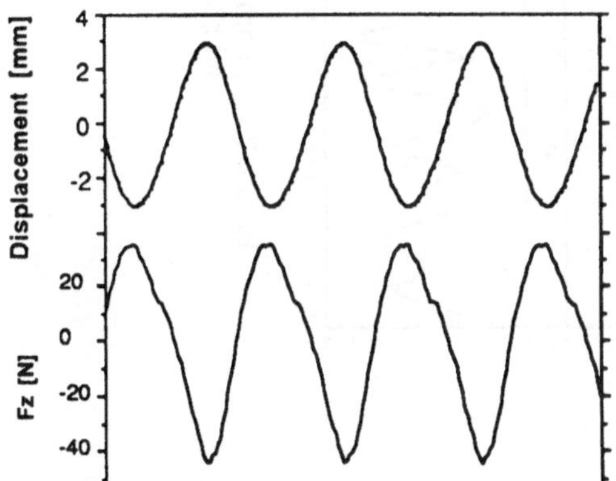

Figure 5 Axial displacement and axial force produced by the BAe 80K cryocooler.

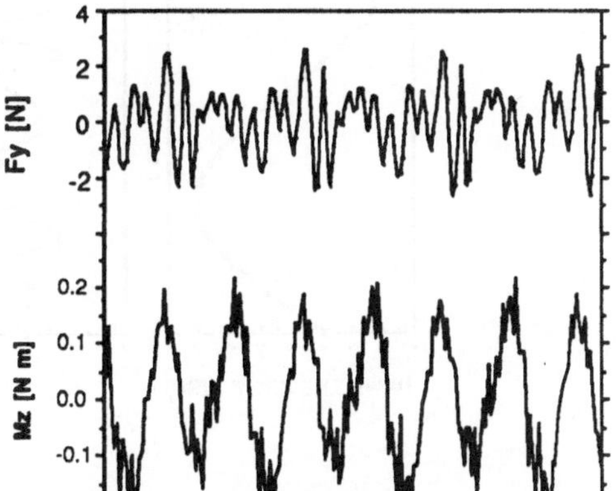

Figure 6 Radial force on cryocooler, and torque about piston axis.

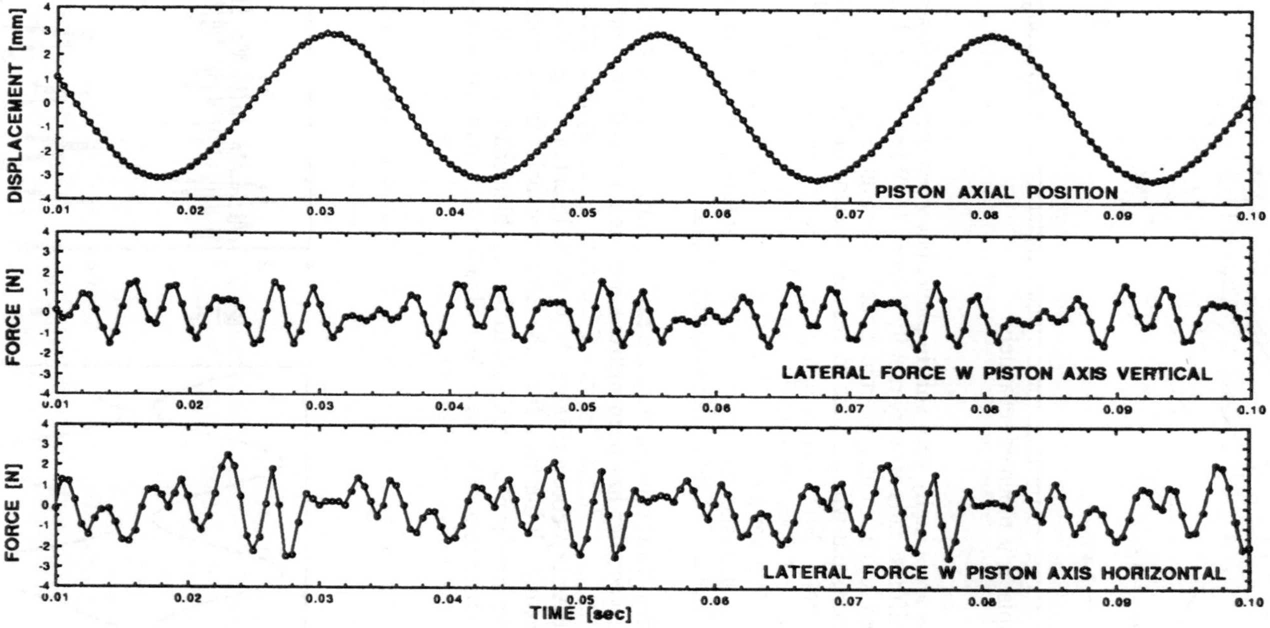

Figure 7   A comparison of the lateral force on the cryocooler when it is running in vertical and horizontal orientations.

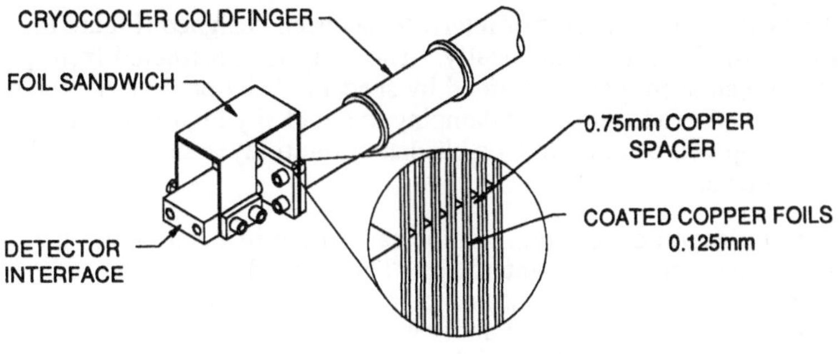

Figure 8 Flexible thermal strap for coldfinger/detector interface.

1991. A 65K cryocooler manufactured by Stirling Technology Inc. will also be instrumented and undergo similar testing. Cryocooler characteristics will be categorized as either generic or design specific. Interface forces from the cryocoolers will be measured for different open-loop and closed-loop control systems, and then compared with the vibration results using the manufacturer supplied cryocooler control electronics. Control systems methodologies to minimize vibration will be evaluated. With this approach, the effects of different operating parameters on the vibration of the machine can be easily examined.

NASA/GSFC's long term goal is to design and to verify a flight control system which will reduce cryocooler fundamental and harmonic vibration levels to acceptable levels over extended mission operational lifetimes. Information and experience gained from the BAe and Stirling Technologies cryocooler characterization studies support this long term goal.

RELIABILITY

NASA needs a longlife mechanical cryocooler to support instrument missions of 5 years or more. To achieve the necessary reliability figures for mechanical systems with moving parts requires that these parts do not touch during cryocooler operation. This is an extremely difficult objective given the 10 to 20 μm clearances between the moving parts. We expect that touch contact will produce a small signature change in the time plots of the cryocooler forces, and plots of motor current versus voltage. The NASA/GSFC testing program will also include thorough thermal vacuum testing with the cryocooler in multiple orientations during operation. Thermal vacuum chamber operational parameters will include transient and steady state environments a spacecraft is expected to see while on orbit. The purpose of this testing is to verify that touch contact, which might invalidate manufacturer supplied reliability predictions, does not occur during temperature transients and that undesirable orientation restrictions are not imposed upon the cryocooler user.

THERMAL INTERFACES

The thermal interface between the cryocooler and the instrument structure is very important in establishing proper operation of the cryocooler. Between 30-50W must be removed from each of these machines while it is operating, with the cryocooler body typically at a temperature between 0-20°C. A thermal-vacuum test facility has been refurbished to test the performance of the cryocoolers under simulated space conditions. Both the temperature profile and the vibration characteristics of the cryocooler will be monitored while it is operating under vacuum.

A second thermal interface, between the coldfinger of the cryocooler and the instrument's detectors, must be constructed with a tradeoff between thermal conductance and structural flexibility. Both the coldfinger and the detector components exhibit changes in dimensions as the components cool down to their operating temperature. The cryocooler coldfingers are typically intolerant to any appreciable side loads, so loads from this thermal contraction, as well as loads from assembly misalignment, must be isolated from the coldfinger.

A flexible thermal strap for a detector interface has been designed (Figure 8), and will be characterized thermally and mechanically. The strap is constructed from pairs of thin copper foils, with adjacent pairs separated by spacers. Friction between the two foils in each pair is expected to add sufficient damping to avoid any high-amplitude resonance of the strap. The strap will be covered by a blanket of multi-layer insulation (MLI) to reduce the parasitic heat load.

Thermal characterization of the strap will be carried out in a liquid nitrogen dewar with simulated cryocooler and detector interfaces. The thermal conduction and parasitic heat load on the thermal strap will be measured in the temperature range of 70-90K. The stiffness of the strap will be measured under vacuum at room temperature as a function of the forcing frequency. The effects of damping in the MLI blanket and from friction between adjacent foils will be examined.

CONCLUSION

Characterization of currently available cryocoolers through analysis and test is important to establish interface characteristics, as well as reliability, of cryocoolers. It is equally important that this information be supplied to cryocooler manufacturers so that cryocooler technology can be improved. NASA/GSFC has started an analysis and test program to achieve these cryocooler characterization objectives, with the goal of improving the performance of cryocoolers for use in space flight instruments.

REFERENCES

1. R. G. Ross, Jr., D. L. Johnson, & R. S. Sugimura, Characterization of Miniature Stirling-Cycle Cryocoolers for Space Application, in: "Proc. Sixth International Cryocooler Conf.," DTRC-91/002, David Taylor Research Ctr, Bethesda, Maryland (1991), p. 27

THERMAL CHARACTERIZATION OF THE BAe 80K STIRLING CYCLE COOLER

W. Burt, R. Orsini

TRW Space & Technology Group
Redondo Beach, California

ABSTRACT

The British Aerospace (BAe) 80 K Stirling Cycle Cooler is unique in the world as the only cooler in production for long-life space applications. The performance of one of these coolers was recently characterized at TRW. This paper reports on the test results and the parametric sensitivities for the measured cooler. Measured data included cooldown rates, heat sink temperature sensitivity, load line parametrics and static conduction losses. In addition, the accrued running hours of flexure bearing coolers at Oxford and TRW are summarized.

INTRODUCTION

The British Aerospace (BAe) 80 K cooler represents the current state of the art in cryogenic coolers for long-life space application. Since its commercial introduction in 1989, its performance database has continued to develop. Previous reports have presented basic performance data on the cooler[1,2] or its Oxford University precursor[3,4] and a recent BAe product specification provides additional representative data.[5] This work reports on further independent measurements performed at TRW on the thermal characteristics of a BAe cooler operating under thermal and vacuum conditions simulating space application and, in addition, summarizes lifetime running experience at TRW,[6] Oxford and BAe.

APPARATUS

The cooler is now under test in an environment designed to simulate the thermal and vacuum environments of space (Fig. 1). To effect this the cooler is mounted in a vacuum chamber maintained at $< 1 \times 10^{-5}$ torr. The external temperature of the cooler is maintained by a circulating glycol coolant loop. The compressor rejection block and displacer mounting flange were conductively coupled to the respective fluid cooling loop heat exchangers which were in series with the displacer's exchanger located upstream. The compressor and displacer bodies were thermally isolated by a G10 insulating support structure.

A commercial Lakeshore DT 470 silicon diode thermometer and a 500-ohm heater were attached to a copper mounting block that bolted to the cooler's cold finger. Low thermal conductance phosphor-bronze 4-wire signal leads were used to minimize heat leak. TRW designed power meters were calibrated and used to measure both true AC power and any DC component to a resolution of 0.05 W for the compressor and 0.005 W for the displacer.

Advances in Cryogenic Engineering, Vol. 37, Part B
Edited by R.W. Fast, Plenum Press, New York, 1992

Fig. 1. Vacuum test apparatus showing the BAe 80 K cooler compressor (left) and displacer (right). Corrugated tubing is part of the ambient heat removal system.

The cooler is driven by the laboratory electronics supplied by BAe (Model CPL122). This electronics operates to maintain constant stroke in the compressor. Compressor and displacer strokes could be monitored on a digital oscilloscope with resolution to 0.005 mm based on the LVDT voltage calibration provided by BAe. Phase and frequency were measured to better than 0.05 deg and 0.05 Hz, respectively. The test facility includes automated data logging of cooler reject temperature, load temperature, input power to the compressor motor, and cold tip temperature.

METHODOLOGY

Steady state thermal characterization of linear resonant Stirling coolers is complicated by the large number of interactive parameters. Figure 2 shows the six parameters or variables that are available for user control. Frequency and fill pressure, which in design are further constrained by cooler dynamics, are not made available for user selection. Of the six adjustable parameters, stroke, offset, reject temperature, and input power can be separately characterized for the compressor and displacer.

If all 12 parameters were characterized, 66 data sets would be needed. To reduce the data collection effort several approaches were considered. An automated data collection process for a complete parametric study was rejected because of the cost of the custom software. In addition, other practical considerations of the current hardware limit the test range, such as the heater power supply limitations on the standard BAe test set.

The approach used for the present work was to judiciously select the data set near the manufacturer's suggested operating point and hence limit the number of data points. Displacer stroke was fixed at 2.6 mm peak-to-peak with 70 deg relative phase shift between compressor and displacer. Displacer motor input power is typically less than 100 mW and is generally determined by the other settings. The nominal compressor stroke was set at 8.0 mm peak-to-peak to characterize the constant stroke data. Despite these limitations, considerable application information is generated from this smaller data set.

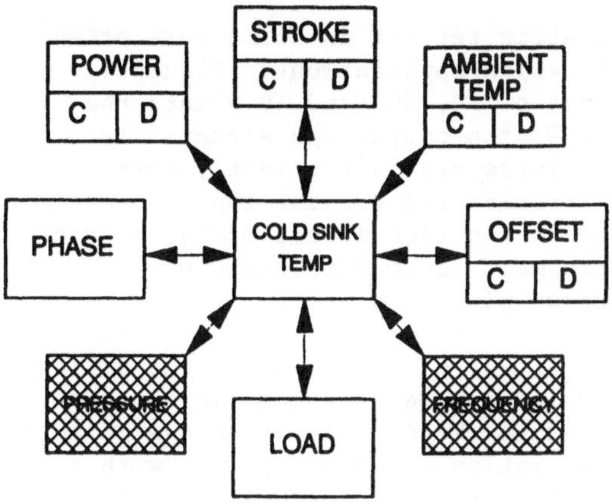

Fig. 2. Parameters determining the cooler operating point. C = compressor, D = displacer.

TEST DATA

Figure 3 shows the constant stroke cooler load line of cold tip temperature versus applied heater load versus input power and is of use to perform scaling comparisons. The data are straightforwardly obtained since all of the input parameters are independently adjustable. Namely, fixing the stroke S, the reject temperature Th and the load Ql determines the temperature of the load, Tl, and the motor input power, Win. The stroke is maintained constant at any set point by a position feedback loop (roughly stable to within $\pm$0.06 mm). As operating conditions change, the DC center positions of the moving piston and displacer vary slightly which was manually compensated during the initial settings.

Figure 3 extends previous published data to loads greater than 2 W and shows that this cooler has a cooling capacity of about 20 mW/K under these conditions. This graph displays one of the interesting characteristics of this device, namely, that it is a constant refrigeration machine. That is, for the same stroke, the cooler will draw more power as the refrigeration temperature is lowered. Furthermore, the cooler's nominal performance of 800 mW net load at 80 K can be obtained well within recommended operating parameters by increasing stroke (up to 9 mm peak-to-peak) or by reducing rejection temperature.

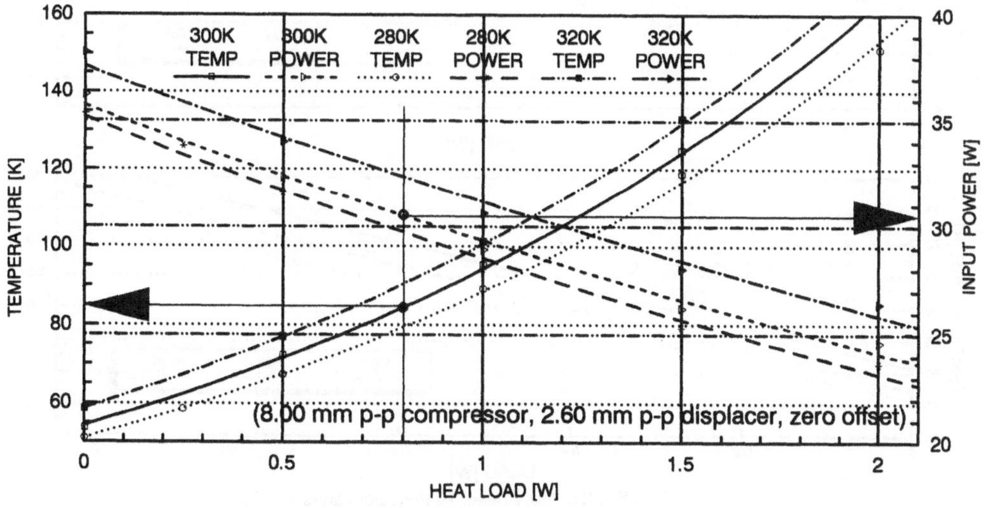

Fig. 3. BAe cooler load lines for constant nominal stroke.

Figures 4 through 7 present parametric results in forms more often of interest to users evaluating point design applications. Figure 4 presents constant input power load lines of 20, 30, and 40 W into the compressor. For applied loads below 1 W, the curves are essentially linear with a slope of about 33 mW/K. These data are slightly more difficult to accumulate than constant stroke data because input power is not independently adjustable but rather a function itself of Ql, S, and Th. However, given a large enough data map, these data could be extracted from a three-dimensional stroke plot.

Figure 5 displays the significant effect of variation of the ambient temperature on performance. At higher input power a change in ambient sink temperature of 5 K change causes approximately 1 K of change in cold tip temperature. Figure 6 shows the variation of load capacity with input power at selected constant temperatures. The load lines of Fig. 7 are particularly useful for the sensor system designer who must optimize

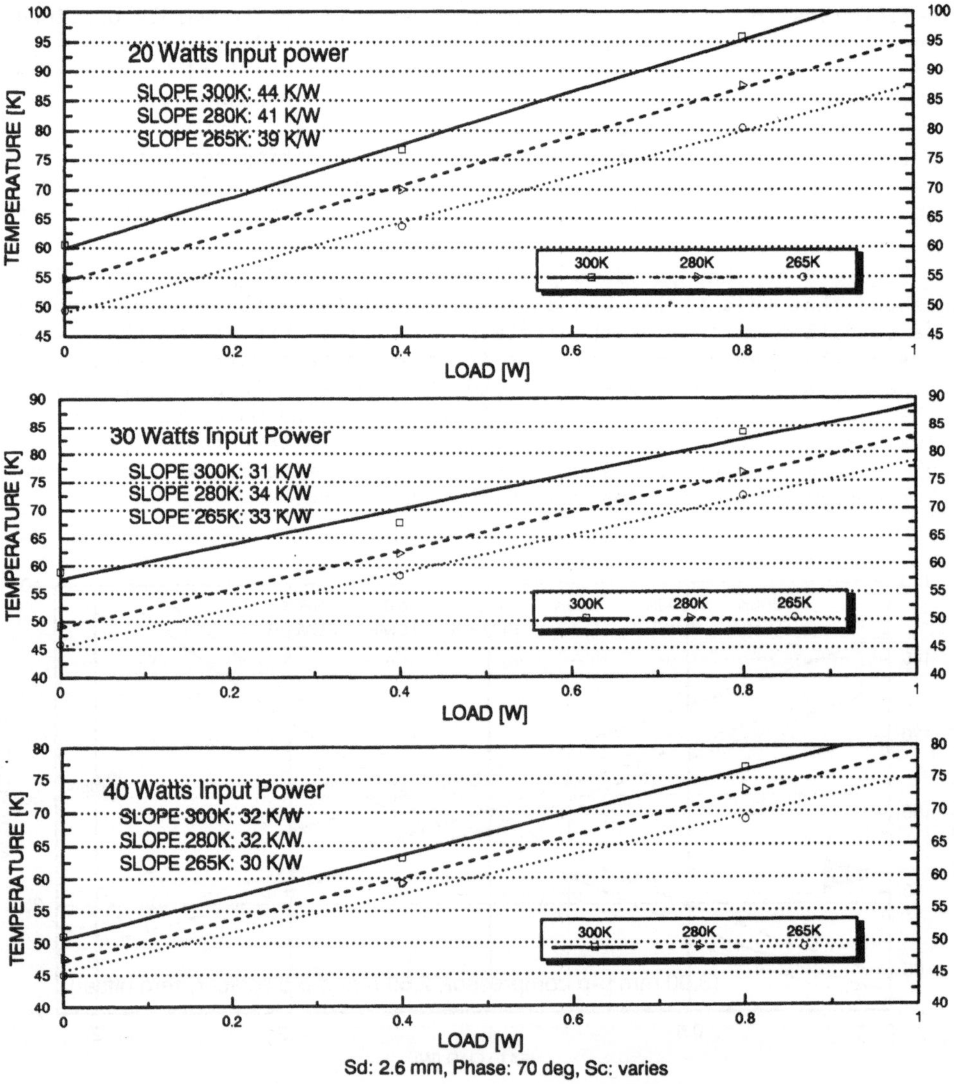

Fig. 4. Constant power load lines.

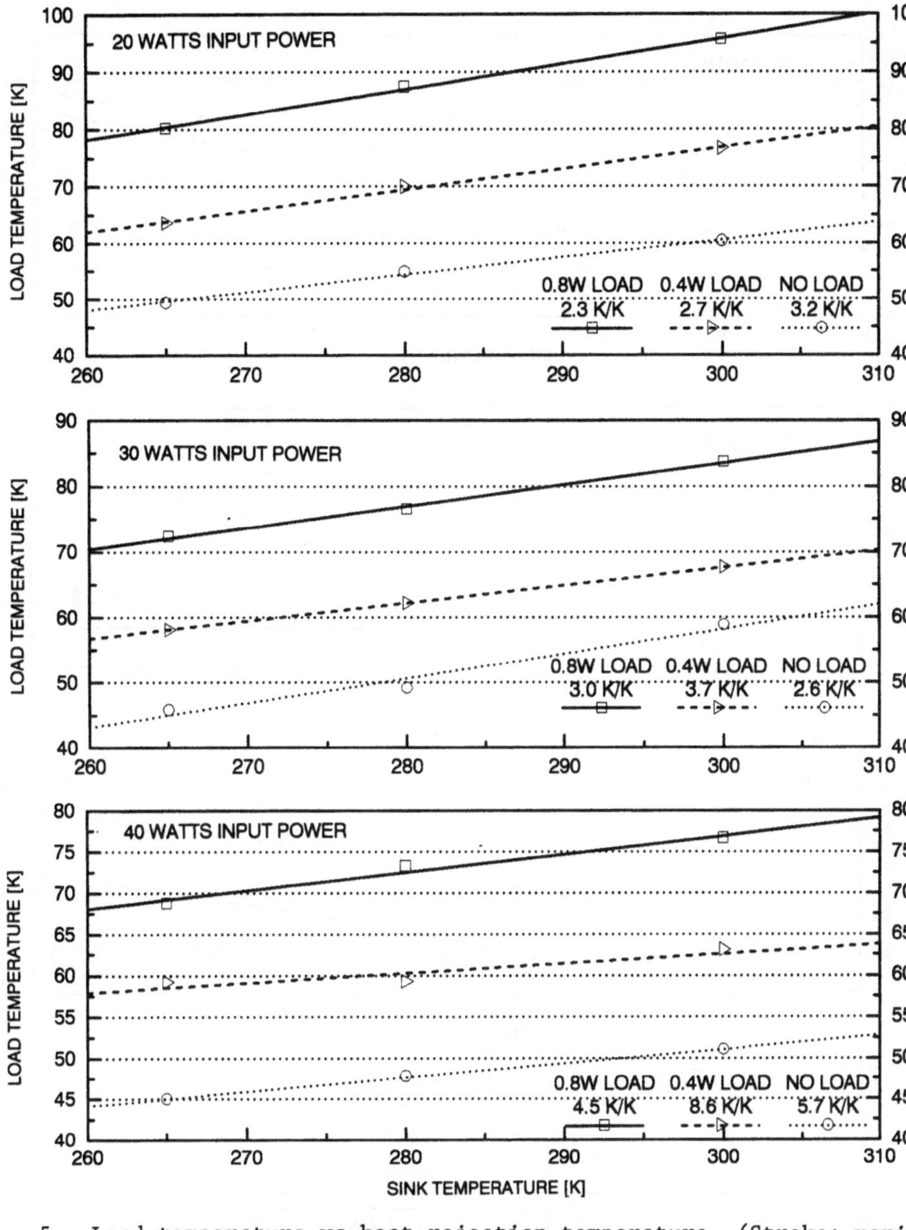

Fig. 5. Load temperature vs heat rejection temperature. (Stroke: variable, phase: 70 deg, offset: zero.)

between temperature and power for a fixed anticipated load. Figure 8 indicates the cooldown capability of the cooler, ranging from 40 minutes at no load to roughly 1 hour with full rated load of 0.8 W applied continuously. Thus little absolute difference exists in the cooldown rate between the loaded and unloaded cooler.

Some applications may require the use of redundant coolers to increase system reliability. In that case, one cooler in the standby or off mode will provide a parasitic heatload to the operational cooler. Figure 9 shows the results of parasitic heat load as function of cold tip temperature using the Oxford "warmup" method[7] in which the heat leak is determined by measuring the time to warmup between two fixed temperatures as a function of heater power. Because this is a transient measurement

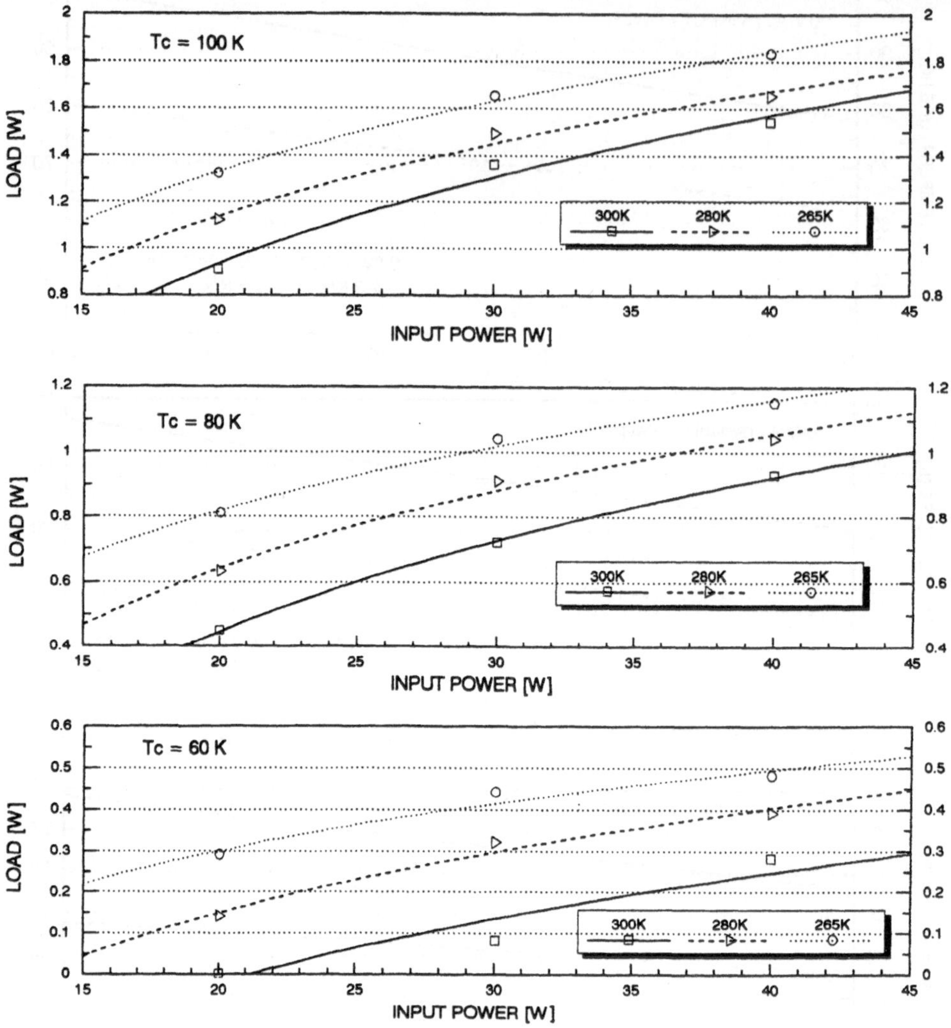

Fig. 6. Input power vs applied load at constant temperature. (Phase: 70 deg, Sd: 2.6 mm, Sc: varies.)

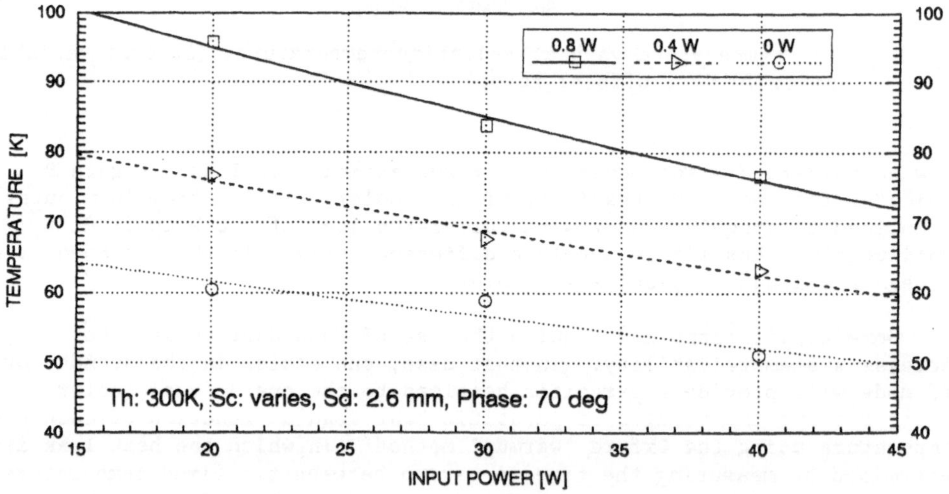

Fig. 7. Input power vs cold tip temperature for constant loads.

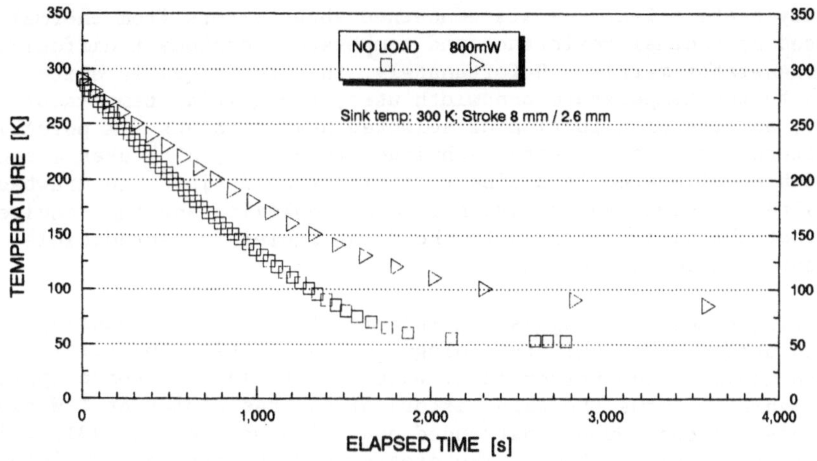

Fig. 8. Characteristic cool down at no load and rated 80 K load.

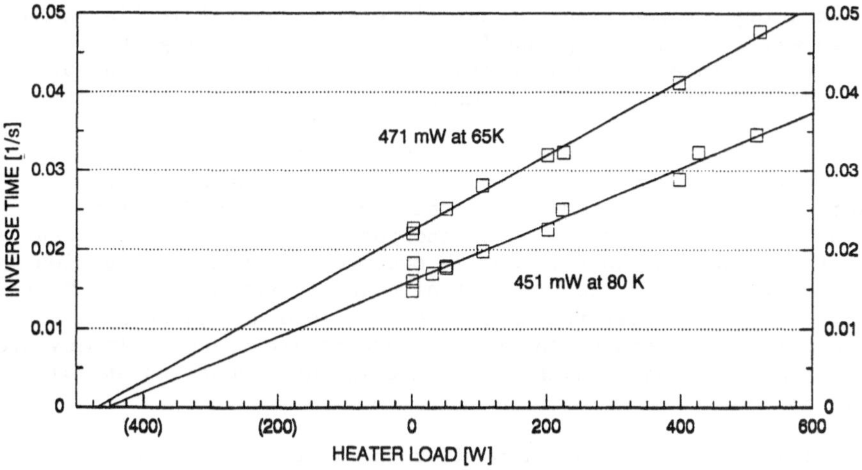

Fig. 9. Parasitic heat load using the warmup method.

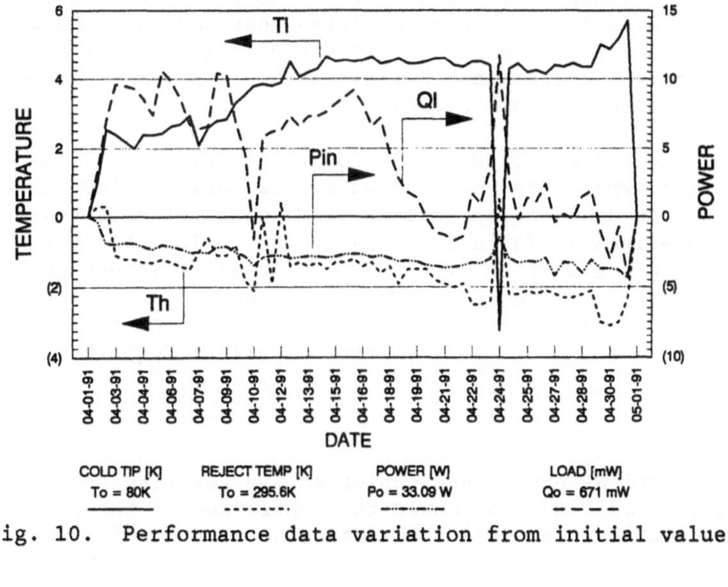

Fig. 10. Performance data variation from initial values.

of a steady-state effect, we are concerned about errors from thermal time lags caused by thermal resistance and temperature dependent diffusivity of the regenerator matrix. Unfortunately, this technique is quite sensitive to the temperature bandwidth used, the initial temperature stability, and the distribution of selected heater values. A potentially better alternative steady state technique proposed by JPL[8] uses a second cooler with a calibrated load line to cool the cold tip. The drawbacks of this alternative technique are hardware complexity and the requirement to know that the parasitic load to the first cooler is reproducible independent of load connection.

The single most distinguishing feature of the flexure bearing BAe cooler and others using this technology is its lifetime potential of years with undiminished performance, which is crucial for space applications. Figure 10 shows the time history of the BAe cooler at TRW since it was placed in continuous unattended operation on 1 April 1991, using the apparatus described above. The greater variability than anticipated is partially attributable to instability in the heater power supply in the cooler test set, temperature sensitivity of the electronics drive (described below), and marginal regulation of the ambient cooling bath. Even accounting for the fluctuations, several Kelvin of initial temperature rise was evident during the first 2 weeks. Following this, cold tip temperature remained steady at a nearly constant value (within $\pm$0.25 K) even in the case of the drifting and off-setting external parameters.

A second interesting feature is the very sharp temperature dip on 24 April which is equivalent to the addition of several hundred milliwatts of load. Unfortunately, data were not logged at rates sufficient to provide instantaneous indications of change but the data suggest that either a control anomaly occurred or that an excursion in ambient conditions affected the control power operating point. There is no reason to believe that the parasitic heat load decreased by the approximately 200 mW equivalent to the temperature change. In any event, cooler operation remained within power set-point boundaries and did not trip off.

A third feature was the trip-off that did occur on 01 May. This was indirectly related to the removal of a defective ion gauge controller from the cooler instrument rack on 30 April. This removal had the unintended effect of substantially cooling the rack interior by providing a large front panel opening. The unexpectedly large thermal coefficients in the test set's control electronics resulted in the cooler's subsequent trip off. This also led to the recognition that ambient temperature stability of this set of nonflight, laboratory drive electronics can significantly affect operation. Following the trip-off, the cooler was restarted and returned to operation.

The experimental duration testing of this cooler at TRW is but one part of a larger experience acquired on similar units. The composite running history to date of the various ISAMS class flexure-bearing coolers is summarized in Table 1. None of the coolers has experienced a shutdown due to a malfunction of the cooler itself although periodic interruptions were experienced in the Oxford and ESTEC testing caused by loss of power, drift in ambient conditions, or operator intervention unrelated to the running tests as we have experienced in this initial testing.

CONCLUSIONS

Characterizations of the BAe cooler at TRW verified reproducibility of the as-built load line and showed that operation of the entire cooler in vacuum has little to no effect on load line characteristics. In

Table 1. Flexure Stirling life test summary for ISAMS class coolers

	Oxford (ISAMS Cooler)	ESTEC (ISAMS Cooler)	TRW (Production Cooler)	BAe (Production Cooler)
Cumulative hours (through 4/29/91)	28,187 (cooler) 34,625 (compressor alone)	28,901	766*	0
Start date	Jan 1986	Jan 1987	April 1991	July 1991 (planned)
Interruptions (planned or environmental)	<50	<200	1	N/A
Test Purpose	Life test qualification	Mechanism validation	Long-term character-ization	Life test qualification
Comments	Cooler exposed to STS loads before test Test curtailed 6/88 Test resumed 10/90	Plans to tear down, examine, and rebuild at 30,000 hrs	Entire cooler in vacuum	Will be exposed to vibration loads before test

*Does not include approximately 400 hours of factory acceptance testing.

addition, sufficient margin exists in the cooler design to accommodate manufacturing variability from unit to unit. Finally, some initial temperature rise was observed during initiation of continuous operation after which performance has been constant within the long-term stability limits of the control electronics.

REFERENCES

1. C. Jewell and B. G. Jones, Mechanical Coolers; An Option for Space Cryogenic Cooling Applications, ESTEC Bulletin 62 (1990), p.79.
2. R. G. Ross, D. L. Johnson and R. S. Sugimura, Characterization of Miniature Stirling-cycle Cryocooler for Space Application, in: "Proc. of the 6th International Cryocooler Conference," p. 27-38, Plymouth, Massachusetts (1990).
3. S. T. Werrett and G. D. Peskett, Development of the Stirling Cycle Coolers for the Improved Stratospheric and Mesospheric Sounder, "Proc. of the 4th International Cryocooler Conference," p. 289-301, Easton, Maryland (1986).
4. S. T. Werrett, et al, Development of a Small Stirling Cycle Cooler for Spaceflight Applications, "Advances in Cryogenic Engineering," Vol. 31, p. 791-800, (1985).
5. Product Specification for Mechnical Coolers, British Aerospace Document, PSP/MCC/A0073/BAe, Issue 1, (1991).
6. TRW Space and Technology Group, Redondo Beach, California, is the commercial representative for the British Aerospace 80 K cryocooler in the United States.
7. A. H. Orlowska and G. Davey, Measurement of Losses in a Stirling Cycle Cooler, in: "Cryogenics," Vol. 27, p. 645-651 (1987).
8. R. Ross, private communication

EXPERIMENTAL TESTS FOR IMPROVING REGENERATOR EFFICIENCY OF A V-M CRYOCOOLER

Shimo Li, Huaiyu Pan, S. Ge, B. Zhao, J. Lin, H. Shao

Cryogenic Lab.
Zhejiang University
Hangzhou, China

Yi Long

Department of Materials
University of Science and Technology Beijing
Beijing, China

ABSTRACT

In this paper, a liquid nitrogen-cooled Vuilleumier cryocooler has been described. Attempts have been made to improve the regenerator efficiency and operating parameters. The cryocooler achieved a no-load temperature of 14. 85 K using lead and phosphor bronze as the regenerator matrix. Magnetic material Er $(Ni_{0.79}Co_{0.21})_2$ whose volumetric specific heat is very high at low temperature was partially substituted for lead, whose specific heat is very small in the temperature region below 15 K. In comparison with the results of the former test, it was found that a terminal temperature under no heat load condition of 11. 72 K was reached this way, and the refrigeration capacity was improved considerably. A comparison between a theoretical evaluation and the experimental data is discussed in this paper.

INTRODUCTION

The Vuilleumier cryocooler is a heat -driven machine, having the potential advantages of long life operation, small vibration, compactness and low weight. In our experiments[1], ambient is regarded as a heat source and liquid nitrogen is used as an intermediate cold source. In this case, it can attain a terminal temperature of 14. 85 K at no load and cooldown time is about 42 minutes (from ambient temperature) and the consumption of liquid nitrogen is 3 L/h. In low temperature region, particularly below 15 K, in conventional material used for the regenerator, lead, only the lattice vibrations and the conduction electrons make a contribution to the specific heat. Its specific heat decreases rapidly with decreasing temperature, so it is impossible to produce satisfactory refrigeration efficiency at these temperature using lead. Some investigations show that the rare earth-3d transition metal intermetallics which undergo magnetic transition or Schottky anomalies at low temperature possess extremely high peaks and broadened large peaks in specific heat. Hence, we used magnetic material Er$(Ni_{0.79}Co_{0.21})_2$ instead of lead in the low temperature side of the regenerator. Under these circumstances, a no-load temperature of 11. 72 K was obtained, and refrigeration capacity was increased considerably at the same temperature.

Advances in Cryogenic Engineering, Vol. 37, Part B
Edited by R.W. Fast, Plenum Press, New York, 1992

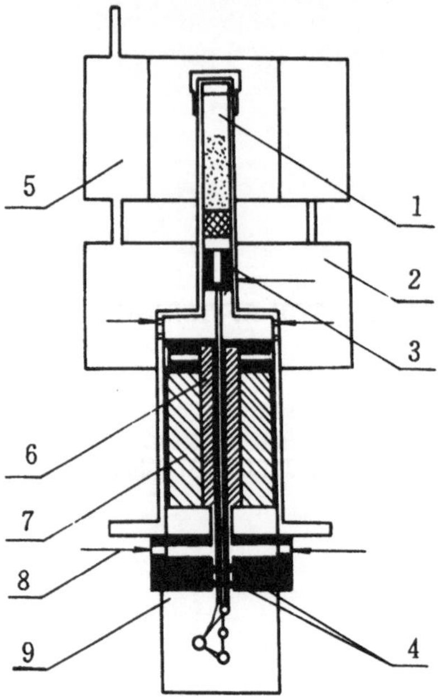

1: cold end regenerator
2: lower liquid nitrogen tank
3: cold displacer
4: seals of displacer rods
5: upper liquid nitrogen tank
6: hot end regenerator
7: hot displacer
8: ambient exchanger
9: crankcase

Fig. 1. Schematic of the VM cryocooler

DESCRIPTION OF EXPERIMENTAL SYSTEM

Figure 1 shows a schematic of the V-M cryocooler which was tested as an experimental machine. The working medium of this cryocooler is helium whose purity is 99. 99% , and charge pressure is 2. 0 Mpa. Helium absorbed heat from ambient and cold end while desorbed heat at liquid nitrogen tank by means of the movement of the cold and the hot displacers at a proper phase angle. The cryocooler was drived by an electric motor whose speed could be adjusted by the electric potential. The reciprocating speed could be varied from 50 to 160 rpm. In the ambient temperature chamber, a manometer was assembled to determine the working pressure.

Both the cold and the hot displacers are made of epoxy glass. The cylinder material is stainless steel. The hot dispacer is 95 mm in diameter, 162 mm in length, and with a stroke 20 mm. The matrix materials of the internal regenerator inside the hot displacer are approximately wire screen disks of 400 mesh stainless steel. Outside the regenerator was filled with glass wool for insulation. We used four seals which are made of glass loaded Teflon to assure gas flow wholly through the regenerator. The cold displacer is 26 mm in diameter, 195 mm in length, and also with a stroke of 20 mm. It was designed with a 0. 1 mm radical clearance between the displacer and cylinder, and at the end of the skirt of the cold displacer there were three seals made of epoxy glass. The regenerator with stainless steel sleeve was located within the cold displacer, its diameter is 23. 8 mm, length is 116 mm. Various matrix materials were stacked in it. In these experiments, a three part repenerator matrix was used, consisting of 400 mesh phosphor bronze wire screens, 0. 2 mm lead shot (whose porosity is 0. 33), and 0. 15-0. 3 mm magnetic material of porosity is 0. 46.

The refrigeration temperature was measured by AuFe-NiCr thermocouples which were calibrated by the Cryogenic Center, Academic Sinica, Beijing. The thermocouple was attached to the cold head. The upper part of the cold finger was wound with enamel manganin wire of 0. 09 mm in diameter by about 29 m in length as heat load for measuring cooling capacity. The electric resistance of the manganin heater is about 1. 9 KΩ. Outside the cold head was wrapped with the aluminized Mylar and glass fabric alternately to form a vacuum multilayer insulation. A liquid nitrogen radiaton shield to reduce heat leak from ambient was also used.

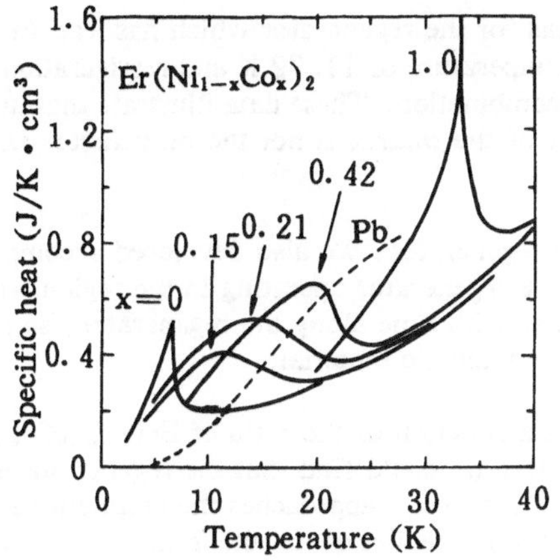

Fig. 2. Temperature dependence of volumetric specific heat of intermetallic compounds $Er(Ni_{0.79}Co_{0.21})_2$

EXPERIMENTAL RESULTS AND THEORETICAL ANALYSIS

As mentioned above, the temperature range across the cold regenerator was 77 K to about 14 K. To improve the regenerator efficiency and thus achieve lower no load temperature below 15 K, a hybrid structure with $Er(Ni_{0.79}Co_{0.21})_2$ grains (0. 15-0. 3 mm in size), lead shot and phosphor bronze wire screens was adopted for the cold regenerator. This is because the specific heat of phosphor bronze is higher than that of lead in the temperature region above 53 K. But as shown in Figure 2, $Er(Ni_{0.79}Co_{0.21})_2$ has larger specific heat than that of lead below 15 K.

In this cold regenerator, $Er(Ni_{0.79}Co_{0.21})_2$ grains occupied the volume of the lower temperature part, phosphor bronze wire screens (400 mesh) occupied the upper part, and lead shot (0. 2 mm in diameter) were filled in the other part of the regenerator.

According to an earlier analysis, the proportion of $Er(Ni_{0.79}Co_{0.21})_2$ in the matrix, which has a great effect on the no-load temperature, must be optimized. In these experiments, we kept the amount of phosphor bronze constant, about 450 disks, and changed the ratio of $Er(Ni_{0.79}Co_{0.21})_2$ to lead. Table 1 gives the speeifications of out test. No-load temperature and refrigeration capacity at 20 K for the primary tests and the theorical evaluation data are shown in Table 2.

Table 1. Composition of the regenerator matrix

	Phosphor Bronze (Screens)	Lead (g)	$Er(Ni_{0.79}Co_{0.21})_2$ (g)
Test 1	About 500	240	0. 0
Test 2	About 450	180	65
Test 3	About 450	150	90
Test 4	About 450	200	45

Table 2. No Load Temperature and Cooling Capacity at 20 K

	No Load Temperature (K)	Cooling Capacity at 20 K (W)
Test 1	14. 85	0. 71
Test 2	13. 90	1. 08
Test 3	14. 50	0. 83
Test 4	11. 72	1. 25
Calculation	10. 70	1. 59

As shown in Table 2. the optimum ratio exists for the regenerator which has lead (about 200 g) and $Er(Ni_{0.79}Co_{0.21})_2$(45 g). A no-load temperature of 11.72 K and refrigeration capacity at 20 K of 1.25 W was obtained with this combination. These data illustrate that in the temperature range around 20 K, the specific heat of the matrix is not the only factor which influence the performance of the cryocooler.

In accordance with the basic equation of the regenerator, we also developed a computer program for roughly calculating the efficiency of the regenerator operating in the region where the specific heat of the matrix and helium gas varies with time along the regenerator, so as to provide the basis for selecting the proportion of the magnetic material.

Figure 3 shows the regenerator efficiency as a function of the ratio of $Er(Ni_{0.79}Co_{0.21})_2$ to lead. In comparison with the experimental results, we could find that the regenerator efficiency of Test 2 is located in the second wavecrest, and Test 4 approaches the optimum value. From the theoretical prediction, with a further proper reduction in the ratio of $Er(Ni_{0.79}Co_{0.21})_2$ to lead, lower no-load temperature could be expected.

As we know, helium gas deviates from the ideal gas below 50 K, because the interaction among the molecules becomes great. If the helium gas is still assumed to be ideal at such low temperature regions, the deviation increases. So we accounted for the compressibility factor for the working fluid. The refrigeration capacity for VM cryocooler is given by the following expression[2].

$$Q_c = \oint p dv_c = nRT_A V_{CM} \int_0^{2\pi} \cos\theta / (a + b\cos\theta + c\sin\theta) d\theta \qquad (1)$$

where $a = V_{HM}(Z_H/Z_A + T_A/T_H)/Z_H + V_{CM}(Z_C/Z_A + T_A/T_C)/Z_C + 2V_V T_A/(Z_V T_V)$, $b = V_{HM}(Z_H/Z_A - T_A/T_H)/Z_H$, $c = V_{CM}(Z_C/Z_A - T_A/T_C)/Z_C$. After calculating Q_C, we should reduce all kinds of loss, including shuttle loss Q_{SH}, pumping loss Q_{PU}, heat transfer through displacer loss Q_D, heat transfer through cylinder wall loss Q_{CY}, regenerator loss Q_R, and intermediate heat exchanger loss Q_{EX}. Here we neglected friction between displacer and cylinder loss due to the existence of radial clearance, therefore, the net refrigeration Q_{CO} is described in the following,

$$Q_{CO} = Q_C - Q_{SH} - Q_{PU} - Q_D - Q_{CY} - Q_R - Q_{EX} \qquad (2)$$

Figrue 4 shows the comparison of the refrigeration capacity dependence of the temperature for the calculation and the experimental results. Figure 5 gives the results of cooldown performance. From Figures 4 and 5, we can find that after using the $Er(Ni_{0.79}Co_{0.21})_2$ instead of lead, lower no load temperature is obtained and refrigeration capacity of the cryocooler is improved.

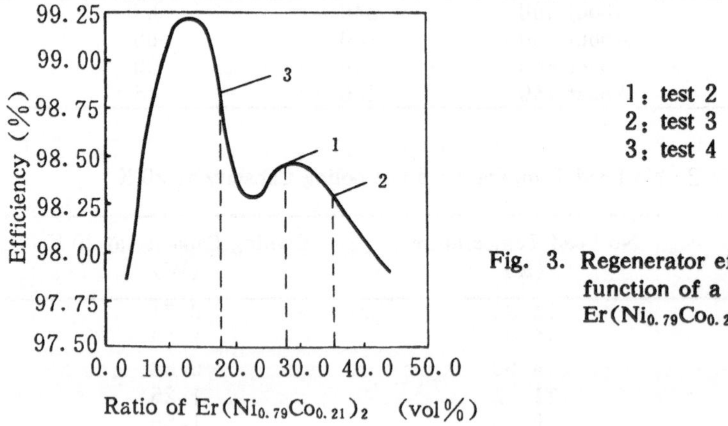

Fig. 3. Regenerator efficiency as a function of a ratio of $Er(Ni_{0.79}Co_{0.21})_2$

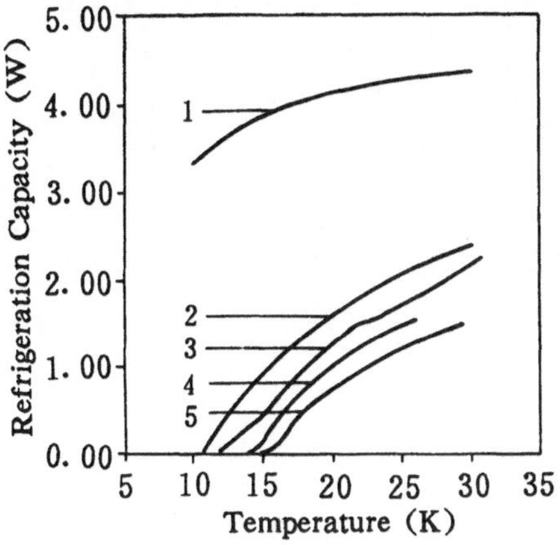

Fig. 4. Refrigeration capacity performance

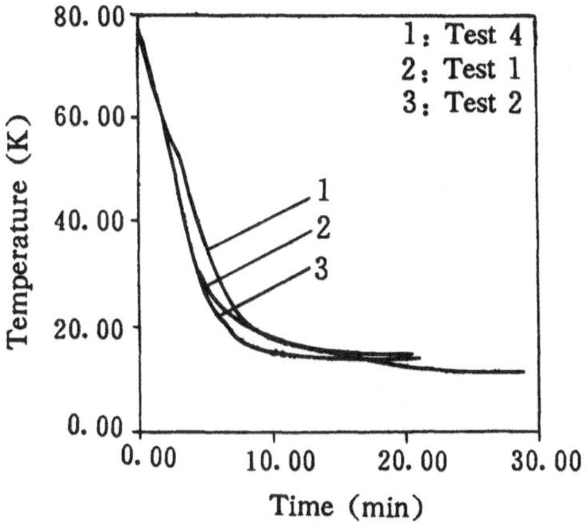

Fig. 5. Coldfinger temperature histories

CONCLUSIONS

The primary test of the V-M cryocooler achieved a no-load temperature of 11.72 K and refrigeration capacity at 20 K of 1.25 W by partially replacing lead with $Er(Ni_{0.79}Co_{0.21})_2$ in the cold regenerator. It shows that no-load temperature and regenerator efficiency below 15 K can be improved by adopting magnetic materials which possess large specific heat in this temperature region instead of the conventional materials e. g. lead, and the no-load temperature with a hybrid material regenerator are greatly influenced by the ratio of the magnetic matrix. material to lead.

REFERENCES

1. F. F. Chellis and W. H. Hogan, Adv. Cryog. Eng., Plenum Press, Vol 9. (1964), p. 545
2. G. Walker, Cryocooler, Part 1 & Part 2, Plenum Press, (1983)
3. R. Li, M. Ogawa, Magnetic intermetallic compounds for cryogenic regenerator, Cryogenics, 30: 521-526 (1990)

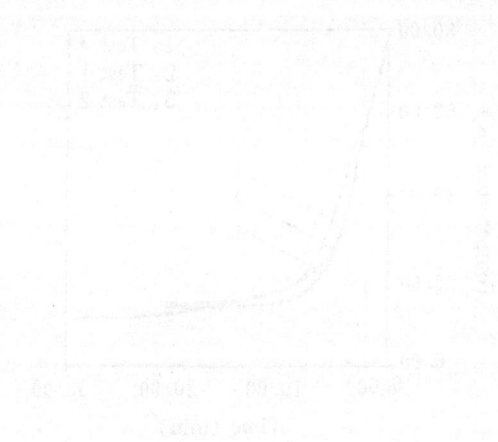

REGENERATOR PERFORMANCE IN A VUILLEUMIER REFRIGERATOR COMPARED WITH A THIRD-ORDER NUMERICAL MODEL*

P. E. Bradley, Ray Radebaugh, and John Gary

National Institute of Standards and Technology
Boulder, Colorado

ABSTRACT

A 3-stage Vuilleumier refrigerator was used to measure the performance of various third stage regenerators. The refrigerator operates between 2.5 and 5.0 Hz and, depending on the material used in the third stage regenerator, achieves temperatures of 8 to 20 K at the cold end of the third stage. This paper presents a comparison of regenerator performance for four regenerator materials: 229 μm diameter spheres of Pb+5%Sb, 229 μm diameter spheres of brass, 216 μm irregular-shaped GdRh powder, and a mixture of 229 μm and 762 μm diameter spheres of Pb+5%Sb. The experimental results are compared with a first-order model that neglects the void volume within the regenerator and with a third-order model that considers the effect of pressure oscillations in the regenerator void volume. Experimental results indicate that regenerator losses are dominated by the pressure oscillation in the void volume rather than the mass flow through the temperature gradient in the regenerator. These results are consistent with the third-order numerical model. This model shows that the heat capacity of the gas in the void space as well as the heat capacity of the matrix influences the regenerator performance.

INTRODUCTION

Cryocoolers such as the Stirling and the Vuilleumier (VM) refrigerators, employing regenerative heat exchangers (regenerators), are commonly used for temperatures down to about 10 K because of the potential of the cryocoolers for high reliability. Lower temperatures are difficult to achieve with these regenerative cryocoolers because of the low matrix heat capacity in this temperature range compared with that of the helium gas passing through the regenerator. Figure 1 shows the volumetric heat capacities of various regenerator materials as well as that of helium gas. Because regenerators operate by storing the heat transferred from the gas during a half cycle, they must have a sufficiently high heat capacity. Qualitatively, materials with higher volumetric heat capacities, such as GdRh,

---

*Research sponsored by NASA/Goddard space Flight Center.

would be expected to lead to a higher regenerator effectiveness, ε. This is the ratio of actual heat transfer divided by the maximum possible heat transfer. The first-order models of Hausen[1] and of Kays and London[2] are often used to give quantitative information regarding the performance of regenerators. These first-order models neglect the void volume in the regenerator. Regenerator performance is evaluated from the ineffectiveness, λ, which is 1 - ε. Figure 2 is from the Kays and London model and shows how the regenerator ineffectiveness varies with the number of transfer units, N_{tu}, and the ratio of matrix heat capacity to heat capacity of the fluid passing through the regenerator, C_r/C_f. Figure 2 shows that C_r/C_f should be large compared with 1 in order to reduce the ineffectiveness to as small a value as possible, but that values much larger than 5 are not necessary. Usually it is not possible to achieve C_r/C_f values as high as 5 at temperatures below about 15 K unless very large regenerator volumes are used. According to the first-order model there is no penalty to pay for the large regenerator void volume except for the larger compressor required to produce the same pressure ratio. The effect of the regenerator void volume is best seen by looking at the equations for conservation of energy describing the thermal behavior of the gas and the matrix in a regenerator (conduction terms are neglected):

$$\text{GAS:} \quad h_t A (T_m - T) = \frac{\partial(\dot{m}h)}{\partial(x/L)} + V_{rg}\frac{\partial(\rho u)}{\partial t}\ , \tag{1}$$

heat transfer   *enthalpy change*   *energy storage*

$$\text{MATRIX:} \quad h_t A (T - T_m) = + C_r \frac{\partial T_m}{\partial t}\ . \tag{2}$$

heat transfer   *energy storage*

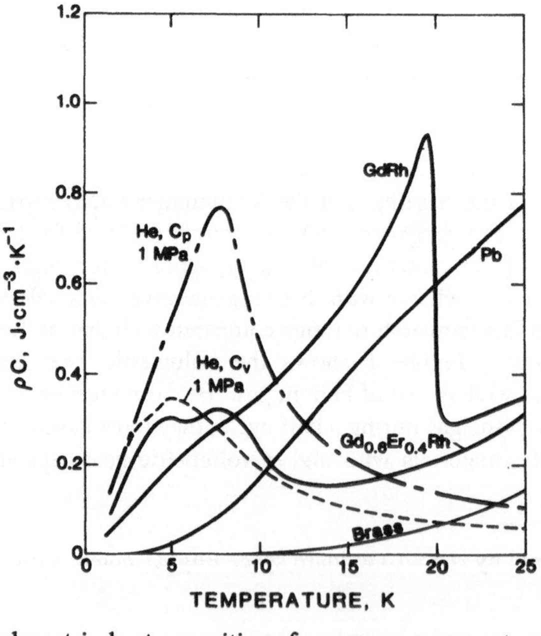

Fig. 1. Volumetric heat capacities of common regenerator materials.

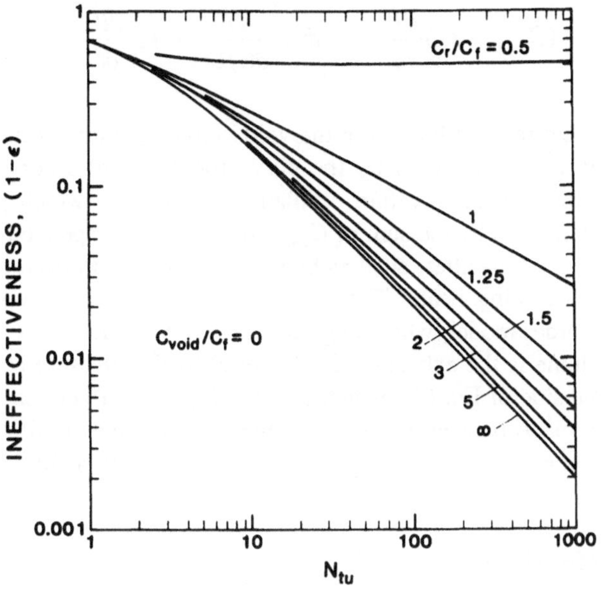

Fig. 2. Regenerator ineffectiveness as a function of N_{tu} and C_r/C_f for a first-order model.

In these equations x is the position along the regenerator in the axial direction, t is time, h_t is the heat transfer coefficient, A is the heat transfer surface area, L is the regenerator length, T_m is the matrix temperature, T is the gas temperature, $\dot{m}$ is the mass flow rate, h is the specific enthalpy, ρ is the gas density, u is the specific internal energy, and P is the pressure. The equations for conservation of mass and conservation of momentum complete the description of the regenerator behavior. Models of Hausen[1] and of Kays and London[2] assume the void volume in the regenerator, V_{rg}, is zero, thus the last term in Eq. (1) is ignored. These are referred to as first-order models in this study. A second-order model is one where V_{rg} is finite but pressure is assumed to be constant. A third-order model, as defined in this study, is one that encompasses oscillations in time of both gas pressure and temperature. Enthalpy change (mass flow through a temperature gradient) is the only gas heat source (sink) in a first-order model, whereas the gas energy storage term in the third-order model gives rise to an additional gas heat source (sink) that can exist even with no temperature gradient. A third-order model similar to that used in this study was described by Gary, Daney, and Radebaugh.[3]

Figure 3 shows how the ineffectiveness varies with C_r/C_f at N_{tu} = 84 for both the first- and third-order models. In most regenerators $C_r/C_f > 2$, in which case the third-order model predicts a higher ineffectiveness than does the first-order model. The higher ineffectiveness is a result of the greater heat flow between the gas and the matrix caused by the changing pressure in the void volume. For $C_r/C_f < 1$, the third-order model predicts a lower ineffectiveness than does the first-order model. The reason for that behavior is best explained by combining Eqs. (1) and (2) with the conservation of mass equation to obtain

$$C_{void}\frac{\partial T}{\partial t} + C_r\frac{\partial T_m}{\partial t} = - \dot{m}C_p\frac{\partial T}{\partial (x/L)} + V_{rg}\frac{\partial P}{\partial t} , \qquad (3)$$

where C_{void} is the heat capacity of the gas in the void volume of the regenerator and C_p is the gas specific heat at constant pressure. For large h_tA and small C_r the matrix temperature follows very close to the gas temperature, i.e., $T_m \approx T$. For that case

$$(C_{void} + C_r)\frac{\partial T}{\partial t} = - \dot{m}C_p\frac{\partial T}{\partial (x/L)} + V_{rg}\frac{\partial P}{\partial t} . \tag{4}$$

For small C_r the heat capacity of the gas in the void volume dominates the behavior of the regenerator ineffectiveness, as shown by the curve for the third-order model in Fig. 3. Equation (4) shows that the temperature oscillation of the gas, which gives rise to the ineffectiveness, is determined by the sum of C_{void} and C_r. Thus, regenerator materials with very low volumetric heat capacities, such as brass at low temperatures, are assisted by the void volume gas in absorbing heat. The resulting ineffectiveness is not as low as that predicted by the first-order model, which neglects the void volume. For $C_r \gg C_{void}$ the only effect of the void volume is the extra heat flow caused by the pressure oscillation (second term on the right hand side of Eq. (4)) which causes a higher ineffectiveness. The purpose of the measurements made here is to show that the void volume significantly affects the performance of a regenerator at low temperatures, and that only a third-order model can explain the results.

MEASUREMENT METHOD

Until recently there has been little opportunity to compare calculated regenerator performance with experimental results from an actual refrigerator operating in the region of 10 K. The refrigerator used for these measurements was a three stage Vuilleumier cryocooler, as described by Walker.[4] It was designed for the cooling of infrared sensors in satellite applications and as a result was fitted with a minimum of sensors. Temperatures and net refrigeration powers at the three stages could be measured, along with the operating speed and the average pressure.

If the various low temperature losses are relatively small, they may be treated as independent (decoupled) losses. The gross refrigeration power can then be expressed as

$$\dot{Q} = \dot{Q}_{net} + \dot{Q}_c + \dot{Q}_s + \dot{Q}_{reg} + \dot{Q}_{rad} + \dot{Q}_h + \ldots, \tag{5}$$

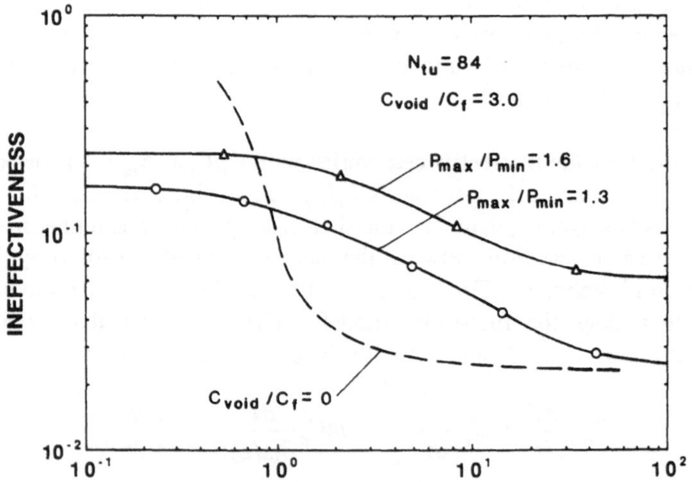

Fig. 3. Ineffectiveness as a function of the ratio of C_r/C_f for the first-(dashed line) and third-order (solid line) models.

Table 1. Regenerator Materials

Material	Particle Size (μm)	Porosity	T_{min} (K)	$\dot{Q}_{net}$ @ 10K (W)
Pb + 5% Sb	229	38.5	9.25	.068
Pb + 5% Sb	60%762 + 40%229	32.8	9.45	.064
GdRh	216	51.7	8.4	.132
Brass	229	47.5	14.75	---

where $\dot{Q}_{net}$ is the net refrigeration power, $\dot{Q}_c$ is the conduction loss, $\dot{Q}_s$ is the shuttle loss due to the oscillating displacer, $\dot{Q}_{reg}$ is the loss due to regenerator ineffectiveness, $\dot{Q}_{rad}$ is the radiative heat flow, and $\dot{Q}_h$ is the loss caused by an excess enthalpy flow through the regenerator brought about by non-ideal-gas effects. In accordance with Eq. (4), $\dot{Q}_{reg}$ can be separated into two components,

$$\dot{Q}_{reg} = \dot{Q}_{reg,T} + \dot{Q}_{reg,P}, \tag{6}$$

where $\dot{Q}_{reg,T}$ is the heat load on the expansion space due to the flow of fluid through a temperature gradient in the regenerator, and $\dot{Q}_{reg,P}$ is the heat load on the expansion space due to the pressure oscillations in the regenerator void volume.

The instrumentation on the VM refrigerator is such that only $\dot{Q}_{net}$ can be measured. Because the geometry and material properties are known, $\dot{Q}_c$ and $\dot{Q}_s$ can be calculated with fair accuracy. It is possible to reduce the temperature at the warm end of the third stage regenerator to that at the cold end by reducing the input power on the second stage. This increases the net refrigeration of the third stage by an amount equal to $\dot{Q}_c + \dot{Q}_s + \dot{Q}_{reg,T}$. According to Eq. (5) the resulting net refrigeration power, which shall be referred to as $Q_{net'}$, is given by

$$\dot{Q}_{net'} = \dot{Q} - (\dot{Q}_{reg,P} + \dot{Q}_{rad} + \dot{Q}_h). \tag{7}$$

A radiation shield attached to the second stage makes $\dot{Q}_{rad}$ negligible for the third stage. Because $\dot{Q}_c$ and $\dot{Q}_s$ can be calculated, the value of $\dot{Q}_{reg,T}$ is determined by

$$\dot{Q}_{reg,T} = \dot{Q}_{net'} - (\dot{Q}_{net} + \dot{Q}_c + \dot{Q}_s) \,. \tag{8}$$

As long as the void volume does not change more than about 50%, $\dot{Q}$ and $\dot{Q}_h$ will not change as various regenerator materials are tried for the third stage regenerator. If all the other operating conditions are kept the same, the change in the net refrigeration power with a zero temperature gradient, for a change in regenerator material, becomes

$$\Delta\dot{Q}_{net'} = \Delta\dot{Q}_{reg,P} \,. \tag{9}$$

This equation shows that it is not possible to measure the value of $\dot{Q}_{reg,P}$, but its change from one material to the next can be measured. This change can then be compared with that predicted by the third-order model. Only the model can provide the absolute value of $\dot{Q}_{reg,P}$.

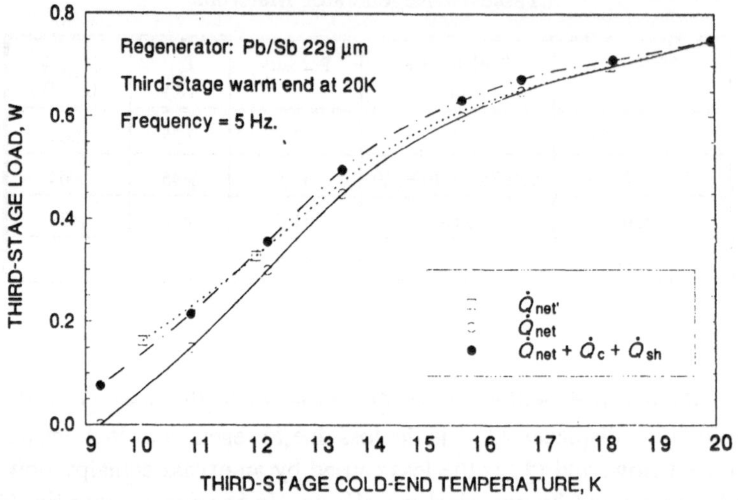

Fig. 4. Net refrigeration for single size Pb+5%Sb with the regenerator warm end at 20K.

EXPERIMENTAL RESULTS

Four different regenerator materials were tested in the third stage of the VM refrigerator. These materials are listed in Table 1. The GdRh consisted of irregularly shaped powder with sizes ranging from 178 μm to 254 μm, whereas the other materials consisted of spherical powder with a size variation of ±25 μm. The brass powder had a few small nodules on the otherwise spherically shaped particles. As a result its porosity was higher than that of the Pb+5%Sb powders. By reducing the porosity of the powder a greater mass can be inserted into a fixed volume regenerator, resulting in a larger C_r. This was accomplished using a mixture of two sizes of the Pb+5%Sb powder. The third stage regenerator had a total cross-sectional area of 1.48 cm^2 and a length of 6.0 cm.

Figures 4 through 7 show the measured $\dot{Q}_{net}$ and $\dot{Q}_{net'}$ for the four different regenerator materials for an operating frequency of 5 Hz and an average pressure of about 3.7 MPa.

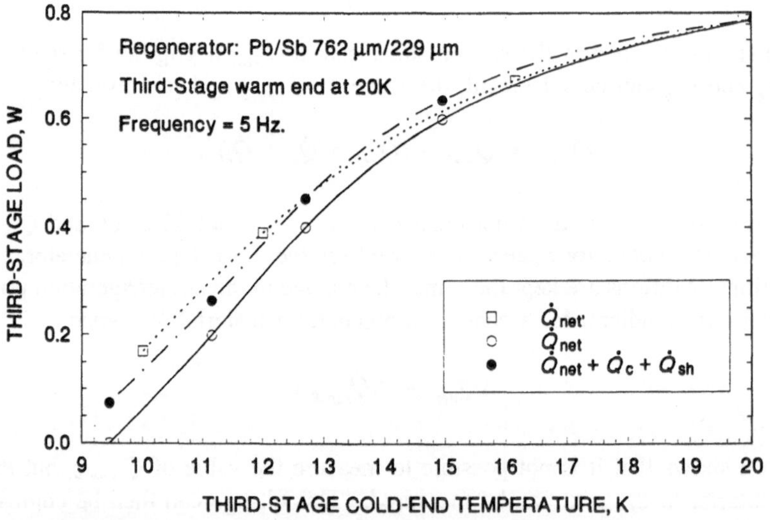

Fig. 5. Net refrigeration for two size Pb+5%Sb with the regenerator warm end at 20 K.

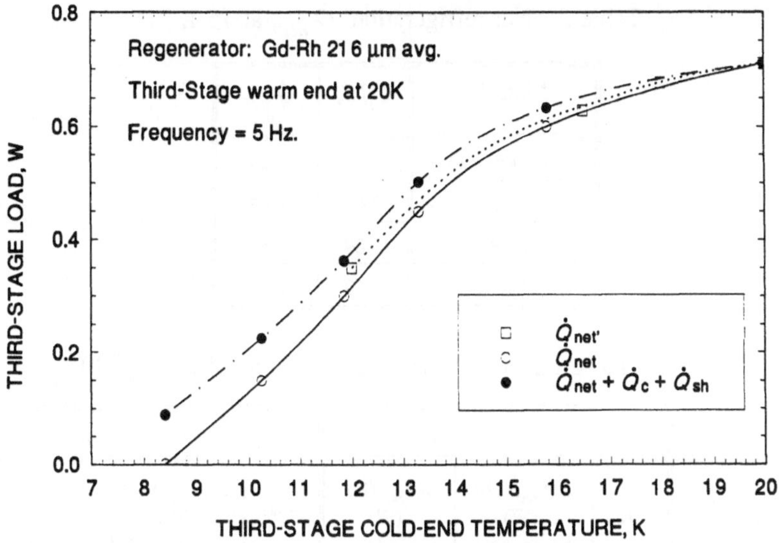

Fig. 6. Net refrigeration for GdRh with the regenerator warm end at 20 K.

The $\dot{Q}_{net}$ curve is for a second stage temperature of 20 K. The third curve in each figure shows the expected $\dot{Q}_{net}$ when the conduction and shuttle losses are absent. According to Eq. (8) these figures show that $\dot{Q}_{reg,T}$ is very small and nearly undetectable within the measurement accuracy. In some cases the curves suggest that $\dot{Q}_{reg,T}$ is negative, but the value is less than the overall experimental uncertainty. Thus, it can be concluded from the figures that $\dot{Q}_{reg}$ is dominated by $\dot{Q}_{reg,P}$ for the temperature gradients used in these measurements.

Table 1 shows the minimum temperature reached, T_{min}, and $\dot{Q}_{net}$ at a cold end temperature of 10 K, while the warm end was maintained at 20 K for the four different regenerator materials. As expected, GdRh achieved the lowest temperature due to its higher volumetric heat capacity below 20 K. Table 2 shows how $\dot{Q}_{net'}$ varies with C_r/C_f for the four materials with a third stage temperature of 15 K and a second stage temperature of 15

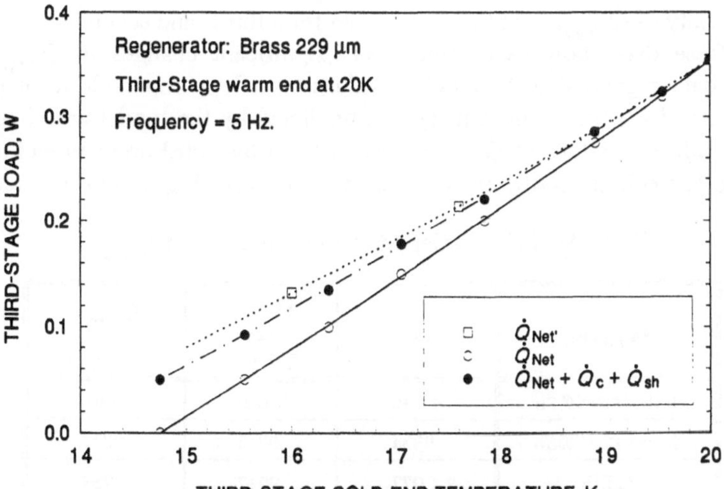

Fig. 7. Net refrigeration for Brass with the regenerator warm end at 20 K.

Table 2. Net refrigeration, $\dot{Q}_{net'}$, at 15 K

Material	C_r/C_f	$\dot{Q}_{net'}$ (W)
1 size Pb+5%Sb	3.29	0.58
2 size Pb+5%Sb	3.61	0.62
GdRh	3.89	0.58
Brass	0.19	0.08

Table 3. Experiment and Model calculations of $\Delta\dot{Q}_{reg}$(Pb)

	$T_c = T_h = 15$ K			$T_c = T_h = 20$ K		
Material	C_r/C_f	$\Delta\dot{Q}_{reg}$ (Pb) (W) Experiment	$\Delta\dot{Q}_{reg}$ (Pb) (W) Model	C_r/C_f	$\Delta\dot{Q}_{reg}$ (Pb) (W) Experiment	$\Delta\dot{Q}_{reg}$ (Pb) (W) Model
1 sz. Pb+5%Sb	3.29	----	----	6.93	----	----
2 sz. Pb+5%Sb	3.61	- 0.045	- 0.075	7.48	- 0.040	- 0.096
GdRh	3.89	0.005	- 0.075	4.66	0.040	0.385
Brass	0.19	0.495	1.12	0.88	0.395	0.986

K. The 2-size Pb shows the greatest refrigeration power, presumably because of its smaller void volume, which reduces $\dot{Q}_{reg,P}$. Conversely, GdRh has a greater void volume than the single size Pb which counteracts the effect of the higher C_r/C_f.

COMPARISON OF EXPERIMENTAL AND THEORETICAL RESULTS

As stated previously, only changes in $\dot{Q}_{reg}$ can be determined from the experimental results. Experimental values for the differences between $\dot{Q}_{reg}$ and $\dot{Q}_{reg}$(Pb) (single size Pb was used as the standard due to its common usage) as a function of C_r/C_r at a temperature of 15 K are shown in Table 3. Shown for comparison are the values calculated from the third-order model. Because the second stage temperature was held at 15 K for these data, $\dot{Q}_{reg}$ consists only of $\dot{Q}_{reg,P}$. Table 3 shows data for a third- and second-stage temperature of 20 K. These data show two points: (1) significant changes in $\dot{Q}_{reg,P}$ from the experimental data suggest that the absolute values must be comparable to or greater than these changes, and (2) the changes in $\dot{Q}_{reg,P}$ predicted by the model are about twice the experimental values. Values of $\dot{Q}_{reg}$ calculated from the third-order model are listed in Table 4. The last column shows the calculated values for $\dot{Q}_{reg,T}$ with the cold end at 15

Table 4. Third-Order Model calculations of $\dot{Q}_{reg}$

Material	$\dot{Q}_{reg}$ (W) T_c=15K, T_h=20K	$\dot{Q}_{reg,P}$ (W) T_c=T_h=15K	$\dot{Q}_{reg,T}$ (W) T_c=15K, T_h=20K
1 size Pb+5%Sb	1.130	1.020	.110
2 size Pb+5%Sb	.9834	.9013	.0821
GdRh	1.077	.9012	.1758
Brass	2.922	2.094	.828

K and the warm end at 20 K. The $\dot{Q}_{reg,T}$ values are only about 10% of the total $\dot{Q}_{reg}$, which is in qualitative agreement with the experimental results. Table 4 shows that the calculated value of $\dot{Q}_{reg,P}$ for the 1-size Pb is 1.0 W and that of brass is only 2.1 times higher even though the C_r/C_f value for brass is about 15 times smaller than that of Pb. This behavior shows that C_{void} is dominating the performance of the brass regenerator, which is expected because C_r/C_{void} = 0.05. The experimental results show an even smaller change in $\dot{Q}_{reg,P}$ for the brass. The difference between the experimental and the theoretical values may be due to the use of perfect isothermal boundary conditions in the model that do not occur in practice. A first-order model is entirely unsatisfactory for these results since it predicts $\dot{Q}_{reg}$ is zero for the case where the two end temperatures are equal.

CONCLUSIONS

A three-stage VM refrigerator was used to compare the performance of four different powdered regenerator materials with varying volumetric heat capacities and varying porosities. The results show that $\dot{Q}_{reg,P}$, due to the pressure oscillations in the void volume, dominates the regenerator performance when the warm end is at about 20 K or below. With low matrix heat capacities, such as brass, the heat capacity of the void volume gas helps to absorb some or most of the heat transferred in a half cycle. These results can be explained only by a third-order model, which considers pressure oscillations in the void volume. The model predicts somewhat larger changes in $\dot{Q}_{reg,P}$ between the various materials than the experimental results show. However, the third-order model used here is a valuable tool for predicting the regenerator performance. A first-order model cannot be relied on to qualitatively predict regenerator performance correctly in this low temperature range.

REFERENCES

1. Hausen, H., in Wärmeübertragung im Gegenstrom, Gleichstrom und Kreuzstrom Springer-Verlag, Berlin, 1950).
2. Kays, W. M., and London, A. L. in Compact Heat Exchangers, McGraw-Hill, New York, 1984.
3. Gary, J., Daney, D. E., and Radebaugh, R., "A Computational Model for a Regenerator," in Proceedings of the Third Cryocooler Conference, National Bureau of Standards Special Publication 698, 1984, p.199.
4. Walker, G., in Cryocoolers Part 1: Fundamentals, Plenum Press, New York, 1983.

AN OVERVIEW OF THE DEVELOPMENT OF SORPTION REFRIGERATION

Lawrence A. Wade

Aerojet Electronic Systems Division
Post Office Box 296
Azusa, California

ABSTRACT

Cryogenic cooling improves sensor performance in any spectral band. The high cost of achieving improved performance has led to a ten-year lifetime requirement, along with already stringent requirements for vibration, size, weight, power and temperature. The development of cryogenic refrigerators capable of meeting these requirements is a challenge that has not yet been met.

Sorption refrigeration has the potential to meet this challenge. Recent advances in sorption cooler technology have resulted in cryocooler designs that offer competitive performance and long life. Current hardware development efforts are expected to lead to a near-term flight experiment.

This paper reviews the development status of sorption coolers, sorption materials and component technologies. Anticipated future developments, including potential commercial applications, are also discussed.

INTRODUCTION

This paper primarily reviews the status of sorption cryogenic refrigerator development efforts for spaceborne applications. The extensive commercial heat pump and solar engine developments are not discussed. Typical requirements for spaceborne applications include: multistage cooling of detectors, optics and baffles; high reliability with lifetimes of up to ten years; very low thermophonics, microphonics and emi; and low system power consumption, mass and volume. To date, few refrigerators, of any type, have come close to meeting these requirements. The on-orbit performance of the few machines flown has been disappointing.

In an attempt to eliminate many of the failure modes associated with mechanical refrigerators, the development of thermally driven sorption refrigerators was initiated. There are two types of sorption processes. Physical adsorption, or physisorption, relies on the relatively weak van der Waals interaction between the refrigerant gas and the sorbent material. Activated charcoals, zeolites, and silica gels are some of the materials used in physisorption compressors.

Advances in Cryogenic Engineering, Vol. 37, Part B
Edited by R.W. Fast, Plenum Press, New York, 1992

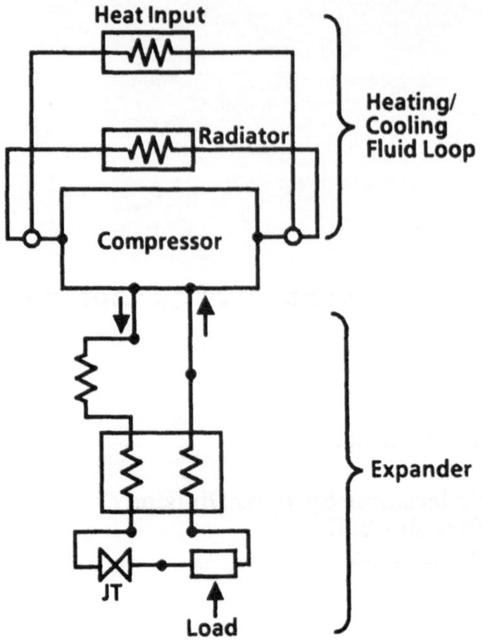

Fig. 1. Sorption Refrigerators Consist of a Sorption Compressor and Expander.

During chemical adsorption, or chemisorption, a chemical bond, usually covalent, is formed between the sorbent and the sorbate. Chemisorption materials include: hydrides (e.g. vanadium and $LaNi_5$) and praseodymium cerium oxide (PCO). The historical development of physisorption and chemisorption systems will be discussed separately.

Sorption refrigerators combine a sorption compressor with an expander (see Fig. 1). Almost any kind of expander (e.g. turbo, Joule-Thomson [J-T], etc.) could be used. In practice, sorption compressors are most efficient when producing the high pressure ratios and low flow rates favored by J-T valves and are, therefore, commonly combined.

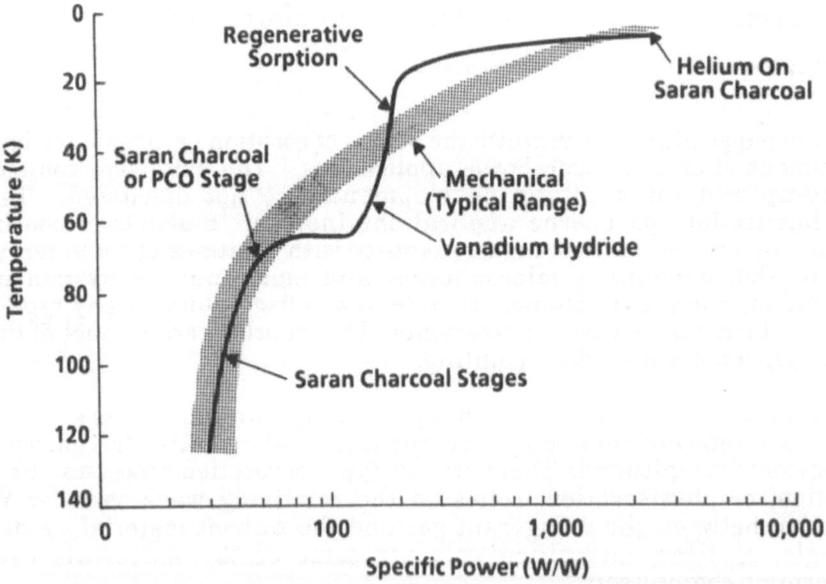

Fig. 2. Typical Sorption Refrigerator Specific Power Requirements as a Function of Cold Tip Temperature Compared to Mechanical Coolers.

Table 1. Typical Operating Conditions for Multistage Sorption Coolers

Stage	1	2	3	3	4	5
Cold Temp. (K)	200-150	150-110	100-70	90-65	60-19	12-5
Sorbent	Saran Charcoal	Saran Charcoal	Saran Charcoal	PCO	Hydride	Saran Charcoal
Sorbate	Xenon	Krypton	Argon or Nitrogen	Oxygen	Hydrogen	Helium
High Pres. (atm)	20-30	12-40	12-40	12-40	60-120	10-25
High Temp. (K)	550	550	550	800	450	80

The sorption compressor is driven thermally, and heating the compressor canister compresses the refrigerant gas. The compressor consists of several canisters, or elements, which are filled with a sorbent material. For space applications, the compressor is connected to a radiator, and, if available, a solar collector. Multistage cooling from 5 to 200 K is achieved using several sorbent/sorbate combinations. Table 1 shows several options for sorption refrigerator stage sorbents, sorbates and typical operation conditions. Fig. 2 shows the specific power (defined as gross input power, including: parasitics, valves, and circulator but not control electronics, divided by net refrigeration at the cold tip) requirements for multistage sorption refrigerators as a function of cold tip temperature. The performance of these coolers will vary considerably depending on the specific operation temperatures chosen.

Sorption refrigerators have the potential to meet the ten-year life requirement due to their relatively simple hardware. The refrigerator primarily consists of plumbing with no close tolerances. The temperatures and pressures the compressor operates over are moderate. Redundancy in the form of extra compressor elements dramatically improves reliability without greatly increasing system mass. Furthermore, sorption coolers using J-T expanders, do not have any cold moving parts.

Sorption refrigerators can be run directly off of a heat input from a solar concentrator or waste heat from a radioisotope power source. Using heat rather than electricity to operate the refrigerator substantially reduces the overall system mass.

Significant advantages during sensor integration are also offered by sorption refrigerators. When using a J-T expander, the cold end can be located directly on the focal plane without inducing vibration. As the low mass flow rate of the system results in low pressure drops, the compressor can be located 50 feet or more from the cold end without significantly affecting performance. Therefore, packaging a sorption refrigerator within a satellite is considerably easier than mechanical refrigerators.

History of Physisorption Cooler Development

Absorption refrigerators (e.g., ammonia/water refrigerators) have been used for industrial and household applications for many years. In addition, adsorption compressors are commonly used for ultrahigh and ultraclean (no oil) vacuum pumping applications. It was not until 1963 that adsorption cryocoolers were first proposed by Vickers.[1] He proposed using silica gel as the sorbent and a J-T expander to provide refrigeration for spaceborne applications.

In 1975, Hartwig[2] initiated an effort to develop an adsorption cryocooler that used zeolites as the sorbent material. He demonstrated the first physisorption cooler.[3] It achieved 3 W and 185 K using N_2O as the refrigerant. In addition, he proposed adsorption systems to operate as low as 4.2 K.[4] Due to the low efficiency of these zeolite-based designs, physisorption research languished until 1986 when Bard[5] demonstrated a 100 K Barneby-Cheney charcoal cooler. While this type of charcoal substantially out-performed zeolites, the specific power of this cooler was still over 300 W/W. Also in 1986, Chan[6] reported on a 27 K hydrogen cooler which incorporated gas gap thermal switches on the compressor.

After studying alternate sorbents, Bard[7] determined that activated saran charcoal offered substantial performance benefits based on isothermal adsorption capacity measurements by Quinn[8]. A saran/krypton compressor[9] was fabricated and has accumulated over 7,000 hours of operation. A continuous operation life test is currently underway[10]. The specific power of this machine is approximately 200 W/W.

Analysis of physical sorption cryocoolers demonstrated that over 98 percent of the input power went towards heating the compressors. This energy was then dumped, in its entirety, to the heat rejection radiator when the elements were cooled down. Regeneration of the waste heat would clearly lead to substantial performance benefits. Several practical methods for regenerating this energy were independently developed. The use of thermal switches to bring opposing compressor pairs into thermal equilibrium can reduce the required input power by nearly half. In addition, for some systems, the waste heat from a chemisorption PCO stage can be used to power physisorption saran carbon stages. A detailed analysis of a cooler incorporating these regenerative techniques was performed by Wade[11] in 1987. Practical concept designs implementing these concepts were developed and analyzed by Bard and Jones[12].

In 1987, Sywulka[13] proposed that an active valved regenerative cooler could substantially improve refrigerator performance. These claims were validated by a regenerative testbed experiment[14] which demonstrated 76 percent regenerative efficiency with a simple compressor design. Based on this work, a 137 K, 2 W net refrigeration cooler was built and is currently undergoing endurance tests[15]. As of 1 June 1991, the compressor has accumulated over 1,800 hours, and as a cooler it has operated over 500 hours. Although this machine uses a relatively low performance Anderson charcoal for the sorbent, it demonstrated 75 W/W specific power. An advanced regenerative cooler using saran charcoal is currently under development at Aerojet and is expected to require less than 30 W/W. This cooler will be discussed later in this paper.

Analytical studies indicate that the regenerative thermal switch compressor system is the most efficient sorption design for low cooling capacity applications. The regenerative fluid loop design is the most efficient for loads in excess of 0.5 W.

Two other sorption coolers of note for very low temperature applications were developed by Roach et al[16] and Duband et al[17]. Both of these systems use charcoal sorption compressors. The system demonstrated by Roach is a dilution refrigerator which achieved temperatures below 200 mK. The cooler demonstrated by Duband is a helium 3 refrigerator which achieved 100 microwatts net cooling capacity at 346 mK.

<u>History of Chemisorption Cooler Development</u>

In 1971, the first chemisorption refrigerator was built at the Phillips Research Laboratories in The Netherlands[18,19]. This chemisorption cooler delivered approximately 1 W of refrigeration at 26 K using hydrogen as the refrigerant and $LaNi_5$ for the sorbent. Studies conducted by Lehrfeld and

Boser[20] indicated that a flight worthy cooler, based on the then available materials and design concepts, would require 2.5 kW to produce approximately one watt of refrigeration at 26 K. In 1979, Steyert[21] conducted a design study to enhance compressor heat transfer.

Jones[22] developed and demonstrated a hydrogen sorption refrigerator with improved compressor heat transfer. By 1983 this refrigerator had accumulated over 1000 hours of operation in the 14 to 29 K temperature range. A similar cooler with improved kinetics was demonstrated by Karperos[23]. Analysis showed that the performance of this machine was very close to theoretical predictions.[24]

Matsubara et al[25] demonstrated a very fast (0.5 Hz) $LaNi_5$ compressor for use in a Gifford-McMahon cycle cooler. Only compressor characterization has been reported to date. This is the only sorption compressor developed for use with an expander other than a J-T.

There were two problems with these $LaNi_5$ hydride based coolers. First, a long term dissociation of the lanthanum into an irreversible lanthanum hydride occurred which would result in severe performance degradation over multiyear missions. Second, the required input power was too high. Research conducted by Bowman et al[26] determined that vanadium had no such intrinsic degradation and even higher capacity than $LaNi_5$. Combining this vanadium hydride with the aforementioned regenerative concepts results in a cooler which should exhibit high performance and long life.

Kumano, et al[27] demonstrated a 16.5 K hydride cooler which incorporated a three-stage compressor to operate over the complete pressure range.

In 1987, Jones[28] proposed using hydrides to achieve cooling to 10 K. A cooler which incorporated a two-stage hydride compressor to achieve the required pressures was studied by Rodriguez and Mills[29]. A proof-of-principle hardware demonstration effort is currently underway.

A second type of chemisorption system is the oxygen/PCO cooler proposed and developed by Bard and Jones[30]. It provides cooling as low as 72 K with a specific power of 585 W/W at 80 K. This compressor has been undergoing life tests for several years accumulating over 17,000 hours and 37,000 cycles to date[31]. Studies indicate that, with a regenerative configuration, specific powers as low as 50 W/W can be achieved at 65 K.

Other chemisorption systems have been examined, to date without yielding significant benefit (e.g., Jones and Lund[32]).

CURRENT AND ANTICIPATED DEVELOPMENT EFFORTS - MATERIALS

The development efforts of JPL[33] have resulted in a well characterized process for the manufacture of saran carbon. In addition, Radebaugh[34] of the National Institute of Standards and Technology (NIST) has done an outstanding job of characterizing a wide variety of sorbents and sorbates in the temperature and pressure ranges needed for sorption cooler design. Table 2 gives an overview of the materials whose isotherms have been characterized. Work is still continuing for broader characterization of saran charcoal properties and the enhancement of its thermal conductivity.

PCO isotherms have been well characterized by JPL. The long-term endurance tests show a slight improvement in capacity and kinetics after 37,000 cycles. It would appear that long-term degradation will not pose a substantial problem for this material.

Table 2. Sorbent/Sorbate Pairs With Characterized Isotherms

	Saran	Anderson AX-31	Barneby-Cheney	Bergbau-Forchung
Helium	x		x	
Nitrogen	x	x	x	x
Methane	x	x		
Ethane		x		
Carbon Monoxide	x			
Krypton	x	x		
Xenon	x	x		

Of the hydride materials, vanadium appears to offer the most promise. It has demonstrated substantial contamination tolerance and has no inherent degradation problems. Bowman is currently conducting compressor and material characterization efforts at Aerojet. In addition he is investigating several other materials for tailored applications. A 10 K proof-of-principle hydride cooler is being developed at JPL for near-term performance demonstrations.

CURRENT AND ANTICIPATED DEVELOPMENT EFFORTS - COMPONENTS/TECHNOLOGIES

While the refrigerator stages developed so far have shown remarkable longevity and stability, there is a great deal to be accomplished before ten-year lifetimes can be achieved. The most critical need is probably in the area of contamination control and ultraclean manufacture technique. Gas phase contamination and refrigerant leakage are thought to have caused the failure of all the refrigerators which have flown to date. Systems which use J-T expansion, even with contamination tolerant designs, have to be particularly careful of contamination if they are to meet end-of-life performance requirements.

Currently, Sematech's gas handling systems used for semiconductor chip fabrication have been certified at less than 10 ppb total contamination at point-of-use[35]. The advanced charcoal refrigerator currently under development at Aerojet will be fabricated, by the vendors who built the Sematech hardware, to their standards. As sorption coolers are primarily plumbing, existing manufacture technique will permit this requirement to be met.

Making use of existing standards and materials is convenient. In the future, advanced compressor designs using super alloys like Inconel 718 will substantially reduce system mass and improve compressor performance over the conventional 316L CRES currently baselined. The development of electropolishing and ultraclean weld techniques with Inconel will become very important once the reliability goals have been met. Further weight savings are available through the incorporation of valves into the compressor body design instead of the currently baselined and existing discrete components.

JPL has also been doing considerable research into the reliability physics of Inconel compressor vessels and heater elements. Tests have demonstrated that compressor creep rate and heater life are consistent with lifetimes in excess of ten years.[31]

It is believed that currently available valve lifetimes greatly exceed even the ten-year life requirement. Wade et al[36] have conducted accelerated valve life tests equivalent to seven-year operation. Futurecraft[37] reports that they have 25 year life flight qualified valves currently available. These valves will be incorporated in the Aerojet advanced charcoal cooler.

Even if the refrigerator is built to exacting standards contamination tolerant J-T valves are likely to be of great import. The phenomenology of J-T plug formation has been studied at Harvey Mudd College[38]. Future work appears likely to lead to improvements in contamination tolerance and the development of quickly defrostable J-T valves. Work is currently underway with General Pneumatics and APD Cryogenics to develop adjustable orifice J-T valves with integral condensers and contamination tolerant heat exchangers and nozzles.

The most important component which requires development is the helium circulator used in the regenerative fluid loop design. Swift and Nutt have analyzed circulator designs and concluded that existing technology designs could meet flight requirements. These circulators typically operate at 600 psia, pumping less than 0.33 g/s across a 6 psia head. There are no efforts currently underway to demonstrate such a circulator.

Mixed gas refrigerators offer some potential for reducing system size and mass. Jones et al[39] is currently involved in the characterization of this type of system.

CURRENT AND ANTICIPATED DEVELOPMENT EFFORTS - REFRIGERATORS

There appear to be no major roadblocks to the development and demonstration of a flight worthy multistage sorption refrigerator. Three single stage systems are successfully undergoing endurance testing. However, no multistage tests have been conducted or are currently planned.

The first flight-like single stage regenerative refrigerator is currently under development at Aerojet for demonstration in 1993. Some of its basic performance specifications are as follows:

Capacity	>1 W
Temperature	130 K
Refrigerant	Krypton
Sorbent	Saran Charcoal
Volume	<1 cu ft
Mass	<40 lb
Power	<40 W/W

As described earlier it will use saran charcoal, flight qualified active valves, and be fabricated using ultrahigh purity standards. Figure 3 shows the advanced charcoal compressor design concept.

Al Johnson of the Aerospace Corp. pointed out the utility of sorption coolers for periodic cooling requirements. Johnson and Jones[40] developed several conceptual designs for this type of application Fig. 4 shows a schematic for a blow down sorption system. This system is suitable for cooling a load for a relatively short time with a longer period for recuperation (e.g., 1 hour cooling every 24-hour period). The advantage of this system is that it is simple. It eliminates the complexity added to the continuous operation sorption designs at the cost of energy efficiency. However, since these systems can spread the energy requirement over a period of hours the power requirement is quite low compared to any other type of refrigerator.

Hybrid cooler designs are likely to be developed in the future. For systems requiring very low temperatures combining a sorption cooler with a cold

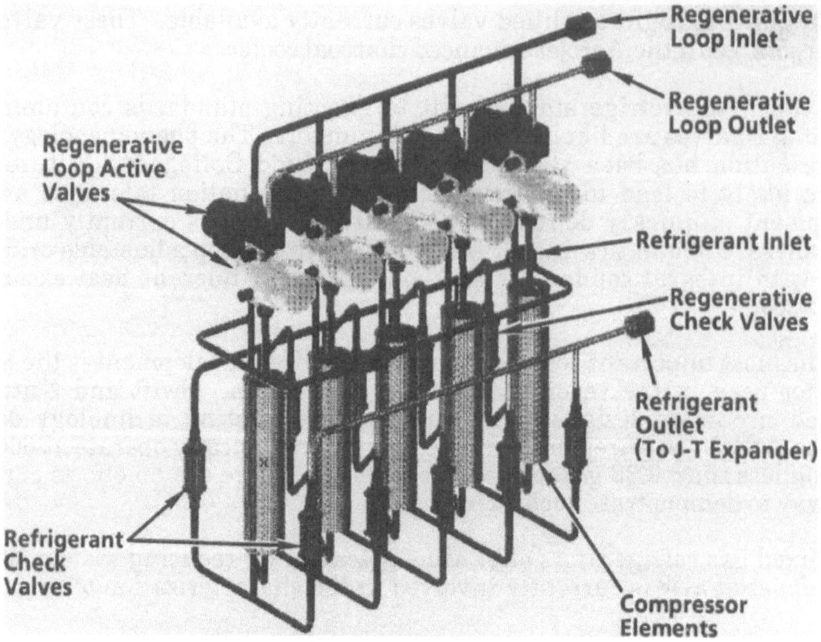

Fig. 3. Advanced Charcoal Compressor Design Concept.

magnetic stage appears to be a viable concept. Sorption coolers are very efficient down to 16 K. Magnetic coolers are good below 20 K. It would appear that for some applications, especially those requiring large capacity, a hybrid system of this sort has merit.

POTENTIAL COMMERCIAL APPLICATIONS OF SORPTION

Commercial applications of sorption systems have a long history. Absorption refrigerators have been around for generations. Cryopumps (sorption compressors) are the standard in ultrahigh vacuum systems. In addition, a great deal of research into sorption heat pumps and sorption fuel storage vessels (saran charcoal/methane and hydrides/hydrogen) is conducted worldwide. With the advent of new high performance materials such as saran charcoal and vanadium hydride it seems likely that new applications will soon be found.

Perhaps the most exciting of these is for CFC free refrigeration systems. There are environmentally safe freon substitutes. But none of these are soluble

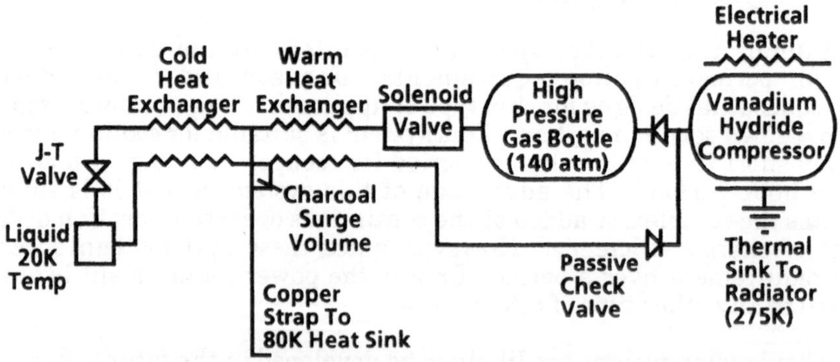

Fig. 4. Schematic of a Typical Blow Down Cooler Concept.

with oil as required by mechanical compressors. Since sorption coolers are oilless, this seems like a natural marriage. While no sorption isotherms have been made with the new refrigerants, it seems likely that the performance of a regenerative sorption cooler will rival that of mechanical ones just as it does in the cryogenic region. Single-stage sorption coolers using xenon as the refrigerant will reject heat at room temperature and provide cooling to 165K for less than 5 W/W. Surely good refrigerants will do considerably better.

If high temperature superconducting electronics are to achieve commercial success, they will need cheap and reliable cooling. The low tolerance, low piece part design of sorption coolers should lead to very inexpensive large-scale manufacturing.

SUMMARY

At present three sorption stages are undergoing long term endurance testing. JPL's PCO compressor has accumulated over 17,000 hours and their saran compressor has accumulated over 7,000 hours. Testing of Aerojet's Anderson charcoal refrigerator has been initiated, and it has accumulated nearly 2,000 hours. The longevity of these prototype laboratory brassboard coolers is highly encouraging.

The demonstration of the advanced charcoal cooler will be a major step in the development of flight qualified sorption coolers. This refrigerator is the first to be built to flight standards.

With the exception of the circulator, component development is well underway. Contamination tolerant J-T valves have been developed and continued improvement seems likely. Long-life valves have been tested individually and system tests are currently underway. Compressor testing of charcoal and PCO elements is currently underway at JPL and Aerojet is currently testing hydride and charcoal compressor elements.

Considerable progress has been made in the past five years in the location and characterization of high performance sorbent materials. Continued research by NIST, JPL and Aerojet is likely to result in improved system performance and the first viable flight cryocooler.

ACKNOWLEDGEMENTS

I gratefully acknowledge the assistance and ideas of S. Bard and J. Jones on numerous conversations over the past five years.

REFERENCES

1. J. M. Vickers, "Intermittent Type Silica Gel Adsorption Refrigerator," U.S. Patent #3,270,512 filed August 1963.

2. W. Hartwig, "Adsorption Pumping Cryogenic Refrigerator Studies", 47th Monthly Progress Report, Contract No. NASA 9-14491, National Aeronautics and Space Administration, Lyndon B. Johnson Space Center, December 1, 1978 - December 31, 1979.

3. W. H. Hartwig, "Cryogenic Refrigeration Concepts Utilizing Adsorption Pumping in Zeolites", Advances in Cryogenic Engineering 23, p. 438, Plenum Publishing Co., N. Y., 1978.

4. W. H. Hartwig, "Requirements for and Status of a 4.2K Adsorption Refrigerator Using Zeolites", Proceedings of the Conference on Refrigeration for Cryogenic Sensors and Electronic Systems, Boulder, Colorado, 1980, NBS SP 607.

5. S. Bard, "Development of an 80-120K Charcoal-Nitrogen Adsorption Cryocooler", Forth Biennial International Cryocooler Conference, Easton, MD, September 1986.

6. C. K. Chan, Performance of Long Life J-T Cryocooler, paper presented at Interagency Meeting on Cryocoolers, Easton, Maryland, 1986.

7. S. Bard, "Improving Adsorption Cryocoolers by Multi-Stage Compression and Reducing Void Volume", Cryogenics 26, 1986.

8. D. F. Quinn et al, "Solid Adsorbents for Storage of CMG for Automotive Use-Saran Carbon", Alternate Energy Conference, Windsor, Ontario, Canada, 1985.

9. H. R. Schember, "Development of an Adsorption Compressor for use in Cryogenic Refrigeration", AIAA-89-0076, 27th Aerospace Sciences Meeting, Reno, Nevada, Jan. 1989.

10. G. Mon et al, "Reliability and Life of Sorbent Materials for Sorption Cryocoolers," in Proc. Sixth International Cryocooler Conference, October 25, 1990, Plymouth, Massachusetts.

11. L. A. Wade, "Parametric Performance Analysis of Sorption/Magnetic Refrigerators, Final Report", Aerojet Electronic Systems Division Report 8936, Nov. 1987.

12. S. Bard and J. A. Jones, "Regenerative Sorption Compressors for Cryogenic Refrigeration", Advances in Cryogenic Engineering 35b, p. 1357, Plenum Press, N. Y. 1990.

13. P. Sywulka, patent application filed June 1990.

14. L. Wade, P. Sywulka, M. Hatter, and J. Alvarez, "High Efficiency Sorption Refrigerator Design," Advances in Cryogenic Engineering 35b, p. 1375, Plenum Press, N.Y. 1990.

15. L. Wade, E. Ryba, C. Weston, and J. Alvarez, "Test Performance of an Efficient 2W, 137K Sorption Refrigerator," submitted to Cryogenics 1991.

16. P. Roach and K. Gray, "Low-Cost, Compact Dilution Refrigerator: Operation from 200 to 20 mK," Advances in Cryogenic Engineering 33, Plenum Press, N.Y. 1988.

17. L. Duband, C. Alsop and A. Lange, "A Rocket-Borne Helium 3 Refrigerator," Advances in Cryogenic Engineering 35b, p 1447, Plenum Press, N.Y. 1990.

18. G. Prast, C.M. Hargreaves, A. Mijnheer, and H.H. Van Mal, "Proceedings of the 13th International Congress on Refrigeration," Washington D.C. 1971.

19. H. H. Van Mal and A. Mijnheer, "Hydrogen Refrigerator for the 20K Region with a $LaNi_5$ Hydride Thermal Absorption Compressor for Hydrogen," Proc ICEC 4, IPC Science & Technology Press, Guildforn, UK, 1972.

20. D. Lehrfeld and O. Boser, "Absorption-Desorption Compressor for Spaceborne/Airborne Cryogenic Refrigerators," Technical Report AFFDL-TR-74-21, Air Force Flight Dynamics Laboratory, Wright-Patterson AFB, Ohio, March 1974.

21. W.A. Steyert, "New Heat Transfer Geometry for Hydride Heat Engines and Heat Pumps," report LA-7822, Los Alamos Scientific Lab., New Mexico, July 1979.

22. J.A. Jones and P.M. Golben, "Design, Life Testing, and Future Designs of Cryogenic Hydride Refrigeration Systems,"Cryogenics 25, p 212, April 1985.

23. K. Karperos, "Operating Characteristics of a Hydrogen sorption Refrigerator, Part 1: Experiment Design and Results," Proceedings of the Fourth International Cryocoolers Conference, Easton, Maryland, September 1986.

24. L. A. Wade, "Operating Characteristics of a Hydrogen Sorption refrigerator, Part II: A Comparison Between a Second Order Analysis and Empirical Data." Proceedings of the Fourth International Cryocoolers Conference, Easton Maryland, September 1986.

25. Y. Matsubara, M. Kaneko, J. Suzuki, and k. Hirosawa, "High Response Hydride Compressor for Regenerative Cryocooler," Proceedings of the Fourth International Cryocoolers Conference, Easton, Maryland, September 1986.

26. R. Bowman, B. Freeman, and R. Phillips, "Application of Vanadium Hydride Compressors for Joule-Thomson Cryocoolers," submitted to Advances in Cryogenic Engineering, 1991.

27. T. Kumano, B. Tada, Y. Tsuchida, Y. Kuraoka, T. Ishige, and H. Baba, "Development of High Pressure Metal Hydride for a Compressor," Metal-Hydrogen Conference, Stuttgart, W. Germany, September 1988.

28. J. A. Jones, Ten Kelvin Hydride Refrigerator, U.S. Patent No. 4 641 499, February 1987.

29. J. Rodriguez and A. Mills, "Development of a Solid Hydrogen Sorption Refrigerator Stage, Phase I, Final Report," JPL D-5070, February 1988, Jet Propulsion Laboratory, Pasadena, California.

30. S. Bard and J. Jones, "Development and Testing of an 80 K Oxide Sorption Cryocooler," Proceedings of the 5th International Cryocooler Conference, Monterey, California, August 1988.

31. S. Bard, C. Blue, and B. Bowlten, "Reliability and Physical Analysis of a Life-Tested Sorption Compressor for Cryogenic Refrigeration," presented at the CEC/ICMC Conference, Huntsville, Alabama, June 1991.

32. J. Jones and A. Lund, "Sorption J-T Refrigeration Utilizing Manganese Nitride Chemisorption," Advances in Cryogenic Engineering 35b, Plenum Press, N.Y. 1990.

33. A. Yavrovrian, private communication, April 1991.

34. R. Radebaugh, private communication, Jan. 1991.

35. D. Hope, et al., "Installing and Certifying Sematech's Bulk-Gas Delivery Systems," Microcontamination, May 1990.

36. See Reference 15.

37. Futurecraft, City of Industry, California.

38. L. A. Wade, C. Donnelly, E. Joham, K. Johnson, R. Phillips, E. Ryba, B. Self, and R. Stanton, "An Investigation into the Mechanics of J-T Valve Plug Formation," Advances in Cryogenic Engineering 33, Plenum Press, N.Y. 1988.

39. J. Jones, S. Petrick, and S. Bard, "Mixed Gas Sorption Joule-Thomson Refrigerator," NASA Tech Briefs, May 1991.

40. A. Johnson and J. Jones, patent application filed 1991.

THE PREDICTION OF THE VAPOR-LIQUID EQUILIBRIUM THERMODYNAMIC PROPERTIES OF N_2-Ar-O_2 MIXTURES INCLUDING AIR

E. W. Lemmon, R. T Jacobsen, and S. W. Beyerlein

Center for Applied Thermodynamic Studies
University of Idaho, College of Engineering
Moscow, Idaho

ABSTRACT

An extended corresponding states model for calculating the vapor-liquid equilibrium (VLE) thermodynamic properties for mixtures of nitrogen, argon, and oxygen has been developed. The model is based on four reference fluids: nitrogen, argon, oxygen, and air as a pseudo-pure fluid. The use of air as one of the reference fluids assures that mixture properties default to those for air when the mixture composition is that of air. Mixture Helmholtz energies are predicted using a Lagrangian interpolation method involving the acentric factors of the component. Modified van der Waals mixing rules containing empirical, mole fraction dependent, binary and ternary interaction parameters are used to determine the pseudo-critical properties of the mixture.

INTRODUCTION

Vapor-liquid equilibrium (VLE) properties describe the thermodynamic state of a mixture of fluids where both liquid and vapor phases are present. Equilibrium states can be defined by various combinations of independent thermodynamic properties including pressure, temperature, liquid composition, and vapor composition. A VLE model for air and air-like mixtures has been developed in this work as a companion to a new wide-range equation of state for air by Jacobsen et al.[1] The VLE model is based on the pure fluid equations of state for three major constituents of air: nitrogen[2], argon[3], and oxygen[4], as well as the equation of state for air[1] as a pseudo-pure fluid. The model was developed for compositions near that of air, but it is theoretically valid for binary and ternary mixtures of nitrogen, oxygen, and argon.

FUNDAMENTAL EQUATIONS OF STATE

The equations of state used in this work are explicit in Helmholtz energy with independent variables of temperature and density,

$$A/RT = \alpha(\delta,\tau) = \alpha^{o}(\delta,\tau) + \bar{\alpha}(\delta,\tau). \qquad (1)$$

In this equation, A is the Helmholtz energy, $\delta = \rho/\rho_c$, $\tau = T_c/T$, ρ is the density, T is the temperature, ρ_c is the critical density, and T_c is the critical temperature.

The term, $\alpha^{o}(\delta,\tau)$, is the ideal gas contribution to the dimensionless Helmholtz energy. For the pure fluid components,

$$\alpha^{o}(\delta,\tau) = \frac{H^{o}_{o}\tau}{RT_c} - \frac{S^{o}_{o}}{R} - 1 + \ln\frac{\delta\tau_o}{\delta_o\tau} - \frac{\tau}{R}\int_{\tau_o}^{\tau}\frac{C^{o}_{p}}{\tau^2}d\tau + \frac{1}{R}\int_{\tau_o}^{\tau}\frac{C^{o}_{p}}{\tau}d\tau, \tag{2}$$

where $\delta_o = P_o/\rho_c RT_o$, $\tau_o = T_c/T_o$, P_o is the reference state pressure and T_o is the reference state temperature. The ideal gas contribution includes the integral of the isobaric heat capacity, which is represented by an equation dependent only on temperature. The term, $\bar{\alpha}$, is the real fluid (compressibility) contribution to the dimensionless Helmholtz energy,

$$\bar{\alpha}(\delta,\tau) = \sum_k N_k \delta^i \tau^j \exp(-\gamma\delta^\ell), \tag{3}$$

where i, j, and ℓ are exponents, and the N_k are the coefficients of the equation determined by fitting experimental data. The coefficient, γ, is zero when ℓ is zero and is one when ℓ is not zero.

CALCULATION OF FUGACITY

In a two-phase, non-reacting mixture of m components, the thermodynamic constraints for VLE are,

$$T' = T'' = T, \tag{4}$$

$$P' = P'' = P, \text{ and} \tag{5}$$

$$f'_i = f''_i, \quad i = 1, 2, ..., m, \tag{6}$$

where the superscripts ' and " refer to the liquid and vapor phases, respectively. The f_i are the fugacities of the ith component. Fugacities can be calculated from a fundamental equation of state explicit in Helmholtz energy,

$$f_i = x_i \rho RT \exp[\partial(n\bar{\alpha})/\partial n_i]_{T,V,n_{j,j\neq i}}. \tag{7}$$

where the x_i are the mole fractions of the specified phase. The derivative, $\partial(n\bar{\alpha})/\partial n_i$, is calculated numerically.

CORRESPONDING STATES THEORY

Corresponding states theory relates properties of one fluid to those of another in reduced form by assuming that the thermodynamic surface for fluids has the same general shape in reduced coordinates. The reducing parameters are generally the critical parameters of the fluid.

The acentric factor indicates the degree to which a fluid diverges from the reduced behavior of a monatomic fluid such as argon. The acentric factor was defined by Pitzer et al.[5] as,

$$\omega = [-\log(P_{\sigma r}) - 1.0], \tag{8}$$

where $P_{\sigma r}$ is the reduced vapor pressure of the fluid evaluated at a reduced temperature of 0.7. Pitzer showed that the compressibility factor can be written as a linear function of the acentric factor,

$$Z = Z^o + \omega Z^r, \tag{9}$$

where Z^o is the contribution from a simple fluid with zero acentric factor, and Z^r is a function that takes into account the departure from spherical geometry of the molecule. The compressibility factors are calculated at the same reduced temperature and pressure.

Corresponding states theory can be extended to use several pure fluid equations of state to predict properties of fluid mixtures. The corresponding states equation of Lee and Kesler[6] is,

$$Z = Z^o + (\omega/\omega^r)(Z^r - Z^o). \tag{10}$$

In Equation (10), r represents a reference fluid that is not monatomic, and o represents a simple fluid. The thermodynamic properties of the reference fluid are generally similar to those of the fluid or mixture being predicted.

The work of Teja et al.[7] replaced the monatomic fluid, r1, with a second complex fluid, r2,

$$Z = Z^{r1} + [(\omega - \omega^{r1})/(\omega^{r2} - \omega^{r1})]\,(Z^{r2} - Z^{r1}). \tag{11}$$

Hwang[8] presented a three-reference fluid model which is equivalent to a quadratic interpolation,

$$Z = Z^{r1}\frac{(\omega - \omega^{r2})\,(\omega - \omega^{r3})}{(\omega^{r1} - \omega^{r2})\,(\omega^{r1} - \omega^{r3})} + Z^{r2}\frac{(\omega - \omega^{r1})\,(\omega - \omega^{r3})}{(\omega^{r2} - \omega^{r1})\,(\omega^{r2} - \omega^{r3})}$$
$$+ Z^{r3}\frac{(\omega - \omega^{r1})\,(\omega - \omega^{r2})}{(\omega^{r3} - \omega^{r1})\,(\omega^{r3} - \omega^{r2})}. \tag{12}$$

LAGRANGIAN INTERPOLATION

The basis for this work is a four-reference fluid model in which nitrogen, oxygen, argon, and air (as a pseudo pure fluid) serve as the reference fluids. A Lagrangian interpolation is used to determine the residual Helmholtz energy of the mixture. The expression for the Helmholtz energy of the mixture, predicted from the reference fluids is,

$$\bar{\alpha}_m = \sum_{i=1}^{4} L_i(\omega)\,\bar{\alpha}_i(T/T_{cm}, \rho/\rho_{cm}), \text{ where} \tag{13}$$

$$L_i(\omega) = \prod_{\substack{j=1 \\ j \neq i}}^{4} \frac{(\omega_m - \omega_j)}{(\omega_i - \omega_j)}. \tag{14}$$

T_{cm}, P_{cm} and ω_m are mixture values determined using mixing rules. The equivalent equation for density is derived by solving for the mixture specific volume, and then inverting the expression,

$$\rho_m/\rho_{cm} = \{\sum_{i=1}^{4} L_i(\omega)/[\rho_i(T/T_{cm},P/P_{cm})/\rho_{ci}]\}^{-1} . \qquad (15)$$

The use of this interpolation model is advantageous in this work because air has been characterized by a pseudo-pure fluid equation of state. For this composition, the predicted values default to the pseudo-pure fluid equation of state for air when the air equation is included as a reference fluid equation.

PSEUDO-CRITICAL MIXING RULES

The pseudo-critical temperature, density, pressure and the acentric factor for the mixture were calculated using van der Waals mixing rules[8]. These rules are shown in the following equations.

$$T_{cm} = [\sum\sum x_i x_j T_{cij} v_{cij}]/v_{cm} \qquad (16)$$

$$v_{cm} = \sum\sum x_i x_j v_{cij} \qquad (17)$$

$$P_{cm} = (0.2905 - 0.085\ \omega_m)RT_{cm}/v_{cm} \qquad (18)$$

$$\omega_m = \sum x_i\ \omega_i \qquad (19)$$

$$v_{cij} = \theta\ \xi_{ij}(v_{ci}^{1/3} + v_{cj}^{1/3})^3/8 \qquad (20)$$

$$T_{cij} = \sigma\ \zeta_{ij}(T_{ci}T_{cj})^{1/2} \qquad (21)$$

$$v_{ci} = (0.2905 - 0.085\ \omega_i)RT_{ci}/P_{ci} \qquad (22)$$

In Equations (20) and (21), ζ_{ij} and ξ_{ij} are binary interaction parameters and θ and σ are ternary interaction parameters discussed in the next section. The index, i, refers to nitrogen, argon, and oxygen, respectively.

For a pure fluid, when the critical volume is given by Equation (22), the mixing rules default to the critical point of that fluid. However, this is not true for air. The pseudo-critical point for air must be calculated using the critical points of nitrogen, argon, and oxygen. Furthermore, the properties at the maxcondentherm are used as reducing parameters in the equation of state for air, rather than those at the critical point. If the critical point determined by the mixing rules is not identical to the maxcondentherm of the equation of state for air, the properties of air mixtures calculated from the model do not agree with the values from the equation of state for air. For this reason, the reducing parameters for air were reassigned the values predicted from the mixing rules at the composition of air and the linear coefficients, N_k, in Equation (3) for air, were adjusted.

BINARY AND TERNARY INTERACTION PARAMETERS

Binary and ternary interaction parameters can be used to adjust VLE models to represent available experimental data by accounting for interactions between molecules of different species. The interactions between similar molecules are accounted for by the interaction parameters set to one. The interaction parameters between unlike molecules account for the change in mixture properties from that of an ideal mixture.

In this work, selected data from Wilson[9] and the data of Hiza[10] were used to determine the interaction parameters. The data of Hiza[10] were used to determine the binary interaction parameters for the N_2-O_2 and N_2-Ar binary mixtures. Selected data of Wilson[9] were used to determine the ternary interaction parameters for the ternary mixture. Both data sets were used to determine the binary interaction parameters for the Ar-O_2 binary.

A non-linear, least squares Marquardt algorithm[11] was used to determine the coefficients in the proposed interaction parameter functions.

$$\xi_{13} = 1.0442663 - 0.0252890[x_{N2}/(x_{N2}+x_{O2})] \quad (23)$$

$$\zeta_{13} = 1.0011516 - 0.0016409[x_{N2}/(x_{N2}+x_{O2})] \quad (24)$$

$$\xi_{12} = 1.0152796 - 0.0075723[x_{N2}/(x_{N2}+x_{Ar})] \quad (25)$$

$$\zeta_{12} = 0.9982822 - 0.0006550[x_{N2}/(x_{N2}+x_{Ar})] \quad (26)$$

$$\xi_{23} = 0.9670891 + 0.0169474[x_{O2}/(x_{Ar}+x_{O2})] \quad (27)$$

$$\zeta_{23} = 0.9965121 - 0.0039081[x_{O2}/(x_{Ar}+x_{O2})] \quad (28)$$

$$\theta = 1 - 0.141689 x_{N2}x_{O2}x_{Ar} \quad (29)$$

$$\sigma = 1 + 0.068842 x_{N2}x_{O2}x_{Ar} \quad (30)$$

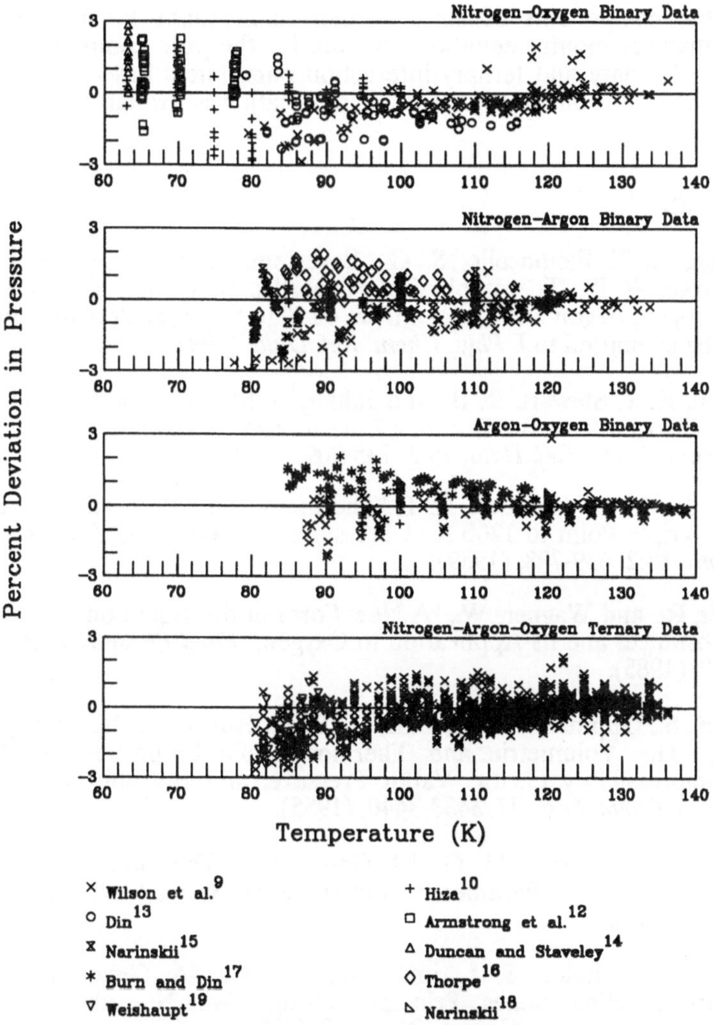

Figure 1. Comparisons of Selected Experimental VLE Data to Predictions from the Four Reference Fluid Model.

ACCURACY ASSESSMENT

The model presented here provides accurate properties for binary and ternary mixtures of nitrogen, argon and oxygen over a wide range of temperatures, pressures and compositions. Figure 1 compares selected experimental VLE data to predictions from the model. The model can be used to predict equilibrium pressures of nitrogen-oxygen binary mixtures to within 1% for temperatures above 90 K, and to within 3% for temperatures below 90 K. For both nitrogen-argon and oxygen-argon binary mixtures, the model predicts equilibrium pressures within 2% for temperatures above 100 K, and within 3% for temperatures below 100 K. The model can be used to predict equilibrium pressures of ternary nitrogen-argon-oxygen mixtures to within 2% for temperatures above 90 K and within 3% for temperatures below 90 K.

CONCLUSIONS AND RECOMMENDATIONS

An extended corresponding states VLE model has been developed as an adjunct to the pseudo-pure fluid equation of state for air developed by Jacobsen et al.[1]. This model allows the calculation of mixture VLE thermodynamic properties for air-like mixtures. It can be extended to include any number of fluids, and is not limited to the components of air. Further applications for hydrocarbon and refrigerant mixtures require equations of state for the pure components as well as optimization of binary and ternary interaction parameters based on experimental data. For many fluid mixtures, sufficient information is available to develop such correlations.

REFERENCES

1. Jacobsen, R. T, Penoncello, S. G., Beyerlein, S. W., Lemmon, E. W., and Clarke, W. P., "Thermophysical Properties of Air and Similar Mixtures of Nitrogen, Argon and Oxygen from 65 K to 873 K at Pressures to 70 MPa," to be submitted to *J. Phys. Chem. Ref. Data*, (1991).

2. Jacobsen, R. T, Stewart, R. B., and Jahangiri, M., "Thermodynamic Properties of Nitrogen from the Freezing Line to 2000 K at Pressures to 1000 MPa," *J. Phys. Chem. Ref. Data*, 15:2, 735-909, (1986).

3. Stewart, R. B. and Jacobsen, R. T, "Thermodynamic Properties of Argon from the Triple Point to 1200 K at Pressures to 1000 MPa," *J. Phys. Chem. Ref. Data*, 18:2, 639-798, (1989).

4. Schmidt, R., and Wagner, W., "A New Form of the Equation of State for Pure Substances and its Application to Oxygen," *Fluid Phase Equilibria*, 19, 175-200, (1985).

5. Pitzer, K. S., Lippmann, D. Z., Curl, Jr., R. F., Huggins, C.M., and Peterson, D. E., "The Volumetric and Thermodynamic Properties of Fluids. II. Compressibility Factor, Vapor Pressure, and Entropy of Vaporization," *J. Am. Chem. Soc.*, 77, 3433-3440, (1955).

6. Lee, B. I., and Kesler, M. G., "A Generalized Thermodynamic Correlation Based on Three-Parameter Corresponding States," *AIChE J.*, 21:3, 510-527, (1975).

7. Teja, A. S., Sandler, S. I., and Patel, N. C., "A Generalization of the Corresponding States Principle Using Two Nonspherical Reference Fluids," *The Chem. Engr. J.*, 21, 21-28, (1981).

8. Hwang, S-C, "Vapor-Liquid Equilibrium Calculations for N_2-Ar-O_2 Mixtures with Modified B-W-R Equation of State and a Corresponding States Principle," *Fluid Phase Equilibria*, 37, 153-167, (1987).

9. Wilson, G. M., Silverberg, P. M., and Zellner, M. G., "Argon-Oxygen-Nitrogen Three Component System Experimental Vapor-Liquid Equilibrium Data," *Adv. in Cryo. Eng.*, 10, 192-207, (1965).

10. Hiza, M. J., "Liquid-Vapor Equilbria Data on Mixtures of Nitrogen, Oxygen, and Argon," unpublished results from the Boulder Laboratories of the Thermophysics Division, Center for Chemical Technology, National Institute of Standards and Technology, (1990).

11. Marquardt, D. W., "An Algorithm for Least Squares Estimation of Nonlinear Parameters," *J. Soc. Indust. Appl. Math.*, 11:2, 431-441, (1963).

12. Armstrong, G. T., Goldstein, J. M., and Roberts, D. E., "Liquid-Vapor Phase Equilibrium in Solutions of Oxygen and Nitrogen at Pressures Below One Atmosphere," *J. Research NBS*, 55:5, 265-277, (1955).

13. Din, F., "The Liquid-Vapour Equilibrium of the System Nitrogen + Oxygen at Pressures up to 10 atm," *Trans. Faraday Soc.*, 56, 668-681, (1960).

14. Duncan, A. G. and Staveley, L. A. K., "Thermodynamic Functions for the Liquid Systems Argon + Carbon Monoxide, Oxygen + Nitrogen, and Carbon Monoxide + Nitrogen," *Trans. Faraday Soc.*, 62, 548-552, Pt. 3, (1966).

15. Narinskii, G. B. , "Liquid-Vapour Equilibrium in the Argon-Nitrogen System I. Experimental Data and Their Verification," *Russ. J. Phys. Chem.*, 40:9, 1093-1096, (1966).

16. Thorpe, P. L., "Liquid-Vapour Equilibrium of the System Nitrogen + Argon at Pressures up to 10 atm," *Trans. Faraday Soc.*, 64:549, 2273-2280, (1968).

17. Burn, I., and Din, F., "Liquid-Vapour Equilibrium of the System Argon + Oxygen at Pressures up to 10 Atmospheres," *Trans. Faraday Soc.*, 58, 1341-1356, (1962).

18. Narinskii, G. B., "Experimental Data on Liquid-Vapor Equilibrium in the O_2-Ar System at 90.5 K, 110 K, 120 K," *Kislorodnyi*, 10:3, 9-16, (1957).

19. Weishaupt, J., "Bestimmung des Gleichgewichtes siedender Stickstoff-Argon-Sauerstoffgemische bei 1000 Torr," *Angew. Chem.*, 20:12, pp. 321-316, (1948).

DEW AND BUBBLE POINT PROPERTIES OF AIR

S. G. Penoncello, R. T Jacobsen and E. W. Lemmon

Center for Applied Thermodynamic Studies
University of Idaho, College of Engineering
Moscow, Idaho

ABSTRACT

Four new ancillary functions for the calculation of pressures and densities of states at the bubble and dew points of air are presented. These functions were developed using experimental data and calculated values. The experimental data for the bubble and dew point pressures and densities of air are summarized and evaluated. In the absence of experimental data at high-pressure phase equilibrium states, a Leung-Griffiths model modified for ternary mixtures was used to calculate pseudo-data. This ternary mixture model was also used to calculate new values for the critical point, maxcondenbar and maxcondentherm for air. The calculated properties at the maxcondentherm were used as reducing parameters in the ancillary functions. Graphical comparisons of the ancillary equations to the experimental data and pseudo-data are presented to justify the estimated accuracies of the new ancillary functions. The equations presented here have been used to calculate dew and bubble point pressures and densities for the determination of the phase boundary for a wide-range equation of state for air treated as a pseudo-pure fluid.

INTRODUCTION

Since air is a mixture, conventional pure fluid characteristics cannot be used in the development of the ancillary functions. Perhaps the most prominent difference is in the critical region. The critical point is not the point of maximum pressure and maximum temperature on the coexistence boundary as it is for a pure fluid. As shown in Figure 1, the critical point of air lies approximately 0.1 K below the point of maximum temperature (maxcondentherm) and approximately 0.01 MPa below the point of maximum pressure (maxcondenbar).

The existing experimental data for the bubble and dew point pressures and densities of air are summarized in Table 1. The data of Blanke[1] and Michels et al.[4] were considered to be the most accurate representations of the phase boundaries available. This conclusion is verified by several other investigators including Rainwater and Jacobsen[6], Sychev et al.[7] and Blanke and Weiss[8]. These experimental data were used in this study to develop the ancillary functions for air.

The critical region for air was modeled using a scaling law approach for mixtures known as the Leung-Griffiths model. The model used in this study was based upon original work of Leung and Griffiths[9] and subsequently modified for binary mixtures by Moldover, Rainwater and co-workers[10-13]. This method was extended to ternary mixtures and applied to the nitrogen-argon-oxygen system by

Advances in Cryogenic Engineering, Vol. 37, Part B
Edited by R.W. Fast, Plenum Press, New York, 1992

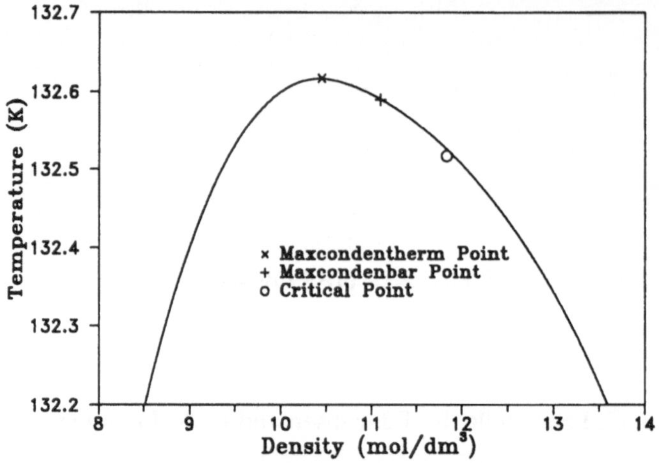

Fig. 1. Critical region phase boundary for air with fixed points determined from the ternary Leung-Griffiths model.

Van Poolen[14]. Calculated "pseudo-data" generated from this ternary model and the experimental data of Blanke[1] and Michels et al.[4] were used in the development of the ancillary functions.

The basic idea of the Leung-Griffiths model is to transform the coexistence boundary from "density-space" to "field-space". The field variables have equivalent values for coexisting liquid and vapor phase, whereas the density variables have different values. Conventional ancillary functions for bubble and dew point density are a mixed representation of both field variables (P or T) and a density variable (ρ).

The Leung-Griffiths model is not constrained to a pre-determined critical point, maxcondentherm or maxcondenbar. These points can be determined once the model is optimized using experimental vapor-liquid equilibrium (VLE) data. In this study, air is treated as a ternary mixture with mole fractions of nitrogen ($x=0.7812$), argon ($x=0.0092$) and oxygen ($x=0.2096$). The critical point lies on the critical surface of this mixture. However, the maxcondentherm and maxcondenbar are on the isopleth representing the composition of air, but removed from the

Table 1
Experimental Data for Air on the Phase Boundary

Author	Year	Temperature Range (K)	Number of data	Data form*
Blanke[1]	1973	60-129	9	bpp
Blanke[1]	1973	67-132.3	11	dpp
Furukawa and McCoskey[2]	1953	60-85	8	dpp
Kuenen and Clark[3]	1917	123-132.5	14	bpp
Michels et al.[4]	1954	118-131.9	3	bpp
Michels et al.[4]	1954	118-132.6	6	dpp
Walker et al.[5]	1966	60-64	5	bpp
Blanke[1]	1973	60-128	9	bpd
Blanke[1]	1973	67-132.3	11	dpd
Kuenen and Clark[3]	1917	123-132.5	11	bpd
Kuenen and Clark[3]	1917	129-132.5	8	dpd
Michels et al.[4]	1954	118-131.9	13	bpd
Michels et al.[4]	1954	118-132.6	18	dpd

* bpp = bubble point pressure   dpp = dew point pressure
bpd = bubble point density   dpd = dew point density

Table 2
Calculated Values of the Critical Point, Maxcondentherm and Maxcondenbar for Air

	Pressure (MPa)	Density (mol/dm^3)	Temperature (K)
Critical Point	3.7860	11.8308	132.5168
Maxcondentherm	3.78502	10.447700	132.61738
Maxcondenbar	3.78909	11.094825	132.58970

critical surface. The values of the critical point, maxcondentherm and maxcondenbar points for air from the optimized Leung-Griffiths model reported by Van Poolen[14] are given in Table 2.

EQUATIONS FOR THE BUBBLE AND DEW PROPERTIES OF AIR

Generally, the critical point properties are used as reducing parameters in ancillary functions representing the phase boundary. However, for air, using the critical point as a reducing parameter would make the dew point density and pressure equations double valued for temperatures between the critical temperature and the maxcondentherm temperature. To eliminate this problem, the maxcondentherm properties are used as reducing parameters in the ancillary functions. The ancillary functions for the bubble point properties also include the dew point values between the critical point and the maxcondentherm.

The data used to develop these ancillary functions include the experimental points of Blanke[1] and Michels et al.[4] and the pseudodata generated from the ternary Leung-Griffiths model. In the critical region where measurements do not exist, the pseudodata were used.

The functional form for the bubble and dew point pressures is

$$\ln(P/P_j) = (T_j/T) \sum_{i=1}^{30} N_i(1-T/T_j)^{i/2} \tag{1}$$

where P_j and T_j are the maxcondentherm pressure and temperatures, respectively. The values of the coefficients, N_i, are given in Table 3.

The bubble and dew point densities were correlated using the function:

$$\ln(\rho/\rho_j) = \sum_{i=1}^{25} N_i(1-T/T_j)^{i/3} \tag{2}$$

In Equation (2), T_j and ρ_j are the maxcondentherm temperature and density, respectively. The coefficients for these bubble and dew point density equations are given in Table 4.

Table 3
Coefficients for the Bubble and Dew Point Pressures for Air - Equation (1)*

Bubble Point Pressure	Dew Point Pressure
N_1 = 0.2095592444	N_1 = -0.1537763029
N_2 = -6.654905539	N_2 = -5.544542064
N_4 = 22.13718815	N_{10} = 312.7182733
N_5 = -84.14553609	N_{11} = -895.9553274
N_6 = 135.9753732	N_{13} = 1834.176566
N_7 = -83.66895082	N_{14} = -1321.892808
N_{12} = 17.97856602	

*Coefficients not listed are zero.

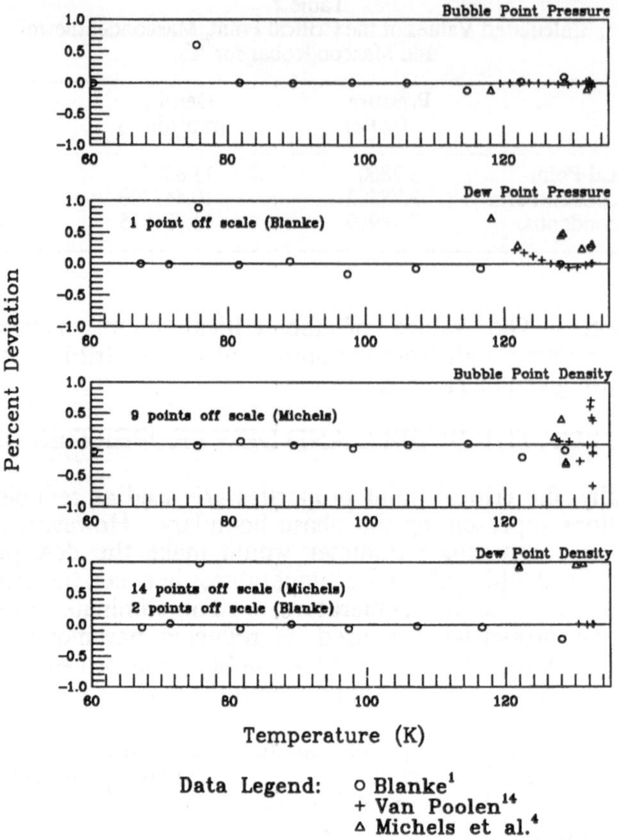

Figure 2. Comparisions of Experimental and Predicted Data to Values Calculated Using Correlated Equations

A graphical comparison of bubble and dew point properties calculated using Equations (1) and (2) to experimental and calculated data are shown in Figure 2. In this figure, the percent deviation in a property, x, is given by

$$\%\Delta x = 100(x_{data} - x_{calc})/x_{data}. \qquad (3)$$

Table 5 shows numerical comparisons in the form of the absolute average deviation (AAD) and the bias (BIAS). These values are defined by,

Table 4
Coefficients for the Bubble and Dew Point Densities for Air-Equation (2)*

Bubble Point Density		Dew Point Density	
N_2	= 19.04163405	N_1	= -0.4260410355
N_4	= -477.1948003	N_2	= -9.794208888
N_6	= 15083.59618	N_3	= 23.87307954
N_7	= -70795.97403	N_6	= -1549.372838
N_8	= 152653.7408	N_7	= 7459.189780
N_9	= -169972.7900	N_8	= -14627.38944
N_{10}	= 82260.28048	N_9	= 11872.26681
N_{13}	= -12399.98758	N_{11}	= -4091.109590
N_{15}	= 3657.873321	N_{15}	= 1782.608401
		N_{17}	= -936.1183672

*Coefficients not listed are zero.

Table 5
Summary of Deviations of Calculated Properties from Selected Experimental and Calculated Data

Form	Michels et al.[4]		Blanke[1]		Van Poolen[14]	
	AAD	BIAS	AAD	BIAS	AAD	BIAS
Bubble point pressure	0.077	-0.062	0.096	0.068	0.021	-0.007
Dew point pressure	0.427	0.427	0.275	-0.076	0.065	0.038
Bubble point density	3.380	-3.297	0.067	0.055	0.290	0.083
Dew point density	1.410	1.410	0.047	-0.035	0.001	0.000

$$AAD = \sum |\%\Delta x|/n, \text{ and} \tag{4}$$

$$BIAS = \sum (\%\Delta x)/n, \tag{5}$$

where n is the number of points in the data set. The AAD reflects the accuracy of the data, and the BIAS is an indication of the precision.

With the exception of the dew point pressure, the data of Blanke[1] show absolute average deviations less than 0.10 percent with low bias values. The higher average deviations for the dew point pressure are caused by a single point at 71.689 K. In general, the data of Michels et al.[4] show larger deviations, especially on the coexisting density lines. These data are in the critical region at temperatures greater than 118 K where experimental uncertainties are larger than in other regions. The dew point density values of Michels et al.[4] are inconsistent with the Blanke[1] data and the ternary Leung-Griffiths model. As a result, these data were not used in the determination of the dew point density equation. The pseudo-data calculated with the ternary Leung-Griffiths model have absolute average deviations less than 0.07 percent with the exception of the bubble point density, where the AAD is 0.29 percent. These deviations are well within the expected accuracy of the model.

CONCLUSIONS

Four new ancillary functions for the representation of the bubble and dew point pressures and densities for air have been developed. These equations were based on selected experimental data of Michels et al.[4] and Blanke[1]. In the critical region, a modified ternary Leung-Griffiths model was used to calculate pseudodata. These pseudodata were included with the experimental data where appropriate in the development of the ancillary functions. The modified Leung-Griffiths model was also used to determine new values for the critical point, maxcondentherm and maxcondenbar. The calculated properties at the maxcondentherm are used as reducing parameters in the ancillary functions. The accuracy of the ancillary equations is estimated to be ±0.10 percent in density or pressure along the bubble point and dew point curves from 60 K to the maxcondentherm.

REFERENCES

1. Blanke, W., Messung der thermischen Zustandgrossen von Luft im Zweiphasengebiet und seiner Umgebung, Ph.D. dissertation, Ruhr University, Bochum, Fed. Rep. Germany (1973).

2. Furukawa, G. T. and McCoskey, R. E., "The Condensation Line of Air and the Heats of Vaporization of Oxygen and Nitrogen," Tech. Note, U.S. NACA No. 2969, (1953).

3. Kuenen, J. P. and Clark, A. L., "Critical Point, Critical Phenomena and a Few Condensation Constants of Air," Commun. Phys. Lab. Univ. Leiden, 150:124, (1917).

4. Michels, A., Wassenaar, T., and Van Seventer, W., "Isotherms of Air Between 0° C and 75° C and at Pressures up to 2000 Atm," *Appl. Sci. Res.*, A4, (1954).

5. Walker, G., Christian, W. J., and Budenholzer, R.A., "The Vapor Pressure of Dry Air at Low Temperatures," *Adv. Cryog. Engr.*, 11, (1966).

6. Rainwater, J. C. and Jacobsen, R. T, "Vapor-Liquid Equilibrium of Nitrogen-Oxygen Mixtures and Air at High Pressure," *Adv. Cryog. Engr.*, 28 (1988).

7. Sychev, V. V., Vasserman, A. A., Kozlov, A. D., Spiridonov, G.A., and Tsymarny, V.A., Thermodynamic Properties of Air, National Standard Reference Data Service of the USSR, Selover, T.B., English Language Edition Editor, Hemisphere Publishing Co., Washington, (1987).

8. Blanke, W. and Weiss, R., "Korrelationsgleichungen fur die thermischen Zustandsgroben der Luft auf der Grenzkurve des Sattigungsgebietes," *PTB-Mitteilungen*, 27:97 (1987).

9. Leung, S. S. and Griffiths, R. B., *Phys. Rev.*, A8 (1973).

10. Moldover, M. R. and Gallagher, J. S., "Phase Equilibria in the Critical Region: An Application of the Rectilinear Diameter and the 1/3 Power Laws to Binary Mixtures," ACS Symposium Series, 60 (1977).

11. Moldover, M. R. and Gallagher, J. S., "Critical Points of Mixtures: An Analogy with Pure Fluids," *AIChE J.*, 24:2 (1978).

12. Rainwater, J. C. and Moldover, M. R., "Thermodynamic Models for Fluid Mixtures near Critical Conditions," Chemical Engineering at Supercritical Fluid Conditions, Paulaitis, M.E., Penninger, J. M. L., Gray, Jr., R. D., and Davidson, P., editors, Ann Arbor Science, (1983).

13. Rainwater, J. C. and Williamson, F. R., "Vapor-Liquid Equilibrium of Near-Critical Binary Alkane Mixtures," *Int. J. Thermophysics*, 7:1 (1986).

14. Van Poolen, L. J., private communication to R. T Jacobsen (1988).

ON THE PREDICTION OF THERMAL CONDUCTIVITY OF GAS MIXTURES AT LOW TEMPERATURES

W. Sheng and B.C.-Y. Lu

Department of Chemical Engineering
University of Ottawa
Ottawa, Ontario, Canada

ABSTRACT

Thermal conductivity of pure gases were correlated by means of an extended form of the modified Enskog theory together with a modified volume-translated Peng-Robinson equation of state at low temperatures and at pressures up to 370 bar. Two different approaches were used in the correlation. A substance and temperature dependent parameter was introduced in both correlations. The pure-component parameters thus obtained were used to predict the thermal conductivity of five binary mixtures (Ar-He, Ar-N_2, Ar-Ne, He-N_2 and N_2-Ne) without using any binary adjustable parameters with various degrees of success.

INTRODUCTION

The thermal conductivity λ of gases and their mixtures represents the proportionality of the local heat flux to the temperature gradient. Industrially, it is useful in the design of heat exchangers. There is a need to reduce the experimental efforts for obtaining this transport property through correlation of λ for pure components and prediction of λ for their mixtures. A number of correlation and prediction methods have been proposed in the literature. Some of these methods have been evaluated by Reid et al.[1]. Recently, an attempt[2] was made to predict shear viscosity of mixtures of cryogenic interest by means of an extended form of the modified volume-translated Peng-Robinson equation of state[3]. In this work, a similar but simpler approach was made to predict thermal conductivity of gas mixtures of cryogenic interest using only the parameters obtained from the correlation of pure-component thermal conductivities. However, two approaches for the correlation of λ were tested and a temperature dependent parameter was introduced in the course of development.

CORRELATION OF λ OF PURE COMPONENTS

According to the approximate kinetic theory of Enskog, the thermal conductivity λ of a dense gas of rigid spherical molecules may be represented by

$$\lambda = \lambda_o b\rho \left[1/b\rho\chi + 1.200 + 0.755\, b\rho\chi\right] \qquad (1)$$

where λ_o is the dilute-gas thermal conductivity, ρ is the molecule density,

$b(=2\pi\sigma^3/3)$ is the co-volume, there σ is the molecular diameter, and χ is the value of the equilibrium radial distribution function at a distance σ from the center of an individual molecule. The virial expansion of χ may be given by[4]

$$\chi = 1. + 0.6250b\rho + 0.2869(b\rho)^2 + 0.111(b\rho)^3 + 0.109(b\rho)^4 + ... \quad (2)$$

For real gases, Enskog suggested that the quantity $b\rho\chi$ should be determined from compressibility experiments, using the "thermal pressure":

$$b\rho\chi = \frac{V}{R}\left(\frac{\partial P}{\partial T}\right)_V - 1 \quad (3)$$

where R is the gas constant, P, V and T are the pressure, volume and temperature, respectively. The quantity χ is a function of density. Its value approaches unity as ρ approaches zero. In this work, the modified volume-translated Peng-Robinson equation of state[3] was used without any additional term for the evaluation of $b\rho\chi$. The equation is of the following form

$$P = \frac{RT}{V - b'} - \frac{a(T)}{V^2 + (2b' - 4c)V + 2c^2 - b'^2} \quad (4)$$

where

$$a(T) = a(T_c)\alpha(T), \quad (5)$$

$$a(T_c) = 0.45724R^2T_c^2/P_c,$$

$$b' = 0.3112RT_c/(2+u)P_c,$$

$$c = (2 + u)b'/4,$$

Adachi et al.[5] observed that the temperature function of α suggested by Soave[6]

$$\alpha(T) = [1 + s(1 - \sqrt{T_r})]^2 \quad (6)$$

is most suitable for the Peng-Robinson equation. The quantity s of equation 6 was obtained simultaneously with u of equation 5 from the PVT data in the supercritical region. The values of s and u obtained for the six gases investigated in this work are reported in Table 1. It should be mentioned that these values are different from those previously reported[2] as no quantum correction was added to equation 4.

Substituting equation 4 into equation 2 yields,

$$b\rho\chi = \frac{V}{R}\left[\frac{R}{V - b'} - \frac{\frac{da(T)}{dT}}{V^2 + (2b' - 4c)V + 2b'^2 - c^2}\right] - 1 \quad (7)$$

Since $\chi \to 1$ as $\rho \to 0$, it can be readily shown that

$$b = \lim_{\rho \to 0} b\chi = b' - (da/dT)/R \quad (8)$$

Hence

$$b\rho = [b' - (da/dT)/R]/V \quad (9)$$

In practice, λ is expressed by $\lambda = \lambda^{tr} + \lambda^{int}$, where λ^{tr} and λ^{int} are the translational and internal contributions to the total thermal conductivity λ.

Following the presentation of Cohen and Sandler[7], we adopted the expression (Approach I)

$$\lambda = \lambda_o b\rho[(1/b\rho\chi) + (1.2/D) + (0.755\ b\rho\chi/D)] \qquad (10)$$

for correlation λ and treating the quantity $D(= 4\lambda_o/15\ k\eta_o)$ substance and temperature dependent. The quantities λ_o, k and η_o are the low density thermal conductivity, the Boltzmann constant and the dilute gas viscosity, respectively.

Another approach (Approach II) was also considered in this work. For simplicity, it was assumed that the virial equation of state truncated to the second virial coefficient could approximately describe the PVT behavior of the gas, leading to

$$\lambda = \lambda_o b\rho\chi[(1/b\rho\chi) + (1.2/D) + (0.755\ b\rho\chi/D)] \qquad (11)$$

with a χ value equal to unity. This approach is identical to that previously adopted by Sheng et al.[8]. Both equations 10 and 11 were used in the correlation of λ in conjunction with equations 3, 4, 5 and 9.

Many correlations are available in the literature for λ_o. In this work, a three-parameter polynomial correlation

$$\lambda = \lambda_o = \lambda_{o1} + \lambda_{o2}T + \lambda_{o3}T^2 \qquad (12)$$

was adopted to describe the temperature dependence. The obtained λ_{oi} values are reported in Table 2. The deviations in the calculated λ_o values over the temperature range investigated in this work is generally around 0.2%. The D values of equations 10 and 11 are further correlated as a linear function of temperature. A typical result of the correlation is shown in Fig. 1 for nitrogen. The coefficients obtained from the correlation for six arbitrarily selected pure substances are presented in Table 3. The correlated λ values for these substances are reported in Table 4. A comparison of the calculated and literature λ values for hydrogen at 280 K is depicted in Fig. 2.

PREDICTION OF λ FOR BINARY MIXTURES

In the calculation of the parameters for mixtures, the following conventional mixing rules were adopted for the parameters a, b' and c of equation 4:

$$a = \sum_i \sum_j x_i x_j a_{ij} \qquad (13)$$

$$b' = \sum_i x_i b_i' \qquad (14)$$

$$c = \sum_i x_i c_i \qquad (15)$$

For the purpose of prediction, it was decided that no binary interaction parameters would be introduced into these mixing rules. The calculated results for five arbitrarily selected binary systems are reported in Table 5. A comparison of the calculated and literature λ values for the nitrogen-neon mixtures at 300 K for three compositions is depicted in Fig. 3.

DISCUSSION AND CONCLUSIONS

While it was expected that equation 10 would yield better correlation of pure component λ values, the limited results obtained in this study indicated that it yields somewhat higher average absolute percentage deviations in most

Table 1 Values of s (for α) and u in the supercritical region determined from PVT data[9]

	T, K	P, bar	Data Points	s	u	AAPD
Ar	180 ~ 320	1 ~ 300	50	0.44737	1.8254	0.66
H_2	100 ~ 300	1 ~ 400	60	0.19807	1.6563	0.26
He	100 ~ 300	0.1 ~ 100	35	0.19485	2.8455	0.03
N_2	150 ~ 350	1 ~ 400	55	0.39991	1.3640	0.62
Ne	100 ~ 300	0.1 ~ 200	60	0.42628	1.7112	0.40
CH_4	200 ~ 300	0.2 ~ 400	56	0.36761	1.5662	1.45

Table 2 Correlation of pure-component λ_o as a function of temperature, $\lambda_o = A + BT + CT^2$

	A	B	C	Ref.
Ar	-0.08920	0.070599	-0.37586×10^{-4}	9
H_2	-8.5504	0.82092	-0.62075×10^{-3}	9
He	22.215	0.53532	-0.34938×10^{-3}	9
N_2	-0.64569	0.10617	-0.58970×10^{-4}	9
Ne	4.4604	0.19223	-0.14074×10^{-3}	9
CH_4	-0.41968	0.10122	0.48276×10^{-4}	10

Table 3 Correlation of the parameter D as a linear function of temperature, $D = D_o + D_1T$

	Equation 10		Equation 11	
	D_0	D_1	D_0	D_1
Ar	1.355	0.2269×10^{-3}	0.873	0.1840×10^{-2}
H_2	0.891	0.1527×10^{-2}	0.777	0.1441×10^{-1}
He	0.949	0.9773×10^{-4}	1.475	0.1132×10^{-1}
N_2	0.556	0.3069×10^{-2}	0.026	0.7128×10^{-2}
Ne	1.676	-0.2023×10^{-2}	2.916	-0.1316×10^{-2}
CH_4	1.723	-0.1883×10^{-2}	0.057	0.5726×10^{-2}

Table 4 Comparison of calculated and literature λ values for six pure components

Substance	T,K	P, Bar	Data Points	AAPD Approach 1	AAPD Approach 2	Ref.
Ar	180	1~300	15	13.03	7.92	9
	200	1~300	15	11.35	5.67	9
	300	1~300	15	3.97	1.50	9
	350	1~300	11	2.74	0.90	9
	174	4~33	10	0.39	0.33	11
	225	3~53	13	0.65	0.43	11
	305	13~91	15	0.45	0.20	11
	223	4~85	15	1.89	1.01	12
	266	4~92	14	0.70	0.20	12
	308	6~62	17	0.27	0.43	12

Table 4 (Continued)

Substance	T,K	P, Bar	Data Points	AAPD Approach 1	AAPD Approach 2	Ref.
H_2	80	1~100	11	1.18	0.90	9
	90	1~150	12	0.93	1.22	9
	100	1~250	14	0.89	3.64	9
	110	1~250	14	0.60	2.91	9
	120	1~250	14	0.47	2.19	9
	130	1~250	14	0.28	1.65	9
	140	1~250	14	0.22	1.63	9
	150	1~300	15	0.39	1.33	9
	160	1~350	16	0.23	1.40	9
	170	1~350	16	0.25	1.02	9
	180	1~350	16	0.33	0.83	9
	190	1~350	16	0.19	0.83	9
	200	1~350	16	0.21	0.78	9
	210	1~350	16	0.45	1.03	9
	220	1~350	16	0.29	0.51	9
	230	1~350	16	0.35	0.56	9
	240	1~350	16	0.30	0.29	9
	250	1~350	16	0.38	0.46	9
	260	1~350	16	0.39	0.52	9
	270	1~350	16	0.38	0.50	9
	280	1~350	16	0.28	0.34	9
	290	1~350	16	0.56	0.73	9
	300	1~350	16	0.79	0.91	9
He	270	1~300	7	0.64	0.26	9
	280	1~300	7	0.68	0.31	9
	290	1~300	7	0.65	0.27	9
	300	1~300	7	0.42	0.15	9
	310	1~300	7	0 64	0.28	9
N_2	177	3.1~86	16	2.91	1.65	11
	233	5~103	16	0.99	0.88	11
	270.5	4~88	16	8.38	0.61	11
	150	1~40	5	1.70	1.13	9
	160	1~50	6	1.94	1.26	9
	170	1~300	15	4.42	2.80	9
	180	1~300	15	8.82	2.68	9
	190	1~300	15	4.56	2.25	9
	200	1~300	15	4.05	2.01	9
	220	1~300	15	4.06	2.59	9
	240	1~300	15	3.43	1.30	9
	260	1~300	15	2.67	0.88	9
	280	1~300	15	2.09	0.62	9
	300	1~300	15	1.97	0.70	9
Ne	275	1~200	7	0.47	0.64	14
	298	1~310	15	0.65	0.41	13
	323	1~330	17	0.38	0.19	13
	348	1~370	14	0.19	0.13	13
CH_4	200	1~20	4	0.51	0.39	10
	210	1~50	5	2.79	2.17	10
	220	1~50	5	1.97	1.43	10
	230	1~50	5	1.62	1.15	10
	240	1~100	6	4.96	3.55	10
	250	1~100	6	3.84	2.61	10
	260	1~100	6	2.96	1.89	10
	270	1~300	8	5.27	2.41	10
	280	1~300	8	4.68	1.94	10
	290	1~300	8	4.18	1.60	10
	300	1~300	8	3.58	1.19	10

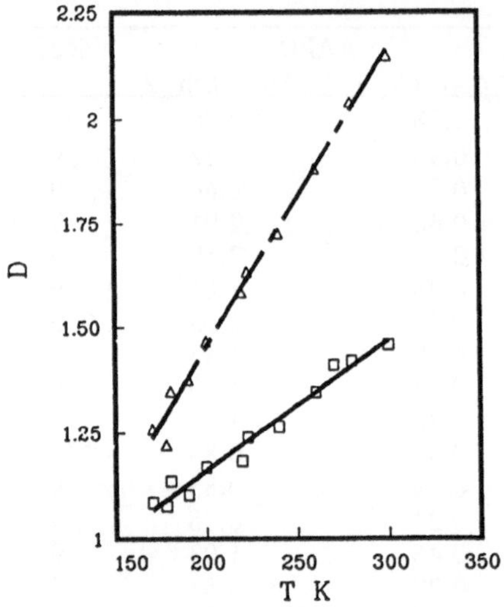

Fig. 1. Correlation of parameter D with temperature for nitrogen. □, values from equation 10, Δ, values from equation 11.

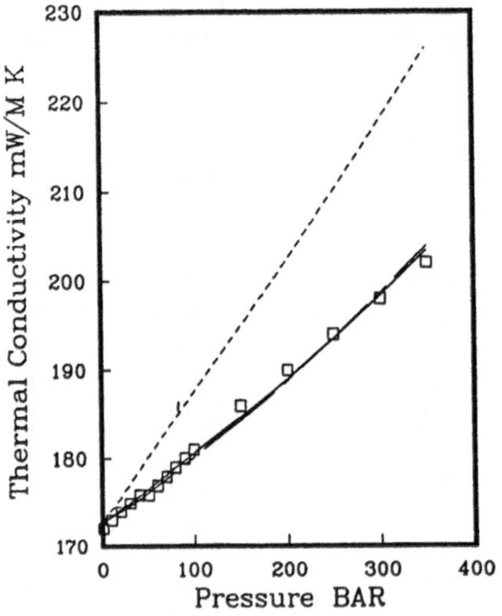

Fig. 2. Comparison of experimental and calculated λ values for H_2 at 280 K. □, data (ref. 9); ——, calculated with optimized D (equation 10); -----, calculated with D = 1.0 (equation 10); — — —, calculated with optimized D (equation 11).

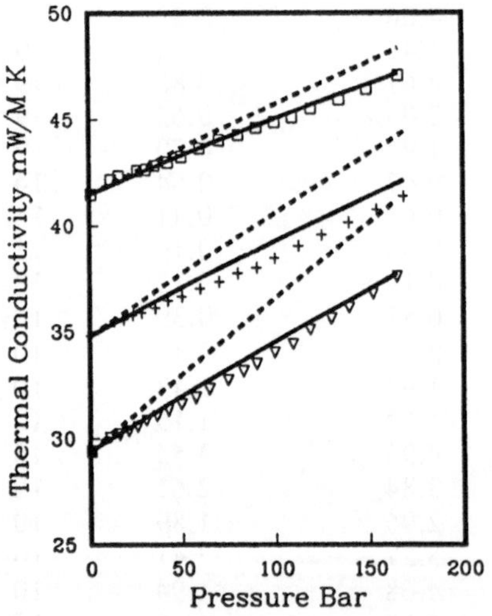

Fig. 3. Comparison of experimental and calculated λ values for the binary system N_2(1) - Ne(2) at 300 K. Experimental data (ref. 16): , ∇, y_2 = 0.2190; +, y_2 = 0.4937; □, y_2 = 0.7637; ——, calculated with equation 10; -----, calculated with D = 1.0.

Table 5 Comparison of predicted and literature λ values for binary mixtures

System	T,K	P, Bar	y_2	Data Points	AAPD Equ. 10	AAPD Equ. 11	Ref
Ar(1)-He(2)	275	1~200	0.1	4	3.40	1.82	14
		1~200	0.2	4	4.95	0.89	14
		1~200	0.3	4	4.87	0.46	14
		1~200	0.4	4	3.92	0.22	14
		1~200	0.5	4	2.68	0.17	14
		1~200	0.6	4	1.37	0.57	14
		1~200	0.7	4	0.25	0.85	14
		1~200	0.8	4	1.30	0.48	14
		1~200	0.9	4	2.44	0.59	14
	325	1~200	0.1	4	2.53	1.91	14
		1~200	0.2	4	4.36	0.67	14
		1~200	0.3	4	4.85	0.55	14
		1~200	0.4	4	3.93	0.58	14
		1~200	0.5	4	1.96	0.11	14
		1~200	0.6	4	0.35	0.68	14
		1~200	0.7	4	1.34	0.82	14
		1~200	0.8	4	2.28	0.92	14
		1~200	0.9	4	2.62	1.06	14
Ar(1)-N_2(2)	301	1~70	0.165	3	0.86	2.88	15
		1~70	0.404	4	0.28	1.73	15
		10~70	0.633	3	2.67	0.80	15
		1~70	0.909	4	0.70	2.08	15
	308	10~70	0.134	4	1.92	3.29	15
		10~70	0.446	4	0.86	1.44	15
		30~71	0.730	3	1.40	1.45	15
		1~70	0.934	4	1.62	3.27	15
	300	7~94	0.2793	12	1.52	0.72	16
		8~89	0.4980	10	1.61	0.80	16
		13~84	0.7644	9	1.96	0.73	16
Ar(1)-Ne(2)	275	1~200	0.1	7	1.29	2.30	14
		1~200	0.2	7	2.29	1.95	14
		1~200	0.3	7	3.02	1.86	14
		1~200	0.4	7	3.64	1.65	14
		1~200	0.5	7	4.07	1.28	14
		1~200	0.6	7	4.15	1.20	14
		1~200	0.7	7	3.65	1.06	14
		1~200	0.8	7	2.42	1.15	14
		1~200	0.9	7	0.51	1.84	14
	325	1~200	0.1	7	0.78	3.14	14
		1~200	0.2	7	1.05	3.07	14
		1~200	0.3	7	1.56	2.33	14
		1~200	0.4	7	1.83	1.99	14
		1~200	0.5	7	1.82	1.73	14
		1~200	0.6	7	1.57	1.59	14
		1~200	0.7	7	1.16	1.58	14
		1~200	0.8	7	1.68	1.72	14
		1~200	0.9	7	0.17	1.71	14
He(1)-N_2(2)	300	10~133	0.1630	16	2.77	1.49	16
		10~125	0.4658	15	0.95	0.15	16
		10~61	0.7798	11	2.80	0.71	16
N_2(1)-Ne(2)	300	10~167	0.2190	19	0.99	2.94	16
		13~170	0.4937	16	1.29	2.19	16
		12~168	0.7637	16	0.30	2.38	16

cases. The treatment of parameter D as temperature and substance dependent reduces the deviations in the correlation of pure component λ values very significantly as shown in Fig. 2. The predicted results indicate that the volume-translated Peng-Robinson equation with the s and u values determine from PVT data is suitable for calculating $b\rho\chi$ values. The predicted λ values for the five binary mixtures are generally acceptable without using any binary interaction parameters in the mixing rules used in the determination of mixture parameters.

ACKNOWLEDGEMENT

The authors are indebted to the Natural Sciences and Engineering Research Council of Canada for financial support.

REFERENCES

1. R.C. Reid, J.M. Prausnitz and B.E. Poling, "The properties of gases and liquids", 4th ed., McGraw-Hill, New York (1987).
2. W. Sheng and B.C.-Y. Lu, Calculation of shear viscosity of mixtures by mean of equation of state, in "Advances in Cryogenic Engineering", Vol. 35, Plenum Press, New York (1990), p. 1533.
3. W. Sheng and B.C.-Y. Lu, A modified volume-translated Peng-Robinson equation with temperature dependent parameters, Fluid Phase Equilib., 56: 71 (1990).
4. C. Chapman and T.C. Cowling, "The mathematical theory of non-uniform gases", 3rd ed., chapter 16, Cambridge Univ. Press, London (1970).
5. Y. Adachi, H. Sugie and B.C.-Y. Lu, Temperature dependence of the cohesion parameter for calculating binary VLE values for systems containing helium and neon, in: "Advances in Cryogenic Engineering". Vol. 33, Plenum Press, New York (1988), p. 1031.
6. G. Soave, Equilibrium constants from a modified Redlich-Kwong equation of state, Chem. Eng. Sci., 27: 1197 (1972).
7. Y. Cohen and S.I. Sandler, The viscosity and thermal conductivity of simple dense gases, Ind. Eng. Chem. Fundam., 19: 186 (1980).
8. W. Sheng, G.J. Chen and H.C. Lu, Prediction of transport properties of dense gases and liquids by the Peng-Robinson (PR) equation of state, Int. J. Thermophys., 10: 133, (1989).
9. N.B.. Vargaftik, "Tables on the thermophysical properties of liquids and gases", 2nd ed., Hemispheres, Washington, DC (1975).
10. D.G. Friend, J.F. Ely and H. Ingham, Thermal physical properties of methane, J. Phys. Chem. Ref. Data 18:583 (1989).
11. J. Millat, M.J. Ross and W.A. Wakeham, Thermal conductivity of nitrogen in the temperature range 177 to 270K, Physica 159A:28 (1989).
12. U.V. Mardolcar, C.A. Nieto de Castro and W.A. Wakeham, Thermal conductivity of argon in the temperature range 107 to 423 K, Int. J. Thermophys. 7:259 (1986).
13. J.V. Sengers, W.T. Bolk and C.J. Stigter, The thermal conductivity of neon between 25-75°C at pressure up to 2600 atm, Physica 30:1018 (1964).
14. K. Stephan and Heckenberger, "Thermal conductivity and viscosity data of fluid mixtures", Dechema Chemistry Data Series, Vol. X., Part 1, Frankfurt (1989).
15. M. Yorizane, S. Yoshimura, H. Masuoka and H. Yoshida, Thermal Conductivity of binary gas mixtures at high pressures: N_2-O_2, N_2-Ar CO_2-Ar and CO_2-CH_4, Ind. Eng. Chem. Fundam. 22:458 (1983).
16. R.D. Fleeter, J. Kestin and R. Paul, The thermal conductivity of mixtures of nitrogen with four noble gases at room temperature, Physica 108 A: 371 (1981)

ADSORPTION OF HELIUM IN COMMERCIALLY AVAILABLE ACTIVATED CARBONS

Ben P. M. Helvensteijn and Ali Kashani
Sterling Federal Systems, Inc.
Palo Alto, California *

Randall A. Wilcox
Trans-Bay Electronics, Inc.
Richmond, California*

ABSTRACT

A newly constructed cryostat has been employed in order to establish the helium adsorption characteristics of several commercially available activated carbons. So far, two carbon samples have been studied. Considered are the steady-state adsorption-pressure-temperature relations of ^3He as well as ^4He (unmixed). Pressures are measured with room-temperature instrumentation. Applying thermo-molecular pressure difference corrections the cold end pressures are obtained. The helium adsorption is derived from straightforward mass balance computations. The measurements span the range of temperatures between 2 K and 50 K for pressures between 1 Pa and 1 MPa. Interpolation allows presentation of the data in the form of adsorption isotherms as well as isobars (only adsorption isotherms are presented here).

INTRODUCTION

Activated carbon is known for its potential to purify many a substance (e.g., gas masks, water filters). This convenient quality is the result of the van der Waals interaction between the solid and the individual molecules and an internal labyrinth providing a relatively large surface area ($\approx 1000\ m^2g^{-1}$). The amount of any given substance adsorbed is a complex function of surface area, temperature, concentration/gas pressure and even carbon morphology. Adsorption generally becomes significant at temperatures on the order of the normal boiling point and below. Not surprisingly, activated carbon is frequently applied in cryogenics to assure a thermally isolating high vacuum. In this type of application the amount

* Mailing address: NASA-ARC, MS 244-10, Moffett Field, CA 94035.

Advances in Cryogenic Engineering, Vol. 37, Part B
Edited by R.W. Fast, Plenum Press, New York, 1992

of carbon is generally not a major concern. However, when carbon is used for its ability to adsorb and desorb large amounts of helium reversibly, efficiency becomes an issue. This is the case in certain types of low temperature refrigerators in which the carbon affects evaporative cooling[1,2] or provides the means to activate heat switches[3]. The efficiency of these carbon controlled devices may be improved significantly by precise knowledge of the adsorption relations of the incorporated carbon.

Data on helium (^4He) adsorption onto activated carbon have been published by Vazquez et al. and others,[4,5] however, these particular carbons are not readily available. Variation in adsorption qualities of nonidentical carbons, difficulties in acquiring already tested material and the scarcity of data published on ^3He adsorption has lead us to the present study. The results to be discussed here are on two samples of activated carbon: 1) Amoco* (Am), 2) Barnebey Cheney PE+ (BCh). Data[6] we have published previously on Am are to be considered erroneous due to the discovery after publication of inadequate thermal anchoring of the thermometry. This problem has since been overcome. The data contained in the present paper have all been recorded with improved thermometry. To our dismay the Am carbon is no longer in production, although this situation is subject to change. Still, the Am data are considered relevant because of the unusually high surface area (2000 m^2/g) which suggests a great potential for helium adsorption. The tests on BCh ($\approx$ 1100 m^2/g) have resulted in adsorption data on a type of carbon which is indeed easily accessible, as has been the intent from the start of this effort. As indicated above, adsorption has been determined both for ^3He and ^4He (unmixed).

CRYOSTAT LAYOUT

The cryostat has been discussed in detail in an earlier publication.[6] Therefore, we will merely recapitulate the layout using Fig. 1. The activated carbon test assembly (ACTA) consists of an annular shell surrounded by an annular vacuum space. The shell/container has a volume of 40 cm^3 and a wall area of 90 cm^2 to which the carbon is bonded with a thin layer of Stycast 2850 FT (for BCh the area was increased twofold by means of a copper screen). In tests with Am carbon 3.50 g has been used; the sample of BCh weighed 5.69 g. A manganin wire heater is epoxied to the shell as well as an Allen Bradley carbon resistance thermometer. The pressure inside the shell is measured through a 1.6 mm inner diameter tube by an external room temperature gauge#. The pressure gauge is fitted with either a 35 kPa (5 psi) or a 1.4 MPa (200 psi) pressure capsule. The gauge volume including the space between valves equals 30 cm^3. For data-acquisition, graphics and computations Macintosh computers have been used.

PROCEDURE

The carbon is taken from a weighed container and glued to the shell after which the container with the remaining carbon is weighed again. The shell is soft-soldered with woodsmetal to its inner wall/coolfin. Then, after leaktesting, the cryostat is prepared for an experiment by purging the shell at 60°C repeatedly with nitrogen gas and at a later stage with helium gas. Each purge is followed by a thorough evacuation using a turbo pump. Once the dewar is filled with liquid helium and the cryostat is cold, the activated carbon is given a first charge of helium gas. The amount admitted is derived from the tank pressure measured before

---

* Courtesy APD Cryogenics Inc., Allentown, PA 18103, USA.

\+ Courtesy Barnebey Cheney, USA.

\# Mensor, Model 100.

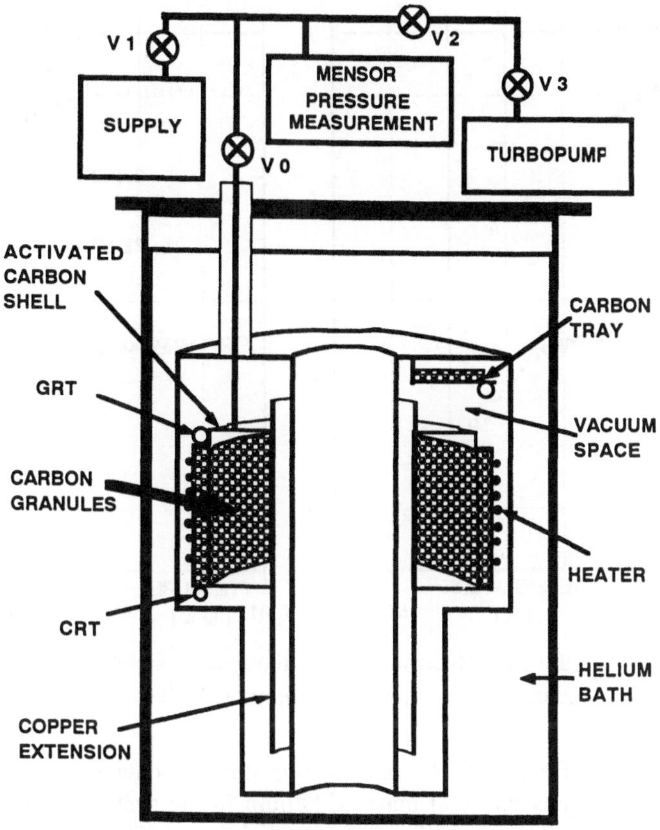

Figure 1. Activated carbon test assembly.

and after passing gas through valve V0. The pressure gauge is pumped out (^{4}He by the turbo; ^{3}He by the carbon sample) after which the helium from the sample is allowed to also enter the pressure gauge.

Being charged with helium, the cryostat is ready for testing. The shell is heated up while monitoring the progression in temperature and pressure readings. Data have been recorded during slow continuous cooldowns and near steady-state stepwise warmups. No significant differences in data values (< 0.5% pressure) have been observed between both modes of data collection. When the pressure and/or temperature range of interest has been traversed, the helium is pumped out of the pressure gauge by the carbon sample. The completion of one set of data at a particular charge of helium to the shell is succeeded by a repetitive series of additional charges of helium to the system and more data collection.

The recorded data is processed correcting for thermo-molecular pressure differences between the room-temperature pressure gauge and the cold shell.[7] The helium contained/adsorbed by the carbon is derived employing a simple mass balance, which considers the helium charge and the amount of helium present in the gas phase inside the gauge and the shell (using the van der Waals equation of state), . Subsequently, the data is interpolated[8] to produce isotherms (and isobars) of adsorption versus pressure (respectively temperature).

RESULTS

The results gathered at various charges of helium to the system, corrected for thermo-molecular pressure differences, are plotted in Figs. 2a-2d. Figures 2a and 2b show the data

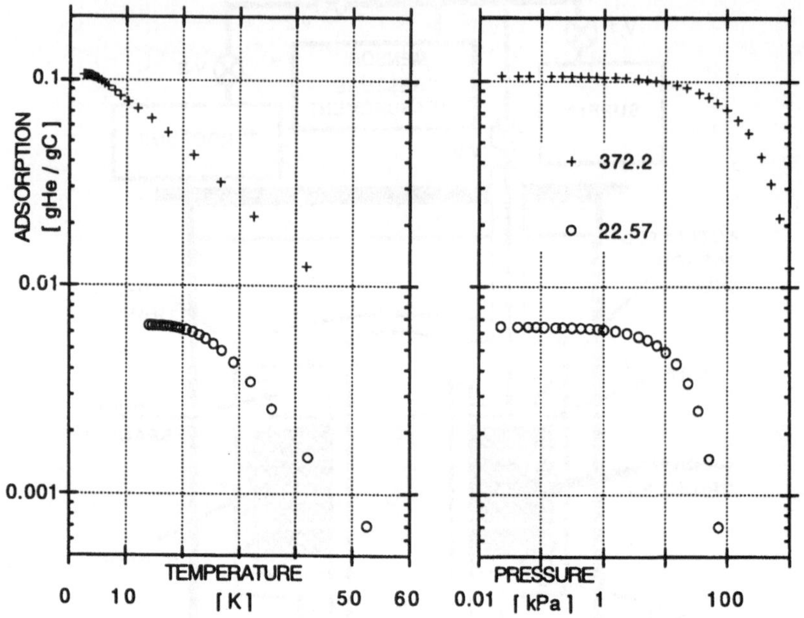

Figure 2a. ^3He adsorption of the Amoco sample.
The legend indicates the number of milligrams of ^3He in the system.

for the Am sample for ^3He and ^4He respectively; the BCh data are contained in Figs. 2c and 2d. The legend shows the total mass of helium in the system for each curve. The measurements are presented as adsorption versus both the measured temperature and pressure in order to be able to contain all information on a particular sample/gas without interpolation of the data. Points on the pressure and temperature traces are correlated through the ordinate giving the adsorption.

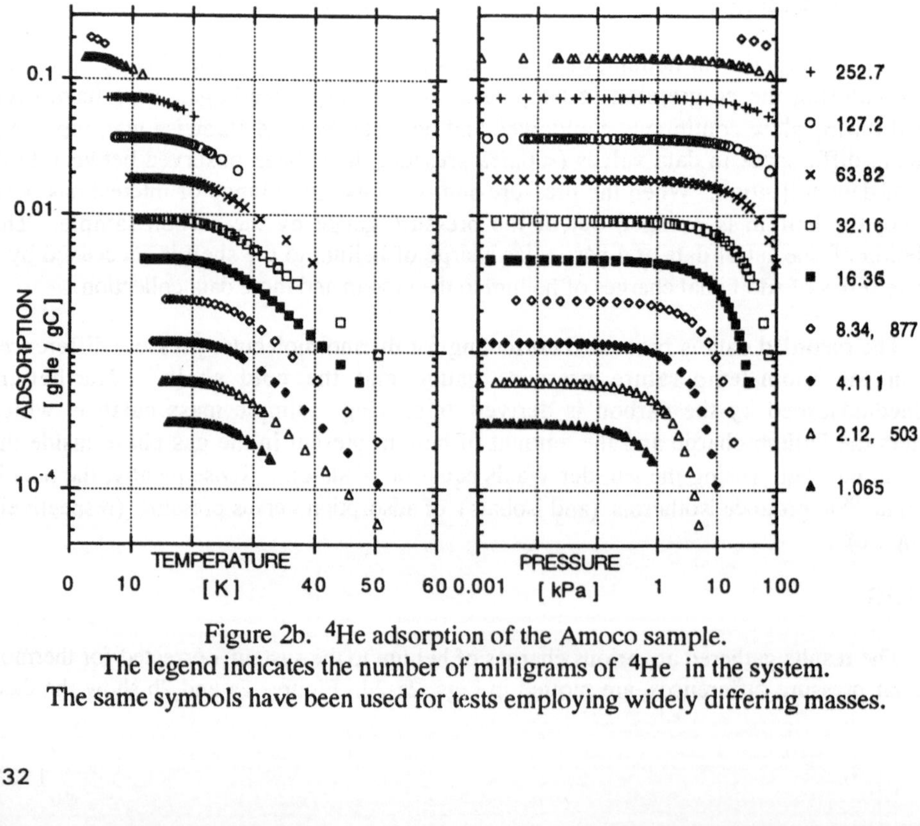

Figure 2b. ^4He adsorption of the Amoco sample.
The legend indicates the number of milligrams of ^4He in the system.
The same symbols have been used for tests employing widely differing masses.

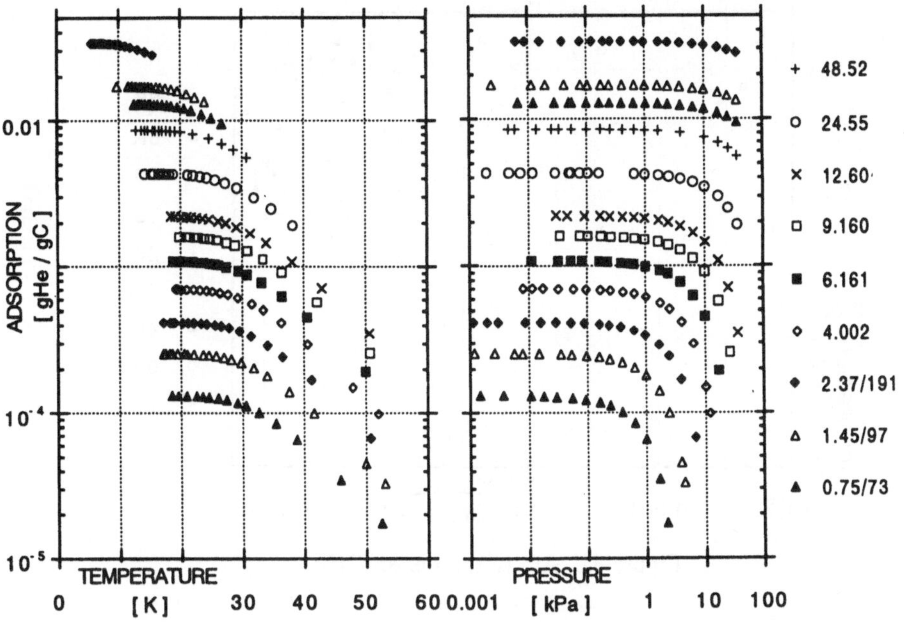

Figure 2c. ³He adsorption of the Barnebey Cheney PE sample.
The legend indicates the number of milligrams of ³He in the system.
The same symbols have been used for tests employing widely differing masses.

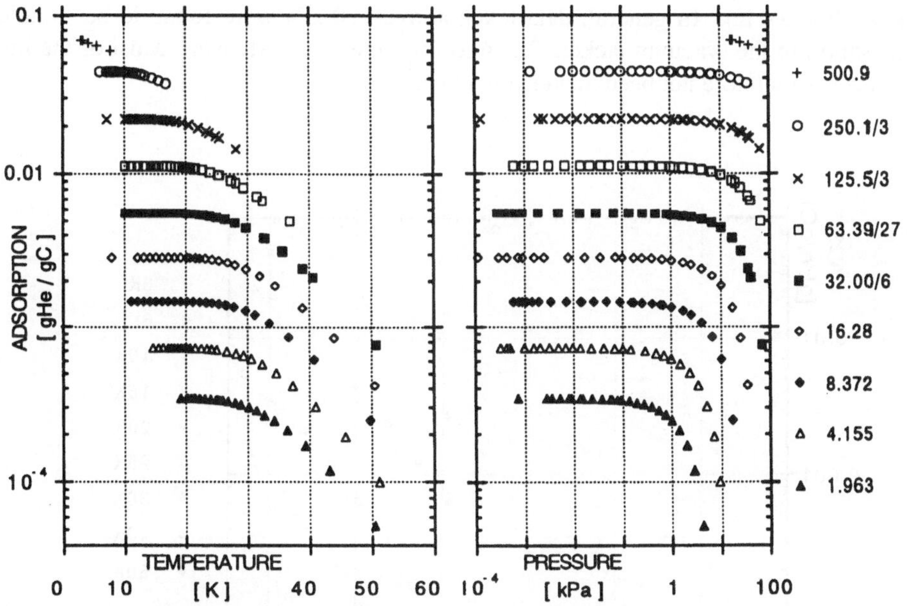

Figure 2d. ⁴He adsorption of the Barnebey Cheney PE sample.
The legend indicates the number of milligrams of ⁴He in the system.
A slash indicates data taken in separate tests with nearly
the same mass except for the last decimal place(s).

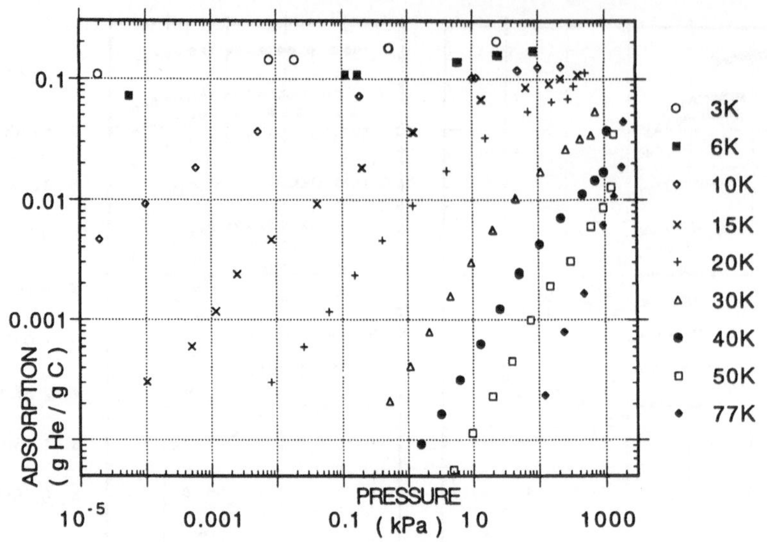

Figure 3a. Am isotherms for ^{4}He.

The data in Figs. 2a-2d point out that at moderate temperatures (3K to 20K pending the charge) the adsorption is nearly independent of temperature and pressure. meaning that nearly all the gas admitted to the system is adsorbed by the carbon. This implies that a high adsorption is attainable at low pressures. However, at a given temperature the pressure does increase with the amount of helium admitted to the system. In order to attain a low pressure at a high charge, the carbon needs to be at a low temperature (< 4.2K at the 503 mg ^{4}He / Am test). Ultimately, the saturation pressure is reached (877 mg ^{4}He / Am trace at $P \approx 24$ kPa, $T \approx 3.0$ K). In general, data taken above 100 kPa were found to be troubled by condensation in the (vacuum jacketed) connecting line to the Mensor. Most of the high pressure data has therefore not been included in the graphs.

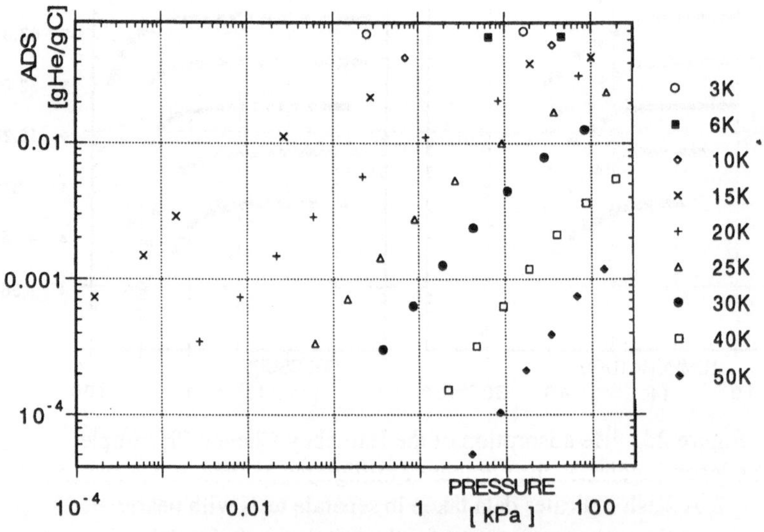

Figure 3b. BCh isotherms for ^{4}He.

Comparing the ^3He and ^4He adsorption is appropriate only in terms of mole/gC, for which one divides the adsorption by the respective molar masses. For instance the Am sample at 1/8 mole charge (≈ 503 mg ^4He; 372 mg ^3He) gives at 100 kPa an adsorption of 0.073 g^{3}He/gC and 0.101 g^{4}He/gC, converting to molar adsorptions of 0.0242 and 0.0252 mole/gC respectively. In general, the difference in molar adsorption for the two helium isotopes is observed to be less than 5%. This holds true for the Barnebey Cheney sample as well.

In order to facilitate comparison between the Am and BCh data, and to data published elsewhere as well, isotherms of adsorption have been generated by interpolation of the data and have been plotted in Figs. 3a and 3b. For the combination of high pressures (> 1 kPa) and low temperatures (< 25 K), the Am sample has the greater adsorption capability, while in all other cases the helium adsorption of the BCh sample is roughly equal to or exceeds the Am adsorption. Therefore, the BCh sample appears to be a better candidate for compression as well as for achieving a low vacuum. For the purpose of affecting evaporative cooling at moderately high pressures the Am carbon appears more suitable. It is found that in comparison to the data published by Vazquez et al., the Am sample has the greater adsorption when both high pressures and low temperatures prevail. The BCh adsorption falls below the Vazquez data.

CONCLUSIONS AND FUTURE WORK

The experiments have demonstrated the capability of the apparatus to provide us the desired data-base on various activated carbons. The Amoco carbon is found to be an excellent adsorber of helium when at low temperatures under high pressures. Barnebey Cheney PE appears more suitable for achieving low vacuum or even compression. The ^3He and ^4He adsorption are found to be nearly identical. More types of carbon are yet to be tested.

REFERENCES

1. P. R. Roach and K. E. Gray, Low-Cost, Compact Dilution Refrigerator: Operation from 200 to 20 mK, in: " Advances in Cryogenic Engineering," Vol. 33, Plenum Press, New York (1988), p. 707.
2. L. Duband, D. Alsop, D. Lange and P. Kittel, A Rocket-Borne ^3He Refrigerator, in: "Advances in Cryogenic Engineering," Vol. 35B, Plenum Press, New York (1990) p. 1447.
3. B. P. M. Helvensteijn and A. Kashani, Conceptual Design of a 0.1W Magnetic Refrigerator for Operation between 10K and 2K, in: "Advances in Cryogenic Engineering," Vol. 35B, Plenum Press, New York (1990), p. 1115.
4. I. Vazquez, M. P. Russell, D. R. Smith and R. Radebaugh, Helium Adsorption on Activated Carbons at Temperatures between 4 and 76K, in: "Advances in Cryogenic Engineering," Vol. 33, Plenum Press, New York (1988), p. 1013.
5. P. Roubeau, Nigohossian and O. Avenel, Adsorption de l'Helium 4 par le Charbon Actif, Colloque International: "Vide et Froid", Grenoble (1969) p. 22.
6. B. P. M. Helvensteijn, A. Kashani and R. A. Wilcox, Activated Carbon Test Assembly, in: "Proc. Sixth Intl. Cryocoolers Conf.," (1990) p. 103.
7. T. R. Roberts and S. G. Sydoriak, Thermomolecular Pressure Ratios for ^3He and ^4He, Phys. Rev., 102:305 (1990).
8. KaleidaGraph. Version 2.1.2. Copyright 1984-1991 by Abelbeck Software

A CRYOGENIC APPARATUS WITH ON-LINE PURIFICATION SYSTEM FOR SCINTILLATION AND IONIZATION STUDIES IN NOBLE LIQUIDS

J. Seguinot, T. Ypsilantis
Collège de France, Paris

M. Bosteels, G. Passardi, J. Tischhauser
CERN, Geneva, Switzerland

Y.Giomataris
World Laboratory, Lausanne, Switzerland

ABSTRACT

The scintillation and ionization properties of liquid argon, krypton and xenon have been investigated by means of a small laboratory apparatus. The setup consists of a cryogenic system which has the capability of cooling any liquid from ambient down to liquid nitrogen temperature and of a pulsed 10 to 100 kV electron gun used as excitation source of the liquids. By changing the electron beam intensity and the acceleration voltage, a total energy ranging from 1 MeV to 100 GeV per pulse can be deposited in the liquid (about a tenth of a liter). A forced circulation system allows continuous purification and recondensation of the evaporated gas. A description of the various purification methods, of the used materials and of the adopted cleaning procedure are given. The achieved degree of purity of liquid argon, krypton and xenon permitted an electron lifetime ≥ 200 μs corresponding to an oxygen equivalent contamination of less than 10 ppb. Measurements of the temperature uniformity of the liquid and its influence on the light scattering are presented. A scintillation yield in liquid xenon (at a wavelength of 175 nm) of 26×10^6 photons/GeV was measured.

INTRODUCTION

Liquid argon calorimeters have found, during the last decade, several applications in high energy physics. Apparatus, using several tons of argon, have been constructed and successfully operated in various laboratories for elementary particle physics. Even larger devices[1] are being designed for the next generation of experiments. All are of the sampling type with sandwiched ionization layers of argon and metallic plates (lead or uranium) of very short radiation length.

More recently, the scintillation and ionization properties of liquid xenon and krypton have been the object of laboratory type[2-8] investigations in view of their possible application (an alternative to argon) for calorimetry. These liquids have a higher density, a shorter radiation length (2.8 cm for xenon instead of 14.3 cm for argon) and a small Fano factor allowing totally active compact calorimeters of superior energy and spatial resolution.

The main purpose of the laboratory test setup was an experimental investigation of the fundamental properties of the noble liquids. We have, nevertheless, devoted particular attention to the technical problems and have tried to adopt solutions of realistic application for large size devices.

EXPERIMENTAL APPARATUS

Cryogenic System

The cryogenic system is based on a two-phase forced circulation scheme (Fig.1). Gas evaporated from the test cell is passed (after warming at ambient temperature) via a small membrane blower[9] through the purification system and a heat exchanger for recondensation. All of the circuit is maintained just above atmospheric pressure. Typical value for the vapor pressure in the test cell is, independent of the liquid used, about 110 kPa corresponding to 88 K for argon, 121 K for krypton and 167 K for xenon. The blower only creates the pressure drop necessary to overcome the flow impedance of the purifier and of the piping.

As coolant for the condenser we use saturated liquid nitrogen at atmospheric pressure provided by a standard CERN dewar. To match the low temperature of liquid nitrogen (77 K) with the higher condensation temperature required for argon, krypton and xenon and to avoid possible freezing, the heat was exchanged from the system to the coolant by means of a massive (thermally inert) stainless steel plate (Fig.2). The plate is kept at constant average temperature (typically 86 K for argon, 163 K for xenon) by admitting a controlled flow of liquid nitrogen. Nitrogen is supplied to one side of the plate from the bottom of a parallel tubing heat exchanger and the noble gas is con-

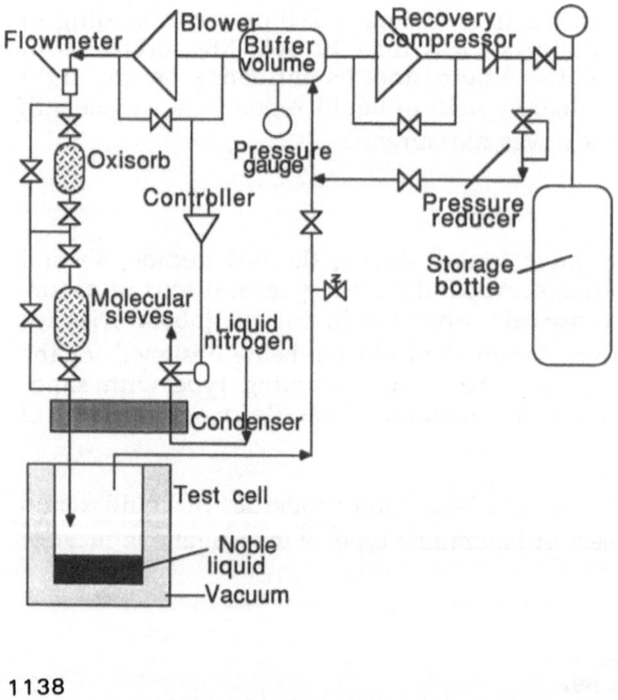

Fig.1. Flow diagram of the cryogenic and purification system.

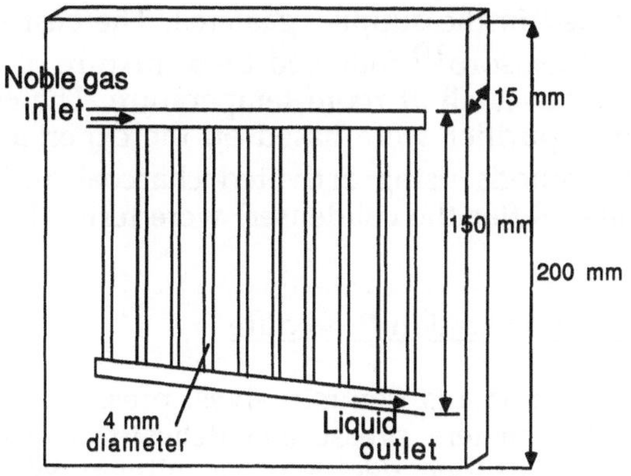

Fig. 2. The stainless steel plate used as condenser.

densed on the other side of the plate from the top of an identical tubing arrangement.

After filling the cell with the desired quantity of liquid (evaluated from the pressure drop in the storage gas volume), the gas supply valve is closed and the system is operated automatically by adjusting the flow of the coolant to the desired pressure of the cell (the controller is pneumatic and drives an analog valve at nitrogen exhaust). A constant mass flow (typically 0.2 g/s) is maintained by the blower during cooldown, filling and operation. During operation only that fraction of the flow which is necessary for keeping the pressure constant is liquefied. The excess gas is separated from the liquid at the top cell entrance (the cell is fed from the top) and returns to the blower. Satisfactory cell saturated bath pressure and, therefore temperature, was achieved (typical oscillations ± 0.5 kPa with a period of a few minutes). A membrane compressor[9] was used for gas recovery in the storage cylinder.

Purification Method and System Cleaning

The response of calorimeters based on the ionization measurements can be severely degraded by the presence of electronegative contaminants because of their capability of trapping the electrons generated in the liquid by the incoming particles. In spite of the fact that the photon absorption cross section is one to two orders of magnitude smaller than the electron trapping cross section, for calorimeters using the scintillation and transmitting the light over long distances, photon absorption by impurities results in an attenuation of the detected light. The achievement and the maintaining of a high degree of purity of the liquid is, therefore, mandatory and constitutes a crucial point for assessing the feasibility of the calorimeters.

To our knowledge, recent experiments have adopted similar solutions to obtain a very high degree of purity. Basically, they use an apparatus whose construction is based on guidelines used in ultrahigh vacuum technology. The noble gas is passed through the purifiers only during the filling and, then, is recondensed directly in the test cell without recirculation through the purifier. We believe that this method (perfect for small scale apparatus) leads to excessive technical difficulties when applied to very large devices.

We have followed a different approach, avoiding the use of ultrahigh vacuum technology and of sophisticated cleaning, mounting and leak tightness test procedures of the system. The intention is of eliminating on-line the possible source of contamination (material outgassing, back diffusion of air via small leaks, impurities contained in the supply gas etc.). The best set of purifier configuration utilized an Oxisorb[10] followed by a mixture of silica gel, 13 X and 4 A type molecular sieves all at room temperature. The choice and the preparation of the second purifier were based on the experience of other experimenters[11]. Other methods using activated charcoal and 4 A molecular sieve on the liquid phase (after the condenser) were tested but did not produce satisfactory results.

Constructing Materials and System Preparation Procedure

We have employed system components (joints, valves, pressure indicators and reducers, safety valves, flow meters, pressure switches) from various suppliers with no special concern as to their degree of cleanliness. We used stainless steel pipes from the CERN stockroom. The blower and the recovery compressor have neoprene membranes and, to prevent back diffusion of air we flushed the external side of the membrane with argon. Teflon and Viton sealings were used on various plant components. Materials in contact with the liquid were: epoxy glue, boron nitride, ceramic electrical insulators and copper wires. The recovery-storage cylinder is a standard gas bottle. After assembling, all the system components are evacuated at room temperature by means of a turbomolecular pump for at least 24 hours. The tightness of the system is controlled by measuring the pressure drop in a given time with the system at its operational pressure (typical leak rate from the system to the atmosphere is 0.3 Pa-L/s).

Test Cell

The pulsed electron beam (total energy 1 Mev to 100 GeV per pulse) enters from the bottom of the cell ionizing the liquid and, at the same, time producing UV scintillation light (Fig.3). The charge signal and the scintillation light are detected by a suitable arrangement of collecting grids and photodetectors. For light, detection various methods have been used: CsI photocathodes, silicon strip photodiode (both immersed in the liquid) and a photomultiplier with a MgF_2 window and CsTe photocathode. The photomultiplier was kept at room temperature and the light transmitted through a CaF_2 cold window (only this configuration is shown in the figure). The individual energy of the electrons is in the range of 10 to 100 keV. The electrons enter the cell via a 12 μm Mylar window (diameter 1 cm) glued internally to the bottom of the cell. This very thin window was capable of withstanding a differential pressure of 100 kPa and avoided an excessive electron energy loss at the cell entrance.

The cell is thermally insulated by means of vacuum and superinsulation. Radiation shields are also mounted inside the neck. The measured heat load, with the cell filled with xenon, is 5 W and can be scaled for argon and krypton approximately as the temperature ratio. The main source of the heat load is of conductive origin (via the large diameter stainless steel neck of the cryostat). Under this thermal load all the liquid mass is passed through the purifier about every two hours.

EXPERIMENTAL RESULTS

The experimental results are mainly presented elsewhere[12], here we report only those which are related to the chosen cryogenic system and its associated purifier.

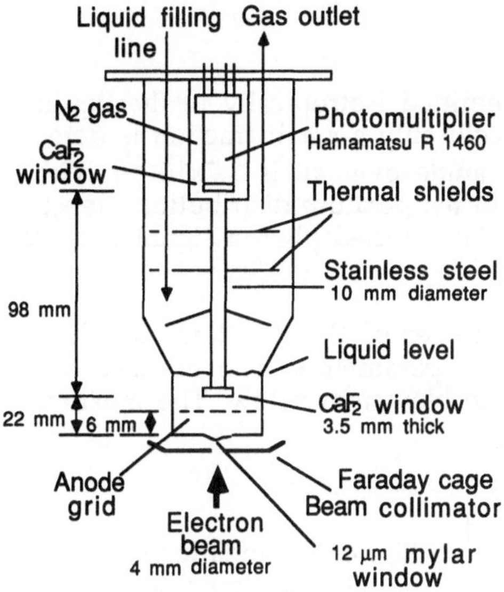

Fig. 3. Schematic view of the ionization and scintillation test cell.

Degree of Purity of the Liquid

Those ionization electrons which escape the initial recombination with positive ions are driven to the collecting anode (typical drift length l=6 mm) by the applied electrical field. All the measurements reported here are for a field gradient ≥ 2 kV/cm (above this value the drift velocity is almost independent of the applied electrical field). During their motion (at a drift velocity v_d) the electrons induce a charge signal Q(t) on the anode. If the concentration of the electronegative impurities is sufficiently low, the electron lifetime becomes much longer than the transit time $t_d = l/v_d$ and all the generated electrons will reach the anode. Q(t) will rise linearly with time (the source of ionization is point-like). Otherwise, the analysis of the shape of Q(t) allows the evaluation of the average electron lifetime before capture which, in turn, is inversely proportional to the concentration of the electronegative impurities.

The Q(t) measurements for xenon are presented in Fig. 4 (a and b), for krypton in Fig. 5 and for argon in Fig. 6. A lifetime for xenon of 200 μs was obtained only after two weeks of continuous circulation through the purifier. For krypton and argon, lifetimes in excess of 200 μs were obtained only after a few hours. Assuming that only the oxygen molecule is present as contaminant, this lifetime corresponds to an impurity level of less than 10 ppb of O_2 (1 ppb of O_2 is defined as one molecule of O_2 per 10^9 atoms of noble liquid). The supply krypton quality was 99.998 %, argon quality 99.9999%[13]. Xenon was pre-purified before use in our system by an existing CERN unit of very high performance[14]. Figure 7 shows the long term time dependance of the collected charge for xenon at a deposited beam energy of 5 GeV. The upper curve refers to the pre-purified gas and the lower curve to the industrial gas of 99.995% purity[13]. The liquid quantity in both cases was 40 cm^3. Increasing the quantity of liquid xenon by a factor of three resulted in a equal increase in the cleaning time.

Scintillation Yield

The scintillation photons are emitted isotropically in the liquid from a point-like source and, in our test cell, only a small fraction is detected (the ratio between the detection solid angle over 4π is 1.7×10^{-4}). With such a geometry, a non uniform bath liquid temperature distribution (dependent on cryogenic parameters) can modify the detected photon signal because of the scattering of the emitted light.

We have observed that the number of the the detected photons was dependent on the cryogenic control parameters. The number of photons tends to increase when reducing the blower overflow. The maximum yield

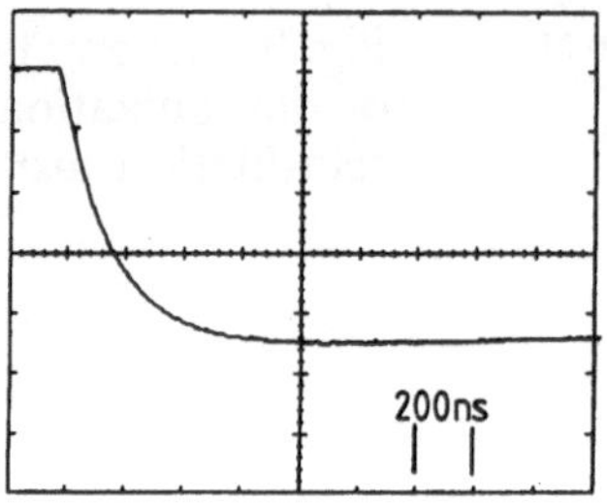

Fig. 4a. Ionization signal Q(t) for liquid xenon after filling the test chamber. The Q(t) shape indicates strong electron absorption by impurities.

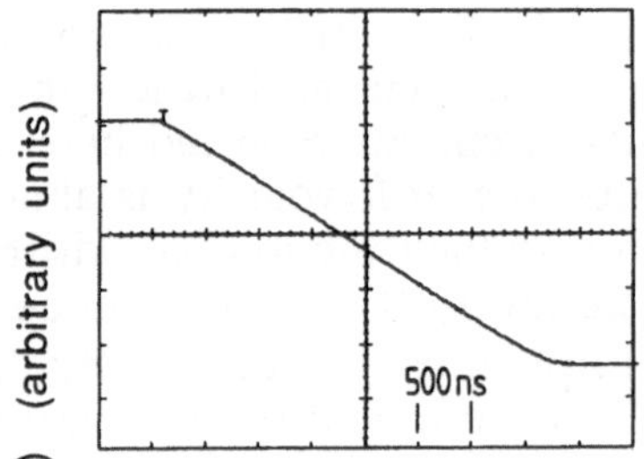

Fig. 4b. Ionization signal Q(t) for liquid xenon after two weeks of on-line purification. The linear signal indicates that no significant electron absorption occurred in the 6 mm drift gap.

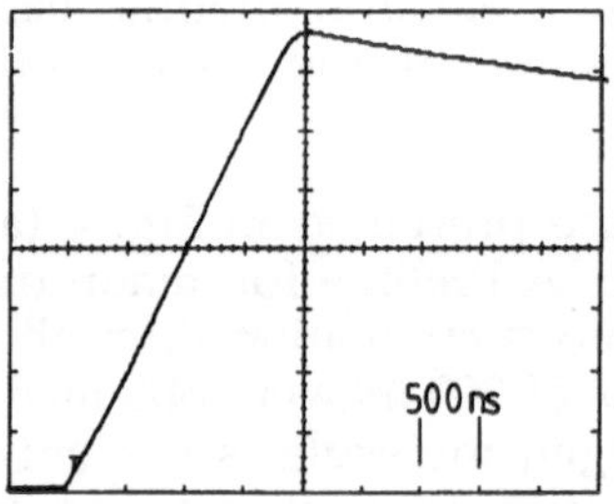

Fig. 5. Ionization signal Q(t) for liquid krypton a few hours after filling the test chamber.

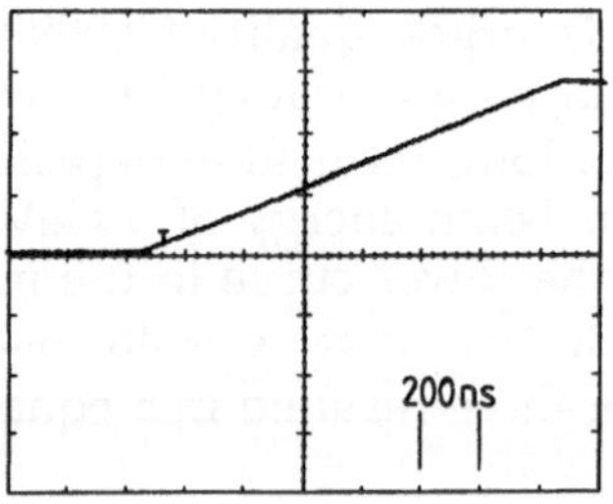

Fig. 6. Ionization signal Q(t) for liquid argon a few hours after filling the test chamber.

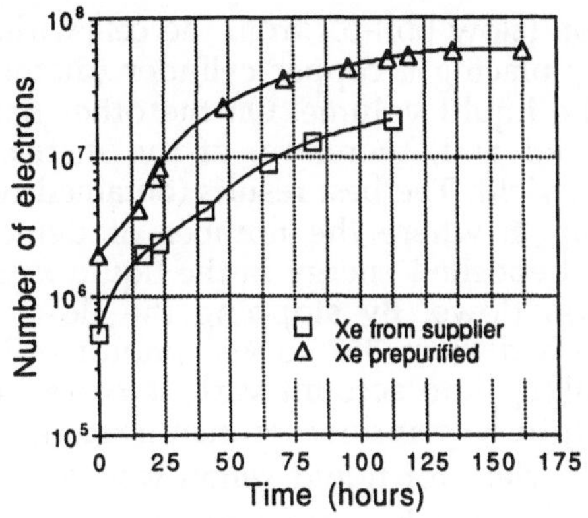

Fig. 7. Dependence of the collected charge on the on-line purification time for liquid xenon. The data refer to a beam energy of 5 GeV and a field gradient of 4 kV/cm.

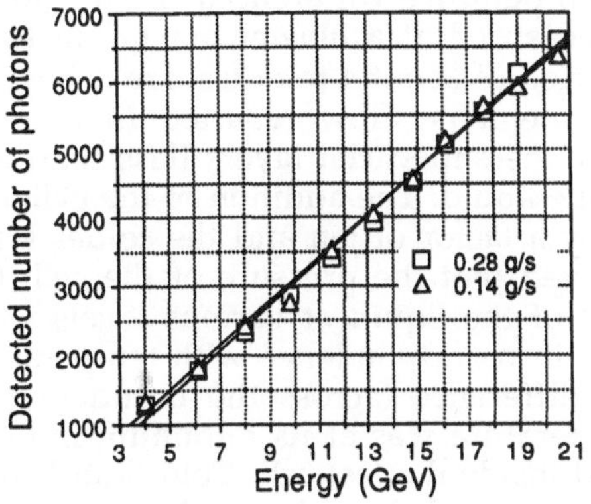

Fig. 8. Dependence of the detected number of photons on the beam energy deposited in liquid xenon.

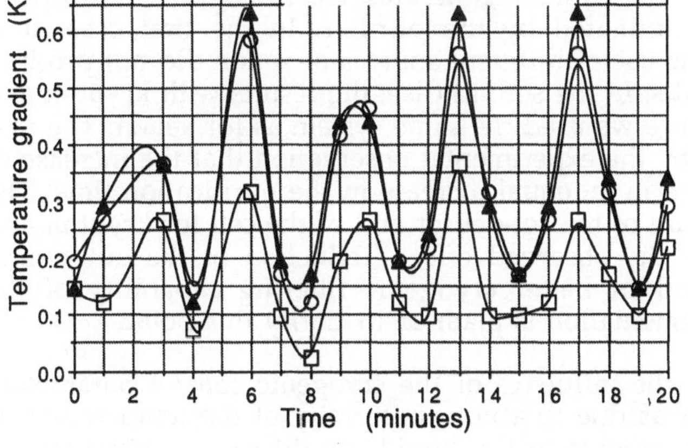

Fig. 9. Variation with time of the temperature gradient across the 4 cm deep volume of liquid xenon.

was obtained by stopping the circulation (slow boil-off from the cell without refilling). This effect was minimized by placing a copper cylinder (diameter int.4 cm, ext.4.2 cm, height 5 cm) in the liquid volume. On the other hand, the increase in the heat load to the cell and, therefore, of the increased turbulence did not affect the scintillation yield. The best results (obtained with the copper cylinder) are shown in Fig. 8 where the number of detected photons is plotted as a function of the deposited energy in the liquid xenon. The data refer to two circulation mass flows. By stopping the flow, the number of the detected photons increased from 357 to 460 photons/GeV. Integrating over the solid angle and taking into account various corrections (quantum efficiency of photomultiplier, window transmission), the maximum measured photon scintillation yield for liquid xenon was 26.3×10^6 photons/GeV (this value is in agreement with other measurements[6]).

A confirmation was obtained by measuring the temperature gradient in the liquid volume filled with xenon and its dependance on mass flow and temperature equalizer cylinder. We used a movable platinum resistor mounted in such a way that temperature of the liquid could be explored both vertically and on a plane of a given radius. The liquid depth was 4 cm. Fig.9 shows the time dependance of the temperature differences in the liquid (bottom level minus indicated level) measured at a blower maximum mass flow of 0.28 g/s. The oscillations of a period of about 4 minutes are driven by the pressure control loop. Reducing the mass flow resulted in a more uniform temperature of the liquid; only the superficial layers (less than one cm deep) were colder than the bulk of the liquid. The addition of the cylinder improved the temperature uniformity by a factor of ten and the colder layer was restricted to the surface. We observed that the pressure of the cell, the temperature of the heat exchanger and of the liquid at various levels were oscillating in phase (pressure minimum in correspondence with temperature minima). The maximum temperature difference (across the liquid layers) occurred when the heat exchanger temperature was at its minimum indicating a convective movement of the liquid inside the test cell. Cold liquid from the condenser first contacts the bath surface and then slowly flows down to the bottom of the cell.

CONCLUSION

Electron lifetimes of more than 200 µs were achieved for argon, krypton and xenon using on-line purification and conventional (not ultrahigh vacuum quality) technology but the purification of xenon required a much longer time than originally expected. There are two possible explanations: either xenon contains or the system generates electronegative impurities which are inefficiently eliminated by the purifier. In the first case, it is difficult to explain why the use of different sources of xenon did not produce significantly different results. In the second case, the results with krypton and argon are contradictory since we used the same system as for xenon. The first explanation is supported by the experimental observation that the increase of the xenon quantity resulted in an equal increase in the purification time. The lower operating temperature of the condenser and of the cell for krypton and argon might also explain the observed effect (reduction of the outgassing from the cold surface and/or better cryogenic filtering capability of the condenser). Further experimentation is planned to clarify this point.

We have shown that the influence of the cryogenic control parameters on the scintillation yield was due to a non uniformity of the temperature of the liquid. Temperature gradients in the liquid are difficult to eliminate in large size devices but their effect should disappear if larger solid angles and/or reflecting walls are used for light collection.

ACKNOWLEDGMENTS

This work was supported by the LAA project leaded by A. Zichichi. The encouragement of M. Froissard, H. Wenninger and J. Schmid is gratefully acknowledged. We thank D. Schinzel, A. Gonidec for helpful discussion and suggestion. We have particularly appreciated the technical skill of J.L. Escourrou in constructing the experimental apparatus.

REFERENCES

1. B. Aubert et al, "Liquid Argon Calorimetry with LHC Performance Specifications," CERN/DRDC/90-3, (1990).
2. T. Doke, Recent Developments of Liquid Xenon Detectors, Nuclear Instr. and Methods 196 : 87 (1982).
3. S. Kubota et al, Liquid and Solid Argon, Krypton and Xenon Scintillators, Nuclear Instr. and Methods 196 : 101 (1982).
4. T. Lindblad et al, On the Development of Liquid Ionization Detectors as Spectroscopic Instruments, Nuclear Instr. and Methods 215 : 183 (1983).
5. D.A. Imel and J.Thomas, Design, Construction and Performance of a Liquid Xenon and Liquid Argon Ionization Chamber, Nuclear Instr. and Methods in Physics Research A273 : 291 (1988).
6. E. Aprile at al, A study of Scintillation Light Induced in Liquid Xenon by Electrons and Alpha Particles, in: "Proceedings of the IEEE 1989 Nuclear Science Symposium," San Francisco, 15-19 January 1990.
7. V.M. Aulchenko et al, Development of the Liquid Krypton Electromagnetic Calorimeter of the KEDR Detector, in: "Proceedings of the Inter. Conf. on Calorimetry in High Energy Physics," Fermilab (Batavia), 29 Oct.-1 Nov.1990.
8. J. Seguinot et al, "Liquid Xenon (Krypton) Calorimetry," CERN/DRDC/90-70, (1990).
9. KNF, Kurt Neuberger France.
10. Messer Griesheim GmbH, Oxisorb Gas Purifying System.
11. E. Buckley et al, A Study of Ionization Electrons Drifting over Large Distances in Liquid Argon, Nuclear Instr. and Methods A275 : 364 (1989).
12. J. Seguinot et al, report in preparation, (1991).
13. Argon, Krypton and Xenon are supplied by Carbagas, Switzerland.
14. The off-line purification unit was provided by D.Schinzel.

CRYOGENIC FUEL TECHNOLOGY AND ELEMENTS

OF AUTOMOTIVE VEHICLE PROPULSION SYSTEMS

W. Peschka

German Aerospace Research Establishment
Stuttgart, Federal Republic of Germany

ABSTRACT

Carbon free fuels will be important in the future because they are environmentally safe. Liquid hydrogen is the most promising carbon free fuel for internal combustion engines. In addition to on board fuel storage and handling the availability of cryogenic high pressure fuel conditioning and injection systems represent the most crucial requirement for developing high performance hydrogen internal combustion engines. Primary pressurization of supercritical hydrogen from 1.5 MPa at 35 K up to 20 MPa is carried out by a compact three-cylinder high-speed cold compressor. Cryogenic high pressure injectors for direct cylinder injection are still under test. Details of the cryogenic fuel injectors and of the cold compressor are provided and its application in compact liquid nitrogen precooled small refrigerators and liquefiers below 60 K are briefly discussed.

INTRODUCTION

The primary argument for the use of hydrogen as fuel lies in obvious future problems regarding the atmospheric pollution with CO_2 as a result of continued unrestrained use of fossil energy. While for the heat released hydrocarbons produce about the same amount of CO_2 during combustion, no considerable improvement can be achieved by selecting fuels with a higher hydrogen content (see Fig. 1). Hydrogen is practically the only carbon-free fuel which is technically feasible. Apart from lubricant combustion products only nitric oxides are critical pollutants in I.C. engines. In this case, they can be specifically reduced without causing any other side effects. Although this requires increased technical complexity in the engine and the mixture formation.

MIXTURE FORMATION WITH CRYOGENIC HYDROGEN

The advantage of liquid hydrogen storage in a vehicle has been known for a long time and demonstrated [1-5,20,21]. There are various ways to use the advantage properties of cryogenic hydrogen. Here it is assumed that for mixture formation hydrogen primarily can

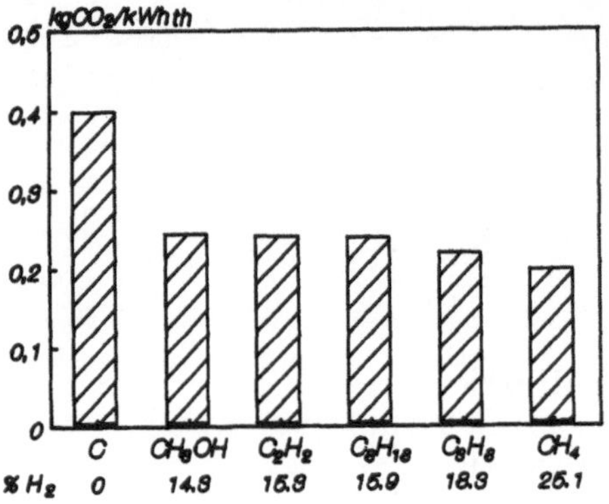

Fig. 1. CO_2 emission of various fuels related to their heat value.

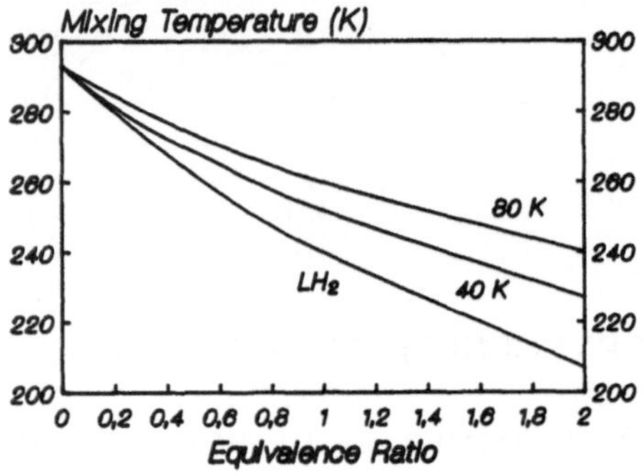

Fig. 2. Mixing temperature of dry air with cryogenic hydrogen.

be used either as a cold low pressure gas or supercritically near the critical temperature. Because of the small mass flow rate and corresponding small evaporation enthalpy it is hardly possible to deliver hydrogen in a liquid state into the combustion chamber of automotive engines. It could be done in stationary large scale engines or ship engines.

EXTERNAL MIXTURE FORMATION

Low pressure injection of cryogenic or liquid hydrogen directly into the intake manifold causes a temperature drop in the fuelair mixture. According to Fig.2, for a stoichiometric fuel-to-air ratio ($\varphi = 1$) a maximum temperature drop of about 80 C can be achieved for homogeneous mixtures. The corresponding increase in mixture mass density is equivalent to supercharging of the engine. Due to the considerably higher density of the cryogenic hydrogen compared to ambient hydrogen, the fuel volume which has to be injected is also considerably smaller. As has been demonstrated for the first time by Furuhama[6,22] loss of power compared to conventional hydrocarbon fuel can be partly compensated. The temperature drop of the mixture also results in the reduction of the mixtures tendency toward uncontrolled preignition. Experience with individual continuous and individual sequential timed intake port injection of cryogenic hydrogen with a 2-liter four-cylinder engine[7] showed that for steady operation at around $\varphi = 1$ the design properties for gasoline operation for brake mean effective torque (b.m.e.t.) and power were achieved under steady operation conditions Fig. 3. However, considerable breaks in the torque curve were obtained under partial load and especially at lower speed. The test engine described[7] did not experience any problems with condensation of the humid air and

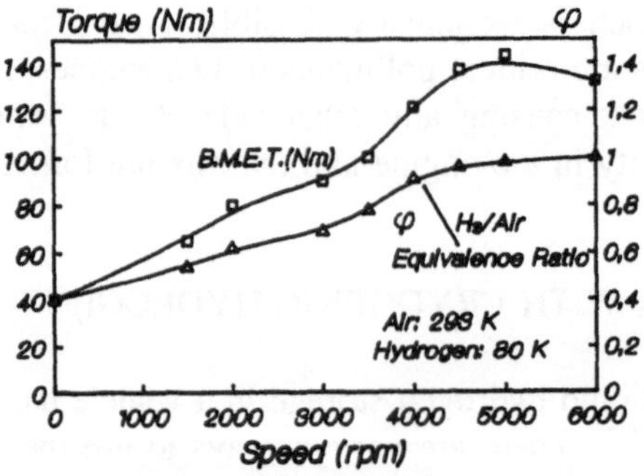

Fig. 3. BMW-M20 engine, torque with cryogenic hydrogen.

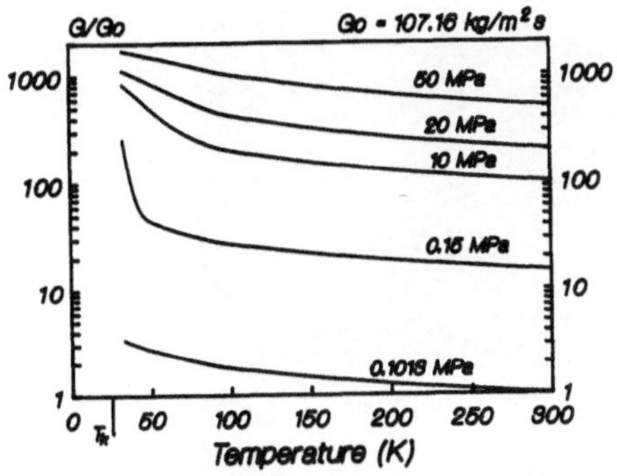

Fig. 4. Critical mass flow density of cryogenic hydrogen.

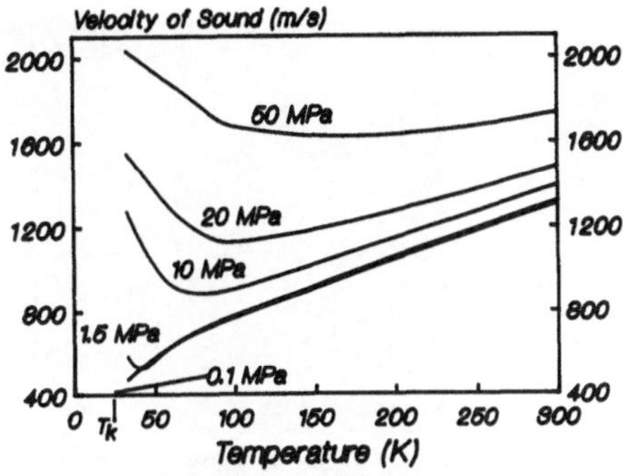

Fig. 5. Sound velocity of cryogenic hydrogen.

the associated ice formation. Evidently, the cryogenic hydrogen-air mixture did not remain in the cylinder intake ports of the air intake manifold much longer than the time required for the formation of condensation nuclei and sublimation zones.

INTERNAL CRYOGENIC MIXTURE FORMATION

Internal cryogenic mixture formation with injection start at the beginning of the compression stroke (early injection) does not have any significant advantages compared to ambient hydrogen. Above all, uncontrolled preignition and thus torque losses under partial load cannot be eliminated. On the other hand, from experience to date, internal cryogenic mixture formation with injection start close to top dead center (late injection, about 5° b.t.d.c.) by means of mechanically or hydraulically actuated cryogenic hydrogen injectors represents the most promising engine design. In terms of road performance this corresponds to conventionally fueled Otto cycle engines and Diesel-engines, however, with substantially lower amounts of harmful exhaust emissions[8].

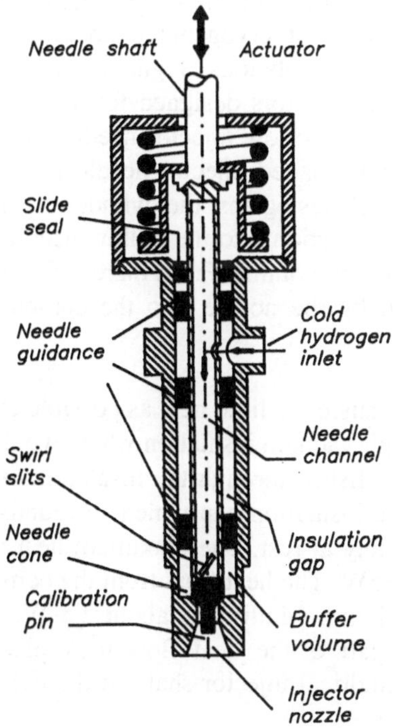

Fig. 6. Principle of the high pressure cryogenic hydrogen injector.

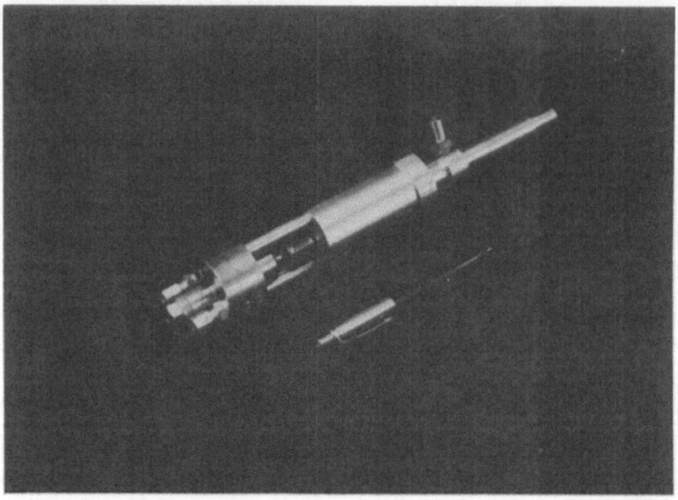

Fig. 7. View of the cryogenic hydrogen injector.

CRYOGENIC FUEL INJECTORS

Due to the short period of time available (about 5 ms at 5,000 rpm.) in high-speed engines, attempts should be made to attain minimum injection periods of about 0.3 ms to 0.5 ms and maximum injection periods up to 3 ms. This variation is equivalent to a ratio of minimum to maximum fuel mass injected of about 1:6 to 1:10. In general the development of these types of externally actuated gaseous fuel injectors is an essential problem in the development of H_2 engines with internal mixture formation[9-13]. The very short periods of time require an injection pressure ranging between 15 MPa to about 20 MPa. On one hand it is necessary to provide a larger initial density and thus larger impulse to the H_2-jet which exits the injector at the local speed of sound. On the other hand, it is necessary to keep the gaseous volume which has to be injected as small as possible in order to enable smaller injector orifices thus moving smaller amounts of mass. Figure 4 and figure 5 show the critical mass flow density and the speed of sound of cryogenic hydrogen[14]. The advantage of cryogenic hydrogen under high pressure is obvious. Figure 6 shows the principle and figure 7 one of the cryogenic hydrogen injectors designed for 15 MPa. It consists of the injector shaft, the injector nozzle and the hollow injector needle which is guided in sleeves made from PTFE bronze. The spring loaded injector needle is pushed into the nozzle seat via a titanium spring retainer and closes against the cylinder pressure. The cryogenic hydrogen flows from the upper injector section into the hollow needle and after passing three diagonal swirl slits into a small storage volume. From there it flows via the open seat of the needle and the calibration pin by the nozzle into the combustion chamber.

The injector configuration was designed to transfer as little heat as possible to the cryogenic hydrogen without using the double wall high-vacuum insulation which would be extremely difficult to use in this case. This is accomplished through the insulation of the hydrogen flowing in the hollow nozzle needle by the insulation gap which surrounds the outside of the needle which contains gas approximately at rest. The maximum heat flow rate to the cryogenic hydrogen measured was about 35W. The heat flow from the combustion space (400 C) via the nozzle and needle seat (1 mm width) was about 2.5 W. This small value was achieved because a considerable part of the heat flow took place in the bypass via the exterior of the nozzle and the cylindrical injector shaft on the cylinder

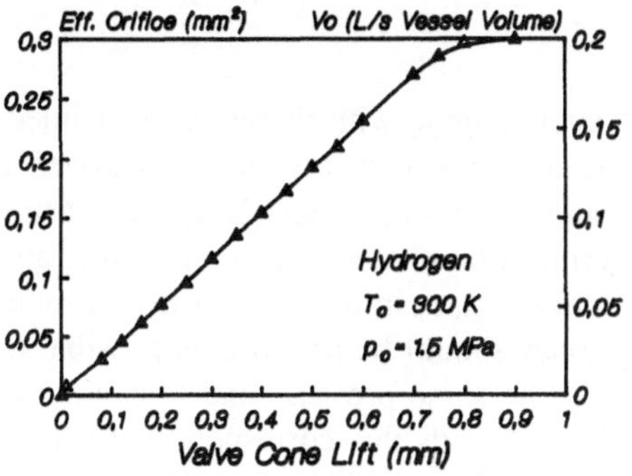

Fig. 8. H_2 mass flow rate of the cryoinjector as a function of the injector needle stroke.

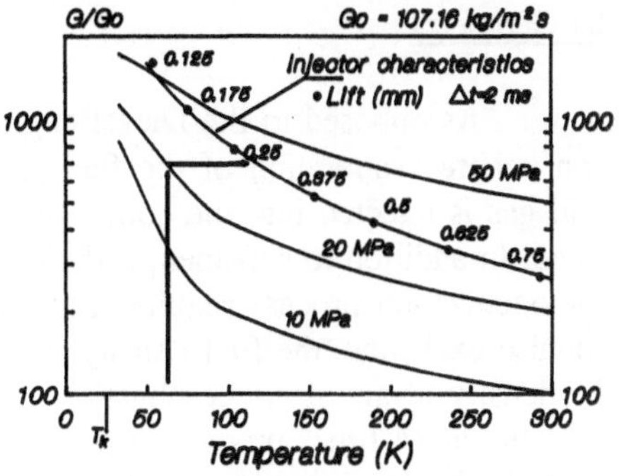

Fig. 9. Mass flow rate of the cryoinjector for cryogenic hydrogen.

head (120 °C). At a H_2 mass flow rate of about 0.8 g/s per cylinder (full load), hydrogen is warmed up by about 10 K. Under partial load this temperature span is correspondigly increased in dependence from the decrease in the fuel mass flow rate.

Figure 8 shows the steady mass flow rate of ambient hydrogen as a function of the needle stroke. Maximum mass flow rate is achieved at about a 0.8 mm stroke. The shortest open period attained was 1 ms, where 0.4 ms is included in each for the opening and closing procedure (needle motion). As figure 4 shows, the mass flow rate increases considerably with cryogenic hydrogen. With an admission pressure of about 20 MPa, the maximum values required by the M30 test engine already can be attained with a 0.25 mm needle stroke and at a 3 ms opening period (see Fig.9). As a result of the shorter stroke the opening and closing procedure decreased to about 0.15 ms each, so that the shortest feasible reproducible opening period was about 0.5 ms. A comparison shows that with hydrogen below 80 K and 20 MPa pressure, the maximum volume (φ= 1) which has to be injected amounts to about 0.57 cm^3 compared to about 0.1 cm^3 with gasoline or Diesel fuels. Thus, high presure cryogenic internal mixture formation leads to an amount of fuel to be injected per stroke of nearly the same order of magnitude.

Injector Actuation

The lower limit of the opening period of actuation via hydraulic pressure pulses with currently available in-line pumps is about 0.5 ms. The elasticity of the pressure lines as well as transientphenomena may have negative effects on the pulse shape. Another desirable reduction of the pulse wich seems to be possible along with the development of appropriate hydraulic systems like distributor pumps or pump elements which are directly actuated by the camshaft of the engine, as well as through the selection of other linear drives. One method which has already been tested with ambient hydrogen is appropriately designed, extremely fast acting electromagnetic linear drives[15,16].

Piezohydraulic Actuation

A potentially interesting procedure for achieving a fast acting linear drive is the electrostriction of piezoceramic materials like Bariumtitanate, with which upper cutoff frequencies up to about 100 kHz can be achieved. The very small attainable strokes of less than 1/10 mm however must be increased to about 0.5 mm. In addition to purely mechanical devices, hydraulic transmissions are particularly well suited for this.

Jet Geometry

As opposed to the Diesel engine or Otto cycle engine with direct cylinder injection, where evaporation of the fuel plays an important role, with cryogenic hydrogen a cold gas is injected into the combustion chamber and accordingly must be rapidly distributed. In addition to sufficient turbulence in the combustion chamber, a short mixing and homogenization process requires maximum jet impulse. Also, in the interest of optimum impulse exchange, the fuel density in the mixing zones should be as close as possible to the density of air in the combustion chamber[17,18]. At the same pressure and temperature, air is about 14 times more dense than hydrogen. With liquids this corresponds to approximately the density ratio of water to mercury. By using cryogenic high pressure hydrogen this situation is considerably improved with regard to density and jet impulse.

Hybrid Mixture Formation

At about 0.4-0.5 MPa injection pressure a lean cylinder charge ($\varphi < 0.6$) is obtained with sequentially timed intake manifold injection of cryogenic hydrogen and ignited at about 40° b.t.d.c. This lean mixture does not exhibit tendency toward uncontrolled preignition or significant NO_x emission. At about 5° b.t.d.c. cryogenic hydrogen at about 20 MPa is injected into the propagating flame front. As a result of flame turbulence and flame front propagation, problems regarding rapid mixing and ignition of the late injected fuel fraction are secondary. Figure 10 shows the low-pressure injectors arranged on the newly developed air intake manifold as well as the cryogenic high-pressure injectors which are mounted directly on the cylinder head (see Figs.6, 7). The injectors are actuated by the Diesel injection pump, serving as a pressure pump, which is located on the bottom right side of the engine. In the foreground in the top right part of the figure the vacuuminsulated cryogenic hydrogen feed line can be seen. This hybrid mixture formation for hydrogen enables stoichiometric combustion at very low NO_x emissions.

Fig. 10. BMW-M30 engine converted for hybrid mixture formation.

Fig. 11. View of the small Cryomec LH_2 pump.

Cryogenic Fuel Conditioning System

With external mixture formation the pressure in the cryogenic fuel tank (0.3-0.4 MPa) is sufficient for fuel injection. Also it could be done by a reciprocating or membrane pump which is integrated in the tank. For internal mixture formation with late injection start more sophisticated systems are necessary which nevertheless can be designed as a highly redundant device[8,18,19]. Pressurization of the fuel fraction for internal mixture formation up to the 20 MPa level can either be made in one step via a LH_2-pump driven by the engine or by a cold compressor which is supplied with supercritical hydrogen at about 1.5 MPa and 35 K via a booster pump which is integrated in the fuel tank. Figure 11 shows a compact LH_2 pump from Cryomec AG with a flow rate of 6 L/min at 450 strokes/min, tested at the DLR with LH_2 up to a pressure of 25 MPa. Caused by losses in the LH^2 high pressure pump or also in the LH^2 booster pump a small fraction of liquid hydrogen becomes vaporized and has to be released into the air. With hybrid mixture formation it can be used in the fuel fraction for external mixture formation. Figs.12 and 13 show the principle and the front view of a fast speed three cylinder "cold compressor" designed for supercritical hydrogen still under development. It is suitable for further miniaturization, uses a stepped piston, whose upper slender part moves in the cylinder by means of a gas film acting as a clearance seal. The cylinder contains a buffer space in its midsection. Its volume is approximately the volume of the gas film of the slender piston part in the cylinder. The part of the cylinder below the buffer space in the area of the cylinder block can be warmed by the flow of a liquid medium at ambient temperature (water for instance). Thus the average gas film temperature of approximately 200 K is considerably higher than the upper part of the cylinder (35-50 K). Calculations as well as experimental results revealed that through the increase of the gas film temperature via heat input the gas leakage of a miniaturized cold compressor (bore 12 mm, stroke 30 mm, speed 3000 rpm., pressure 20 MPa, mass flow rate per cylinder approximately 2-3 g/s) can to about 5% of the mass flow rate by selecting a radial piston clearance in the cylinder of

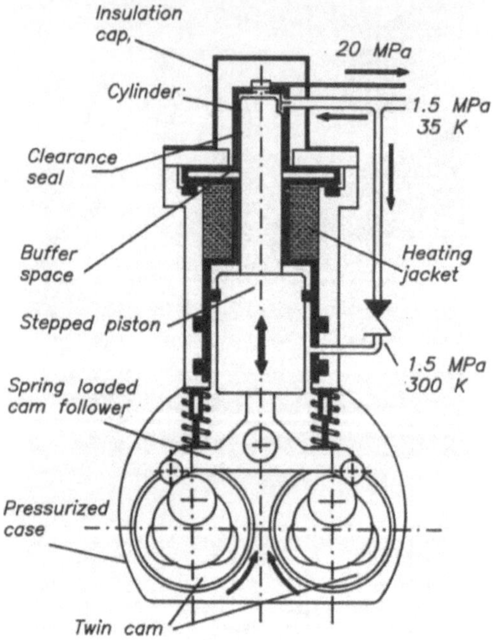

Fig. 12. Principle of the cold compressor for cryogenic hydrogen.

about 5 . 10^{-3} mm. The total length of the cylinder is then only 70 mm. With a cold gas film, for comparable gas leakage the total length of the cylinder would have to be more than 700 mm. This is contrary to miniaturization and practically impossible from technical reasons especially with regard to the narrow film gap.

However with the desired small total length the heat input which has to be supplied for the gas film in the area below the buffer space would be about 400 Watt per cylinder. This would be difficult to achieve with respect to the heat transfer in this very compact design. For this reason, the leakage of the piston-cylinder gap is further reduced by the lower part of the piston (increased diameter) together with an appropriate expansion of the lower cylinder section (gas film for piston sealing or piston sleeves with dry lubrication as shown in Fig. 12). By the input of a small amount of warm hydrogen under admission pressure (1.5 MPa) at bottom dead center (piston sleeve position below the intake slot) via

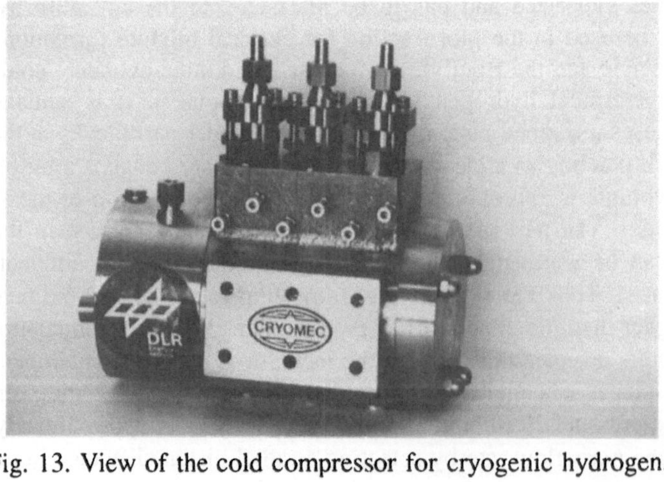

Fig. 13. View of the cold compressor for cryogenic hydrogen.

the one-way valve the lower cylinder section above the piston sleeve is filled up. Therefore the lower piston section around the piston sleeve serves as a supercharger for the gas film of the upper, slender piston section. Therefore during the compression stroke a gas film flow is achieved in a direction towards the high-pressure section. In this manner, through appropriate matching of the buffer space and film volume, a gas film flow directed from the cold high-pressure section towards the warm low pressure end during the compression stroke is prevented and excessive cooling of the midsection of piston and cylinder that it causes is eliminated even with a small total length of piston and cylinder. Therefore, in order to maintain a piston/cylinder temperature of about 200-300 K in the midsection only heat conduction losses of each cylinder of about 20-30 Watt have to be offset in the piston, cylinder wall and cylinder block. This can be achieved by means of a suitable liquid flow in the cylinder jacket. In this manner ambient temperature is maintained in the crank case which is under admission pressure of the compressor of approximately 1.5 MPa so that a "cold drive" and its problems can be avoided.

Apart from use in vehicles this type of cold compressor can be used as a cycle compressor in refrigeration or liquefaction processes such as for instance in the Linde-Thomson process or Claude process. The advantage is a considerably smaller size, correspondingly smaller shaft power demand and no oil or hydrocarbons in the compressed gas since the cold compressor has to be designed to run with dry lubrication. As a heat sink for the cold compressor a liquefied gas like liquid nitrogen for example can be used. Thus this type of refrigeration equipment might be applicable where compact design is required, that is small space and weight requirements, as well as where only limited electrical power is available and where affordable liquid nitrogen is available.

SUMMARY

Considerable improvement can be obtained in internal combustion engine operation by cryogenic hydrogen fuel. Liquid hydrogen the preferred fuel storage option onboard the vehicle, additionally satisfies the prerequisite for the advantageous use of the cryogenic properties of hydrogen. In the case of external mixture formation this leads to an increase in power to displacement ratio and reduced NO_x emissions as a result of cylinder charge cooling. Substancial improvements are obtained with internal mixture formation with high pressure hydrogen injection. With a tenable amount of effort the fuel pressurization can only be achieved with LH_2. Cryogenic properties of hydrogen such as large density and considerable heat sink capacity favor fuel injection, mixture formation and thus the combustion process. As an "inconvenient fuel", hydrogen leads to a relatively sophisticated fuel conditioning system. According to recent state of development in cryo-technology it can be designed and carried out to high redundancy. Hybrid cryogenic mixture formation which is an advantageous combination of external and internal cryogenic formation could be very attractive with respect to power output and torque as well as satisfactory transient operation.

REFERENCES

1 W. F.Stewart, Hydrogen as a Vehicular Fuel, in: "Recent Developments in Hydrogen Technology" K.D., Williamson, Jr., Edeskuty, F.J., eds., CRC Press Cleveland Ohio, (1986), Vol.2, pp.69-146.

2 W. Peschka, "Fluessiger Wasserstoff als Energietraeger", Springer- Wien New York, (1984). Engl. Neuauflage (1991)

3 W. Peschka, Liquid Hydrogen for Automotive Vehicles, Experimental Results, in: ASME-Paper No. 81-HT-83 (1981).

4 W. Peschka, Liquid Hydrogen for Automotive Vehicles, Status and Development in Germany, in: "Cryog. Proc. and Equipment", ASME, Plenum Press, New York, (1984),p.97.
5 W. Peschka, Hydrogen Combustion in Tomorrow's Energy Technology., Int. J. Hydrogen Energy, 12:481(1987).
6 S. Furuhama, M.Hiruma, Y.Enemoto, Development of a Liquid Hydrogen Car, Int. J. Hydrogen Energy 3:61(1978).
7 W. Peschka, W.Nieratschker, Experience and Special Aspects on Mixture Formation of an Otto-Engine Converted for Hydrogen Operation, Int.J.Hydrogen Energy, 11:653 (1986).
8 W. Peschka, Fluessiger Wasserstoff als Motorenkraftstoff der Zukunft, Maschinenwelt und Elektrotechnik, 43:1 (1988).
9 C. A.McCarley, W.D.Van Vorst, Electronic Fuel Injection Techniques for Hydrogen Powered I.C. Engines, Int. J. Hydrogen Energy 5:179 (1980).
10 C. A.McCarley, Development of a High Speed Injection Valve for Electronic Hydrogen Fuel Injection, in: (Proc., 3rd World Hydrogen Energy Conf.), Vol. 2, Pergamon Press, New York(1980), p.1119.
11 T. Krepec, T. Giannacopoulos, D. Miele, New Electronically Controlled Hydrogen-Gas Injection Development and Testing, Int. J. Hydrogen Energy, 12:12 (1987).
12 K. S.Varde, G.M.Frame, A Study of Combustion and Engine Performance Using Electronic Hydrogen Fuel Injection, Int. J. Hydrogen Energy, 9:327 (1984).
13 K. S.Varde, G.A.Frame, Development of a High-Pressure Hydrogen Injection for SI Engine and Results of Engine Behaviour, Int. J.Hydrogen Energy, 10:743 (1985).
14 R. D. McCarty, J.Hord, H.M.Roder, in:"Selected Properties of Hydrogen," NBS Monograph 168 (1981).
15 A. H.Seilly, Colenoid Actuators- A New Concept in Extremely Fast Acting Solenoids. SAE-paper 810 462, (1981)
16 T. Krepec, T.Giannacopoulos, D.Miele, New Electronically Controlled Hydrogen-Gas Injector Development and Testing, Int. J. Hydrogen Energy, 12:855 (1987).
17 S. Furuhama, T.Fukuma, T.Kashima, Liquid Hydrogen Fuel Supply System for Hot Surface Ignition Turbocharged Engine. in: "Cryogenic Proc. and Equipment", ASME, Plenum Press, New York, (1984) p. 105.
18 W. Peschka, Liquid Hydrogen Reciprocating Pumps for Automotive Application, in: "Adv. Cryog. Eng.," Vol. 35 B, Plenum Press, New York (1990) p.1783.
19 W. Peschka, Liquid Hydrogen Pumps for Automotive Application, Int.J.Hydrogen Energy, 15:817, (1990).
20 W. Peschka, Liquid Hydrogen-Cryofuel in Ground Transportation, in:"Adv. Cryog. Eng.,"Vol.31, Plenum Press, New York (1986) p.1035.
21 W. Peschka, The Status of Handling and Storage Techniques for Liquid Hydrogen in Motor Vehicles, Int. J. Hydrogen Energy, 12:753 (1987).
22 S. Furuhama, T.Fukuma, Liquid Hydrogen Fueled Diesel Automobile with Liquid Hydrogen Pumps, in:"Adv. Cryog. Eng.", Vol. 31, Plenum Press, New York (1986) p.1047.

SLUSH HYDROGEN QUANTITY GAGING AND MIXING FOR THE NATIONAL AEROSPACE PLANE

R. S. Rudland
I. M. Kroenke
A. R. Urbach

Ball Electro-Optics/Cryogenics Division
Boulder, Colorado

ABSTRACT

The National Aerospace Plane (NASP) design team has selected slush hydrogen as the fuel needed to power the high-speed ramjet/scramjet engines. Use of slush hydrogen rather than normal hydrogen provides significant improvements in density and cooling capacity for the aircraft. The loading of slush hydrogen in the NASP tank must be determined accurately to allow the vehicle size and weight to be kept to a minimum. A unique sensor developed at Ball to measure the slush density will be used in each region of the hydrogen tank to accurately determine the total mass of fuel loaded in the vehicle.

The design, analysis, and test configuration for the mixing system is described in this paper. The mixing system is used to eliminate large-scale disturbances in the fluid produced by the large heat flux through the wall. The mixer also provides off-bottom suspension of the solids to create a more uniform slush mixture. The mixer design uses a pump to supply flow to an array of jets that produce mixing throughout the tank. Density sensors will be used in the test configuration to evaluate the mixing effectiveness.

The location and sizing of the jet array are determined by using CFD analysis with a slush mixing algorithm developed to model solid settling and agglomeration. Preliminary jet array selection is accomplished using a two-dimensional code, followed by a more detailed analysis using the three-dimensional code "FLOW-3D" which can model the slush and free surface effects. This design and analysis will be verified by test at Ball's test facility.

INTRODUCTION

The National Aerospace Plane (NASP) challenges the capabilities of current vehicle design in the areas of structural design for lightweight, thermal protection systems for control of high heating rates and temperatures, and design of system components to withstand the high thermal environments. The hypersonic speeds that the NASP will traverse to achieve orbit require active cooling of significant areas of the outer surface, where heating rates and temperatures exceed uncooled material capabilities.

Advances in Cryogenic Engineering, Vol. 37, Part B
Edited by R.W. Fast, Plenum Press, New York, 1992

Also, the fuel of choice for the high-speed propulsion (scramjet engines) is liquid hydrogen, a very low-density propellant but excellent high heat flux cooling media.

Hydrogen Densification Requires Slush Hydrogen Measurement

Since liquid hydrogen is the propellant of choice for the propulsion system, ways of densification were evaluated to reduce the relatively large storage volume due to the low density of the liquid hydrogen. Liquid hydrogen densification allows a smaller vehicle structure to be used for the same impulse value of the contained fuel.

By densifying the liquid hydrogen (LH_2), slush hydrogen is created which is a mixture of solid hydrogen and LH_2 at the triple-point temperature. For an assumed mixture of 50 percent solids and 50 percent triple-point hydrogen, the temperature of the triple-point mixture is approximately 13.8 K and the vapor pressure is 7 KPa. The density of the slush mixture is 15.2 percent greater, thereby lowering the volume by approximately 13 percent.

Assuming a hypothetical tank with a maximum pressure of 140 KPa, heat absorption comparisons can be made. The heat required to raise one kilogram of LH_2 to 140 KPa vapor is 450 KJoules. Triple-point LH_2 can absorb 12 percent more heat, and 50 percent slush hydrogen can absorb 18 percent more heat than NBP LH_2. It is absolutely essential to mission success that an adequate quantity of slush hydrogen be loaded onboard the vehicle.

Measurement of the loaded slush hydrogen is complicated by the settling of the solids. This requires that as the propellant is loaded, homogeneity must be created and maintained to achieve an adequate assessment of the actual slush status on the vehicle. A slush hydrogen density sensor has been developed, built, and tested to provide this assessment of slush status.[1] To combine this density sensor into a NASP fuel tank will require a mixing system to thoroughly mix the solids in the tank so that a homogenious mixture is maintained during loading measurements.

This paper will describe the density array/mixer systems design, analysis, and testing effort to be performed in 1991 and 1992 to support NASP development.

CALIBRATION OF DENSITY SENSOR

Capacitance measurements allow one to determine the dielectric constant of the slush hydrogen. This, in turn, is a measure of the density. Previous studies showed that the Clausius-Mossotti equation fit the data over a wide range of densities. However, all data were for homogeneous liquid hydrogen at various pressures. Thus, it was found that no actual data existed relating the density of slush hydrogen with dielectric constant. During the course of our investigation we have shown that the dielectric constant of slush hydrogen does in fact behave as one would expect. The total is equal to the weighted average of the volume fractions of solid and triple-point liquid hydrogen dielectric constants.

To meet the configuration requirements of the hypersonic vehicle– namely light weight, ±0.5 percent accuracy, and large distances between the sensor and support electronics – a novel capacitance sensor was developed.

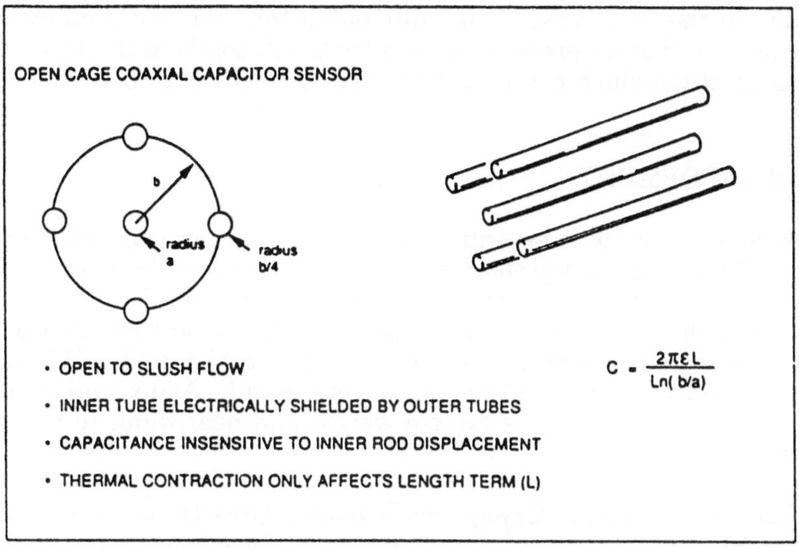

Fig. 1. Density sensor measurement volume geometric configuration.

It allows free flow of the solids around the sensor, is lightweight (a few pounds), is insensitive to vibration, and sensor location may be hundreds of feet from support electronics without cable variations entering into the measurement. The sensor's measurement volume is shown in Fig. 1.

The calibration equipment has been used to develop the density sensor design with its attached electronics. This glass dewar was used to demonstrate better than 0.4 percent error with slush hydrogen, as shown in Fig. 2. The array/mixer system will use the calibration glass dewar with an improved slush mixer to obtain sensor calibrations of better than 0.2 percent. The density sensor design has also been improved to simplify fabrication and

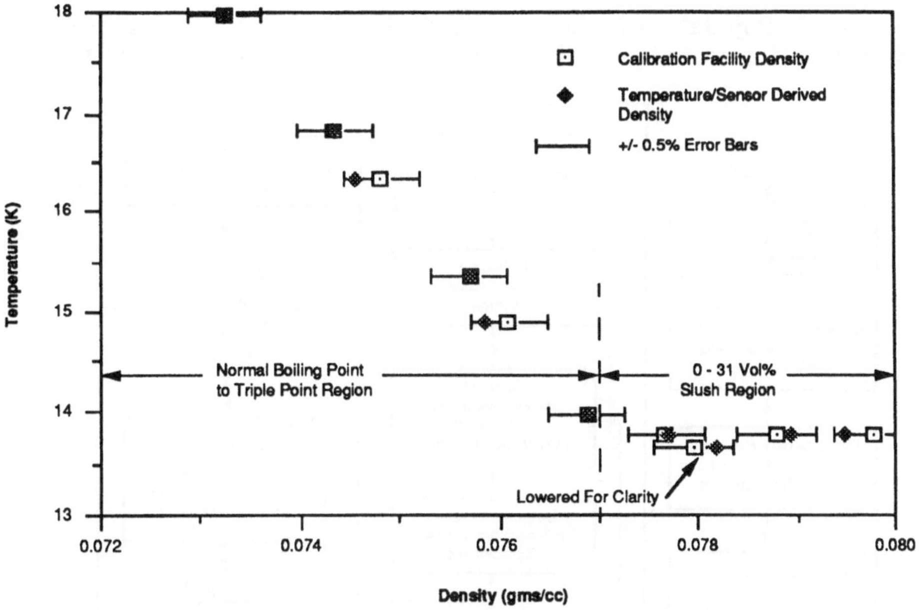

Fig. 2. Liquid/slush hydrogen accuracy test calibration using precision temperature sensor.

installation in the test tank. It is anticipated that the array/mixer testing will demonstrate that by properly mixing the slush solids in the tank, the average density of the slush can be determined to better than 0.5 percent accuracy.

MIXER DEVELOPMENT

A mixing system is essential for a hypersonic vehicle fueled by slush hydrogen. Functional requirements for this system may be summarized as keeping solids in suspension and maintaining the proper fluid level in each tank during flight and ground loading phases. The solid hydrogen particles must be kept in suspension to ensure accurate density and solid mass determination, and to avoid stratification of the liquid. Additional system requirements include minimizing system weight and heat input to reduce fuel requirements.

Ball Electro-Optics/ Cryogenics Division's (BECD) approach to the selection, design, and analysis of a mixing system for a hypersonic vehicle using slush hydrogen is shown in Fig. 3. This includes selecting the mixer system, predicting the mixer performance using FLOW-3D, performing tests, validating the analysis model, and predicting future NASP density gage performance.

Mixing Effectiveness

Studies at BECD and NBS (NIST) have determined that the primary objective of a slush hydrogen mixing system for a space flight vehicle is to provide complete, off-bottom suspension and mobility of the solid particles in the slush during loading, ground hold, and transfer operations. This objective applies to one-g operations and requires the solids to be suspended and moving up off the bottom of the tank. The term "mixing effectiveness" has been chosen to quantify the mixing system objective.

Solids Population is the technique selected for measuring mixing effectiveness. This technique is based on the time rate of change of the solid

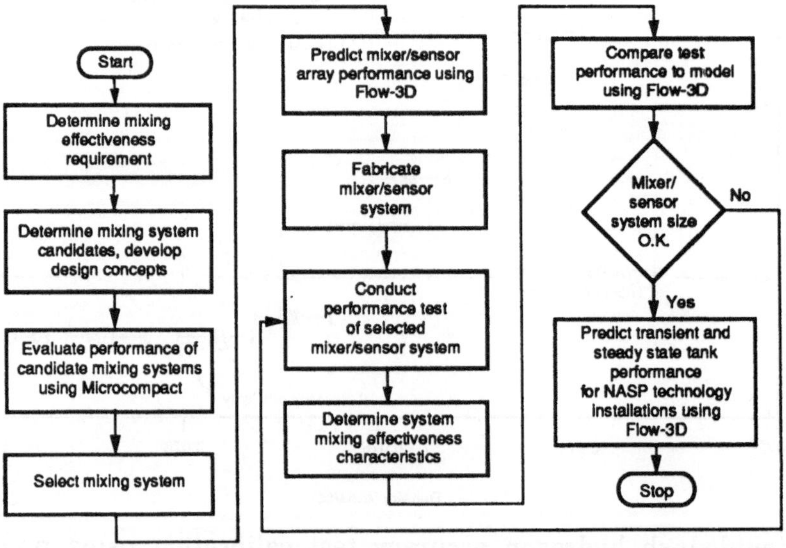

Fig. 3. Systematic approach to design, analysis, and test.

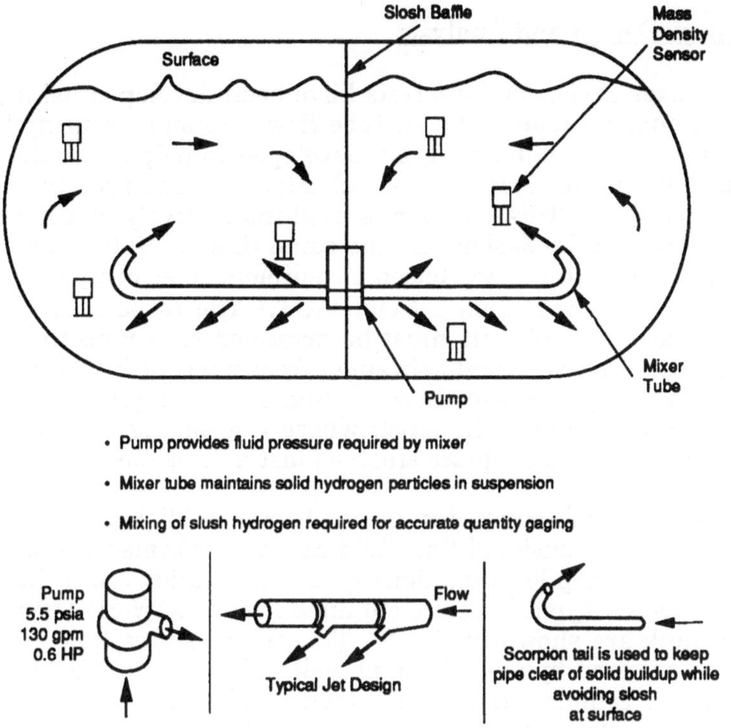

Fig. 4. Concept for mixing of integrated test tank.

hydrogen particle population within a mixture through a given control volume. Since density is proportional to the solid fraction within the triple-point liquid, a series of local densimeters can be used to determine mixing effectiveness locally. The sensor data is then averaged over the entire volume under consideration.

Baseline Mixing System Selection

Several candidate mixing systems have been studied and analyzed for cryogenic systems at BECD and outside sources. The candidate systems that have or are capable of being used with slush hydrogen include the jet tube, axial jet, radial jet, ducted propeller, high-shear impeller, unshrouded propeller, axial turbine, radial turbine, and paddles. Initial trade study results, specific to a hypersonic vehicle, narrowed the field of candidates to the three jet mixing systems and the ducted propeller concepts. In addition to the mixing requirement, these trade studies were concerned with weight, reliability, and heat input to the tank.

For a typical hypersonic vehicle tank, additional analysis using the microCOMPACT computational fluid dynamics (CFD) program revealed that a single jet orifice, representative of the axial jet, radial jet, and ducted propeller, would not adequately mix slush hydrogen. Significant settling occurred in areas of the tank, even with extremely high flow velocities.

This data resulted in the selection of the jet tube configuration as the baseline mixing system. The jet tube mixer is a multiple jet configuration consisting of an electric, motor-driven pump feeding a perforated tube with an end nozzle. A schematic diagram of the baseline mixing system is shown in Fig. 4.

Jet Tube Mixer Design and Analysis

The mixer design and analysis have been developed using a variety of design and analysis tools. The jet tube flow rate and jet sizing have been determined using a computer program developed to help select the jet diameter, number of jets, jet location, jet velocity, jet flow, pressure drop, and pump power required. This program is used interactively by the designer to arrive at a basic jet tube design. At the same time, a preliminary flow field study is performed using a PC-based two-dimensional computational fluid dynamic code (CFD) such as microCOMPACT.[2] The two-dimensional analysis will show how multiple jets must be arranged to stir up the solids and prevent settling. Typical results obtained from microCOMPACT for mixing in a tank using a single jet are shown in Fig. 5. This figure shows how well heat enters the tank and melts slush where the mixing velocity is low, so that slush mixing does not replace slush as fast as it is melted away.

Once a baseline mixer design has been established, then a detailed study of the three-dimensional flow field can be performed to assess the influence of baffles, pump location, density sensor locations, wall heating, and free surface effects. Typical results obtained from FLOW-3D for multiple jet mixing in a tank are shown in Fig. 6. This figure shows how multiple jets in a manifold powered by a small pump will induce a flow field in the tank. This flow field is complex and three-dimensional due to the location of the jets, pump, and walls. Once baffles and a free liquid surface are added, the flow will become more complex. Solid settling, mixing, and agglomeration can be added to the FLOW-3D source code, as necessary, to study the off-bottom suspension of solids by the jet tube mixer. This modeling will be supported by tests with slush hydrogen to verify the mixer design.

Development Testing

Ball has designed a test system to produce slush hydrogen and perform tests on jet tube mixer systems developed for use on NASP. The test

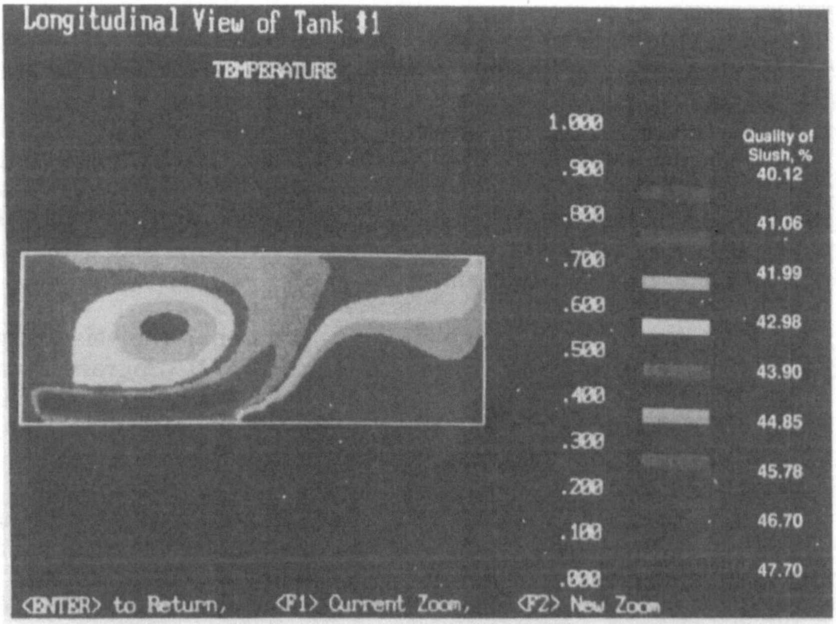

Fig. 5. View of slush hydrogen quality locations.

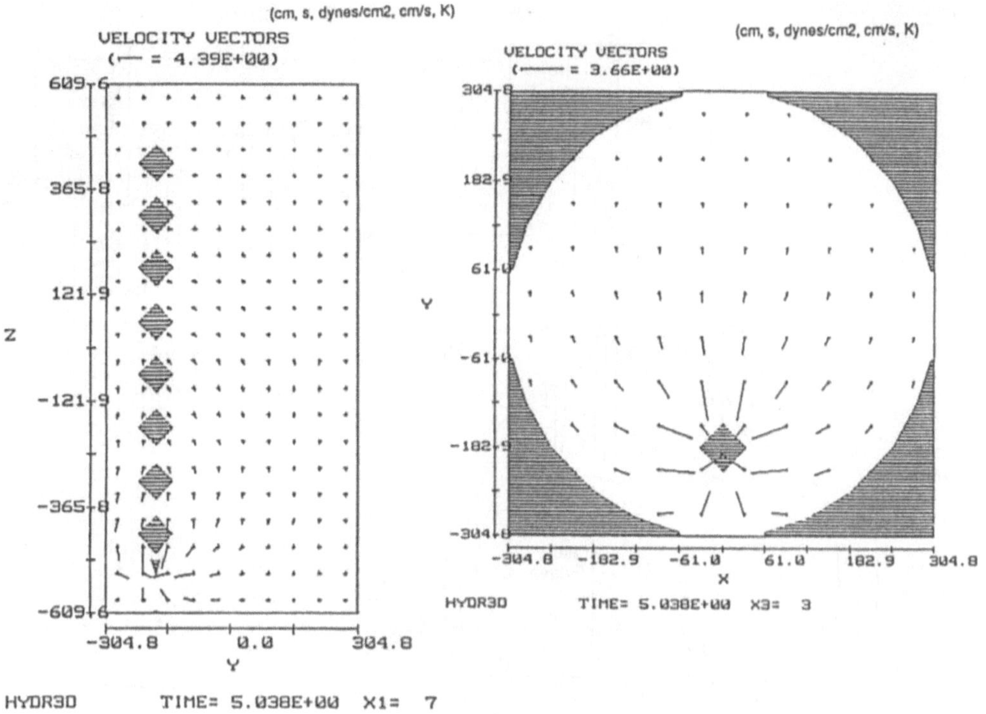

Fig. 6. Model with mass sources/sink.

tank is able to produce slush using the freeze-thaw method and uses calibrated density sensors to measure the mixing effectiveness by measuring the variation in density throughout the tank. The test schematic is shown in Fig. 7. Pressurization can be performed with cold or warm hydrogen or helium gas, as needed. The test tank is able to cool down, load hydrogen, produce slush, top off, and prepare for testing in one 8-hour shift. This allows many tests to be performed in sequence once slush is produced. Wall heaters will be positioned in the tank to simulate environmental heating that will tend to melt the slush solids. Two large windows are located at the top of the tank to watch the slush mix and check the slush production and aging.

Array/Mixer System

The array/mixer system will be able to measure the average density by measuring local density in several locations, combined with a mixer that has good mixing effectiveness as determined by analysis and test. The current projection of six sensors, one pump, and a jet tube designed to thoroughly mix the slush will be refined by analysis and test to arrive at the optimum design. Since each density sensor includes a precision silicon diode temperature sensor, this signal is also processed to verify the accurate operation of the sensor as needed. Other pressure and temperature sensors are used in test to monitor the fluid behavior.

CONCLUSIONS

Ball has shown in this paper that:

- A slush hydrogen sensor with appropriate accuracy has been demonstrated.

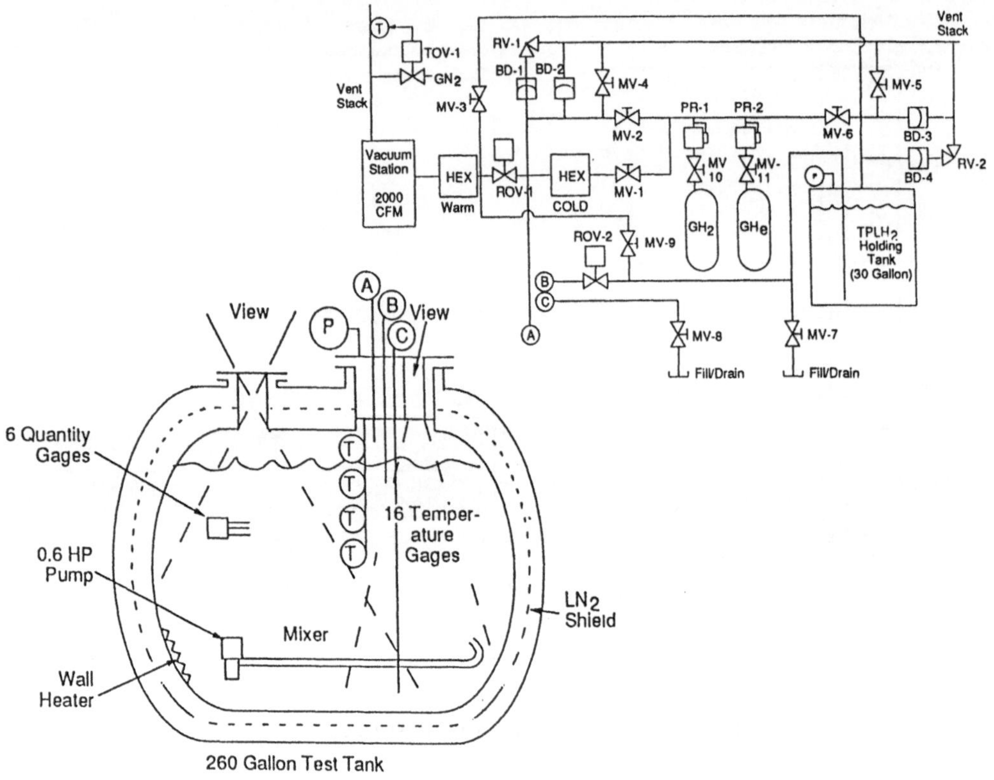

Fig. 7. Test schematic shows flexibility in testing.

- A preliminary mixer design and modeling have been accomplished.
- A development test fixture is state-of-the-art.
- A mixer/slush density sensor system can be demonstrated in the required time frame for NASP.

ACKNOWLEDGEMENTS

This work has been supported in part by Rockwell International Corporation.

REFERENCES

1. W. Horsely, I. Kroenke, F. Chandler, "Slush Hydrogen Density Gage Operation in Extreme Environments," AIAA-90-5235, AIAA 2nd International Aerospace Planes Conference, October 29-31, 1990, Orlando, Florida.

2. A. Dreher, R. Bell, T. Flaska, "Flow Field Analysis of Slush Mixing," AIAA-90-5215, AIAA 2nd International Aerospace Planes Conference, October 29-31, 1990, Orlando, Florida.

LOW TEMPERATURE RESEARCH FACILITIES FOR SHUTTLE AND BEYOND

P. Mason, U. Israelsson, T. Luchik, W. Owen, D. Petrac, and D. Strayer

Jet Propulsion Laboratory, Pasadena, California

ABSTRACT

This paper describes the Low Temperature Research Facility program now under development at the Jet Propulsion Laboratory. This program will develop the experiments and cryogenic facility to fly experiments requiring temperatures in the liquid helium range, from the superfluid range (1.6 K) to the critical point (5.2 K). A shuttle facility is planned for 1998. It will be flown about every two years after that time. A free flyer is also being evaluated for the end of the decade or early in the next century. A brief discussion of the quality of micro-gravity available in various carriers is also given.

THE LOW TEMPERATURE RESEARCH PROGRAM

Background

The Superfluid Helium Experiment(SFHE) was flown on and the shuttle as part of Spacelab 2 in 1985[1,2]. The cryostat

Fig. 1. The Lambda Point Experiment facility in its magnetic shield. The experiment SQUID electronics is mounted on the head of the cryostat on a copper heat sink.

facility was intended for multiple flights and will be flown again as the cryogenic facility for the Lambda Point Experiment (LPE)[3,4] in 1992. The cryostat, enclosed in a magnetic shield and with the experiment in place, is shown in Fig. 1. The LTRF cryostat will be similar in appearance, but somewhat larger.

Program Description

The Microgravity Science and Applications Office of the National Aeronautics and Space Administration intends to sponsor a continuing program of low temperature research. New facilities will be built by JPL to accommodate these experiments. These low temperature research facilities (LTRF-I AND LTRF-II) are described in this paper. Both are superfluid helium cryostats with associated control and command and data electronics. LTRF I is to be a shuttle-based facility with a vibration isolation system and a lifetime of 20 days. LTRF-II is to be a free flyer with a 100 day lifetime.

The Low Temperature Research Facilities will combine operation at superfluid helium temperatures, (bath temperatures of 1.3 to 2.17 K), a microgravity vibration environment (static acceleration less than 10^{-5} g, g-jitter less than 10^{-3} g, and a very stable thermal environment (temperature fluctuations less than 10^{-6} K). If lower acceleration levels are required, a passive vibration isolation system can be added to reduce the vibration levels by a factor of 10. In addition, ambient magnetic fields at the experiment can be reduced by a factor of 100, to allow operation of superconducting quantum interference detectors at full sensitivity.

A tentative schedule of the program is shown in Fig. 2.

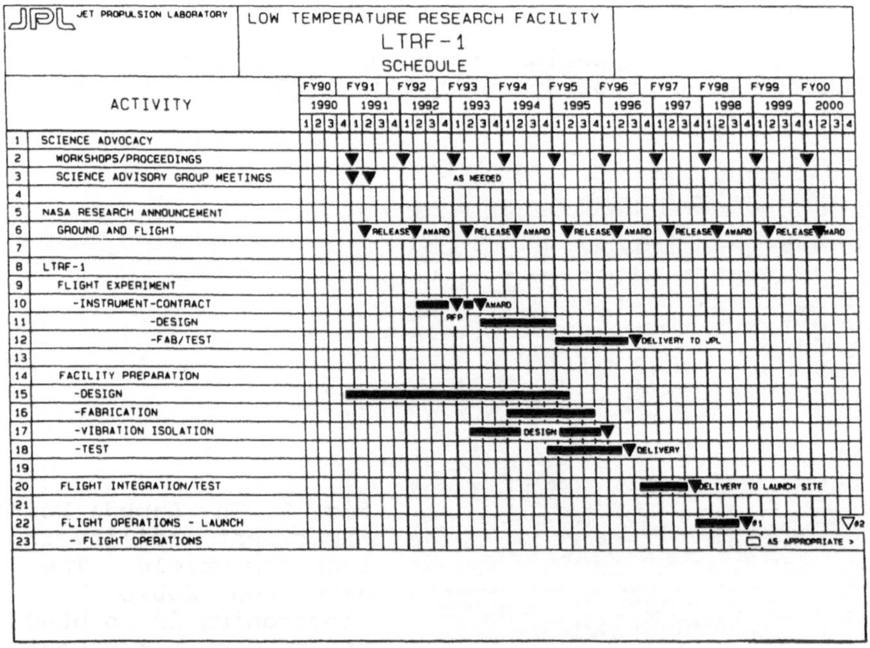

Fig. 2. LTRF-I schedule

Development of Experiments

In developing the Low Temperature Research Facility program, JPL first had to show that a need, based on science, existed for such a facility. Once it was concluded that there was a science need, the capabilities of the facility to meet the science needs were to be determined.

JPL held its first workshop in 1989 in Oregon to address the question of the need for the facility. This first workshop was attended by some 40 scientists and engineers specializing in a variety of low temperature disciplines. The results of the workshop showed that there was a class of superfluid helium and critical point experiments that could use a common facility based on the Space Shuttle or Space Station-based low temperature facility, where the expected acceleration levels are of the order 10^{-3} to 10^{-5} g. However, gravitation and relativity experiments, although sharing the need for low temperatures in common with the above experiments, have such stringent requirements on the residual acceleration level (10^{-12} g) as to not make them feasible for a shuttle or station-based facility. It was also identified in this workshop that other supporting technologies, such as SQUID technology, active refrigeration and magnetic shielding, should be supported by a long range μ-gravity program.

The final action of the workshop was the formation of a Science Advisory Group (SAG) on low temperature research in space. The group was chaired by Professor Russell Donnelly of the University of Oregon and members included an international collection of experimental and theoretical physicists.

In the year that followed the first JPL workshop, several SAG meetings were held. These meetings were held to identify and prioritize specific low temperature research areas for near-term experiments. The SAG was also asked to review and comment on an initial 10 year plan developed by JPL for the Low Temperature Research Facility (LTRF) program. The group emphasized the need for microgravity, identifying the measurement limitations due to hydrostatic, finite size and surface effects. They further commented that a focus should be placed on experiments which cannot be done in earth-based laboratories. The SAG's main comment on the 10 year plan for LTRF was the need to shorten the time between ground-based feasibility and flight development such that an experiment might be flown in a PhD student's academic career (normally 5-6 years). JPL responded with a program that included this constraint.

A second workshop on low temperature research was held in Washington D.C. in January, 1991. This workshop attendees not only included low temperature scientists and technologists, but also included JPL management and the management of the funding agency, NASA Microgravity Science and Applications Division (MSAD). The workshop served two purposes. It identified to NASA directly that there was sufficient low temperature science that required the low gravity environment of space and that there was support within the science community. The science community was shown clearly NASA's commitment to the development of a low temperature research facility in space. The workshop resulted in a clear definition of "important" science from the scientist's perspective and it helped in the definition of the LTRF program.

The initial portion of the program involves the release of a NASA Research Announcement (NRA). This announcement includes all of the fundamental science discipline and is due for release sometime in the summer of 1991 (we cannot mention exact dates due to legalities). Proposals selected from this announcement will be funded to perform initial ground-based definition studies and flight development work. This research will also be used to define specific science requirements for the Low Temperature Research Facility (LTRF-1). Currently only general requirements have been specified in the conceptual design phase of LTRF. The NRA will also be used to help identify low temperature supporting technology development which will significantly enhance the facility.

Once implemented, the LTRF-1 program is currently expected to fly an experiment every other year on the Space Shuttle for at least ten years. An average of 4 to 5 ground-based experiments per year along with supporting technology developments are expected to be funded by the program.

As shown in Fig. 2, annual workshops will be an integral part of the LTRF program. They are intended to inform scientists of opportunities in low temperature research in space, to exchange information about the science and technology, and to develop advocacy or the program.

PROPOSED EXPERIMENTS

A number of experiments have been discussed in the workshops. These are summarized in Table 1. Some of these can be performed only in space, while others can be performed in a one g, but with much lower resolution or accuracy. Typical of the latter is the measurement of the lambda point, which can be performed to a resolution of about 10^{-6} K in the laboratory, but to 10^{-10} K in space, where the transition is not broadened by the pressure dependence of the lambda transition.

SHUTTLE LOW TEMPERATURE RESEARCH FACILITY (LTRF-I)

The first low temperature research facility (LTRF-I) is now in the conceptual design stage. It is planned for flight as early as 1996 to 1998. It is an improved version of the cryostat flown on Spacelab 2 with the Superfluid Helium Experiment and the Lambda Point Experiment.

If required by the experiment, LTRF-I will be mounted on a vibration isolation system to reduce the vibrations seen by the shuttle by a factor of 10 above 10 Hz. Its lifetime will be increased from 10 to 20 days to achieve operation for the full Shuttle flight time of 7 to 10 days even after a two day launch delay. To achieve this, the liquid helium complement will be doubled to 200 liters and the parasitic heat leaks will be reduced by several means. A third vapor cooled shield will be added. As shown in Fig. 3, the configuration of the support straps will be changed to provide increased length to area ratio while maintaining adequate stiffness and fatigue

Table 1. Candidate low gravity low temperature experiments

CRITICAL POINT EXPERIMENTS	
Transport Properties	Static Properties
• Thermal conductivity	• Specific heat
• Viscosity	• Expansion
• Lambda transition in a heat current density	• Superfluid
• Low frequency sound attenuation near T_λ	• Order parameters
• Non-linear sound near T_λ	• Finite size and surface effects
• Mutual friction	
• Tri-critical point in ^3He-^4He mixtures	
• Conversion of first sound to second sound at interfaces	
FINITE SIZE AND SURFACE EFFECTS	
• Boundary free fluid flow (Levitated helium drop)	
• Static: expansion, superfluid density, specific heat	
• Dynamic: thermal conductivity	
OTHERS	
• Boundary free fluid flow (levitated helium drop)	
• Freezing/melting dynamics	
• Third sound - thick films, persistent currents	
• Colloidal suspensions	

life; and the wires and plumbing necessary to support the experiment will be heat sunk more effectively to the vapor cooled shields.

The cryostat is designed to accommodate an experiment of the volume and weight of the Lambda point experiment. The cold experimental volume is 8.5" x 28" and the allowable mass is 35 kg. Experiment electrical and gaseous interfaces are accommodated by suitable wiring and plumbing. The experiment can dissipate up to 100 mW to the helium bath while meeting its lifetime requirement of 20 days.

The temperature of the bath can be as low as 1.3 K or as high as about 2.3 K. Higher experiment temperatures, for example helium critical-point experiments at 5.2 K can be accommodated by using a thermal control system. An active thermal control system can also be used to provide a high degree of temperature regulation. The five-stage thermal control system developed for the Lambda Point Experiment, for example, regulates temperature to about 10^{-9} K.

The command and data system currently provides 512 bytes/sec of data downlink Three 18-byte commands per second can be transmitted during a 5 minute window every 45 minutes. However this rate is subject to change as needed by the experimenter.

The LTRF-I concept is shown in Fig. 4, mounted on the cross-bay MPESS structure. The MPESS provides power and

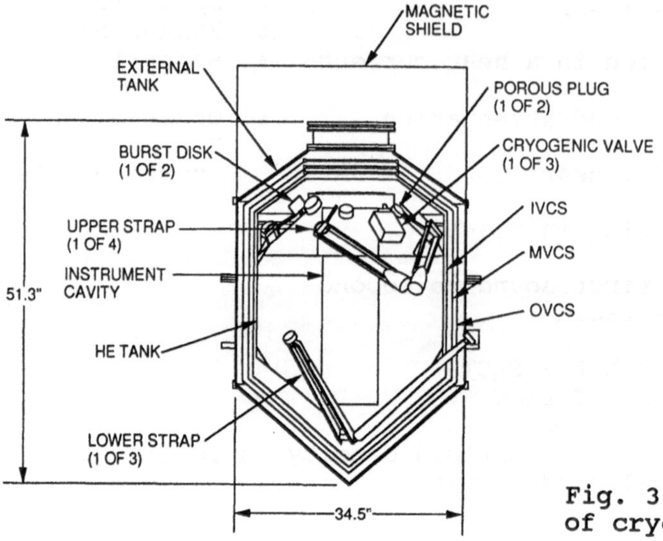

Fig. 3. Cross section of cryostat

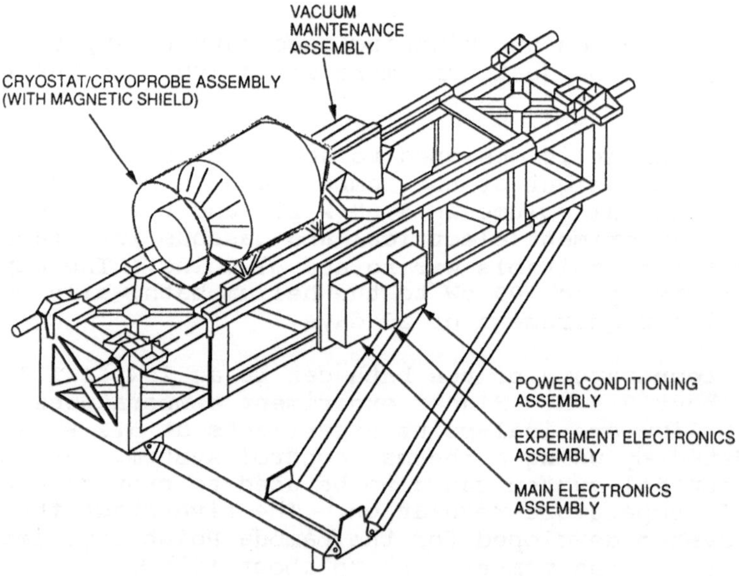

Fig. 4. LTRF-I mounted in the Shuttle bay.

command and data system interfaces. A Shuttle Acceleration Measurement system is capable of measuring accelerations with a sensitivity of 10^{-5} g.

FREE FLYER LOW TEMPERATURE RESEARCH FACILITY (LTRF II)

The second low temperature research facility, now in the conceptual stage, is intended for free flight in order to provide static and dynamic accelerations in the range of 10^{-6} g. It will be recoverable to allow resupply or replacement of experiments. It will have a lifetime of at least 100 days in order to allow improvement of signal to noise ratio by long term averaging.

The carrier might be a shuttle-launched and recoverable free flyer such as the MBB EURECA, which offers up to six months of orbital lifetime.

THE ZERO-G ENVIRONMENT

Since the key reason for going to space is to conduct experiments in near-zero gravity, we include a brief description of the acceleration environment as experienced on various carriers.

The static acceleration is limited primarily by residual atmospheric drag and gravity-gradient effects caused by the finite size of the carrier. The gravity gradient effects are zero at the center of gravity, but it is seldom possible to operate at the CG in the Shuttle or Space Station. Dynamic accelerations are caused by operating equipment and crew activity.

Table 2. Quality and duration of zero gravity for various carriers

CARRIER	STATIC ACCELERATION	DYNAMIC ACCELERATION	DURATION
DROP TOWER	10^{-5} G	NOT AVAILABLE	5 SECONDS
KC-135	10^{-3} G	NOT AVAILABLE	20 SECONDS
SHUTTLE	10^{-5} TO 10^{-6} G	10^{-2} TO 10^{-3} G	7-10 DAYS
SHUTTLE WITH VIBRATION ISOLATION	10^{-5} TO 10^{-6} G	10^{-3} TO 10^{-4} G	7-10 DAYS
FREE FLYER	10^{-5} TO 10^{-6} G	$<10^{-3}$ G	MONTHS TO YEARS
DRAG COMPENSATED FREE FLYER	10^{-10} TO 10^{-11} G	10^{-7} G	MONTHS TO YEARS

As shown in Table 2, drop towers operating in vacuum achieve excellent quality, of the order of 10^{-5} g, but are limited in time to about 5 seconds. This time can be doubled as in the NASA-Lewis drop tower by launching the payload vertically from the bottom towers, but in practice the disturbances caused by launch are so large that the capability is seldom used. Zero-g aircraft, such as the NASA Johnson Space Center KC-135 can achieve 10^{-3} g for up to 20 seconds, but it is difficult to release the test package without acceleration and drift which limit both the quality and duration of the weightlessness

The shuttle static acceleration is in the range of 10^{-5} to 10^{-6} g, but the dynamic accelerations are in the range of 10^{-2} to 10^{-3} g, mostly as a result of mechanisms such as antennas and crew motions. Free fliers without drag compensation are much better, perhaps as good as 10^{-6} g, while the ultimate of 10^{-10} to 10^{-11} g is obtained when a proportional thrusting system is used to obtain a drag-free environment.

ACKNOWLEDGEMENTS

The research described in this paper was carried out at the Jet Propulsion Laboratory, California Institute of Technology, under contract to the National Aeronautics and Space Administration. The Low Temperature Research Facility program is sponsored by the NASA Microgravity Science and Applications Division.

REFERENCES

1. P. V. Mason, D. Petrac, D. D. Elleman, T. Wang, H. W. Jackson, D. J. Collins, P. Cowgill, and J. R. Gatewood, "The Scientific Results of the Spacelab 2 Superfluid Helium Experiment", Proceedings of the International Cryogenic Engineering Conference, West Berlin, April 22-25, 1986. Butterworth, Guilford, Surrey, 1986.

2. D. Petrac and P. Mason, "Cryogenic Performance of the Spacelab 2 Superfluid Helium Cryostat," Proceedings of the International Cryogenic Engineering Conference, West Berlin, April 22-25, 1986. Butterworth, Guilford, Surrey, 1986.

3. J. A. Lipa, T. C. P. Chui, and D. Marek, "Lambda Point Experiment in Zero Gravity", Aerospace Century 21 Advances in the Astronomical Sciences (1987), 64, 1245

4. D. Petrac, U. Israelsson, D. Otth, L. Simmons, J. Staats, and A.Thompson, "Technical Challenges Involved in Supporting the Lambda Point Experiment", Proceedings of the International Cryogenic Engineering Conference, Beijing, China, April 24-27, 1990, Butterworth, Guilford, Surrey, 1986. C. S. Hong, Ed.

THE CONE PROGRAM - AN OVERVIEW

William J. Bailey

Martin Marietta Astronautics Group
Civil Space & Communications
Denver, Colorado 80201

Hugh Arif

National Aeronautics and Space Administration
Lewis Research Center
Cleveland, Ohio 44135

ABSTRACT

Subcritical cryogenic fluid management (CFM) has long been recognized as an enabling technology for future space applications such as Space Transfer Vehicles (for near-earth, lunar and interplanetary missions) and On-Orbit Cryogenic Fuel Depots. Space Station Freedom may also derive benefits from an understanding of CFM. This CFM technology contains many elements which are required to successfully support future system development and mission goals associated with the cost of system operation and reuse. Subcritical liquid storage and supply, however, have never been demonstrated on-orbit. In-space demonstration of this technology using liquid nitrogen (LN2), with a few well defined areas of focus, would provide the confidence level required to implement low-gravity subcritical cryogen use and is a first step towards the more far reaching issue of cryogen transfer and tankage resupply.

The Cryogenic Orbital Nitrogen Experiment (CONE) is a LN2 cryogenic storage and supply system demonstration placed in orbit by the National Space Transportation System (NSTS) Orbiter and operated as an in-bay payload whose objective is to demonstrate needed critical components and technologies. A conceptual approach has been developed by Martin Marietta and an overview of the CONE program is described which includes the following: (1) a definition of the background and scope of the technology objectives being investigated, (2) a description of the payload design and operation, major features and rationale for the experiments being conducted, and (3) the justification for CONE relating to potential near-term benefits and risk mitigation for future systems.

INTRODUCTION

The CONE (shown in Figure 1) is an integrated experimental payload designed to investigate specific subcritical cryogenic fluid management technologies in a low-gravity space environment. Data and criteria will be provided to correlate in-space performance with analytical and numerical modeling of cryogenic fluid management systems, as well as demonstration data for the mitigation of design risk. CONE results are tailored to provide increased confidence for the use of subcritical cryogen storage and supply for various applications including Space Station Freedom growth options and space propellant storage.

Technical objectives of the CONE mission are highly focused and are divided into priority experiments and demonstrations which are considered a secondary set of cryogenic

Advances in Cryogenic Engineering, Vol. 37, Part B
Edited by R.W. Fast, Plenum Press, New York, 1992

technologies. Collectively they form the CONE Experiment Set which provide both experimental and demonstration data for future space missions, providing fluid management technology in the following areas of emphasis: 1) cryogenic liquid storage, 2) liquid nitrogen supply, 3) pressurant bottle recharging, 4) active pressure control, and 5) liquid acquisition device performance (expulsion efficiency). Active pressure control is the highest priority of scientific investigation and is the only experiment category of test. All others are demonstrations where the technical objectives tend to be less scientific in nature.

The CONE payload will provide an opportunity to demonstrate the feasibility of combining various methods of integrating pressure control, liquid acquisition, and tank fluid outflow into a subscale experimental tank design that will provide subcritical LN2 storage and outflow characterizations. Controlling tank pressure and supplying single-phase liquid to accomplish transfer and resupply/topoff of tankage is essential to having a space-based operational capability. Using LN2 as the test fluid will provide data acceptable for both oxygen and nitrogen subcritical systems and provides for extrapolation to LH2. In addition, LN2 allows collection of needed cryogenic data without the safety implications associated with liquid hydrogen (LH2) while operating in the more restrictive post-Challenger era of payload safety within the Orbiter cargo bay.

LOW-GRAVITY CRYOGENIC FLUID MANAGEMENT TECHNOLOGIES

Subcritical cryogenic fluid management technology for cryogens stored at low pressure contains many elements which are required to successfully support future space system development and mission goals associated with the cost of system operation and reuse. One of the technical elements is a knowledge of low-g transfer phenomena which is critical since in-space systems are currently end-of-life limited by depletion of on-board consumables. A more basic element of the technology, however, is an understanding of the storage and tank outflow requirements for subcritical cryogenic fluids.

Existing cryogen storage and supply systems utilize well characterized supercritical techniques that are outdated and weight intensive due to the high pressure required to assure that a single phase is maintained in the tank. Substantial weight savings can be realized by low pressure subcritical storage where two phase fluid is stored and liquid only is provided to a user by a surface tension liquid acquisition device (LAD). CONE on-orbit testing in the

Fig. 1. CONE payload configuration on the Hitchhiker- M carrier

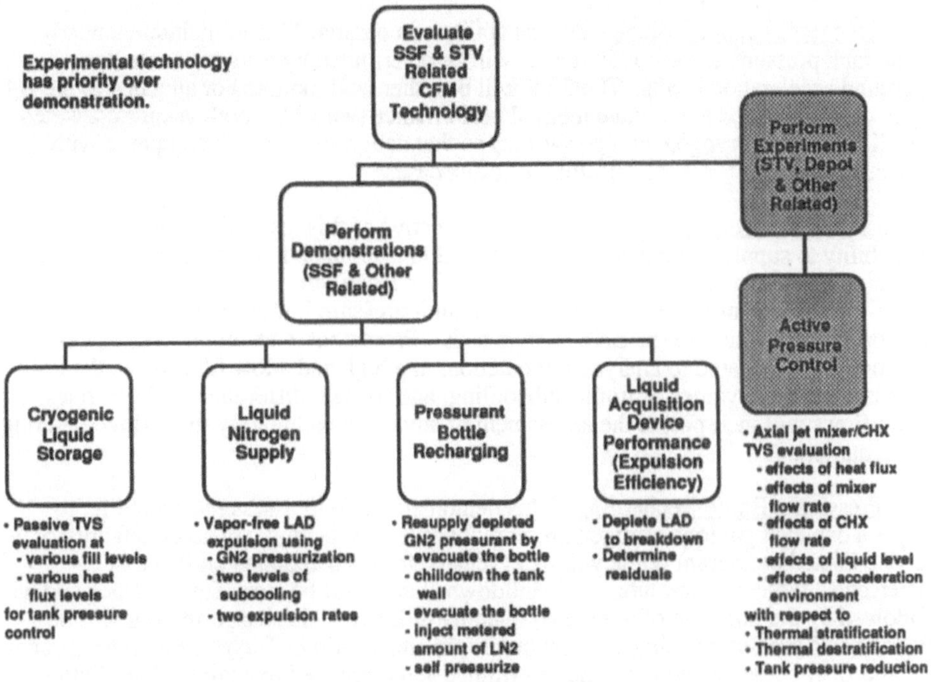

Fig. 2. CONE experiment science technical objectives

areas of liquid storage using both active and passive techniques to control tank pressure, along with the assessment of supplying vapor-free LN2 from the tank by use of a LAD is the first step towards the implementation of future in-space subcritical cryogen use.

Over the years, a three step approach to cryogenic fluid management technologies has been defined starting with individual component and hardware development, progressing to subsystem element ground based testing and finally to in-space experimentation. All result in the establishment of a cryo data base and to the development of refined analytical models which make use of the available data for validation and correlation purposes. CONE is the first precursor experiment to implement a cost effect, subscale on-orbit test approach focused on the near term needs of the space community.

CONE EXPERIMENT SET OVERVIEW

The experimentation and demonstration test categories comprise the primary and secondary technical mission objectives. Tests in each category include low-g fluid and thermal process investigations, demonstrations of performance capabilities, as well as technology evaluations to achieve an overall test mix that provides for a maximized technological return of data over the duration of the mission. Figure 2 shows these objectives and the subelements of each. A brief listing of individual experiment objectives is as follows:

Tank Active Pressure Control - Investigating the phenomena associated with the control of cryogenic storage tank pressure using an axial jet induced mixer coupled with a compact thermodynamic vent system (TVS) heat exchanger is the specific objective of this test series. A parametric assessment will be conducted to investigate the effects of tank heat flux, mixer flow rate, compact heat exchanger flow rate, tank liquid level and the acceleration environment on: 1) thermal stratification, 2) thermal destratification by mixing and , 3) tank pressure reduction during heat exchanger operation. Results will be compared with analytical predictions to provide partial verification of of the models which describe the physical processes involved.

The remaining technical objectives are lower priority system demonstrations consisting of the following:

Cryogenic Liquid Storage - The capability of a passive TVS to maintain a nearly constant tank pressure at various fill levels will be determined by tests conducted at background acceleration levels. The TVS will be either wall mounted or attached to the LAD. Heat flux will be varied to simulate thermal performance typical for both vacuum jacketed and foam/MLI insulated cryogen storage systems so that data results can be compared with analytical predictions for these different heat flux cases.

Liquid Nitrogen Supply - The specific objective of this test series is to demonstrate the capability to supply subcooled vapor-free LN2 to a simulated user by expelling tank liquid using a total communication capillary type fine mesh screen LAD. Expulsion will be provided by gaseous nitrogen (GN2), stored in high pressure bottles, and regulated to 30 psia prior to introduction to the LN2 storage tank. Pressurant consumption rates will be determined and compared to analytical predictions at a high and a low fill level in the LN2 storage tank, for two values of liquid subcooling, and for two different expulsion rates. Tank outflows will incorporate the assessment of advanced technology mass flow metering instrumentation.

Pressurant Bottle Recharging - This demonstration will assess the capability to resupply a depleted gaseous nitrogen pressurant bottle by injecting a metered quantity of LN2. A depleted pressurant bottle will be evacuated to space and then chilled to a predetermined target temperature. The chilldown charge will be evacuated to space and will be followed by the injection of a metered quantity of LN2. This charge amount will be allowed to self-pressurize using ambient environmental heating. Target temperature, charge mass, and final bottle pressure will be determined and compared to analytical predictions.

LAD Performance - Investigations of the capability of the LN2 storage tank LAD to provide vapor-free liquid to the point of breakdown will determine the expulsion efficiency for the device and will provide a single data point on the LAD retention capability for comparison to analytical models describing the LAD performance .

EXPERIMENT SUBSYSTEM DESCRIPTION

The CONE Experiment Subsystem is composed of the following three major elements: liquid nitrogen storage and supply tank, LN2 storage tank valve module, and GN2 pressurization & LN2 bottle recharge module. Figure 3 shows a schematic of the inter-relationship of the experiment subsystem elements.

Definition of Major Subsystem Elements

LN2 Storage Tank. This tank has a pressure vessel of 0.226 m^3 (8 ft^3) capable of holding 171 kg (375 lbs) of LN2 at 95% full and is completely contained by a vacuum jacket (VJ). To maintain a shape relationship to typical space based transfer vehicle tankage and in order to provide for a stable fluid interface, a cylindrical tank shape was selected. The overall tank VJ has a diameter of 68.6 cm (27 in) and a length of 121.9 cm (48 in) to provide the largest size tank that could be mounted to the front face of the Hitchhiker-M carrier. The pressure vessel (PV) has a diameter of 58.4 cm (23 in) and contains a total communication LAD with an outlet at the top of the tank. A vent/pressurization penetration which feeds directly into the tank via a diffuser is located at the top end so that contact with ullage is maintained for both horizontal and vertical positions of the tank. Pressurant is introduced into the tank by this line. An axial jet spray system is provided through which liquid can be introduced from a mixer pump to provide mixing of the bulk fluid. Mixer pump fluid is passed through a compact heat exchanger (CHX) where energy can be removed from the tank to actively control or reduce tank pressure. An internal thermodynamic vent system (TVS) heat exchanger (HX) routed on the LAD is provided to cool the bulk fluid and passively control tank pressure. Thermal control heaters uniformly cover the pressure vessel and are used to vary the heat flux for pressure control and stratification experiments. A layer of foam insulation covers the heater blanket to provide protection against a loss of annular vacuum and limit the venting potential for this off-nominal case. Multi-layer insulation (MLI) is located over the foam insulation for nominal on-orbit control of tank heat leak. All plumbing penetrations from the PV are routed internal to the VJ and exit at the girth ring area. An outlet heat exchanger (OHX) mounted to the VJ girth allows for subcooling of the tank

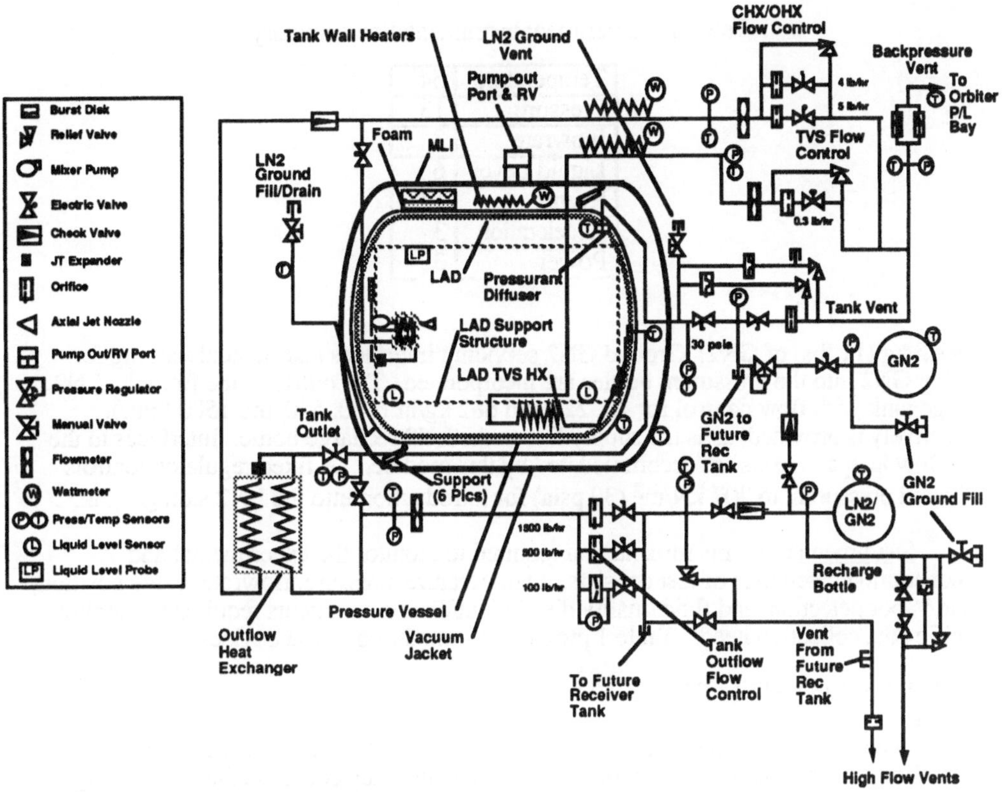

Fig. 3. Experiment subsystem simplified schematic

outflow. The PV connects to the VJ with a strap suspension system. Figure 4 shows the details of the mixer/CHX and the tank outlet.

<u>LN2 Storage Tank Valve Module</u>. All components that interface with the back pressure vent and provide LN2 storage tank pressure relief or isolation for TVS, CHX, OHX flow rate control and tank venting/pressure introduction, as well as instrumentation for flow monitoring are accommodated by this module. Redundant mechanical burst disk/relief assemblies provide overpressure protection for the tank at 345 kN/m ^{2} (50 psia).

<u>GN2 Pressurization & LN2 Bottle Recharge Module</u>. Components associated with pressurant storage, ground servicing, regulation and distribution, as well as the LN2 storage tank outflow line and the LN2 recharge bottle are assembled into a self contained module. Pressurant storage is provided by two spherical 35.6 cm (14 in) diameter 0.023 m^3 (0.83 ft^3) tanks pressurized to 20670 kN/m^2 (3000 psia) on the ground prior to flight. Each tank

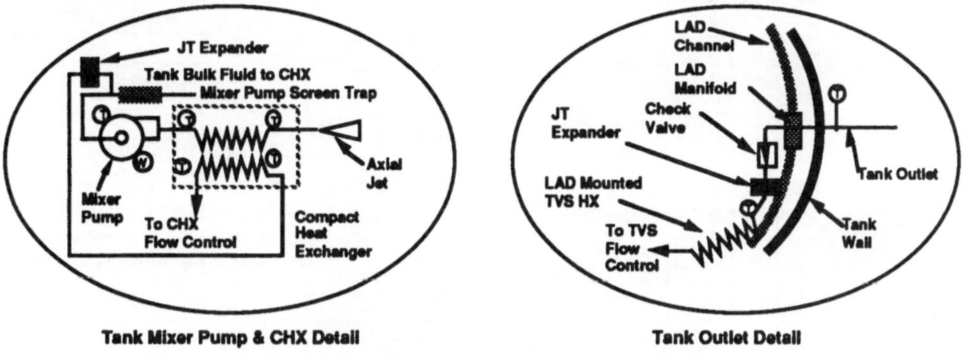

Fig. 4. LN2 storage tank internal details

Table 1. Experiment instrumentation summary

Temperature	54
Pressure	13
Flowrate	3
Liquid Level	6
Events	17
Acceleration	3
Power	3

stores 5 kg (11 lbs) of GN2. Ground GN2 servicing interfaces and manual valving for loading GN2 into the pressurant bottles are incorporated. The outflow line from the LN2 storage tank with flow control for 45, 227 and 682 kg/hr (100, 500 and 1500 lb/hr), respectively is provided, as is the supply line to the LN2 recharge bottle. Interfaces to the high flow in-space vents are accommodated by this module. A fixed regulator controls delivered pressurant to 207 kN/m^2 (30 psia) for introduction into the LN2 storage tank.

Instrumentation. Instrumentation required to monitor the experimental and demonstration categories of test consists of temperature, pressure, flowrate, valve position, liquid/vapor detection, and fluid quality discrimination measurements required to monitor specific data collection needs. Table 1 provides a summary of these devices.

SUPPORT SUBSYSTEM DESCRIPTION

The support subsystems augment the experiment subsystem in the accomplishment of the mission science and operational objectives. Together they comprise the payload flight element of the CONE System. The following provides a brief description of the support subsystems which are functionally depicted with the experiment subsystem in Figure 5. Figure 6 views the CONE mounted to the HH-M carrier.

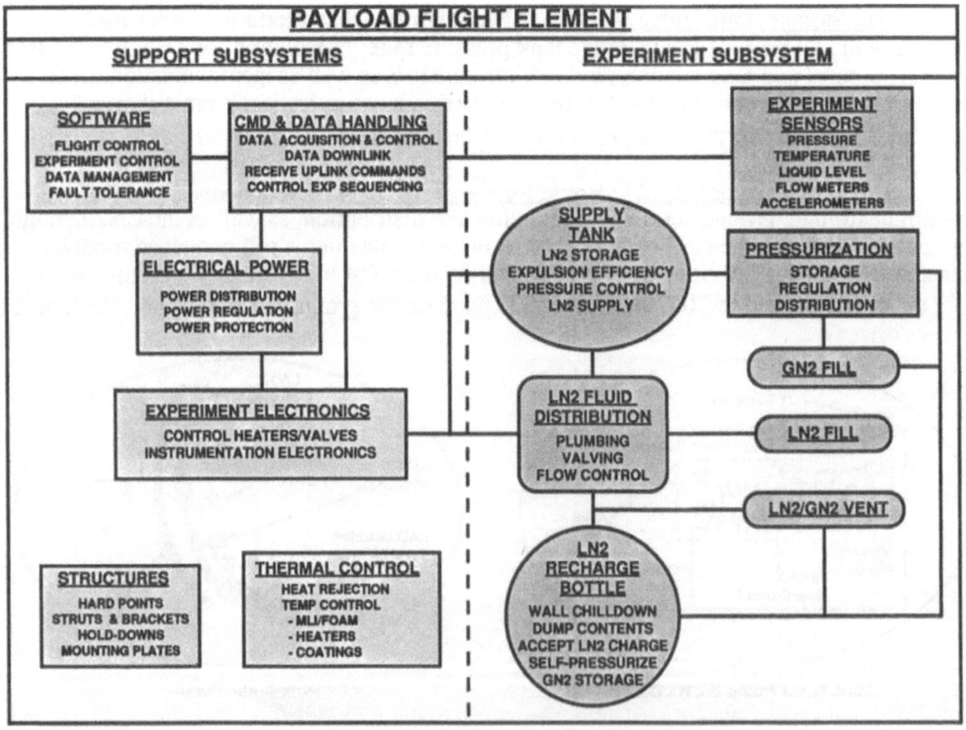

Fig. 5. Payload flight element subsystem functional block diagram

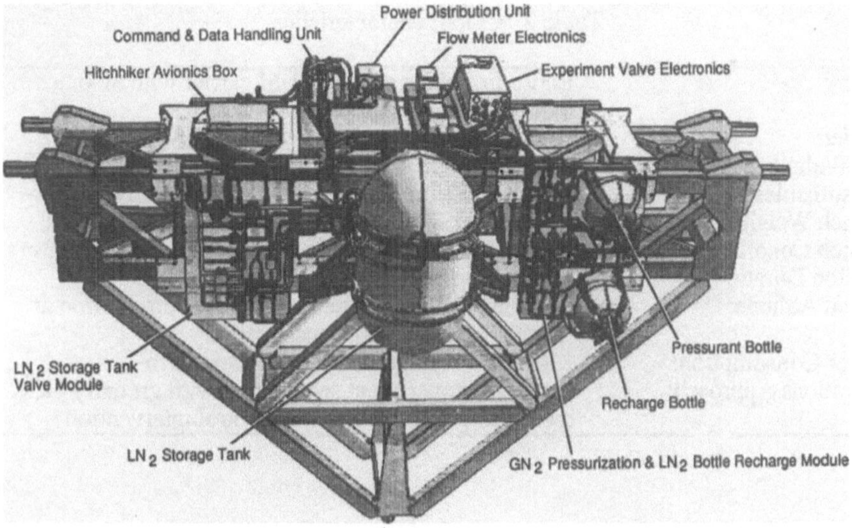

Fig. 6. CONE payload element definition

Major Support Subsystem Elements

Structures Subsystem. This subsystem provides the structural mounting for all support and experiment subsystem elements to the HH-M carrier, as well as attachment of CONE subsystem elements to one another. The supports and struts connecting the LN2 storage tank to the HH-M truss comprise the major elements of this subsystem. All component mounting brackets, clamps,base plates and attachment structure complete the definition of this subsystem.

Electrical Power Subsystem (EPS). The EPS controls, conditions, distributes, monitors and provides power bus isolation and protection. A Power Distribution Unit (PDU) provides protection, control and distribution of 28 vdc electrical power from the HH-M avionics unit to electrical elements on the payload. The PDU also provides power to the experiment mixer pump, heaters and valves via the Experiment Valve Electronics (EVE) unit.

Command & Data Handling (C&DH) Subsystem. The C&DH provides for formatting and transmission of housekeeping and experiment data and the capability for the decoding and distribution of commands to operate the payload during the mission. The C&DH also provides for the transmission of data downlink and the acceptance of ground command uplink via standard Orbiter communication links that interface with the HH-M avionics unit. Off-the shelf hardware is utilized and contains elements that accomplish control and monitoring of the sensors, valves, and other components of the experiment subsystem. Control of the experiment sequencing functions is handled by the on-board computer (OBC). Communications between the C&DH and the other subsystems is via a multiplex data bus.

Experiment Electronics Subsystem.This subsystem contains the experiment valve electronics (EVE) unit which controls the application of power commands to heaters and valves in the experiment subsystem. It provides the interface between the C&DH and the components that have to be controlled in the experiment subsystem. Also all instrumentation requiring unique electronics for signal conditioning or power regulation have these units located in this subsystem.

Thermal Control Subsystem (TCS). The TCS consists of thermal coatings, insulation, sensors, heaters and associated thermostats, and radiative surface configurations required to control and monitor the thermal environment to within proper operating ranges for all payload hardware. This design approach provides passive thermal balance of the entire payload using the above specified techniques, as appropriate.

Table 2. CONE characteristics

Size:	Entire front face of the HH-M carrier with avionics mounted on two standard plates
Carrier:	Hitchhiker-M
Dry Weight:	393 kg (865 lb)
Consumables:	171 kg (375 lb) LN2 and 10 kg (22 lb) GN2
Launch Weight:	574 kg (1262 lb)
Launch Condition:	Powered down with LN2 storage tank TVS operating
Mission Duration:	Six days nominal
Orbital Attitude:	Random - with fixed periods of low perturbation at maximum Orbiter drag acceleration
Power Consumption:	250 watts average - 500 watts maximum
Operations Approach:	On-board experiment sequencing with ground POCC monitoring and contingency control intervention

Vehicle Flight Software. Vehicle flight software consists of computer programing which will be resident in the OBC that will perform operations and computations in support of experiment and health monitoring functions, manage data collection, control the telemetry and experiment subsystem operation, and provide necessary system management and fault protection function.

Table 2 lists the general top-level characteristics of the CONE payload.

GROUND SEGMENT AND INTERFACES DESCRIPTION

The CONE Ground Segment provides for associated pre-flight, in-flight, and post-flight support functions. Figure 7 provides a definition of the CONE System and the functional interfaces required to operate the system. Internal payload and ground support equipment interfaces (both mechanical and electrical) are also shown.

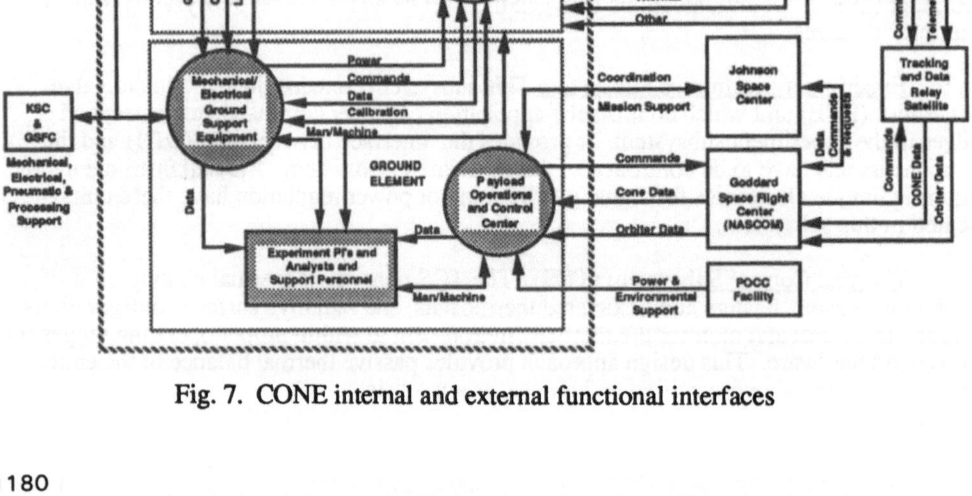

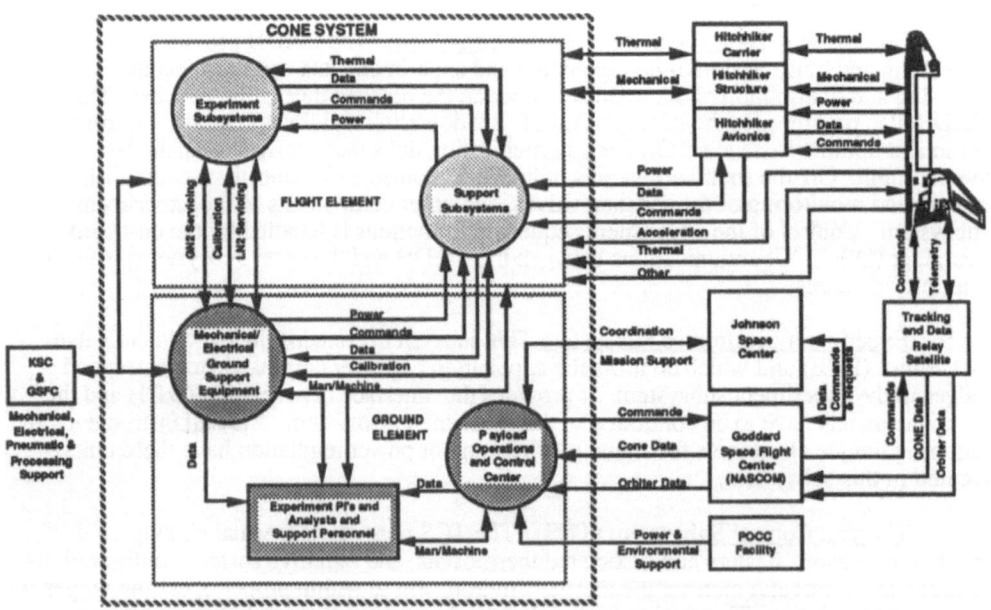

Fig. 7. CONE internal and external functional interfaces

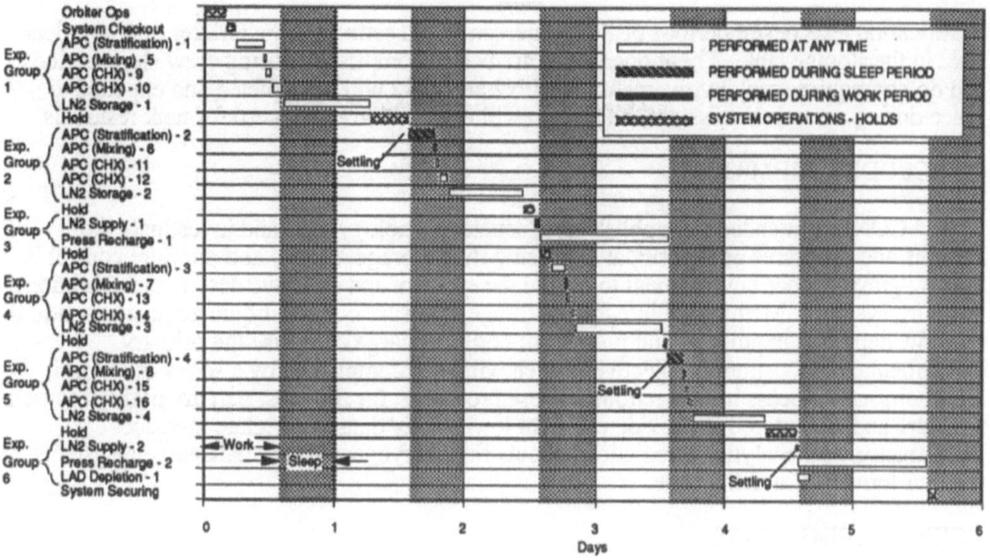

Fig. 8. CONE experiment/demonstration science on-orbit timeline

Major Ground Segment Elements

Mechanical Ground Support Equipment (MGSE). The MGSE provides ground servicing, handling support, transportation support, and maintenance functions for the payload. The major MGSE structural hardware items include a protective cover for transport, handling and rotation dolly, handling/lifting slings & strongback, holding fixtures and installation tools. Support equipment for the experiment subsystem includes a LN2 servicing/deservicing system, GN2 pressurant servicing system, experiment system leak check kits, fluid support equipment and miscellaneous calibration equipment.

Electrical Ground Support Equipment (EGSE). The EGSE provides command, control, calibration, simulation, ground 28 Vdc power to the payload, and data management of the support and experiment subsystems during ground test and checkout operations. A major portion of the EGSE hardware is comprised of the payload C&DH support equipment comprised of a mission sequence and command generation system, EEPROM programmer, power supplies, monitoring system, display equipment and printer. In addition, a power distribution system, payload electrical power subsystem support equipment and electronics integration test set are provided.

Payload Operations Control Center (POCC). A POCC located at NASA Goddard Space Flight Center provides in-flight command, control and management of flight data (both realtime and recorded) for CONE. Operations software is included in the POCC. The POCC includes a router, telemetry preprocessor, data management work stations, mission planning and scheduling work stations, and personal computers for experiment commanding, monitoring and data processing.

MISSION DESCRIPTION

The CONE on-orbit mission was developed so that the experiments and demonstrations could be accomplished on a nominal seven day Orbiter mission. Figure 8 shows the mission timeline that was assembled to accommodate all of the tests within this time constraint while allowing for periods of system operations holds to provide for coordination with the crew and other operations that may impact the flow of desired CONE operations. Tests are arranged into six groups with the first three occurring at the high fill level for LN2 in the storage tank and the last three at a lower level resulting from the first expulsion (supply test). Active pressure control (APC) assessments comprise the majority of the mission events. The ordering of events had to accommodate crew availability to support orbiter operations for attitude and thruster firings for fluid positioning/settling. Two stratification tests desire periods of fluid quiescence and settling using Orbiter drag to orient LN2 in the storage tank. These operations are best accomplished during crew sleep periods and do not require crew involvement. All LN2 and GN2 will be depleted and expelled to space during a nominal mission; the system will return without concern for tank residuals.

CONCLUDING REMARKS

CONE is intended to establish an adequate technology base and investigate the systems and processes of subcritical cryogenic fluid storage, supply and pressure control in the low-gravity space environment to enable the efficient and effective design and operation of future systems. Methods of integrated storage, pressure control, liquid acquisition, liquid outflow, depletion expulsion, and pressurant bottle recharge comprise the primary experimental/demonstration objectives which will be accomplished by a series of on-orbit liquid nitrogen process investigations. These processes, for the most part are highly focused in nature and are tailored to provide essential low-gravity data to correlate in-space performance with analytical predictions of subcritical cryogenic fluid management systems for near term STV and SSF use.

ACKNOWLEDGMENTS

This feasibility study was performed under Contract NAS3-25063 with the NASA Lewis Research Center (LeRC), Cleveland , Ohio 44135 under the direction of this paper's coauthor Hugh Arif - Project Manager. The NASA Principal Investigator was John C. Aydelott and the Science Advisor was Naseem Saiyed.. Eugene P. Symons provides overall program management and is the Chief of the Cryogenic Fluids Technology Office (CFTO) at NASA LeRC.

FINAL CRYOGENIC PERFORMANCE REPORT FOR THE NASA COSMIC BACKGROUND EXPLORER (COBE)

S.M. Volz, M.J. DiPirro, S.H. Castles, M.G. Ryschkewitsch

NASA/Goddard Space Flight Center
Greenbelt, Maryland

R. Hopkins

Ball Electro-Optics/Cryogenics Division
Boulder, Colorado

ABSTRACT

The cryogenic operation of NASA's Cosmic Background Explorer (COBE) ended on September 21, 1990, with the depletion of the liquid helium cryogen. The COBE had successfully completed more than 10 months of dewar and instrument operation. We report on the cryogenic performance of the COBE dewar and of the two cryogenic instruments throughout the mission lifetime. We discuss the steady state dewar performance, and the dewar and instrument response to a variety of transient thermal phenomena, including external radiation (from the earth and the sun) and instrument power variation. We present the effectiveness of using approximate mass gauging techniques in determining the liquid helium content. Finally we discuss the dewar behavior during the depletion of the helium, and the expected thermal performance of the dewar cryogen tank and the cryogenic instruments as they approach final thermal equilibrium.

INTRODUCTION

The first phase of the science mission of the Cosmic Background Explorer (COBE) came to an end on Friday, September 21, 1990, after 306 days of cryogenic operations as the last of the superfluid helium contained within the dewar was consumed. The final depletion of the helium cryogen marked the successful end of scientific operations for one of the three COBE instruments, the Far Infrared Absolute Spectrophotometer (FIRAS), and for a portion of a second instrument, the Diffuse Infrared Background Experiment (DIRBE). A substantial portion of the DIRBE will continue to make useful and accurate observations of the sky. The third instrument, the Differential Microwave Radiometer (DMR), is completely unaffected by the depletion of the liquid cryogen and has successfully operated without interruption throughout the full 18 months that the spacecraft has been on orbit.

In this paper we present a summary of the dewar cryogenic performance throughout the COBE mission lifetime. Following a brief description of the COBE spacecraft and dewar, we present an overall summary of the dewar performance through the period of liquid cryogen operations. Included is a discussion of the dewar and instrument sensitivities to internal and external perturbations, and the effectiveness of the dewar thermal math model in correlating the performance with the actual on orbit environment and operations. Next is a detailed description of the FIRAS and DIRBE cryogenic performance. We focus on such issues as the temperature stability of the instruments, the amount of thermal cross-talk between them, and the instrument power dissipation. Our ability to determine the helium boil off rate and to predict the lifetime of the liquid cryogen was important in the planning of scientific observations. We discuss the several methods employed to estimate the helium

Advances in Cryogenic Engineering, Vol. 37, Part B
Edited by R.W. Fast, Plenum Press, New York, 1992

flow rate, the effectiveness of these methods, and present our best estimates of the helium lifetime and flow rate throughout the mission. The final depletion of the liquid helium is presented next, and projections of the anticipated long term equilibrium condition of the dewar and the two cryogenic instruments.

SPACECRAFT AND DEWAR DESCRIPTION

The COBE was launched November 18, 1989 from the Western Space and Missile Center aboard an expendable Delta rocket. The satellite was injected into a circular (890±9 km), sun-synchronous orbit of 99° inclination. While in orbit the satellite is three axis stabilized with the aft end pointing generally towards the earth. The spacecraft is spinning at approximately 8.59×10^{-2} rad/sec (0.82 RPM) about the long axis (the X axis).

A schematic of the COBE dewar and instrument module is shown in Figure 1. The helium is contained in a toroidal 664 liter cryogen tank (CT). The FIRAS and DIRBE instruments are assembled together as the Cryogenic Optical Assembly (COA). Primary components of the FIRAS include the Mirror Transport Mechanism (MTM), located at the base of the FIRAS, and the External Calibrator (XCAL) situated at the top of the instrument with a good view of deep space. The COA is structurally and thermally anchored to the cryogen tank through a bolted interface at the COA Mounting Flange (COAMF), located at the base of the CT. The COAMF serves as the primary thermal interface between the instruments and the CT, carrying more than 95% of the total instrument generated heat to the helium. Additional thermal contact is provided by two separate 99.999% pure copper straps running from the top of the COA to the top of the CT (one strap is for DIRBE and one is for the FIRAS XCAL). Gaseous helium is vented through a porous plug[4] and past three vapor cooled shields before exiting the dewar. Outside the dewar approximately 15 feet of additional plumbing carries the helium to the final overboard vent. The exhaust vent is located on the spacecraft aft end, along the spin axis of symmetry.

DEWAR CRYOGENIC PERFORMANCE

At launch the dewar cryogen tank contained 616 L helium at 1.67 K. The vent valves opened as scheduled on ascent and over the next two weeks the dewar cooled to its equilibrium on orbit operating temperature[1]. At the same time as the dewar was approaching thermal equilibrium the spacecraft and instruments underwent thorough check out and performance testing. Of the several anomalies discovered[2], two had a direct impact on the performance and operation of the dewar. First, one of the three spacecraft gyros failed shortly after launch. COBE has operated successfully on two gyros since that time, but the loss of the one gyro meant that the spacecraft was more vulnerable to a temporary failure of the attitude control system. As a safety measure the spacecraft roll angle (the angle between the instrument aperture plane and the sun) was maintained at 4° or greater for most of the mission. The second anomaly concerned the FIRAS MTM, and a detailed description of the

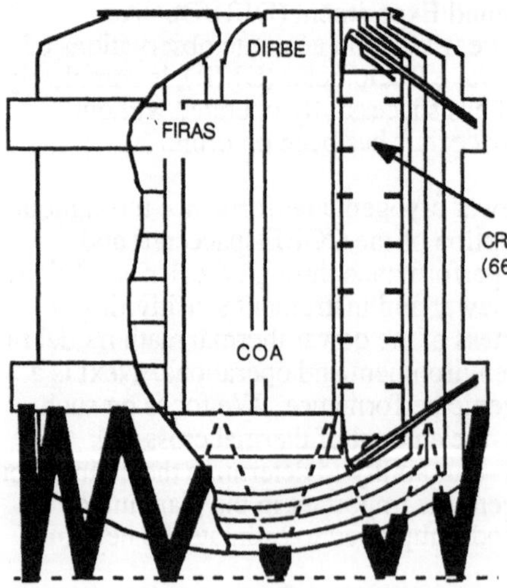

Figure 1: Cutaway view of the liquid helium dewar of the COBE satellite. The primary cryogen components are identified.

Table 1: COA on-orbit operating power budget (all power in mW)

MET (days)	COAMF	FIRAS additional	DIRBE additional	Instrument Total
20	25	10	3	38
133	25	7	3	35
150	27	7	3	37
180	30	10.5	5	45.5
250	27	8.5	3	38.5

failure and corrective actions initiated is given below in the discussion of instrument performance and in reference 1.

By twenty days into the mission a normal operating procedure had been established and the instruments had begun full time operations. During the on orbit checkout period measurements were made of the thermal conductance paths between the COA and the CT to determine an instrument operating power budget for the normal operating mode. The resulting heat budget is shown in Table 1. The "COAMF" lists the heat passing through the COA Mounting Flange at the base of the COA. This path includes most of the heat generated by the instruments during every day operations. The "FIRAS" and "DIRBE additional" lists extra heat produced by those instruments either dissipated through the Cu thermal straps at the top of the cryogen tank or produced by transient instrument phenomena. The powers listed in the Tables are time averaged over many orbits. The power budget was continuously updated throughout the mission lifetime as the operating modes changed and the dewar on-orbit environment changed.

Shown in Figures 2 and 3 are plots of the temperatures of the dewar helium bath and the three vapor cooled shields (VCS), respectively. The first feature that is clearly evident are the temperature spikes in the helium bath and dips in the VCS occurring approximately every 30 days. These are the monthly FIRAS calibration measurements. During these events, for a duration of one to three days the FIRAS instrument generated between 1 and 10 kJ, sometimes doubling the overall heat load to the helium. In response the helium temperature rose and the helium mass flow increased. The increased flow provided more cooling and so the VCS cooled. This combination of events was a clear sign of an increase in internally generated power.

At a Mission Elapsed Time of 155 days (MET 155) a dramatic change is apparent in the temperature profile of the dewar. Immediately following an XCAL operation on MET 155, the spacecraft was rolled to a new sun angle of 1.75° instead of the 4° maintained up to that time. Over the next two weeks the helium bath temperature increased from 1.410 K to

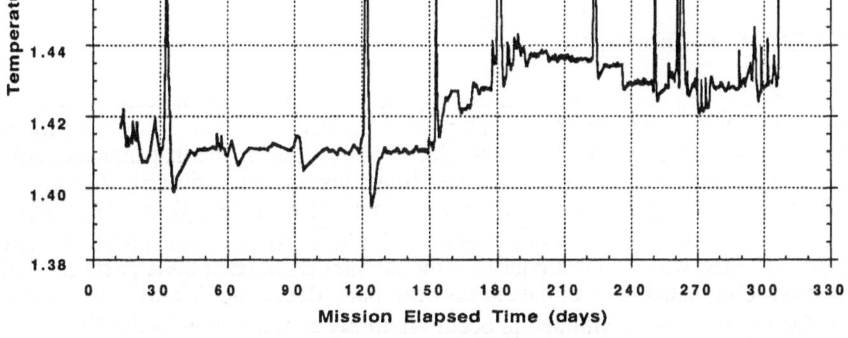

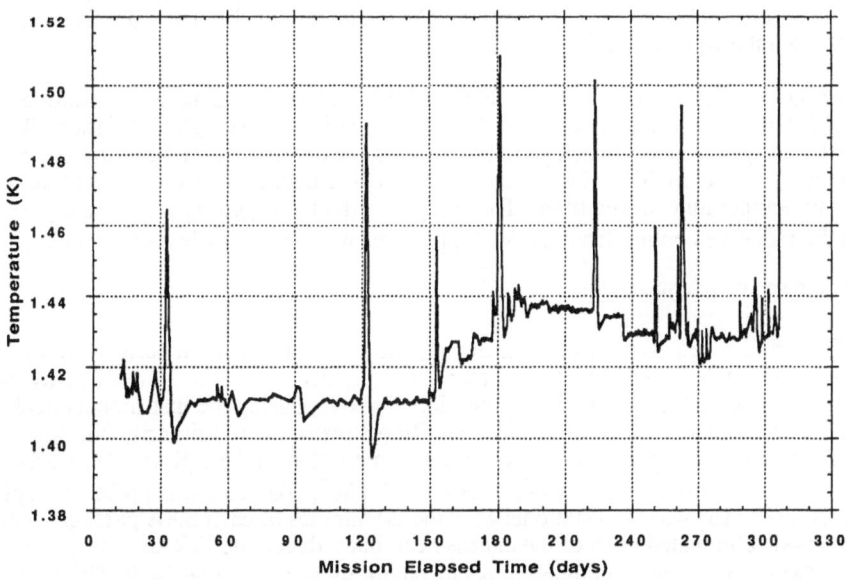

Figure 2: Dewar helium bath on-orbit operating temperature.

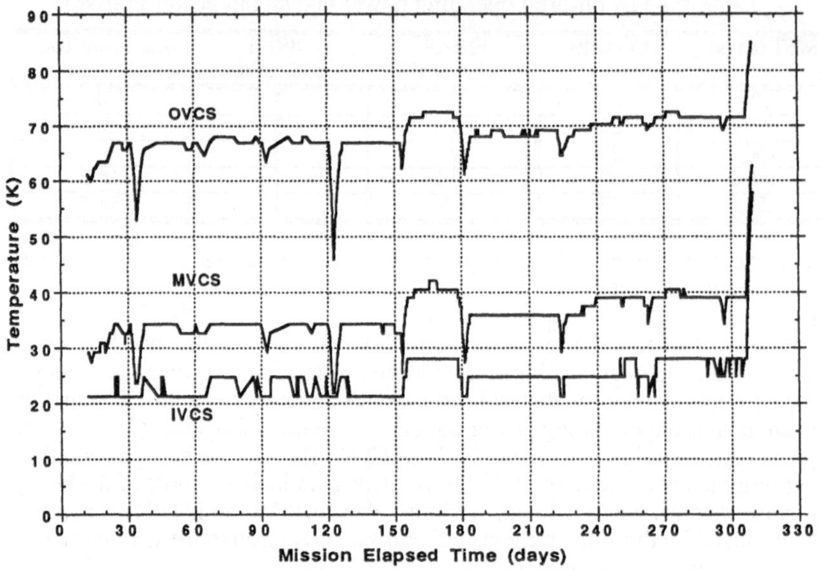

Figure 3: Dewar Vapor Cooled Shield on-orbit operating temperatures.

1.427 K. At the same time the VCS temperatures were also increasing. During this period the instruments had not changed their operating mode, and in any case the temperature profile was not indicative of an increase in internally generated heat. It is suspected that sunlight was being reflected onto the VCS primarily, and onto the COA secondarily. This heat (amounting to approximately 150 mW spread among the three VCS and about 2 mW additional to the COA) warmed the VCS. This increased the parasitic heat load on the CT but also warmed the vent line plumbing and hence the flow impedance. As the flow impedance rose the helium bath back pressure increased and the tank temperature rose. While we were never able to absolutely confirm the cause of the effect, this "sun glint" heating was present for the remainder of the mission[2].

Beginning around MET 180 the summer eclipse season began to affect the dewar and instruments. Radiation from the sunlit north pole of the Earth added a significant amount of heat directly into the instrument apertures and secondarily into the VCS. Symptomatic of an increase in COA generated power, the VCS temperature remained relatively constant while the bath temperature increased. The net increase in power dissipated to the helium bath during this period was about 10 mW from the COA, and an unknown amount from increased parasitics from the warmer VCS.

At the conclusion of the earth limb season we expected the tank temperature to return to around 1.410 K, the value seen before the first appearance of sun glint radiation. The spacecraft was again rolled 4° back from the sun and the instruments were in the same operating mode as before MET 150. Instead, the VCS temperatures were again increasing and the bath temperature stayed high. The sun glint effect was getting worse, and possibly in addition the radiative cooling from the VCS apertures was becoming less efficient[2].

INSTRUMENT PERFORMANCE

The primary requirements for the dewar, and the actual on-orbit performance are shown below in Table 2. The low temperature and stability requirements were easily met for the 10% power variations specified. As mentioned above, during the instrument check out period a major anomaly was discovered in the FIRAS instrument. Primarily during spacecraft passages through the South Atlantic Anomaly (SAA) the FIRAS MTM was malfunctioning and producing excess power of 400 mW[2]. To avoid this problem, beginning on MET 20 the MTM was held stationary for the 20 minutes of each SAA passage. This operating mode eliminated most of the excess heat, but reduced the FIRAS duty cycle to ≤90%. MTM malfunctions continued to occur randomly at other times in the COBE orbit. Figure 4 shows examples of two such events. It is evident that the failures in the MTM

Table 2: Primary Dewar Performance Requirements

Area	Requirement	On-Orbit Actual
Helium Lifetime	Six month minimum, with a design goal of one year	306 days total, 303 days after aperture cover ejection
Instrument thermal sink	FIRAS and DIRBE heat sink temperature shall be less than 1.6 K	FIRAS 1.520-1.528 K DIRBE 1.523 K
Thermal sink stability	Heat sink temperature shall vary by less than ±0.05 K for transient instrument power variations of less than ±10% (≤2mW)	For power variations of ≈ 4 mW, $\Delta T \leq 0.00025$ K

strongly affected the temperature stability of the DIRBE. Both the DIRBE and FIRAS instruments are equipped with separate thermal rods connecting the detectors directly to the base of the COA. They are, however, thermally linked to each other through the COA Mounting Flange and then connected to the cryogen tank. As a result, as heat is generated in the MTM it passes to the COAMF and then both into the cryogen tank and up into the DIRBE detectors The much larger heat capacity of the helium absorbs the heat without any apparent change in temperature. This was true even within days of the final depletion of liquid helium.

HELIUM MASS FLOW ESTIMATES

Knowledge of the helium mass flow rate was essential to the COBE science team for mission operations planning. Pre-launch estimates were based on limited ground tests and on correlation of the dewar thermal math model. To permit accurate on orbit determination of the helium mass contained within the dewar, COBE was equipped with a calorimetric mass gauge system[3]. The system was ground tested, and was intended for use after one complete sky survey was completed by all three instruments (after approximately six months observing time). Once the COBE satellite was in the sky, the COBE Project management team made the decision not to employ the heat pulse calorimeter unless its use was specifically required by the project scientists and necessary to properly plan COBE Observatory activities. Arguments for the use of the system based on engineering requirements, such as the desire to know the helium mass and flow rate precisely and to gauge the access the accuracy of the thermal math model, and the need to test the mass gauge calorimeter for future missions and for COBE, were not sufficient.

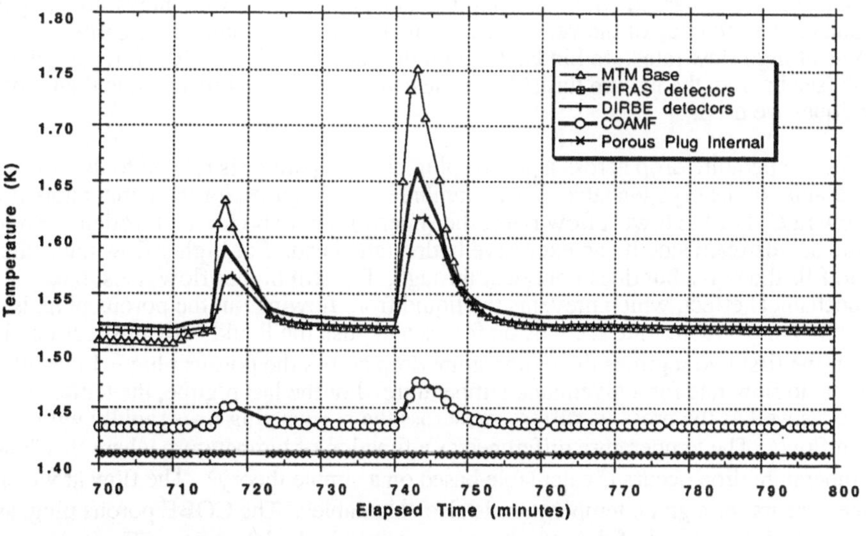

Figure 4: Response of DIRBE and FIRAS base temperature and dewar to MTM malfunction.

Without the mass gauge system, alternative methods for gauging the helium flow rate or mass of helium within the dewar were attempted. Our methods included; one, projections of flow rate from the ground developed and tested thermal math model; two, measurements of the spacecraft thrust produced by the vented helium; three, using the cryogen tank porous plug as a flow meter, and four, the use of high power generating instrument (FIRAS) activities as heat pulses for heat pulse mass gauge calorimetry. We will discuss the success of each of these methods in turn.

At launch the COBE dewar contained 615 L (92.7% full) of liquid helium. The helium boil off rate during the first few weeks was much larger than during normal operations, as the helium bath first pumped down to operating temperature and as the instruments went through initial check out tests. We estimate the helium loss during the initial 30 days to be 10-12% of the total dewar volume. With this initial loss, the time averaged flow rate for the remainder of the mission would be about 3.25 mg/sec. The actual helium flow rate was highly irregular, varying daily, weekly and seasonally as the instrument and external heat load on the cryogen changed. During the first 30 days on orbit the dewar was slowly approaching equilibrium conditions. Attempts to model the transient dewr behavior during this period were unsuccessful. A fuller discussion of the transient model is given in Reference 8. Between MET 30 and 150 the dewar appeared to be operating in a very consistent and reasonably well understood fashion. Only very minor, and physically reasonable adjustments needed to be made to the steady state thermal math model to reproduce the actual temperature profiles seen in the cryogen tank and the dewar VCS[4]. Under these operating conditions the thermal model calculated a helium flow rate of 2.3 mg/sec. The model was less successful in modelling the dewar behavior for mission operations after MET 150. With heat coming from sun glint radiation, earth limb heating, and variable instrument operations we were unable to adequately correlate the model with the dewar performance for the remainder of the mission. The thermal model, by itself, was not an adequate gauge of dewar performance.

Six days after launch, following the failure of one control gyro, COBE was commanded to a negative pitch attitude as a safety measure. The resulting helium venting force component in the direction of motion was significantly higher than anticipated. The direct effecton the mission was a higher than predicted orbital decay rate. This increased decay rate did not impact the mission science operations in any way. Once the cryogen was totally depleted, the orbital behavior of COBE resembled that of a normal non-venting spacecraft. The combined orbit and attitude behavior was analyzed by the Flight Dynamic Facility at Goddard to monitor the total helium venting force. During the first 100 mission days the flow rate calculated by this was 3.8 mg/sec. At the time this value of flow was considered far in excess of what was possible, and the estimate was discounted. For the entire mission time the rate was estimated at 3.6 mg/sec. Because the early thermal math model data strongly disagreed with these numbers, and because we were never able to place an estimate on the accuracy of the vent thrust calculations, we discounted the early predictions of high flow rates. In hindsight, with the exception of the XCAL mass gauge technique (see below), the helium thrust flow rate estimates were more accurate than most of the techniques we used.

The temperature drop across a porous plug phase separator is related to the flow rate and temperature of the cryogen tank. There are three main regimes for the temperature drop versus flow rate. For the lowest flow rates, the superfluid may be forced through the porous plug to the downstream side by an excessive hydrostatic head. For higher flow rates, the superfluid fills the pores but does not break through. For still higher flow rates, the thermomechanical effect, which prevents the liquid from flowing out the porous plug, is strong enough to overcome surface tension forces and push the liquid back upstream in the plug[6]. For the first two regimes the temperature drop across the porous plug is low and is proportional to flow rate for a given tank temperature. For the last regime, the temperature difference, as well as the pressure difference across the porous plug rises rapidly with increasing flow[7]. The temperature difference as a function of temperature where this "knee" in the temperature drop occurs is calculable based on a simple theory[5]. The flow at which this "knee" occurs for a given temperature is also calculable[6]. The COBE porous plug, as well as porous plug material of the same type, was tested in the laboratory. These data are shown in Figure 5. Each set of data represents a different vent line impedance. As the flow

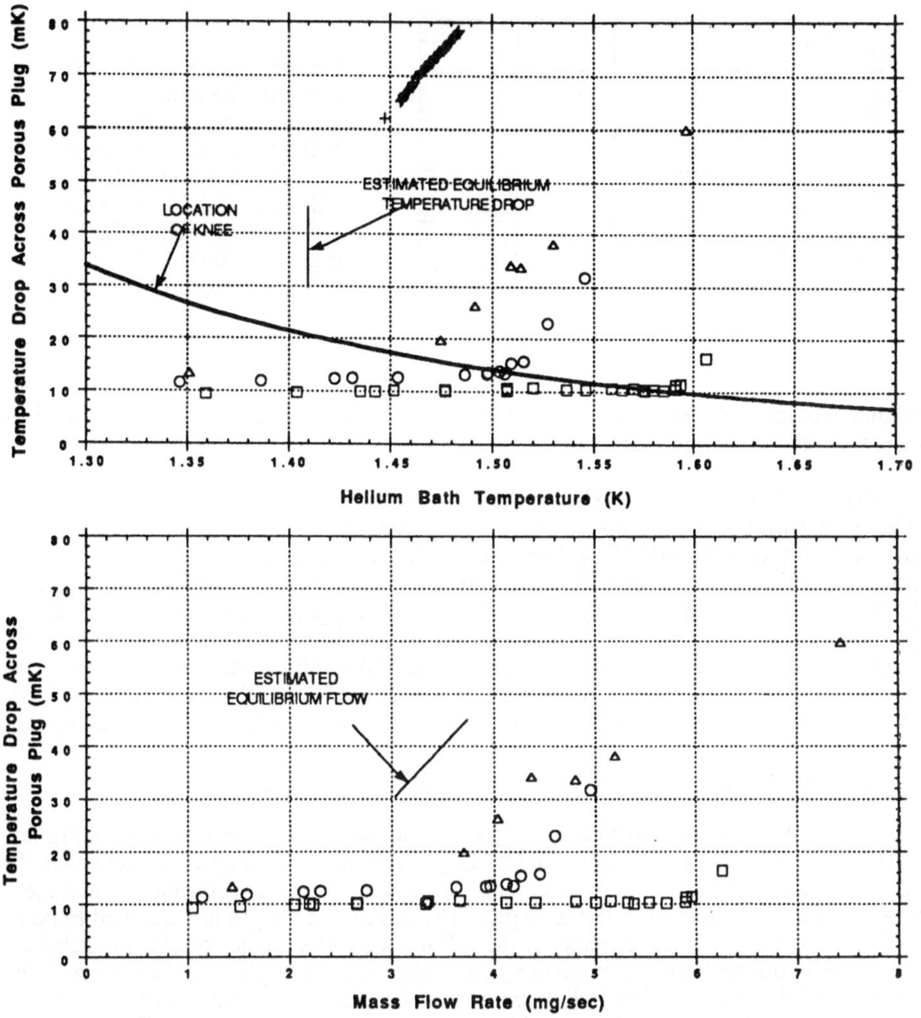

Figure 5. The COBE porous plug temperature drop as a function of tank temperature and as a function of flow rate. (+) are on-orbit data taken during the intial cooldown (time MET<30), (Δ), (O), and ( ) are laboratory data taken at three increasing vent line impedances. The COBE on orbit vent line impedance is lower than these. The calculated location of the "knee" is also shown.

rate increased, the cryogen tank temperature increased and the knee was reached. Also shown in Figure 5 is the on orbit COBE data. One can see that all the on-orbit data show the porous plug to be in the regime where the mass flow is a very strong function of the porous plug temperature gradient. Note that the lowest COBE data was so low that the downstream side of the porous plug was slightly lower than 1.388 K, colder than the readable temperature output of our one external porous plug GRT. From the progression to lower impedance seen in Figure 5, one can also see that the actual COBE vent line impedance is lower than the test impedances used in the lab. From the equations in Reference 5 one can calculate that the flow rate at the location of the knee, and for COBE's actual conditions ($\Delta T \approx 21$ mK at a cryogen tank temperature of 1.41 K), is 2.6 mg/sec. This would be a lower bound on the COBE flow rate since the actual temperature drop is at least 22 mK.

To estimate the actual temperature drop across the porous plug, one may look at the temperature drop as a function of tank temperature when the external porous plug temperature is on scale. One may then extrapolate these data to lower tank temperatures. This method yields a temperature drop of between 30 and 46 mK at a tank temperature of 1.41 K, depending on whether the extrapolation is taken as being quadratic in pressure or linear in pressure. This leads to an estimate of the flow rate, as computed from the slope of the lowest

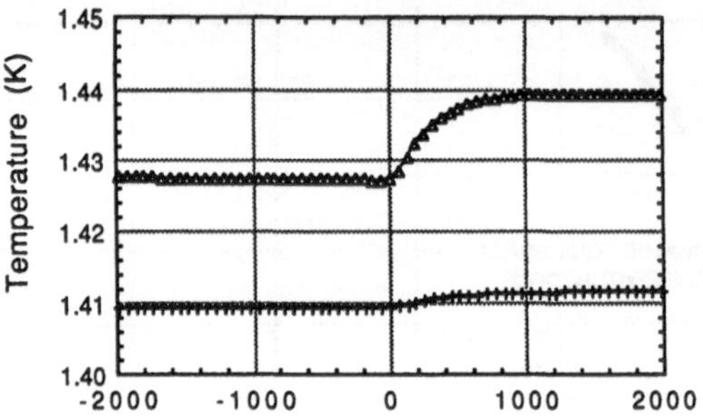

Figure 6: Two samples of the liquid helium temperature response to FIRAS XCAL heat pulses. (+) is for a motion on MET 60, and (Δ) is for an operation on MET 298, eight days before the end of the mission.

impedance lab data, of between 0.4 and 1.1 mg/second above the flow computed above. We thus have a worst case upper limit of 3.7 mg/sec.

The last mass gauge method employed the monthly calibration measurements of the FIRAS. To perform the calibration the FIRAS External Calibrator (XCAL)was rotated from a stowed position into the FIRAS skyhorn (see Figure 1). The motor to move the XCAL generates a large heat input (between 1 and 3 W) for 56 seconds. This heat is partially dissipated in the COA and in the cryogen tank, and partially radiated into space from the COA. A sample of the dewar response to two XCAL motions is shown in Figure 6. We used this XCAL motor as a mass gauge heater. Under ideal circumstances, the heater applies a quantity of heat Q_{TOTAL} directly to the CT. This heat is adsorbed by the CT, the liquid and gaseous helium in the CT, the COA, and to a degree by the boil-off gas vented into space.

$$Q_{TOTAL} = Q_{HE} + Q_{VENT\ GAS} + Q_{COA} + Q_{CT} \qquad (1)$$

On the LHS of equation (1), $Q_{TOTAL} = Q_{HEATER} * \Delta t$, where Δt is the heater operating time. The RHS of the equation (1) is dominated by Q_{HE}. The heat shows up in the helium in three ways; an increase in enthalpy in the liquid, an increase in the enthalpy of the gas, and the conversion of a small amount of liquid to gas. The liquid and the gas are assumed to be in thermal equilibrium. The heat capacity of the CT and the COA are negligible compared to that of the helium liquid and vapor, so Q_{COA} and Q_{CT} may be neglected. For the very short heater time and the relatively low helium vent rate the amount of cooling from the vent gas is also small compared to the tank heat capacity terms, and may be neglected. The helium mass (or volume) can be determined by solving equation (2) for the helium fill fraction f.

$$Q_{XCAL}=\{r_{liq}*f*C_{p,liq}+r_{gas}*(1-f)*C_{pgas}+(1-f)*L*(\partial r_{gas}/\partial T)\}*V_{CT}*(dT_{He}/dt) \qquad (2)$$

The temperature rise rate, dT_{He}/dt, is measured, and the other terms are known from the helium temperature. The two unknowns in equation (2) are the XCAL heater power,

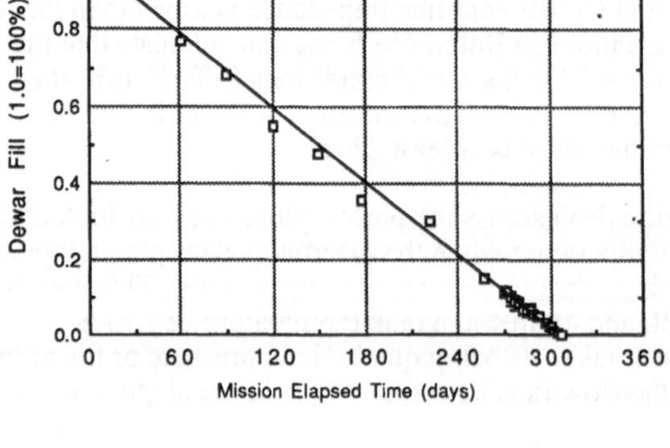

Figure 7: Predicted COBE dewar liquid fill level versus time as determined by the XCAL mass gauge calculations.

Q_{XCAL}, and the helium fill, f. To determine a fill level at a given time we needed to know the fill at a different time. To furnish the starting point we estimated the fill at the time of the first FIRAS calibration, on day MET 31, to be 80%. Figure 7 shows the results of using the XCAL "mass gauge." These data reflect the best estimates of the liquid level versus time up to the final XCAL operation, performed on MET 303. Using the XCAL method the depletion of helium from the dewar was predicted for MET 299, seven days before the actual event. In comparison, the thermal math model consistently estimated the helium flow rate ≈1/3 lower than the actual amount. The accuracy of the porous plug flow meter was hindered by the lack of the lowest on-orbit temperature data, and we had no confidence in the predictions made by the helium vent thrust calculations.

FINAL DEPLETION OF LIQUID HELIUM

The temperature and stability of the helium bath continued to meet all mission requirements up until less than five minutes of the final end of life (EOL) of the helium. As our lifetime predictions could not predict the precise time of the EOL, the instruments were performing normal observations up until the end. Figure 8 shows the temperatures of the instruments and the helium bath at the EOL. As the last of the bulk helium within the dewar evaporated, at t=3540 seconds, the COAMF and the instruments began to warm. Four minutes later the helium film within the CT has evaporated and the CT thermometer started warming. The internal porous plug thermometer was still cold as liquid was retained within the plug and the wire wick leading to the plug. At six minutes into the warming the automatic over temperature control protection circuit sensed that the FIRAS and DIRBE temperatures had exceeded 1.65 K and the FIRAS instrument was automatically powered off. After eight minutes the GRT on the inside of the plug was uncovered, and began to warm. A large temperature gradient then appeared across the plug. The porous plug dried completely after eleven minutes. All of these events occcurred while the spacecraft was out of contact with ground control. When the mission operations crew next contacted the spacecraft, 25 minutes later, the dewar temperature was about 2 K and was warming at a rate of 0.5 K/hour.

Ten days later the helium tank had warmed to 32 kelvin. The FIRAS bolometers became insensitive within the first days and the FIRAS was shut down. Figure 9 shows the long term warming trend of the dewar and the DIRBE instrument. The thermal math model of the dewar predicted an equilibrium temperature of 80 K for the dewar , measured at the cryogen tank. However the model did not adequately incorporate the radiative cooling provided by the FIRAS and DIRBE instrument apertures and the COA aperture closeout. As shown in Figure 9, the DIRBE instrument is running substantially colder than expected (the FIRAS may also be much colder, but when the instrument was turned off the data from the GRTs became unavailable). Six of ten of the DIRBE detectors continue to supply useful data,

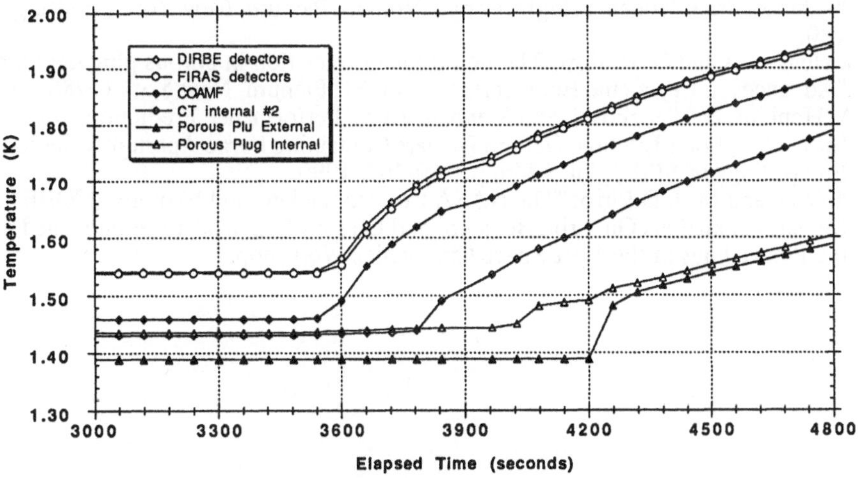

Figure 8: Final depletion of liquid helium in the COBE dewar.

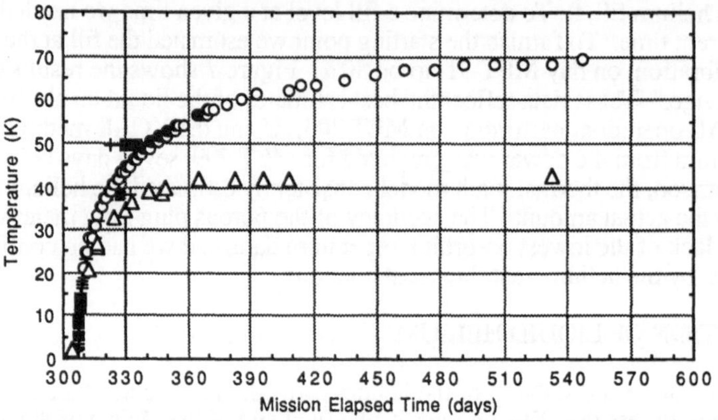

Figure 9: Long term warm up of dewar cryogen tank and DIRBE instrument. The data plotted are (O) COA Mounting Flange PRT, (Δ) average of 10 DIRBE detector temperatures, and (+) COA Mounting Flange GRT.

typical S/N ratios for these six are about 10/1 at 42 K, down from about 100/1 at 1.8 K. There are expected to continue to provide valuable scientific data as long as they are less than 70 K. Current trends indicate they may not exceed 50 K at final equilibrium.

REFERENCES

1 S M. Volz, et al., Cryogenic on-orbit performance of the NASA Cosmic Background Explorer (COBE), Proceedings of SPIE, Vol. 1340, p. 268-279, 1990.

2 S. M. Volz and M. J. DiPirro, Anomalous on-orbit behavior of the NASA Cosmic Background Explorer (COBE) dewar, to be published in the Proceedings of the 1991 Space Cryogenics Workshop.

3 S. M. Volz, M. J. DiPirro, and M. G. Ryschkewitsch, A calorimetric mass gauge system for the Cosmic Background Explorer (COBE), "Advances in Cryogenic Engineering," Vol. 35, Plenum Press, Los Angeles (1989), p. 1703 - 1710.

4 R. A. Hopkins, S. Nieczkoski, and Stephen Volz, Performance predictions for spaceborne, long-lifetime helium dewars containing large aperture telescopes, Proceedings of SPIE, Vol. 1340, p. 260-267, 1990.

5 M. J. DiPirro, F. Fash, and D. McHugh, Precision measurements on a porous plug for use in COBE, "Proc. 1983 Space Helium Dewar Conf.", University of Alabama Press (Huntsville), 1984, p. 121.

6 M. J. DiPirro and J. Zahniser, The liquid/vapor phase boundary in a porous plug," "Advances in Cryogenic Engineering," Vol. 35, Plenum, New York (1989), p. 173-180.

7 J. B. Hendricks and G. R. Karr, The liquid-vapor transition in porous plug operation, "Advances in Cryogenic Engineering," Vol. 31, Plenum, New York (1986), p. 861.

8. R. A. Hopkins, S. Nieczkoski, and Stephen Volz, "Performance Predictions for Spaceborn, Long-Lifetime Helium Dewars Containing Large Aperture Telescopes," Proceedings of SPIE, Vol. 1340, p. 260-267, 1990.

9 S. M. Volz and M. J. DiPirro, "The NASA Cosmic Background Explorer (COBE) Dewar: Anomalous On-Orbit Behavior and Lessons Learned," to be published in the Proceedings of the 1991 Space Cryogenics Workshop.

THE X-RAY SPECTROMETER--A CRYOGENIC INSTRUMENT ON THE ADVANCED X-RAY ASTROPHYSICS FACILITY

Susan R. Breon

NASA/Goddard Space Flight Center
Greenbelt, Maryland

Richard A. Hopkins and Stephen J. Nieczkoski

Ball Electro-optics/Cryogenics Division
Boulder, Colorado

ABSTRACT

The X-ray Spectrometer (XRS) is an instrument on the Advanced X-ray Astrophysics Facility (AXAF), the third of NASA's Great Observatories scheduled for launch in 1998. The XRS detectors have a resolution of approximately 10 eV over the range 0.3 - 10 keV. To achieve this resolution, the detectors are maintained at or below 0.1 Kelvin using an adiabatic demagnetization refrigerator inside a superfluid helium dewar. In addition, split-Stirling-cycle mechanical coolers are used to extend the anticipated on-orbit helium lifetime to a minimum of 4 years. This paper describes the challenges of developing this hybrid cryogenic system and presents an overview of the current design of the system.

INTRODUCTION

The XRS is one of four focal plane instruments which will be launched with the AXAF, shown in Figure 1. The AXAF scientific objectives are to (1) determine the nature of celestial objects from normal stars to quasars, (2) understand the nature of physical processes which take place in and between astronomical objects, and (3) understand the history and evolution of the universe[1]. In order to meet the scheduled launch in 1998, the XRS will be delivered to the Marshall Space Flight Center for calibration in the spring of 1996. The elements described herein will be completed and ready for integration at the Goddard Space Flight Center in early 1995.

X-ray-emitting plasmas produce spectral lines which are often very nearly the same energy. The XRS detectors are sensitive over nearly the entire AXAF spectral bandwidth: the detector resolution is approximately 10 eV over the range 0.3 - 10 keV. An energy resolution of approximately 10 eV will allow XRS to distinguish between spectral lines that are very close together. These lines are used as diagnostics for the source electron temperature, ionization temperature, density, mass motion, and abundance of various elements. Depending on the model used to interpret the diagnostics, it may be possible to infer the history of the plasma from its current state.

As is discussed below, a hybrid cryogenic system consisting of an adiabatic demagnetization refrigerator (ADR) and a superfluid helium dewar is needed to allow the detectors to meet the 10 eV resolution requirement. A second requirement--an on-orbit

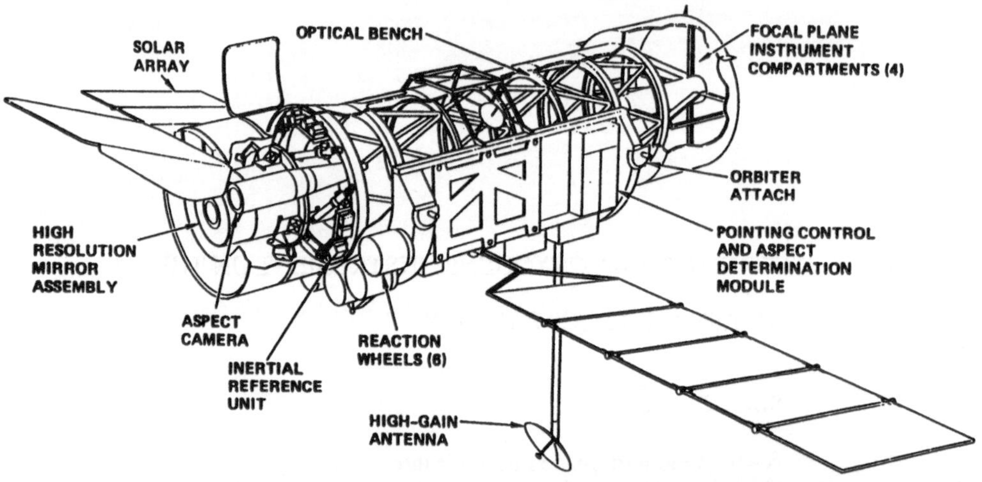

Fig. 1. The Advanced X-ray Astrophysics Facility (AXAF).

lifetime of at least 4 years--is also a major factor determining the hybrid configuration. The AXAF Observatory will be in orbit for 15 years. Approximately every 5 years AXAF will be serviced: the primary servicing mode is to replace the focal plane instruments, although on-orbit helium resupply is not precluded in the present design. In order to meet the lifetime requirement within the constraints of weight and volume, mechanical coolers are used to cool the outermost vapor-cooled shield of the superfluid helium dewar, thus increasing the helium lifetime by a factor of 3. This paper describes the factors which were considered and the interdependencies between the various XRS elements which were traded against each other to arrive at the basic cryogenic system design.

THE XRS CRYOGENIC SYSTEM

The Detectors

The XRS uses a calorimetric measurement to determine the energy of incident x-rays. There are a total of 32 detectors, each comprised of silicon implanted with a resistor which is used as a thermometer, and backed by an absorber such as mercury telluride. The full array of detectors covers an area approximately 3 mm square. When an x-ray strikes a detector, the energy of the x-ray is deposited in the mercury telluride. This raises the temperature of both the mercury telluride and the silicon to which it is bonded, thereby causing the resistance of the thermometer to change. By measuring the change in temperature and knowing the heat capacity of the detector, the energy of the x-ray can be determined.

It is important to lower the temperature of the detectors below 0.1 K for two reasons. First, lowering the temperature decreases the heat capacity of the detectors. The smaller the heat capacity, the larger the temperature rise--and therefore signal--for a given x-ray energy. Second, reducing the temperature of the detectors also lowers the phonon and Johnson noise. To minimize the amount of infrared and ultraviolet radiation striking the detectors, 5 thin filters made of aluminum and Lexan™ are placed in the dewar aperture[2].

The detector signal is amplified as close to the detectors as possible using Junction Field Effect Transistors (JFETs), which are trans-impedance amplifiers, as shown in Figure 2. The JFETs operate at up to 120 K--to reduce noise--and are located within 5 cm of the 0.1 K detectors. To intercept the heat, the JFETs are suspended inside a box which is thermally connected to the innermost vapor-cooled shield at a temperature of approximately 26 K. This box is suspended inside a second box which is mounted on a 1.5 K surface just outside the cryogen tank. The JFET temperature is maintained partially by self-heating and partially by an additional heater which is controlled to maintain the temperature stable to within ±1 K. All JFET wires, including those that run to connectors on the warm dewar girth ring, are heat sunk at both the 26 K and the 1.5 K stations. Although this results in more heat being deposited on the vapor-cooled shield and the

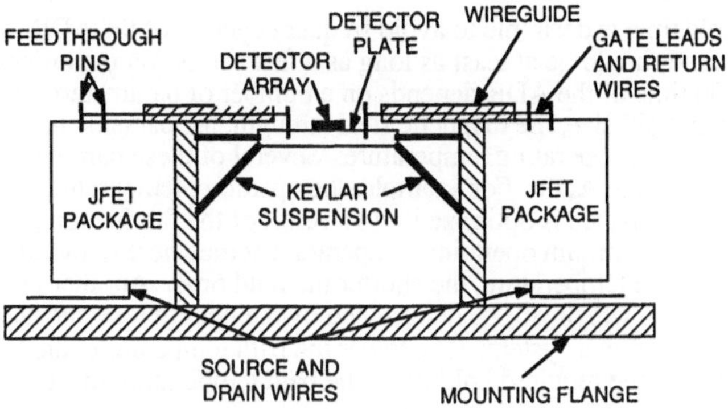

Fig. 2. Schematic diagram of the XRS detector and JFET mount.

cryogen tank, it simplifies the wiring greatly. The wires running from the JFETs to the detectors must be short and taut to minimize noise pick-up. A thermal analysis of the JFET package indicates that all the requirements can be met, although a full-scale package has not yet been built and tested.

The Adiabatic Demagnetization Refrigerator (ADR)

For ground-based experiments a temperature of 0.1 K can be achieved either with a He^3-He^4 dilution refrigerator or with an ADR. Given the complexity of attempting to operate a dilution refrigerator in zero gravity, the XRS Project decided to pursue the development of a flight-worthy ADR. A detailed description of the XRS ADR is provided by Serlemitsos et al.[3,4]

The major components of the ADR are the paramagnetic salt pill, the heat switch, and the magnet, as shown in Figure 3. There are also support structure and a thermal bus to connect the ADR to the detectors. The type of salt used depends on the desired operating temperature range. Over the range of 0.065 to 0.100 K, a reasonable choice for the salt is ferric ammonium alum ($FeNH_4(SO_4)_2(12H_2O)$). Once the salt has been chosen, the design of the ADR is developed based on the required hold time; i.e., the time the detectors are maintained at the operating temperature; the weight; the efficiency; and the reliability. As is true in most flight programs, attempts to minimize weight compete directly with attempts to increase the hold time, the efficiency, and the reliability.

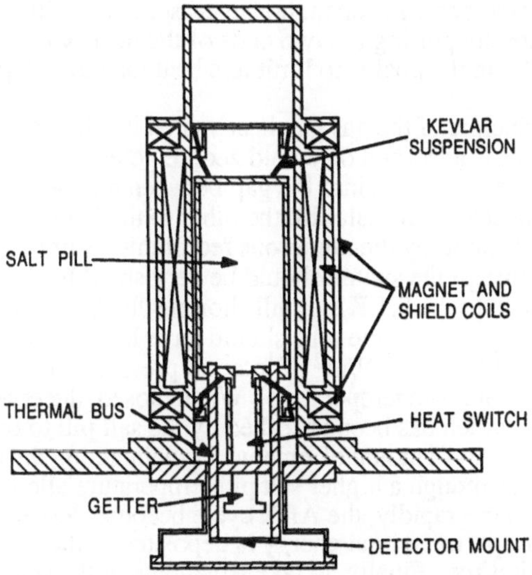

Fig. 3. The XRS adiabatic demagnetization refrigerator (ADR).

A long hold time is desirable to avoid frequent cycling of the ADR. As a practical matter, the hold time should be at least as long as a few orbits--on the order of 250 to 300 minutes. The hold time of the ADR depends on a number of parameters. The most important are the salt pill size, the magnetic field strength, the parasitic heat load, the helium bath temperature, and the operating temperature. Several of these parameters are imposed by elements outside of the ADR. For example, the operating temperature is determined so that the detector performance is optimized. As a result of the fabrication process, the detectors will have an optimum operating temperature somewhere in the range of 0.065 to 0.100 K. The lower the temperature, the shorter the hold time. Another external parameter is the bath temperature, which will be determined by the helium loss rate and the design and fabrication of the dewar vent system. The lower the bath temperature, the longer the hold time. In order to meet the required hold time, the bath temperature must be $\leq$1.5 K.

The salt pill size and magnetic field strength are two factors which are under the direct control of the ADR design. Actually, the two are heavily interrelated and have a direct bearing on the weight of the system. The salt pill is located in the bore of the magnet, a superconducting solenoid. A weight optimization study performed in 1986[5] showed that, while the salt pill should fill the bore as much as possible radially, axially the center of the salt should coincide with the center of the magnet, where the field is greatest, and the salt should not extend beyond the point where the field drops below about 0.8 Tesla. Between 0.6 Tesla and 0.8 Tesla the salt provides negligible cooling. Below 0.6 Tesla the salt would fail to reach the operating temperature and would act as a parasitic heat load. Therefore, for a given magnet the maximum salt pill volume is determined. In the present design, the salt pill is 7 cm in diameter and 16 cm long. To increase the hold time, either the magnetic field strength must be increased or both the salt pill and magnet size must be increased. Both options increase the overall weight of the system.

The magnet is shielded by three superconducting coils, one located at each end of the main magnet and one outside the main coil extending over its full length. The coils are all wired in series, with the shield coils wound counter to the main coil. The purpose of the shielding is to lower the stray magnetic field outside of the ADR. XRS may not produce a field which interferes with the facility or with other instruments. The shielding, though necessary, has two undesirable features: it adds weight and it decreases the field at the salt pill. The field strength at the center of the shielded magnet is 1.9 Tesla.

Another factor within the control of the ADR design is the parasitic heat load. Heat is conducted to the salt pill along two paths: through the supports and the heat switch. The heat from the detectors is negligible. The ADR is supported by tensioned Kevlar™ fibers. The fibers are pretensioned to keep them taut when the ADR is cooled down. The fibers must be strong enough to withstand this pretensioning and the launch loads. The heat load through the Kevlar™ is <10 μW. The heat switch is designed to be conductive when closed. Even when the heat switch is open, however, some parasitic heat is conducted through the Vespel™ tube supporting the two ends of the heat switch. Although the design of the switch is still evolving the goal is to limit this heat load to <10 μW.

The operating principle of the gas gap heat switch is relatively simple. When the switch is open, helium gas is adsorbed on a cold zeolite getter. To close the switch, the zeolite is heated and the helium flows into the gap between the two ends of the switch. One end of the switch is connected to the salt pill, the other to the helium bath. The design of the heat switch is made difficult by the numerous requirements imposed upon it. When open, the conductance through the switch should be very small because any heat flowing from the 1.5 K helium bath to the 0.1 K salt pill shortens the hold time. Since the heat is conducted through the support tube, the tube should have low conductance but still be strong enough to support 1.01×10^5 N/m^2 of internal pressure, survive launch loads, and be impermeable to helium at room temperature. When closed, the conductance should be large to allow the energy which has been absorbed by the salt pill to be deposited in the helium bath in a relatively short amount of time while maintaining the temperature of the salt to rise above about 2 K. Although a higher salt pill temperature allows the heat to be conducted to the helium more rapidly, the ADR cycle becomes less efficient. As the efficiency of the cycle drops, more total energy is deposited in the helium bath, thus decreasing the mission lifetime. Finally, to take advantage of the occultation periods when

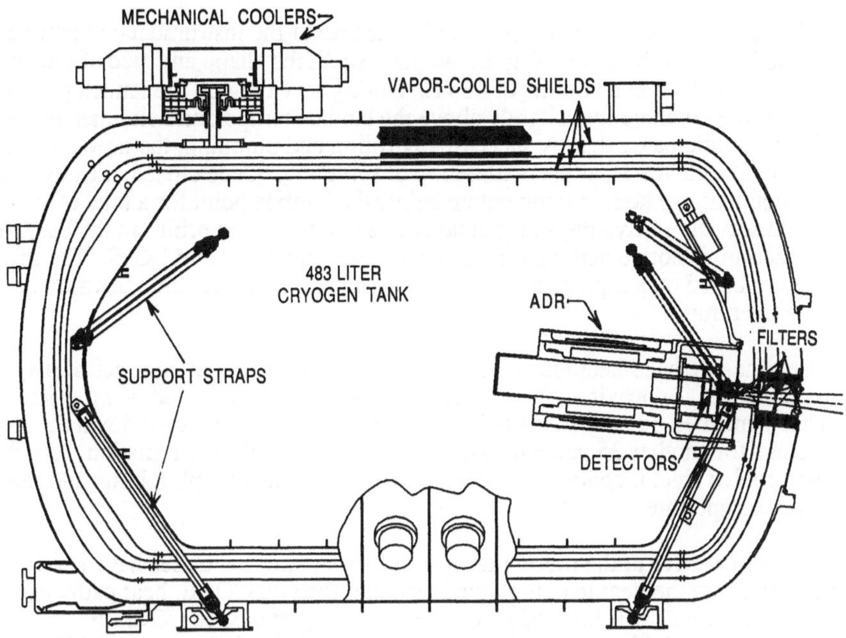

Fig. 4. The XRS dewar with the ADR and mechanical coolers.

the XRS would not be observing anyway, the total time to recycle the ADR should be less than 20 to 30 minutes.

A mechanical heat switch has been considered as an alternative to the gas gap heat switch. Although some very preliminary work has been done on a mechanical heat switch for XRS, questions about the reliability of repeated operation during ground tests and over a 5-year period on-orbit have kept this work at a low priority. It may also be difficult to develop sufficient force in the closed position to allow the heat to be exhausted to the helium bath in a short period of time. The main advantage of a mechanical heat switch would be a significant reduction in the parasitic heat load, which would allow for a much lighter system.

The Dewar

Two long-life superfluid helium dewars have flown to date. IRAS, launched in 1983, and COBE, launched in 1989, both achieved a lifetime of 10 months. The XRS dewar, shown in Figure 4, is required to have a lifetime greater than or equal to 4 years with a goal of 5 years. Since the lifetime is such an important issue and there may be some confusion caused by different numbers appearing in different documents, a short discussion is in order. Until very recently, the AXAF Project has not had a stated requirement on the helium lifetime. All instruments, including the XRS, must be designed for 5 years in orbit, consistent with the AXAF servicing interval. However, the helium was specifically excluded from this requirement and the helium lifetime was not addressed by AXAF. The XRS Project, in recognition of the tremendous technological advancement required to meet a multi-year lifetime requirement, established 5 years as the helium lifetime goal. Based on the results of parallel studies conducted in 1987-88 by the Goddard Space Flight Center, Ball Aerospace Systems Division (now Ball Electro-optics/Cryogenics Division), and Lockheed Research and Development Division, the XRS Project concluded that a lifetime of 4 years represented a challenging but achievable requirement. Within the past month the AXAF Project has initiated a change which will establish a helium lifetime requirement of 3 years and a goal of 5 years. This change notwithstanding, the XRS Project is maintaining the requirement at 4 years with a goal of 5 years.

A detailed description of the XRS dewar is given by Nieczkoski et al.[6] The helium lifetime is a function of the helium volume and the heat load on the cryogen tank. The dewar main shell volume is constrained both by the size of the instrument compartment and by the allowed mass of the XRS. Within the main shell, the shape and size of the cryogen tank has been optimized at a volume of 483 L. During servicing on the launch pad, the dewar will be filled with helium slightly above the lambda temperature and then pumped superfluid. It is anticipated that by successively filling and pumping, the dewar will be about 97 percent full, 469 L, at the completion of servicing. Once closed, the dewar is required to maintain the helium temperature below the lambda point for a total of six days after the final servicing: five days in ground hold and one day on orbit. In the baseline design, the helium is not vented on ascent as was done on COBE and IRAS. The lifetime predictions are based on a 3 percent loss of helium during on-orbit pumpdown to the operating temperature.

The time-averaged heat load on the helium is 8.9 mW, about an order of magnitude less than on COBE. Approximately 50 percent of the heat is conducted down the support straps. Heat deposited directly in the helium by the ADR and conducted down the JFET wires accounts for another 25 percent. The remaining heat paths are radiation from the inner vapor-cooled shield, conduction down the dewar wiring and plumbing, and radiation down the dewar aperture.

In order to arrive at the present design, numerous lifetime trades have been performed. Because the heat load is so much smaller than on COBE, heat paths which were insignificant on COBE have taken on increased importance for XRS. For example, a fourth vapor-cooled shield was added to intercept the radiation between the 45 K shield and the cryogen tank, increasing the dewar weight. Another trade found that the thermal connection between the support straps and the vapor-cooled shields has a significant impact on lifetime. Thermally attaching the two innermost shields closer to the cryogen tank resulted in a substantial improvement in lifetime. However, this complicates the mounting of the vapor-cooled shields.

One significant change from COBE will be the XRS dewar wiring. The coaxial cables used on COBE are being replaced by ribbon cables, which should be easier install in the dewar. On COBE, in order to attach the fine wires in the cables to connectors, the ends of the cables were cut open and the conductors were soldered to transition boards. Heavier wires leading to the connectors were then soldered at the opposite end of the transition boards. This was a very labor-intensive process. On XRS, connectors will be installed by the cable manufacturer and the ribbon cables and shields will be left intact, which should improve the electrical performance and will definitely simplify the installation of the cables. It is very important that the ribbon cables be properly heat sunk: if the wires are poorly heat sunk or not heat sunk at all, the lifetime will be reduced by 20 to 50 percent. Even the choice of conductor material is critical. Stainless steel wires provide a 7 percent lifetime improvement over manganin, although the electrical performance of the stainless steel and the ability to meet the electromagnetic interference requirements have not yet been analyzed.

Within the scope of the dewar design, only limited trades can be done on the major sources of heat--the support straps and the ADR/JFET assembly. The support straps must provide adequate stiffness and be strong enough to support the suspended mass, 170 kg, during launch and landing. Although AXAF is presently baselined for a Space Transportation System (STS) launch, a switch to a Titan IV launch has been discussed. The straps will be sized to accommodate either launch vehicle. When the AXAF focal plane instruments are replaced on-orbit, they will be returned to earth in the Shuttle orbiter. On landing, XRS may be placed in either of two orientations, one rotated 90° from the other about the longitudinal axis. The straps will be sized to accommodate both landing positions. The straps will be made of fiberglass/epoxy. Alumina/epoxy was initially considered, but as a result of an increase in the launch loads the fiberglass/epoxy was found to have a lifetime advantage.

There are two important dewar parameters which affect the amount of heat deposited in the helium by the ADR. The first is the temperature of the innermost vapor-cooled shield, which serves as a heat sink to the JFET. As the shield temperature gets higher, more heat is conducted down the detector wires, which are heat sunk at the cryogen tank. The second parameter is the helium bath temperature. The ADR becomes less efficient at higher bath temperatures, shortening the hold time and resulting in more energy being dissipated in the helium bath. The bath temperature is required to be ≤1.5 K; currently the analytical model predicts a temperature of 1.3 K. The bath temperature is determined by the pressure drop in the dewar vent line, which in turn is a function of the helium loss rate and the size and geometry of the vent line.

The Mechanical Coolers

The key to obtaining the required helium lifetime is the use of mechanical coolers. The coolers lower the temperature of the outermost vapor-cooled shield to about 85 K, which in turn lowers the heat load on the helium and raises the lifetime by a factor of 3. In the baseline design, two pairs of British Aerospace split-Stirling-cycle coolers are mounted on the dewar girth ring. The weight of the coolers, 29 kg for all four coolers with electronics, is small relative to the lifetime improvement they offer. The power required is 40 W per cooler, including electronics. Although this makes the coolers a major user of XRS power, it is still well within the power budget for the instrument. Potentially, the most serious drawback for using the coolers is the induced vibration[7]. Steps are being taken to minimize the vibration, including operating the coolers only in balanced pairs and providing vibration isolation mounts for the compressors. Additionally, it may be possible to fine tune both the mechanisms and the electronics to further reduce vibrations. The extent of the problem and possible solutions are currently being evaluated.

There are three areas regarding cooler operations which have been examined for options to improve the overall performance of XRS. The first is the compressor temperature. As the compressor temperature drops the cooler performance improves[8], allowing more heat to be removed from the shield. The primary mode of heat rejection for the compressors is by radiation to a shroud over the instrument compartment. In order to lower the compressor temperature, the compressors must be thermally strapped to the dewar main shell which is at approximately 250 K. Although this raises the main shell temperature slightly which would tend to increase the helium boiloff rate, the boiloff rate actually decreases because the outer vapor-cooled shield temperature drops. However, the British Aerospace coolers are not rated for operation at these low temperatures and their ability to perform under these conditions is unknown at the present time.

The second area of consideration is the expected reliability of the coolers. Life test data on the coolers are extremely limited. A precursor to the British Aerospace cooler, built at Oxford University, has been on a life test for nearly 3 years. No life test data have yet been published on the design built by British Aerospace, which has been modified relative to the Oxford cooler. Although the lifetime of the XRS dewar would greatly exceed the 5-year goal if four coolers were operated continuously over the life of the mission, it is risky to assume that the coolers will not fail in that period of time. If all coolers were operating and failed after 2.5 years, XRS may not even meet its 4-year lifetime requirement. To be conservative the XRS Project has decided to operate only one pair of coolers for the first two years in orbit. After 2 years, the second pair of coolers will be operated also. The current model predicts that a single pair of coolers operated continuously (or two pairs operated sequentially, if one pair fails) will allow the XRS to meet the 5-year lifetime goal.

The third area that is presently being evaluated is the use of a heat switch to disconnect inoperational coolers. A cooler that is either turned off or failed acts as a parasitic heat load on the outer vapor-cooled shield. A heat switch which relies on differential thermal contraction to make the thermal contact is under consideration. Preliminary analysis shows that the lifetime benefit of a realistic switch design is relatively small, although not negligible. However, work is proceeding to better define a suitable switch design and to better analyze the thermal performance of the cooler interface with and without a switch before a final decision is made.

SUMMARY

The XRS cryogenic system design is driven by numerous requirements which are often in conflict with each other. The overriding concern is to produce an instrument which will perform the scientific observations for which it is intended. Within this context, options involving the performance of the ADR, the dewar, and the mechanical coolers must be weighed within the constraints of mass, volume, complexity, and cost. Although XRS is still early in the preliminary design stage, the work that has been done to date provides confidence that the requirements of the XRS cryogenic system can be met.

ACKNOWLEDGMENTS

The authors would like to thank Dr. Richard Kelley for discussions of the XRS detectors and science requirements. We also wish to acknowledge the assistance of Mr. Marcelino SanSebastian in providing the ADR drawing.

REFERENCES

1. C. E. Winkler, C. C. Dailey, and N. P. Cumings, Advanced X-Ray Astrophysics Facility (AXAF) instrumentation, in: "Proc. SPIE Intl. Sym. Opt. Engr. and Photonics in Aero. Sensors," Orlando, Florida (1991).
2. R. A. M. Keski-Kuha, X-ray metal film filters at cryogenic temperatures, App. Optics, 7:2965 (1989).
3. A. T. Serlemitsos et al, The AXAF/XRS ADR: engineering model, presented at: Cryo. Engr. Conf., Huntsville, Alabama (1991).
4. A. T. Serlemitsos, M. SanSebastian, E. Kunes, Design of a space worthy ADR, presented at: Space Cryo. Workshop, Cleveland, Ohio (1991).
5. S. R. Breon, "Summary of Adiabatic Demagnetization Refrigerator Optimization," Technical Report 86-037, Swales & Associates, Greenbelt, Maryland (1986).
6. S. J. Nieczkoski, R. A. Hopkins, and S. R. Breon, 5-year lifetime superfluid helium dewar for the AXAF X-ray Spectrometer (XRS), presented at: Cryo. Engr. Conf., Huntsville, Alabama (1991).
7. R. G. Ross, Jr., D. L. Johnson, R. S. Sugimura, Characterization of miniature Stirling-cycle cryocoolers for space application, in: "Proc. Sixth Intl. Cryocooler Conf.," Plymouth, Massachusetts (1990).
8. C. A. Lewis, "British Aerospace Cryogenic Coolers for Space Applications," TP 9073, British Aerospace Public Limited Company, Stevenage, Hertfordshire, UK (1987).

CRYOGENIC MODELLING OF THE ISOCAM EXPERIMENT FROM TEST RESULTS

L. De Sa, B. Collaudin

AEROSPATIALE
Cannes la Bocca
France

ABSTRACT

ISOCAM is an infra-red camera experience to be mounted aboard the ISO (Infra-red Space Observatory) satellite, which shall be launched in 1993. The mission of the satellite is to collect an important amount of data on near infrared wavelength astronomy (2 to 200 µm) during its 18 months lifetime. The operating temperature is in the range 2.4 to 3.4 K.

A previous article[1] has described the general thermal analysis to be performed and the cryogenic test facilities. This article overviews the main cryogenic test results and associated fundamental measurements. Model fitting has provided an evaluation of thermal conductances (through fixation screws, through ball bearings) and of the heat distribution in the rotor and stator of a cryogenic stepper motor when functioning.

Special phenomena has been put into evidence such as the "chopped" functioning of motors, or the two-time temperature rise after a continuous functioning of the motors.

Model correlation with the test results has been led to an important level which ensures a high degree of confidence in the model. ESATAN software was the thermal modelling tool used here.

INTRODUCTION

Previous analysis have enabled a good approach to the thermal processes within the ISOCAM experiment[1] (see also for a full description of the camera, satellite and mission). However, these remained mostly first order evaluation of phenomena which could in fact be quite complicated. It is the case for stepper motors which have a peculiar thermal behavior at first sight but quite explainable after a thermal analysis.

As an end goal, we were set out to correctly thermally model the experiment in order to:

1) Verify the requirements verification in a nominal functioning mode.
2) Ensure a good follow-up of the thermal history of a series of processes resulting from a functioning mode on-flight.
3) Detect possible critical areas and limitations to the use of different functioning modes.
4) Provide full thermal interface data with respect to the telescope (therefore the spacecraft).

So the work of modelling had to provide : first, a good explanation of test results together with adequate fits, second, good predictive capabilities when forced into a representative process.

The presentation of this article first recalls the thermal requirements. It then sketches how the experimental results were exploited to enable model fitting. Then it extracts from the test results the necessary data which could not be obtained via database research, for instance the contact conductances between the base plate and the Optical Support Structure (OSS), the conductance through the ball bearings or the thermal behavior of motors. By this time, some important new data will have been highlighted. Then some words on the nominal cycle simulation and on how the performances are compliant with the specifications.

SPECIFICATIONS, CRITICAL AREAS

The set of thermal specifications on the camera were:

1) Parasitic light : at any operational moment, no point of the camera must reach or overcome 10 K.
2) Detector stability : at any operational moment, the temperature fluctuation of the detector cannot be greater than 0.1 K (after calibration).
3) Average dissipation : shall not exceed 10 mW.

The first specification is valid at any point, even those that are apparently hidden from the detectors (e.g. inside the motors, inside the ball bearing assembly, etc...). The detector stability must be kept during all the acquisition phase which can last for nearly 3000 seconds (both short and long term stabilities are thus required). The average dissipation requirement comes from the ISO satellite allowance to each of the four experiments mounted aboard in order to comply to the design lifetime of the spacecraft, i.e. 18 months.

MAJOR CRYOGENIC TEST RESULTS - THEORETICAL APPROACH

Temperature measurements were guaranteed by Silicon diodes from Lake Shore Cryotronics or Southampton and carbon resistors of the C10 type (LINDE supplier, 10 Ω at room temperature). Differences existed between measurements at zero dissipation which most probably came from the

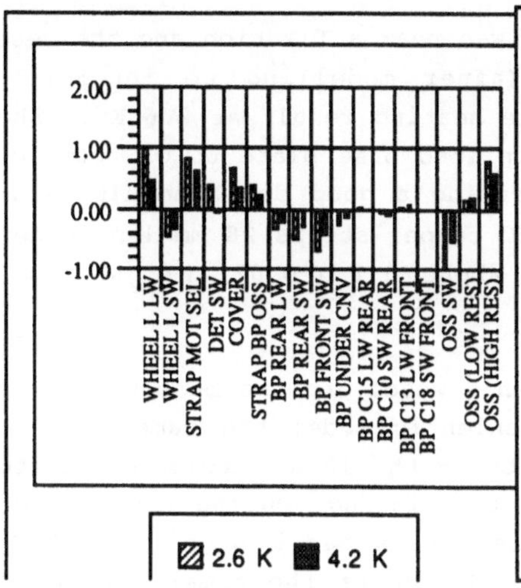

Fig. 1 : Additive corrections to sensors

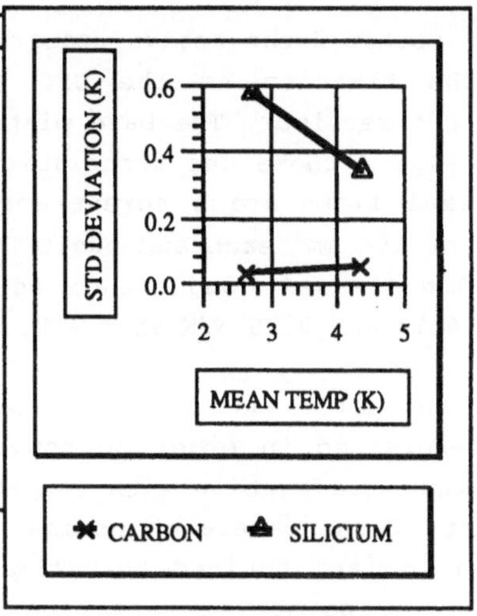

Fig. 2 : Standard deviation for sensors

calibration accuracy of each temperature probe. Indeed, diodes were provided with a unique general calibration curve whereas carbon resistors showed a better behavior one with respect to another because calibration curves were dedicated. Probe locations, corrections and standard deviations are reproduced in figs. 1 and 2.

DETERMINING BASIC CONDUCTANCES - BALL BEARINGS AND CONTACTS

The first conductance to be determined which is a driver in the instrument thermal concept is the base plate-OSS conductance. It can be determined either from the temperature difference between the base plate and the OSS or from the temperature increase in the base plate after heating up. Two methods were used : method 1 consisted in a central heating (middle of the base plate, 50 and 100 mW) and sensing the temperatures near the fixation points of the camera (attachments to the

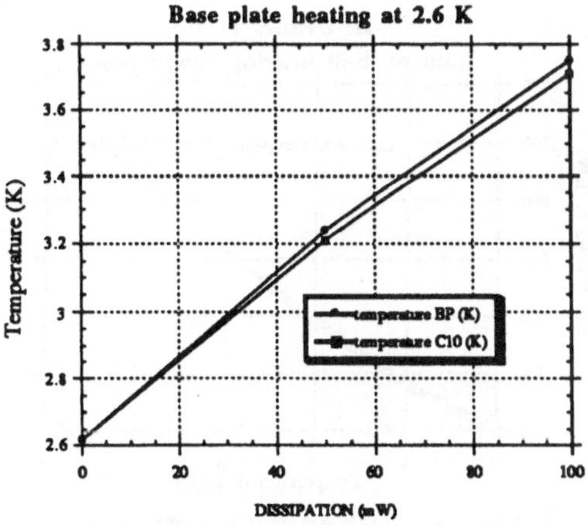

Fig. 3 : Base plate heating at 2.6 K

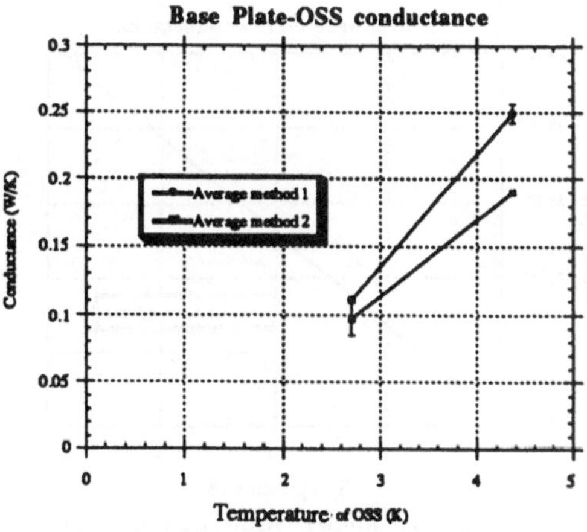

Fig. 4 : Base plate- OSS conductance

OSS); in method 2 the active heater was near a fixation and the detection after the fixation on the OSS (after modelling to interprete the temperature results). The base plate heating result at 2.6 K is shown in fig. 3. Fig. 4 shows the translation into base plate to OSS conductance. The thermal links are 5 screws ensuring a good dry aluminium-aluminium contact of 154 mm^2 each and two OFHC copper straps 28 mm length per 1 mm thick. The thermal conductance is about 0.1 (±0.02) W/K at 2.6 K and between 0.19 and 0.26 W/K at 4.4 K.

The thermal conductivity between a wheel and the base plate had also to be determined in order to conveniently model the camera and ensure requirement verification with respect to the 10 K maximum temperature in any point of the camera at nominal functioning. During test cooldown it had been in fact noticed that these wheels had a high thermal inertia thus determining the overall time constant of the camera. A heater had been placed on a wheel aside a temperature sensor and another temperature sensor at the base plate, near the closest wheel axis support. The measured temperature gradient (with sensor correction) was used to determine the thermal conductivity. It would have been useful to determine how a wheel would heat up after a continuous functioning but sensor implementation was not compatible with this functioning mode. The results of this measurement are shown in fig. 5. This thermal conductivity between the wheel and the base plate is very largely driven by the ball bearings that enable wheel axis rotation. It was tempting to compare this single point measurement to a set of measurements on ball bearings alone[2]. The result presented in this paper is in good agreement with an extrapolation of the presented results[2] (see fig. 6), in which were used the same type of ball bearing : stainless steel, about 4 mm diameter.

Another useful thermal conductivity is the one between a detector and the base plate. Again the heating and sensing approach was used here. The results of this measurement are shown in fig. 7.

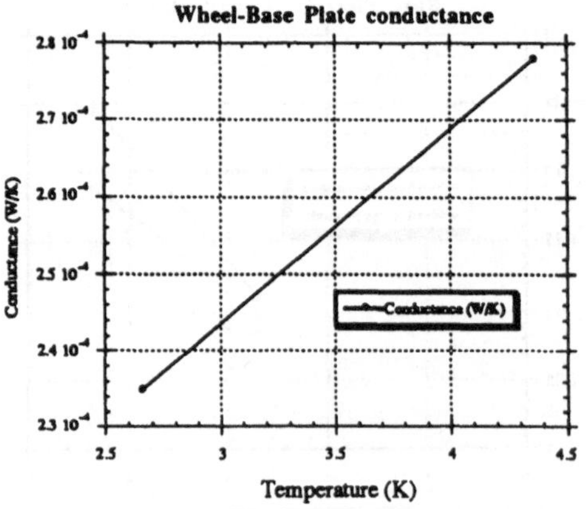

Fig. 5: Wheel-vase plate conductance

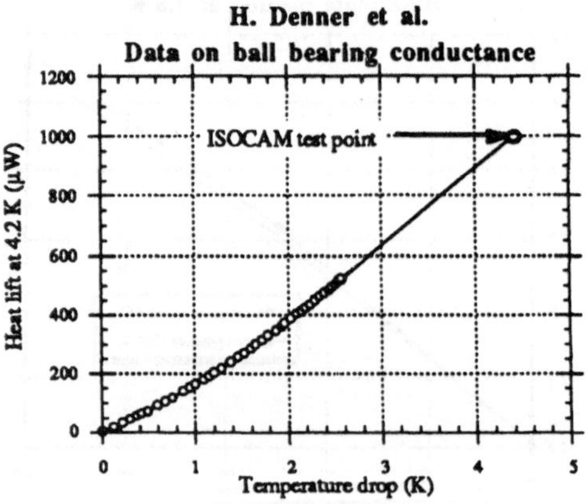

Fig. 6: ISOCAM data compared to other papers

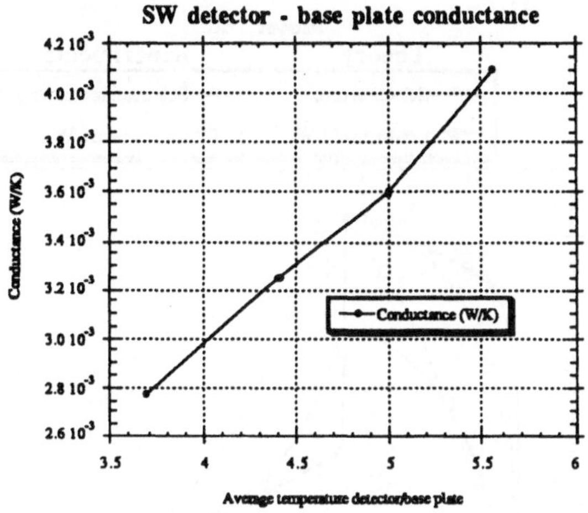

Fig. 7 : SW detector-base plate conductance

The thermal conductivity is nearly linear and can be described by the relation:

$$K = 0.0001747 + 0.0006978*T \text{ in W/m/K}$$

Here the thermal conductivity is mainly dominated by the contact conductances in the heat path.

CRYOGENIC STEPPER MOTORS AND HEAT PROCESSES ASSOCIATED

One of the most critical thermal aspects of the camera are the cryogenic stepper motors and their intrinsic functioning. These motors are used to actuate the cogged wheels via a pinion connected to the rotor of the motor. Tests showed that a two-step temperature rise could be sensed at a motor stator (outer shell) when it was ON (see fig. 8). The explanation of this phenomena is of utmost importance since it was suspected that the heat generated by the functioning was spreaded between the rotor and the stator of the motor (no access to the rotor from outside) and that two different time-constants were in fact being displayed. The fit was meant to give a reasonable approach to the thermal processes that took place anywhere that could be reached by a sensor (motor shell included) in a medium time (long w.r.t. the nominal cycle motor actuation but not yet the real thermal equilibrium of the motor). This adjustment to the test results can be seen in fig. 8.

The parameters used then were :

1) Motor dissipation : 50 mW
2) Stator/rotor dissipation ratio :50%
3) Rotor to stator conductance : 1.5 10^{-5} T in W/K until 20 K and constant above (3 10^{-4} W/K)
4) Rotor-stator radiative exchange : 7.8 10^{-3} (rotor surface)

Aside from the nominal functioning cycle for the camera, operational circumstances could lead to a much more frequent use of the motors. As a

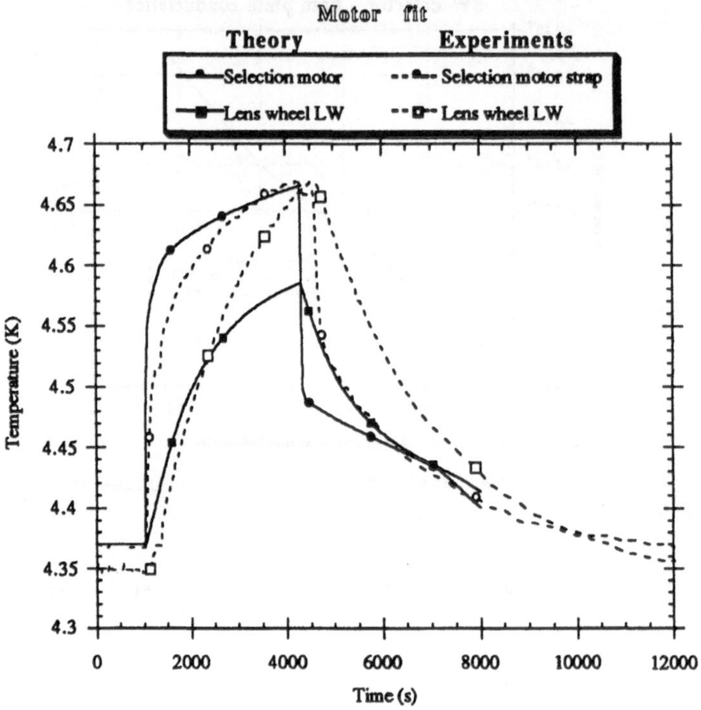

Fig. 8: Fit of motor functioning

long time-constant exists within the motor, the thermal inertia can introduce a stepwise increase of the average camera temperature which could at length become contradictory with the requiments. In order to define the limited use of the motors, a theoretical motor functioning period versus stator and rotor temperature was performed. The latent periods between two motor fixed functioning time-length are 1000, 500,

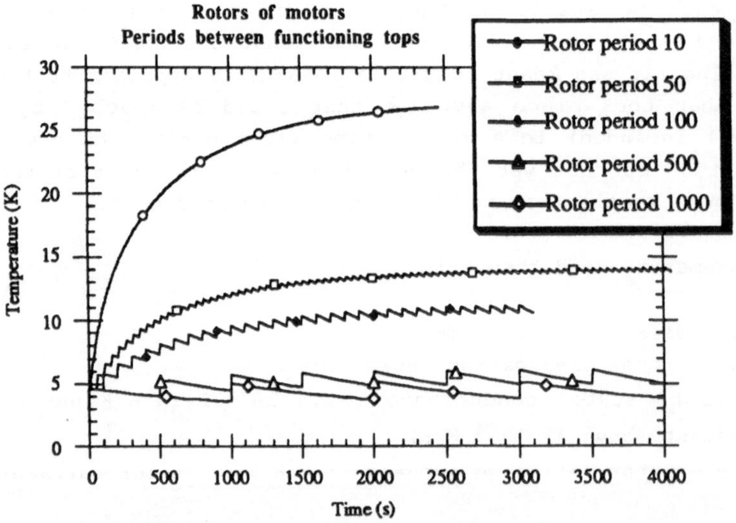

Fig.9: Thermal response of a rotor at different functioning periods

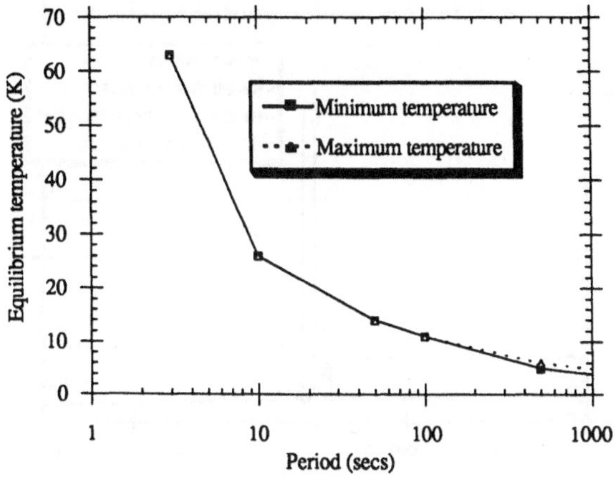

Fig. 10 : Final equilibrium temperature at rotor periodic functioning

100, 50 and 10 seconds. The fixed time-length corresponds to a full wheel turn of the largest wheel (3 seconds). As the latent time diminishes, the top-like temperature effects on the stator become less and less perceptible. The temperature profile of the rotor is saw-like (fig. 9) and the mean rotor temperature over a motor cycle tends to a limit which increases with the decreasing latent period.

The set of limits as a function of the latent period is reproduced in fig. 10. To verify the 10 K parasitic light specification, the latent period cannot be smaller than 100 seconds.

NOMINAL OPERATIONAL THERMAL CYCLE FOR ISOCAM

The theoretical thermal dissipation cycle for ISOCAM is depicted in fig.11. All the required data for a model has been either extracted from a database (and fitted correctly the experiment results) or calculated as it was previously described. Some nodes have been represented in fig. 12 for the SW side (the LW side being quite identical qualitatively). With

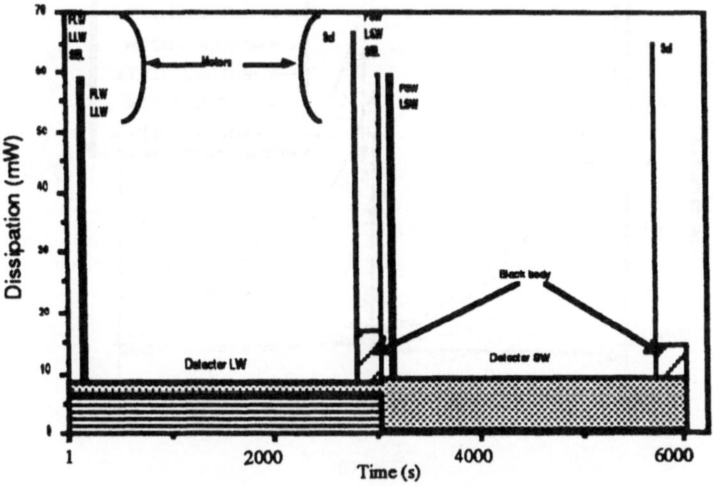

Fig. 11: Nominal theoretical dissipation cycle

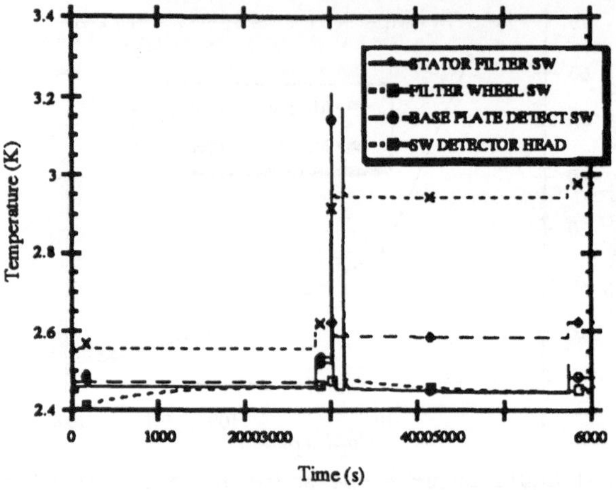

Fig. 12 : Nominal cycle - SW side

what was collected as supplementary data (motors, for instance) it is now possible to determine an average dissipation for the camera. With fig. 11 nominal cycle, this dissipation is 9.8 mW thus below the 10 mW maximum allowed by the requirements (ISO satellite allowance).

Fig. 13 shows how the rotors behave during the nominal cycle as a result of the previous analysis on motors. The 10 K maximum temperature for parasitic light is verified during the whole cycle.

Fig. 14 is a detail of what goes on in the immediate vicinity of detector switch-on (stand-by to ON in fact). The 0.1 K temperature stability is very locally not verified, in a 10 to 40 seconds lapse of time. But this feature only occurs as calibration starts just before a measurement cycle and reflects the thermal equilibrium of the detector itself. No further unrequested instability occurs in the operational mode. The total unavailability is at a maximum of about 2 % of the total cycle time.

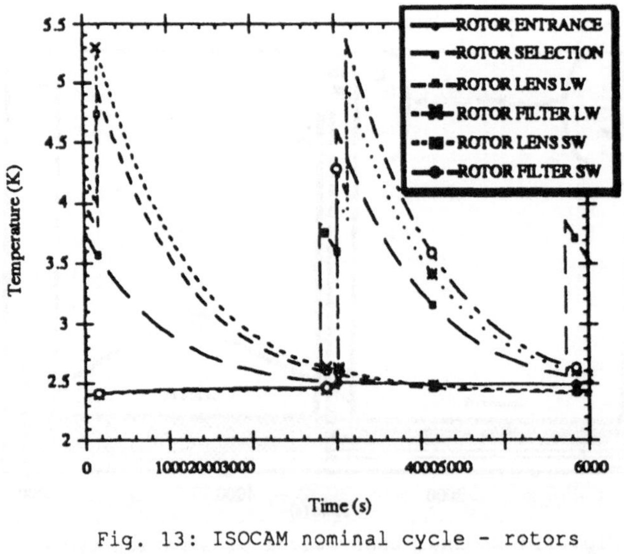

Fig. 13: ISOCAM nominal cycle - rotors

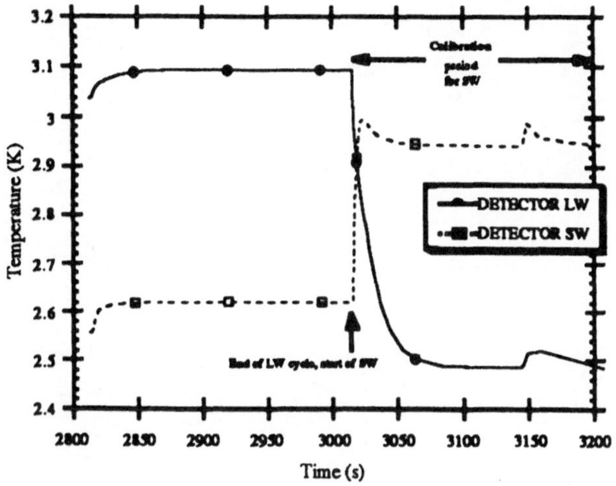

Fig. 14 : Transition LW to SW channel

CONCLUSIONS

All requirements have been verified so that the camera is compatible with the scientific measurements and as a payload with the telescope and cold source facility.

The flight model has been delivered to ISO and there is a good confidence in the thermal model so that fully representative thermal schemes can be expected to conveniently support the ISOCAM project team throughout the operational mission.

The authors would like to thank D. AUTERNAUD, ISOCAM project manager at **aerospatiale**, for all the project support. They are also obliged to the Service d'Astrophysique at the C.E.A.-C.E.N.S. for allowing the output of these results, with particular emphasis to J.F. BONNAL with whom they had useful exchange of ideas. Special thanks also to the efficient work accomplished at the cryogenic test laboratory where ISOCAM tests were performed, the test team being led by S. TARIDE.

REFERENCES

1. L. DE SA, S. TARIDE, ISOCAM, the ISO's satellite infra-red camera. Cryogenics, 13th ICEC proceedings, 30, September Supplement 1990, p. 480.

2. X. LI et al., Static and dynamic thermal conductance of ball bearing at LHe temperatures. Cryogenics, 13th ICEC proceedings, 30, September Supplement 1990, p. 499.

DESIGN OF THE CRYOGENIC SYSTEM FOR THE SAFIRE INSTRUMENT

J. H. Lee, D. A. Payne, and R. D. Averill[†]

Ball Electro-Optics/Cryogenics Division
Boulder, Colorado

[†]NASA Langley Research Center
Hampton, Virginia

ABSTRACT

One of the primary goals of the NASA Mission to Planet Earth is to improve understanding of the ozone chemistry of the atmosphere over an extended period of time. The Spectroscopy of the Atmosphere using Far Infrared Emission (SAFIRE) instrument is being developed to conduct, for the first time, global measurements of the key ozone chemistry constituents in both the mid- and far-infrared spectral regions. Such remote, long-term observations are made possible by the recent development of compact, long-life, hybrid cryogenic dewars which are necessary to cool the sensitive detectors to the 3-4 K range. The success of this hybrid concept is based on the use of long-life, Stirling cycle cryocoolers to intercept parasitic heat from the internal radiation shields of the superfluid helium dewar. Extensive system trade studies are required to optimize the mass, power, and lifetime of these space-borne cryogenic systems. The SAFIRE Cryogenic Subsystem (CSS) is described, including the thermal performance trades leading to the chosen system configuration and the important dewar/cryocooler interface issues. SAFIRE is an international program involving contributions from the United Kingdom, Italy, and France, with the United States responsible for overall instrument integration.

INTRODUCTION

The NASA Mission to Planet Earth has been planned to understand both natural and human induced global change effects over an extended 15-year period.[1] The Earth Observing System (Eos) is one of the major initiatives in support of the global change program. Eos consists of two series of sun-synchronous polar-orbiting platforms designed for remote, global observations of the Earth's environment. SAFIRE uses, simultaneously, a broadband radiometer in the mid-infrared (MIR) (6 to 17 μm) and a high resolution Fourier transform spectrometer in the far-infrared (FIR) (25 to 125 μm) to measure ozone chemistry parameters in the middle atmosphere.[2] SAFIRE has the capability to study diurnal variations in the atmosphere, ozone decline in the Antarctica region and local dynamic atmospheric processes, wherever they may occur. These SAFIRE science objectives uniquely address the important goals of the second Eos mission to conduct measurements of the global distribution of the atmospheric chemicals, dynamic processes, and solar energy input that control ozone concentrations.

SAFIRE REQUIREMENTS

The Eos platforms are planned to transport and support a selected group of synergistic experiments into a sun-synchronous polar orbit for a mission length of five years. The instruments will be attached to the platform in individual Earth-facing bays with appropriate spacecraft power, command and data interfaces. The SAFIRE instrument specifications are given in Table 1. The configuration to accommodate the Eos instrument envelope and interface requirements is shown in Fig. 1. The instrument optical design is divided into four optical modules. The Front End Optics (FEO) module contains a scan mirror which images the instrument field of view through either a forward or aft viewing aperture. The optical beam passes through the FEO where the field of view is split at the telescope focus into the MIR and FIR paths. The FIR beam enters the Fourier Transform Interferometer (FTI) module which is developed by the Italian partners. The emerging FIR beam is then transferred to the Cold Optics and Detector Module (CODM) where it is spectrally split and refocused onto three parallel focal planes maintained at a temperature of about 3 K. The CODM is a joint international effort, with the French producing the detectors, British developing the Cold Optics Subsystem (COS) and the two-stage Stirling cycle cryocoolers, and the United States providing the Cryogenic Subsystem (CSS).

The CODM requirements are particularly severe due to the extended mission life and because of the envelope constraints and mass and power limitations imposed by the Eos platform. The CODM consists of the CSS and the COS. Table 1 contains the system budgets allocated to the CODM as well as the performance requirements for the CSS. A vacuum shell temperature of 265 K is predicted for the CSS superfluid helium dewar. The efficiency of the cryocooler is affected by the compressor and displacer case temperature. The predicted compressor and displacer temperature for SAFIRE is about 265 K. Table 1 also shows the COS center-of-mass misalignment requirement. Misalignment requirements for the optical components and focal plane assemblies within the COS will be determined later.

CODM BASELINE CONFIGURATION

The baseline configuration of the CODM is shown in Fig. 2. The 124-liter cylindrical helium tank is supported by ten dual-link straps from the aluminum vacuum shell.

Table 1. SAFIRE Instrument and CODM Specifications

PARAMETER	SPECIFICATION
Instrument Type	Passive spectrometer-radiometer, limb scanning (Fore or Aft)
MIR Radiometer	7 channels, 6-17 μm, 105 detectors, HgCdTe (PV and PC), split-Stirling cryocooler
FIR Fourier Transform Spectrometer	7 channels, 25-125 μm, 48 detectors, Ge:Ga (stressed and unstressed), split-Stirling cryocoolers/SfHe dewar
Instrument	
Envelope	Eos double pallet size: 1.6 m x 1.6 x 1.8 m height
Life	5-year mission
Mass	407 kg
Power	465 W
Thermal Rejection	Passive radiation
Cooler/dewar Radiator	258 K nominal
Electronics Radiator	295 K nominal
CODM	
Envelope	74 cm x 74 cm x 135 cm
Mass	<115 kg
Power	<168 W
Vacuum Shell Temperature	≤265 K
Cryocooler Case Temperature	253 K ≤ T ≤ 310 K
Focal Plane Assembly Temperature	3-4 K
Cold Optics Subsystem (COS) Temperature	≤30 K
Cold Preamplifiers Temperature	80 K
COS Center of Mass Alignment	±0.2 mm (displacement) ±20 arcsec (tilt)

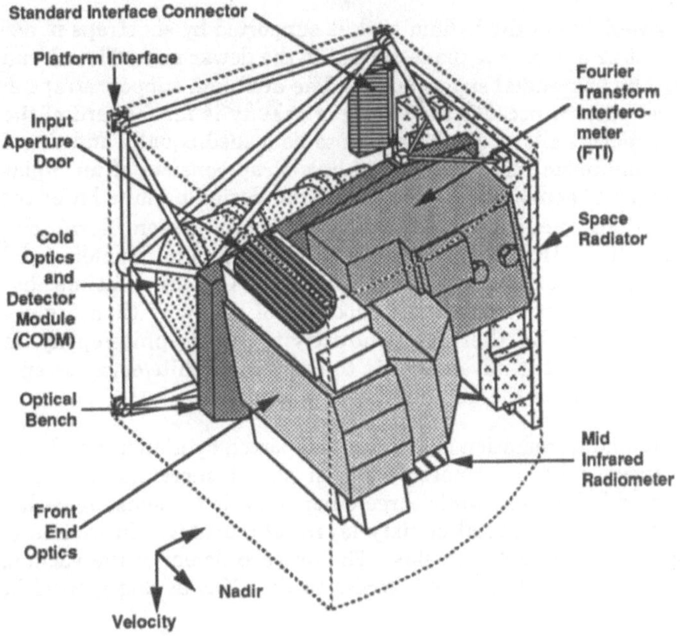

Fig. 1. SAFIRE instrument configuration.

The forward girth ring accommodates the mounting of the cryocoolers, electrical connectors, and fluid management external manifold. The girth ring internal space shares a common vacuum with the dewar except in areas where the cryocooler displacers are mounted; each displacer cold finger is contained in a separate vacuum compartment.

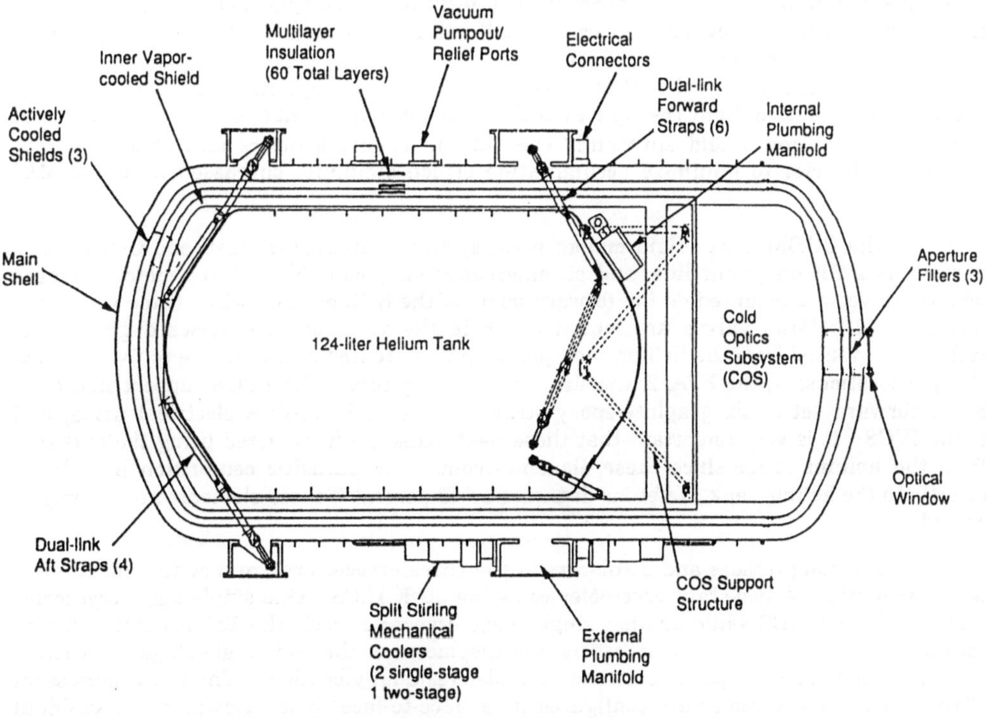

Fig. 2. SAFIRE/CODM 124-liter superfluid helium dewar.

The forward end of the helium tank is supported by six straps in a zig-zag pattern which are angled at 22.5° from the plane normal to the dewar axis. The aft end of the helium tank is supported by four radial straps at 35°. The dual-link support straps are arranged in an unsymmetrical pattern because the center of gravity is far forward of the center of the tank. This arrangement allows all the straps to be loaded equally and therefore all can be the same size to minimize cost. Each dual-link strap consists of an S-glass/epoxy strap connected to a graphite/epoxy strap. The dual strap design permits lower conduction parasitic heat leak than the single S-glass/epoxy strap design flown on the Infrared Astronomical Satellite (IRAS)[3] and the Cosmic Background Explorer (COBE).[4] An all S-glass/epoxy support strap approach degrades CODM lifetime by 25 percent. Graphite/epoxy has greater strength and modulus of elasticity than S-glass/epoxy and at temperatures below 30 K, thermal conductivity of graphite/epoxy is lower than S-glass/epoxy. Therefore, it is advantageous to use the graphite/epoxy strap on the inboard section (helium tank end) of the dual-link strap assembly.

Four aluminum radiation shields are supported by aluminum blocks bonded to the S-glass/epoxy straps. An innermost radiation shield surrounding the helium tank is cooled by the vented helium gas while three other radiation shields are cooled by three split Stirling-cycle cryocoolers. A total of sixty layers of multilayer insulation (MLI) is placed between the three actively-cooled shields. The forward domes of the vacuum shell and radiation shields are removable to allow access to the COS during ground integration and testing.

The COS, which includes the optics, focal plane assembly (FPA) and the detector preamplifiers, is mounted to the forward end of the helium tank by a graphite/epoxy tube truss structure. The COS is cooled to about 30 K by a thermal strap attached to the inner actively-cooled shield (IACS). The preamplifiers and feedback resistors are cooled to about 80 K by a thermal strap connected to the middle actively-cooled shield (MACS). The FPA is cooled by two thermal links. One is attached to the helium tank, and the other is attached to the helium vent line. The enthalpy available from the vented helium gas absorbs about 80 percent of the heat load from the FPA, while 20 percent of the heat load is conducted to the helium tank. If there is a cryocooler failure, the extra helium enthalpy generated from the rise in helium boil-off can accommodate the increase in parasitic heat load absorbed by the FPA, thus maintaining the FPA temperature at about 3 K. The optical beam enters the COS through the optical window on the vacuum shell and three thermal blocking filters mounted on the removable end dome of the radiation shields. The thermal blocking aperture filters have spectral characteristics that maximize signal throughput, but limit thermal background radiation from the optical train. An aluminum tube which is cantilevered from each radiation shield surrounds each of the aperture filters and acts as a contamination barrier to minimize accumulation of contaminants outgassed from the MLI blankets.

The CODM dewar fluid management system is similar to IRAS and COBE. The dewar has a fill line, vent line, and an emergency vent manifold. Three motor-operated bellows valves are mounted on the forward dome of the helium tank, while three warm motor-operated bellows valves are mounted inside the vacuum shell forward girth ring. Helium gas vented from the helium tank porous plug is routed in a vent line to five heat exchanger locations: 1) the FPA, 2) the six graphite/epoxy tubes of the COS support structure, 3) the forward set of six graphite/epoxy straps, 4) the JFET-to-FPA electrical wires, and 5) the IVCS. It is very important that these heat loads are intercepted using the enthalpy from the helium vapor since these elements contribute parasitic conduction heat loads directly to the helium tank. Lifetime degrades by 45 percent if these elements are not vapor-cooled.

The compressors and displacers of the three cryocoolers are mounted on the forward girth ring. A two-stage cryocooler cools the 30 K IACS. One single-stage cryocooler cools the 80 K MACS while another single-stage cryocooler cools the 135 K outer actively-cooled shield (OACS). The compressors and displacers of the two single-stage cryocoolers are mounted as opposed pairs to attenuate cooler residual vibrations. The two compressors of the two-stage cryocooler are configured as a "face-to-face" pair. However, the residual force from the displacer of the two-stage cooler is not balanced. A resultant force and moment may be induced onto the CODM or transmitted to other SAFIRE modules. This is discussed in a later section.

For preliminary dynamic analysis, we have assumed that the CODM is attached to the SAFIRE optical bench via a kinematic mount. A spherical bearing on the forward girth ring restricts the translation of the CODM along three orthogonal directions, while an aluminum tube truss arrangement on the aft girth ring prevents rotation of the CODM about the three orthogonal axes.

CRYOCOOLERS CONSIDERATION

SAFIRE instrument success will depend greatly on consistent and reliable performance of the cryocoolers over the extended mission life of the instrument. The unprecedented requirement for long-life (>5 years) cooling capability is being addressed by the cryocooler community. Recent work done at the Jet Propulsion Laboratory to characterize the vibration performance, electromagnetic characteristics, and thermal performance of existing Stirling-cycle coolers has been published.[5]

The SAFIRE team addressed the cryocooler issue very early in the instrument conceptual phase and selected the Oxford-type, split Stirling-cycle coolers produced by British Aerospace as our baseline.[6] Early system studies indicated that the single-stage coolers alone, with cooling capacity in the 50 K and above temperature range, would not provide the thermal performance needed for the SAFIRE instrument. The Rutherford Appleton Laboratory (RAL) in England, a SAFIRE team partner, developed a two-stage split Stirling-cycle cooler to produce cooling in the 20 K and above temperature range which meets the SAFIRE cooling requirements. It is a further extension of the Oxford cooler heritage, using two compressors in parallel driving a single two-stage displacer.[7] The two-stage cooler was initially developed and tested at RAL, and a prototype model has been built and is being tested by British Aerospace in Bristol.

Cooler Induced Vibration

The dewar/cryocooler interface issues[8] have been addressed in the SAFIRE CSS design. Residual vibration induced by the cryocoolers can cause unacceptable microphonics coupling in the detector readout or jitter in the optical system. The BAe split-Stirling single-stage cryocoolers have been found to have a strong primary reaction at the operating frequency and many secondary reactions (harmonics) at integral multiples of the operating frequency.[5] The primary reaction acts only in the direction of the cooler axis and can be compensated out to about the level of the secondary reactions by mounting the cooler compressors and displacers as back-to-back pairs and driving them in phase. The secondary reactions, however, act in all three axes and cannot be compensated out. They can only be minimized by improving the design and precision of the cooler mechanisms and by refining the drive electronics. Currently, for each pair of BAe single-stage cooler compressors or displacers, we expect less than 0.44 N (0.1 lb) reactions in all three axes at each harmonic from the operating frequency up to about 400 Hz. This result was obtained by Ross, et al.[5] using low-distortion laboratory drive electronics.

Since the operating frequencies will be accurately controlled, we assume that all the harmonics will occur in very narrow frequency bands so that the vibration input to the dewar is expected to be many very narrow spikes spread over the frequency range of the harmonics. We also expect the dewar transmissability to exhibit many narrow spikes over this frequency range because at low amplitudes, the dewar structure possesses low damping. Thus, the analytical problem becomes one of trying to predict how many resonances will occur between cooler harmonics and dewar modes. This cannot be handled by simply driving a finite-element model with the assumed input since no model is accurate enough to predict exact resonances at high frequencies. Instead, we have used probability theory to predict the worst-case resonance conditions. Assuming that the resonances occur at the worst modes, we find that the allowable COS center-of-mass displacement (0.2 mm) and line-of-sight jitter (20 arcsec) correspond to allowable cooler reactions at all harmonics and in each axis of 0.98 N (0.22 lb).

SAFIRE uses two single-stage coolers operating at 40 Hz and a two-stage cooler operating at 35 Hz. Using Ross's results[5], the combined residual vibration force at 40 Hz and its harmonics from the two back-to-back mounted single-stage cooler compressors and displacers is about 0.89 N (0.2 lbf). The RAL-developed two-stage cooler compressors have similar construction as the single-stage cooler compressor, and they are configured as a "momentum-compensated" pair. But the sole two-stage cooler displacer is unbalanced and generates 5.3 N (1.2 lbf) residual vibration at 35 Hz. However, since this large primary reaction is well below the dewar minimum resonance frequency of 40 Hz, the residual vibration force at 35 Hz and its harmonics from the two-stage cooler compressors and displacer is also about 0.89 N (0.2 lbf). Based on the measured BAe cryocooler performance, the allowable SAFIRE COS jitter and displacement requirements can be met but without margin. It should be noted, however, that the COS center-of-mass acceleration which corresponds to the allowable line-of-sight jitter and displacement is about 1 g. This is currently being investigated from a detector microphonics point of view. Preliminary assessment shows that the 35 Hz to 500 Hz disturbance generated by the cryocoolers may not cause any significant microphonics problem for the CODM because the CODM science data readout frequency is on the order of 1.3 kHz to 6 kHz.

Aside from improving the cooler design itself, one can obtain some reduction in vibration levels by changing the location and method of mounting the coolers. For this study we assumed hard-mounted coolers and located them on the forward girth ring as close to the main spherical bearing kinematic mount as possible. This reacts as much of the cooler vibration as possible into the SAFIRE optical bench and was found to reduce optics vibration about 20 percent versus other mounting locations on the dewar vacuum shell. It is also possible to mount the compressors on vibration isolators which act as low-pass filters to attenuate high-frequency residual vibrations. The displacers must be hard-mounted because of their delicate thermal interface to the dewar radiation shields. As the COS and CODM designs become more mature, and the system requirements on optics and FPA jitter become better defined, we will evaluate options to reduce mechanical coolers residual vibration to provide more performance margin, and to evaluate the effects of vibrations transmitted to the SAFIRE optical bench.

<u>Cooler Displacer/Dewar Shields Interface</u>

The thermal link between the displacer cold tip and the dewar radiation shield is designed to minimize the thermal gradient along this link. The baseline displacer cold finger concept used for both the single-stage or the two-stage cryocoolers is shown in Fig. 3. The cold finger consists of two sets of thin copper foils attached to a copper rod. The two sets of copper strips accommodate radial as well as axial displacement of the shields; the dewar radiation shields deflect due to launch excitation and during cooldown from 300 K to its operating temperature. At the baseline operating condition, the overall thermal conductances of the cold finger from the cold tip to the radiation shield are 470 mW/K, 190 mW/K, and 170 mW/K for the IACS, MACS, and OACS, respectively. The thermal link is designed to transmit no more than about 0.5 N of side load to the cold tip. An S-glass/epoxy truss configuration supports the cold finger for launch. This stiff S-glass/epoxy truss transmits most of

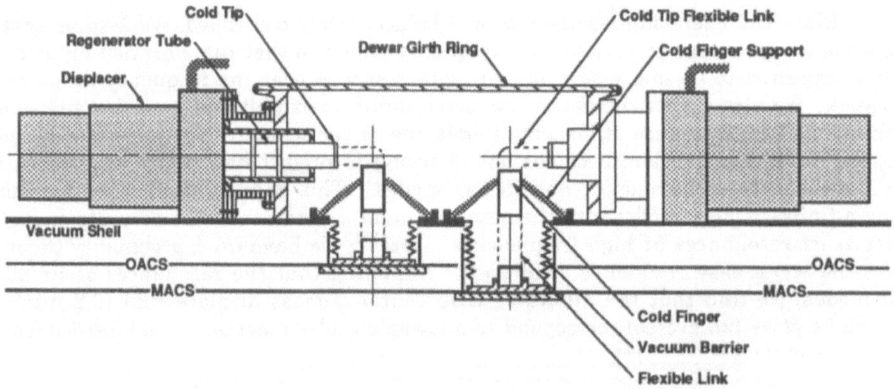

Fig. 3. Cooler displacers/dewar shields interface.

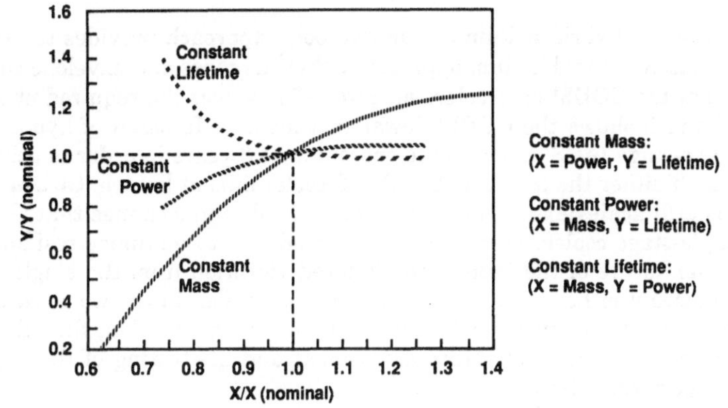

Fig. 4. CODM system performance tradeoffs; nominal values are: CODM mass = 115 kg, lifetime = 7.7 years, and cooler power = 160 W.

the cold finger disturbance load to the dewar vacuum shell rather than to the displacer cold tip. About 20 to 30 mW of parasitic heat is conducted from the vacuum shell to the displacer cold tip. Parasitic heat load due to radiation from the girth ring and vacuum shell surfaces is minimized by wrapping MLI around the displacer regenerator tube, cold tip, and cold finger/thermal link assembly.

CODM PERFORMANCE

Cryogenic subsystem lifetime of 7.7 years (5 years + 54 percent margin) is predicted for the CODM. The tank support straps contribute 45 percent of the total helium tank parasitic heat load while the IVCS radiation contributes 25 percent. About 160 W of power is required from the Eos spacecraft bus to operate the cryocoolers. The two-stage RAL-developed cryocooler provides 330 mW of cooling at the 30 K IACS, while the two single-stage BAe cryocoolers provide 384 mW of cooling at the 80 K MACS, and 1,367 mW of cooling at the 135 K OACS, respectively.

The critical CODM system parameters (mass, lifetime, and cooler power) have been analyzed to determine the performance tradeoffs. Results are shown in Fig. 4. As the amount of helium is increased above the nominal value of 18 kg, the dewar size and mass also increase. The helium tank support strap cross-sectional area has to increase in order to meet the same structural design criteria as in the nominal case. This increases the amount of parasitic heat leak to the helium tank. As a result, for constant cooler power, lifetime does not increase significantly as the mass of CODM or helium is increased. Likewise, for constant lifetime, the required cooler power does not decrease significantly as the mass of CODM or helium is increased.

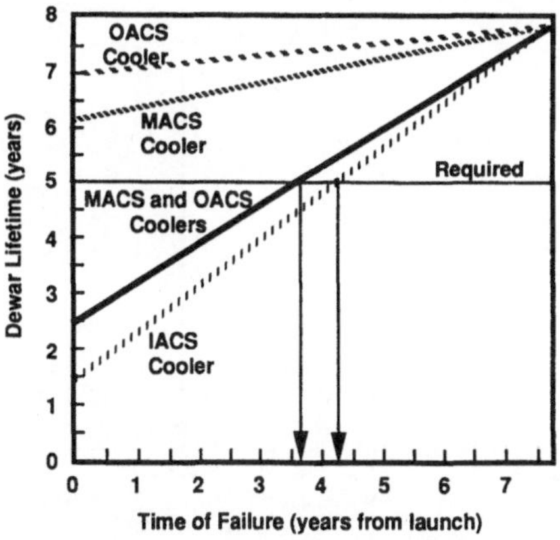

Fig. 5. Impact of cooler failure on dewar lifetime.

The baseline hybrid helium dewar/cryocooler approach provides ten times greater lifetime than a passive stored helium approach within the mass and envelope constraints. It is imperative that the CODM cryocoolers operate reliably over the required mission lifetime of 5 years. Figure 5 shows the CODM dewar lifetime as a function of hypothetical cooler failures. The two-stage IACS cooler has to operate for 4.3 years in order to achieve a 5-year CODM lifetime. If either the MACS or the OACS cooler fails at launch, CODM lifetime of at least 6 years is still achievable. Our current baseline design assumes that the MACS and OACS BAe single-stage coolers operate as an opposed pair to minimize residual vibrations. Depending on the nature of the failure, the induced vibration from the single-stage coolers may not be balanced if either the MACS or OACS cooler fails. In the worst case, if both single-stage coolers have to be turned off at the beginning of the mission, then the CODM lifetime decreases to 2.5 years. Reliability assessment and life-testing of the cryocoolers are currently being performed by BAe.

Several alternative concepts have been evaluated to minimize the impact of the two-stage cooler failure on the mission. These options can be grouped into four general approaches: 1) increase helium mass, 2) use different radiation shield cooling schemes with two-stage coolers, 3) include a redundant two-stage cooler, and 4) add solid hydrogen or neon. There is no obvious favorable option because each option impacts system mass, volume, power, and lifetime. Performance of the CSS using a redundant two-stage cooler is shown in Fig. 6. Note that if the original two-stage cooler does not fail during the mission, dewar lifetime compared with the baseline of 7.7 years is shorter by 48 percent and 14 percent for the cases without thermal switch and with thermal switch, respectively. In order to achieve the baseline lifetime of 7.7 years, both the original and redundant two-stage coolers have to be operating simultaneously after 1.75 years from launch or after 4.75 years from launch, respectively, for the cases without thermal switch or with thermal switch. As a result, CODM mission average cooler power consumption increases by 50 percent or 25 percent, correspondingly. More detailed risk assessment studies are planned as the SAFIRE instrument design progresses.

CONCLUSIONS

The SAFIRE CODM hybrid cryogenic system combines proven superfluid helium storage technology with split-Stirling single-stage and two-stage cryocoolers to achieve

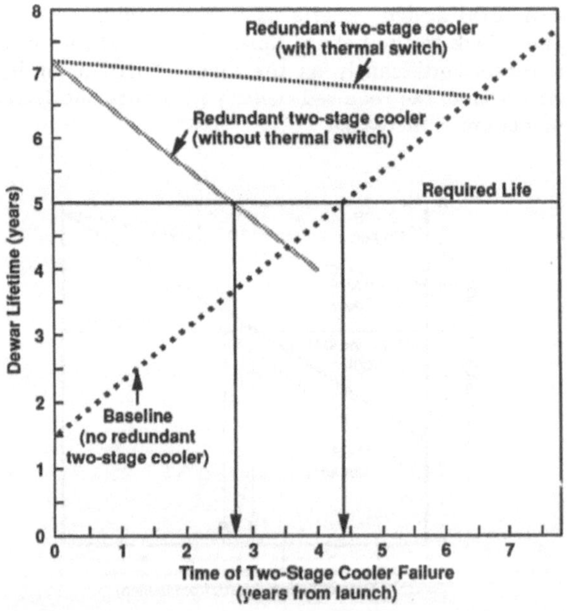

Fig. 6. Effect of redundant two-stage cooler on mission lifetime.

greater than 5 years on-orbit lifetime. Critical interface issues between the dewar and cryocoolers have been addressed in the design. Preliminary analysis shows that the allowable misalignment of optics and focal plane assemblies due to cryocooler residual vibrations is achievable, but with no margin. There is ongoing effort on cryocoolers and the SAFIRE system design to improve the performance margin. Premature failure of the two-stage cryocooler poses the greatest risk in curtailing SAFIRE mission lifetime. Failure of the two-stage cryocooler at launch would reduce mission lifetime to 1.5 years. Several options, including implementation of a redundant two-stage cooler to alleviate this impact, have been considered. But these options either increase system mass, volume, and power consumption or decrease baseline lifetime. Future system trades will benefit from the results of life-testing and reliability assessment of our baseline RAL-developed, two-stage cryocooler currently being performed by British Aerospace.

ACKNOWLEDGEMENTS

The NASA Langley Research Center is funding development of the SAFIRE instrument at Ball Aerospace under contract NAS1-19129. The principal author expresses his gratitude to R. Hopkins, C. Downey, and R. Reinker for their valuable suggestions and comments made during the SAFIRE Cryogenic Subsystem Conceptual Design Study phase.

REFERENCES

1. "Eos-A Mission to Planet Earth," Eos Program Office, NASA Headquarters (Code EE), Washington, DC 20546 (1990).
2. J. M. Russell III, "An Overview of the Spectroscopy of the Atmosphere Using Far-infrared Emission Experiment (SAFIRE)," in Proc. International Symposium on Optical Engineering and Photonics in Aerospace Sensing, SPIE meeting, April 1-5, 1991, Orlando, Florida.
3. A. R. Urbach and P. V. Mason, "IRAS Cryogenic System Flight Performance Report," in Advances in Cryogenic Engineering, Vol. 29, Plenum Press, New York (1984).
4. S. M. Volz, et al., "Cryogenic On-orbit Performance of the NASA Cosmic Background Explorer (COBE)," in Proc. of SPIE, Vol. 1340, (1990).
5. R. G. Ross, Jr., D. L. Johnson and R. S. Sugimura, "Characterization of Miniature Stirling-cycle Cryocoolers for Space Application," in Proc. 6th Intl. Cryocooler Conf., DTRC-91/002, David Taylor Research Center, (1991), p. 27.
6. G. Davey and A. H. Orlowska, "Miniature Stirling Cycle Cooler," Cryogenics, 27:645 (1987).
7. A. H. Orlowska, T. W. Bradshaw, and J. Hieatt, "Closed Cycle Coolers for Temperatures Below 30 K," Cryogenics, 30:246 (1990).
8. R. A. Hopkins, et al., "Long-lifetime Stored Cryogen Systems Using Refrigerators to Reduce Parasitic Heat Input," in Proc. 6th Intl. Cryocooler Conf., DTRC-91/002, David Taylor Research Center, (1991) p. 153.

NUMERICAL SIMULATION OF THE HELIUM GAS SPIN-UP CHANNEL PERFORMANCE OF THE RELATIVITY GYROSCOPE

Gerald R. Karr and Josephine Edgell

University of Alabama in Huntsville
Huntsville, Alabama

Burt X. Zhang

Air Products and Chemicals, Inc.
Gardner Cryogenics Department
Lehigh Valley, Pennsylvania

ABSTRACT

The dependence of the spin-up system efficiency on each geometrical parameter of the spin-up channel and the exhaust passage of the Gravity Probe-B (GPB) is individually investigated. The spin-up model is coded into a computer program which simulates the spin-up process. Numerical results reveal optimal combinations of the geometrical parameters for the ultimate spin-up performance. Comparisons are also made between the numerical results and experimental data. The experimental leakage rate can only be reached when the gap between the channel lip and the rotor surface increases beyond physical limit. The computed rotating frequency is roughly twice as high as the measured ones although the spin-up torques fairly match.

INTRODUCTION

In a process of evaluating the gas spin-up system used on the gyroscope of the GPB, a computer program simulating the performance of the gas spin-up process is developed. The dependencies of the spin-up torque on the geometrical parameters of the spin-up channel, the rotor housing cavity, as well as the exhaust passage are investigated.

For the numerical modeling purpose, the gas spin-up system is simplified to avoid complexity and is schematically shown in Figure 1. The gas leaked through the gap between the spin-up channel lip and the rotor surface is represented by the flow through passage A. Passage B denotes the exhaust holes and slots between the rotor housing cavity and the vacuum chamber. The Mach number at the spin-up channel entrance is first calculated using the following isentropic relationship

$$\frac{W_s}{A_s} = \sqrt{\frac{k}{R}}\frac{P_o}{\sqrt{T_o}} M_{si}\left(1+\frac{k-1}{2}M_{si}^2\right)^{\frac{k+1}{2(k-1)}} \tag{1}$$

where W_s is the mass flow rate, A_s is the cross-sectional area of the spin-up channel, k is the ratio of the specific heats of helium gas, R is the gas constant, P_o and T_o are the stagnation pressure and temperature respectively, and M_{si} is the entrance Mach number. The pressure drop across the channel is then computed using the relationship between the entrance Mach number and the stagnation pressure

$$\Delta P = P_o\left(1+\frac{k-1}{2}M_{si}^2\right)^{-\frac{k}{k-1}}\left[1 - M_{si}\sqrt{\frac{2\left(1+\frac{k-1}{2}M_{si}^2\right)}{1+k}}\right] \tag{2}$$

Advances in Cryogenic Engineering, Vol. 37, Part B
Edited by R.W. Fast, Plenum Press, New York, 1992

The pressure at the entrance can also be calculated using

$$P_{si} = P_o\left(1+\frac{k-1}{2}M_{si}^2\right)^{\frac{k}{1-k}} \tag{3}$$

In order to determine the pressure in the cavity, consider the relationship between the pressure drop across a channel and the shear force acting on the channel wall

$$\frac{\Delta PA}{LC} = \frac{1}{2}\rho V^2 f \tag{4}$$

where ΔP is the pressure drop under discussion, A and C are the cross-sectional area and perimeter length of the channel respectively, L is the length of the channel, ρ is the density of helium gas, V is the local gas velocity, and f is the local coefficient of friction. For laminar flow,

$$f \approx \frac{16}{Re} = \frac{16A\mu}{WD} \tag{5}$$

where μ is the dynamic viscosity, Re is the Reynolds number, W is the mass flow rate, D is the hydraulic diameter of the channel.

An expression for the local average velocity can be obtained from the equation

$$\frac{W}{A} = \sqrt{\frac{k}{R}}\frac{P}{\sqrt{T_o}}M\sqrt{1+\frac{k-1}{2}M^2} \tag{6}$$

as

$$V = \sqrt{kR}\frac{T}{\sqrt{T_o}}M\sqrt{1+\frac{k-1}{2}M^2} \tag{7}$$

where the relationship

$$\frac{W}{A} = \rho V \tag{8}$$

and the perfect gas law have been applied. Substituting Equations 5 and 7 into Equation 4, one gets

$$M = \frac{\Delta PAD}{8LC\mu}\frac{\sqrt{T_o}}{T}\frac{1}{\sqrt{kR}}\left(1+\frac{k-1}{2}M^2\right)^{-\frac{1}{2}} \tag{9}$$

This equation can be applied to passages A and B individually resulting the expressions for the entrance Mach numbers of these passages

$$M_{ai} = \frac{(P_{bi}-P_{ai})A_aD_a}{8L_aC_a\mu}\frac{\sqrt{T_o}}{T_a}\frac{1}{\sqrt{kR}}\left(1+\frac{k-1}{2}M_{ai}^2\right)^{-\frac{1}{2}} \tag{10}$$

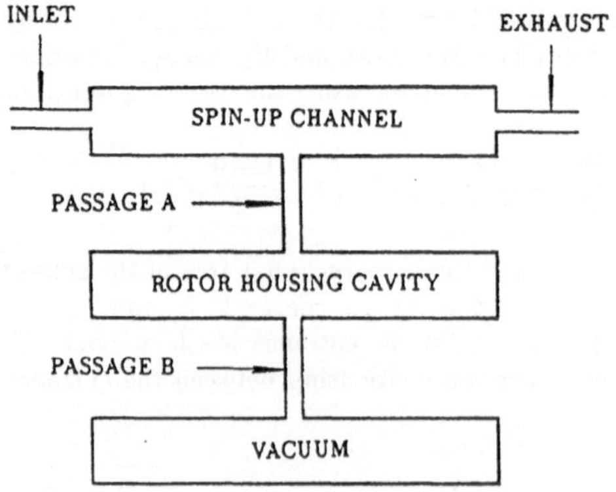

Figure 1: Schematic of gas spin-up system

$$M_{bi} = \frac{(P_v - P_{bi})A_b D_b}{8L_b C_b \mu} \frac{\sqrt{T_o}}{T_o} \frac{1}{\sqrt{kR}} \left(1 + \frac{k-1}{2} M_{bi}^2\right)^{-\frac{1}{2}} \tag{11}$$

where the subscripts a and b denote passages A and B respectively. Equations 10 and 11 can be combined such that

$$\frac{M_{ai}}{M_{bi}} \sqrt{\frac{1 + \frac{k-1}{2} M_{ai}^2}{1 + \frac{k-1}{2} M_{bi}^2}} = \frac{P_{bi} - P_{ai}}{P_v - P_{bi}} \frac{A_a D_a L_b C_b T_b}{A_b D_b L_a C_a T_a} \tag{12}$$

Another expression for the quantity on the left-hand side of Equation 12 can be derived by applying the continuity equation (Equation 6) to passages A and B. Such a procedure produces

$$A_a \sqrt{\frac{k}{R}} \frac{P_{ai}}{\sqrt{T_o}} M_{ai} \sqrt{1 + \frac{k-1}{2} M_{ai}^2} = A_b \sqrt{\frac{k}{R}} \frac{P_b}{\sqrt{T_o}} M_{bi} \sqrt{1 + \frac{k-1}{2} M_{bi}^2} \tag{13}$$

or

$$\frac{M_{ai}}{M_{bi}} \sqrt{\frac{1 + \frac{k-1}{2} M_{ai}^2}{1 + \frac{k-1}{2} M_{bi}^2}} = \frac{A_b}{A_a} \frac{P_{bi}}{P_{ai}} \tag{14}$$

Equating Equations 12 and 14, an expression relating the pressure in the spin-up channel with the pressures in the cavity and the vacuum chamber is obtained

$$\frac{P_{ai}}{P_{bi}} \frac{P_{bi} - P_{ai}}{P_v - P_{bi}} = \frac{A_b^2 D_b L_a C_a T_a}{A_a^2 D_a L_b C_b T_b} \tag{15}$$

This equation can be readily solved for P_{bi} which equals to the cavity pressure P_c. The Mach numbers at the entrances of passages A and B can be consequently obtained by solving the following equation

$$M_i = \frac{P_e}{P_i} \sqrt{\frac{k+1}{2\left(1 + \frac{k-1}{2} M_i^2\right)}} \tag{16}$$

where P_i and P_e denote pressures at the entrance and exit of the passage. Once these parameters are determined, the leaking mass flow rate can be easily computed using Equation 6. Finally, the local spin-up torque is given by

$$\Gamma_L = F_L r = \frac{\Delta P A_s}{\Delta L_s C_s} S_s r \tag{17}$$

where Γ_L is the local spin-up torque, r is the rotor radius. The tangential force acting on an infinitesimal area of the rotor surface is given by

$$F_L = \tau S_s \tag{18}$$

where S_s denotes the infinitesimal area of the rotor surface exposed to the spin-up channel. This area spans the entire width of the spin-up channel and has a length of ΔL in the flow direction which is an arbitrarily small quantity. Mathematically, this area is defined as

$$S_s = W_s \times \Delta L_s \tag{19}$$

The local shear force τ in Equations 17 and 18 is given by

$$\tau = \frac{\Delta P A_s}{\Delta L_s C_s} \tag{20}$$

Substituting Equation 19 for S_s into Equation 17, the local spin-up torque becomes

$$\Gamma_L = \frac{\Delta P A_s W_s r}{C_s} \tag{21}$$

The total spin-up torque acting on the rotor can be obtained by integrating the local spin-up torque over the entire spin-up channel length

$$\Gamma_T = \frac{1}{L_s} \int_0^{L_s} \Gamma_L dl \tag{22}$$

The maximum achievable spin speed is given by the expression

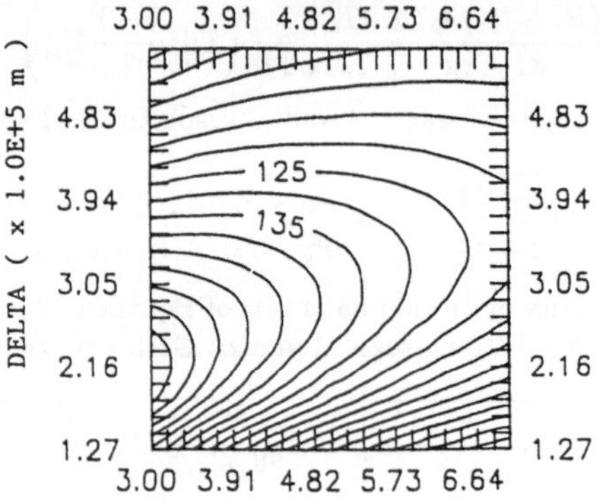

Figure 2: Dependence of rotating frequency on channel lip width and gap between lip and rotor surface

$$\omega_{max} = \Gamma_T \left(\frac{\mu S_L r^2}{\delta} + \frac{\pi P_c r^4}{\sqrt{2RT}} \right)^{-1} \tag{23}$$

where S_L is the surface area of the channel lip, δ is the gap between the channel lip and the rotor surface, and P_c is the pressure in the cavity.

RESULTS

Numerical results indicate that when the stagnation pressure P_o and the pressure in the vacuum chamber P_v are kept constant, the rotating frequency of the gyro can be altered by changing the geometry of passages A and B. Figures 2 and 3 show the dependence of the frequency on δ and w, which are the gap between the spin-up channel lip and the rotor surface, and the width of the lip respectively. It is seen that corresponding to each fixed value of w, there is a single value of δ which offers the optimal system performance. For example, the maximum frequency for $w = .5\ mm$ is obtained at $\omega = 89.58\ Hz$ by setting the gap value at $\delta = .01852\ mm$. It is also obvious that within a certain range of δ values, the frequency will increase as the lip width is reduced. However, the frequency will drop sharply to a very low value when w is reduced to below $.2\ mm$.

The geometrical characteristics of passage B is far more complicated. The passage consists holes and slots of different sizes. In order to model the flow characteristics at passage B, a geometrical parameter B is defined as

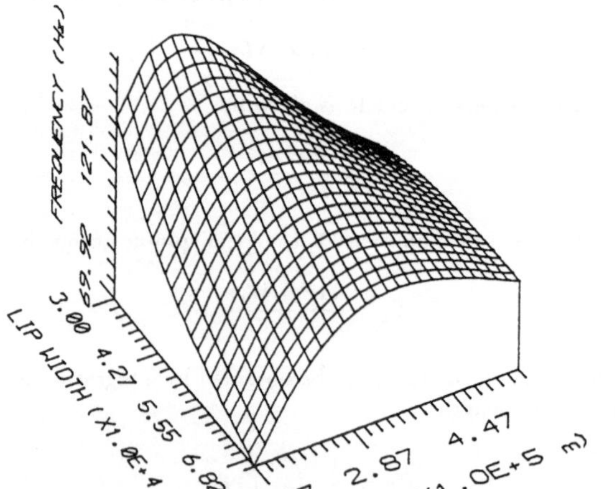

Figure 3: Topological presentation of geometrical dependence of rotating frequency

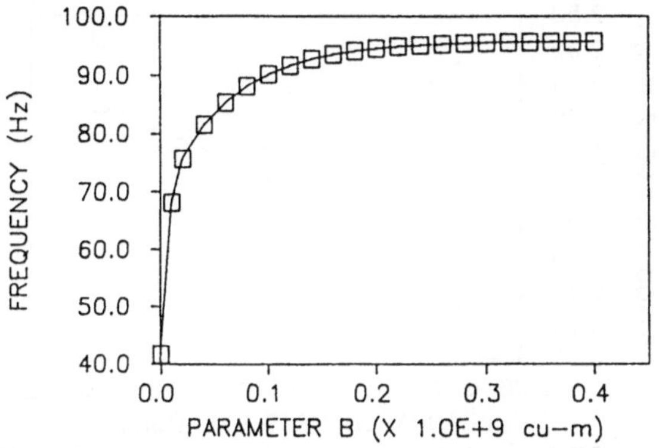

Figure 4: Dependence of rotating frequency on parameter B ($W = 5.7E-9\ kg/s, w = 5E-4\ m, \delta = 1.27E-5\ m, P_o = .092\ torr$)

$$B = \sum_{i=1}^{n} \frac{A_i^3}{L_i C_i^2} \tag{24}$$

where A, C, and L are the cross-sectional area, the wetted perimeter, and the depth of each of the holes and slots comprising passage B individually, and n denotes the total number of holes and slots. Figure 4 shows the maximum achievable frequency at different values of parameter B when the mass flow rate is fixed at $\dot{m} = 1.22 \times 10^{-9}\ kg/s$. As depicted in this figure, the frequency increases rapidly as the value of B increases at the lower end. When B is sufficiently large, the rotating frequency will be much less sensitive to any variation of B.

The numerical results are also compared with experimental data acquired by researchers at Stanford. As shown in Figure 5, the computed frequencies at different gas flow rates are much higher than those experimentally measured. The leakage rate in each case is found to be approximately 9% of the total gas flow rate, which is much lower than the measured value (approx. 50% of the total gas flow rate). In order to correlate the numerical results with the experimental data, the value of δ is gradually increased until the computed leakage rate matches the measured leakage rate. This process is completed when δ reaches a value which is about 5 times greater than the upper limit of the designed value ($.254 \times 10^{-4}\ m$). The computed spin-up torques at different gas flow rates are then compared to the experimentally measured ones (see Figure 6). The computed rotating frequencies are also compared to the experimentally measured frequencies. As depicted in Figure 7, they are about twice as high as those experimentally measured. Apparently, a higher stagnation pressure drives s larger gas flow rate which in turn produces a larger spin-up torque. Nevertheless, it is also responsible for a larger leakage rate resulting a higher background pressure P_c which contributes to the resistance to the spin-up process. Figure 8 shows this kind of resistance in terms of Γ/ω_{max} which increases as the stagnation pressure is raised.

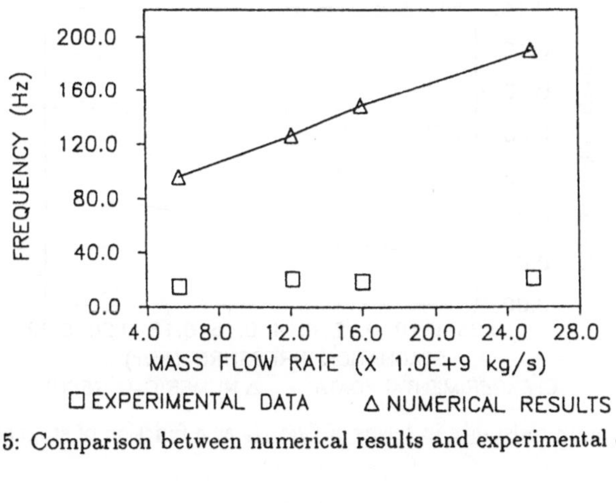

Figure 5: Comparison between numerical results and experimental data

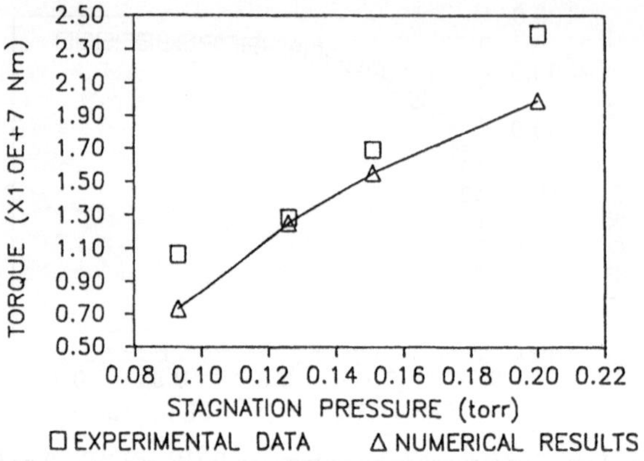

Figure 6: Comparison between computed and measured spin-up torques

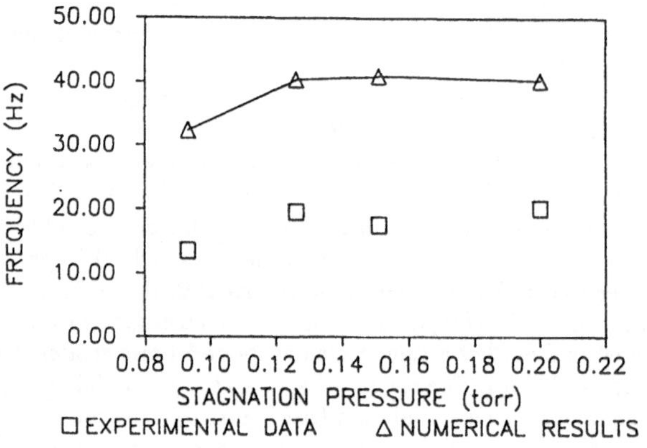

Figure 7: Rotating frequency as a function of stagnation pressure

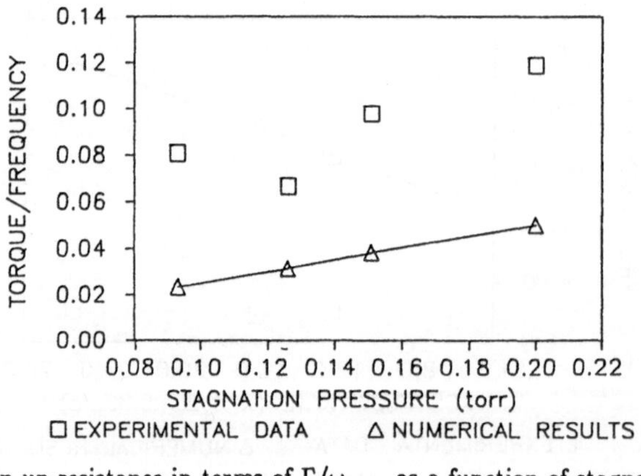

Figure 8: Spin-up resistance in terms of Γ/ω_{max} as a function of stagnation pressure

CONCLUSION

The efficiency of the spin-up system for the gyroscope of the GPB has been investigated. The influence from each relevant geometrical parameter is individually analyzed. For a fixed width of the spin-up channel lip, there is an optimum dimension for the gap between the lip and the rotor surface, which corresponds to a highest achievable rotating frequency.

It is also concluded that besides the dependence of the rotating frequency on various geometrical parameters, this frequency is primarily determined by P_v, which is the pressure in the vacuum chamber, and the helium gas flow rate which in turn is a function of P_v and the stagnation pressure P_o.

ACKNOWLEDGEMENTS

The research was conducted at the University of Alabama in Huntsville and supported by NASA/Marshal Space Flight Center under Contract # NAS8-36955.

REFERENCE

1. Liepmann, H.W. and Roshko, A. *Elements of Gasdynamics*, John Wiley & Sons, New York, 1957.

2. Shapiro, A.H. *The Dynamics and Thermodynamics of Compressible Fluid Flow*, Vol. 1, Ronald Press, New York, 1953.

SHOOT PERFORMANCE TESTING

M. J. DiPirro, P. J. Shirron, S. M. Volz, and M. E. Schein

Code 713, NASA/Goddard Space Flight Center
Greenbelt, Maryland

ABSTRACT

The Superfluid Helium On-Orbit Transfer (SHOOT) Flight Demonstration is a shuttle attached payload designed to demonstrate the technology necessary to resupply liquid helium dewars in space. Many SHOOT components will also have use in other aerospace cryogenic systems. The first of two SHOOT dewar systems has been fabricated. The ground performance testing of this dewar is described. The performance tests include measurements of heat leak, impedances of the two vent lines, heat pulse mass gauging accuracy, and superfluid transfer parameters such as flow rate and efficiency. A laboratory dewar was substituted for the second flight dewar for the transfer tests. These tests enable a precise analytical model of the transfer process to be verified. SHOOT performance is thus quantified, except for components such as the liquid acquisition devices and a phase separator which cannot be verified in one gravity.

INTRODUCTION

The SHOOT flight program has two primary objectives. The first is to demonstrate the technology to resupply superfluid helium in space and the second is to develop components that will be useful for other superfluid helium or other cryogenic payloads as well. SHOOT is currently manifested for flight as a shuttle attached payload in January, 1993.

A series of component developments have taken place over the last few years[1-4]. The components were integrated into the first of two flight dewars which was recently tested. The performance of this dewar along with associated electronic boxes and software is described in this paper. The objectives of this first performance test were to measure the parasitic heat leak into the dewar, the gas flow impedance of the vent lines, the impedance to liquid flow of the transfer line, the efficiency of a transfer, and to evaluate the accuracy of the flowmeters and mass gauging system. Due to the fact that the test was conducted in one gravity, no information was expected on the performance of the liquid acquisition devices.

SHOOT has two dewars each consisting of a 207 liter liquid helium tank surrounded by two vapor cooled shields (VCSs), all housed in a vacuum shell. The VCSs and liquid tank are suspended from the vacuum shell by 6 fiberglass-epoxy straps[5]. The low flow vent line is anchored to the inner vapor cooled shield (IVCS) and the outer vapor cooled shield (OVCS) to provide the vapor cooling. A high flow vent line, which is much straighter, shorter and of larger diameter (1.80 cm inner diameter versus 1.17 cm for the low flow vent) is weakly anchored to the VCSs through stainless steel brackets, provides a low impedance vent path to be used during transfer operations. The transfer line which connects the two dewars is also only weakly anchored to the VCSs to minimize heat input to the transferred liquid while keeping the parasitic heat leak to an acceptable level. Over 400 manganin and fine copper wires enter the dewar main shell and terminate on or in the liquid helium tank.

HEAT LEAK

The normal vent path out of the dewar during ground operation and standby between transfers is through the low flow vent line. This stainless steel line is interrupted by two copper sections which are anchored to the two VCSs. These relatively short sections provide nearly perfect heat exchange at the low flow rates at equilibrium. The high flow vent is valved off for heat leak tests.

A thermal model using the SINDA program was developed. This model predicted a heat leak in the range of 126 mW with a main shell at room temperature (296 K). A schematic is shown in Figure 1. The actual parasitic heat at 4.2 K measured by repeated weighing of the dewar, by vented gas rate, and by superconducting level detector readings was 152± 6 mW. Excellent agreement was obtained between the calculated volumes obtained from liquid level measurements and the weighed liquid and gas. Note that one must make a substantial correction for the gas remaining in the dewar after boiling when at a temperature around 4.2 K. The saturated vapor has a density of .01649 g/cm^3 compared to 0.1254 g/cm^3 for the liquid. Thus as the liquid boils, 13% remains in the dewar, instead of venting.

The first test of the flight dewar culminates in a qualification vibration. Low temperature accelerometers are mounted on the cryogen tank to monitor this vibration and verify the structural model. These accelerometers require special vacuum feedthroughs out of the main vacuum shell. These feedthroughs were found to leak at a combined level of 10^{-3} standard cubic centimeters per second (sccs). This leak, which will be eliminated when the dewar is warmed after vibration, is a likely contributor to the higher than expected heat load. While this much air directly condensed onto the liquid helium tank only will release 0.7 mW, the air must pass through a number of layers of MLI on its way to the tank. This results in a residual pressure especially between the OVCS and the cryogen tank main shell. This pressure was measured to be 0.03 Pa by a cold cathode gauge just inside the main shell. This extra heat load on the OVCS would result in a higher temperature, leading to a higher heat load to the IVCS and ultimately to the cryogen tank. One other difference in the first dewar test is the presence of atmospheric helium in the high flow vent line. Conduction in the helium gas in this line could be as much as 10-20 mW. This line would normally be valved off by Valve D in Figure 2 preventing direct communication with the liquid in the dewar. This drawback is very significant when the dewar is placed on its side - the configuration it will be in during all activities while integrated with the carrier up to a couple weeks before launch. In the horizontal orientation, the low temperature end of the high flow vent is higher than the warm end of the tube. This leads to a severe convection problem. The heat leak in this orientation is about 2.2 W! This considerably limits the hold time in this position. Once again, however, the flight configuration will have this line valved off and it can even be evacuated to eliminate all convection.

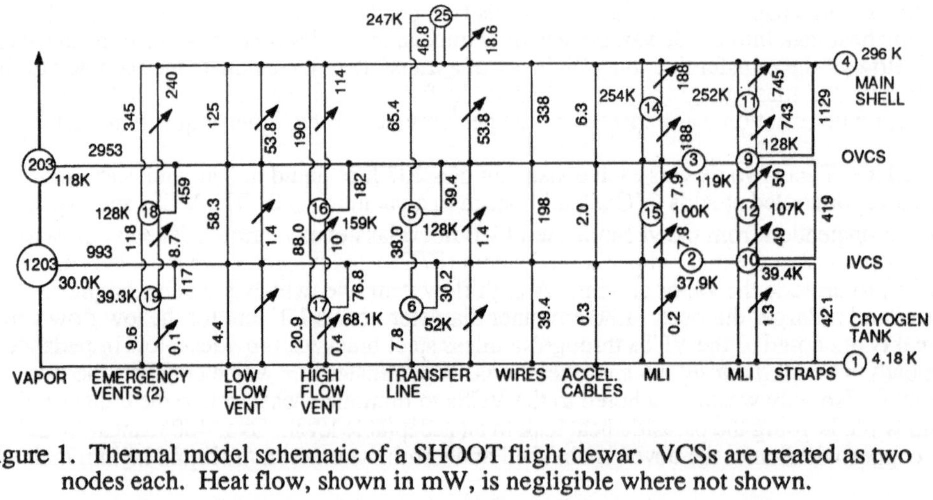

Figure 1. Thermal model schematic of a SHOOT flight dewar. VCSs are treated as two nodes each. Heat flow, shown in mW, is negligible where not shown.

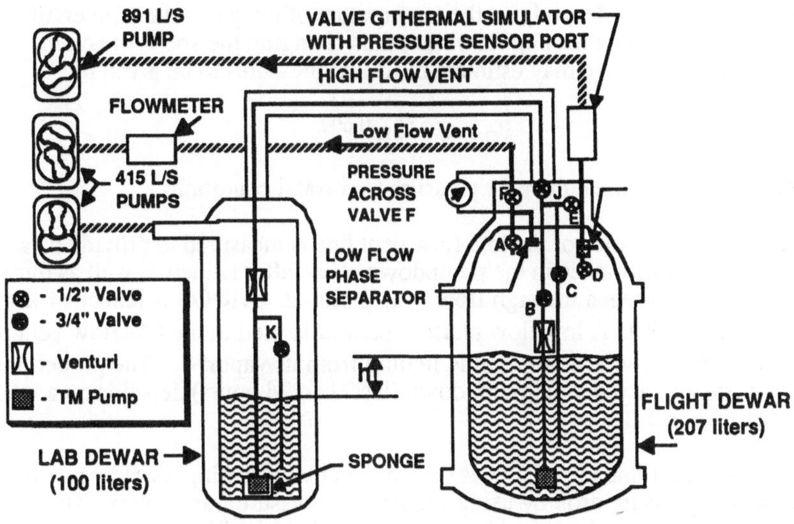

Figure 2. Schematic of the SHOOT dewar performance test apparatus.

VENT LINE IMPEDANCES

During transfers large amounts of heat are dissipated in the supply dewar. To maintain a low temperature during a transfer, a low impedance vent path is required. On SHOOT the high flow vent serves this purpose. To predict the transfer performance on orbit it is important to measure the impedance of this line. The primary reason that Valve D is not in place in the high flow vent is to measure the impedance of the vent itself to gas flow. In normal operation in space Valve D and the high flow phase separator in series with the high flow vent line will be immersed in liquid. The main pressure drop will be in the gas venting out the high flow vent. On the ground the liquid is below the valve and phase separator making these components have a much higher pressure drop than the vent line.

The impedance was measured by flowing helium at a known rate through the vent while measuring the pressure at the inlet (cryogen tank) and outlet. (Refer to Figure 2.) The flow is achieved by heating the superfluid within the cryogen tank at a measured rate. The boil off gas passes through either the low flow vent where it is measured, or out the high flow vent. The flow rate out the high flow vent is therefore the boil off rate minus the flow out the low flow vent. At constant heater power, the flow rate and temperature of the superfluid in the dewar come to equilibrium and one data point is taken. Some of the data are repeated with a heater on the outlet of the high flow vent line simulating the heat input from the uninsulated globe valve at this point. Various powers are used to test the sensitivity of the overall impedance to this parameter. The temperature of the exiting gas was measured using a Platinum Resistance Thermometer (PRT) and Germanium Resistance Thermometer (GRT).

Pressure and flow data were obtained for flows between 0.05 and 0.9 grams per second. The data were fit with the following:

$$P_t^2 - P_o^2 = 0.4746\dot{m}^{1.091} \tag{1}$$

where P_o is the back pressure at the outlet of the vent and is controlled on the ground by the pumping speed of the vacuum pump, P_t is the pressure in the cryogen tank, both in kPa, and $\dot{m}$ is the flow rate in grams per second.

On orbit, the impedance in the high flow vent due to the finite pumping speed of the vacuum pump will not be present. To correct for this the temperature of the exiting gas is computed from the vent rate and expected parasitic heat input, and by assuming that in space choked flow exists at the exit. In choked flow the gas velocity at the outlet is at the speed of sound. The speed of sound depends on the temperature of the gas. The temperature of the exiting gas decreases with increasing flow rate, thus changing the speed of sound. From our data and the above analysis we may estimate the outlet pressures to be given by:

$$P_o = 0.259\dot{m}^{0.607} \tag{2}$$

and using (1) we may calculate the tank pressure for orbital conditions.

Similarly, the impedance of the low flow vent line is measured to provide data for the prediction of dewar performance in the pumpdown phase after launch as well as during standby periods on orbit when the high flow vent is closed. SHOOT is launched with liquid above the lambda point[6-8]. A low flow phase separator is used in the low flow vent line to phase separate normal as well as superfluid helium from its vapor[4,7]. The proper phase separation performance and a rapid pumpdown (less than 24 hours) depends on a relatively low impedance vent line.

The low flow vent line impedance is determined by measuring the flow rate using a gas flow meter and the pressure drop by using Baratron™ pressure transducers. The pressure transducers' taps were located just inside and just outside Valve F on the warm end of the low flow vent. The majority of the pressure drop in the line was expected to occur there. In addition, another Baratron™ was reading the pressure in the high flow vent, which, when there is no flow in this line, measures the pressure in the cryogen tank. At temperatures below the lambda transition, the tank temperature is used to determine the saturated vapor pressure.

The majority of the pressure drop occurred across Valve F. Using data for pressures below 7 kPa and flows above 30 mg/s the following relation was obtained for the pressure across Valve F:

$$P_i^2 - P_o^2 = -0.156 + 12.9\dot{m} + 22.7\dot{m}^2 \tag{3}$$

where P_i and P_o are the valve inlet and outlet pressures in kPa and $\dot{m}$ is the mass flow rate in grams per second. This will prove too high an impedance to obtain rapid pumpdown on orbit[7,8], and this valve will be replaced with one of negligible impedance in the refurbishment of this dewar. For flows between 10 and 50 mg/sec. the remainder of the low flow vent pressure drop is fit by

$$P_t^2 - P_i^2 = 0.128\dot{m} + 30.8\dot{m}^2 \tag{4}$$

where P_t is the tank pressure. This impedance is very low compared to the pressure drop expected across the low flow phase separator except at the lowest flow rates[7].

TRANSFERS

A laboratory dewar and engineering unit transfer line were combined with the flight dewar to allow superfluid transfers to be demonstrated. (See Figure 2). The transition between the transfer line and the flight dewar was made using a vacuum insulated connection. Both the transfer line and coupler used had a higher heat leak than that expected in the flight transfer line and dewar couplers (10W versus about 5 W) so transfers in this set up had larger losses, particularly at low flow rates. Vacuum pumps were hooked up to the laboratory dewar, the flight dewar low flow vent line and the high flow vent line. Transfers could be made in either direction: from the flight to the lab dewar and vice versa. The transfer line is vacuum jacketed from the thermomechanical pump in the flight dewar to the venturi flow meter just above Valve K in the lab dewar. The transfer line is nominally 1.17 cm inner diameter and approximately 5 meters long. The center 2 meters of the line is made of nominal 1.3 cm inner diameter bellows. The line is blanketed with varying numbers of MLI layers. The transfer path from the flight dewar to the laboratory dewar starts at the TM pump and extends up the line through the venturi flow meter and Valve B, through Valve J, the external transfer line and into the lab dewar through Valve K. A transfer from the lab dewar goes up from the TM pump through the external transfer line, and into the flight dewar

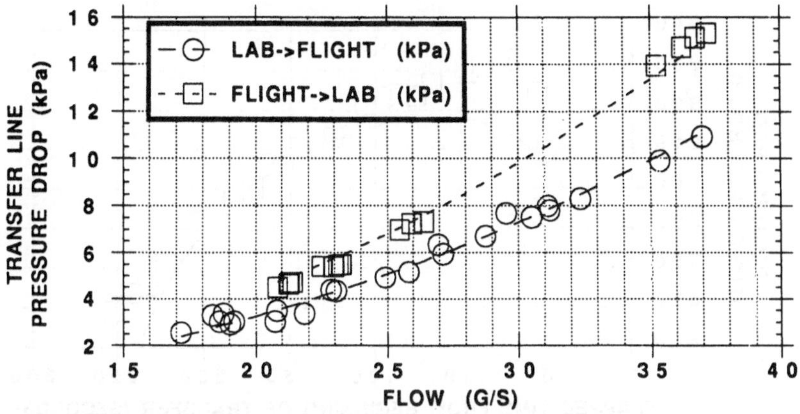

Figure 3. Pressure drop in the transfer line as a function of flow rate. The two sets of data for flows in opposite directions differ due to the increased impedance of Valve K. The lines represent fits to the data based on fully turbulent flow.

through Valves J and C. During precool Valve C is closed and the cold gas is routed through Valve E and out the high flow vent. Even though the valves in the transfer line are the larger, lower impedance version, the majority of the impedance in the transfer line is calculated to be in the valves.

The pressure drop in the transfer line was calculated from measurements of the TM pump pressure, supply and receiver dewar pressures and hydrostatic head difference between the two dewars (h in Figure 2). The results are shown in Figure 3. The dashed lines are best fits to the function

$$\Delta P = A\dot{m}^2 \tag{5}$$

The fits are very good, indicating that the flow is fully turbulent in these transfers. There is a difference between the two directions of transfer mainly due to the extra valve in line when transferring from the flight dewar to the lab dewar. In this direction there are three valves (B, J, and K in Figure 2), as opposed to two valves (J and C) when transferring in the other direction. In addition Valve K has a smaller orifice and higher impedance than the other 3/4 inch valves. With this in mind, the measured pressure drop is within 25% of the expected pressure drop in the transfer line and valves, calculated and extrapolated from room temperature flow impedance measurements respectively.

The lab dewar experienced a high heat leak under certain conditions. In the temperature range above 1.6 K heat leaks of up to 20 W were measured during pumpdown to operating temperatures. This problem was traced to a resonance (not a thermoacoustic oscillation) in the lab cryostat driven by the roots pump through the gas in the vent line. Hence at low pressures the effect is greatly diminished. This problem led to greatly reduced overall efficiency in the transfer process. Still, the overall losses per transfer ranged from 5 to 10% depending on flow rate and supply dewar temperature.

A very simple model of the transfer process appears to work very well, where the transfer rate is calculated from the heat input to the TM pump heater, the impedance of the transfer line, and the pressures in the supply and receiver dewars. This model has also been adapted to work in transient situations where the temperatures (and hence pressures) of the supply and receiver dewars are allowed to vary over time due to heat inputs from the pump and cooling through the vent lines. These changes in turn cause the flow rate to change similarly to that observed in these transfer tests. An example of the flow rates calculated in this way is given in Figure 4. The calculation is performed for one of the flight dewar to lab dewar transfers using equations (1), (3-5), and (8) and the initial conditions as measured. The flow rate as determined by the TM pump flow meter (see next section) is also shown. The calculated curve *has no adjustable parameters*.

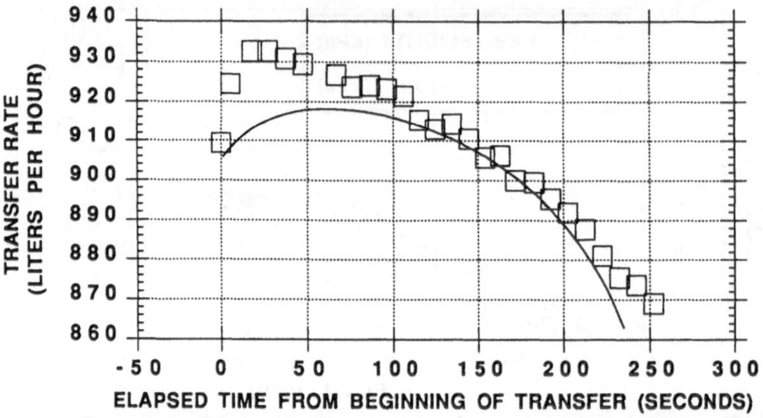

Figure 4. Transfer rate as a function of time for a typical flight dewar to lab dewar transfer. The solid line is calculated from a simple model of the transfer process using the actual initial conditions. The data are flow rates calculated using equation (8) and actual TM pump temperatures. The decreasing flow rate is due to the inability of the laboratory vacuum pump to keep the temperature low. This did not occur for flow rates of 600 liters per hour or less.

TEMPERATURE MEASUREMENT, MASS GAUGING AND FLOW METERING

A problem involving GRT calibration shift was discovered during testing. The flight dewar cryogen tank contains 8 GRTs immersed in the liquid that were read out often. All of these GRTs were calibrated both in a vacuum and immersed in superfluid. Seven of eight of the GRT crystals were suspended in copper barrels that were not sealed, but allowed the liquid or vapor to come in intimate contact. The other GRT was in the more traditional sealed container with gaseous helium. Upon cooldown of the flight dewar it was noted that disagreements of up to 7 mK occurred between sensors even though they were all immersed in an isothermal superfluid bath. A calibration check was performed by allowing the dewar to drift through the lambda point at 2.1768 K. The only thermometer which read the correct temperature was the sealed one. All others deviated by amounts ranging from 1 to 7 mK both high and low. The readings were made both with a laboratory readout which uses a DC technique and the flight electronics which uses an AC technique[9]. Both systems agreed to within 0.1 mK and both accurately read out fixed resistors to better than 0.01%. The problem appears to be related to the crystal being exposed to air for a period of about one year. The mechanism for this calibration shift, which was repeatable over the month long testing at superfluid temperatures, is not known. This shift, which is very regular in temperature, can either be calibrated out by in situ measurements, or the GRTs may be replaced with the sealed variety.

The heat pulse mass gauging technique was used a few times in the lab and flight dewars. Typical heat pulse responses are given in Figure 5 (a) and (b). The accuracy is better than 3%. The inaccuracy in measurements like 5 (b) is mostly due to thermal stratification of the gas in the ullage. The correction due to this gas varies in magnitude with temperature and fill level; the higher the temperature and lower the fill, the larger the correction. For the heat pulse shown in 5 (a), the correction term represents 2.4% of the total. The liquid volume is given by:

$$V_L = \frac{(1+(1-\varepsilon))}{2} \frac{\{Q - V_T[(\rho_{gf}H_{gf} - \rho_{gi}H_{gi}) - (\rho_l H_{li} - \rho_{gi}H_{gi})\delta]\}}{(\rho_l H_{lf} - \rho_{gf}H_{gf} - (1-\varepsilon)(\rho_l H_{li} - \rho_{gi}H_{gi}))}$$

with

$$(1-\varepsilon) = \frac{\rho_l - \rho_{gf}}{\rho_l - \rho_{gi}} \quad \text{where } \varepsilon << 1 \quad \text{and} \quad \delta = \frac{\rho_{gf} - \rho_{gi}}{\rho_l - \rho_{gi}} \quad \text{where } \delta << 1$$

where V_L is the liquid volume, V_T is the tank volume, ρ_l is the liquid density, ρ_g is the gas density, Q is the total heat applied, H_l is the liquid enthalpy, and H_g is the gas enthalpy. The

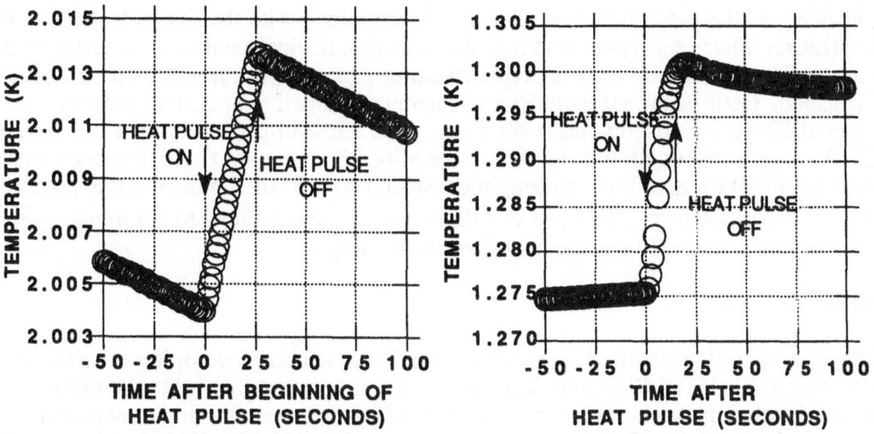

Figure 5. Typical mass gauging results. (a) Typical sharp equilibration time. (b) Typical results when heater is not immersed in liquid.

subscripts i and f stand for before and after the heat pulse. While this equation looks complex, it really only relies on a lookup table for liquid and gas density and enthalpy at saturated conditions. Note that when a heater which was not immersed in the liquid was used to provide the heat pulse, the heat soaked into the liquid over a few seconds and the sharp step was not observed as in 5 (b). This made the correction for temperature drift more difficult. Such a problem was not observed in the COBE ground test due to use of the steady state heat flow technique[10]. This problem should not be present in orbit since it is expected that a thick helium film will coat the walls where the heaters are placed.

Two types of flow meters are to be used on SHOOT. A venturi will be used immediately downstream of the pump. The venturi is designed and positioned so that the potential for cavitation will be minimized. The venturi pressure drop is given by:

$$\Delta P = Bv^2 + Cv \tag{7}$$

where C is the loss term which is negligible in our case. The venturi has a tapered portion downstream of the throat to recover approximately 85-90% of the pressure drop. The pressure drop measured across the two venturis followed equation (7) to within 2% which is the accuracy of the flow measurement by other means.

The second type of flowmeter is unique to superfluid being pumped by a TM pump. The temperature just downstream of an ideal pump follows the following relation[4]:

$$\dot{m} = \frac{\dot{Q}}{TS} \tag{8}$$

where $\dot{m}$ is the mass flow rate, T is the temperature on the warm side of the pump, S is the entropy of superfluid at that temperature, and Q is the heater power applied to the TM pump. The entropy is a well measured, tabulated quantity. GRTs accurate to one mK (except for the previously noted problem) are used to measure the downstream temperature. The TM pumps used in SHOOT have nearly ideal thermal behavior, thus equation (8) should hold. The TM pump flowmeter results were integrated and compared with volume readings obtained from the liquid/vapor discriminators and the superconducting level detectors. The volume readings are adjusted for the losses due to boil off during the transfer which is typically a few liters per hour. The agreement of the TM pump flowmeter and the volume measurements is within 1.5%, which is within the error in temperature and volume measurements.

FUTURE TESTING

The first SHOOT dewar has completed a series of performance tests. These tests successfully demonstrated the operating characteristics of the TM pump, the flowmeters, valves, heaters and thermometers as a complete system along with the flight electronics and ground software. Phase separators and liquid acquisition devices could not be tested at the system level. The dewar will undergo a qualification vibration test. It will then be warmed and refurbished. Refurbishment includes replacement of failed temperature sensors, installation of low and high flow phase separators, and a low impedance vent valve to replace Valve F. After refurbishment, this dewar will be joined by a second flight dewar and tested along with the flight transfer line. Among the tests to be performed will be a low pressure topoff and stratification test[4]. This test will demonstrate ideas on how to get longer ground hold times before launch without the use of vacuum pumps.

ACKNOWLEDGMENTS

The authors wish to thank S. Sutherland and C. Mosier for developing the SINDA thermal model of the SHOOT dewar. The authors are especially grateful to the group of engineers and technicians who assembled the SHOOT dewar system with great care allowing the first test to be so successful. This work is supported by NASA's Office of Space Flight.

REFERENCES

1. M.J. DiPirro, E.R. Quinn, and R.F. Boyle, Tests of a nearly ideal, high rate thermomechanical pump, "Proc. of ICEC 12", p. 646, Butterworths, Southampton, UK (1988).
2. M.J. DiPirro and A.T. Serlemitsos, Discrete liquid/vapor detectors for use in liquid helium, "Adv. Cryo. Eng." Vol. 35, (1990) p. 1617.
3. M.J. DiPirro, Liquid acquisition devices for superfluid helium transfer, Cryogenics Vol. 30, (1990), p. 193.
4. M.J. DiPirro, et al., The SHOOT cryogenic components: testing and applicability to other flight programs, "Proc. SPIE 1340 Cryogenic Optical Systems and Instruments IV", 1990, p. 291.
5. M.J. DiPirro, et al., The SHOOT cryogenic system, "Superfluid Helium Heat Transfer", J.P. Kelly and W.J. Schneider, eds., ASME HTD-Vol. 134, (1990), p. 29.
6. M.J. DiPirro, D.C. McHugh, and J. Zahniser, Phase separators for normal and superfluid helium, "Proc. of ICEC 12", Butterworths, Southampton, UK, (1988), p. 681.
7. P.J. Shirron and M.J. DiPirro, A liquid/gas phase separator for He I and He II", to be published in "Adv. Cryo. Eng." Vol. 37.
8. M.J. DiPirro and P.J. Shirron, The SHOOT orbital operations, to be published in Cryogenics.
9. C.E. Woodhouse, High precision, rapid readout of cryogenic temperature sensors in the space shuttle environment, IEEE Trans. Instr. and Meas. Vol. 39, (1990), p. 279.
10. S.M. Volz, M.J. DiPirro, and M.G. Ryschkewitsch, A calorimetric mass gauge system for the Cosmic Background Explorer (COBE), "Adv. Cryo. Eng." Vol. 35, (1990), p. 1703.

DETAILED MODELING OF THE NO-VENT FILL PROCESS

D. H. Beekman and T. A. Martin

Civil Space & Communications Systems
Martin Marietta Astronautics Group
Denver, Colorado

ABSTRACT

An analytical model has been developed to simulate the process of filling a non-vented cryogenic tank. The analysis considered data gathered during no-vent fill testing with LH_2 of a 538 L (19 ft^3), 69 kg (152 lb_m), aluminum tank. The model consists of single liquid and vapor nodes and multiple wall, insulation, and penetration nodes and uses the initial tank wall and fluid temperature distribution, the inlet fluid pressure and temperature, and the inlet line flow resistance from test data. The model simulates several thermal and thermodynamic interactions, including heat and mass transfer phenomena at the bulk liquid/vapor and spray/vapor interfaces and boiling heat transfer on the tank wall. The model adequately predicts the tank pressure, liquid fill level, and inlet flow rate histories when loading through the fill and drain line. When the axial spray system was activated, however, the model was found to be deficient. Reasons for the deficiency are postulated. The paper summarizes the significant portions of the model, compares simulation results with experiment data, and presents an evaluation of uncertainties in several parameters. The sensitivity analysis indicated that simulation results are quite sensitive to inlet fluid temperature.

INTRODUCTION

The transfer of fluids between tanks is straightforward in the presence of a settling acceleration because vapor can be readily vented as a consequence of the known liquid position. Back pressure relief is necessary to sustain the fluid transfer and is particularly important when dealing with cryogens because the fluid transfer is typically accompanied by rapid boiloff in the receiver tank which produces an increase in receiver tank pressure. In the absence of a settling acceleration, as during an orbital fluid transfer operation, the fluid transfer process is more complicated because surface forces may dominate body forces, resulting in an uncertain liquid position and decreased probability of venting vapor. One method to overcome this difficulty is to thermally condition the receiver tank to limit the back pressure buildup and then simply fill the receiver tank while its vent is closed, assuming that non-condensable vapors are not present. The essential part of the no-vent fill is condensation of the vapor so that the receiver tank pressure does not increase to a level which would inhibit the fluid transfer process. Several approaches using atomizing sprays and liquid positioning jets have been proposed to promote vapor condensation, and numerous experimental investigations have evaluated the effectiveness of the no-vent fill procedures[1,2,3]. Martin Marietta Astronautics Group investigated the no-vent fill process under an Independent Research and Development project to evaluate the sensitivity of fill level to variations in various parameters[4]. This paper presents the results of an analysis of the Martin Marietta no-vent fill experimentation using the integrated fluid and thermal cryogenic analysis program

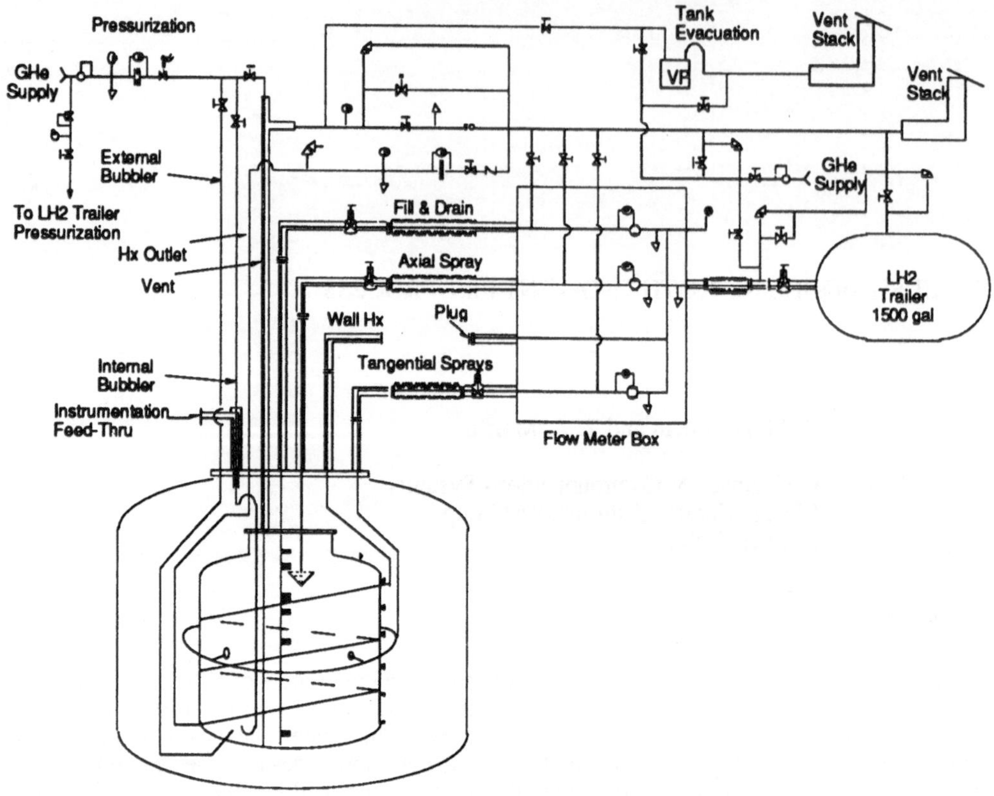

Figure 1 No-Vent Fill Test Apparatus Schematic

MMCAP (Martin Marietta Cryogenic Analysis Program), a general purpose cryogenic analysis program which has been used extensively on the Cryogenic Onorbit Liquid Depot Storage and Transfer (COLD-SAT), Cryogenic Orbital Nitrogen Experiment (CONE), and Space Transfer Vehicle (STV) projects. An overview of the test program, modeling approaches, and simulation results are presented below.

TEST PROGRAM OVERVIEW

The test apparatus, illustrated in Figure 1, consisted of a receiver tank placed in a vacuum chamber. The tank was suspended from the the vacuum chamber lid which also provided the fluid line penetrations to the receiver tank. Outside the chamber, the transfer lines were connected to an evacuated flowmeter enclosure which, in turn, was connected to the LH_2 supply trailer. The apparatus permitted two types of no-vent fill operations to be performed. The method most commonly used was to fill the tank from the bottom with the downward pointing fill/drain line. The second method of filling the tank was to inject fluid into the ullage region with an atomizing axial spray fixture located at the top of the receiver tank. The tank wall initial temperature was varied by performing charge-hold-vent cycles until the desired temperature was achieved. A total of 14 tests were performed in 1990, and a second series of tests was performed in 1991. Details of the apparatus design and testing are presented elsewhere[4].

MODELING APPROACH

The analysis was performed to reconstruct test results and also to validate a general purpose no-vent fill model. Therefore, the analysis was performed with the intent of minimizing the number of specific code modifications required to match the test results. The receiver tank was modeled as a fluid and thermal network with several boundary and initial conditions as summarized in Figure 2. The principal initial conditions were the thermal

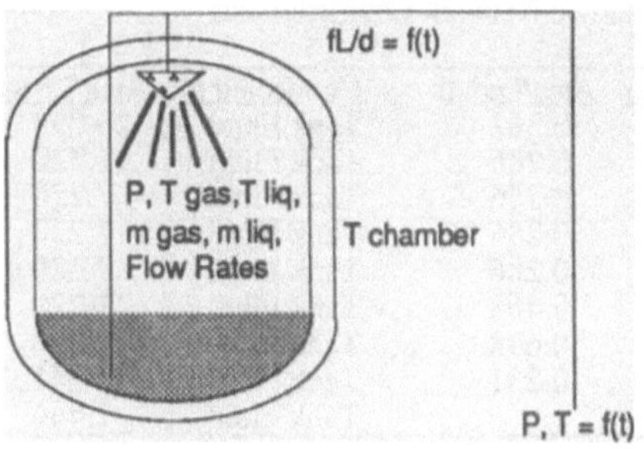

Figure 2 Simulation Overview

network temperatures, fluid masses, and tank pressure. The principal boundary conditions were the pressure and temperature of the inlet liquid and the flow resistance of the fill & drain and axial spray lines. Details of the significant models are described below.

Thermodynamics and Fluid Properties

The thermodynamic models assume that the liquid and vapor portions of the tank behave as a simple compressible substance. This is a non-restrictive assumption which permits the same differential equations for state, conservation of energy, and enthalpy to be used for either liquid or vapor. The modified Benedict-Webb-Rubin equation of state was used to produce all fluid properties derivable from the equation of state[5]. Transport properties were calculated from curvefits which are accurate to within 5% of published properties. The thermodynamic models were solved simultaneously with a 4-term Runge-Kutta method. The combination of thermodynamic models which account for non-ideal behavior, accurate fluid properties, and an efficient solution method minimize thermodynamic solution uncertainties.

Thermal Models

Interfacial heat and mass transfer is a significant parameter in modeling the no-vent fill process because vapor condensation on the liquid surface is necessary to inhibit the buildup of receiver tank back pressure. The interface was modeled as a massless control volume separating the liquid and vapor portions of the tank. Conservation of energy at the interface yields the following expression for interfacial mass transfer.

$$\dot{m}_e = \frac{q_{gs} - q_{sl}}{h_{fg}}$$

The relative magnitudes of heat transfer from the vapor to the surface, q_{gs}, and surface to liquid, q_{sl}, determines if evaporation ($dm_e/dt > 0$) or condensation ($dm_e/dt < 0$) will occur. Thus, the mass transfer at the interface is governed by the heat transfer rates from the ullage to the surface and from the surface to the liquid. Free convection heat transfer was assumed for both heat transfer rates at the interface. A model for a submerged jet impinging on the liquid surface was not included because the fill and drain line points toward the bottom of the tank. The following flat plate free convection heat transfer correlation was used to describe the gas to surface and surface to liquid heat transfer processes[6].

$$Nu = 0.27\ Ra^{0.25}$$

Heat and mass transfer to spray droplets is another process to promote vapor condensation by enhancing heat transfer from the vapor to the liquid. The spray system was designed to produce 1.2 mm (0.0475") diameter droplets by ensuring that the stream velocity exceeded the minimum value for liquid column atomization. The spray Reynolds number was calculated from the droplet terminal velocity, which was generally less than 0.7 m/s. A height volume table was included in the model so that the droplet residence time in the ullage region based on droplet velocity and distance could be calculated. Spray heat and mass

Table 1 Tank Thermal Network Characteristics

Node	Mass (kg)	Connected to	Area/Length	Connected to	Area/Length
Tank 1	13.1	Tank 2	0.362	Tank Fluid	25.098
Tank 2	8.5	Tank 3	0.286	Tank Fluid	7.720
Tank 3	8.5	Tank 4	0.286	Tank Fluid	7.720
Tank 4	8.5	Tank 5	0.286	Tank Fluid	7.720
Tank 5	8.5	Tank 6	0.286	Tank Fluid	7.720
Tank 6	8.5	Tank 7	0.463	Tank Fluid	7.720
Tank 7	10.8	Tank 8	0.698	Tank Fluid	9.026
Tank 8	2.3	Tank 9	0.231	Tank Fluid	2.941
Tank 9	0.8			Tank Fluid	7.446

transfer was modeled by considering heat transfer from the ullage to the droplet surface and from the droplet surface to the interior. The net condensation or evaporation rate at the droplet surface was based on the average heat transfer rates during the droplet residence time in the ullage. An analysis of transient spherical conduction by Brown[7] was used to describe heat transfer from the surface to the droplet interior. The following correlation was used to describe heat transfer from the ullage to the droplets[8].

$$Nu_d = 2 + \left(0.4\, Re_d^{0.5} + 0.06\, Re_d^{0.67}\right) Pr^{0.4} \left(\mu_\infty/\mu_s\right)^{0.25}$$

The receiver tank volume was 538 L (19 ft^3), weighed 69 kg (152 lb$_m$), and was fabricated from aluminum. The receiver tank wall was modeled with 7 nodes corresponding to the 7 thermocouple locations on the test article plus 2 nodes for the top flange, resulting in a total of 9 wall nodes. The wall nodes were thermally connected axially with each other and radially to the tank contents and the 15 layer multilayer insulation (MLI) blanket located on the outside surface. The MLI blanket was radiatively connected to the vacuum chamber wall. All fluid transfer lines, cabling, and support straps were modeled as conductors between the vacuum chamber and receiver tank walls. The conductors between the wall and fluid nodes were automatically adjusted as the liquid level increased. The simulation immediately covered the bottom dome of the tank with liquid after filling began because the test data indicated a rapid quench of the bottom dome, presumably as a consequence of flashing and splattering of the incoming superheated flow stream. All other wall nodes were covered with liquid when the liquid reached the appropriate level. Table 1 summarizes the conduction and convection characteristics of the tank thermal network. The first two columns present the node name and mass and the following four columns summarize the nodes and geometries for the nodes connected to the entry in the first column.

Fluid Transfer Models

Fluid transfer was simulated with input flow resistances, which can be specified as functions of time, and the pressure difference between the supply line and the receiver tank. Flow resistance may be expected to remain relatively constant with single phase flow but may vary widely with transient two-phase flow. Two-phase flow occurred during flow initiation because the receiver tank was initially evacuated to a pressure lower than the saturation pressure of the supply trailer and also because it was not possible to pre-chill the transfer line downstream of the flow meter box. Two-phase flow may also have occurred near the end of the tests because the degree of subcooling decreased as a consequence of the decaying flow rate as the pressure difference across the transfer line decreased. Instances of two-phase flow at the start and end of the runs were evident from the flow meter data. The initial variation in flow characteristics was accommodated in the model by ramping the flow resistance down to a steady state value during the first minute of the simulation. The steady state value was determined by averaging the calculated flow resistances for several tests, neglecting the flow resistances corresponding to two-phase flow. Figure 3 presents the flow resistances as functions of the pressure difference across the transfer line, which indicates three distinct flow resistance groups. The first group occurs when only the fill & drain line was used to fill the tank without any bleed flow to thermally precondition the flow meter. The second group occurs when the bleed flow was used with the fill & drain line. The third

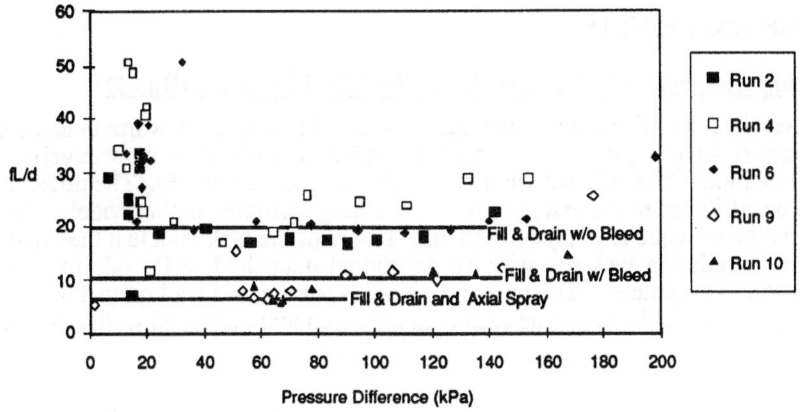

Figure 3 Average Flow Resistance

group occurs when the fill and drain and axial spray systems were used simultaneously. The change in flow resistance when the spray system was activated is understandable because an additional flow path was added to the network. The change in flow resistances for the cases with and without the bleed flow is more difficult to explain. Fill and drain line isolation was accomplished with a manual valve, and the personnel conducting the experiment may have opened the valve to different positions for different experiments. Also, thermal conditioning by the bleed flow may have significantly changed the flow characteristics of the equipment inside the the flow meter enclosure. Regardless of the cause, the change in flow resistance was implemented in performing the analysis, so three resistances were used depending on the simulation case.

The fluid transfer model also requires specifying the temperature of the transfer line fluid so that the enthalpy of the incoming flow stream can be calculated. This parameter was not measured accurately in the first test series, however, so the temperature of the fluid in the transfer line was not known. To overcome this difficulty, it was assumed that data from the second test series could be used to determine the temperatures from the first test series. The transfer line temperature was accurately measured in the second test series, and the results show that the fluid remained nearly saturated for low flows but became increasingly subcooled at higher flow rates as shown in Figure 4. A linear regression of the subcooling was incorporated into the model so that the transfer line temperature could be calculated from flow rate and transfer line saturation temperature. The temperature calculation in the model also decreased the subcooling calculated by the linear regression by 0.8 K to account for heating downstream of the transfer line temperature measurement.

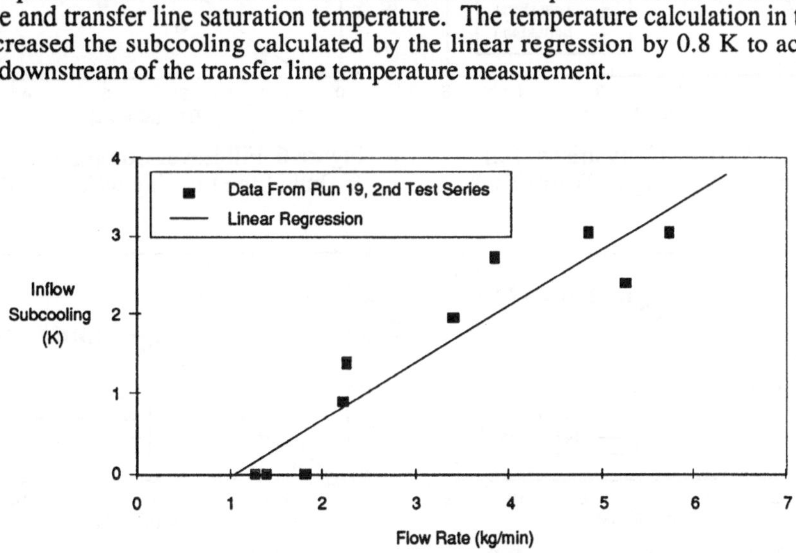

Figure 4 Supply Line Fluid Temperature Adjustment

MODELING RESULTS

Loading Through the Fill & Drain Line - Tank Initially Warm (Run 1)

The results for a fill and drain line loading case with an initially warm tank are shown in Figures 5 and 6, which compare tank pressure and fill level histories, respectively. The tank pressures compare reasonably well and the fill levels agree within 5%. The difference in fill levels can be attributed to the extent of wall quenching permitted in the model. The test data indicate that the wall quench may have included more of the tank than just the bottom dome, and another simulation was run with an additional quenched wall node to evaluate the sensitivity to this parameter. The pressure histories for the two wall quenching cases were quite similar but the fill level for the case with the additional quenched wall was much closer to the test data.

Loading Through the Fill & Drain Line - Tank Initially Cold (Run 4)

The results for a fill and drain line loading case with an initially cold tank are shown in Figures 7 and 8, which compare tank pressure and fill level histories, respectively. The tank pressures and fill levels compare quite well. The computed tank pressure experienced an inflection earlier than was observed in the test, and this may have resulted from the inlet fluid temperature used in the simulation being slightly colder than actually occurred during the test. Comparison with the previous case indicates the importance of thermally preconditioning the tank prior to testing to achieve a high fill level. If the tank wall is relatively warm, then the back pressure will rapidly increase and the flow will stop. These filling simulations using the fill and drain line indicate that the modeling approach without sprays works well.

Loading Through the Axial Spray Line - Tank Initially Warm (Run 3)

The tank pressure histories for a spray loading case with an initially warm tank are shown in Figure 9. Fill level is not shown because essentially no liquid accumulated during

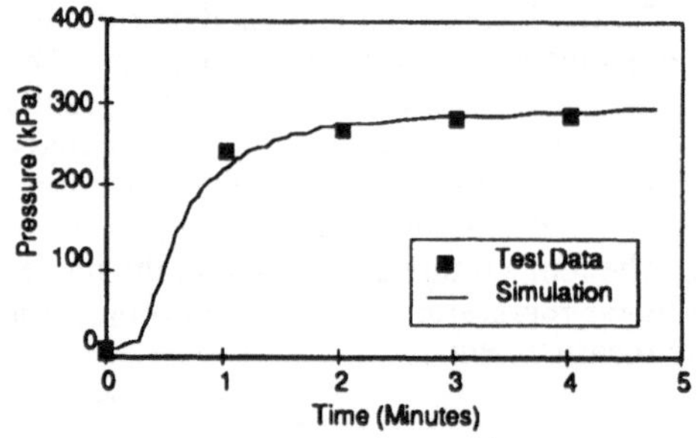

Figure 5 Pressure Comparison, Fill & Drain Line Loading w/ Warm Tank

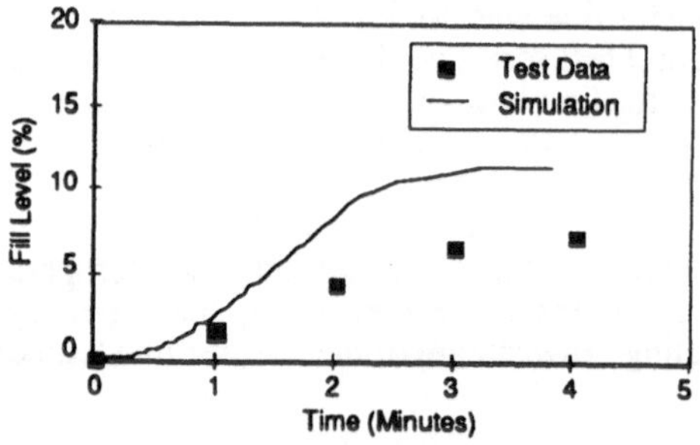

Figure 6 Fill Level Comparison, Fill & Drain Line Loading w/ Warm Tank

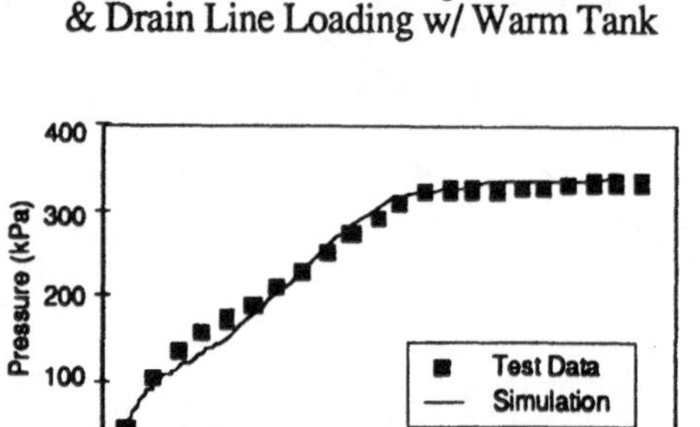

Figure 7 Pressure Comparison, Fill & Drain Line Loading w/ Cold Tank

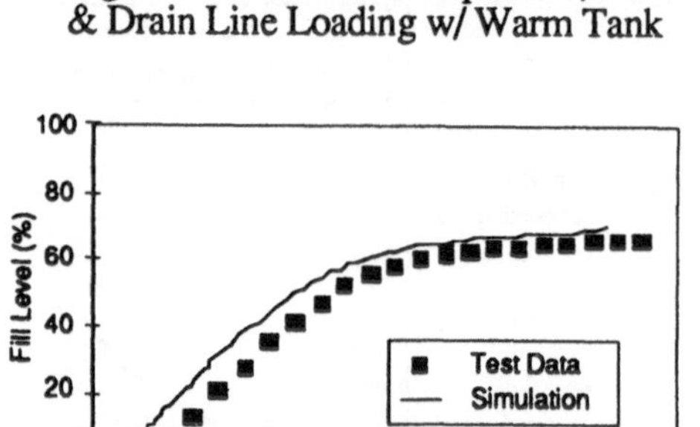

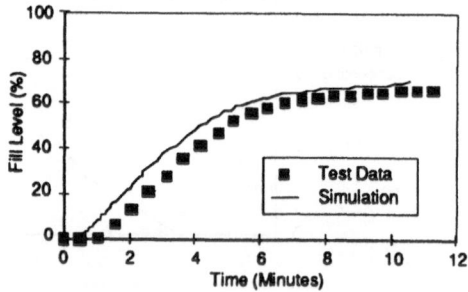

Figure 8 Fill Level Comparison, Fill & Drain Line Loading w/ Cold Tank

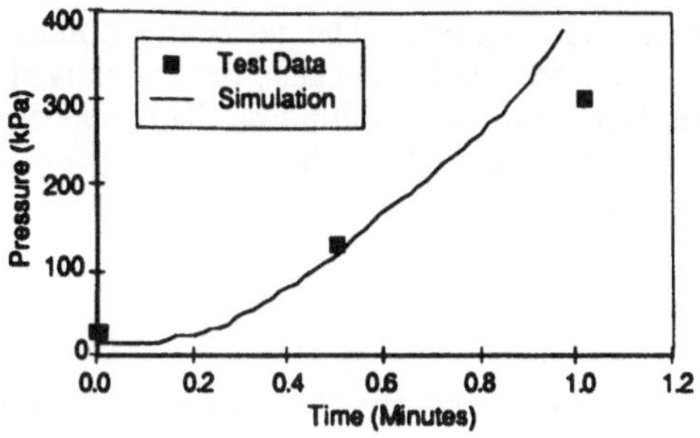

Figure 9 Pressure Comparison, Axial Spray Loading w/ Warm Tank

the fill. The tank pressures compare reasonably well, although the simulation indicates somewhat higher pressure increase at the end of the simulation.

Loading Through the Fill & Drain and Spray Lines - Tank Initially Cold (Run 10)

The results for a fill and drain line loading with axial spray case with an initially cold tank are shown in Figures 10 and 11, which compare tank pressure and wall temperature histories, respectively. The tank pressures compare quite well until the axial spray was activated at 3 minutes. After spray initiation, the predicted pressure initially stabilized as the ullage was cooled but then increased rapidly. As a consequence of the increasing back pressure, the fill rate decreased so that the final predicted fill level was approximately 70% whereas the test data indicated a final fill level of 95%. The cause of the pressure increase was not immediately apparent because the spray model exhibited adequate capacity to cool the ullage and the tank wall as shown in Figure 11. The ullage temperature, which is not shown, closely tracked the wall temperature. Thus, the spray model could not produce adequate condensation to inhibit back pressure buildup even though it could reduce and maintain the ullage temperature . Possible causes for inadequate condensation are discussed below.

The assessment of condensation mechanisms consisted of determining the total condensation rate based on test data and then evaluating the condensation rates at the liquid/vapor interface and to the sprays. The total condensation rate was determined from the differential equation of state for the ullage assuming ideal gas and a constant ullage temperature, yielding the following expression for the total condensation rate as a function of ullage volume, pressure, temperature, and change rates for ullage volume and pressure.

$$\dot{m}_{ct} = -\frac{\dot{P}V + \dot{V}P}{RT}$$

The maximum condensation rate to the spray was determined from a steady state energy balance of the spray from the spray nozzle to the liquid/vapor interface, assuming no heat

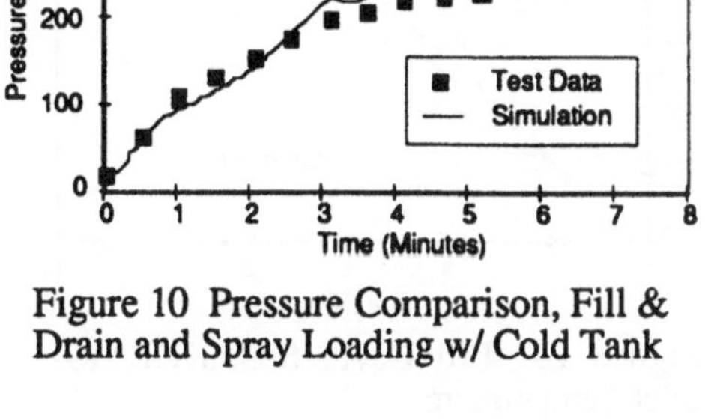

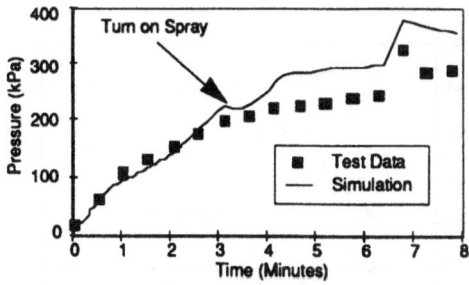

Figure 10 Pressure Comparison, Fill & Drain and Spray Loading w/ Cold Tank

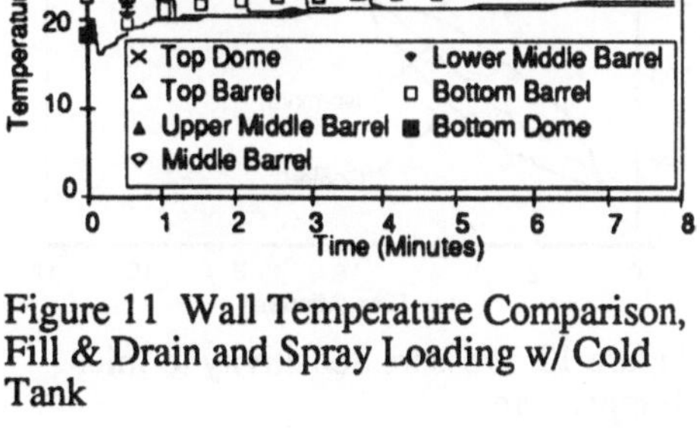

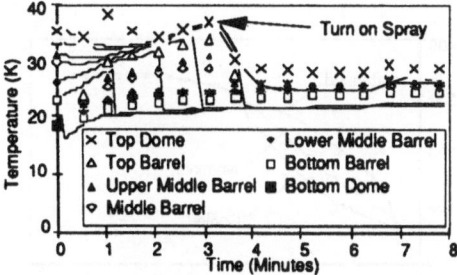

Figure 11 Wall Temperature Comparison, Fill & Drain and Spray Loading w/ Cold Tank

transfer from the ullage to the droplets. The adiabatic condition assured maximum condensation to the spray and is reasonable because the test data show that the ullage was chilled nearly to saturation following spray initiation. This approach yields the following expression for maximum condensation rate to the spray as a function of spray flow rate, subcooling, and fluid properties.

$$\dot{m}_{cs} = \dot{m}_s \frac{C_{pl} \Delta T_l}{h_{fg}}$$

These expressions were evaluated for the following conditions which occurred following spray initiation:

Ullage Volume:	119 L (4.2 ft^3)
Pressure:	224 kPa (32.5 psia)
Temperature:	25 K (45 R)
Volume Change Rate:	-65 L/min (-2.3 ft^3/min)
Pressure Change Rate:	10 kPa/min (1.45 psi/min)
Spray Flow Rate:	1.36 kg/min (3.0 lbm/min)
Spray Subcooling:	0.9 K (1.6 R)

The total required condensation rate was 0.130 kg/min (0.286 lbm/min) and the maximum condensation rate to the spray was 0.028 kg/min (0.061 lbm/min). The simulation indicated a maximum interface condensation rate of 0.021 kg/min (0.046 lbm/min), so the spray and interface condensation rates comprised only 38% of the total required condensation rate. Since no additional vapor could condense on the spray, the additional condensation must have occurred on the liquid/vapor interface. In order to match the required condensation rate, the interfacial rate would have had to increase by at least a factor of 5, probably as a consequence of forced convection due to surface agitation by the sprays. Heat transfer to the spray droplets would diminish the condensation capacity of the spray, however, so the interfacial condensation rate would have to be increased by more than a factor of 5 to account for heat transfer to the spray droplets. It must be emphasized that the necessity for increased mass transfer at the interface assumes the same filling rates and inlet temperatures as used throughout the rest of the analysis. The interfacial mass transfer would not have to be increased as much if the inlet temperature was colder or if the filling rate was lower. A colder spray would have additional condensation capacity whereas a slower fill rate would require a lower total condensation rate. Neither of these possibilities can be disregarded as the source of the modeling discrepancy, but future experimentation may help resolve the uncertainties.

SENSITIVITY ANALYSIS RESULTS

The modeling activity required making several assumptions, and it was desirable to evaluate the sensitivity of the results to uncertainties in several parameters. The first analysis involved an assessment of the sensitivity to uncertainty in inlet liquid temperature. The inlet temperature was not accurately measured during the 1990 tests, so the temperature was

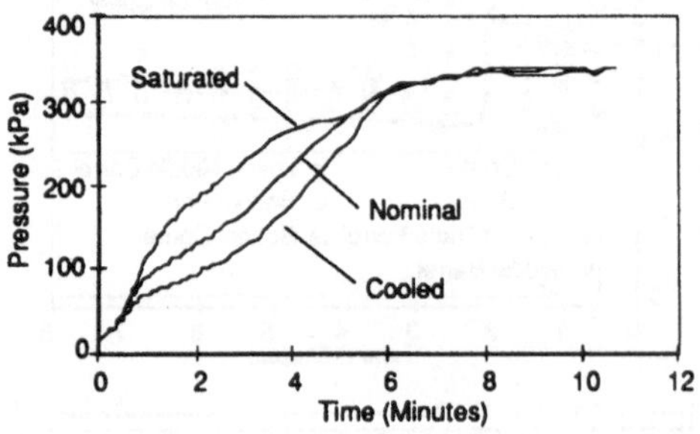

Figure 12 Pressure Sensitivity to Inlet Temperature

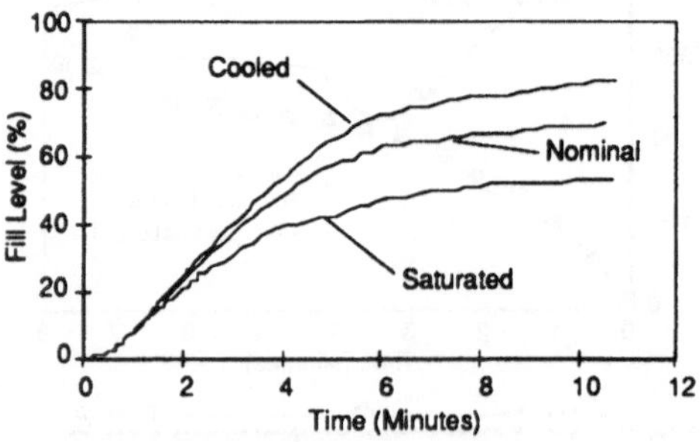

Figure 13 Fill Level Sensitivity to Inlet Temperature

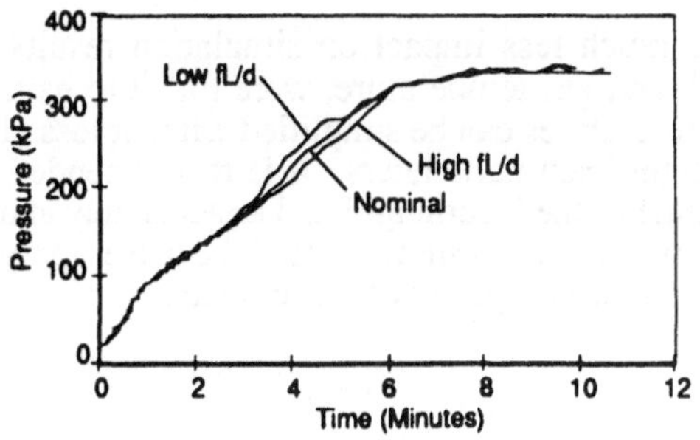

Figure 14 Pressure Sensitivity to Transfer Line Flow Resistance

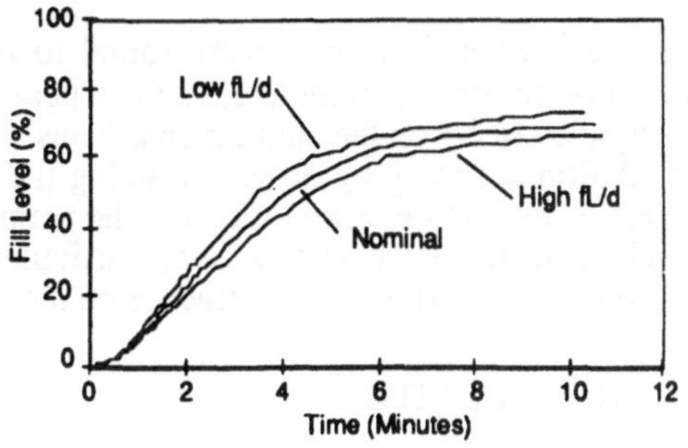

Figure 15 Fill Level Sensitivity to Transfer Line Flow Resistance

inferred from the liquid flow rate as determined in the 1991 tests described above. Two cases were considered. The first case assumed no subcooling of the inlet liquid whereas the second case shifted the linear regression in Figure 4 so that it passed through the origin, reducing the inlet liquid temperature by 1.6 K. The results of this analysis are shown in Figures 12 and 13, which illustrate the effect of inlet temperature uncertainty on tank pressure and fill level. The tank pressure was quite sensitive to liquid inlet temperature, which is consistent with the analytical and experimental results reported by Chato[2,3]. The inlet temperature sensitivity also affected the final fill level because flow rate was proportional to the square root of the pressure difference between the supply line and the receiver tank.

A second sensitivity analysis considered the effect of ±25% variations in transfer line flow resistance. This was done because the assessment of the experiment results revealed an uncertainty in transfer line flow characteristics as illustrated in Figure 3 The results of this analysis are shown in Figures 14 and 15 which compare the pressure fill level histories for the three cases. The tank pressure was minimally affected and the fill level was affected only slightly more. Thus, the simulation results were not very sensitive to uncertainties in the transfer line flow resistance.

Several other parameters were considered in the sensitivity analysis. These included ±11 K initial gas temperature variations, ±4 K initial wall temperature variations, and ± 1 order of magnitude variations in heat transfer from the vacuum chamber wall to the MLI surface. In all cases, the variations were found to have an insignificant effect on simulation results. Large variations in wall temperature, however, have a significant effect on the no-vent fill process as illustrated in Figures 5-8.

CONCLUSIONS

This analysis demonstrates that the no-vent process can be adequately simulated with relatively simple models if the tank is being filled with a fill and drain line pointing away from the liquid vapor interface. In such a situation, free convection at the liquid/vapor interface is adequate to describe interfacial heat and mass transfer. If a spray system is used, however, then the modeling becomes more difficult due to the interaction of the spray droplets, ullage, and liquid/vapor interface. The complexity of the spray modeling was identified in the axial spray simulations which demonstrated sufficient ullage and tank wall temperature reduction capability but inadequate condensation capability. It was found that the spray condensation capability in the model was limited by the degree of inlet spray subcooling, so the deficiency in the total condensation rate was attributed to heat and mass transfer at the liquid/vapor interface, which is the only other location in the system where condensation can occur. The deficiency could also be explained by a greater degree of spray subcooling or a lower filling rate, so the modeling difficulties with the spray system cannot be resolved without further test data. The sensitivity analysis indicated that the dominant parameter in performing a no-vent fill simulation is the temperature of the incoming liquid which, in this analysis, was inferred from subsequent testing. Variations in transfer line flow resistance, however, were found to have much less impact on simulation results. Variation in other parameters, such as heat leak and gas temperature, were found to have negligible effects on the simulations. Thus, future analyses can be simplified without loss of simulation accuracy by de-emphasizing the insignificant parameters. It is recommended, however, that future testing measure the temperature of the incoming liquid as accurately and as close to the tank as possible, particularly if a spray system is used because spray subcooling has a significant influence on the condensation capacity of the droplets.

NOMECLATURE

Variable	Description	Units
C_{pl}	droplet specific heat at constant pressure	kJ/kg-K
fL/d	transfer line flow resistance	
h_{fg}	heat of vaporization	kJ/kg
m_{cs}	maximum condensation mass to the spray droplets	kg
m_{ct}	total required condensation mass	kg
m_e	evaporation mass at the liquid/vapor interface	kg
m_s	spray mass	kg
Nu	Nusselt number for heat transfer at the liquid/vapor interface	
Nu_d	Nusselt number for heat transfer to the spray droplets	
P	ullage pressure	kPa
Pr	Prandtl number of the ullage gas	
q_{gs}	heat transfer rate from the ullage to the interface	kJ/s
q_{sl}	heat transfer rate from the interface to the liquid	kJ/s
R	gas constant for hydrogen	kJ/kg-K
Ra	Rayleigh number at the liquid/vapor interface	
Re_d	Reynolds number of the spray droplet	
T	ullage temperature	K
ΔT_l	spray subcooling	K
V	ullage volume	m^3
μ_∞	ullage gas viscosity, evaluated at bulk conditions	kg/m-s
μ_s	ullage gas viscosity, evaluated at droplet surface conditions	kg/m-s

REFERENCES

1) D. J. Chato, Thermodynamic Modeling of the No-Vent Fill Methodology for Transferring Cryogens in Low Gravity, AIAA Paper 88-3403, (1988).
2) D. J. Chato et al, Initial Experimentation on the Nonvented Fill of a 0.14 m^3 (5 ft^3) Dewar With Nitrogen and Hydrogen, AIAA Paper 90-1681, (1990).
3) D. Vaughan and G. Schmidt, Analytical Modelling of No-Vent Fill Process, AIAA Paper 90-2377, (1990).
4) J. E. Anderson et al, No-Vent Fill Testing of Liquid Hydrogen, CEC Presentation, (1991).
5) R. T. Jacobsen et al, Thermophysical Properties of Nitrogen from the Fusion Line to 3500 R [1944 K] for Pressures to 150,000 psia [10342 x 10^5 N/m^2], Nat. Bur. Stand. (U. S.) Tech Note 648, (1973).
6) F. P. Incropera and D. P. DeWitt, "Fundamentals of Heat Transfer", John Wiley & Sons, New York, p 445, (1981).
7) G. Brown, Heat Transmission by Condensation of Steam on a Spray of Water Drops", Inst. Mech. Engr. General Discussion on Heat Transfer, p. 49-52, (1951).
8) S. Whitaker, AIChE J., 18, p. 361, (1972).

NO-VENT FILL TESTING OF LIQUID HYDROGEN

John E. Anderson
Paul M. Czysz
Dale A. Fester

Martin Marietta Civil Space & Communications
Denver, Colorado

ABSTRACT

A receiver tank test system has been fabricated and tested with liquid hydrogen to evaluate tank filling processes including chilldown and no-vent fill. The test system and the approach employed in conducting the test program is described. Results are presented for a series of chilldown and no-vent fill tests. The effectiveness of spray systems to promote vapor condensation and tank filling was demonstrated. A comparison of the results with and without spray system operation clearly showed the need for condensing hydrogen vapor from the ullage to accomplish complete filling of the tank.

INTRODUCTION

Future long duration space programs such as those dedicated to orbital operations and deep space exploration will use cryogenic fluids in numerous applications including propulsion, thermal control and life support. In addition to the requirement for long term storage of these fluids, there will be a need to provide periodic replenishment of those fluids consumed during normal operations. The resupply of these fluids will be accomplished in a low gravity environment. Because of this low gravity field, the ullage gas volume will not always be located in the vicinity of the tank vent. As a result, conventional methods of filling a tank by continuously venting to eliminate pressure buildup cannot be used, since significant loss of liquid could occur if the tank were vented during the filling process. A promising method for preventing this loss and completing the liquid transfer in low-g is called the no-vent fill process. In this process, the tank vent is closed while the liquid is transferred into the tank. As the liquid volume in the tank increases, the ullage volume is compressed tending to increase the tank pressure. If sufficient heat and mass transfer can be maintained within the ullage and at the liquid-vapor interface to condense vapor that is being compressed by the entering liquid, the tank pressure increase can be controlled so as to permit complete filling of the tank. One method of enhancing the condensation process is to inject or spray liquid droplets into the ullage region. The atomized spray provides an increased surface area upon which the ullage vapor can condense and slow the rate of pressure rise in the tank. The purpose of the test program reported in this paper was to investigate the no-vent fill method and determine the effectiveness and requirements of the spray system for controlling the tank pressure rise during loading.

The first no-vent fill testing performed by Martin Marietta was in 1968 as part of a company funded research and development program to investigate handling of fluorine propellants[1]. The primary objective of this program was to demonstrate the feasibility of the no-vent loading concept when handling extremely toxic fluids such as fluorine based propellants. Satisfactory testing of the concept was accomplished first with liquid nitrogen and then fluorine in 0.63 m^3 (165 gallon) tanks. The next series of no-vent fill tests of a cryogenic tank conducted at Martin Marietta was under an inhouse research and development program in 1982. The test article employed was a small, 0.33 m (13-inch) diameter spherical tank installed in a vacuum chamber and insulated with multilayer insulation (MLI). Testing was conducted with both liquid nitrogen and liquid hydrogen. The results of this test program indicated as in the previous vent free fluorine program that the test tank could be completely filled with liquid nitrogen using the no-vent fill approach without overpressurizing the tank. A method for condensing ullage vapors was not required to fill the tank. However, the situation was entirely different when liquid hydrogen was used. The maximum level that could be obtained was only 60% of the fully loaded tank, indicating that fluid characteristics and properties have a significant influence on the no-vent fill process. A method for condensing the ullage vapor was not incorporated in the 0.33 m (13-inch) test tank. The results of these tests together with analytical investigations of the no-vent fill process indicated the need for increasing interface heat transfer between the vapor and the incoming liquid and promoting condensation of the ullage vapor. One way to increase the interface heat transfer is to induce mixing of the tank contents and causing forced convection between the bulk liquid and the interface. A test program[2] was conducted by Sam Dominick at Martin Marietta to investigate condensation rates produced by using a liquid jet submerged in the bulk liquid to promote forced convection at the liquid-vapor interface. The jet could be rotated through an angle of 90 degrees so that the jet could be directed normal to or parallel to the fluid interface The test fluid used was Freon 113. The results of this test program indicated that the jet mixing was most effective when the jet was directed at or normal to the interface. As the direction of the jet was moved away from the normal to the interface, the mixing created by the jet became less effective in producing the desired condensation. Another test program to investigate no-vent filling in a small scale test facility and using Freon 114 as the test fluid was conducted by Doctors Stephen Traugott and Henry Obremski at Martin Marietta Laboratories in Baltimore. The test tank was a 0.33 m (13-inch) plexiglass sphere in which two spray systems were employed to promote mixing and condensation of the ullage vapor. The first of these systems employed nozzles which were free to rotate as result of the reactive force of the spray. The second spray system employed a base ring in the bottom of the tank and a vertical pipe extending upward into the tank. Both the ring and the pipe had holes drilled to provide the spray. No-vent fill test with and without spray operation were conducted. The best performance was observed with the ring and vertical pipe system in which the tank could be filled to 95% of the total tank volume.

It was the results of these tests that prompted an investigation in the possible use of spray systems to control tank pressure. At Martin Marietta, Mr. John Gille conducted a thorough analysis of fluid transfer in orbit[3] A part of this analysis was devoted to an evaluation of the effect of fluid properties in the heat transfer and condensation of ullage vapor during no-vent filling. By combining heat transfer and condensation equations, he established a fluid transfer parameter (FTP) based solely on fluid properties. This parameter is defined as follows:

$$\mathrm{FTP} = \frac{\rho_l^{3/2} K_l^{1/2} C_{pl}^{1/2}}{h_{evap}\, \rho_v} \left(\frac{dT_{sat}}{dP}\right)$$

where

ρ_l = liquid density
ρ_v = vapor density
K_l = liquid thermal conductivity
C_{pl} = liquid specific heat
h_{evap} = heat of vaporization
$\left(\frac{dTsat}{dP}\right)$ = change in saturation temperature with pressure

Calculation of the FTP was made for various fluids. The results are presented in Table 1. Comparison of the FTP values indicates a significant variation from the very cold cryogens, helium and hydrogen, to the earth storable fluids such as water and hydrazine. This

TABLE I FLUID TRANSFER PARAMETER COMPARISON

FLUID	FTP*
Helium	2.74×10^{-4}
Hydrogen	8.39×10^{-4}
Nitrogen	2.29×10^{-2}
Oxygen	4.02×10^{-2}
Fluorine	5.06×10^{-2}
Nitrogen Tetroxide	6.52×10^{-2}
Freon 114	7.20×10^{-2}
Hydrazine	1.58×10^{-1}
Water	6.690

*Units of FTP are Kg/Pa-hr-m4

variationis an indication of the difficulty in attempting to no-vent fill the fluid, with helium and hydrogen being most difficult. Previous testing at Martin Marietta and NASA-LeRC[4] both have indicated that no-vent filling with liquid nitrogen can be achieved withoutmuch difficulty. Therefore, it would seem that any fluid in Table 1 with an FTP value greaterthan liquid nitrogen would also be easy to fill by the no-vent fill technique. It also indicates that liquid nitrogen would be a reasonable fluid for simulation of liquid oxygen no-vent filling because of their proximity in the table. For liquid hydrogen, no other fluids other than helium would be representative. Therefore,we concentrated our test program on the no-vent filling of liquid hydrogen believing that the test results will provide better information for design of tank filling techniques than would test results with referee fluids.

TEST SYSTEM DESCRIPTION

An aluminum receiver test tank was designed and fabricated for experimental evaluation of the no-vent fill processes using liquid hydrogen as the test fluid. The test tank was a 0.76 m (30-inch) cylinder with a 0.76 m (30-inch) barrel section and $\sqrt{2}$ elliptical domes providing a total volume of 0.54 m^3 (19.13 cubic feet). It was constructed of 6061 aluminum with a wall thickness of 0.64 cm (0.25 inch). This heavy wall construction was employed as a safety precaution in providing a test bed tank for tank pressurization tests as well as no-vent filling evaluations. It was recognized that the thick wall of the tank would introduce problems in subsequent chilldown testing because of its high mass-to-volume ratio. A bolted flange is located in the top dome to provide access for modifying or changing the internal test equipment and instrumentation. The tank is insulated with 10 layers of MLI. The drawing and photograph, presented in Figure 1, show the details of the receiver tank design. Filling and draining of the tank is accomplished through a dip tube extending from the top to the bottom of the tank . This fill and drain line also serves as an instrumentation support for temperature and liquid level sensors.

Two ullage spray systems were incorporated in the tank design. The first was an axial spray system attached to the top flange of the tank. It was designed to direct the spray into the ullage in a manner similar to a conventional shower nozzle. This spray nozzle was attached to the fluid inlet line with conventional tube fittings to facilitate removal and replacement for evaluating different configurations. The cone nozzle configuration currently being tested is shown in Figure 2. The second ullage spray system was a tangential system as identified in Figure 1. It consisted of two nozzles located 180 degrees apart on the inner surface in the upper one-third of the tank. These nozzles were fabricated from 0.32 cm (1/8-inch) diameter tubes bent 90 degrees so as to inject tangentially to the tank wall. The end of the tube was flattened to produce a fan-shaped spray parallel to the wall. These nozzles were also attached to a manifold with mechanical fittings so that removal and replacement could be easily accomplished. Both axial and tangential spray systems can be used for tank chilldown prior to no-vent filling as well as for promoting condensation of the ullage vapor during no-vent fill. The tank was also provided with a wall heat exchanger or cooling coil for tank chilldown. This coil is a 0.64 cm (1/4-inch) internal diameter tube that is welded directly to the tank wall.

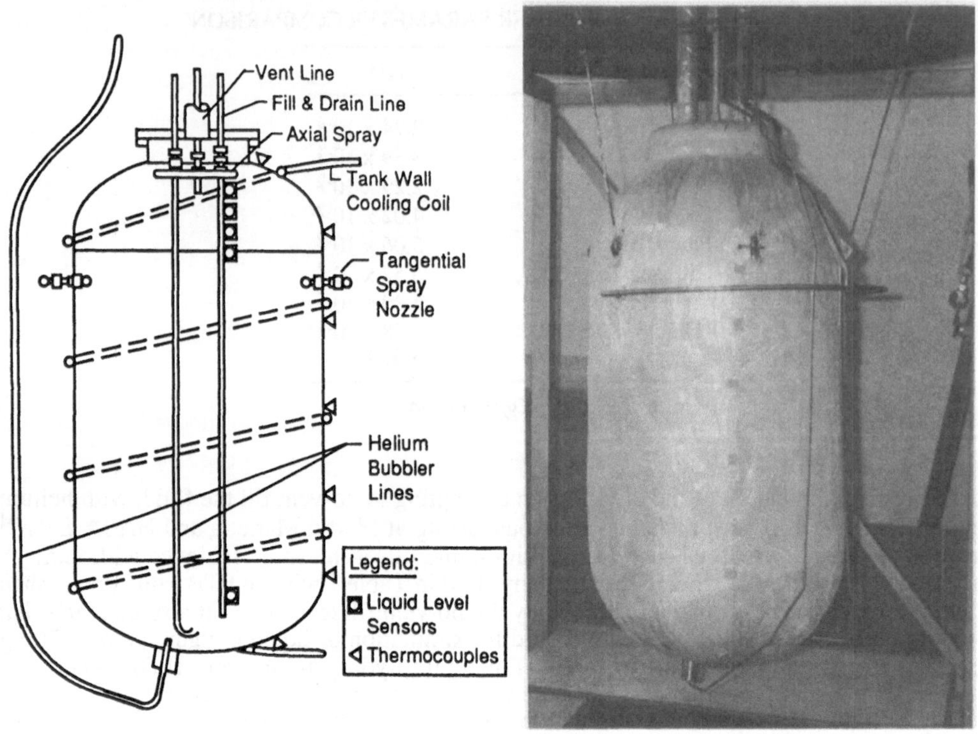

Figure 1 Martin Marietta's Liquid Hydrogen Receiver Tank

In addition to the chilldown and no-vent filling capabilities, the tank design also incorporates features for pressurization system performance evaluation. Tank pressurization may be accomplished by three methods. The first is by direct injection of the pressurant into the ullage in a conventional manner. The second and third methods are concerned with bubbling the pressurant through the bulk liquid into the ullage. The first of these methods transports the pressurant from the top of the tank down to the bottom through a tube in which the pressurant is cooled to the liquid temperature before bubbling up into the ullage. The third method provides direct injection of ambient temperature pressurant into the bottom of the tank. Pressurization studies are not presently being conducted.

The no-vent fill testing is being conducted in Martin Marietta's Liquid Hydrogen Laboratory. The test system schematic is shown in Figure 3. The receiver tank is installed in a 1.83 x 3.05 meter (6 by 10 foot) vertical vacuum chamber capable of providing a vacuum pressure of 10^{-6} Torr. Liquid hydrogen is transferred from a storage trailer through a series of vacuum jacketed lines to the test chamber. Separate flow lines are provided for fill and drain, axial spray, tangential spray, and the wall cooling coil. Venturi meters are used to measure liquid flow rates in the fill and drain, axial spray, and tangential spray lines. Flow rate for the cooling coil is measured by a gas orifice meter located at the exit line from the vacuum chamber. In addition to the flow meters, instrumentation consists of silicon diodes for measurement of fluid temperatures inside the receiver tank, carbon resistors for liquid level sensing within the receiver tank, and chromel-constantan thermocouples attached to the receiver tank wall for tank cooldown measurement. A capacitance probe is also installed in the receiver tank to determine liquid level during the filling process. Pressure measurements consist of flow meter inlet and differential pressures and the tank ullage pressure.

TEST PROCEDURE

In conducting no-vent fill tests with liquid hydrogen, several preparatory steps are required. The first is to pressurize the test tank and all lines with gaseous helium to check all fittings and connections for leaks. The system is then evacuated with a vacuum system and repressurized with helium to remove condensible gases that would freeze and block lines and

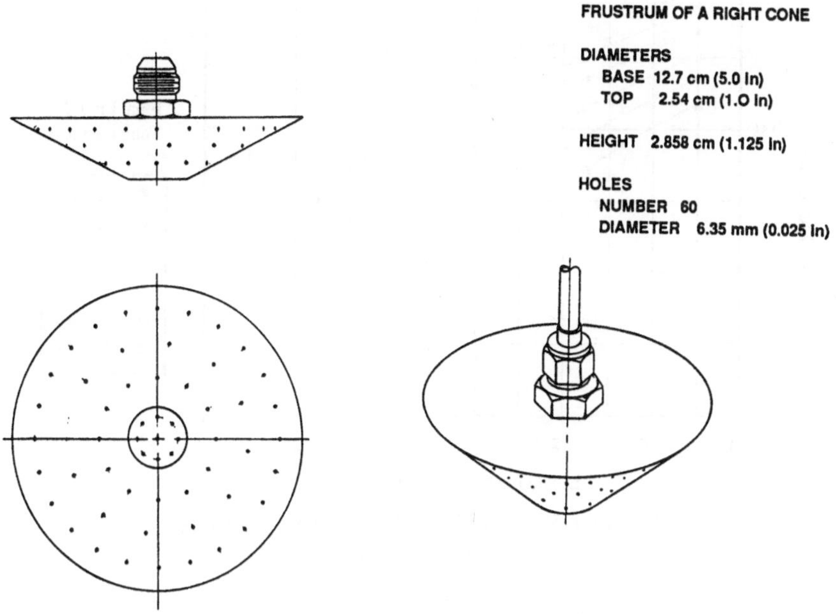

Figure 2 Axial Spray Nozzle

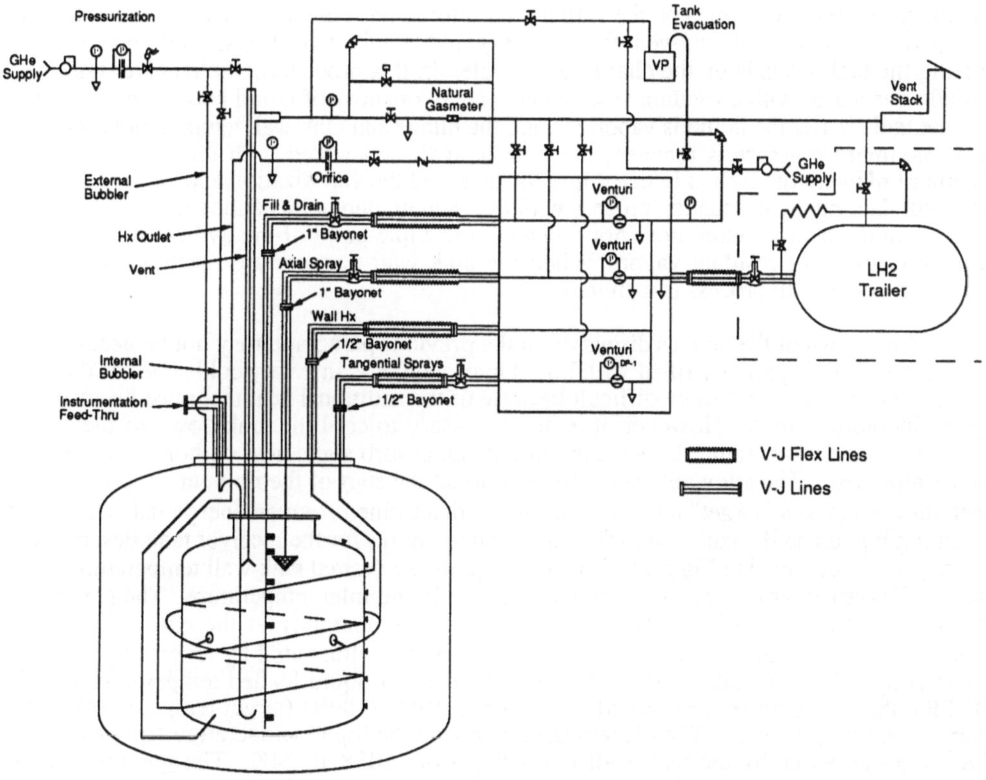

Figure 3 LH_2 Receiver Tank No-Vent Fill System Schematic

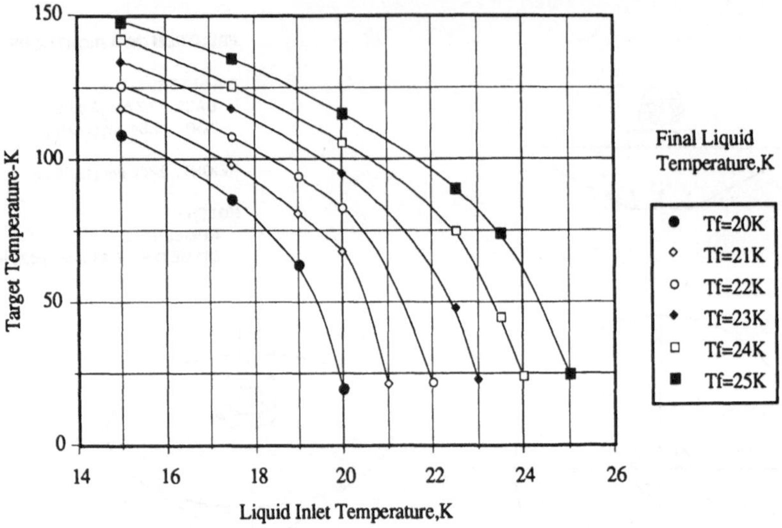

Figure 4 Receiver Tank Target Temperatures
$M_T/V_T=7.9$

fittings. This purge process is repeated several times. After purging of the system is complete, cooldown of the transfer lines from the storage trailer to the vacuum test chamber is initiated by opening bleed valves. When the transfer lines had been cooled so that liquid is observed in the bleed lines, chilldown of the test tank is then initiated. Two methods are available to accomplish this process. The first is to open the appropriate inlet valve to the tank wall heat exchanger coil to initiate liquid hydrogen flow. As the flow through the coil continues, the thermocouples on the wall are monitored until the desired wall temperature is achieved. At that time, the no-vent fill process would be initiated. The second procedure for cooling the tank walls is by the charge-vent cycle. In this procedure, the receiver tank was initially evacuated with a vacuum system and a small quantity of liquid hydrogen is injected into the tank. After the liquid is vaporized and the ullage and tank wall temperatures stabilize or a maximum pressure is reached, the tank is again evacuated with the vacuum pump. Another liquid charge is then injected into the tank and the vaporization and venting process repeated. During the evacuation process with the vacuum pump, precaution must be taken to assure that the tank pressure does not go below the triple point of the hydrogen in order to prevent freezing of residual vapor. When the tank wall temperature reaches the desired value, the no-vent fill process is then initiated.

Chilldown of the tank as discussed in the previous paragraph may not be necessary if the receiver tank is partially filled with liquid hydrogen or is in a very cold state. If the tank is warm, no-vent filling is more difficult because of the additional heat that must be absorbed by the incoming liquid. However, it is not necessary to cool the tank down to the actual incoming liquid temperature because the liquid can absorb some heat without an excessive temperature rise. The allowable tank temperature at the start of the no-vent fill process has been referred to as a "target" temperature[5] and is determined from an energy balance on the incoming liquid and the tank wall. The target temperatures for the receiver tank described in this paper are presented in Figure 4. This figure plots the desired tank wall temperature at the start of fill (target temperature) as a function of the liquid inlet temperature. The parameter temperature on the plot is the final liquid and wall temperature at the end of fill. The significance of the data presented in Figure 4 can best be illustrated by an example. If the liquid supplied to the tank is 20K (36R) and the desired liquid loaded temperature is 24K (43.2R), the tank temperature should be cooled to 105K (189R) (target temperature) at the start of the filling process. The 4K temperature rise in the liquid temperature corresponds to the energy given up by the tank wall in cooling from 105K to 24K. The maximum tank pressure would be the saturation pressure at 24K. If the tank wall temperature were cooled to approximately 84K for the same 20K liquid inlet temperature, the final loaded liquid temperature would be 22K.

After the tank wall is cooled to the desired temperature, no-vent filling is initiated. Two approaches can be followed. The first is to introduce liquid into the evacuated tank through the fill and drain line. Some flashing of the liquid will occur as it enters the tank. As the liquid volume increases, compression of the ullage vapor will increase the tank pressure until it is sufficient to terminate the liquid inflow at an accumulated liquid volume significantly less than a full tank. This test forms a basis for evaluating the effectiveness of the spray system. The next no-vent fill test is performed in a similar manner except that the spray system is activated at some predetermined pressure level in the tank. The ability of the spray system to promote condensation and control the filling process will be demonstrated by the reduced pressure level maintained during filling and the increased liquid accumulation in the tank. The spray system flow rate may be varied to determine its effect on tank pressure level.

TEST RESULTS

Testing to evaluate receiver tank chilldown and no-vent filling with liquid hydrogen was initiated at Martin Marietta in 1989 and is continuing into this year. The first problem to be investigated during the test program was that of tank chilldown. Testing to evaluate chilldown has been primarily concerned with the charge vent procedure in which small quantities are injected into the tank and allowed to vaporize. Both the axial spray and the tangential nozzles were evaluated. Because of the high mass-to volume ratio of the tank as indicated above, difficulty was encountered in lowering the tank wall temperature to the desired target temperature. Both nozzle systems were designed to provide spray for condensing ullage vapor during no-vent fill and lacked the capacity required for tank chilldown. Several hours were required to reach the desired value. One observation from this test indicated that the tangential nozzles provided more uniform temperature distribution of the tank wall during cooldown than the axial nozzle. This condition is attributed to the fact that the tangential nozzles directed the injected flow along the wall so that the resulting forced convection provided more efficient cooling. The axial spray nozzle flow depends more on natural convection to cool the wall. Testing of the charge-vent process was discontinued in order to proceed with no-vent filling. Tank chilldown for subsequent tests was accomplished by filling the tank normally, allowing temperature stabilization, and then draining the tank back to the supply trailer. The no-vent fill test was then initiated.

A total of 14 no-vent fill tests have been performed. The first series of tests conducted was to determine how much liquid could be loaded through the normal fill and drain line before the maximum ullage pressure of 345 KPa (50 psia) was reached. At this pressure, the inlet flow rate would terminate. The maximum quantity of liquid hydrogen that

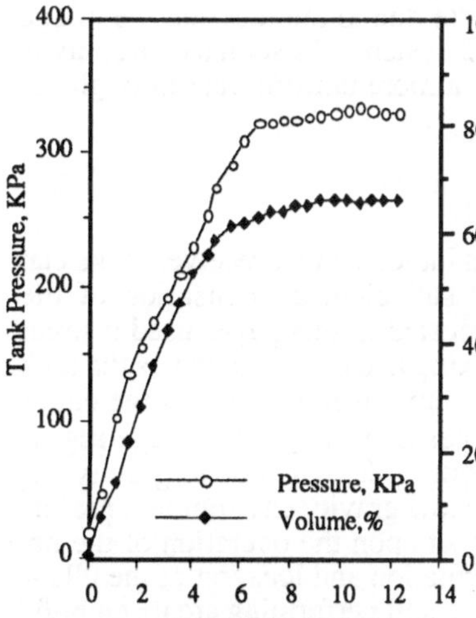

Figure 5 Tank Pressure and Liquid Volume for Run 4 w/o Sprays

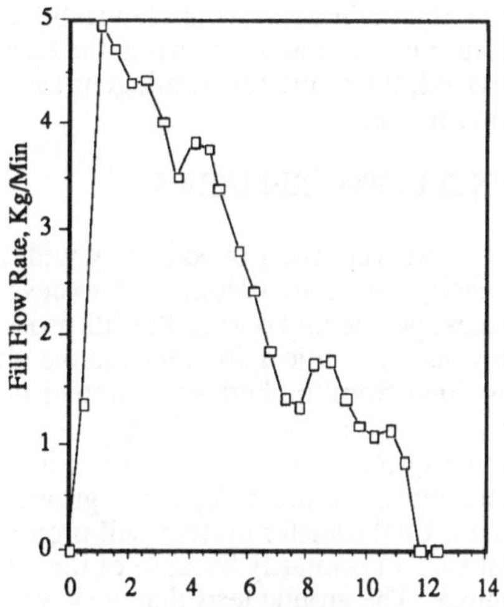

Figure 6 Main Fill Flow Rate for Run 4 w/o Sprays

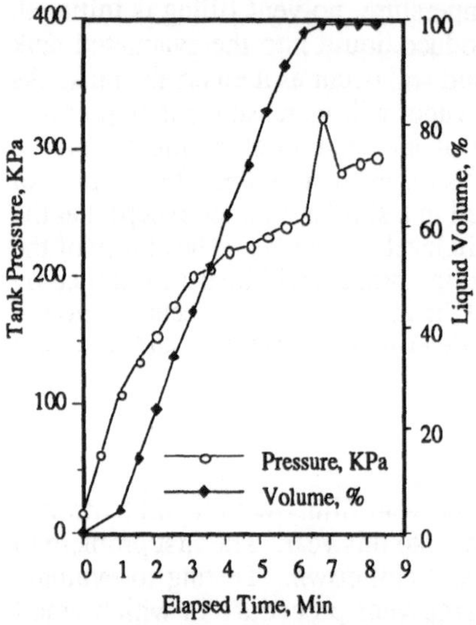

Figure 7 Tank Pressure And Liquid Volume for Run 10 w/ Axial Spray

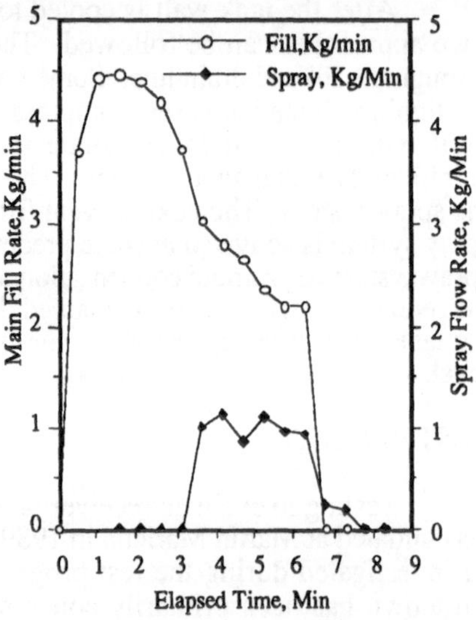

Figure 8 Main Fill and Axial Spray Rates for Run 10

could be loaded under these conditions varied from 65 to 75% of the full load. Figure 5 presents a typical tank pressure history during a no-vent fill without a method for increasing ullage condensation. At approximately seven minutes the ullage pressure was sufficient to significantly reduce fluid transfer into the tank. At this time, the liquid volume was approximately 65% of the fully loaded tank. Figure 6 shows the main fill flow rate during this test. The next series of tests employed the axial spray system in parallel with the normal fill and drain loading. Using this spray system, sufficient condensation of the ullage vapor was provided so that the tank could be loaded to the 100% level. Figure 7 illustrates a typical run using the axial sprays in conjunction with the main fill and drain. The tank filling was initiated using the main fill line. When 207KPa (30 psia) was reached in the tank, the axial spray system was activated At this time, the rate of rise of the tank pressure is seen to decrease allowing the liquid volume to reach 100%. The pressure spike at the end of the run is partly attributed to the liquid level reaching the spray nozzle so that part of the spray was injected directly into the liquid rather than the ullage. Another factor influencing the generation of the pressure spike is the rapidly reducing ullage volume in the top of the tank. Figure 8 presents both the main fill flow and the injected spray flow. Prior to activating the spray, the main fill flow is seen to be reducing at a rapid rate due to the fact that the ullage pressure is rising and increasing the back pressure on the fill system. As soon as the spray is activated, the reduction in ullage pressure rise rate permits a more uniform rate through the main fill line.

CONCLUDING REMARKS

No-vent fill ground testing with liquid hydrogen has indicated that receiver tanks can be easily filled to volumes in excess of 90% provided sufficient condensation of the ullagevapor occurs during the fill process. The testing reported in this paper used a small spray nozzle to inject droplets into the ullage to promote mixing and condensation as the tank was filled from the bottom. Another method that has been employed is to inject the entire flow through a nozzle in the top of the tank. The liquid passes through the ullage region providing the condensation surface as it settles to the bottom of the tank due to gravity. In either approach, the influence of gravity is significant. The low gravity environment present in an orbital transfer system will provide a significant impact upon the operation of the no-vent system primarily because of the uncertainty in the orientation and location of the ullage volume. The ground tests that we have accomplished and are still performing are meaningful to identify problems and establish operational criteria. However, an orbital transfer experiment is an absolute necessity to provide final verification of the tank loading procedures.

Results of the charge-vent testing indicated that tank chilldown can best be accomplished by injecting the coolant tangentially to the tank wall. The inherent forced convection provides more efficient cooling.

REFERENCES

1 Fester, D.A., Page, G.R., and Bingham, P.E.,"Liquid Fluorine No-Vent Loading Studies," Journal of Spacecraft and Rockets, Vol. 7, No. 2, February 1970, pp. 181-185

2 Dominick, S.M.,"Mixing Induced Condensation Inside Propellant Tanks," AIAA Paper 84-0514, AIAA 22nd Aerospace Sciences Meeting, Reno, Nevada,January 1984

3 Gille, J.P.,"Analysis and Modeling of No-Vent Transfer of Cryogens in Orbit," AIAA Paper 86-1252, AIAA/ASME 4th Joint Thermophysics and Heat Transfer Conference, Boston, Massachusetts, June, 1986

4 Chato,D.J., Moran, M.E., and Nyland,T.W.,"Initial Experimentation on the Nonvented Fill of a 0.14m3 (5 ft3) Dewar with Nitrogen and Hydrogen," AIAA Paper 90-1681, AIAA/ASME 5th Joint Thermophysics and Heat Transfer Conference, Seattle, Washington, June 1990

5 DeFelice, D.M., and Aydelott, J.C., "Thermodynamic Analysis and Subscale Modeling of Space-Based Orbit Transfer Vehicle Cryogenic Propellant Resupply," AIAA Paper 87-1764, 23rd Joint Propulsion Conference, San Diego, California, June 1987

HYDROGEN NO-VENT FILL TESTING IN A 34 LITER (1.2 CUBIC FOOT) TANK

Matthew E. Moran and Ted W. Nyland

NASA Lewis Research Center
Cleveland, Ohio

Susan L. Driscoll

NASA Marshall Space Flight Center
Huntsville, Alabama

ABSTRACT

Experimental results of no-vent fill testing with liquid hydrogen in a 34 liter stainless steel tank are presented. More than 40 tests were performed with various liquid inlet temperatures, inlet flowrates, initial tank wall temperatures, and liquid injection techniques. Maximum pressure within the receiver tank was limited to 0.207 MPa (30 psia), and fill levels equal to or exceeding 90 percent by volume were achieved in 40 percent of the tests. Three liquid injection techniques were employed; top spray, upward pipe discharge, and bottom diffuser. Effects of each of the varied parameters on the tank pressure history and final fill level are evaluated. The final fill level is found to be indirectly proportional to the initial wall and inlet liquid temperatures and directly proportional to the inlet liquid flowrate. Furthermore, the top spray is the most efficient no-vent fill method of the three configurations examined. The success of this injection method is primarily due to condensation of the ullage vapor onto the incoming liquid droplets. Ullage condensation counteracts the tank pressure rise resulting from energy exchange between the fluid and the warmer tank walls, and ullage compression.

INTRODUCTION

The test series described in this paper is part of an ongoing effort to advance cryogenic fluid handling technologies for space based applications. The ground test program being conducted by the Cryogenic Fluids Technology Office of the NASA Lewis Research Center (LeRC) is focused on verifying analytical models which simulate key cryogenic fluid handling processes.

One of these key processes is the transfer of liquid hydrogen between tanks using the no-vent fill technique. This technique involves chilling the receiver tank to a predetermined average wall temperature and then filling the tank without venting. The no-vent fill method is well suited to microgravity cryogen handling since it eliminates the need for tank venting during the fill process. Venting of a tank in a microgravity environment risks substantial loss of cryogen due to the uncertain positioning of liquid and vapor within the tank.

No-vent fill testing with both nitrogen and hydrogen has been performed at the LeRC Cryogenic Components Laboratory, Cell 7 (CCL-7). Preliminary results of earlier testing with a 142 liter receiver tank have been reported[1]. This paper discusses a more extensive array of test runs recently completed with a smaller 34 liter receiver tank. These tests incorporate three different liquid injection techniques; top spray, upward pipe discharge, and

Advances in Cryogenic Engineering, Vol. 37, Part B
Edited by R.W. Fast, Plenum Press, New York, 1992

bottom diffuser. The fill technique used dictates the heat and mass transfer process which takes place at the liquid-vapor interface. As a result, each of these configurations generates a unique tank pressure profile during the fill process.

EXPERIMENTAL RIG DESCRIPTION

A simplified piping schematic of the test rig is shown in Fig. 1. The facility is designed to accommodate both nitrogen and hydrogen testing. A detailed description of the test rig and its operation has been documented previously[1].

Fluid handling tests in the CCL-7 facility are performed with a supply dewar and two interchangeable receiver dewars. The supply dewar is a vacuum jacketed stainless steel tank that contains multi-layer insulation (MLI) within the vacuum annulus. The dewar is cylindrical, with an internal height of 137 cm and an inside diameter of 56 cm. Internal volume of the supply tank is approximately 306 liters.

The receiver dewars are of similar construction as the supply dewar. The large receiver, with an internal height of 71 cm and an inside diameter of 56 cm, has an internal volume of approximately 142 liters. The small receiver has an internal height of 51 cm and inside diameter of 32 cm, resulting in an internal volume of 34 liters. The lids of both receiver tanks are composed of a flat flange which supports a short cylindrical section with an inverted dome bottom. The space between the flange and cylindrical section is evacuated and insulated with MLI to minimize heat transmission through the dome from the environment. With the lid in place, the interior walls of the assembled receiver tanks form a cylindrical storage volume with domed ends.

Heat transfer from the environment is a function of liquid fill level for the supply and receiver tanks. This is due to the disproportionate heat flux entering from the tank top as a result of various lid mounted penetrations and the coupling of the lid walls to ambient temperatures at the tank flange. The overall heat flux for all three tanks was experimentally determined, and ranges from 3 to 32 W/m^2 (1 to 10 Btu/hr·ft^2) depending on the fill level and test fluid (nitrogen or hydrogen).

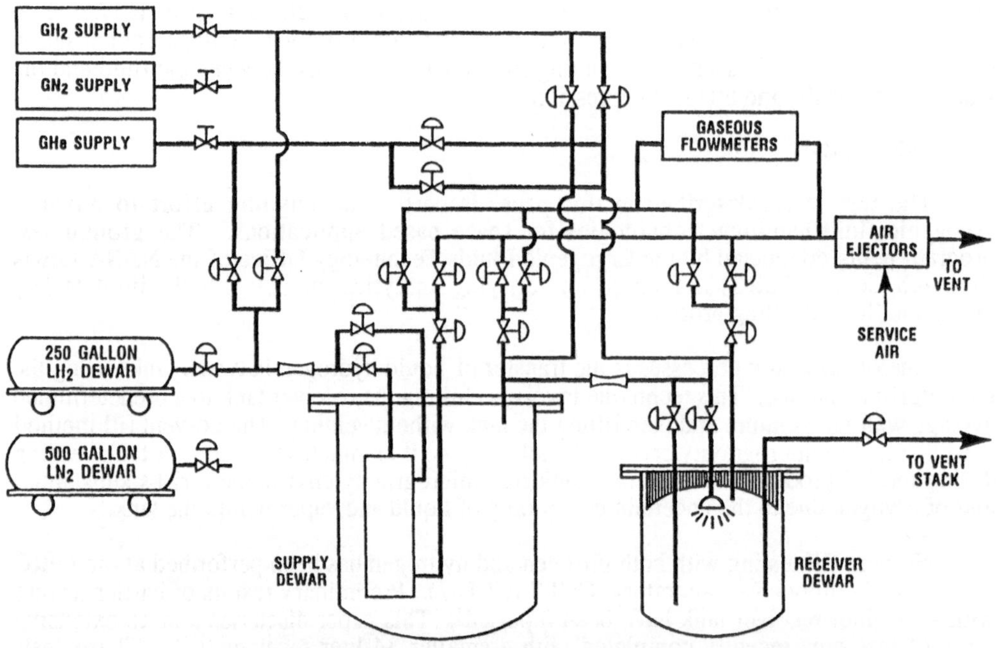

Fig. 1. Simplified piping schematic of the CCL-7 test rig

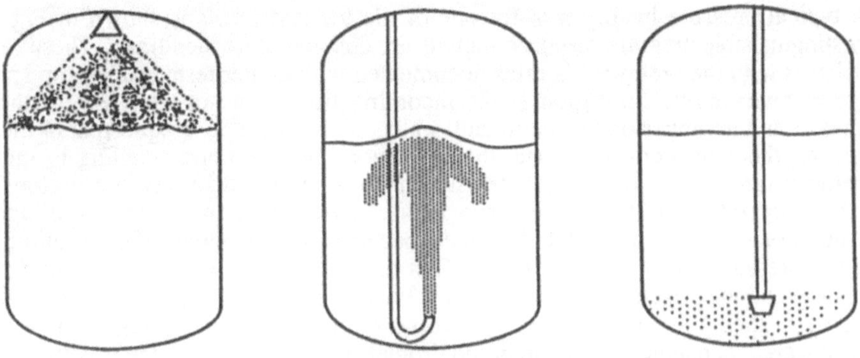

Fig. 2. Fill configurations used for no-vent fills at CCL-7.

Temperature sensors are positioned throughout the rig and on all tank walls, selected lines, and components. Temperatures are measured with type T (copper-constantan) thermocouples and silicon diodes, and thermistors are utilized to indicate the presence of liquid or vapor. Tank wall sensors are located in the annular vacuum space of the supply and receiver tanks, and are mounted to the inner tank wall. Within each tank is an instrument tree with silicon diodes and thermistors attached at various heights. This tree is in direct contact with the tank contents, whether liquid or vapor. Silicon diode sensors are accurate to within ± 0.1 K, whereas, the thermocouples are accurate to within ± 1 K.

Transducers provide continuous pressure measurement throughout the system with an estimated accuracy of ± 0.5 percent. An air ejector system provides sub-atmospheric pressure control in the vent lines to as low as 0.014 MPa (2 psia). Pressurization with helium, hydrogen, or nitrogen is available for the supply and receiver tanks. Each tank is equipped with a capacitance type level probe which is used to calculate the liquid fill level. The level probe in the small receiver tank was calibrated in liquid hydrogen against point sensors (thermistors) and found to agree within 2 cm for liquid levels greater than 10 percent.

N0-VENT FILL TEST PROCEDURE

Performance of a no-vent fill test at CCL-7 involves five sequential steps. First, the system is pressurized to 0.172 MPa (25 psia) with gaseous helium and checked for leaks. The helium is then vented through the air ejectors. This purge cycle is repeated a total of four times. Second, the supply dewar is filled from the roadable dewar with enough liquid to perform the planned test. With the supply tank filled, the liquid is thermally conditioned by controlling the tank pressure with the air ejector system. Third, with the cryogen conditioned to the desired temperature, the supply tank is pressurized for liquid transfer. The transfer line and associated components (e.g. valves, fittings, etc.) are then prechilled with a low flowrate of liquid. In the fourth step, the receiver tank pressure is reduced below atmospheric with the air ejectors. A charge of liquid is then loaded into the receiver tank with the vent valve closed. The vent remains closed while the liquid vaporizes, thus removing heat from the tank walls. When the tank pressure reaches a predetermined maximum or stabilizes, the vent valve is opened. Additional cooling is achieved as the tank pressure is once again brought below one atmosphere using the air ejector system. The resulting charge-hold-vent cycle is repeated until the tank wall temperature is reduced to the desired starting condition. In the fifth and final step, the liquid cryogen is transferred from the supply to the receiver tank until the receiver is filled to the desired level or until the pressure reaches a predetermined maximum value.

RESULTS AND DISCUSSION

A total of 42 no-vent fills were performed with the small, 34 liter receiver tank: 19 top spray, 7 upward pipe discharge, and 16 bottom diffuser tests. Figure 2 illustrates the three fill configurations tested.

A typical pressure history plot for one of the top spray fills is shown in Fig. 3(a). Three distinguishable pressure response regions are denoted by dashed lines. These regions are consistent with the pressure histories documented with earlier tests using a top spray in the large receiver tank[1]. In region 1, the incoming liquid flashes as it enters the low pressure tank and impinges on the warm tank walls. This results in a rapid rise in the tank pressure. As the walls cool down, and the saturation pressure corresponding to the inlet liquid temperature is approached, the pressure history curve transitions into region 2. In region 2, vaporization of the liquid cryogen decreases, and the effect of ullage gas condensation onto the incoming liquid droplets becomes more evident. The magnitude and sign of the pressure curve slope in region 2 is dictated by the competing processes of condensation and vaporization within the tank. Finally, in region 3, the pressure begins to rise sharply as the spray nozzle starts to become submerged by the rising liquid. As the nozzle is covered, condensation on the liquid droplets ceases and the ullage is compressed.

Internal tank temperatures recorded by the instrument tree during the same no-vent fill drop rapidly as the inlet spray fills the tank volume. This behavior is consistent with previous no-vent fill tests performed with the large receiver tank using the top spray configuration. On the tank wall, all temperature sensors except the top one remain at liquid hydrogen temperatures throughout the run. The top sensor, which is not exposed to the inlet liquid spray, cools gradually over most of the test run. Toward the latter portion of the test, the top sensor rises in temperature as the remaining tank ullage is compressed and the spray nozzle becomes submerged.

A pressure history plot for one of the upward pipe discharge no-vent fills is given in Fig. 3(b). As with the top spray configuration, the pressure response curve displays three distinguishable regions. In region 1, the pressure rises sharply as the liquid enters the low pressure tank and flashes. However, this initial pressure rise is less dramatic for the upward pipe discharge configuration. Region 2 is characterized by a gradual increase in tank pressure as the liquid level rises in the tank. Condensation at the liquid-vapor interface is enhanced due to fluid circulation induced by the upward liquid motion. This circulation reduces the thermal stratification within the bulk liquid, and brings the cooler fluid to the interface. Lastly, an increase in the slope of the pressure curve is evident in region 3 as the effect of ullage compression becomes more pronounced, and the liquid interface rises into the upper dome region. Once again, the pressure rise in region 3 is less dramatic than for the spray configuration since the condensation process is not altered significantly (i.e. as compared to submergence of the spray nozzle for the spray fills).

Internal tree temperatures for the same test show the two upper silicon diodes dropping in temperature rapidly in the initial moments of the fill, whereas, the lower sensors experience a sudden transitory peak in temperature and then drop to liquid hydrogen temperatures. The rapid drop shown by the top sensors indicates that the liquid is gushing to the upper portions of the tank early in the test before the pipe becomes submerged in the liquid. The behavior of the lower silicon diodes is assumed to be caused by circulation of the initially warm upper ullage vapor as the liquid is injected upward into the tank at the start of the fill. Later in the test, the uppermost sensor slowly heats up as the ullage is compressed and then cools as the interface approaches and eventually submerges the sensor. Meanwhile, the tank wall sensors cool to liquid hydrogen temperatures within the first few seconds of the test, indicating that the injected liquid fountain initially reaches the top of the tank and impinges on the side walls. Approximately one third of the way into the run the top wall sensor begins to rise in temperature rapidly and then tapers off while continuing to rise during the remainder of the test due to ullage compression.

The pressure history of one of the tests using the bottom diffuser configuration is shown in Fig. 3(c). No-vent fills using this injection technique do not exhibit the three pressure response regions typical of the other two fill configurations. Pressure rise at the start of this test is gradual, as the liquid is injected into the bottom of the small receiver tank through a porous plug diffuser. A moderate tapering of the pressure rise in the tank is evident in the middle portion of the run with a slight increase in the pressure curve slope toward the end of the test. This fill configuration results in the smallest and most quiescent liquid interface of the three configurations tested and, therefore, the lowest condensation

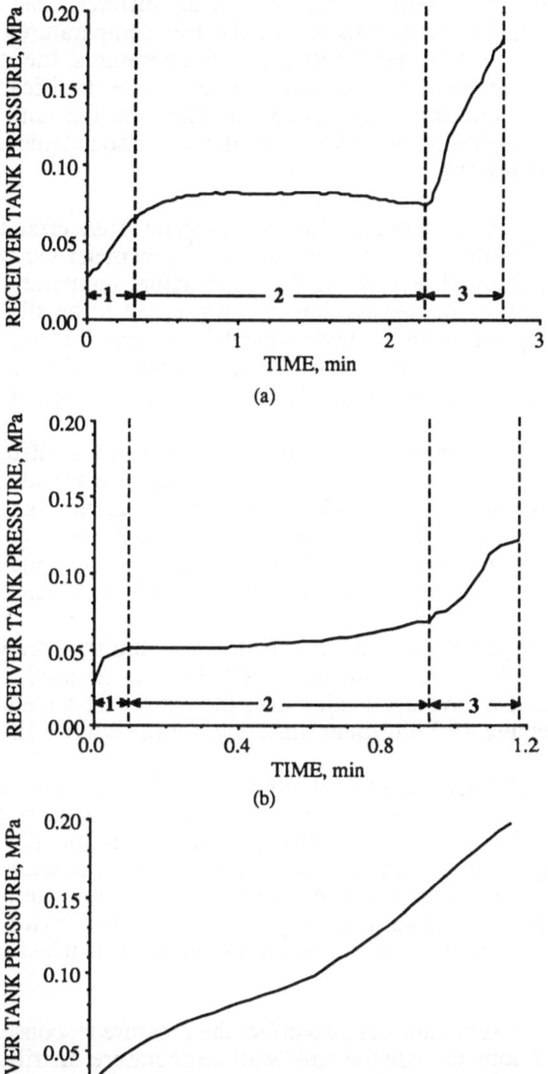

Fig. 3. Tank pressure as a function of time for hydrogen no-vent fill tests with (a) top spray, (b) upward pipe discharge, and (c) bottom diffuser fill configurations.

rate. In addition, energy from the tank wall is absorbed by the cryogen in a more gradual manner as the liquid level rises in the tank.

Thermal stratification of the vapor is present at the initiation of the test shown in Fig. 3(c) and remains throughout the run. All of the internal instrument tree sensors rise in temperatures gradually until they are submerged by the liquid interface. The upper most sensor is the only one to remain in the vapor space throughout the test. Response of the internal temperatures for this configuration is markedly different than for the top spray and upward pipe discharge configurations where the tree temperatures drop rapidly at the initiation of the tests. As with the internal tree temperatures, the wall sensors indicate thermal stratification in the tank at the start of the test. The stratification persists until the liquid interface passes the individual sensor locations on the tank wall. Temperature indicators in the vapor region of the tank wall continue to rise in temperature throughout the test due to ullage compression.

Liquid inlet temperature is one of the primary parameters effecting the tank pressure response. This temperature is measured at the venturi flowmeter located in the transfer line between the supply and small receiver tanks. The actual temperature of the liquid as it enters the receiver tank is not sensed, but it is presumed to be slightly higher than the venturi flowmeter temperature due to environmental heat leak into the transfer line. Liquid temperature rise in the transfer line from the supply tank to the venturi flowmeter was found to range from less than 0.5 K to nearly 1.7 K for most of the tests.

The effect of liquid inlet temperature for the bottom diffuser configuration is illustrated in Fig. 4(a). Pressure-versus-fill level response for all three configurations indicates a lower pressure rise in the tank as the inlet temperature is decreased. This trend was also observed in initial no-vent fill tests performed with the large receiver tank[1]. The test runs depicted in Fig. 4(a) were selected to match the inlet flowrate and equivalent initial tank wall temperature as closely as possible, with the widest variance of inlet temperature.

Another primary parameter is the liquid flowrate into the receiver tank. For higher flowrates, the tank pressure as a function of fill level is reduced. The liquid flowrate specified for the tests is an averaged value over the entire test run and is derived from the tank fill level data. Figure 4(b) illustrates this effect for the bottom diffuser.

The effect of initial equivalent wall temperature on the pressure response was more difficult to isolate for this series of tests. Initial equivalent wall temperature is a weighted average temperature that accounts for variable wall mass and the strong temperature dependence of the specific heat of stainless steel at cryogenic temperatures. Generally, most of the no-vent fill tests were initiated with large axial temperature gradients in the tank wall. Figure 4(c) illustrates the variance in pressure response for two bottom diffuser tests with different equivalent initial wall temperatures and well matched inlet temperature and flowrate.

Other secondary test parameters also effect the pressure response during a no-vent fill test. These test conditions include the tank wall temperature distribution, the inlet flow-versus-time profile, and the sensible heat gain in the liquid as it flows through the transfer line. Careful attention was paid to the secondary parameters when matching test runs for comparison.

CONCLUDING REMARKS

A key result of the testing performed is the demonstration of the no-vent fill technique for hydrogen in a repeatable fashion under normal gravity conditions. With a maximum allowable pressure of 0.207 MPa (30 psia) in the receiver tank, approximately 40 percent of the hydrogen no-vent fill tests performed resulted in final fill levels equal to or greater than 90 percent by volume. Variable test parameters and liquid injection configurations provide some insight into the conditions necessary for a successful transfer of liquid hydrogen by this method. Analytical work being conducted with this and other test data will further quantify the no-vent fill process[2,3].

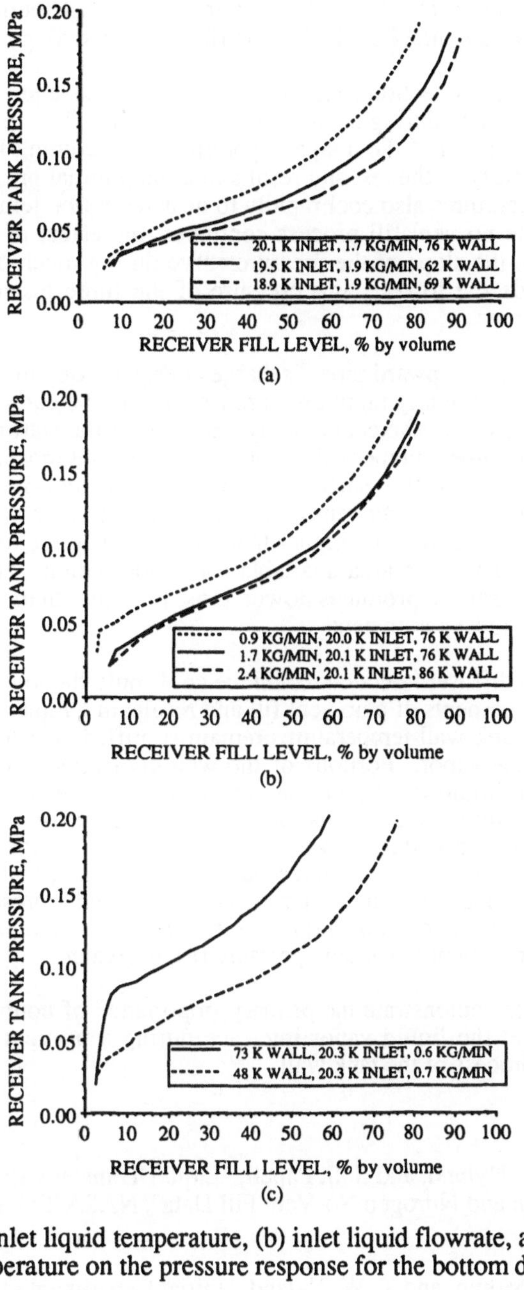

Fig. 4. Effect of (a) inlet liquid temperature, (b) inlet liquid flowrate, and (c) equivalent initial wall temperature on the pressure response for the bottom diffuser configuration.

The effect of several test parameters is illustrated by the comparison of pressure histories as a function of tank fill level for carefully selected test runs. The magnitude of the receiver tank pressure profile is found to be directly proportional to the inlet liquid temperature and equivalent initial wall temperature and indirectly proportional to the inlet liquid flowrate (i.e. higher liquid inlet and wall temperatures, and lower inlet flowrates, produce higher receiver tank pressures). Since the tank pressure is limited to a maximum allowable value, the final fill level is indirectly proportional to the tank pressure response. Secondary test conditions, such as initial wall temperature distribution, flowrate history, and sensible heat gain of the inlet liquid, also contribute to the tank pressure response.

Three different liquid injection techniques were employed, and each displayed unique responses in the receiver tank during the no-vent fill process. The top spray configuration, which promotes condensation of the ullage vapor onto the incoming liquid droplets, cools the tank walls rapidly early in the test and results in a steep initial pressure rise in the tank. The internal tank temperatures also cool rapidly to near saturation temperature as the spray enters the tank. As the no-vent fill process continues, the effect of ullage condensation results in a decrease in the slope of the pressure curve during much of the remaining test. This fill method yields the lowest tank pressure of the three tested configurations for comparable test conditions.

Much like the top spray, the upward pipe discharge configuration initially cools most of the tank wall and results in a similarly rapid pressure rise during the early moments of the fill. The internal tank temperatures also cool quickly. Soon after the start of the test, the outlet of the discharge pipe becomes submerged by the rising liquid interface, and condensation of the ullage vapor is limited to this surface. The bulk fluid motion induced by the pipe discharge arrangement enhances condensation at the interface as cooler fluid is circulated upward. This fill method results in a higher tank pressure than the top spray method due, primarily, to the reduced surface area available for condensation. However, the upward pipe discharge configuration produces lower tank pressure than the bottom diffuser configuration for similar test parameters.

In contrast, the bottom diffuser injection technique cools only the lower portion of the tank wall during the initial moments of a no vent fill and results in a rather gradual pressure rise at the start of a test. Tank wall temperatures remain stratified near the starting conditions while in contact with the vapor. Portions of the wall cool rapidly when in contact with liquid hydrogen as the liquid level increases in the tank. The internal instrument tree temperatures react in a similar manner, and both the wall and internal temperatures increase gradually while in the vapor due to vapor compression. As the fill progresses, the submerged diffuser does not produce much bulk liquid motion, and the liquid near the vapor interface remains close to saturated temperature at the tank pressure. This condition results in much less condensation at the liquid-vapor interface as compared to the other two configurations, and consequently, the tank pressure rise is greater.

All three configurations demonstrate the primary importance of controlling the mass and heat transfer process at the liquid-vapor interface during a no-vent fill, and to a lesser extent, the cooldown process of the tank wall.

REFERENCES

1. M. E. Moran, T. W. Nyland, and S. S. Papell, "Liquid Transfer Cryogenic Test Facility - Initial Hydrogen and Nitrogen No-Vent Fill Data", NASA TM-102572, March, 1990.

2. D. J. Chato, M. E. Moran, and T. W. Nyland, "Initial Experimentation on the Nonvented Fill of a 0.14 m^3 (5 ft^3) Dewar With Nitrogen and Hydrogen", NASA TM-103155, 5th Joint Thermophysics and Heat Transfer Conference, AIAA-90-1681, June, 1990.

3. W. J. Taylor and D. J. Chato, "Improved Thermodynamic Modelling of the No-Vent Fill Process and Correlation With Experimental Data", 26th Thermophysics Conference, Honolulu, Hawaii, June 24-27, 1991.

ANALYSIS OF PULSED INJECTION FOR MICROGRAVITY RECEIVER TANK CHILLDOWN

Scott C. Honkonen, Joe R. Pietrzyk, and John R. Schuster

General Dynamics Space Systems Division
San Diego, California

ABSTRACT

The dominant heat transfer mechanism during the hold phase of a tank chilldown cycle in a low-gravity environment is due to fluid motion persistence following the charge. As compared to the single-charge per vent cycle case, pulsed injection maintains fluid motion, and the associated high wall heat transfer coefficients, during the hold phase. As a result, the pulsed injection procedure appeared to be an attractive method for reducing the time and liquid mass required to chill a tank. However, for the representative conditions considered, no significant benefit can be realized by using pulsed injection, as compared to the single-charge case.

A numerical model of the charge/hold/vent process was used to evaluate the pulsed injection procedure for tank chilldown in microgravity. Pulsed injection results in higher average wall heat transfer coefficients during the hold, as compared to the single-charge case. However, these high levels were not coincident with the maximum wall-to-fluid temperature differences, as in the single-charge case. For representative conditions investigated, the charge/hold/vent process is very efficient (i.e., little mass savings can be realized since the mass used to chill the tank is only 20% greater than the lower theoretical limit, but more than a factor of three less than the theoretical upper limit). A slightly shorter chilldown time (10%) was realized by increasing the number of pulses (from 1 to 7).

INTRODUCTION

Long-term storage and fluid transfer in the microgravity environment of space gives rise to many concerns, particularly for cryogens where thermal effects play a critical role. The technology needed to successfully manage these cryogens has been studied under COLD-SAT (Cryogenic On-Orbit Liquid Depot — Storage, Acquisition and Transfer Satellite) funded by the NASA Lewis Research Center[1]. COLD-SAT is a free-flying orbital experiment that would demonstrate enabling technologies for the development of on-orbit systems and provide an engineering data base to correlate analytical tools used for cryogenic system design and analyses.

Two of the challenges of microgravity cryogenic fluid management are the efficient chilldown and filling of receiver tanks. Charge/hold/vent chilldown to the target temperature followed by a no-vent fill[2] is a desirable method of filling a tank in a microgravity environment since vapor cannot be preferentially vented with liquid present. The chilldown consists of a series of charge/hold/

vent cycles. During the charge phase, liquid is injected into the tank through nozzles where it flashes to vapor as it strikes the tank walls and absorbs energy. Following this injection, the vapor is held in the tank to absorb additional energy from the tank walls. This hold phase continues until the pressure rises to a predetermined maximum or until the walls are cooling at an unpractically slow rate. The vent phase is then initiated and the tank contents are vented to space and the cycle is repeated. The time required to reach the target temperature (the tank wall temperature that allows liquid filling of the tank with the vent closed) and the liquid mass used during the chilldown are the quantities of most interest.

The GDNVF (General Dynamics No-Vent Fill) computer program[3] models the tank chilldown and no-vent fill processes in low-g environments. During the charge/hold/vent cycle the dominant heat transfer regime changes: forced convection is used to model the relatively short charge phase; during the hold phase the actual heat transfer coefficient is expected to be higher than that predicted using free convection correlations due to fluid motion persistence from the earlier charge phase; as fluid motion ceases, the heat transfer coefficient will approach a lower limit, which can be no less than that provided by pure molecular conduction. The decay to this lower limit continues into the vent phase.

This paper presents the results of a numerical investigation of using pulsed injection during the charge phase in an effort to maintain fluid motion and the associated high heat transfer levels throughout the hold phase. Since the hold phase can account for as much as 90% of the predicted chilldown time, model refinements for this phase will have the greatest impact on reducing the predicted chilldown time.

FLUID MOTION PERSISTENCE

The characteristics of fluid motion persistence and the corresponding convective heat transfer during the hold phase of a tank chilldown process depend on the manner in which fluid is injected into the tank during the charge phase. Three different nozzle configurations were proposed for the COLD-SAT study: tangential, axial, and radial. Tangential injection at the tank wall will generate a swirling mean flow pattern that is expected to persist longer than flow established with the other nozzle configurations. Radial injection towards the tank walls will generate the least "organized" flow field in the tank. Large turbulent eddies are expected to be the dominant flow structures in the tank following radial injection. These eddies are expected to decay rapidly in the absence of a significant mean shear layer.

In previous work[4], a numerical analysis of a decaying swirling flow in a circular tank was used to simulate fluid motion persistence following tangential injection. A theoretical analysis of large eddy decay was used to simulate fluid motion persistence following radial injection. These two configurations represent limiting cases for fluid motion persistence during the hold phase of a tank chilldown cycle in low-g. Fluid motion persistence models, based on these results, were implemented into the GDNVF computer program.

Pulsed injection was expected to increase the efficiency of tank chilldown by maintaining fluid motion and a corresponding high wall heat transfer coefficient during the hold phase. Previous work[4] indicated that the flow field following radial injection is chaotic, and the duration of fluid motion during the hold is too short to result in significant convective heat transfer from the wall. The flow field following tangential injection is more organized and the fluid motion decays more slowly, making it the better candidate for pulsed injection. Only tangential injection results are presented in the present paper.

PULSED INJECTION CHARACTERISTICS

The GDNVF computer program used to predict the results presented in this paper can model a wide range of tank configurations. However, all results presented in this paper pertain to a single representative case, the COLD-SAT 1.1-meter-diameter receiver tank. This tank is designed of 2219 aluminum and weighs 34 kg. It is chilled, in a low-g environment ($g/g_e = 10^{-7}$), from 250 K to 55.6 K using liquid hydrogen at 20.3 K. The maximum working pressure for the tank is 276 kPa.

The typical charge/hold/vent cycle for the single-charge (per vent cycle) case is apparent from the tank pressure and wall temperature histories shown in Figure 1. During the charge the tank pressure rapidly rises and the wall temperature rapidly drops as energy is transferred at a high rate from the warm tank wall to the cool fluid. Following the charge, fluid motion within the tank decays, as does the heat transfer rate between the wall and the fluid. As a result, the tank pressure rise rate decreases and the wall temperature decreases less rapidly during the hold as compared to the charge. Following the hold, the fluid is vented, causing a rapid drop in tank pressure. This cycle is repeated until the tank is chilled to the target temperature.

Similar trends can be seen in the results for the pulsed case shown in Figure 2. However, the effects of the multiple charges per vent cycle are also apparent. The tank pressure rises rapidly during each short charge, followed by a less rapid pressure rise rate during each short hold period

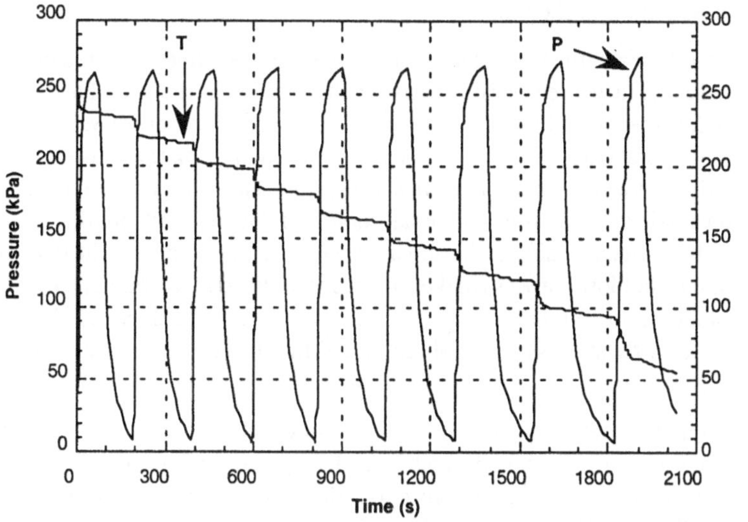

Figure 1. Pressure and tank wall temperature for single-pulse injection.

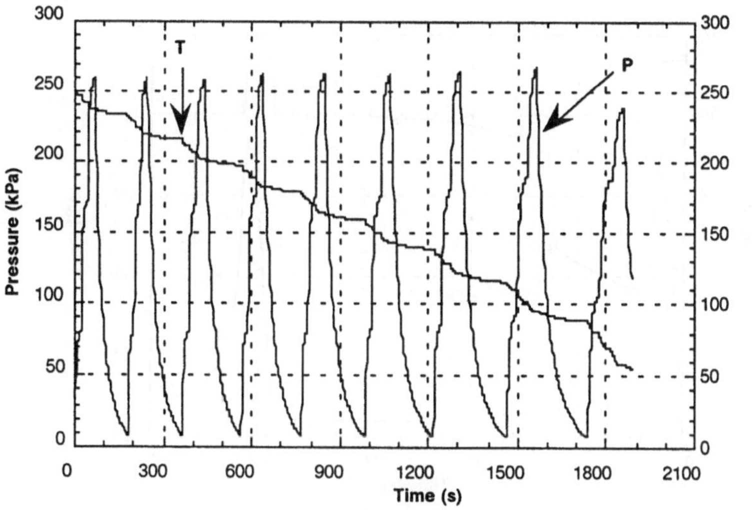

Figure 2. Pressure and tank wall temperature for three-pulse injection.

between pulses. A potential drawback of the pulsed injection procedure is that each successive charge (during a vent cycle) must be injected against increasingly higher tank pressures. For the single-charge case, the entire charge is injected into the tank initially when the pressure is lowest, before all heat transfer between the wall and fluid has occurred.

The wall temperature history for the single-charge case decays in a stepwise manner (see Figure 1). As the number of pulses per vent cycle increase, the number of wall temperature "steps" increase, and the size of the temperature "steps" decrease.

The last cycle may significantly decrease the efficiency of the chilldown process. Should the target temperature be reached during the charge phase of the last cycle, a significant portion of mass with substantial cooling capability would be lost, or the tank temperature would fall below the target temperature if the last cycle were completed. For maximum efficiency, an integral number of complete vent cycles should be completed to chill the tank to the target temperature. However, the exact number of cycles cannot be determined a priori. It depends on a number of interacting parameters and processes[5]: charge mass, vent criterion, heat transfer rates, etc. For the conditions imposed, nine vent cycles were required to reach the target temperature for both the single-charge and pulsed cases, as shown in Figures 1 and 2. For both of these cases, the target temperature was reached during the vent phase; consequently no significant cooling capacity was lost with the remaining mass in the tank.

The time required to reach the target temperature may also be affected by the last cycle. For the single-charge case, the target temperature was reached almost at the end of a vent. The total chilldown time was 2,030 seconds. For the pulsed case the target temperature was reached in the middle of the vent, at a total elapsed time of 1,880 seconds. Comparing these two numbers indicates that the pulsed chilldown was completed in 8% less time. However, it would take additional time to complete the final vent with little additional cooling of the tank wall, but the resulting time difference between the pulsed and single-charge cases would be less than 8%. The discrete nature of the charge/hold/vent chilldown process makes it somewhat difficult to obtain an exact comparison of different chilldown procedures. Interpretation of results must be done in light of the effect of the last cycle, particularly in those cases where little difference exists between different chilldown procedures.

The wall heat transfer coefficient histories for the single-charge case and the pulsed case (three pulses per vent cycle) are shown in Figures 3 and 4, respectively. The mass flowrate into and out of the tank is shown at the top of each figure to indicate the different phases of the charge/hold/vent cycle. Only the first 250 seconds are shown for clarity. The duration and average magni-

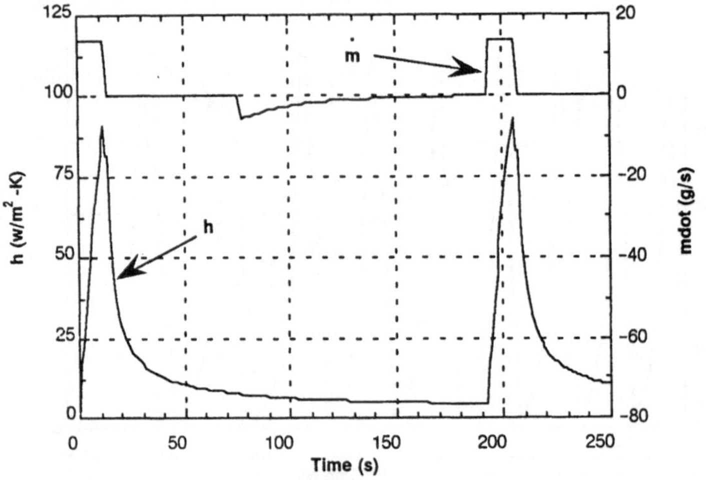

Figure 3. Heat transfer coefficient and mass flowrate for single-pulse injection.

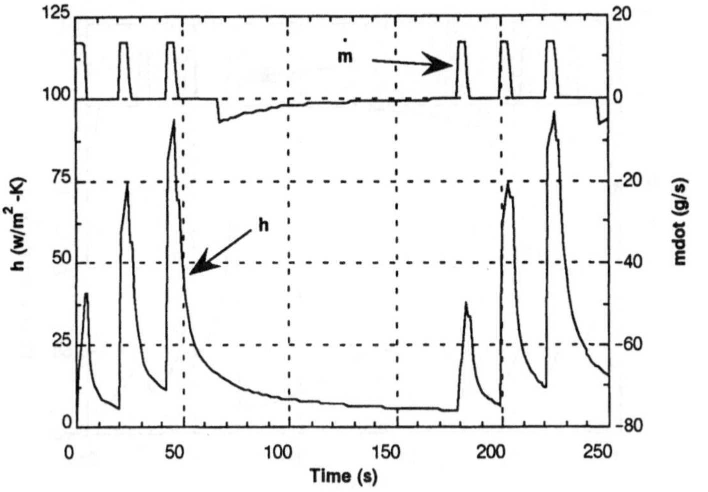

Figure 4. Heat transfer coefficient and mass flowrate for three-pulse injection.

tude of the wall heat transfer coefficient during the charge for the single-charge case is not significantly different from the total duration and average magnitude of all the smaller charges for the pulsed case. For the pulsed case, the maximum wall heat transfer coefficient increases for each successive pulse. This trend is due primarily to an increase in the fluid thermal conductivity with temperature, and is also apparent in the results for the single-charge case.

The potential benefit of the pulsed chilldown procedure is that the wall heat transfer coefficient is maintained at a higher average level during the total hold period by fluid motion persistence following each pulse. As indicated by a comparison of Figures 3 and 4, the single long-duration hold for the single-charge case results in a lower average wall heat transfer coefficient than the three shorter holds for the pulsed injection case. The wall heat transfer coefficient remains low during the vent for both cases.

The heat transfer rate from the wall to the fluid is not only dependent on the wall heat transfer coefficient, but also on the wall-to-fluid temperature difference: $q'' = h(\Delta T)$. These results are shown in Figures 5 and 6 for the single-charge and pulsed cases, respectively. For the

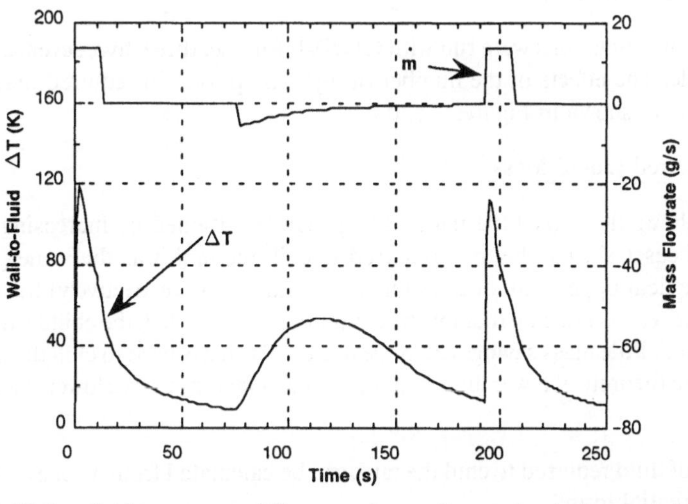

Figure 5. Wall-to-fluid temperature difference and mass flowrate for single-pulse injection.

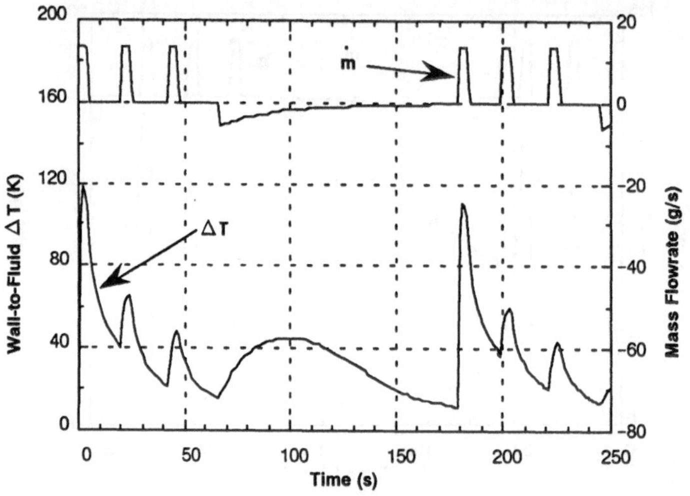

Figure 6. Wall-to-fluid temperature difference and mass flowrate for three-pulse injection.

single-charge case, the highest temperature difference occurs nearly coincident with the highest wall heat transfer coefficient (i.e., during the single-charge at the start of each cycle). This coincidence results in the highest possible heat transfer rates.

For the pulsed case, the highest wall heat transfer coefficient is not coincident with the largest temperature difference (see Figure 6). Each successive pulse occurs after one or more hold periods, during which time additional energy is transferred from the wall to the fluid. Consequently, the wall-to-fluid temperature difference decreases for each successive pulse. The wall-to-fluid temperature difference decreases slowly during each hold, but increases sharply during each charge due to the addition of cold fluid into the tank.

For both the single-charge and pulsed cases, the wall-to-fluid temperature difference history during the vent exhibits the same interesting trend. The temperature difference initially increases as the fluid in the tank cools due to expansion. As the vent continues, the fluid mass in the tank decreases until the expansion cooling becomes less significant than the wall heating. As a result, the wall-to-fluid temperature difference decreases during the later portion of the vent.

SUMMARY OF RESULTS

A number of simulations were run with GDNVF for one, three, five, seven, and ten injection pulses per cycle. The effects of the number of injection pulses on required mass of liquid and chilldown time are shown in Figures 7 and 8.

Effect on Required Liquid Mass

Figure 7 shows that very little mass savings can be attained by increasing the number of pulses. For all cases shown, the time required to chill the tank was the same (~2,000 s). Also shown are practical upper and lower bounds (9.7 and 2.1 kg, respectively) for the liquid mass required. These results indicate that for the conditions investigated, the chilldown process is very efficient (i.e., very little mass savings can be realized). The mass used to chill the tank is only 20% greater than the theoretical lower limit, but more than a factor of three lower than the theoretical upper bound.

The mass of fluid required to chill the tank can be calculated from an energy balance, shown below in differential form:

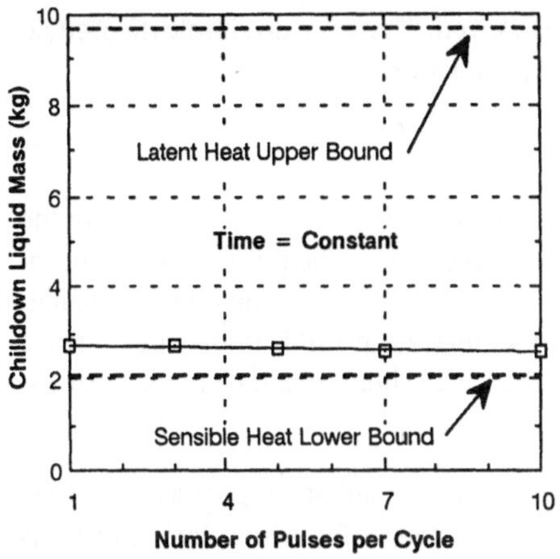

Figure 7. Childown liquid mass as a function of number of pulses per cycle.

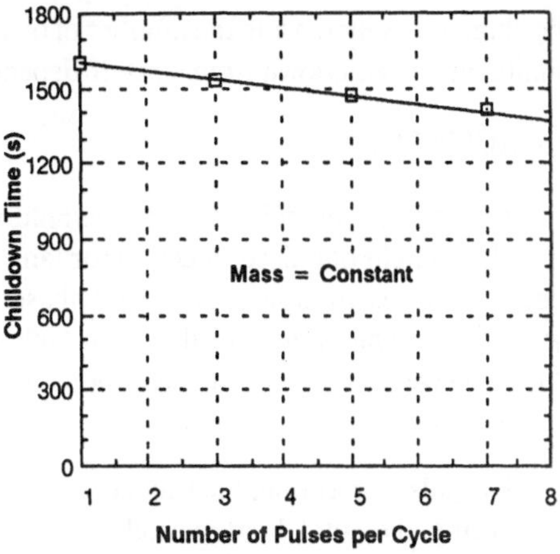

Figure 8. Childown time as a function of number of pulses per cycle.

$$(M\ c)_{tank}\, dT = (h_{in} - h_{out})\, dM_{fluid} \tag{1}$$

where M is mass, c is specific heat of the tank, T is temperature, and h is the enthalpy of the fluid. The tank is assumed to be evacuated prior to and following the childown (i.e., no initial or final mass within the tank is considered in the energy balance). The heat leak from the surroundings into the the tank are considered to be negligible relative to the change in energy of the tank walls. Both these assumptions are valid for the chilldown of a well-insulated tank in space. However, the heat leak term and the energy change of an initial and final mass within the tank could be easily incorporated into Equation 1. After integrating and rearranging, Equation 1 can be expressed as:

$$M_{fluid} = M_{tank} \int_{T_{final}}^{T_{initial}} \frac{c\ dT}{(h_{in} - h_{out})} \tag{2}$$

A theoretical lower bound for the mass of fluid required to chill the tank can be obtained by assuming the fluid is vented at the tank temperature, which decreases during the chilldown process. This is analogous to assuming an infinitely high heat transfer rate or an infinitely long time for heat transfer between the tank and fluid. The integral in Equation 2 was evaluated using third-order polynomial curve fits to specific heat data for 2219 aluminum and enthalpy of gaseous hydrogen.

This analysis is similar to that presented by Barron[6]. However, Barron assumes that the fluid is to be vented at the initial tank temperature. Although this assumption makes the integral in Equation 2 easier to calculate (since h_{out} is constant), it results in a physically unobtainable lower bound.

A theoretical upper bound was calculated by assuming only the latent heat of the fluid is available to chill the tank (i.e., the fluid is vented as saturated vapor). For this case the integral in Equation 2 is simpler, since the exit enthalpy is constant, $h_{out} = h_g$.

Effect on Chilldown Time

Figure 8 indicates that the chilldown time can be reduced by increasing the number of pulses per cycle while using the same liquid mass to chill the tank. A time savings of more than 10% can be obtained by increasing the number of pulses from one to seven. There are no simple theoretical limits for the chilldown time since it depends on heat transfer rates.

CONCLUSIONS

A numerical model of the charge/hold/vent process was modified and used to evaluate the possible benefit of pulsed injection for tank chilldown in microgravity. The pulsed injection procedure involves multiple charge/hold phases prior to each vent. Models based on numerical analyses of decaying tangential flow in a tank were used to simulate the fluid motion persistence following each pulse. For the conditions considered, no significant benefit can be realized by using pulsed injection as compared to the single-charge case.

- For pulsed injection, the wall heat transfer coefficient is maintained at a higher average level during the total hold period due to fluid motion persistence following each pulse, as compared to the single-charge case.
- For the single-charge case, the maximum heat transfer coefficient occurs nearly coincident with the maximum wall-to-fluid temperature difference, resulting in the maximum heat transfer rate. For the pulsed case, the maximum heat transfer coefficient is not coincident with the maximum wall-to-fluid temperature difference. The high heat transfer coefficient associated with each successive pulse occurs in conjunction with a decreasing wall-to-fluid temperature difference.
- The pulsed injection procedure results in very little fluid mass savings, as compared to the single-charge case having equal chilldown time.
- A small time savings can be realized by using pulsed injection. The chilldown time decreases by more than 10% as the number of pulses is increased from 1 to 7.
- Pulsed injection has the disadvantage of having to inject successive charges against increasingly higher pressures. Pulsed injection would require pumps or higher supply tank pressure, as compared to the single-charge case.

REFERENCES

1. J.R. Schuster, E.J. Russ, and J.P. Wachter, "Cryogenic On-Orbit Liquid Depot Storage, Acquisition, and Transfer Satellite (COLD-SAT)," *Feasibility Study Final Report*, NASA CR-185249, July 1990.
2. F. Merino, M.H. Blatt, and N.C. Thies, "Filling of Orbital Fluid Management Systems," NASA CR-159404, CASD-NAS-78-010, General Dynamics Convair Division, July 1978.
3. S.C. Honkonen, F.O. Bennett, and H.K. Hepworth, "An Analytic Model for Low-Gravity Tank Chilldown and No-Vent Fill: The General Dynamics No-Vent Fill Program (GDNVF)," AIAA Paper 91-1380, June 1991.
4. J.R. Pietrzyk, S.C. Honkonen, and J.R. Schuster, "Fluid Motion Persistence in Microgravity Receiver Tank Chilldown," International Astronautical Federation Paper IAF-90-349, October 1990.
5. S.C. Honkonen *et al.*, "Analysis of Cryogenic Fluid Systems in Low-g," Report No. GDSS-ERR-89-406, General Dynamics Space Systems Division, December 1989.
6. R.F. Barron, *Cryogenic Systems*, Second Edition, Oxford University Press, New York (1985), pp 418-422.

AUTOGENOUS PRESSURIZATION OF CRYOGENIC VESSELS USING SUBMERGED VAPOR INJECTION

Robert J. Stochl
Neil T. Van Dresar
Raymond F. Lacovic

National Aeronautics and Space Administration
Lewis Research Center
Cleveland, Ohio

ABSTRACT

Experimental results are reported for submerged injection pressurization and expulsion tests of a 4.89 m^3 liquid hydrogen tank. The pressurant injector was positioned near the bottom of the test vessel to simulate liquid engulfment of the pressurant gas inlet; a condition that may occur in low-gravity conditions. Results indicate a substantial reduction in pressurization efficiency, with pressurant gas requirements approximately five times greater than ideal amounts. Consequently, submerged vapor injection should be avoided as a low-gravity autogenous pressurization method whenever possible. The work presented herein validates that pressurant requirements are accurately predicted by a homogeneous thermodynamic model when the submerged injection technique is employed.

INTRODUCTION

Future space flight will require the transfer of cryogenic liquids under low-gravity (low-g) conditions for use in chemical and nuclear propulsion, life support, and thermal control. Conventional pressurization of settled cryogenic tanks utilizes hardware that diffuses the pressurant flow within the ullage in a manner which minimizes impingement of the pressurant on the liquid-vapor interface or tank walls. In the low-g environment, the distribution of liquid and vapor phases may not be well defined and it becomes difficult to ensure that the pressurant is injected directly into the tank ullage. It is possible that direct injection of the pressurant into the bulk liquid will occur during liquid reorientation, sloshing, or even static conditions. For all of these conditions, the pressurant-liquid interaction may lead to either evaporation of the liquid or condensation of the pressurant gas, depending upon the complex heat and mass transfer processes involved.

Previous studies of submerged gas injection include an experimental investigation of helium gas injection into liquid hydrogen (LH_2) by Johnson[1]. For this situation, interaction of the non-condensible helium gas and LH_2 leads to vaporization of a portion of the LH_2, which in some cases can reduce the required amount of pressurant gas. However, when a condensible pressurant is used for tank pressurization, the potential for pressurant condensation (collapse) is high. This is noted in experiments performed by DeWitt and McIntire[2] with liquid methane. When the pressurant was directly injected into the ullage, liquid sloshing increased the pressurant requirement for a condensible pressurant (methane) and decreased the pressurant requirement for non-condensible pressurants (helium and hydrogen). Finally, the interaction of the pressurant with the liquid frequently results in undesirable liquid heating.

Advances in Cryogenic Engineering, Vol. 37, Part B
Edited by R.W. Fast, Plenum Press, New York, 1992

The results reported herein are concerned with autogenous tank pressurization and expulsion of liquid hydrogen (LH_2) in a normal-g test environment. The gaseous hydrogen (GH_2) pressurant was injected into the tank well below the liquid level in order to simulate the increased interaction with the liquid cryogen that may occur in low-g applications. Data was obtained in a LH_2 tank having characteristics similar to propellent tanks of future spacecraft.

EXPERIMENTAL APPARATUS

The test facility (see Fig. 1) consists of a 7.6 m diameter vacuum chamber containing a 4.0 m diameter cylindrical shroud that in turn encloses the LH_2 test tank. The shroud was maintained at a constant temperature of 295 K by electrical resistance heating to obtain a constant heat input to the test article. This heat input includes penetration heat leaks through the insulation system and is determined from boil-off data[3]. Vacuum chamber pressure during the test series was on the order of 10^{-4} kPa. The test tank is suspended by fiberglass composite struts and all instrumentation lines and flow lines other than the pressurant line are routed though a LH_2 cold guard (not shown) to minimize conductive heat transfer to the test article. Pressurant (normal-GH_2) is supplied from outside high pressure storage bottles. A steam heat exchanger is used to heat the pressurant to 330 K, or the heat exchanger may be bypassed to provide ambient temperature gas at 275 K. Pressurant flow rate is calculated from pressure drop measurements across a square edged orifice (1.15 cm in diameter) placed in the pressurant line. The orifice is instrumented with high and low range differential pressure transducers as well as upstream pressure and temperature transducers. A tank bypass line allows the pressurant line to be thermally conditioned prior to a test when heated pressurant is required. Expelled LH_2 from the test article is returned via a transfer line to an outside storage dewar. The transfer line is instrumented with a venturi flow meter to measure expulsion flow rate. Pressure and temperature transducers, as indicated in Fig. 1, are located on the outflow line to measure the thermodynamic state of flow out of the test vessel.

The LH_2 test tank is approximately an ellipsoidal volume of revolution having a major-to-minor axis ratio of 1.2, a major diameter of 2.2 m, a volume of 4.89 m^3, and a mass of 149 kg. It is constructed of 2219 aluminum. Most of the wall is 2.08 mm thick except for the thick bolted flange and lid at the top, thickened lands for support lugs, and a thickened equatorial region. It is insulated with 34 layers of double aluminized Mylar separated by silk netting. The tank insulation, size, and lightweight construction (other than the lid) are representative of a cryogenic storage tank that could be used in future spacecraft. Pressurant is fed into the test article through a "j-tube" which directs the gas into the LH_2. The j-tube's inside diameter is 1.2 cm and it has a 0.95 cm full cone spray nozzle attached at the outlet.

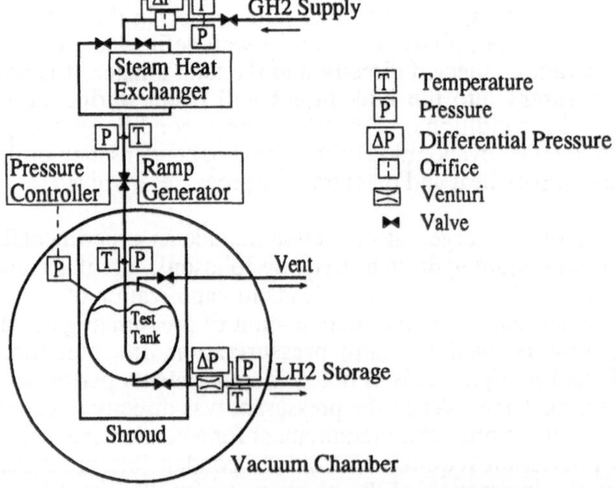

Fig. 1. Schematic diagram of test facility.

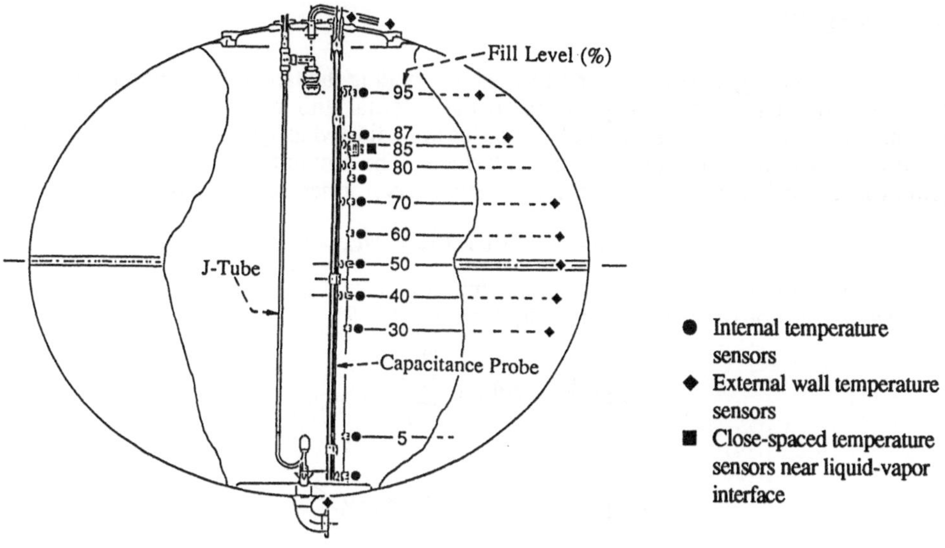

Fig. 2. Schematic diagram of test tank and instrumentation.

The total length of the j-tube from the attachment point at the tank lid to the outlet is approximately 2 m.

Figure 2 is a schematic diagram that indicates the location of the j-tube and various temperature sensors. The j–tube exit is approximately 25 cm from the tank bottom. Liquid fill level in the tank is measured by a capacitance probe, and liquid-vapor temperatures are measured by silicon diode transducers. The external wall temperature distribution is measured by a number of wall-mounted silicon diode transducers. Tank pressure is measured by pressure transducers in direct communication with the tank ullage. Liquid-vapor temperature measurements inside the tank are accurate to ± 0.3 K, while wall temperatures are accurate to ± 0.6 K. An in situ calibration increases the accuracy of liquid-vapor temperature measurements to ± 0.1 K by adjusting the individual sensor readings to known saturation conditions. Tank pressure measurements are accurate to ± 0.01 kPa. Capacitance probe readings are accurate to ± 1.9 cm, translating to a maximum error of ± 1.5 percent fill at the 50 percent fill level (by volume). Pressurant gas flow rate measurements have an estimated accuracy of ± 0.18 and ± 0.40 kg/hr for the large orifice using the low and high range differential pressure transducers, respectively. Liquid outflow could not be properly measured due to cavitation in the venturi; instead it was determined from liquid level change in the tank. Data is sampled by an automated data acquisition system at selected intervals (15 to 60 sec) throughout the duration of the experiments.

TEST PROCEDURE

The tank is prepared for a test by filling to the desired fill level while the tank pressure is maintained at least 15 kPa above atmospheric pressure. If heated pressurant is used, the tank bypass line is opened and the pressurant line is thermally conditioned until the temperature transducer near the j-tube inlet indicates the desired gas temperature. Next the tank is vented to the atmosphere to induce substantial bulk boiling of the tank liquid which produces nearly isothermal conditions within the tank. A venting period of approximately 15 min is necessary to obtain saturated liquid temperatures throughout the tank. A test is initiated by closing the vent line valves and opening the pressurant line valves. In the first portion of a test, a preset tank pressure ramp rate is maintained by controlling the pressurant flow control valve with an automatic ramp generator. After the maximum tank pressure is attained, a 2 min hold period follows during which control of the pressurant flow valve is switched to an automatic pressure controller. The tank pressure is kept constant by addition of pressurant during liquid expulsion. Liquid outflow is regulated by remote operation of flow control valves in the outflow line. Expulsion is stopped at a nominal 5 percent fill level. Data is automatically recorded at regular intervals throughout the duration of the test.

DATA ANALYSIS

Mass and energy balances are performed by dividing the tank interior volume and wall into horizontal segments corresponding to the internal and wall-mounted temperature sensors. At any given time, segment boundaries are adjusted as necessary to accommodate the variable location of the liquid-vapor interface. The amount of vapor condensation or liquid evaporation is determined from a mass balance performed on the ullage volume:

$$\pm M_{t,i\rightarrow f} = M_{u,f} - M_{u,i} - M_{G,i\rightarrow f} \tag{1}$$

A positive value indicates net evaporation. Initial and final ullage masses are obtained by numerical integration of the density profiles where $\rho=f(T,P)$:

$$M_{u,i} = \int_{V_{u,i}} \rho dV \cong \sum_{n=1}^{N_i} \rho_{n,i} V_{u_{i,n}} \tag{2}$$

$$M_{u,f} = \int_{V_{u,f}} \rho dV \cong \sum_{n=1}^{N_f} \rho_{n,f} V_{u_{f,n}} \tag{3}$$

The mass of injected pressurant is calculated by numerical integration of instantaneous flow rate measurements obtained from the calibrated orifice in the pressurant line.

A thermodynamic analysis was performed by applying the first law to the wall and the liquid and vapor contents of the tank. No external work is performed and if kinetic and potential energy terms are neglected, the tank energy balance is:

$$\Delta U_{T,i\rightarrow f} = \int_{t_i}^{t_f} \dot{M}_G h_G dt + \int_{t_i}^{t_f} \dot{Q} dt - \int_{t_i}^{t_f} \dot{M}_L h_L dt \tag{4}$$

Energy added to the tank consists of energy input by the pressurant plus heat leak from the environment minus energy of the liquid outflow from the tank. The various integrals on the right hand side were numerically calculated. The enthalpy of the pressurant was obtained using the measured gas temperature at the tank inlet. Enthalpy of the liquid outflow was based on average values of measured temperature near the tank outlet at the constant expulsion pressure. An average heat leak rate of 28 W times the test duration gives the total heat leak. The energy input results in thermal heating of the vapor, liquid, and tank wall:

$$\Delta U_{T,i\rightarrow f} = \Delta U_{u,i\rightarrow f} + \Delta U_{L,i\rightarrow f} + \Delta U_{w,i\rightarrow f} \tag{5}$$

The quantities on the right hand side of Eq. 5 were calculated as follows:

$$\Delta U_{u,i\rightarrow f} \cong \sum_{n=1}^{N_f} \rho_u \left(h_u - \frac{P}{\rho_u}\right) V_{u_{f,n}} - \sum_{n=1}^{N_i} \rho_u \left(h_u - \frac{P}{\rho_u}\right) V_{u_{i,n}} \tag{6}$$

$$\Delta U_{L,i\rightarrow f} \cong \sum_{n=1}^{N_f} \rho_L \left(h_L - \frac{P}{\rho_L}\right) V_{L_{f,n}} - \sum_{n=1}^{N_i} \rho_L \left(h_L - \frac{P}{\rho_L}\right) V_{L_{i,n}} - \int_{t_i}^{t_f} \dot{M}_L h_L dt \tag{7}$$

$$\Delta U_{w,i\rightarrow f} \cong \sum_{n=1}^{N_w} M_w \int_{T_{i,n}}^{T_{f,n}} C_w dT \tag{8}$$

where ρ and h are functions of temperature and pressure and C_w is the specific heat of the tank wall material. Dropping the i,f subscripts, Eq. 5 may be rearranged as:

$$1 = \frac{\Delta U_L}{\Delta U_T} + \frac{\Delta U_u}{\Delta U_T} + \frac{\Delta U_w}{\Delta U_T} \tag{9}$$

The overall energy balance is then utilized to analyze the resulting distribution of the added energy to the liquid, vapor, and tank wall regions.

Experimentally determined pressurant requirements may be compared to two simple analytical models. The first model gives the so called "worst case" pressurant requirement. It assumes that the pressurant attains thermal equilibrium with the tank contents, i.e. a homogeneous thermodynamic state. Under cryogenic conditions, the energy increase of the tank wall may be neglected. Solutions for the thermal equilibrium prediction are obtained by combining the mass and energy balances applied to the tank contents:

$$M_G = \frac{M_f U_f - M_i U_i + (M_i - M_f) h_L}{h_G - h_L} \tag{10}$$

The second model assumes no energy or mass transfer occurs between the pressurant and the tank or the initial tank contents. This model provides the so called "ideal" pressurant requirements. It is formulated assuming that the initial ullage mass is isentropically compressed during the ramp process. The remaining portion of the initial ullage volume plus the volume vacated by the liquid during expulsion is assumed to be occupied by added pressurant which undergoes an isentropic expansion from its supply condition. The ideal mass requirement for specified initial and final fill levels is:

$$M_G = \rho_{G,f} V_t \left[1 - F_f - (1 - F_i)\frac{\rho_{u,i}}{\rho_{u,f}} \right] \quad \text{where } \rho_{G,f} = f(P_f, s_G) \tag{11}$$

TEST RESULTS

A series of experiments were performed in which the effects of ramp duration, expulsion time, and pressurant gas temperature were investigated. A test summary is provided in Table 1. All tests began with the tank vented to atmospheric pressure (99 to 107 kPa) followed by a ramp pressurization to approximately 275 kPa. Initial liquid fill levels were 84 percent. Liquid expulsions were limited to 4 tests where the final tank fill level was 7 percent (5 percent for Test No. 6).

The tank pressure history for Test No. 5R is shown in Fig. 3 which consisted of a ramp pressurization process followed by a 2 min hold period and then a constant pressure expulsion of the liquid. During the ramp process, the tank pressure was increased from 99 to 275 kPa. Next was the hold period, followed by liquid expulsion from the 84 to 7 percent fill level. The expulsion occurred at constant tank pressure except for a small pressure drop (13 kPa) experienced when the liquid outflow valve is first opened. Pressure histories for the other ramp and expulsion tests listed in Table 1 are similar to that shown in Fig. 3 except for differences due to the parametric variation of the ramp and expulsion rates.

Representative internal tank temperatures (measured near the vertical tank axis) are shown in Fig. 4 for Test No. 5R. Three of the measurement locations were initially below

Table 1. Test Summary

Test No.	Pressurant Temperature (K)	Ramp Duration (min)	Expulsion Duration (min)	Pressurant Consumption Ramp (kg)	Pressurant Consumption Expulsion (kg)	Pressurant Consumption Total (kg)	Evaporated Mass Ramp (kg)	Evaporated Mass Expulsion (kg)
1	275	27	none	3.0	n/a	3.0	-2.4	n/a
2	275	20	none	3.0	n/a	3.0	-2.2	n/a
3	275	18	none	2.9	n/a	2.9	-2.1	n/a
5	275	21	15	3.0	1.6	4.7	-2.3	7.5
5R	275	13	15	3.1	1.7	4.8	-2.0	7.6
6	275	13	25	3.0	1.6	4.6	-1.9	6.2
9	330	12	15	2.4	1.4	3.7	-1.3	7.2

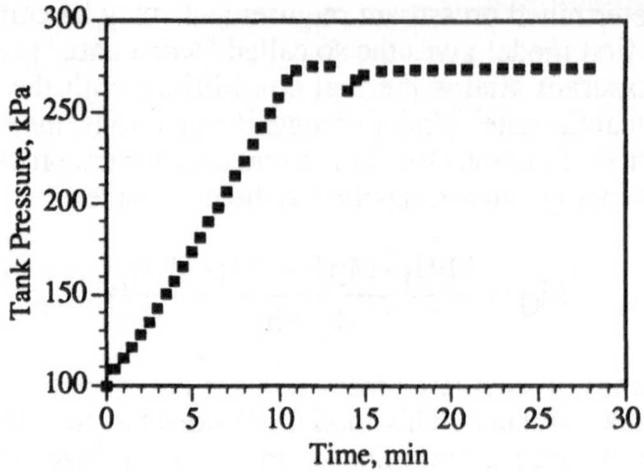

Fig. 3. Tank pressure versus time for Test No. 5R.
Ramp pressurization followed by liquid expulsion.

the liquid level while the remaining location was at all times in the vapor region. All of the liquid temperatures are in close agreement during the ramp process; increasing with time. At the end of the ramp period the liquid temperatures are approximately 0.3 to 0.4 K less than the saturation temperature of LH_2 at 275 kPa. The uppermost temperature was slightly above the saturation temperature during the initial portion of ramp process and then rapidly increased thereafter except for a brief temperature drop attributed to the sudden pressure drop at the start of the expulsion period. Within a few minutes after outflow began, the liquid temperatures reached the saturation temperature corresponding to the expulsion pressure. Two of the temperature sensors became exposed to the ullage during the expulsion and exhibited a steady temperature rise for the remainder of the test as the surrounding vapor becomes superheated. In all of the tests, substantial liquid heating occurred due to the submerged injection of the pressurant gas.

Ramp duration, ranging from 12 to 27 min, did not have a significant effect on the pressurant energy input for the ramp pressurization tests. As shown in Table 1, for gas temperatures of 275 K, the amount of injected pressurant was approximately 3 kg, while at the hotter gas temperature of 330 K, the pressurant mass was 2.4 kg. Total energy input for all ramp tests was 11,500 kJ ± 4 percent.

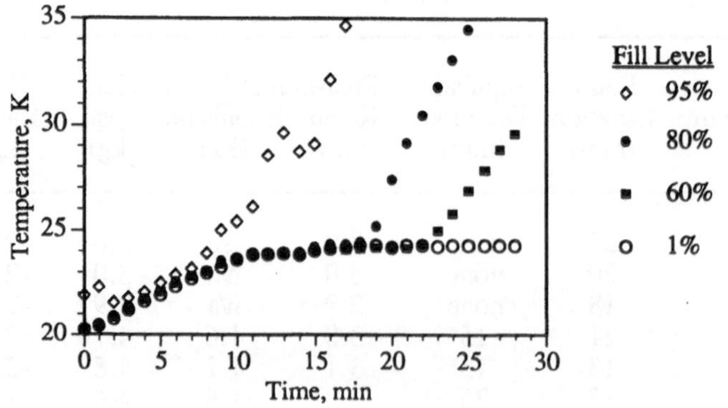

Fig. 4. Internal tank temperature histories for Test No. 5R. Ramp pressurization at 84 percent fill followed by expulsion to 7 percent fill.

Liquid expulsion time also did not have a significant effect on the pressurant energy input. Results were obtained for expulsion durations of approximately 15 and 25 min. As was the case with ramp pressurization, less pressurant mass was needed when the pressurant temperature was increased, with the energy input remaining the same for the two gas temperatures. Total energy input for the expulsion tests, including the ramp period, was 18,100 kJ ± 3 percent.

Mass balances on the vapor region indicate that 63 to 80 percent of the 275 K pressurant gas and 54 percent of the 330 K pressurant gas condenses during ramp pressurization. For the combined ramp and expulsion processes, the mass analysis indicates that net evaporation of the liquid occurs, with the amount of evaporated mass being of the same order as that of the injected pressurant.

Energy balances applied to the combined ramp and expulsion tests were found to balance to within 3 percent. Less than 0.4 percent of the total energy input (ΔU_T) was due to the tank heat leak. Using the analysis described above, it was found that approximately 89 percent of the incoming energy went into liquid heating ($\Delta U_l/\Delta U_T$), 10 percent into vapor heating ($\Delta U_u/\Delta U_T$), and 1 percent was absorbed by the tank wall ($\Delta U_w/\Delta U_T$). This distribution is in good agreement with thermal equilibrium calculations from the homogeneous model. Energy balances applied to only the ramp process were found to be in error by as much as 30 percent, with the calculated liquid heating exceeding the energy supplied by the pressurant. It is theorized that the error is due to the existence of radial temperature gradients in the liquid, with liquid heating away from the central vertical axis lagging that near the axis where measurements were obtained. The radial temperature gradients are thought to be most significant in the liquid region at the end of the ramp period.

Comparison of the experimental results with pressurant requirements predicted by the homogeneous model (Eq. 10) for the combined ramp and expulsion processes shows agreement within 6 percent, with the measured values generally exceeding predictions, as shown in Table 2. For ramp pressurization only, it is seen that the predicted pressurant mass is more than the experimentally measured values. This result is plausible if the liquid has radial temperature gradients.

The last two columns in Table 2 list the ideal pressurant requirements calculated from Eq. 11. The measured total pressurant consumption (for combined ramp and expulsion) exceeds the ideal amounts by a factor of approximately five. For other initial and/or final fill levels, this factor will vary. For the ramp pressurization process only, the factor ranges from 41 to 46. Actual values of this "collapse factor" for direct ullage pressurization fluctuate according to diffuser design and numerous other conditions. Generally, well designed direct ullage pressurization systems have collapse factors that are substantially less than five.

Table 2. Comparison of Measured and Predicted Pressurant Gas Requirements

Test No.	Pressurant Temperature (K)	Measured Ramp (kg)	Measured Total (kg)	Homogeneous Ramp (kg)	Homogeneous Total (kg)	Ideal Ramp (kg)	Ideal Total (kg)
1	275	3.0	n/a	3.2	n/a	0.066	n/a
2	275	3.0	n/a	3.2	n/a	0.065	n/a
3	275	2.9	n/a	3.2	n/a	0.068	n/a
5	275	3.0	4.7	3.2	4.5	0.066	0.90
5R	275	3.1	4.8	3.2	4.5	0.067	0.90
6	275	3.0	4.6	3.1	4.4	0.065	0.92
9	330	2.4	3.7	2.6	3.7	0.058	0.75

CONCLUSIONS

Normal-g pressurization and expulsion tests from the 84 to 7 percent fill level were conducted to simulate tank pressurization in a low-g environment. Autogenous pressurization of LH_2 by the submerged vapor injection technique produces substantial liquid heating. For ramp durations less than 30 min, the liquid heating does not appear to be radially uniform. After the expulsion process begins, liquid in the tank reaches the saturation temperature and approaches the homogeneous thermal state. Measured pressurant requirements for the combined ramp and expulsion processes were predicted by the thermal equilibrium analysis (homogeneous model) to within seven percent. Pressurant gas requirements exceed ideal requirements by a factor of approximately five. For spacecraft design, submerged pressurant injection should be avoided whenever possible due to the excessive amount of pressurant needed and the undesirable liquid heating. If submerged injection cannot be precluded, the thermal equilibrium model should be used to determine pressurant requirements.

NOMENCLATURE

C	specific heat
F	fill level (liquid volume/total volume)
h	specific enthalpy
M	mass
$\dot{M}$	mass flow rate
N	number of segments
P	pressure
$\dot{Q}$	heat leak rate
s	specific entropy
T	temperature
t	time
U	internal energy
V	volume
ρ	density

Subscripts

f	final
G	gas (pressurant)
i	initial
L	liquid
n	summation index
T	total
t	transfer due to phase change
u	ullage
w	wall

ACKNOWLEDGEMENTS

The authors would like to acknowledge the support of the many people who participated in the K-Site Phase 1B test program at Plum Brook Station.

REFERENCES

1. W. R. Johnson, "Helium Pressurant Requirements For Liquid-Hydrogen Expulsion Using Submerged Gas Injection," NASA TN D-4102 (1967).
2. R. L. DeWitt and T. O. McIntire, "Pressurant Requirements For Discharge of Liquid Methane From a 1.52 meter (5 ft) Diameter Spherical Tank Under Both Static and Slosh Conditions," NASA TN-D 7638 (1974).
3. R. Stochl and R. Knoll, "Thermal Performance of a Tank Applied Multi-Layer (34 Layers) Insulation System at Hot Boundary Temperatures of 630, 530, and 140 °R," AIAA paper 91-2400 (1991).

SLOSH WAVE EXCITATION OF CRYOGENIC LIQUID HELIUM IN GRAVITY PROBE-B ROTATING DEWAR

R. J. Hung*, C. C. Lee* and F. W. Leslie+

*The University of Alabama in Huntsville
Huntsville, Alabama

+NASA/Marshall Space Flight Center
Huntsville, Alabama

ABSTRACT

The dynamical behavior of fluids, in particular the effect of surface tension on partially-filled rotatinng fluids (cryogenic liquid helium and helium vapor) in a full scale Gravity Probe-B Spacecraft propellant dewar tank imposed by various frequencies of gravity jitters have been investigated. Fluid stress distribution, caused by the excitation of slosh waves and their associated large amplitude disturbances on the liquid-vapor interface, exerted on the outer and inner walls of rotating dewar container also have been investigated. Results show that fluid stress distribution exerted on the outer and inner walls of rotating dewar are closely related to the characteristics of slosh waves excited on the liquid-vapor interface in the rotating dewar tank.

INTRODUCTION

The Gravity Probe-B (GP-B) Spacecraft is a relativity gyroscope experiment to test two extraordinary, unverified prediction of Albert Einstein's general theory of relativity[1,2]. The requirement for an operational lifetime approaching one year means that a large quantity of cryogenic liquid helium must be used, and that it will be gradually depleted over the lifetime of the experiment. This varying amount of liquid helium gives rise to the possibility of several problems which can degrade the GP-B experiment. The potential problems could be due to asymmetry in the static liquid helium distribution or to perturbations in the free surface.

The equilibrium shape of free surface is governed by a balance of capillary, centrifugal and gravitational forces. In contrasting the effects of surface tension to the effects of gravitational forces on the free surface of liquid, it was found that the surface tension force for most liquids is greater than the gravitational force in a 10^{-8} g_0 level and lower[3-5]. The instability of liquid surface can be induced by the presence of longitudinal and lateral accelerations, vehicle vibration, and rotational fields of spacecraft in a microgravity environment. Slosh waves are, thus, excited which produces high and low frequency oscillations in the liquid propellant. The sources of the residual accelerations range from the effects of the Earth's gravity gradient, atmospheric drag on the spacecraft, and spacecraft attitude motions to the various frequency ranges of "g-jitter" arising from machinery vibrations, thruster firings, and crew motions. Recent study[6-8] suggests that the high frequency accelerations may be unimportant in comparison to the residual motions caused by low frequency accelerations.

Cryogenic liquid propellant is positioned over the tank outlet by using small auxiliary thrusters (or idle-mode thrusters from the main engine) which provide a thrust parallel to the tank's major axis in the direction of flight[9].

Time-dependent dynamical behavior of surface tension in both low gravity and microgravity environments was carried out[5,10-14]. At the interface between the liquid and the gaseous fluids, both the kinematic surface boundary condition, and the interface stress conditions for components tangential and normal to the interface, were applied[10-14]. The experiments carried out by Mason et al[20] showed that the classical fluid mechanics theory is applicable for cryogenic liquid helium in large containers[15,16].

In this study, time-dependent computation have been carried out to investigate the dynamical behaviors of cryogenic liquid helium and helium vapor in the geometry of full scale GP-B propellant dewar container under a microgravity environment. The computation also extends to the study of sloshing waves induced by the various frequencies of gravity-jitters and variations in different rotating speeds of GP-B dewar container and different background gravitational fields. In this study, slosh wave induced stress distribution on the external and internal walls of propellant dewar container will be investigated.

DYNAMICAL BEHAVIORS OF LIQUID-VAPOR INTERFACES

Detailed description of time-dependent mathematical formulation, initial and boundary conditions, together with computational algorithm applicable to cryogenic fluid management under microgravity environment are given in our earlier studies[5,7,8,10-14].

Under microgravity environment, gravity jitters are produced by spacecraft attitude motion, machinery (turbine, pump, engine, etc.) vibrations, thruster firing, thruster shutdown, etc.[6-9,12,14]. Vibration of the gravity environment (gravity-jitters) is governed by the following equation:

$$g = g_B \left[1 + \frac{1}{2} \sin(2\pi f t) \right]$$

where g_B denotes the background gravity environment, and f(Hz) stands for the frequency of gravity-jitters.

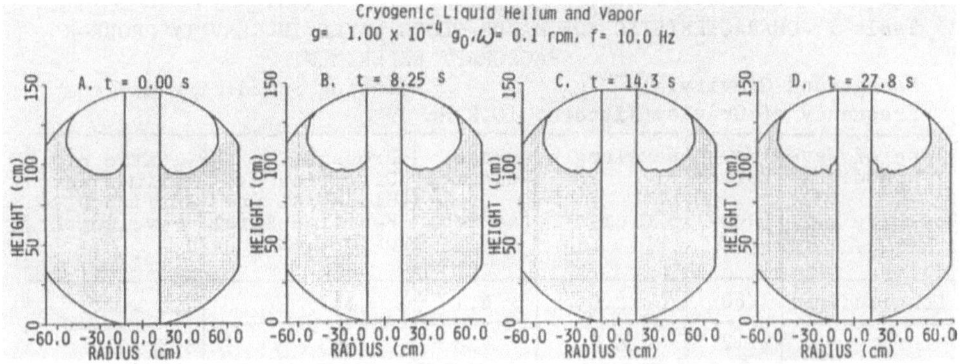

Figure 1(A-D) Time sequence of dynamical evolutions.

Figure 1 shows the time sequences evolution of the dynamical behaviors of liquid-vapor interface oscillations driven by the restoring force fields of gravity jitters with frequency of 10 Hz.

SLOSH WAVE EXCITATION IN A PARTIALLY FILLED ROTATING DEWAR DUE TO GRAVITY JITTERS

Time series of wave amplitudes for liquid-vapor interface fluctuations can be obtained from the numerical simulations of the time sequence interface oscillations accomplished in the Figure 1.

Wave period of various modes of slosh waves can be determined from the Fourier spectral analysis of time series at z_1, z_2, and z_3 while wavelength, phase velocity and propagation direction of slosh waves can be deduced from the cross-correlation analysis of the combinations of any two time series selected out of the three time series[11-20]. Figure 2 shows a sample Fourier power spectral

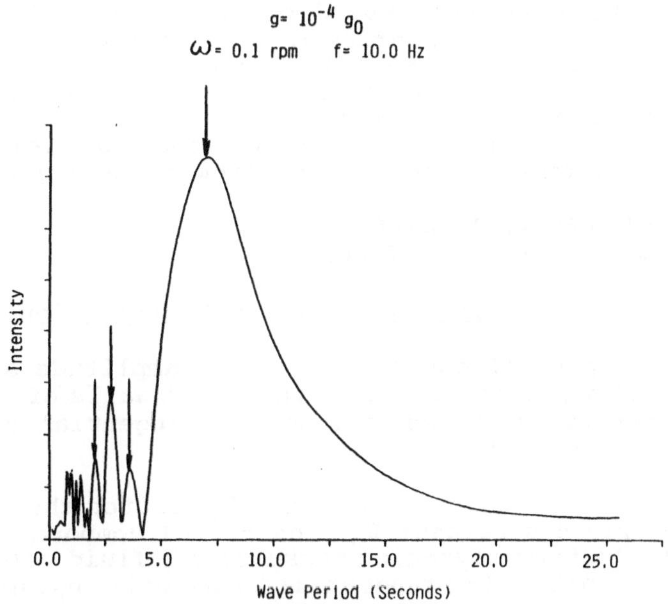

Figure 2 Sample Fourier power spectral analysis of time series.

Table 1 CHARACTERISTICS OF MAJOR SLOSH WAVES IN GRAVITY PROBE-B SPACECRAFT EXPERIMENT

Background Gravity: 10^{-4} g_0 Rotating Speed: $\omega = 0.1$ rpm

Frequency of Gravity Jitters: 10.0 Hz

Type of Wave Mode	Wave Period (sec)	Wavelength (cm)	Phase Velocity (cm/sec)	Propagation Direction (Clockwise from Positive Axial Direction) (Deg)	Ratio of Maximum Wave Amplitude to Wavelength (A/ λ)
Longitudinal	3.60	22.0	6.1	147	2.5×10^{-3}
Transverse	2.12	16.5	7.8	-63	3.8×10^{-3}
	2.65	39.5	14.9	86	2.9×10^{-3}
	6.58	13.2	2.0	101	2.32×10^{-2}

analysis of time series at z_1. A filter shall be properly chosen to apply certain ranges of window for separating these four wave nodes from each other in three time series for the determination of wavelength, phase velocity and propagation direction of slosh waves[17-20].

Table 1 shows the characteristics of major slosh waves caused by the 10 Hz gravity-jitter frequency under $10^{-4}g_0$ background gravity environment and 0.1 rpm rotating speed. Propagation direction (in degree) is measured clockwise from positive axial direction. The ratio of maximum wave amplitude to wavelength, Max (A/λ), [where A stands the wave amplitude; and λ, the wavelength of slosh wave] for each slosh wave is calculated based on the following procedures: (a) determine each wave mode based on the peaks of power intensity shown in the power spectral density analysis in the Fourier domain (see Figure 2); (b) apply proper window of filter to separate each wave mode shown on the peaks of power intensity; (c) calculate wavelength, phase velocity and propagation direction of each wave mode from the cross-correlation analysis based on the separated wave modes through filtering the time series in Fourier domain at locations z_1, z_2, and z_3; (d) reverse Fourier transform the separated wave mode through filtering the time series in Fourier domain to time domain and obtain the amplitude of each mode of slosh waves; and (e) compute the ratio of maximum wave amplitude to wavelength [Max (A/λ)] for each mode of slosh waves from items (c) and (d).

SLOSH WAVE INDUCED FLUID STRESS DISTRIBUTIONS ON THE DEWAR WALLS

(A) Mathematical Formulation of STress Distributions

For the purpose of considering large amplitude slosh wave modified fluid stresses acting on the solid walls of dewar, the fluid stresses are decomposed into the tangential and normal components to the walls.

We introduce symbols as follows: Π_t denotes the tangential component of fluid stresses; Π_n, the normal component of fluid stresses; P, the thermodynamic pressure; u_α, fluid velocity in α direction; t_α, unit vector tangential to the wall; n_β, unit vector normal to the wall; μ, the molecular viscosity coefficient of fluid; and $\delta_{\alpha\beta}$, the Dirac delta function. Subscripts α and β imply the directions of flow fields.

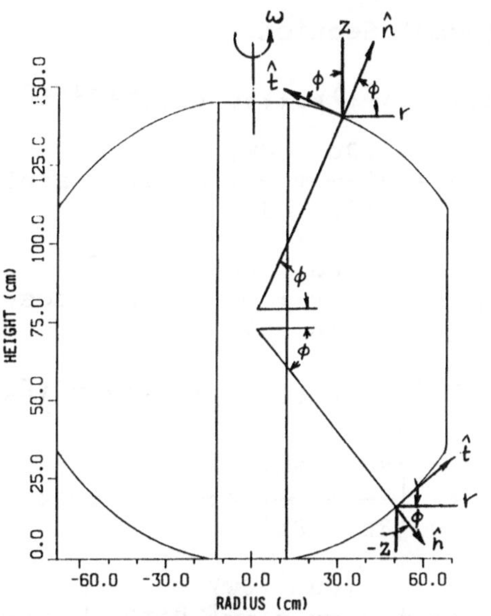

Figure 3 Geometry of Gravity Probe-B Spacecraft dewar propellant tank with coordinate system.

Figure 3 shows the geometry of GP-B dewar propellant tank. In order to make the computation of fluid stresses matching the geometry of dewar tank, mathematical formulations have been divided into three sections: (A) Top wall (dome) section, (B) Bottom wall (dome) section, and (C) Cylindrical section. We define ϕ as the azimuth angle of the dome. Velocity components in cylindrical coordinate of (r, θ, z) are shown as (u, v, w).

Fluid stresses exerted on Top Wall, Bottom Wall and Cylindrical Sections can be shown as follows:

(a) Top Wall (Dome) Section:

$$(\Pi_{ax})_{Top\ wall} = (\Pi_t)_{Top\ wall} \sin\phi + (\Pi_n)_{Top\ wall} \cos\phi$$

$$= \mu\left(\frac{\partial u}{\partial z} + \frac{\partial w}{\partial r}\right) \cos 2\phi \cdot \sin\phi$$

$$+ \left[P - \mu\left(\frac{\partial u}{\partial z} + \frac{\partial w}{\partial r}\right) \sin 2\phi\right] \cos\phi \qquad (1)$$

$$(\Pi_{ra})_{Top\ wall} = (\Pi_t)_{Top\ wall} \cos\phi + (\Pi_n)_{Top\ wall} \sin\phi$$

$$= \mu\left(\frac{\partial u}{\partial z} + \frac{\partial w}{\partial r}\right) \cos\phi \cdot \cos 2\phi$$

$$+ \left[P - \mu\left(\frac{\partial u}{\partial z} + \frac{\partial w}{\partial r}\right) \sin 2\phi\right] \sin\phi \qquad (2)$$

(b) Bottom Wall (Dome) Section:

$$(\Pi_{ax})_{\text{Bottom wall}} = (\Pi_t)_{\text{Bottom wall}}\sin\phi - (\Pi_n)_{\text{Bottom wall}}\cos\phi$$

$$= \mu\left(\frac{\partial u}{\partial z} + \frac{\partial w}{\partial r}\right) \cos 2\phi \cdot \sin 2\phi$$

$$+ \left[-P + \mu\left(\frac{\partial u}{\partial z} + \frac{\partial w}{\partial r}\right) \sin 2\phi \right] \cos\phi \tag{3}$$

$$(\Pi_{ra})_{\text{Bottom wall}} = (\Pi_t)_{\text{Bottom wall}}\cos\phi + (\Pi_n)_{\text{Bottom wall}}\sin\phi$$

$$= \mu\left(\frac{\partial u}{\partial z} + \frac{\partial w}{\partial r}\right) \cos\phi \cdot \cos 2\phi$$

$$+ \left[-P + \mu\left(\frac{\partial u}{\partial z} + \frac{\partial w}{\partial r}\right) \sin 2\phi \right] \cos\phi \tag{4}$$

(c) Cylindrical Section:

$$(\Pi_{ax})_{\text{Cylindrical}} = (\Pi_t)_{\text{Cylindrical}}$$

$$= \mu\left(\frac{\partial u}{\partial z} + \frac{\partial w}{\partial r}\right) \tag{5}$$

$$(\Pi_{ra})_{\text{Cylindrical}} = (\Pi_n)_{\text{Cylindrical}}$$

$$= P \tag{6}$$

where Π_{ax} denotes axial components of stress, and Π_{ra} expresses radial component of stress.

Total stress and direction of stress at any location of interests can be expressed as

$$\Pi = [(\Pi_{ax})^2 + (\Pi_{ra})^2]^{1/2} \tag{7}$$

$$\phi = \tan^{-1}\left(\frac{\Pi_{ax}}{\Pi_{ra}}\right) \tag{8}$$

In this study, flow fields of partially liquid filled rotating dewar have been numerically computed and shown. With the computed results of P, $\partial u/\partial z$, and $\partial w/\partial r$ at each grid point along the wall, one can seek stress distributions at every location along internal and external walls of rotating dewar tank.

(B) Slosh Wave Induced Fluid Stress Distributions

Fluid stress distribution at the interfaces of liquid-solid, vapor-solid and liquid-vapor under the dynamical impact of slosh wave excitation can be computed. In this study, we are particularly interested in the fluid (liquid and vapor) stresses exerted on the solid walls of dewar container, including outer and

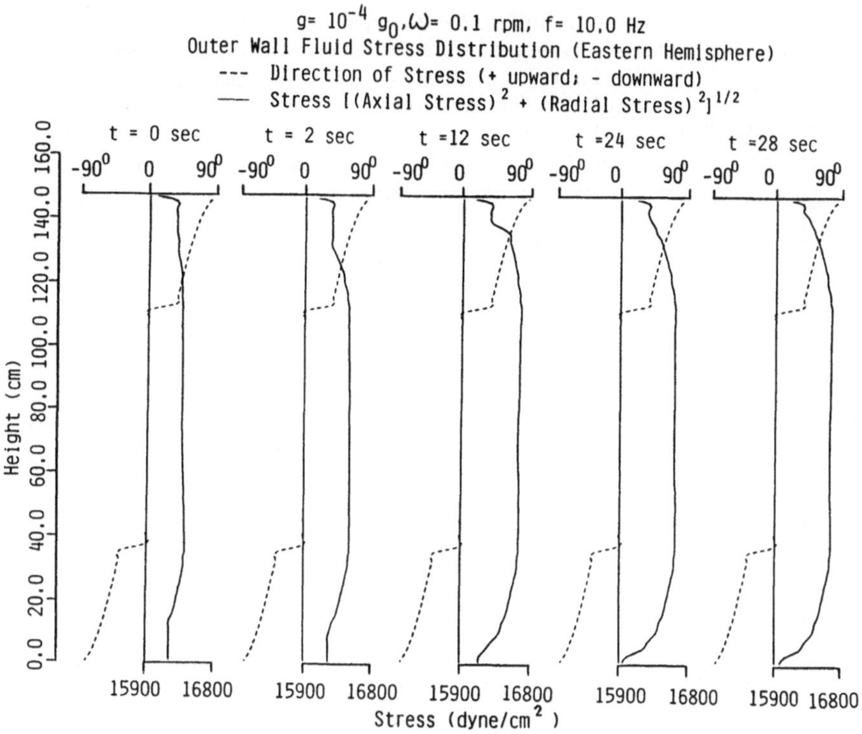

Figure 4 Time evolution of slosh wave disturbance-associated fluid stress distribution exerted on the outer wall.

inner walls of dewar tank. Figure 3 shows that the outer wall contains three sections (top dome, cylindrical, and bottom dome); while the inner wall is in a profile of cylinder.

Figure 4 shows time sequence distributions of fluid stresses exerted on the outer wall, while Figure 5 illustrates time sequence distributions of fluid stresses exerted on the inner wall. Solid lines in these figures show the magnitude of fluid stresses with the unit of dyne/cm^2 (scale shown in the lower side of horizontal axis), and dotted lines illustrate the direction of fluid stress (scale shown in the upper side of horizontal axis with positive angle indicating upward direction; negative angle, the downward direction; and zero angle, the horizontal direction). Vertical axis, in these figures, shows the values at corresponding location of axial coordinate measured from the bottom wall of dewar. Vapor pressure of helium keeps the saturation vapor pressure at 1.8 K which is 16,625 dyne/cm^2, or 12.47 Torr.

Characteristics of slosh waves caused by the 10 Hz gravity jitter frequency under $10^{-4}\ g_0$ background gravity environment and 0.1 rpm rotating speed are illustrated in Table 1. It is also obvious to say that the maximum fluid stress distributions are always located in the top dome and cylindrical sections of the outer wall in this case.

In comparison of fluid stress distribution exerted on between inner and outer walls, it also shows that the contribution

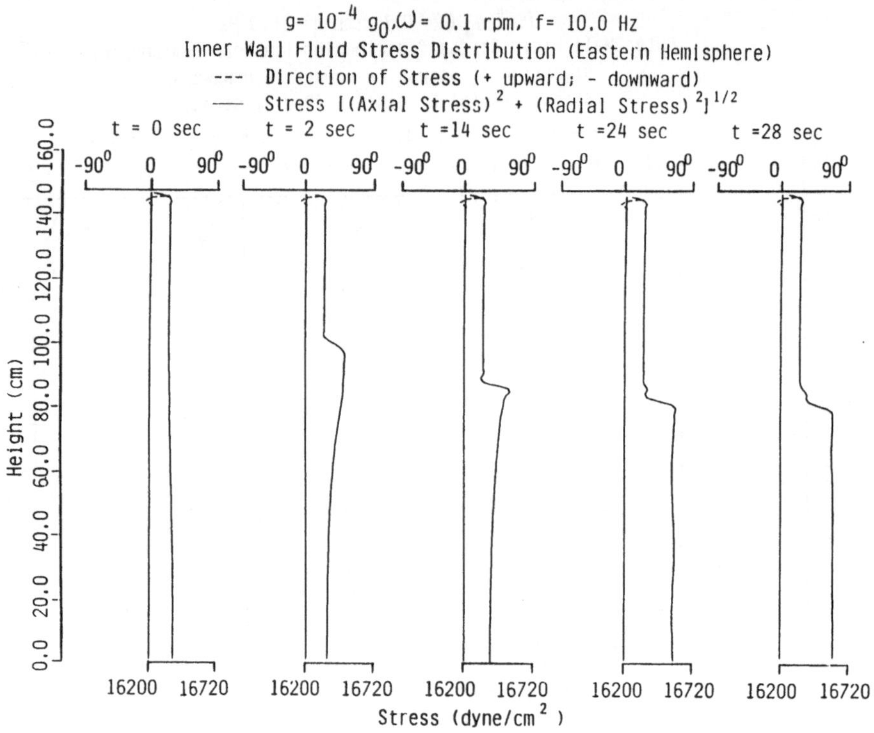

Figure 5 Time evolution of slosh wave disturbance-associated fluid stress distribution exerted on the inner wall.

on the outer wall is much greater than that on the inner wall due to the effect for the contribution of centrifugal force.

CONCLUSION

Time evolution of slosh wave induced fluid stress distribution exerted on the outer and inner walls of rotating dewar have been studied. Results show that fluid stress distribution exerted on the outer wall is always greater than that exerted on the inner wall. It also shows that there are time dependent fluctuations of fluid stress distribution at top dome and upper cylindrical sections of outer wall dependent upon the location of liquid-vapor interface which excite large amplitude slosh waves disturbing the entire flow field of rotating dewar.

In this study, we have demonstrated that the computer algorithm presented, through the execution of supercomputer CRAY X-MP, can be used to simulate the fluid behavior in a microgravity environment, in particular, the production of uneven distribution in fluid stress exerted on the dewar walls. The results obtained will provide the information and suggestion to be used in handling and managing the cryogenic liquid propellant, and the attitude control and guidance of the GP-B Spacecraft propulsion system.

ACKNOWLEDGEMENT

The authors appreciate NASA Grants NAG8-035 and NAG8-129. They would like to express their gratitude to Richard A. Potter of

NASA/Marshall Space Flight Center for the stimulating discussions during the course for the present study.

REFERENCES

1. Wilkinson, D. T., Bender, P. L., Eardley, D. M., Gaisser, T. K., Hartle, J. B., Israel, M. H., Jones, L. W., Partridge, R. B., Schramm, D. N., Shapiro, I. I., Vessort, R. F. C., and Wagoner, R. V., "Gravitation, Cosmology and Cosmic-Ray Physics," Phys. Today, Vol. 39, (1986), p. 43.
2. "Stanford Relativity Gyroscope Experiment (NASA Gravity Probe B)", Proc. Soc. Photo-Optical Instru. Eng., Vol. 619, (1986), p. 1.
3. Hung, R. J., and Leslie, F. W., "Bubble Shapes in a Liquid-Filled Rotating Container Under Low Gravity", J. Spacecraft Rockets, Vol. 25, (1988), p. 70.
4. Hung, R. J., Tsao, Y. D., Hong, B. B., and Leslie, F. W., "Bubble Behaviors in a Slowly Rotating Helium Dewar in Gravity Probe-B Spacecraft Experiment," J. Spacecraft Rockets, Vol. 26, (1988), p. 167.
5. Hung, R. J., Tsao, Y. D., Hong, B. B., and Leslie, F. W., "Dynamical Behavior of Surface Tension on Rotating Fluids in Low and Microgravity Environments", Int. J. Microgravity Res. App., Vol. 11, (1989), p.81.
6. Kamotani, Y., Prasad, A. and Oastrach, S., "Thermal Convections in an Enclosure due to Vibrations Aboard a Spacecraft," AIAA Journal, Vol. 19, (1981), p. 511.
7. Hung, R. J., Lee, C. C., and Shyu, K. L., "Reorientation of Rotating Fluid in Microgravity Environment With and Without Gravity Jitters" J. Spacecraft Rockets, Vol. 28, (1991), p. 71.
8. Hung, R. J., Tsao, Y. D., Hong, B. B., and Leslie, F. W., "Time Dependent Dynamical Behavior of Surface Tension on Rotating Fluids under Microgravity Environment", Adv. Space Res., Vol.8 (12), (1989), p.205.
9. Hung, R. J., Tsao, Y. D., Hong, B. B., and Leslie, F. W., "Axisymmetric Bubble Profiles in a Slowly rotating Helium Dewar Under Low and Microgravity Environments", Acta Astronautica, Vol. 19, (1989), p. 411.
10. Hung, R. J., Lee, C. C., and Leslie, F. W., "Effect of G-Jitters on the Stability of Rotaing Bubble Under Microgravity Environment," Acta Astronautica, Vol. 21, (1990), p. 309.
11. Hung, R. J., Lee, C. C., F. W. Leslie, "Response of Gravity Level Fluctuations on the Gravity Probe-B Spacecraft Propellant System," J. Propulsion Power, Vol. 7, (1991), in press.
12. Hung, R. J., and Shyu, K. L., "Cryogenic Liquid Hydrogen Reorientation Activated By High Frequency Impulsive Reverese Gravity Acceleration of Geyser Initiation," Microgravity Quarterly, Vol 3., (1991), in press.
13. Hung, R. J, Lee, C. C. and Leslie, F. W., "Slosh Wave Excitation in a Partially Filled Rotating Tank Due to Gravity Jitters in a Microgravity Environment", Acta Astronautica, Vol. 22, (1991), in press.
14. Hung, R. J. and Shyu, K. L., "Space-Based Cryogenic Liquid Hydrogen Reorientation Activated by Low Frequency Impulsive Reverse Gravity Thruster of Geyser Initiation", Acta Astronautica, Vol. 22, (1991), in press.
15. Mason P., Collins, D., Petrac, D., Yang, L., Edeskuty, F., Schuch, A., and Williamson, K., "The Behavior of Superfluid

Helium in Zero Gravity," Proc. 7th Int. Cryogenic Eng. Conf., Surrey, England, Science and Technology Press, (1978).
16. Hung, R. J., "Superfluid and Normal Fluid Helium II in a Rotating Tank Under Low and Microgravity Environments", Proc. Nat. Sci. Council, Series (A), Vol. 14, (1990), p. 289.
17. Hung, R. J., Phan T., and Smith, R. E., "Observation of Gravity Waves During the Extreme Tornado Outbreak of April 3, 1974." J. Atmos. Terres. Phys., Vol. 40, (1978), p.831.
18. Hung, R. J., and Smith, R. E., "Ray Tracing of Gravity Waves as a Possible Warning System for Tornadic Storms and Hurricanes", J. App. Meteor., Vol. 17, (1978), p. 3.
19. Hung, R. J., and Kuo, J. P., "Ionospheric Observation of Gravity Waves Associated with Hurricane Eloise", J. Geophys., Vol. 45, (1978), p. 67.
20. Hung, R. J., Phan, T. and Smith, R. E., "Coupling of ionosphere and Troposphere During the Occurence of Isolated Tornados of November 20, 1973", J. Geophys. Res., Vol. 84, (1979), p. 1261.

GRAVITY JITTER EXCITED CRYOGENIC LIQUID SLOSH WAVES IN MICROGRAVITY ENVIRONMENT

R. J. Hung*, C. C. Lee* and F. W. Leslie+

*The University of Alabama in Huntsville
Huntsville, Alabama

+NASA Marshall Space Flight Center
Huntsville, Alabama

ABSTRACT

The dynamical behavior of fluids, in particular the effect of surface tension on partially-filled rotating fluids (cryogenic liquid helium and helium vapor) in a full-scale Gravity Probe-B Spacecraft liquid helium container tank without probe imposed by various frequencies of gravity jitter have been investigated. Results disclose the conditions for the excitation of large amplitude slosh waves which should be avoided in the design of cryogenic liquid propellant system.

INTRODUCTION

The Gravity Probe-B (GP-B) spacecraft is a relativity gyroscope experiment to test two extraordinary, unverified predictions of Albert Einstein's general theory of relativity[1]. The approaches to both cooling and control involve the use of superfluid liquid helium. The boil-off from the cryogenic liquid helium dewar will be used as a propellant to maintain the attitude control and drag-free operation of the spacecraft. In contrasting the effects of surface tension to the effects of gravitational forces on the free surface of liquid, it was found that the surface tension force for most liquids is greater than the gravitational force in a 10^{-8} g_0 level and lower[2-4]. The equilibrium shape of free surface is governed by a balance of capillary, centrifugal and gravitational forces. Determination of bubble profiles based on computational experiments can uncover details of the flow which can not be easily visualized or measured experimentally under microgravity environment[5]. Recent study[6-8] suggests that the high frequency accelerations may be unimportant in comparison to the residual motions caused by low frequency accelerations.

Time-dependent dynamical behavior of surface tension on partially-filled rotating fluids in both low gravity and microgravity environments was carried out by numerically

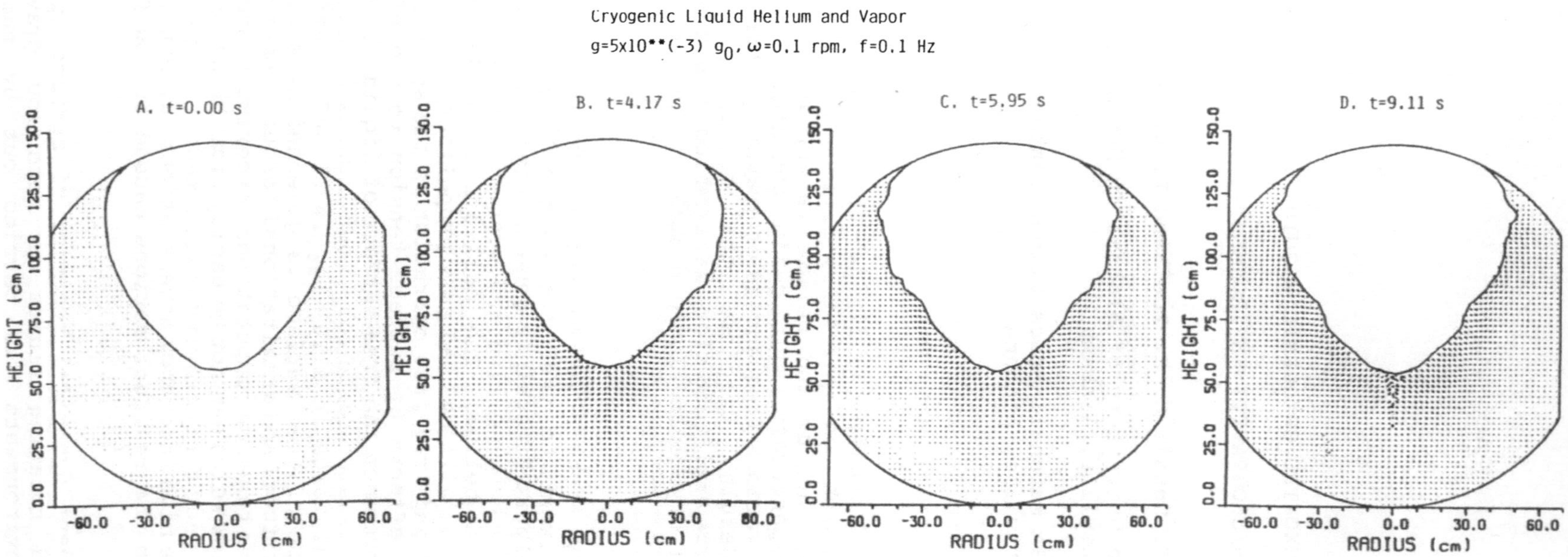

Figure 1(A-D) Time sequence of slosh wave evolutions for interface between liquid helium and helium vapor

computing the Navier-Stokes equations subjected to the initial and the boundary conditions[4,6-11]. At the interface between the liquid and the gaseous fluids, both the kinematic surface boundary condition, and the interface stress conditions for components tangential and normal to the interface, were applied[6-11]. The experiments carried out by Mason et al[12] showed that the classical fluid mechanics theory is applicable for cryogenic liquid helium in large containers[13]. In this study, temperature of cryogenic helium is 1.8K. The following data were used in the numerical simulation: liquid helium density = 0.1457 g/cm^3, helium vapor density = 0.00147 g/cm^3, fluid pressure = 1.6625×10^4 $dynes/cm^2$, surface tension coefficient at the interface between liquid helium and helium vapor = 0.353 dyne/cm, liquid helium viscosity coefficient = 9.609×10^{-5} cm^2/s; and contact angle = 5^0.

The equilibrium configurations of the helium bubble in a rotating dewar were studied[3,11]. In this study, time-dependent computation have been carried out to investigate the dynamical behaviors of cryogenic liquid helium and helium vapor in the geometry of full scale GP-B propellant container without probe under a microgravity environment. In GP-B spacecraft, four gyroscopes and a reference telescope are stored inside the probe surrounded by the superfluid helium II. The purpose of this study is not aimed at the simulation of the real operation conditions of GP-B spacecraft. The computation extends to the study of sloshing waves induced by the various frequencies of gravity-jitter and variations in different rotating speeds of GP-B container and different background gravitational fields.

NUMERICAL SIMULATION OF CRYOGENIC LIQUID-VAPOR INTERFACE

The present study examines time-dependent fluid behavior, in particular the dynamics of interface between cryogenic liquid helium and helium vapor; and the excitation of slosh waves due to gravity jitter. Time-dependent axial symmetry mathematical formulation is adopted[4,6-10]. Under microgravity environment, gravity jitter is produced by spacecraft attitude motion, machinery (turbine, pump, engine, etc.) vibrations, thruster firing, thruster shutdown, etc.[6-8]. Vibration of the gravity environment (gravity-jitter) is governed by the following equation:

$$g = g_B \left[1 + \frac{1}{2} \sin(2\pi f t) \right]$$

where g_B denotes the background gravity environment, and f(Hz) stands for the frequency of gravity-jitter.

In this study, cases of low and high background gravity (5×10^{-4} and 5×10^{-3} g_0), low and high rotation speeds (0.1 rpm and 1.0 rpm), and low, medium and high frequencies of gravity jitters (0.1, 1.0 and 10 Hz) are investigated. Because of the page limitation, selected examples for each cases are given for illustration.

Figures 1(A-D) show the evolution of liquid-vapor interface oscillations caused by gravity-jitter frequency of 0.1 Hz under rotating speed of 0.1 rpm and background gravity environment of 5×10^{-3} g_0. Figure 2 shows the time series of wave amplitudes for

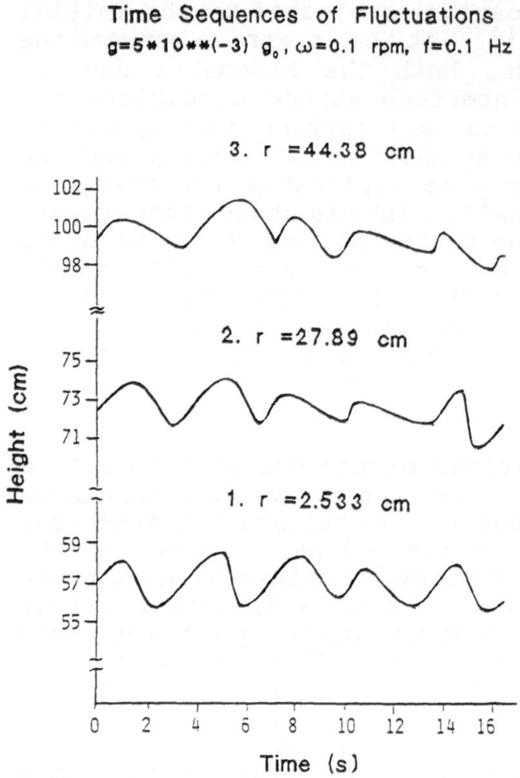

Figure 2 **Time series of wave amplitude fluctuations for the interface under the conditions of $g = 5 \times 10^{-3}\ g_0$, $\omega = 0.1$ rpm, f = 0.1 Hz.**

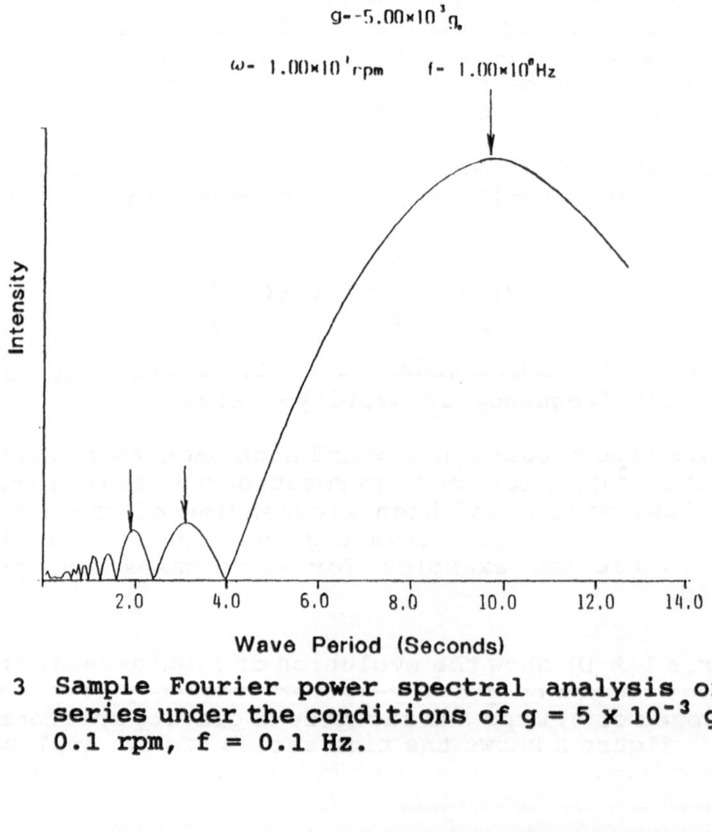

Figure 3 **Sample Fourier power spectral analysis of time series under the conditions of $g = 5 \times 10^{-3}\ g_0$, $\omega = 0.1$ rpm, f = 0.1 Hz.**

Table 1 Characteristics of Major Slosh Waves Caused by Gravity Jitter
Background Gravity: $g = 5 \times 10^{-3}\ g_0$ Rotating Speed: $\omega = 0.1$ rpm
Frequency of Gravity Jitters: $f_0 = 0.1$ Hz

Wave Period (s)	Wavelength (cm)	Phase Velocity (cm/s)	Propagation Direction (Clockwise from Positive Axial Direction) (Deg)	Ratio of Maximum Wave Amplitude to Wavelength (A/λ)	Location of Major Driving Force Field
1.93	42.1	21.8	172	4.56×10^{-3}	From Positive Axial Direction
3.15	13.5	4.3	171	1.64×10^{-3}	From Positive Axial Direction
9.40	55.5	5.9	-13	3.03×10^{-2}	From Negative Axial Direction

liquid-vapor interface fluctuations at $r_1 = 2.533$, $r_2 = 27.89$, and $r_3 = 44.38$ cm in radial coordinate measured from the major rotating axis of the container.

SLOSH WAVE EXCITATION

Figure 3 shows a sample Fourier power spectral analysis of time series for slosh mode wave period analysis with background conditions corresponding to Figures 1(A-D) and 2. This figure clearly indicates that there are three major peaks of wave nodes corresponding to wave periods of 1.93, 3.15, and 9.40 seconds.

Table 1 shows the characteristics of major slosh waves induced by the restoring force field of gravity jitter. There are three major modes of slosh waves: high frequency (wave period of 1.93 sec), medium frequency (wave period of 3.15 sec),

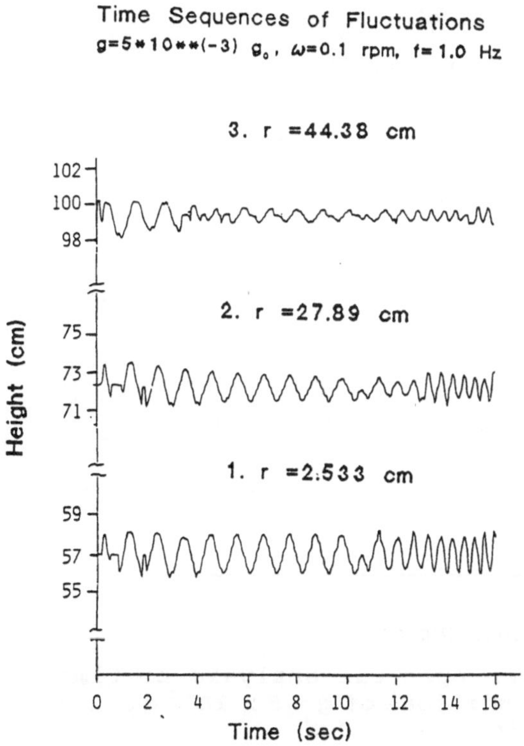

Figure 4 Time series of wave amplitude fluctuations for the interface under the conditions of $g = 5 \times 10^{-3}\ g_0$, $\omega = 0.1$ rpm, $f = 1.0$ Hz.

Table 2 Characteristics of Major Slosh Waves Caused by Gravity Jitter

Background Gravity: $g = 5 \times 10^{-3}\ g_o$ Rotating Speed: $\omega = 0.1$ rpm

Frequency of Gravity Jitters: $f_o = 1.0$ Hz

Wave Period (s)	Wavelength (cm)	Phase Velocity (cm/s)	Propagation Direction (Clockwise from Positive Axial Direction) (Deg)	Ratio of Maximum Wave Amplitude to Wavelength (A/ λ)	Location of Major Driving Force Field
1.49	42.9	28.8	-37	2.39×10^{-3}	From Negative Axial Direction
2.39	13.6	5.7	164	4.12×10^{-3}	From Positive Axial Direction
9.39	75.1	8.0	-30	2.75×10^{-2}	From Negative Axial Direction

and low frequency (wave period of 9.40 sec) modes resulted from this analysis. Propagation direction is measured clockwise from positive direction of major rotating axis. The ratio of maximum wave amplitude to wavelength [Max (A/λ), where A stands the wave amplitude; and λ, the wavelength of slosh wave] for each slosh wave is calculated based on the following procedures: (a) determine each wave mode based on the peaks of power intensity shown in the power spectral density analysis in the Fourier domain (see Figure 3); (b) apply proper window of filter to separate each wave mode shown on the peaks of power intensity; (c) calculate wavelength, phase velocity and propagation direction of each wave mode from the cross-correlation analysis based on the separated wave modes through filtering the time series in Fourier domain; (d) reverse Fourier transform the separated wave mode through filtering the time series in Fourier domain to time domain and

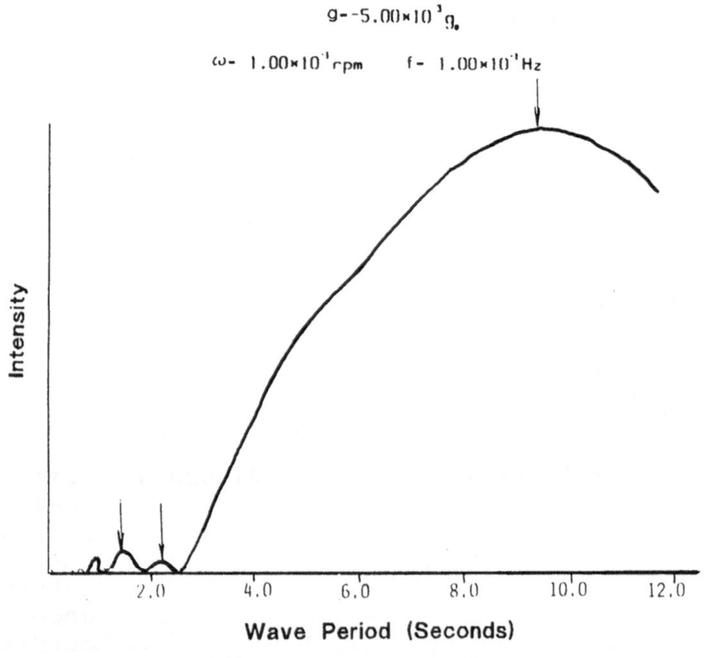

Figure 5 Sample Fourier power spectral analysis of time series under the conditions of $g = 5 \times 10^{-3}\ g_0$, $\omega = 0.1$ rpm, $f = 1.0$ Hz.

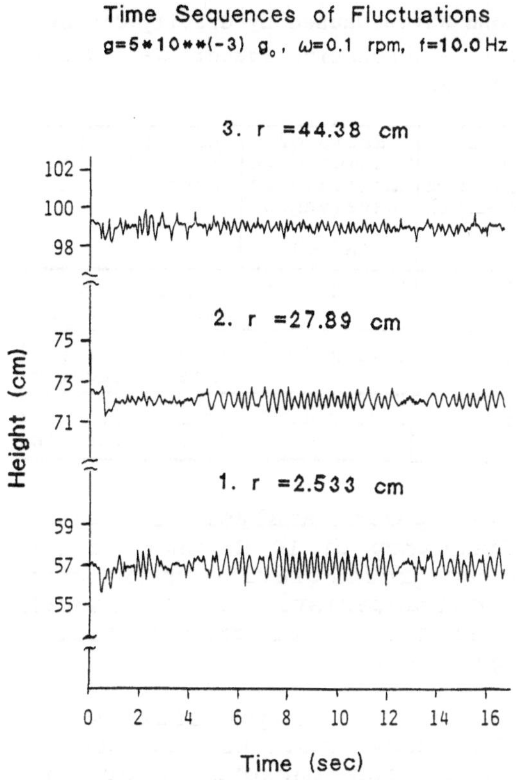

Figure 6 **Time series of wave amplitude fluctuations for the interface under the conditions of g = 5 x 10^{-3} g_0, ω = 0.1 rpm, f = 10 Hz.**

obtain the amplitude of each mode of slosh waves; and (e) compute the ratio of maximum wave amplitude to wavelength [Max (A/λ)] for each mode of slosh waves from items (c) and (d).

Figures 4 shows the time series of wave amplitudes for liquid-vapor interface fluctuations at three locations, as that shown in Figure 2; while Figure 5 shows a sample Fourier spectral analysis of time series for slosh mode wave period analysis for waves caused by a medium gravity-jitter frequency of 1.0 Hz under a low rotating speed of 0.1 rpm and a high background gravity environment of 5 x 10^{-3} g_0. Table 2 shows the characteristics of major slosh waves induced by the restoring force field of gravity jitter applied to the case shown in Figure 4 and 5. Figure 6 shows another set of the time series of wave amplitudes for liquid-vapor interface fluctuations at three locations, as that shown in Figure 2; while Figure 7 shows a sample Fourier Spectal analysis

Table 3 Characteristics of Major Slosh Waves Caused by Gravity Jitter
Background Gravity: g = 5 x 10^{-3} g_o Rotating Speed: ω = 0.1 rpm
Frequency of Gravity Jitters: f_o = 10.0 Hz

Wave Period (s)	Wavelength (cm)	Phase Velocity (cm/s)	Propagation Direction (Clockwise from Positive Axial Direction) (Deg)	Ratio of Maximum Wave Amplitude to Wavelength (A/ λ)	Location of Major Driving Force Field
0.46	47.8	103.9	-36	4.66 x 10^{-4}	From Negative Axial Direction
0.62	29.8	48.1	149	5.97 x 10^{-3}	From Positive Axial Direction
1.65	7.4	4.5	155	2.34 x 10^{-2}	From Positive Axial Direction

Table 4 Characteristics of Major Slosh Waves Caused by Gravity Jitter

Background Gravity: $g = 5 \times 10^{-3}\ g_o$ Rotating Speed: $\omega = 1.0$ rpm

Frequency of Gravity Jitters: $f_o = 0.1$ Hz

Wave Period (s)	Wavelength (cm)	Phase Velocity (cm/s)	Propagation Direction (Clockwise from Positive Axial Direction) (Deg)	Ratio of Maximum Wave Amplitude to Wavelength (A/ λ)	Location of Major Driving Force Field
2.77	55.6	20.1	166	3.95×10^{-3}	From Positive Axial Direction
4.28	59.1	13.8	-24	2.45×10^{-3}	From Negative Axial Direction
9.88	70.1	7.1	-33	2.21×10^{-2}	From Negative Axial Direction

of time series for slosh mode wave period analysis for waves caused by high gravity-jitter frequency of 10 Hz under a low rotating speed of 0.1 rpm and high background gravity environment of 5 x 10^{-3} g_0. Table 3 shows the characteristics of major slosh waves induced by the restoring force field of gravity jitter applied to the case shown in Figure 6 and 7.

Table 4 shows the characteristics of major slosh waves caused by a low gravity jitters frequency of 0.1 Hz under a high rotating speed of 1.0 rpm and a high background gravity environment of 5 x 10^{-3} g_0. Table 5 shows the example for the characteristics of major slosh waves caused by a low gravity jitter frequency of 0.1 Hz under a low rotating speed of 0.1 rpm and a low background gravity environment of 5 x 10^{-4} g_0. Table 6 shows the characteristics of major slosh waves caused by a low

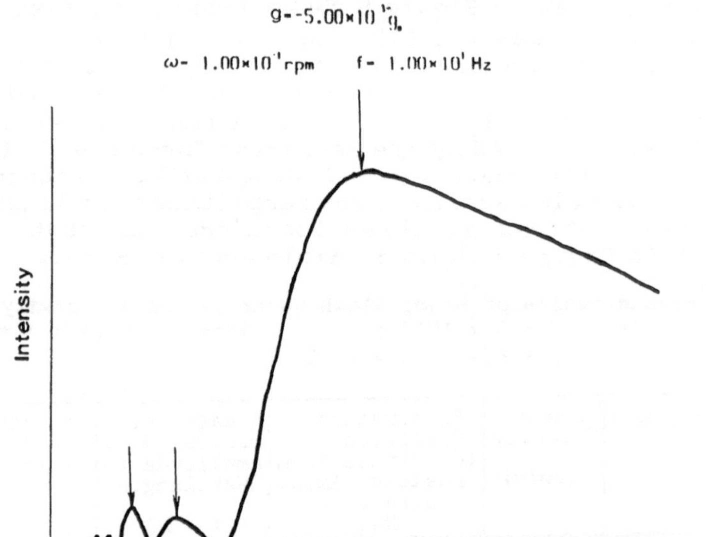

Figure 7 Sample Fourier power spectral analysis of time series under the conditions of $g = 5 \times 10^{-3}\ g_0$, ω = 0.1 rpm, f = 10 Hz.

Table 5 Characteristics of Major Slosh Waves Caused by Gravity Jitter
Background Gravity: $g = 5 \times 10^{-4}\ g_0$ Rotating Speed: $\omega = 0.1$ rpm
Frequency of Gravity Jitters: $f_0 = 0.1$ Hz

Wave Period (s)	Wavelength (cm)	Phase Velocity (cm/s)	Propagation Direction (Clockwise from Positive Axial Direction) (Deg)	Ratio of Maximum Wave Amplitude to Wavelength (A/ λ)	Location of Major Driving Force Field
1.22	53.6	43.9	-27	1.04×10^{-2}	From Negative Axial Direction
1.90	47.1	24.8	-21	4.66×10^{-3}	From Negative Axial Direction
5.57	22.8	4.1	161	4.13×10^{-2}	From Positive Axial Direction

gravity jitter frequency of 0.1 Hz under a high rotating speed of 1.0 rpm and a low background gravity environment of $5 \times 10^{-4}\ g_0$. Table 7 shows the characteristics of major slosh waves caused by a medium gravity jitter frequency of 1.0 Hz under a high rotating speed of 1.0 rpm and a low background gravity environment of $5 \times 10^{-4}\ g_0$. Table 8 shows the characteristics of major slosh waves caused by a high gravity jitter frequency of 10 Hz under a high rotating speed of 1.0 rpm and a low background gravity enviornment of $5 \times 10^{-4}\ g_0$.

CHARACTERISITCS OF SLOSH WAVES EXCITATED BY THE GRAVITY JITTER

Characteristics of slosh waves are studied based on the frequencies of slosh waves excited, frequencies of gravity-jitter imposed on the cryogenic propellant system, levels of background gravity environment, and the rotating speeds of propellant tank.

Wave Characteristics Based on the Lowest Frequency of Slosh Waves Excited

Cases of slosh waves excited by the gravity-jitter have been investigated. Reviews of each case for the values of Max (A/λ) ratio of slosh waves excited based on Tables 1 to 8, and Figures 3, 5 and 7, for the intensities of each slosh waves excited, it is

Table 6 Characteristics of Major Slosh Waves Caused by Gravity Jitter
Background Gravity: $g = 5 \times 10^{-4}\ g_0$ Rotating Speed: $\omega = 1.0$ rpm
Frequency of Gravity Jitters: $f_0 = 0.1$ Hz

Wave Period (s)	Wavelength (cm)	Phase Velocity (cm/s)	Propagation Direction (Clockwise from Positive Axial Direction) (Deg)	Ratio of Maximum Wave Amplitude to Wavelength (A/ λ)	Location of Major Driving Force Field
1.12	55.6	49.6	-22	3.13×10^{-4}	From Negative Axial Direction
1.91	23.7	12.4	150	2.45×10^{-3}	From Positive Axial Direction
5.77	50.8	8.8	-11	2.52×10^{-2}	From Negative Axial Direction

Table 7 Characteristics of Major Slosh Waves Caused by Gravity Jitter
Background Gravity: $g = 5 \times 10^{-4}\ g_o$ Rotating Speed: $\omega = 1.0$ rpm
Frequency of Gravity Jitters: $f_o = 1.0$ Hz

Wave Period (s)	Wavelength (cm)	Phase Velocity (cm/s)	Propagation Direction (Clockwise from Positive Axial Direction) (Deg)	Ratio of Maximum Wave Amplitude to Wavelength (A/ λ)	Location of Major Driving Force Field
1.09	23.6	21.7	173	1.49×10^{-2}	From Positive Axial Direction
1.55	13.2	8.5	-16	5.77×10^{-3}	From Negative Axial Direction
4.67	23.8	5.1	167	2.39×10^{-2}	From Positive Axial Direction

found that the slosh wave with lowest wave frequency for each case is always associated with the highest intensity and the highest value of Max (A/λ) ratio wave mode.

Wave Characteristics Based on the Frequencies of Gravity-Jitter Imposed on the Propellant System

Wave characteristics of slosh wave excited by low, medium and high frequencies of gravity-jitter imposed on the propellant system have been investigated. It is shown that the value of Max (A/λ) ratio slosh waves excited by the low frequency (0.1 Hz) gravity-jitter is always greater than that of slosh waves excited by the medium frequency (1.0 Hz) gravity-jitter which, in turn, is greater than that of slosh waves excited by the high frequency (10 Hz) gravity jitter.

Wave Characteristics Based on the Levels of Background Gravity Environment

Wave characteristics of slosh waves excited by gravity jitter under low ($5 \times 10^{-4}\ g_0$) and high ($5 \times 10^{-3}\ g_0$) reduced gravity environment have been studied. It can be concluded from these studies that the values of Max (A/λ) ratio of slosh waves excited by the gravity-jitter associated with lower levels of

Table 8 Characteristics of Major Slosh Waves Caused by Gravity Jitter
Background Gravity: $g = 5 \times 10^{-4}\ g_o$ Rotating Speed: $\omega = 1.0$ rpm
Frequency of Gravity Jitters: $f_o = 10.0$ Hz

Wave Period (s)	Wavelength (cm)	Phase Velocity (cm/s)	Propagation Direction (Clockwise from Positive Axial Direction) (Deg)	Ratio of Maximum Wave Amplitude to Wavelength (A/ λ)	Location of Major Driving Force Field
1.88	49.1	26.1	-2	1.74×10^{-3}	From Negative Axial Direction
2.59	56.5	21.8	-17	5.75×10^{-3}	From Negative Axial Direction
4.55	43.2	9.5	171	1.81×10^{-2}	From Positive Axial Direction

background gravity environment are always greater than that of slosh waves associated with higher levels of background gravity environment. In other words, higher levels of background gravity environment has a tendency to suppress the excitation of greater amplitude slosh waves. This also indicates that the excitation of greater amplitude slosh waves is much easier under the lower level reduced gravity environment than the higher level reduced gravity environment.

Wave Characteristics Based on Rotating Speeds of Propellant Tank

Wave characteristics of slosh waves excited by gravity jitter under low (0.1 rpm for normal value of spacecraft rotation for GP-B Spacecraft operation) and high (1.0 rpm of spacecraft rotation for instrument calibration of GP-B Spacecraft) rotating speeds of propellant tank have been investigated. It can be summarized from these studies that the values of Max (A/λ) ratio of slosh waves excited by the gravity jitter associated with lower rotating speeds of propellant tank are always greater than that of slosh waves excited by the gravity jitter associated with higher rotating speeds propellant tank. In other words, higher rotating speeds correspond to higher centrifugal forces which tend to suppress greater amplitudes of slosh waves excited by the gravity jitter.

DISCUSSION AND CONCLUSION

The time evolution of slosh waves excited by the gravity jitter have been studied. Results show that there is a group of wave trains with various frequencies and wavelengths of slosh waves generated by the restoring force field of gravity jitter in this study. The following conclusions have been drawn from the present study: (1) slosh waves with the lowest wave frequency are always associated with the highest intensity and the highest values of Max(A/λ) ratio wave modes; (2) the values of Max(A/λ) ratio slosh waves excited by the lower frequency gravity jitter imposed on the propellant system are always greater than that of the slosh waves excited by the higher frequency gravity jitter imposed on the spacecraft propellant system; (3) the values of Max(A/λ) ratio slosh waves, associated with lower levels of background gravity environment, excited by the gravity jitter are always greater than that of the slosh waves, associated with higher levels of background gravity environment, excited by the gravity jitter; and (4) the values of Max(A/λ) ratio of slosh waves, associated with lower rotating speeds of propellant tank, excited by the gravity jitter are very often greater than that of the slosh waves, associated with higher rotating speeds of propellant tank, excited by the gravity jitter.

ACKNOWLEDGEMENT

The authors appreciate NASA Grant NAG8-129. They would like to express their gratitude to Richard A. Potter of NASA/Marshall Spacae Flight Center for the stimulating discussions during the course for the present study.

REFERENCES

1. Stanford Relativity Gyroscope Experiment (NASA Gravity Probe B), Proc. Soc. Photo-Optical Instru. Eng., 619:1 (1986).
2. R. J. Hung and F. W. Leslie, Bubble Shapes in a Liquid-Filled Rotating Container Under Low Gravity, J. Spacecraft Rockets, 25:70 (1988).
3. R. J. Hung, Y. D. Tsao, B. B. Hong, and F. W. Leslie, Bubble Behaviors in a Slowly Rotating Helium Dewar in Gravity Probe-B Spacecraft Experiment, J. Space. Roc., 26:167 (1988).
4. R. J. Hung, Y. D. Tsao, B. B. Hong, and F. W. Leslie, Dynamical Behavior of Surface Tension on Rotating Fluids in Low and Microgravity Environments, Int. J. Microgravity Res. Appl., 11:81 (1989).
5. F. W. Leslie, Measurements of Rotating Bubble Shapes in a Low Gravity Environment, J. Fluid Mech., 161:269 (1985).
6. R. J. Hung, C. C. Lee, and F. W. Leslie, Effect of G-Jitters on the Stability of Rotating Bubble Under Microgravity Environment, Acta Astronautica, 21:309 (1990).
7. R. J. Hung, C. C. Lee, and F. W. Leslie, Response of Gravity Level Fluctuations on the Gravity Probe-B Spacecraft Propellant System, J. Propulsion Power, 7:556 (1991).
8. R. J. Hung, C. C. Lee, and K. L. Shyu, Reorientation of Rotating Fluid in Microgravity Environment With and Without Gravity Jitters, J. Spacecraft Rockets, 28:71 (1991).
9. R. J. Hung, Y. D. Tsao,, B. B. Hong, and F. W. Leslie, Time Dependent Dynamical Behavior of Surface Tension on Rotating Fluids Under Microgravity Environment, Adv. Space Res., 8(12):205 (1989).
10. R. J. Hung and K. L. Shyu, Cryogenic Liquid Hydrogen Reorientation Activataed by High Frequency Impulsive Reverse Gravity Acceleration of Geyser Initiation, Microgravity Quarterly, 1(2):81 (1991).
11. R. J. Hung, Y. D. Tsao, B. B. Hong, and F. W. Leslie, Axisymmetric Bubble Profiles in a Slowly Rotating Helium Dewar Under Low and Microgravity Environments, Acta Astronautica, 19:411 (1989).
12. P. Mason, D. COllins, D. Petrac, L. Yang, F. Edeskuty, A. Schuch, and K. Williamson, The Behavior of Superfluid Helium in Zero Gravity, Proc. 7th Int. Cryogenic Eng. Con., Surrey, England, Science and Technology Press, (1978).
13. R. J. Hung, Superfluid and Normal Fluid Helium II in a Rotating Tank Under Low and Microgravity Environments, Proc. Nat. Sci. Council, Series (A), 14:289 (1990).
14. R. J. Hung, T. Phan, and R. E. Smith, Observation of Gravity Waves During the Extreme Tornado Outbreak of April 3, 1974, J. Atmos. Terres. Phys., 40:831 (1978).
15. R. J. Hung, and R. E. Smith, Ray Tracing of Gravity Waves as a Possible Warning System for Tornadic Storms and Hurricanes, J. Appl. Meteor., 17:3 (1978).
16. R. J. Hung, and J. P. Kuo, Ionospheric Observation of Gravity Waves Associated with Hurricane Eloise, J. Geophys., 45:67 (1978).
17. R. J. Hung, T. Phan, and R. E. Smith, Coupling of Ionosphere and Troposphere During the Occurence of Isolated Tornadoes of November 20, 1973, J. Geophys. Res., 84:1261 (1979)

SLOSH WAVE EXCITATION DUE TO CRYOGENIC LIQUID REORIENTATION IN SPACE-BASED PROPULSION SYSTEM

R. J. Hung, K. L. Shyu and C. C. Lee

The University of Alabama in Huntsville
Huntsville, Alabama

ABSTRACT

The objective of the cryogenic fluid management of the spacecraft propulsion system is to develop the technology necessary for acquistion or positioning of liquid and vapor within a tank in reduced gravity to enable liquid outflow or vapor venting. In this study slosh wave excitation induced by the resettling flow field activated by 1.0 Hz medium frequency impulsive reverse gravity acceleration during the course of liquid fluid reorientation with the initiation of geyser for liquid filled levels of 30, 50, and 80% have been studied. Characteristics of slosh waves with various frequencies excited are discussed.

INTRODUCTION

Operation of cryogenic propellant on space-based spacecraft requires the delivery of vapor-free propellant before full restart. However, as the spacecraft coasts in orbit prior to engine firing, liquid fluid in propellant tank may position itself away from outlet ports because of the adverse acceleration and drag imposed by tenuous gases in the upper atmosphere under microgravity environment. To solve this problem, it is necessary to have the liquid settle with no bubbles near the tank outlet so that the initial flow of propellant will not carry vapor to the pump or engine.

For venting of cryogenic propellant, it is probably necessary that virtually all bubbles be displaced from the bulk liquid so that a two-phase mixture is not vented. Propellant transfer requires that the liquid be completely settled with virtually no bubbles. Outflow of a liquid near the tank outlet can result in the premature ingestion of gas while a significant amount of liquid is still in the tank under microgravity environment.

The process of positioning the liquid over the tank outlet by firing auxiliary thruster is known as impulsive reorientation or settling. Since impulsive reorientation requires the

Advances in Cryogenic Engineering, Vol. 37, Part B
Edited by R.W. Fast, Plenum Press, New York, 1992

expenditure of propellant, it is important to optimize the process to minimize the associated propellant requirements[1]. If the thrust level is too low, the propellant may not reposition. If it is too high, a large geyser may form and vapor pockets may be trapped in the pool. Proper spacecraft design and operation requires a good understanding of the process and parameters which control it.

Recently Leslie[2] was able to measure and to numerically compute the bubble shapes at various ratios of centrifugal force to surface tension force in 2, 4 and 6.3 cm deep cylinders in the microgravity environment. The results showed excellent agreement between model computation and measurements. Hung et al.[3-5] further extended the work to include rotating speeds which resulted with bubbles intersecting and/or without intersecting the top, bottom and side walls of the cylinder.

An analysis of time-dependent dynamical behavior of surface tension on partially-filled rotating fluids in both low gravity and microgravity environments was carried out by numerically solving the Navier-Stokes equations subjected to the initial and the boundary conditions[4,6-8]. Numerical simulation of positive liquid acquisition was attempted by introducing reverse gravity acceleration, resulting from the propulsive thrust of small auxiliary engines which exceeds the critical value for geyser initiation[9].

In this study, time-dependent computations have been carried out to investigate the dynamical behavior of the characteristics of slosh waves excitation during the course of fluid reorientation or resettling of propellant prior to main engine firing for spacecraft restart at the impulsive reverse gravity acceleration with a medium frequency of 1.0 Hz for the activation of the liquid reorientation.

SIMILATION OF SLOSH WAVE EXCITATION

The present study examines time-dependent fluid behaviors, in particular the dynamics of interface between cyrogenic liquid hydrogen and hydrogen vapor; and the excitation of slosh waves during the course of liquid reorientation at medium frequency impulsive reverse gravity acceleration of 1.0 Hz which is great enough to introduce geyser initiation. Propellant tank of liquid filled levels of 30, 50, and 80% are considered in this study. Time-dependent axial symmetry mathematical formulation are adopted. Detailed description of mathematical formulation, initial and boundary conditions suitable for the analysis of cryogenic fluid management under microgravity environment are given in our earlier studies[1,4,6,10,11]. The initial profiles of liquid-vapor interface are determined from computations based on algorithms developed for the steady state formulation of microgravity fluid management[3-7].

Model size is L = 4.23672 cm and D = 4.2672 cm, as shown in Figure 1(A). If the spacecraft had been coasting for a long time, aligned with its direction of motion, the most significant force, drag, would be axial and with acceleration of $10^{-5}g_0$ along upward direction. The hydrogen vapor is, thus, originally positioned at the bottom of the tank. The requirement to settle or to position liquid fuel over the outlet end of the spacecraft propellant tank prior to main engine restart poses a microgravity

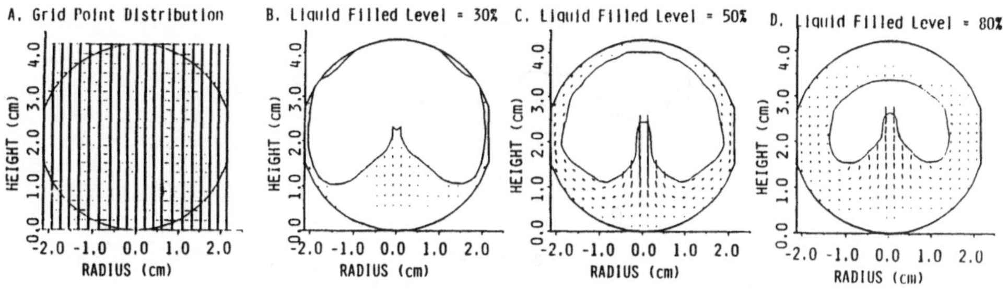

Figure 1 **Grid point distribution and selected flow profiles of fluid reorientation with geyser initiation activated by 1.0 Hz medium frequency impulsive reverse gravity acceleration.**

fluid behavior problem[1,9]. In the recent study of computer simulation,[9] a small value of reverse gravity acceleration (downward direction) was provided by the propulsive thrust of small auxiliary engine to initiate the reorientation of liquid propellant. This small value of reverse gravity acceleration of propulsive thrust increased gradually till reaching the critical value on which initiation of geyser was detected during the course of fluid resettlement[9]. We term this reverse gravity acceleration of propulsive thrust, which is capable to initiate geyser, as "geyser initiation gravity-level". Liquid filled levels of 30, 50, and 80% were considered in our recent study[9]. Cryogenic liquid hydrogen at temperature of 20K was considered. Hydrogen density of 0.071 g/cm^3; surface tension coefficient at the interface between liquid hydrogen and hydrogen vapor of 1.9 dyne/cm; hydrogen viscosity coefficient of 1.873×10^{-3} cm^2/s; and contact angle of 0.5^0 were used in the computer simulation[9].

It was found that these geyser initiation gravity levels are 5.5×10^{-2}, 6.52×10^{-2}, and 8.2×10^{-2} g_0 for liquid filled levels of 30, 50, and 80%, respectively[9]. Figures 1(B), 1(C), and 1(D) show the selected flow profiles with geysering motion activated by medium frequency impulsive thrust of 1.0 Hz during the courses of fluid reorientation for cryogenic hydrogen with liquid filled levels of 30, 50, and 80%, respectively.

In this study, impulsive reverse gravity acceleration, g_i, with frequency of f Hz is defined as follows:

$$g_i = g_{i0} \left(1 + \frac{1}{2} \sin(2\pi ft)\right)$$

where g_{i0} stands for geyser initiation reverse gravity acceleration for corresponding liquid filled levels.

Various modes of slosh waves excitation for liquid filled levels of 30, 50, and 80% activated by corresponding geyser initiation reverse gravity acceleration have been investigated.

<u>(II-A) Slosh Waves Associated with Liquid Filled Level of 30%</u>

Slosh waves excitation during the course of liquid reorientation and resettlement is investigated. Figure 2 shows

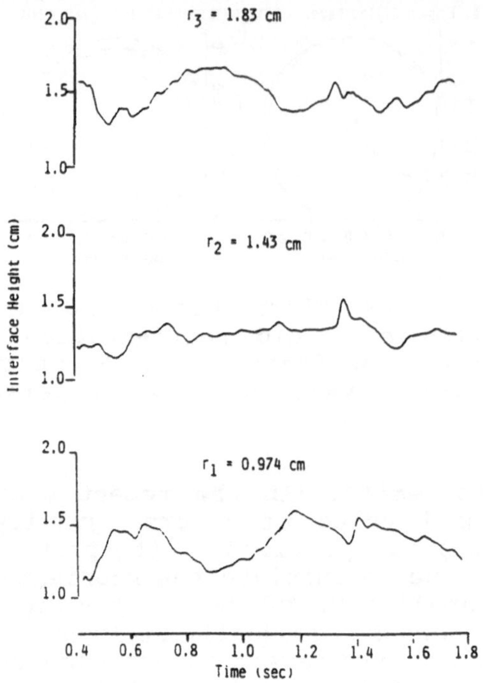

Figure 2 Time series of wave amplitude fluctuations for the liquid-vapor interface activated by 1.0 Hz medium frequency impulsive reverse gravity acceleration with liquid filled level of 30%.

the time series of wave amplitudes for liquid-vapor interface fluctuations. Figure 3 shows a sample Fourier power spectral analysis of time series at r_1. This figure clearly indicates that there are three major peaks of wave nodes corresponding to wave periods of 0.14, 0.21, and 0.50 seconds. A filter shall be properly chosen to apply certain ranges of window for separating these three wave nodes from each other in three time series at r_1, r_2, and r_3 shown in Figure 2, for the determination of wavelength, phase velocity and propagation direction of slosh waves[12-15].

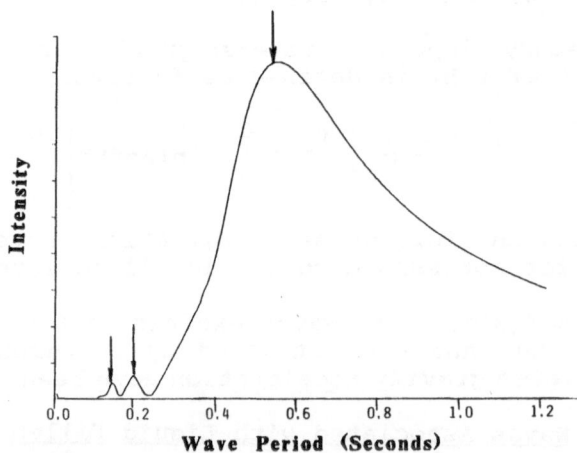

Figure 3 Sample Fourier power spectral analysis of major peaks of wave modes for liquid filled level of 30%.

TABLE 1 CHARACTERISTICS OF MAJOR SLOSH WAVE CAUSED BY RESETTLEMENT OF LIQUID REORIENTATION

1.0 Hz Medium Frequency Impulsive Thrust with Geyser Initiation
Reverse Gravity Acceleration = $5.5 \times 10^{-2} g_o$
Liquid Filled Level = 30%

Wave Period (sec)	Wavelength (cm)	Phase Velocity (cm/sec)	Propagation Direction (Clockwise from Positive Axial Direction) (Deg)	Ratio of Maximum Wave Amplitude to Wavelength (10^{-2})
0.14	0.39	2.82	(±) 6	0.156
0.21	0.16	0.78	0	0.273
0.50	8.00	15.99	0	3.711

Table 1 shows the characteristics of major slosh waves induced by the resettling flow field during the course of liquid fluid reorientation activated by 1.0 Hz medium frequency impulsive reverse gravity acceleration associated with liquid filled level of 30%.

Study of slosh wave excitation induced by fluid reorientation associated with liquid filled levels of 30% draws the following conclusions: (1) A series of longitudinal wave trains of slosh waves which propagate more or less along the axial direction of container; (2) Slosh waves with lowest frequency are associated with the highest value of Max (A/λ) ratio wave modes; (3) Slosh waves with lowest frequency contains highest wave

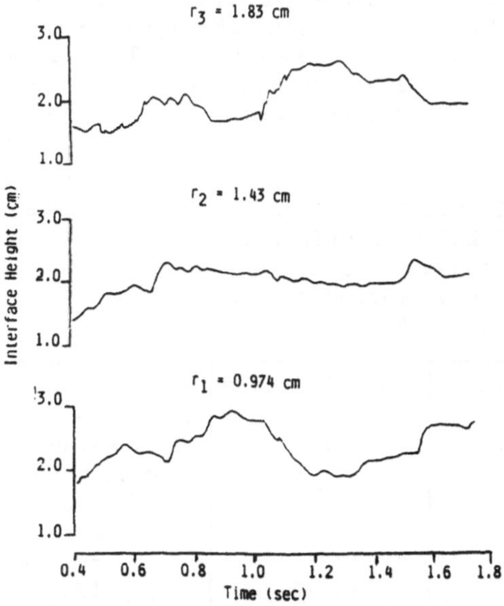

Figure 4 Time series of wave amplitude fluctuations for the liquid-vapor interface activated by 1.0 Hz medium frequency impulsive reverse gravity acceleration with liquid filled level of 50%.

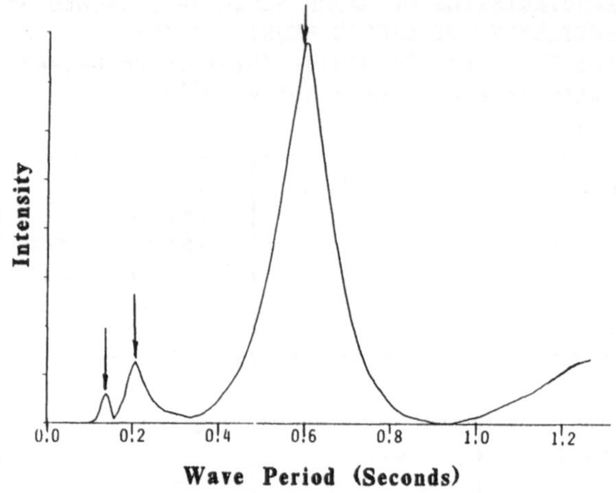

Figure 5 Sample Fourier power spectral analysis of major peaks of wave modes for liquid filled level of 50%.

intensity (see Figure 3) which is also equivalent to the highest wave energy; and (4) Impulsive reverse gravity acceleration along the axial direction is responsible for the excitation of slosh waves with various frequencies.

(II-B) Slosh Waves Associated with Liquid Filled Level of 50%

Figure 4 shows the time series of wave amplitude for liquid-vapor interface fluctuations for 50% liquid filled level. Figure 5 shows a sample Fourier power spectral analysis of time series associated with liquid filled level of 50% at location r_1. This figure shows that there are three major peaks of wave modes corresponding to wave periods of 0.15, 0.21, and 0.57 seconds. Table 2 shows the characteristics of major slosh waves induced by the resettling flow field activated by 1.0 Hz medium frequency impulsive reverse gravity acceleration during the course of liquid fluid reorientation associated with liquid filled level of 50%.

TABLE 2 CHARACTERISTICS OF MAJOR SLOSH WAVES CAUSED BY RESETTLEMENT OF LIQUID REORIENTATION

1.0 Hz Medium Frequency Impulsive Thrust with Geyser Initiation
Reverse Gravity Acceleration = $6.52 \times 10^{-2} g_o$
Liquid Filled Level = 50%

Wave Period (sec)	Wavelength (cm)	Phase Velocity (cm/sec)	Porpagation Direction (Clockwise from Positive Axial Direction) (Deg)	Ratio of Maximum Wave Amplitude to Wavelength (10^{-2})
0.15	1.79	11.96	(±) 176	0.195
0.21	0.30	1.42	(±) 176	0.508
0.57	0.37	0.65	(±) 176	3.125

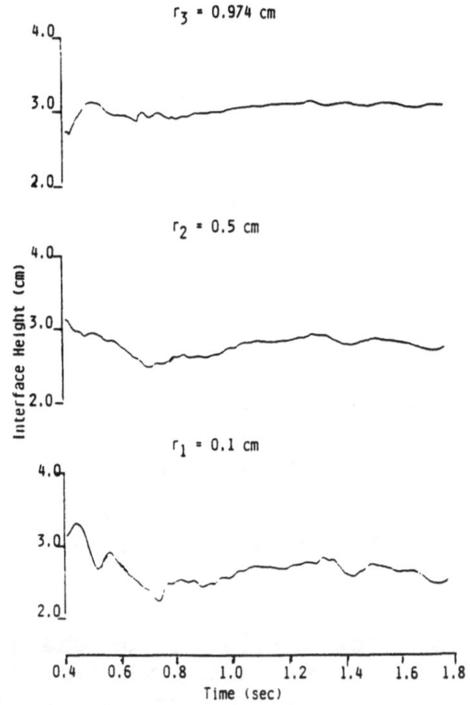

Figure 6 **Time series of wave amplitude fluctuations for liquid-vapor interface activated by 1.0 Hz medium frequency impulsive reverse gravity acceleration with liquid filled level of 80%.**

The conclusions drawn according to slosh wave excitations associated with liquid filld level of 50% are as follows: (1) Conclusions obtained from that associated with liquid filld level of 30% stand basically true; (2) Impulsive reverse gravity acceleration needed to activate the liquid reorientation with an initiation of geyser requires 6.52×10^{-2} g_0 for 50% liquid filled level, and that takes 5.5×10^{-2} g_0 for 30% liquid filled level; and (3) Characteristics of slosh waves associated with 30% and 50% liquid filled levels are basically no prominence differences.

(II-C) Slosh Waves Associated with Liquid Filled Levels of 80%

Figure 6 shows the time series of wave amplitudes for liquid-vapor interface for 80% (r_1 = 0.1, r_2 = 0.5 and r_3 = 0.974 cm) liquid filled level. Figure 7 shows the sample Fourier power spectral analysis of time series associated with liquid filled level 80%. This figure shows that there are three major peaks each of wave modes corresponding to wave periods of 0.09, 0.14 and 0.48 seconds for liquid filled level of 80%. Table 3 shows the characteristics of major slosh waves induced by the resettling flow field activated by 1.0 Hz medium frequency impulsive reverse gravity acceleration during the course of liquid fluid reorientation associated with liquid filled level of 80%.

In addition to the conclusions drawn for slosh waves associated with liquid filled levels of 30 and 50%, it is further confirmed that the reverse gravity acceleration required to activate the liquid reorientation with the initiation of geyser shall increase up to 8.2×10^{-2} g_0 for liquid filled level of 80%.

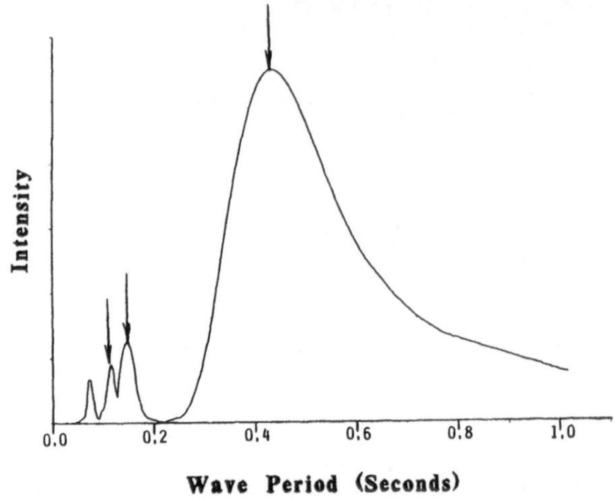

Figure 7 Sample Fourier power spectral analysis of major peaks of wave modes for liquid filled level of 80%.

There is no geysering flow initiated for the reverse gravity acceleration below the corresponding values of liquid filled levels mentioned earlier. Also, there is no prominence difference in the characteristics of slosh wave excitations for the variations of liquid filled levels among 30, 50, 80%. It is true that slosh waves with lowest wave frequency excited by the fluid reorientation for each liquid filled levels contain highest wave intensity longitudinal modes which propagate along the axial direction of the propellant tank.

DISCUSSION AND CONCLUSIONS

Recently Hung and Shyu[16] evaluated the performance between impulsive reverse gravity acceleration and constant reverse gravity acceleration[9] on the effectiveness of liquid fluid reorientation. The criteria for the performance evaluation are based on how efficient the impulsive reverse gravity acceleration can activate the flow parameters in comparison with that of

TABLE 3 CHARACTERISTICS OF MAJOR SLOSH WAVES CAUSED BY RESETTLEMENT OF LIQUID REORIENTATION

1.0 Hz Medium Frequency Impulsive Thrust with Geyser Initiation

Reverse Gravity Acceleration = $8.2 \times 10^{-2} g_0$

Liquid Filled Level = 80%

Wave Period (sec)	Wavelength (cm)	Phase Velocity (cm/sec)	Propagation Direction (Clockwise from Positive Axial Direction) (Deg)	Ratio Maximum Wave Amplitude to Wavelength (10^{-2})
0.09	4.47	49.69	(±) 179	0.273
0.14	1.74	12.42	(±) 179	0.313
0.48	3.41	7.10	(±) 179	2.148

constant reverse gravity acceleration at the same background thrust acceleration. Results show that 1.0 Hz medium frequency impulsvie reverse gravity acceleration is superior than constant reverse gravity acceleration based on the following flow parameters: (A) a higher maximum flow velocity, (B) a shorter time period for flow to reach maximum velocity, (C) a shorter time period for flow to reach tank bottom (tank outlet) for fluid resettling, etc. Comparison between the performance of impulsive thrust and that of the constant thrust, it shows that impulsive thrust, regardless of frequency range, is always superior than the constant reverse thrust in terms of efficient operation of fluid reorientation[16]. As to the impulsive thrust in various frequency ranges from high to low frequencies, 1.0 Hz medium frequency impulsive gravity acceleration can perform better than the other frequency ranges in terms of efficient operation of fluid reorientation[16].

Slosh wave excitation induced by the resettling flow field during the course of liquid fluid reorientation activated by 1.0 Hz medium frequency impulsive reverse gravity acceleration with the initiation of geyser for liquid filled levels of 30, 50, and 80% have been studied. Results show that there is a group of wave trains with various frequencies and wavelengths of slosh waves generated by the resettling flow field activated by the 1.0 Hz medium frequency impulsvie reverse gravity acceleration. Following conclusions have been drawn from the present study: (1) A series of longitudinal wave trains of slosh waves which propagate along the direction of thrust force that activate the reorientation of cryogenic liquid fluid toward the outlet end of tank; (2) Slosh waves with lowest frequency are always associated with the highest values of Max (A/λ) ratio wave modes; (3) Slosh waves with lowest frequency contain highest wave intensity (see Figures 3, 5, 7) which are equivalent to highest wave energy; (4) Impulsive reverse gravity acceleration along the axial direction is responsible for the excitation of slosh waves with various frequencies; (5) Among the impulsive reverse thrusts with various frequency ranges from high to low frequencies, slosh waves excited by 1.0 Hz medium frequency impulsive reverse thrust are always associated with the lowest values of Max (A/λ) ratio wave modes; (6) Impulsive reverse gravity acceleration required to activate the liquid reorientation must increase their thrust forces accordingly to maintain the momentum for the initiation of geyser as the liquid filled levels increase; and (7) There is no distinctive differences in wave characteristics for the slosh waves excited with the various liquid filled levels of propellant tank.

In this study, we have demonstrated that the computer algorithm presented, through the execution of supercomputer CRAY-MP, can be used to simulate the fluid behavior during the course of liquid resettling and reorientation, in particular, the excitation of slosh waves due to resettling flow field activated by the impulsive reverse gravity acceleration.

ACKNOWLEDGEMENT

The authors appreciate NASA Grant NAGW-812, and NASA Marshall Space Flight Center through the NASA Contract NAS8-36955/Delivery Order No. 69.

REFERENCE

1. Hung, R. J., Lee, C. C., and Shyu, K. L., "Reorientation of Rotating Fluid in Microgravity Environment with and without Gravity Jitters," J. Spacecraft Rockets, Vol. 28, (1991), p. 71.
2. Leslie, F. W., "Measurements of Rotating Bubble Shapes in a Low Gravity Environment," J. Fluid Mech., Vol. 161, Dec. (1985), p. 269.
3. Hung, R. J., and Leslie, F. W., "Bubble Shape in a Liquid Filled Rotating Container Under Low Gravity," J. Spacecraft and Rockets, Vol. 25, (1988), p. 70.
4. Hung, R. J., Tsao, Y.D., Hong, B. B., and Leslie, F. W., "Time Dependent Dynamical Behavior of Surface Tension on Rotating Fluids under Microgravity Environment," Adv. Space Res., Vol. 8, No. 12, (1988), p. 205.
5. Hung, R. J., Tsao, Y. D., Hong, B. B., and Leslie, F. W., "Bubble Behaviors in a Slowly Rotating Helium Dewar in Gravity Probe-B Spacecraft Experiment," J. Spacecraft and Rockets, Vol. 26, (1989), p. 167.
6. Hung, R. J., Tsao, Y. D., Hong, B. B., and Leslie, F. W., "Dynamical Behavior of Surface Tension on Rotating Fluids in Low and Microgravity Environments," Int. J. Microgravity Res. Appl., Vol. 11, (1989), p. 81.
7. Hung, R. J., Tsao, Y. D., Hong, B. B., and Leslie F. W., "Axisymmetric Bubble Profiles in a Slowly Rotating Helium Dewar Under Low and Microgravity Environments," Acta Astronautica, Vol. 19, (1989), p. 411.
8. "Stanford Relativity Gyroscope Experiment (NASA Gravity Probe-B)," Proc. Soc. Photo-Optical Instru. Eng., Vol. 619, Society of Photo-Optical Instrumentation Engineers, Bellingham, WA, (1986), p. 1.
9. Hung, R. J., and Shyu, K. L., "Cryogenic Hydrogen Reorientation Activated by Constant Reverse Gravity Acceleration of Geyser Initiation," AIAA Paper, No. 90-3712, (1990), p. 10.
10. Hung, R. J., Lee, C. C., and Leslie, F. W. "Effects of G-Jitters on the Stability of Rotating Bubble Under Microgravity Environment,"Acta Astronautica, Vol. 21, (1990), p. 309.
11. Hung, R. J., Lee, C. C., and Leslie, F. W., "Response of Gravity Level Fluctuations on the Gravity Probe-B Spacecraft Propellant System," J. Propulsion Power, Vol. 7, (1991), in press.
12. Hung, R. J., Phan, T., and Smith, R. E., "Observation of Gravity Waves During the Extreme Tornado Outbreak of April 3, 1974," J. Atmos. Terres. Phys., Vol. 40, (1978), p. 831.
13. Hung, R. J., and Smith, R. E., "Ray Tracing of Gravity Waves as a Possible Warning System for Tornadic Storms and Hurricanes," J. Appl. Meteo., Vol. 17, (1978), p. 3.
14. Hung, R. J., and Kuo, J. P., "Ionospheric Observation of Gravity Waves Associated with Hurricane Eloise," J. Geophys., Vol. 45, (1978), p. 67.
15. Hung, R. J., Phan, T., and Smith, R. E., "Coupling of Ionosphere and Troposphere During the Occurence of Isolated Tornadoes of November 20, 1973," J. Geophys. Res., Vol. 84, (1979), p. 1261.
16. Hung, R. J., and Shyu, K. L., "Initiation of Geyser During the Resettlement of Cryogenic Liquid Activated by the Impulsive Reverse Gravity Acceleration in Microgravity Environment," AIAA Paper, No. 91-0108, (1991), p. 11.

CRYOGENIC LIQUID RESETTLEMENT ACTIVATED BY IMPULSIVE THRUST IN SPACE-BASED PROPULSION SYSTEM

R. J. Hung and K. L. Shyu

The University of Alabama in Huntsville
Huntsville, Alabama 35899

ABSTRACT

The purpose of present study is to investigate most efficient technique for propellant resettling through the minimization of propellant usage and weight penalties. Comparison between the constant reverse gravity acceleration and impulsive reverse gravity acceleration to be used for the activation of propellant resettlement, it shows that impulsive reverse gravity thrust is superior to constant reverse gravity thrust for liquid reorientation in a reduced gravity environment. Comparison among impulsive reverse gravity thrust with 0.1 (low), 1.0 (medium) and 10 (high) Hz frequencies for liquid filled level in the range between 30 to 80 %, it shows that the selection of a medium frequency of 1.0 Hz impulsive thrust over the other frequency ranges of impulsive thrust is most proper based on the present study.

INTRODUCTION

In spacecraft design, the requirements for a settled propellant are different for tank pressurization, engine restart, venting, or propellant transfer. During the prepressurization of a cryogenic propellant in microgravity, significant heat and mass transfer will occur if the liquid interface is disturbed. Interface disturbances may result from (a) impingement of the gas on the liquid surface at a mass flow rate sufficient to cause Kelvin-Helmholtz instability, (b) globule formation from breaking waves caused by wave motion over baffles or internal hardware, (c) globule and surface froth formation resulting from movement of bubbles through the liquid to the surface, and (d) surface froth formation because of gas impingement.

Recently Leslie[1] was able to measure and to numerically compute the bubble shapes at various ratios of centrifugal force to surface tension force in 2, 4 and 6.3 cm deep cylinders in the microgravity environment. Hung et al.[2,3,4] further extended the work to include rotating speeds which resulted with bubbles intersecting and/or without intersecting the top, bottom and side walls of the cylinder.

Advances in Cryogenic Engineering, Vol. 37, Part B
Edited by R.W. Fast, Plenum Press, New York, 1992

An analysis of time-dependent dynamical behavior of surface tension on partially-filled rotating fluids in both low gravity and microgravity environments was carried out by numerically solving the Navier-Stokes equations subjected to the initial and the boundary conditions[3,5,6]. At the interface between the liquid and the gaseous fluids, both the kinematic surface boundary condition, and the interface stress conditions for components tangential and normal to the interface were applied.

An efficient propellant settling technique should minimize propellant usage and weight penalties. This can be accomplished by providing optimal acceleration to the spacecraft such that the propellant is reoriented over the tank outlet without any vapor entrainment, any excessive geysering, or any other undesirable fluid motion.

Instead of applying constant reverse gravity acceleration as we described in Paper I[7], this paper adopts impulsive reverse gravity acceleration with a medium frequency of 1.0 Hz for the activation of fluid reorientation with liquid filled levels of 30, 50, 65, 70 and 80%.

NUMERICAL SIMULATION OF LIQUID HYDROGEN REORIENTATION WITH GEYSER INITIATION AT MEDIUM FREQUENCY IMPULSIVE REVERSE GRAVITY ACCELERATION OF 1.0 HZ

Detailed description of mathematical formulation, initial and boundary conditions suitable for the analysis of cryogenic fluid management under microgravity environment are given in our earlier studies[3,5,8-10]. Model size is height L = 4.2 cm and diameter D = 4.3 cm. Cryogenic liquid propellant shall be positioned over the tank outlet by using auxiliary thrusters (or idle-mode thrusters from the main engine) which provide a thrust parallel to the tank's major axis in the direction of flight.

A small value of reverse gravity acceleration (downward direction) is provided by the propulsive thrust of small auxiliary engine to initiate the reorientation of liquid propellant. This small value of reverse gravity acceleration of propulsive thrust increases gradually until reaching the critical value on which initiation of geyser is detected during the time period of fluid resettlement. We term this reverse gravity acceleration of propulsive thrust, which is capable to initiate geyser, as "geyser initiation gravity level". Table 1 shows some basic geometries and characteristics of cryogenic liquid hydrogen resettlement activated by reverse gravity acceleration at geyser initiation gravity level. Average liquid height $\bar{h}$, and maximum liquid height h_m are shown in Figure 1. Average free fall velocity $\bar{V}_f$, average free fall time $\bar{t}_f$, and free fall velocity from maximum liquid height V_{fm} are computed from the following equations:

$$\bar{V}_f = (2 g_{io} \bar{h})^{1/2} \tag{1}$$

$$\bar{t}_f = \left(\frac{2\bar{h}}{g_{io}} \right)^{1/2} \tag{2}$$

$$V_{fm} = (2 g_{io} h_m)^{1/2} \tag{3}$$

TABLE 1 Some Basic Geometries and Characteristics of Cryogenic Liquid Hydrogen Reorientation

Liquid Filled Level (%)	30	50	65	70	80
Geyser Initiation Gravity-Level g_{io} (10^{-2} g_o)	5.5	6.52	6.6	6.7	8.2
Geyser Initiation Acceleration a_g (cm/s^2)	53.9	63.9	64.7	65.7	80.4
Average Liquid Height, $\bar{h}$ (cm)	2.67	2.05	1.52	1.35	0.93
Average Free Fall Flow Velocity $\bar{V}_f$ (cm/s)	17.0	16.2	14.0	13.3	12.5
Maximum Liquid Height, h_m (cm)	3.41	2.79	2.26	2.09	1.67
Free Fall Velocity from Maximum Liquid Height V_{fm} (cm/s)	19.2	18.9	17.1	16.6	16.4

where g_{io} denotes geyser initiation reverse gravity acceleration.

Figures 2, 3, 4, 5, and 6 show time evolution of fluid reorientation activated by geyser initiation impulse reverse gravity acceleration with a frequency of 1.0 Hz for liquid filled levels of 30, 50, 65, 70 and 80 %, respectively. Each figure contains four sub-figures. Subfigure (A) is initial profile of liquid-vapor interface at the moment of the starting of fluid reorientation at time t = 0; subfigure (B), the flow profile during the course of fluid reorientation before the initiation of geysering motion; subfigure (C), the flow profile with geysering motion; and subfigure (D), the flow profile after the ending of geysering motion.

Examples of selected sequences of time evolution of fluid reorientation illustrate following flow behaviors: (1) The liquid starts to flow in an annular sheet along the solid wall of tank when the fluid reorientation starts; (2) The thrust force gradually pushes the vapor toward the central portion of the lower dome of tank as the net acceleration is applied toward the downward direction of the tank's major axis; (3) The thrust force, reversing the direction of gravity field, is provided by the small

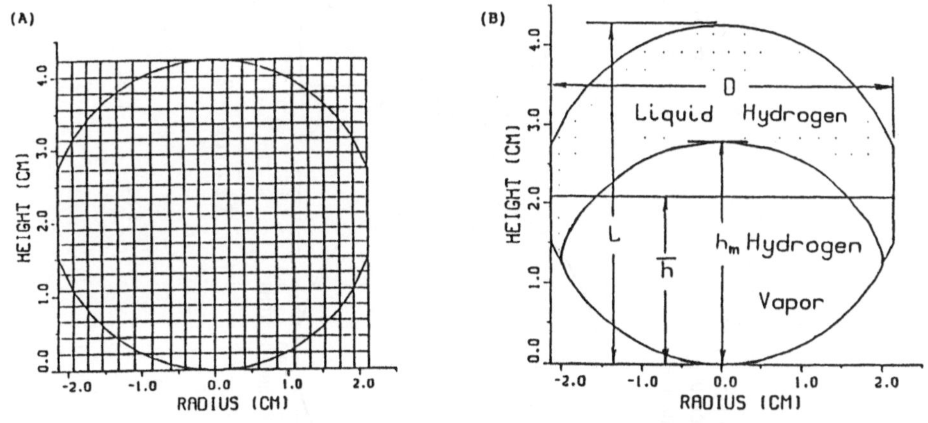

D-4.26720 CM, L-4.23672 CM

Figure 1. (A) Grid points in the radial-axial plane (B) Model size propellant tank.

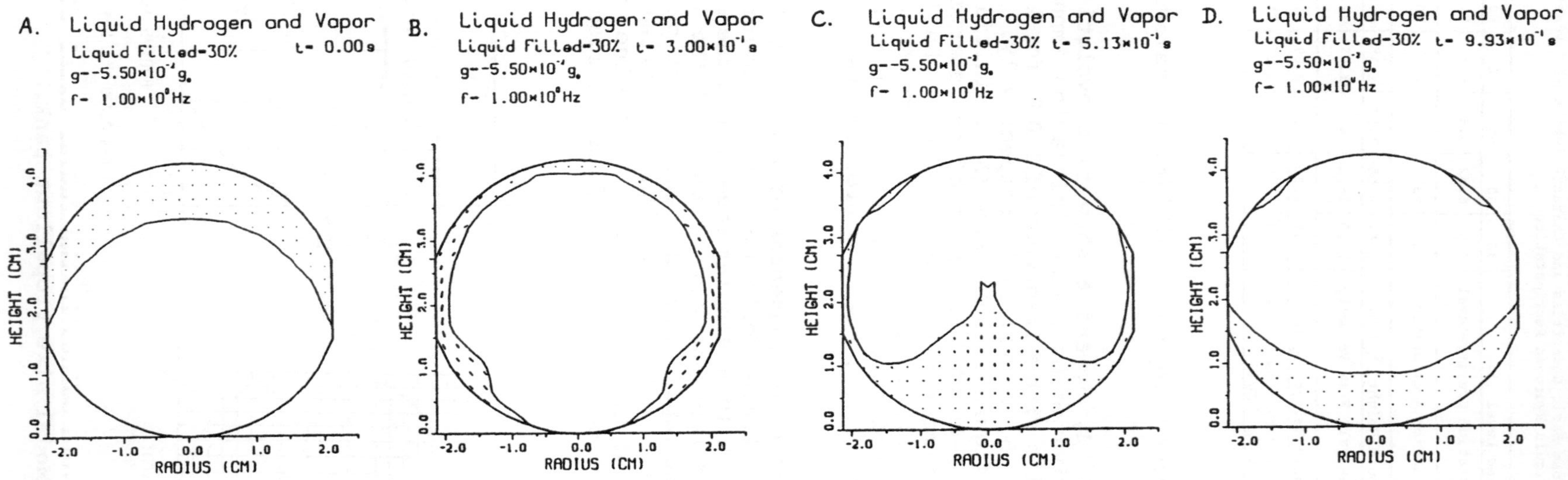

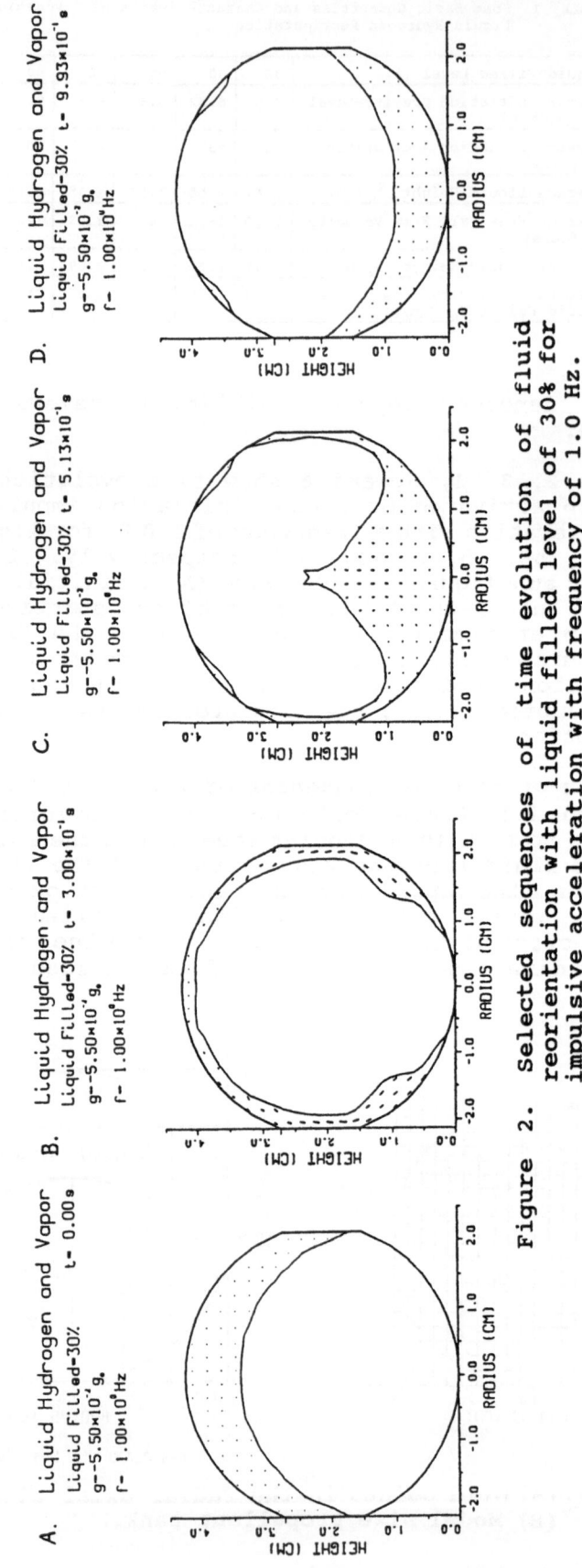

Figure 2. Selected sequences of time evolution of fluid reorientation with liquid filled level of 30% for impulsive acceleration with frequency of 1.0 Hz.

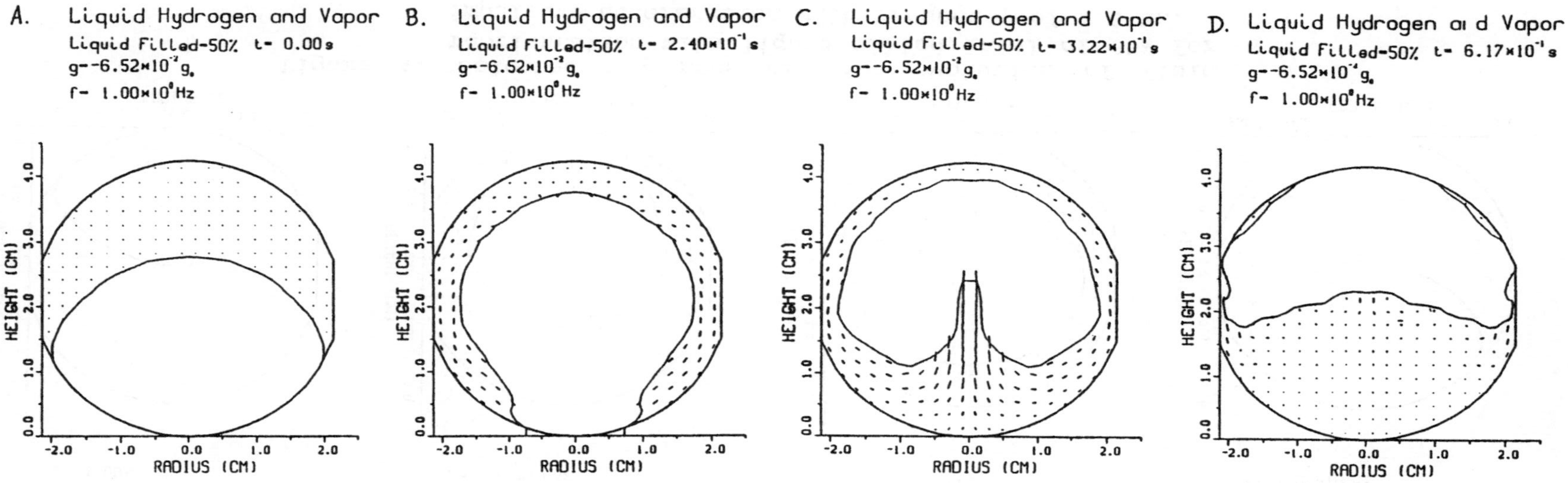

Figure 3. Selected sequences of time evolution of fluid reorientation with liquid filled level of 50% for impulsive acceleration with frequency of 1.0 Hz.

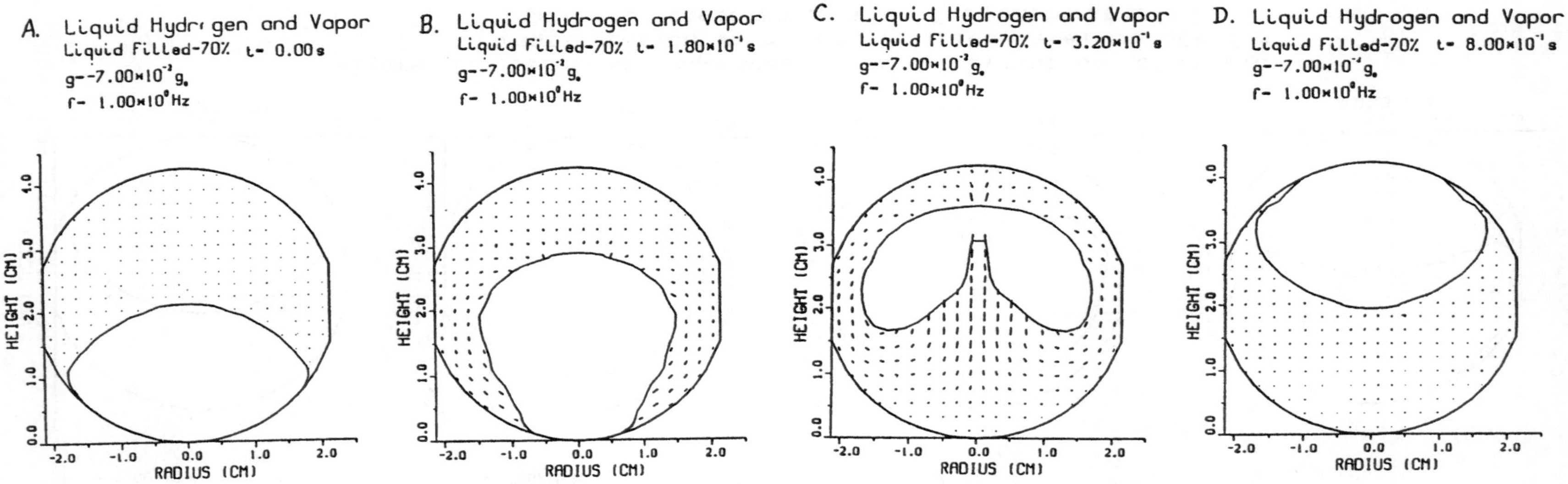

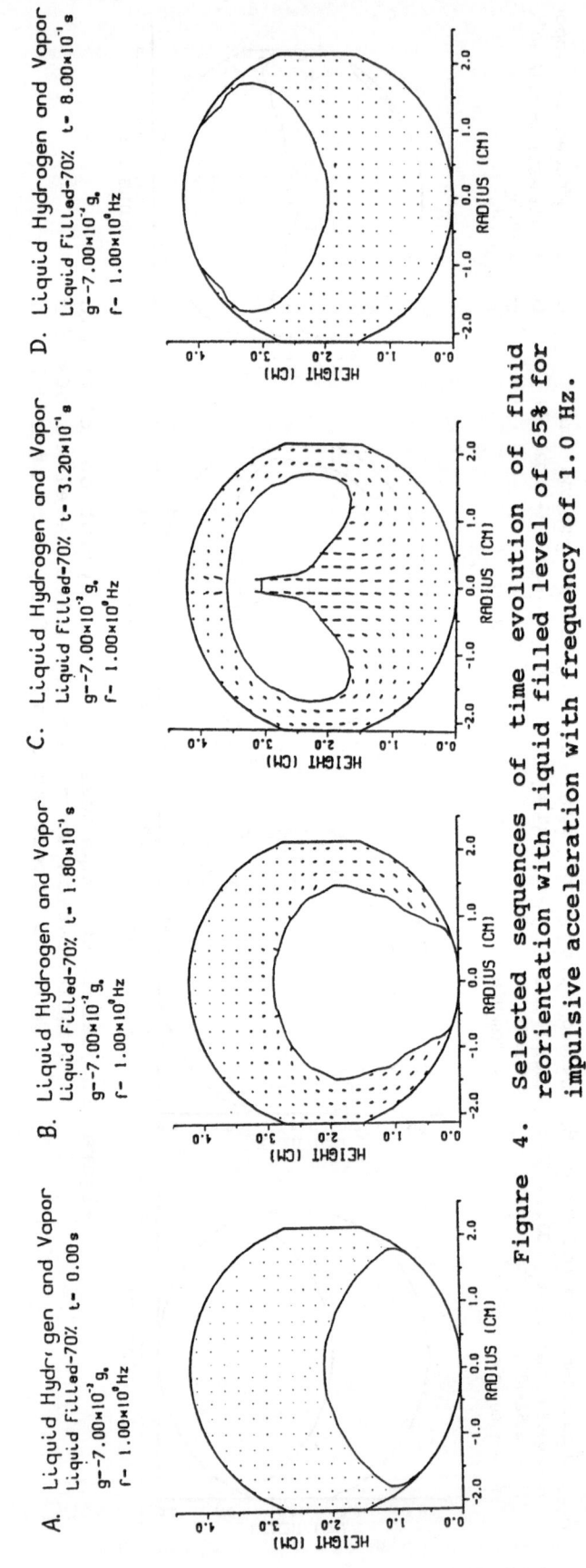

Figure 4. Selected sequences of time evolution of fluid reorientation with liquid filled level of 65% for impulsive acceleration with frequency of 1.0 Hz.

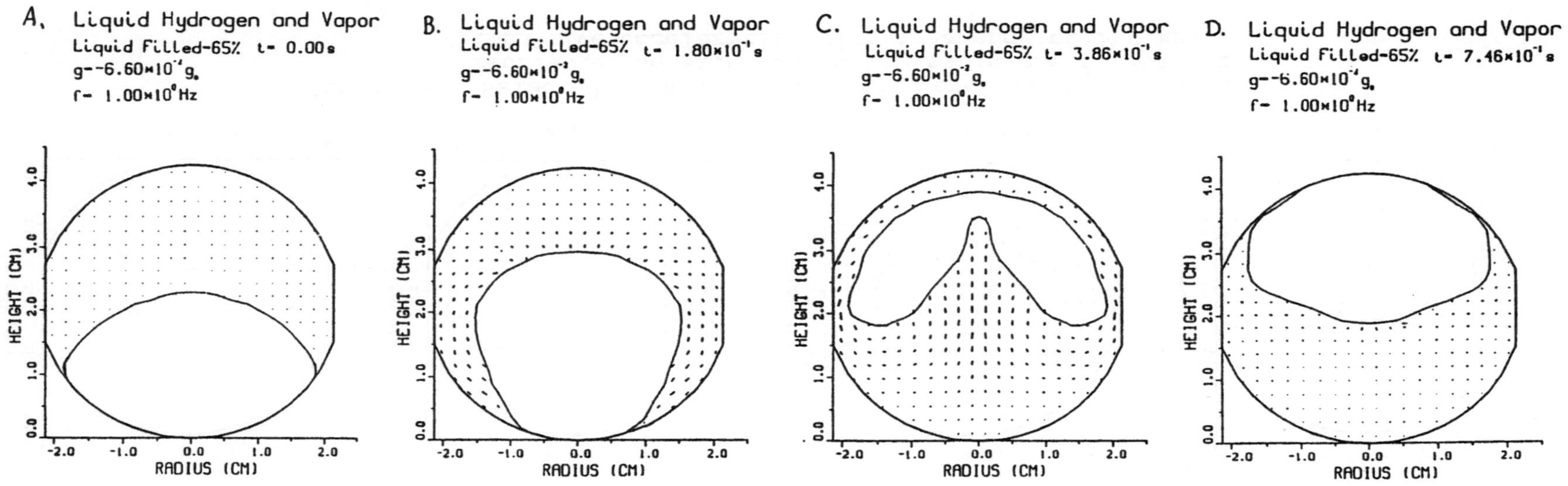

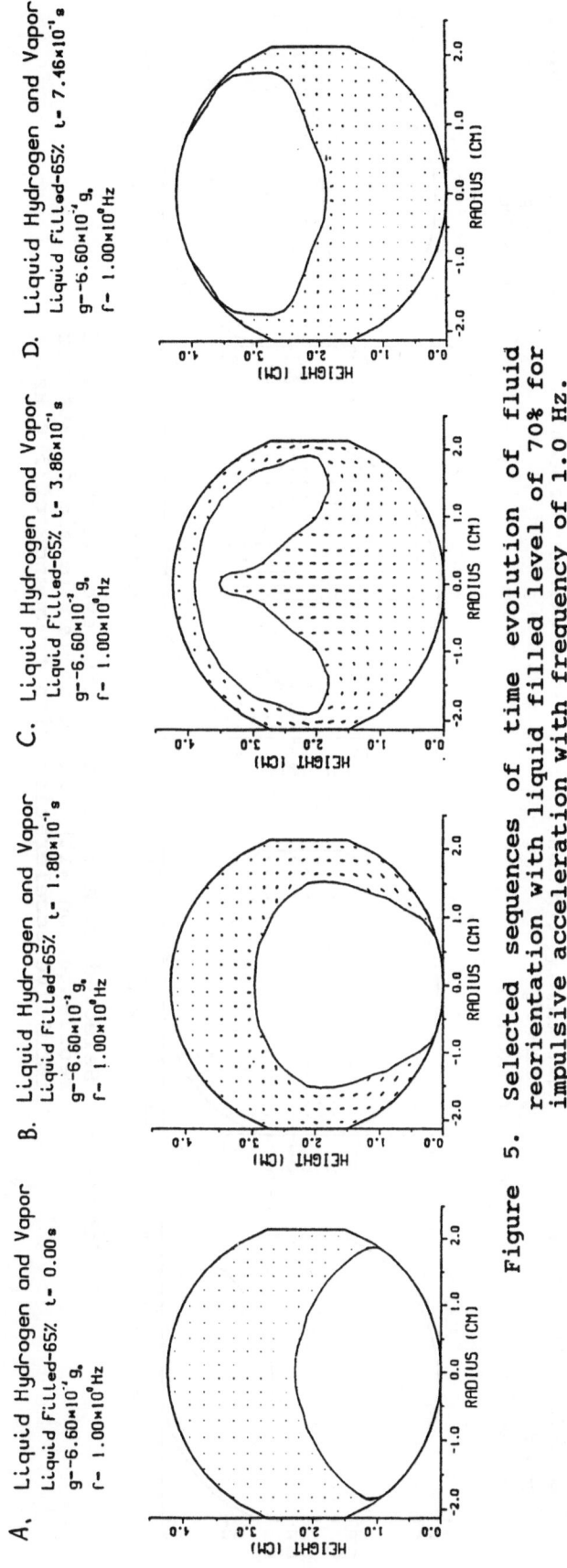

Figure 5. Selected sequences of time evolution of fluid reorientation with liquid filled level of 70% for impulsive acceleration with frequency of 1.0 Hz.

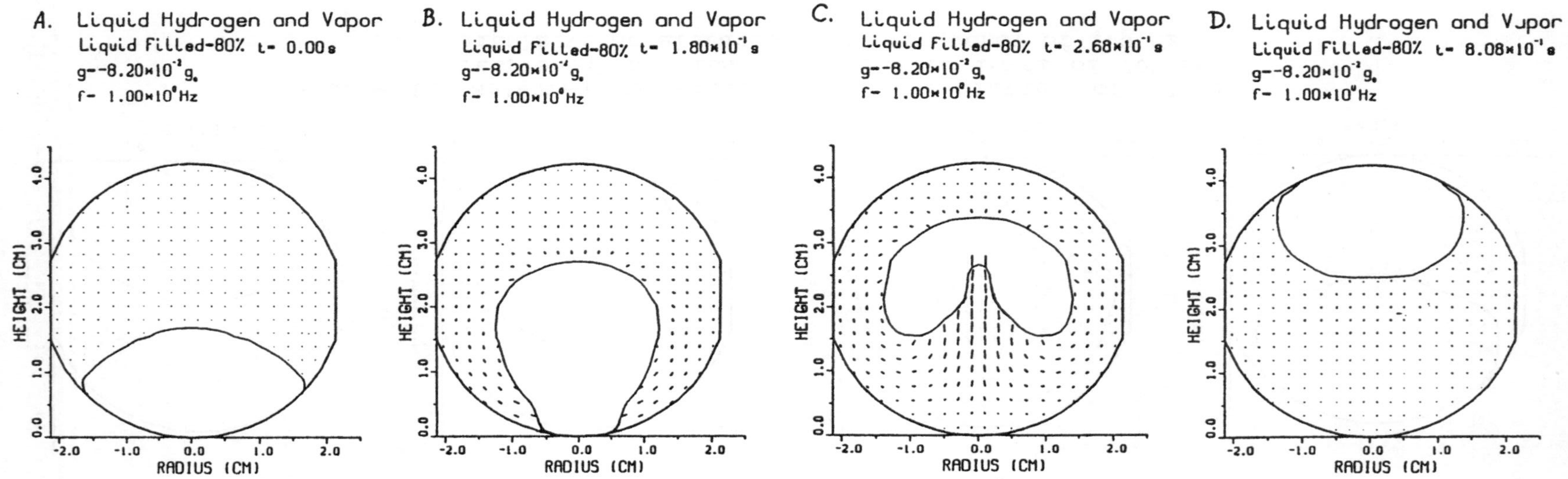

Figure 6. Selected sequences of time evolution of fluid reorientation with liquid filled levels of 80% for impulsive acceleration with frequency of 1.0 Hz.

TABLE 2 Characteristics of Cryogenic Hydrogen Reorientation (Frequency of Impulsive Acceleration = 1.0 Hz)

Liquid Filled Level (%)	30	50	65	70	80
Maximum Flow Velocity, V_m (cm/s)	112.2	107.0	94.8	89.5	82.5
$V_m/\bar{V}_f$	6.6	6.6	6.7	6.7	6.6
Liquid Reaching Bottom Time, t_R (s)	0.35	0.28	0.24	0.22	0.19
$t_R/\bar{t}_f$	1.1	1.1	1.1	1.1	1.2
Time for Observing Maximum Velocity, t_m (s)	0.35	0.30	0.24	0.22	0.18
Scale Length of Maximum Flow Velocity, L_m (= $V_m t_m$) (cm)	39.3	32.1	22.8	19.7	14.8
$L_m/\bar{h}$	14.7	15.6	15.0	14.6	15.9
$t_m/\bar{t}_f$	1.1	1.2	1.1	1.1	1.2
Scale Flow Acceleration Associated with Maximum Velocity, a_m (= $V_m t_m$) (cm/s^2)	320	356	395	407	458
a_m/a_g	5.9	5.7	6.1	6.1	5.7
V_m/V_{fm}	5.8	5.7	5.6	5.6	5.2

auxiliary thrusters; (4) As the downward fluid annular sheet along the tank wall reaches the central bottom dome side of the tank, a geysering flow is observed; and (5) The vapor is thus pushed upward centrally into the liquid and the geysering disappears.

Based on the computer simulation, flow field parameters such as: maximum flow velocity V_m, time for observing maximum flow velocity t_m, and time for reorienting liquid reaching the bottom of propellant tank t_R, are obtained and illustrated in Table 2 for reverse gravity acceleration with impulsive frequency of 1.0 Hz. Scale length of maximum flow velocity L_m, and scale flow acceleration associated with maximum velocity a_m, can be computed from the following parameters:

$$L_m = V_m t_m \tag{4}$$

$$a_m = \frac{V_m}{t_m} \tag{5}$$

Following dimensionless parameters are introduced: $V_m/\bar{V}_f$, $t_R/\bar{t}_f$, $t_m/\bar{t}_f$, a_m/a_g, $L_m/\bar{h}$ and V_m/V_{fm} where a_g stands for geyser initiation acceleration with a unit of cm/s^2, corresponding to geyser initiation gravity level with a unit of Earth gravity g_{io}. Impulsive reverse gravity acceleration, g_i with frequency f Hz is defined as follows:

$$g_i = g_{io}\left(1 + \frac{1}{2}\sin 2\pi ft\right) \tag{6}$$

Figure 7(A) shows the ratio of maximum flow velocity to average free fall flow velocity $V_m/\bar{V}_f$ and its associated parameters of V_m and $\bar{V}_f$ in terms of liquid filled levels for impulsive acceleration with frequency of 1.0 Hz. Figure 7(B) shows the ratio of liquid reaching bottom time to average free fall time $t_R/\bar{t}_f$ and its associated parameters of t_R and $\bar{t}_f$ in terms of liquid filled levels for impulsive acceleration with frequency

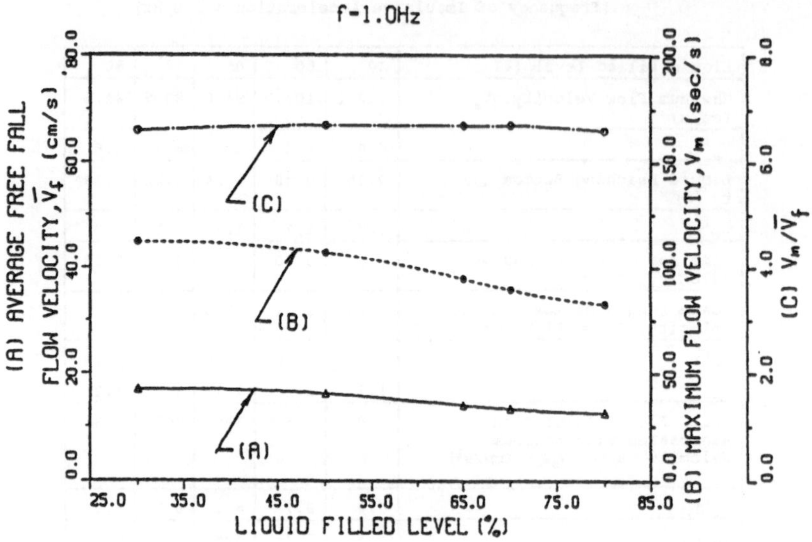

Figure 7(A). Ratio of $V_m/\bar{V}_f$ and its associated parameters.

of 1.0 Hz. Figure 8(A) shows the ratio of time for observing maximum flow velocity to average free fall time $t_m/\bar{t}_f$ and its associated parameters of t_m and $\bar{t}_f$ in terms of liquid filled levels. Figure 8(B) shows the ratio of scale flow acceleration associated with maximum velocity to geyser initiation acceleration for corresponding geyser initiation gravity level a_m/a_g and its associated parameters of a_m and a_g in terms of liquid filled levels for impulsive acceleration with frequency of 1.0 Hz. Figure (9A) shows the ratio of scale length of maximum flow

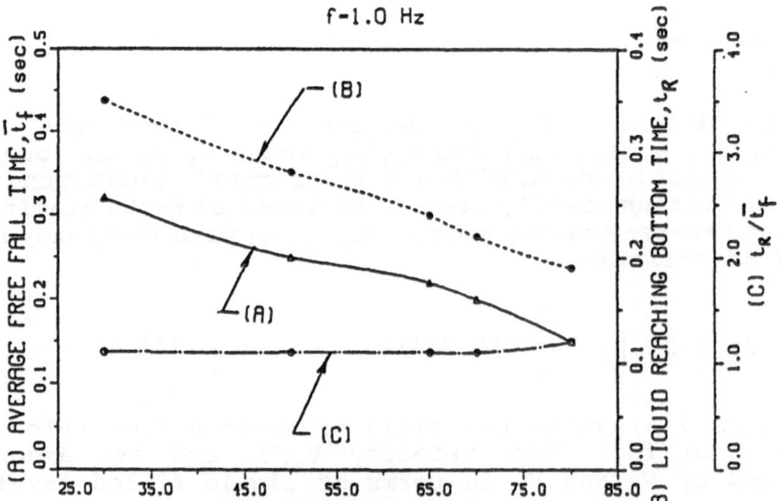

Figure 7(B). Ratio of $t_R/\bar{t}_f$ and its associated parameters.

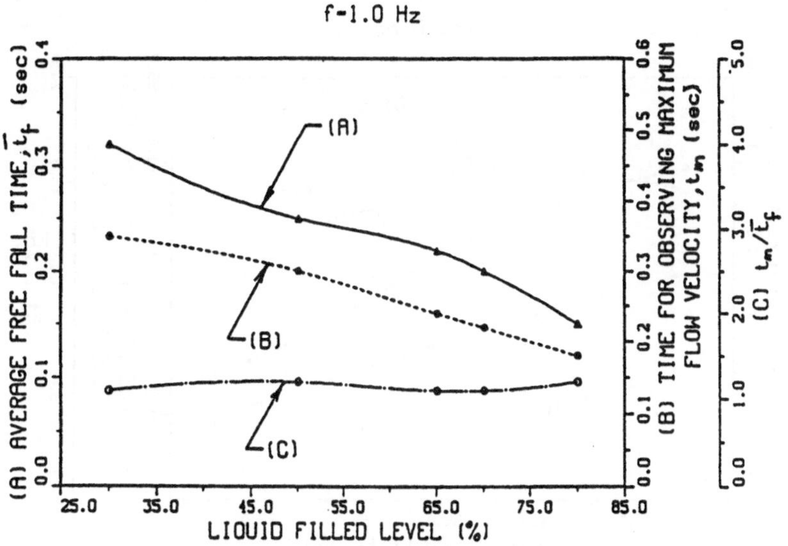

Figure 8(A). Ratio of $t_m/\bar{t}_f$ and its associated parameters.

velocity to average liquid height $L_m/\bar{h}$ and its associated parameters of L_m and $\bar{h}$ in terms of liquid levels for impulsive acceleration with frequency of 1.0 Hz. Figure 9(B) shows the ratio of maximum flow velocity to free fall velocity from maximum liquid height V_m/V_{fm} and its associated parameters of V_m and V_{fm} in terms of liquid filled levels for impulsive acceleration with frequency of 1.0 Hz.

DISCUSSION AND CONCLUSIONS

Evaluation of performance is based on how efficient the impulsive reverse gravity with various frequencies can activate

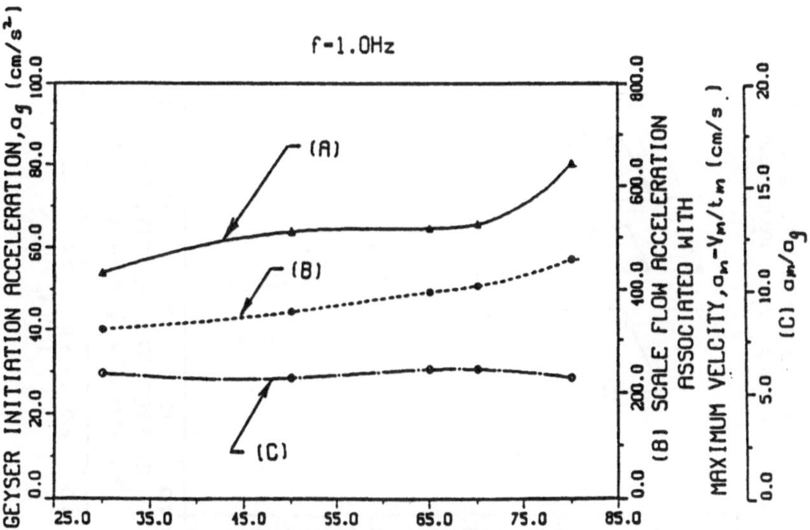

Figure 8(B). Ratio of a_m/a_g and its associated parameters.

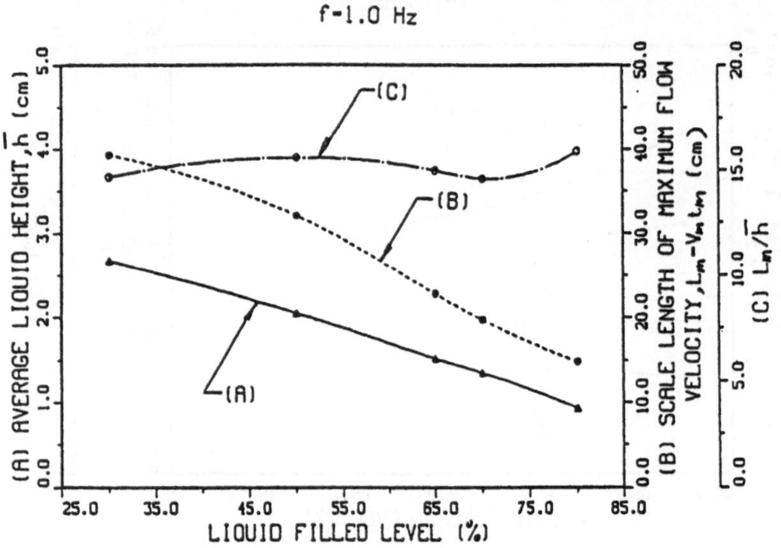

Figure 9(A). Ratio of $L_m/\bar{h}$ and its associated parameters.

following flow parameters at the same background thrust accelerations: (A) a higher maximum flow velocity, (B) a shorter time period for flow to reach maximum velocity, (C) a shorter time period for flow to reach tank bottom (tank outlet) for fluid resettling, (D) a larger length scale of maximum flow velocity, and (E) a higher flow acceleration associated with maximum flow velocity. In summary, it shows that auxiliary engine with various frequencies of impulsive reverse gravity thrust is the better choice of operation for the purpose of activating fluid reorientation. Comparison between the results of present study for impulsive thrust with various frequencies and that of constant thrust, shown in Paper I[7], shows that impulsive thrust,

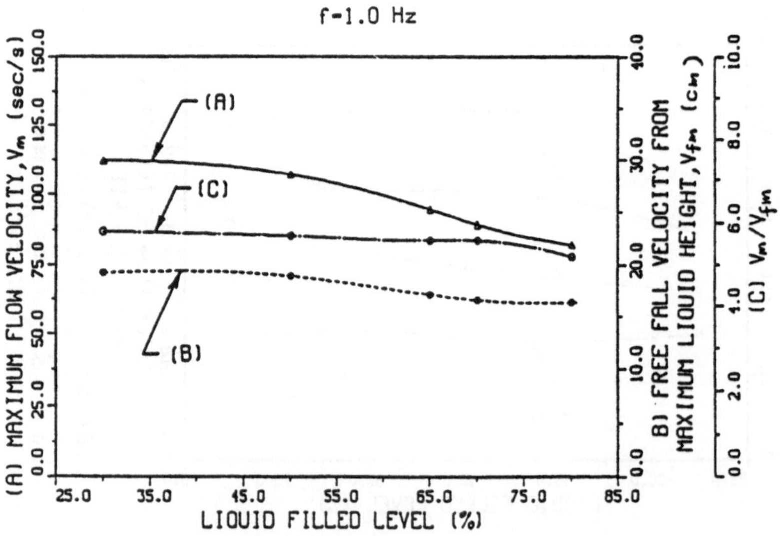

Figure 9(B). Ratio of V_m/V_{fm} and its associated parameters.

regardless of frequency range, is always superior to the constant reverse thrust in terms of efficient operation of fluid reorientation.

Based on the computer simulation of flow fields during the course of fluid reorientation, six dimensionless parameters are presented both in Paper I[7] and this study. It is shown that these parameters hold near constant values through the entire ranges of liquid filled levels during the course of fluid reorientation activated by the reverse gravity acceleration in both constant and impulsive thrusts great enough to initiate geyser. As the denominators of these dimensionless parameters are either predetermined from the geometry of liquid filled levels, as shown in Table 1, or can be deduced from the corresponding calculations associated with the geyser initiation gravity levels, one can predict the values of these flow parameters.

To conclude, we have demonstrated that, the computer algorithm presented, can be used to simulate fluid behavior in a microgravity environment, in particular the development of technology necessary for acquisition or positioning of liquid and vapor within a tank to enable liquid outflow or vapor venting through active liquid acquisition by the creation of a positive acceleration environment resulting from propulsive thrust. Better understanding of the full pictures of flow fields in both constant and impulsive thrusts, during the course of fluid reorientation can provide the proper design techniques for handling and managing the cryogenic liquid propellants to be used in on-orbit spacecraft propulsion. It is important to emphasize that impulsive reverse gravity thrust is superior to constant reverse gravity thrust for the activation of liquid necessary for the resettlement of liquid in a reduced gravity environment. It is also worthwhile to mention that the selection of 1.0 Hz frequency impulsive thrust over the other frequency ranges is most proper based on the present study.

ACKNOWLEDGEMENT

The authors appreciate the support received from the National Aeronautics and Space Administration Headquarters through the NASA Grant NAGW-812, and NASA Marshall Space Flight Center through the NASA Contract NAS8-36955/Delivery Order No. 69.

REFERENCE

1. Leslie, F. W., Measurements of Rotating Bubble Shapes in a Low Gravity Environment, J. Fluid Mech., 161:269 (1985).
2. Hung, R. J., and Leslie, F. W., Bubble Shape in a Liquid Filled Rotating Container Under Low Gravity, J. Spacecraft and Rockets, 25:70 (1988).
3. Hung, R. J., Tsao, Y.D., Hong, B. B., and Leslie, F. W., Time Dependent Dynamical Behavior of Surface Tension on Rotating Fluids under Microgravity Environment, Adv. Space Res., 8(12):205 (1988).
4. Hung, R. J., Tsao, Y. D., Hong, B. B., and Leslie, F. W., Bubble Behaviors in a Slowly Rotating Helium Dewar in Gravity Probe-B Spacecraft Experiment, J. Spacecraft and Rockets, 26:167 (1989).

5. Hung, R. J., Tsao, Y. D., Hong, B. B., and Leslie, F. W., Dynamical Behavior of Surface Tension on Rotating Fluids in Low and Microgravity Environments, Int. J. Microgravity Res. Appl., 11:81 (1989).
6. Hung, R. J., Tsao, Y. D., Hong, B. B., and Leslie F. W., Axisymmetric Bubble Profiles in a Slowly Rotating Helium Dewar Under Low and Microgravity Environments, Acta Astronautica, 19:411 (1989).
7. Hung, R. J., and Shyu, K. L., Constant Reverse Thrust Activated Reorientation of Liquid Hydrogen with Geyser Initiation, J. Spacecraft Rockets, 29:January (1992).
8. Hung, R. J., Lee, C. C., and Shyu, K. L., Reorientation of Rotating Fluid in Microgravity Environment with and without Gravity Jitters, J. Spacecraft and Rockets, 28:71 (1991).
9. Hung, R. J., Lee, C. C., and Leslie, F. W. Effects of G-Jitters on the Stability of Rotating Bubble Under Microgravity Environment, Acta Astronautica, 21:309 (1990).
10. Hung, R. J., Lee, C. C. and Leslie F. W., Response of Gravity Level Fluctuations on the Gravity Probe-B Spacecraft Propellant System, J. Propulsion and Power, 7:556 (1991).

THE ISO - CRYOSTAT; IT´S QUALIFICATION STATUS

A. Seidel, Th. Paßvogel

Messerschmitt-Boelkow-Blohm GmbH
Deutsche Aerospace
Ottobrunn, Germany

ABSTRACT

The Iso-satellite consists of a Service and a Payload Module, the latter being essentially made up of the Cryostat. The present paper describes the current qualification status of this cryostat. The general structural and thermal layout, the He-flow schematic and safety system and the general cover layout are described together with an overview on the existing hardware. Some qualification test results are as well reported.

INTRODUCTION

The Infrared Space Observatory (ISO) project was officially set to Phase C/D by the European Space Agency (ESA) on March 15, 1988. The satellite will be launched in 1993 by an Ariane IV launcher to a highly elliptical 24 hours orbit of 1000 km perigee, 70572 km apogee and an inclination of 5.25 °. ISO carries 4 scientific instruments (FPU´s) with overlapping wavelength ranges, a Camera (3 - 17 µm), an Imaging Photopolarimeter (3 - 200 µm), a Short Wavelength Spectrometer (3 - 45 µm) and a Long Wavelength Spectrometer (45 - 180 µm). The Ritchey-Chrétien telescope has an effective aperture of 60 cm, an overall f-ratio of 15 and a 20 arc min total field of view.

The ISO-satellite consists of a Service Module (SVM) and a Payload Module (PLM) which is made up by the Cryostat (carrying the cold parts of the instruments), the Cryostat Cover and the Optical Subsystem. This paper describes the ISO-cryostat design and hardware status and reports results from the various tests meanwhile performed with the Structural/Thermal Model (STM) of the PLM to qualify the design.

REQUIREMENTS

The ISO-Cryostat shall fulfill the following primary requirements that drive the PLM-design:

• Cryogenic Lifetime	18 months after start of scientific operat. on orbit
• HeII-Bath Temperature on Orbit	1.8±0.1 K
• Launch Site Autonomy without servicing	≥ 72 hours
• Optical Support Structure Temperature (instruments mounting basis)	2.4 - 3.4 K
• Total Average Instruments Dissipation	23 mW
• Maximum Main Baffle Temperature	7 K
• PLM-Overall Structural Frequency	
Axial	≥ 46 Hz
Lateral	≥ 20 Hz

Advances in Cryogenic Engineering, Vol. 37, Part B
Edited by R.W. Fast, Plenum Press, New York, 1992

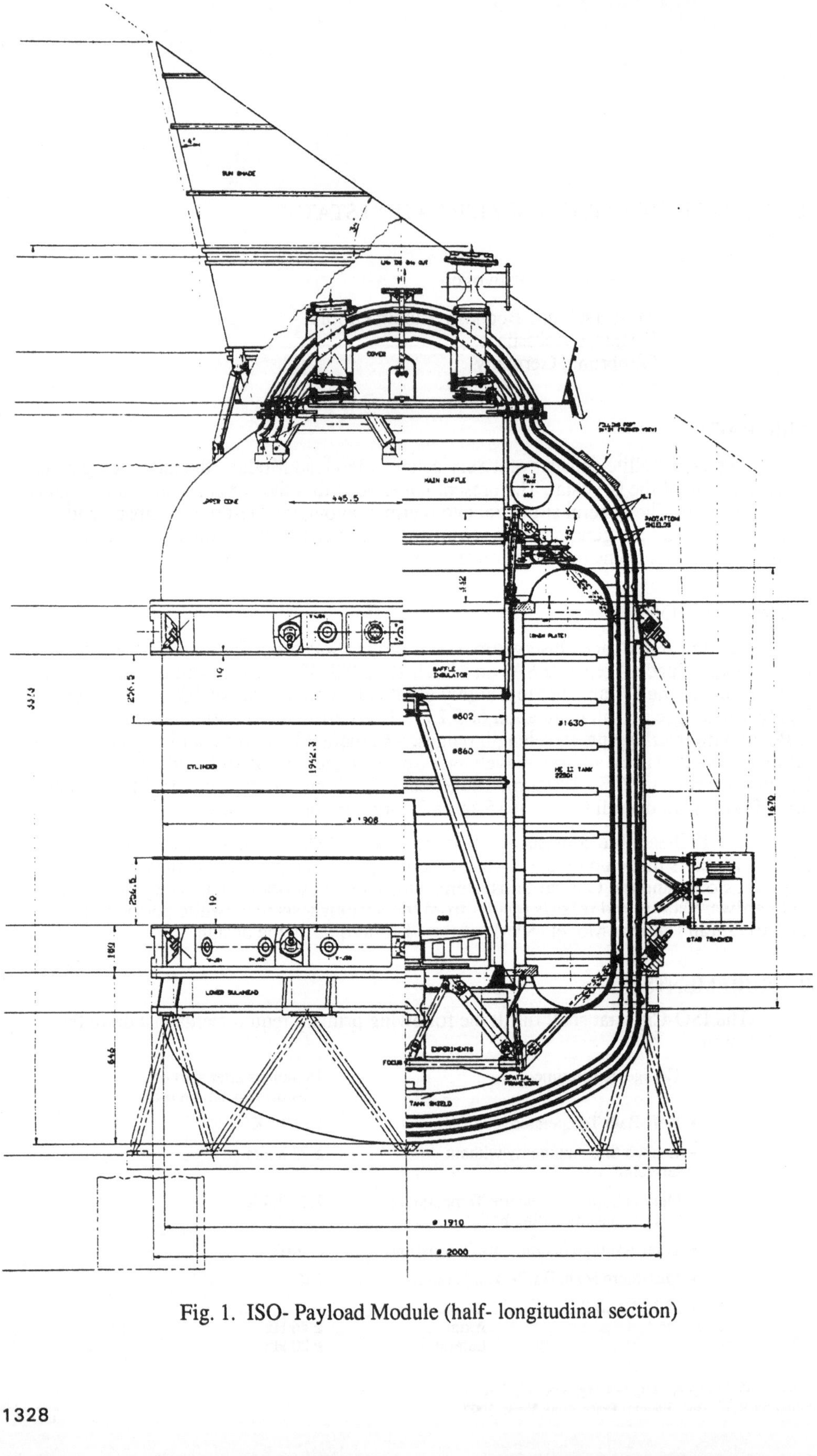

Fig. 1. ISO- Payload Module (half- longitudinal section)

Apart from lifetime and stiffness the cryostat design is driven by:

- thermal loads from the SVM and the startrackers to the PLM, approximately 1100 internal cables supplying the Focal Plane Units (FPU´s), a quadrant star sensor and the housekeeping sensors as well as from external cabling and conducting fixtures between the SVM and the PLM
- mechanical loads from the sunshield, the sunshade, the startrackers and a preamplifier unit fixed to the cryostat vacuum vessel (CVV) as well as SVM/PLM-interface loads
- Minimization of thermoelastic deformation between CVV and startrackers

CRYOSTAT DESIGN AND FUNCTION

a) Overall Description:

The ISO-PLM is shown in a longitudinal half-section by Fig. 1. The main cryostat constituents are:

- The cryostat vacuum vessel (CVV) from AlMg4.5 Mn (5083) alloy wrapped with outside MLI with non-charging white coating
- 16 GFC/CFC tank support straps, and strap tensioning devices
- Upper and lower Spatial Framework (SF) carrying the He-tanks
- The main 2300 l HeII-tank, made from 5083-alloy, with related valves, safety devices, HeII-phase separator and tubing
- The 60 l LHe-auxiliary tank for launch autonomy, also from 5083-alloy
- 3 vapour-cooled shields plus 1 tank attached shield covering the FPU´s, all equipped with MLI (multilayer insulation)
- External He-piping with valves and nozzles for momentum-free He-exhaust

The Optical Support Structure (OSS) carries the FPU´s, the Pyramidal Mirror, the Quadrant Star Sensor and the Telescope. The OSS and the main baffle are thermally decoupled from the HeII-tank and fixed to it by CFC-blade springs. The 16 chain-like tank support straps carry the vapour cooled shields on their stainless steel interconnection bolts. The outer chain elements are of GFC (glass fiber compound) and the inner elements are of CFC (carbon fiber compound)-material to minimize the overall thermal conductance. The SF, consisting of a quadratic frame and 16 struts, each, is made of different types of CFC-material.

In order to reach the longitudinal (axial) fundamental frequency of ≥ 46 Hz while maintaining the cryogenic lifetime a simultaneous structural and thermal optimization of each individual chain and SF-element was performed by the MBB-programme "LAGRANGE".

The optimization resulted in larger cross sections for the lower chain elements than for the upper ones.

b) Cryostat Operation and Safety System:

The cryostat flow schematic as shown in Fig. 2 allows for the various required cryogenic operations e.g.

- Cooldown and LHe filling
- HeII conversion and superfluid top-up
- Launch autonomy without cryoservicing for appr. 100 hours
- Transient behaviour from ground to orbit conditions
- Safety relief functions during abnormal operation on ground

The ISO-PLM He-subsystem, its function and safety system as well as some components are described in more detail in References 1,2 &3.

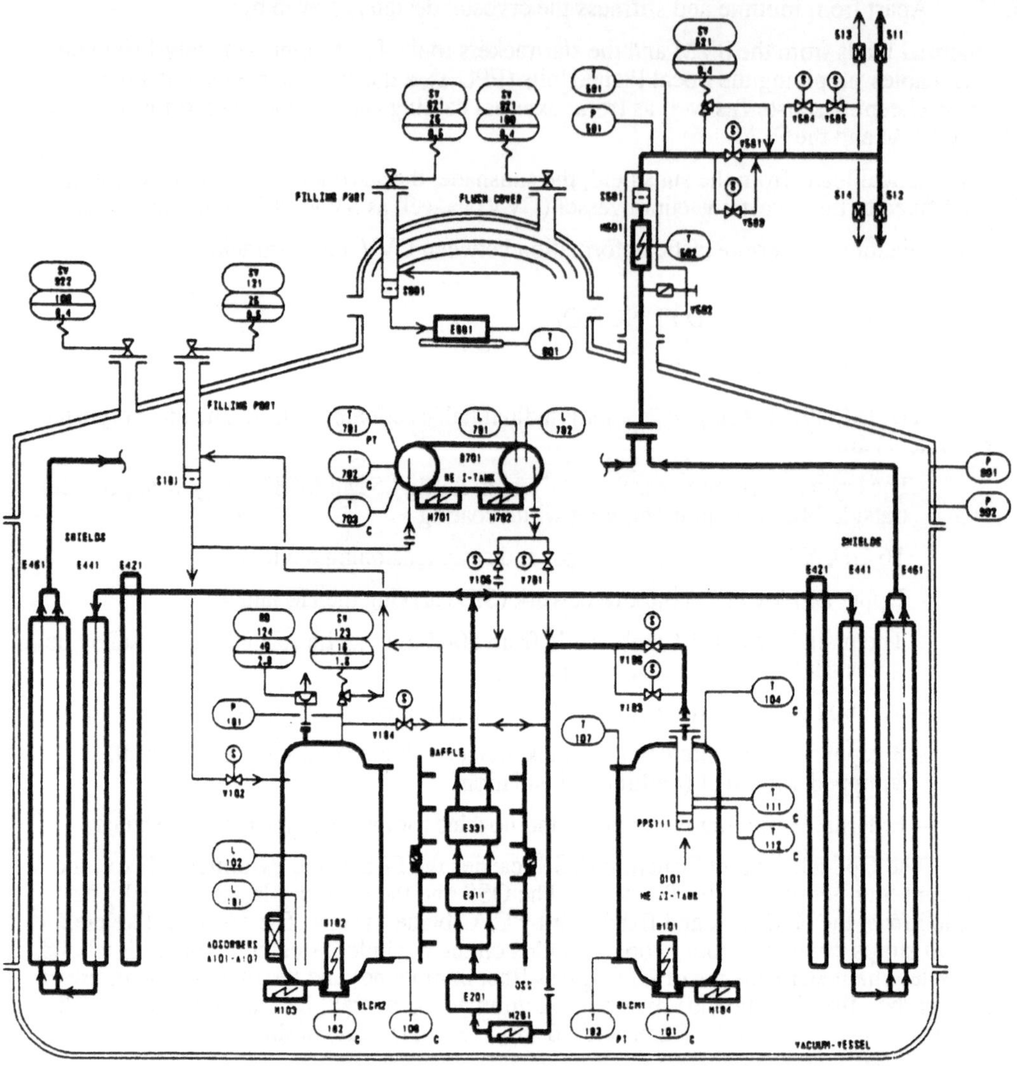

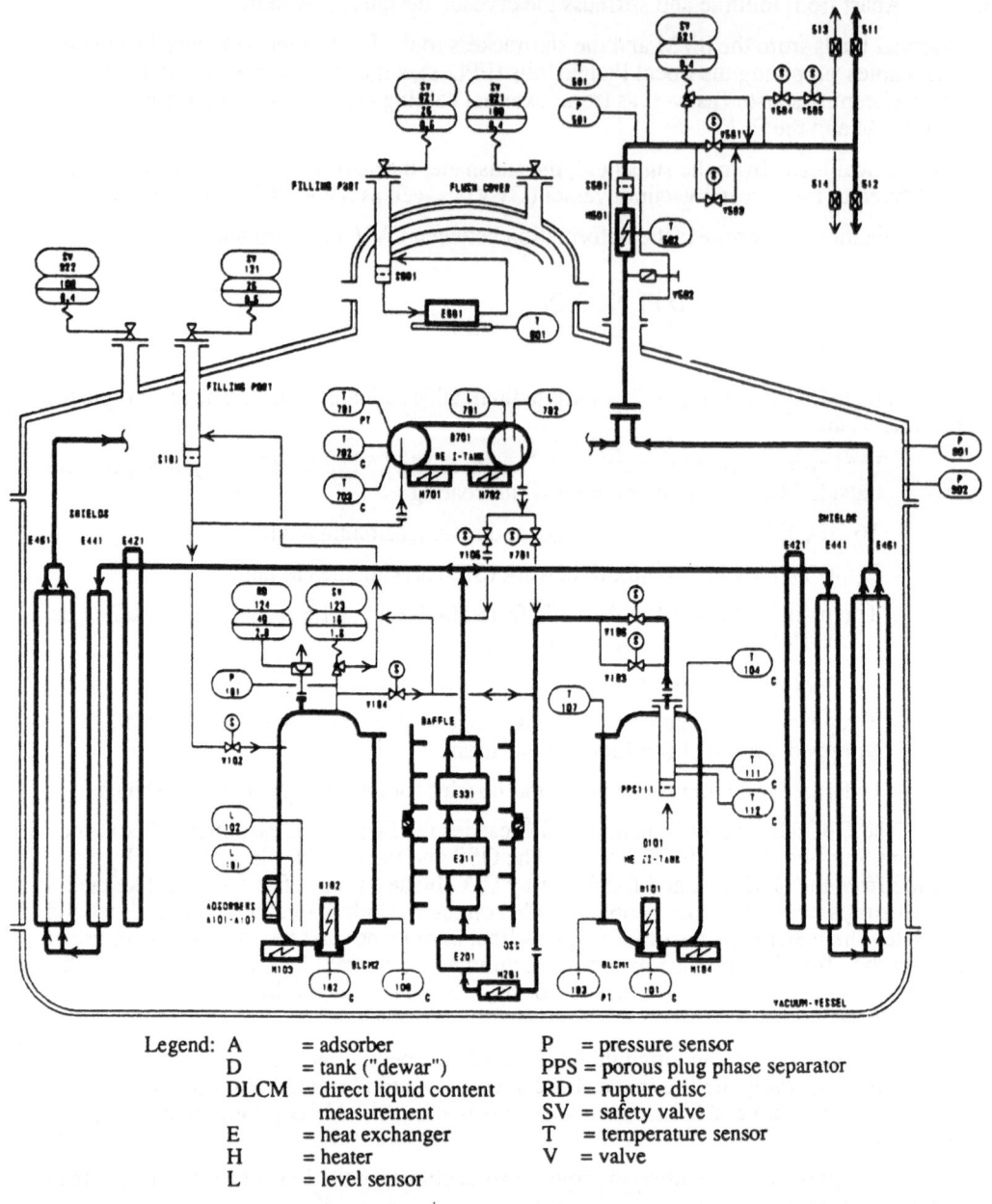

Legend: A	= adsorber	P	= pressure sensor
D	= tank ("dewar")	PPS	= porous plug phase separator
DLCM	= direct liquid content measurement	RD	= rupture disc
		SV	= safety valve
E	= heat exchanger	T	= temperature sensor
H	= heater	V	= valve
L	= level sensor		

Fig. 2. ISO-Cryostat Flow Schematic

c) Thermal Design:

The ISO-PLM thermal design was performed with a 380 nodes model in ESATAN. This model includes all essential thermal couplings (conductive and radiative) and the heat capacities of essential items involved and is used for steady state and transient behaviour calculations. This thermal mathematical model was, after availability of the "ground life-time" (GL) - and "thermal vacuum" (TV) - test results, correlated with these test data. Corrections resulting from the correlation effort were then actually introduced for the effective conductivities of the tank supports and the SF as well as for the effective emissi vity and the thermal conduction of the internal MLI. With the correlated/corrected thermal model transient predictions for the first 40 mission days and steady state ground and orbital calculations were performed.

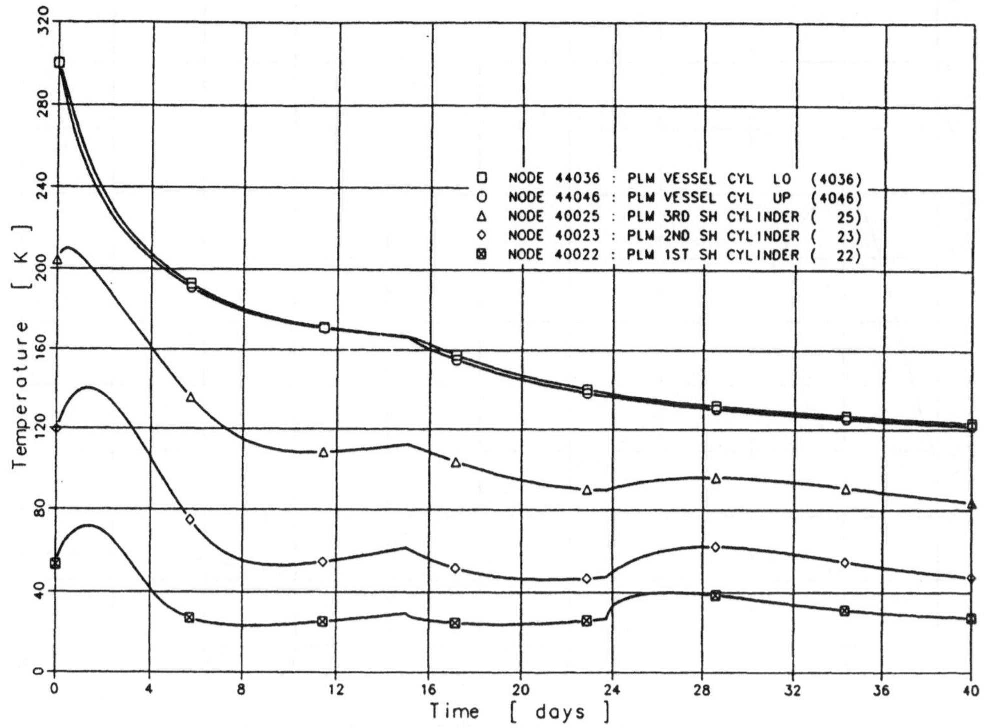

Fig. 3. Cryostat Vacuum Vessel and Heat Shields Temperature Evolution during First 40 Mission Days

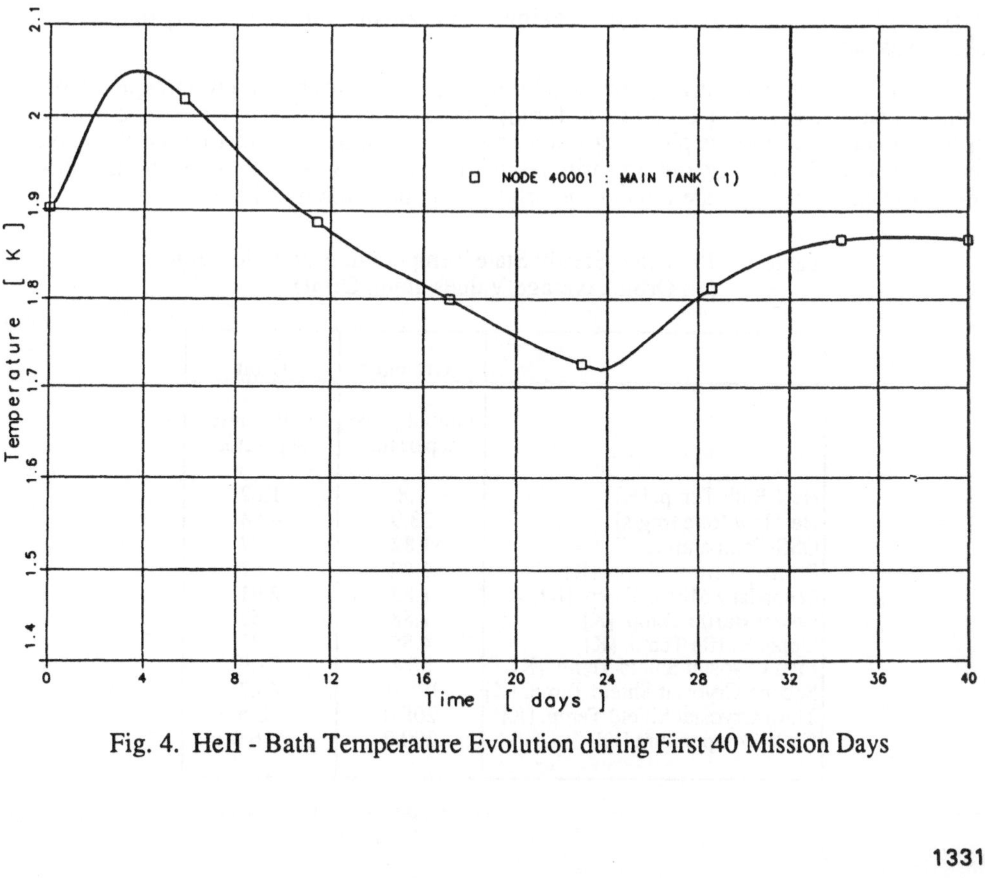

Fig. 4. HeII - Bath Temperature Evolution during First 40 Mission Days

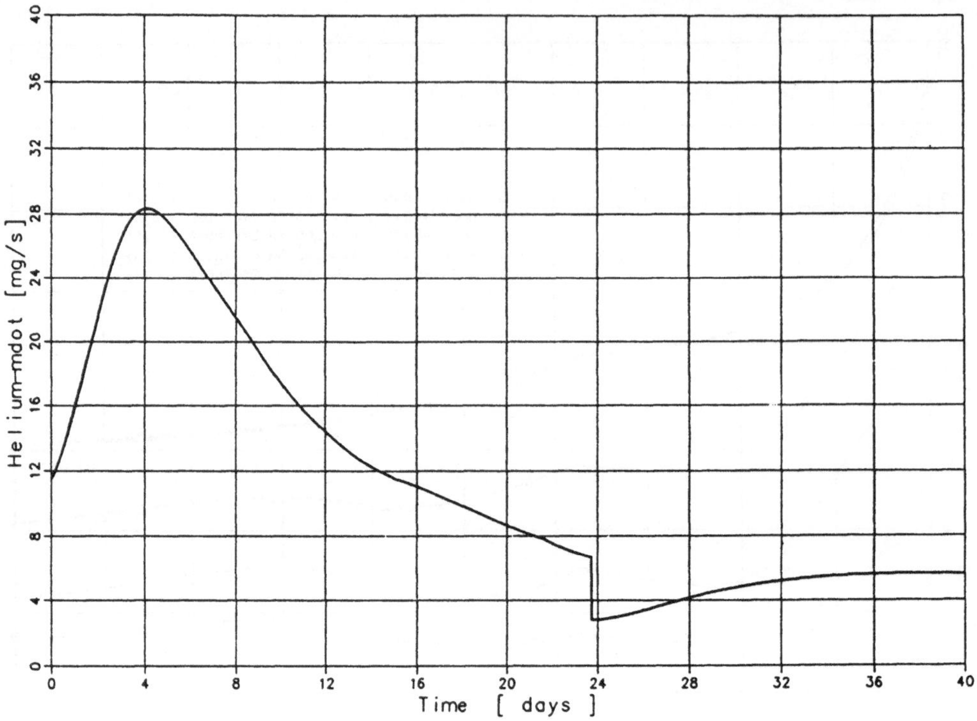

Fig. 5. He - Flowrate Evolution during First 40 Mission Days

Typical transient prediction results are shown by Figure 3 (cryostat heat shield temperatures versus time), Figure 4 (HeII-bath temperature versus time) and Figure 5 (He-flowrate evolution versus time). The unsteady changes in the curve slopes are caused by the closing of the external He-valves V504/505 to shut off the He-venting through the larger vent nozzles 513/514 after the maximum He-flowrate during the early transient phase was passed.

The He-vent rate varies between 33.9 mg/s during equilibrium ground operations without phase separator and 4.85 mg/s during final orbital equilibrium conditions with phase separator. Due to the phase separator and ventline impedance restrictions the maximum transient He-flowrate during early mission is appr. 28.3 mg/s. Essential temperatures predicted for steady state ground and orbital conditions are listed in table 1.

Table 1. Predicted Steady State Temperatures and Flow Rates (on Orbit: Average Values along Orbit)

	Ground	Orbit
	without phase separator	with phase separator
HeII-Bath Temp. [K]	1.8	1.82
He-Flow Rate [mg/s]	33.9	4.84
OSS-Temperature [K]	1.82	2.87
Primary Mirror Temp. [K]	1.82	2.87
Secondary Mirror Temp. [K]	1.83	2.91
Lower Baffle Temp. [K]	1.88	3.52
Upper Baffle Temp. [K]	4.85	6.37
First Cryostat Shield Temp. [K]	58.1	27.9
Second Cryostat Shield Temp. [K]	128.0	45.7
Third Cryostat Shield Temp. [K]	208.0	78.5
CVV-Temperature [K]	300.0	116.0

Fig. 6. Fully Equipped HeII - Tank Head with Spatial Framework

The thermal design of the ejectable Cryostat Cover is governed by 4 radiation shields which are passively cooled via spring loaded thermal contacts transferring the cover shield heat flows to the 3 actively cooled cryostat heat shields and the main baffle. Four thermal contacts are applied per cryostat shield and eight contacts on the main baffle. The thermal conductances of these contacts were separately measured and the results were reported in Ref. 4. The thermal cryostat cover design takes also into account the 4 GFC-bars which support the 4 cover heat shields and the 2 optical channels necessary for optical alignment measurement.

CRYOSTAT INTEGRATION AND TESTS

a) Hardware Integration:

The ISO-cryostat structural/thermal model (STM) integration was started in early August 1989 and was completed just before Christmas 1989. The following figures show some essential hardware during the integration process.

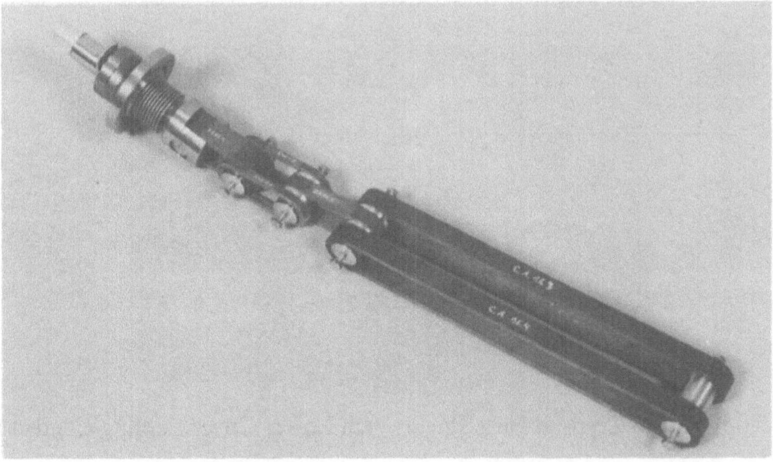

Fig. 7. Tank Suspension Strap with Pretensioning Device (left)

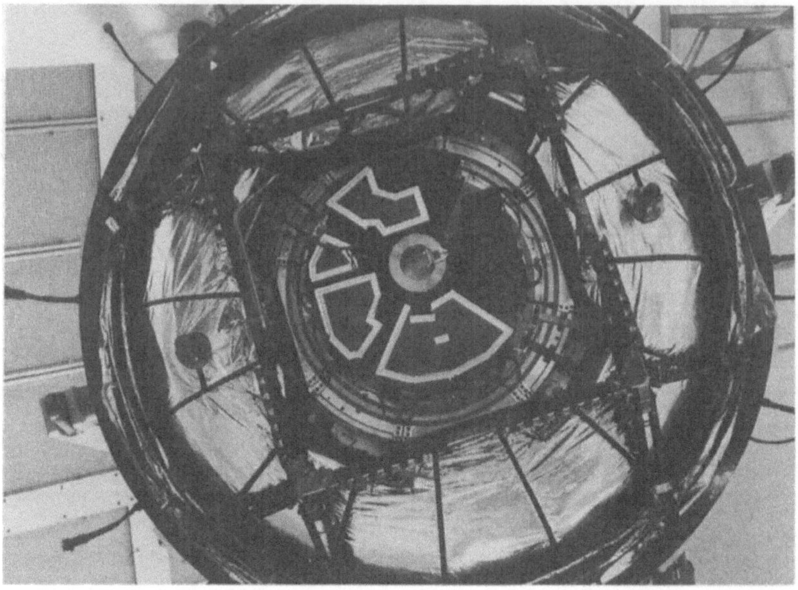

Fig. 8. Lower HeII - Tank Side (with MLI), Mounted together with Heat Shields into the CVV

Fig. 9. Top View to Cryostat Heat Shields with Lower Cover Cooling Contact Parts

Fig. 10. Completely Integrated Cryostat Cover (bottom view)

Fig. 11. Complete ISO - Cryostat

Figure 6 shows the fully equipped main HeII-tank head being already mounted to the Spatial Framework. Figure 7 shows one complete tank suspension strap with the strap pre-tensioning device. Figure 8 shows the lower side of the main HeII-tank with its MLI, being already mounted together with the heat shields into the CVV.

Figure 9 shows the top view of the cryostat heat shields with the lower cooling contacts on them. Figure 10 shows a bottom view of the completely integrated Cryostat Cover. Figure 11 shows the completely integrated and closed cryostat.

After the completion of the cryostat integration with experiment and telescope dummies (mechanical, thermal, mass) in it the insulation system was evacuated and the lowest vacuum achieved prior to start of cooldown was $3 \cdot 10^{-5}$ mbar.

b) Thermal Tests:

The ISO-cryostat had two thermal test campaigns in 1990:

- Ground Lifetime (GLT) Test with normal and superfluid He
- Thermal Vacuum (TV) Test with HeII and low vacuum vessel temperature

The cryostat cooldown started with the cooldown and filling of the auxiliary LHe-tank and the heat shields, followed by the cooldown and filling of the main HeII-tank with normal LHe up to approximately 60% content. The cooldown from room temperature to approximately 4.2 K took 5 hours with a total LHe-consumption of 1440 L.

After the cryostat filling with normal LHe a first near equilibrium test was performed over a duration of approximately 10 days.

Following this test a 100% filling with normal LHe was performed prior to the start of superfluid He production in the main HeII-tank by pumping on it. The lowest HeII-temperature achieved for the time being was 1.66 K.

After the HeII-production and pump-down to this lowest temperature a superfluid He-transfer was conducted using a HeII-transfer line with a Joule-Thomson valve and a LHe-precooling upstream of the J/T-valve by the counterflow of the pumped off cold He-gas from the HeII-tank. A 100% HeII-filling was finally reached with a HeII-bath temperature of 2.05 K before a repumping to 1.77 K with 95% filling prior to the start of the launch autonomy test with He-venting only from the auxiliary LHe- tank and the main HeII-tank completely shut off. During a period of 64 hours with a flowrate of appr. 31mg/s from the auxiliary tank the HeII-bath heated only from 1.82 K to 1.94 K. The insulation vacuum during the HeII-testing was in the range of 2 to $3 \cdot 10^{-8}$ mbar.

The preliminary end of the ground testing in laboratory environment was the so-called ground lifetime (GLT) test with a final thermal equilibrium flow rate of 31.5 mg/s from the main HeII-tank. Essential measured temperatures and flow rates at the end of the tests with normal LHe and HeII in the main tank are listed in table 2.

Table 2. Measured He-Flow Rates and Temperatures during Ground Tests (Equilibrium)

	Normal LHe	HeII
He-Bath Temp. [K]	4.23	1.84
He-Flow Rate [mg/s]	27.7	31.5
OSS-Temperature [K]	-	3.47
Upper Baffle Temp. [K]	8.2	7.8
First Cryostat Shield Temp. [K]	51.5	49.0
Second Cryostat Shield Temp. [K]	119.7	109.6
Third Cryostat Shield Temp. [K]	206.5	199.3

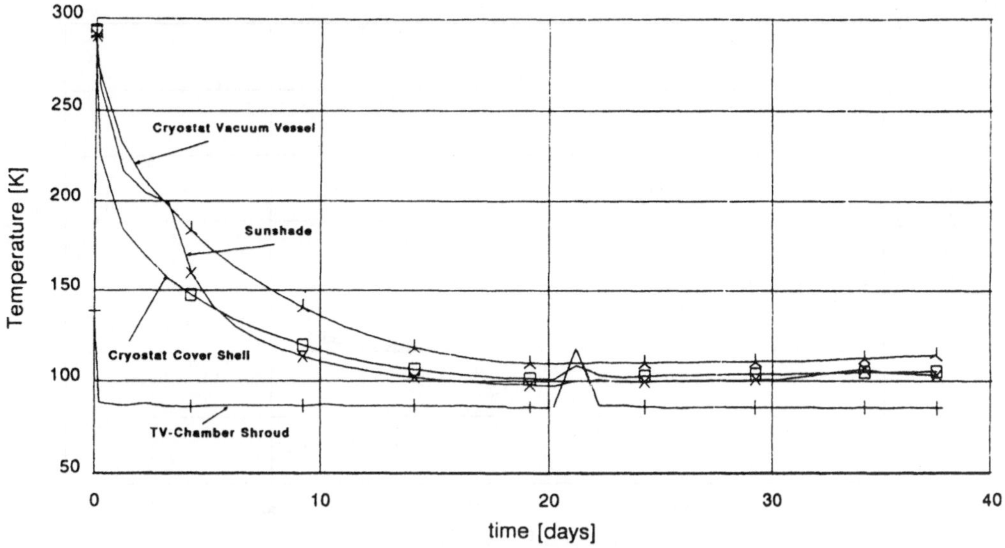

Fig. 12. Cryostat Cooldown during TV - Test

After the completion of the GLT-test the whole PLM was equipped with its external MLI, the sunshade and the startracker housing with thermal startracker dummies. The PLM was then topped up again with HeII and prepared for the thermal vacuum test in the space simulation chamber of IABG at Ottobrunn. During the TV-test the space simulation chamber cold shroud was cooled to an average temperature of approximately 85 K to allow the cryostat-CVV to cool to its expected orbital temperature.

The PLM was brought to a slightly tilted position (approximately 15 ° from vertical) to allow the HeII-phase separator to be just immersed into the HeII-bath. The tilting position was readjusted every day at the beginning of the TV-test and every 4th day at its end to hold the phase separator immersed in the liquid. Figure 12 shows the cool down of the cyostat vacuum vessel (CVV), the cover shell and sunshield over the first 22 TV-test days with the cold shroud kept at appr. 85 K. Figure 13 shows the transient He-flowrate evolution during the whole TV-test period. On the 19th test day the valve V505 in the external ventline to the large vent nozzles 513/514 was closed in order to decrease the

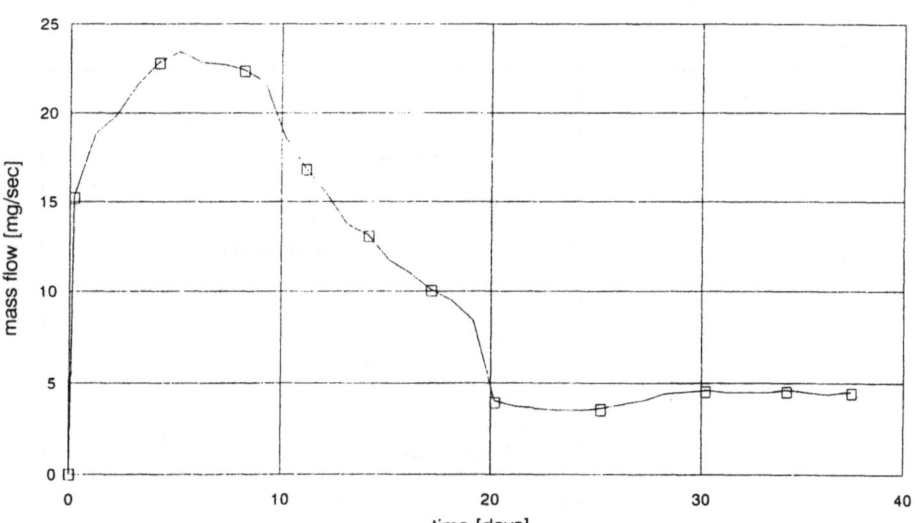

Fig. 13. Transient TV-Test He-Flowrate

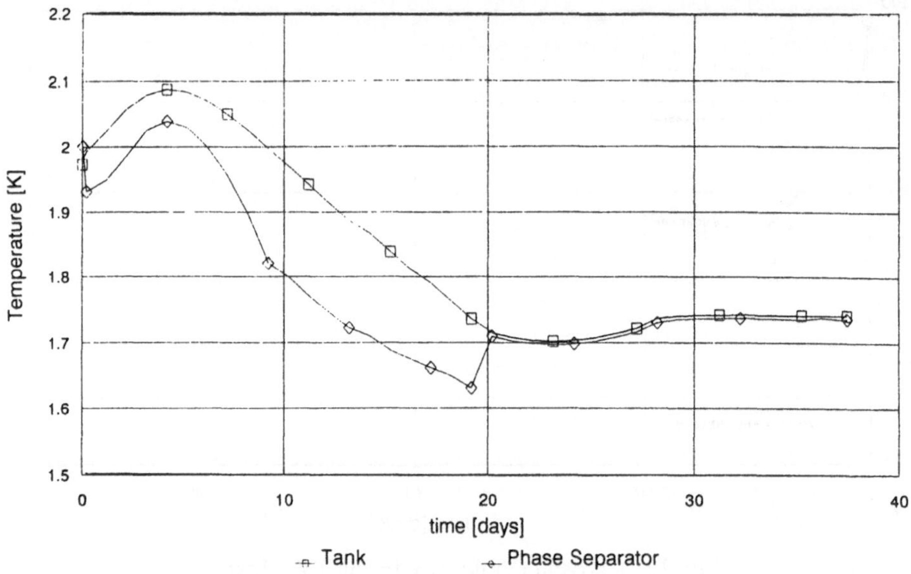

Fig. 14. Transient TV-Test HeII-Bath and Phase Separator Exit Temperatures

flowrate. Figure 14 shows the evolution of the HeII-bath and phase separator exit temperatures over the same period.

Upon ventline switching the He-flowrate is considerably decreased and therefore the Δp across the phase separator as well which causes an increase of the phase separator exit temperature, as shown by Fig. 15. This exit temperature staying below the HeII-bath temperature indicates that the phase separator worked correctly with its exit being dry.

During the TV-test period the following activities were also completed successfully:

- Direct Liquid Content Measurements (DLCM) Test with the "heat pulse" method
- Thermal testing of the scientific experiment thermal dummies
- Startracker housing thermal control to keep the startrackers within their operational limits
- Sunshade heating during 72 h for decontamination (outgassing)

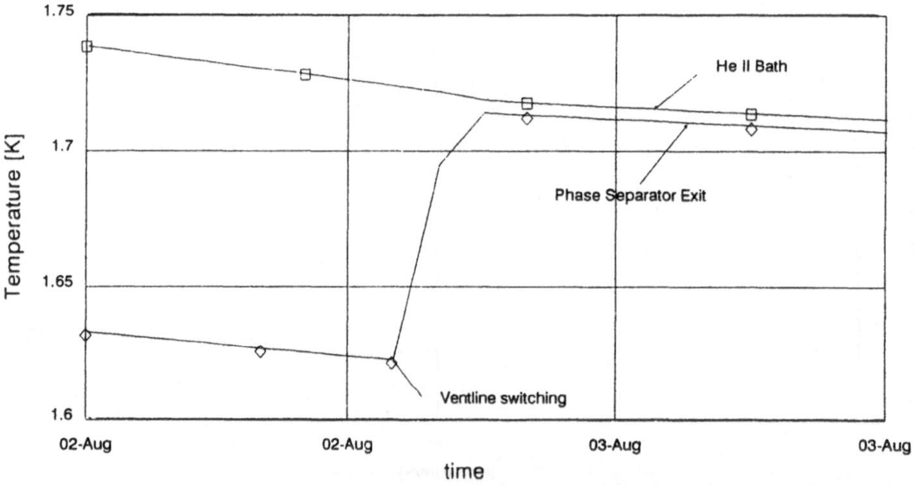

Fig. 15. Tank and Phase Separator Temperatures during Ventline Switching

Table 3. Quasi-Equilibrium Conditions Achieved during TV-test
(* no heat from experiments)

He-Flowrate	4.4 mg/sec
HeII-Bath Temperature	1.742 K
OSS-Temperature *	1.88 K
Upper Baffle Temperature	3.52 K
Innermost Cover Shield Temp.	6.89 K
External Cover Temperature	104.9 K
CVV-Cylinder Temperature	113.0 K
Sunshade Temperature	106.0 K

The external thermal loads from units staying at nearly room temperature (SVM, sunshield, startrackers) were simulated by electrical heaters on the CVV during the TV-test. At the end of the TV-test, which had a total duration of 37.25 days the "quasi-equilibrium" conditions as listed in table 3 were obtained.

c) Structural Tests:

The ISO-PLM was up to now structurally tested in a low-level vibration test program with the HeII-tank completely filled with normal LHe, as well as with the tank empty but cold, and in a full (high) load vibration test program with the cryostat empty and warm.

It was demonstrated in all such tests that the PLM meets its global stiffness requirement in longitudinal and lateral direction. On the other hand it was detected that the HeII-tan alone did not show sufficient stiffness in axial direction. The reason for this stiffness lack was clearly identified and lead to a re-stiffening action for the actual flight model tank which is currently being conducted.

d) Cover Release Tests:

The cryostat cover release was successfully tested under normal laboratory environment as well as in a thermal vacuum chamber with the cover shell, the clampband and the 2 non-contaminating actuators (NCA´s) actually cooled to their expected ejection event temperatures on orbit of approximately 110 K.

The release verification was done with the actuation of both NCA´s as well as with a single NCA-actuation only.

The ejection velocity achieved under cold conditions was 0.4 m/sec.

LIFETIME PREDICTION

As mentioned above the thermal mathematical ESATAN-model of the ISO-PLM was correlated with the thermal test results and corrected accordingly. With this corrected model the ground and orbital equilibrium consitions as well as the transient behaviour from the end of the launch autonomy phase through the first 40 mission days were recalculated (typical result: Fig. 3-5).

From the HeII-content of 310.5 kg at launch, the total He-consumption during the first 40 mission days and the steady state orbital flowrate 4.85 mg/sec an operational lifetime of 650 days (22 months) is predicted, which appears to show sufficient margin over the requirement of 548 days (18 months).

REFERENCES

1. A. Seidel, M. Wanner, A. M. Davidson, " The Current ISO-Cryostat Design" Proc. 12th Intern. Cryogenic Engineering Conf., July 12 - 15, 1988

2. F. Fuchs, A. Seidel, "Directly Operated Valve for Space Cryostat," Cryogenics 27 page 15 (1987)

3. M. Wanner, "Direct Liquid Content Measurement Applicable for HeII-Space Cryostat", Advances in Cryogenic Engineering, Vol. 33, page 917

4. R. Schaellig, A. Seidel, "Very Low Force Cooling Contacts for the ISO-Cryostat Cover", Cryogenics, Vol. 30, March 1990

THE AXAF/XRS TEST DEWAR: A VERSATILE DESIGN

Karl F. Weintz, Aristides T. Serlemitsos, Marcelino SanSebastian and Evan S. Kunes

NASA/Goddard Space Flight Center
Greenbelt, Maryland

ABSTRACT

The X-Ray Spectrometer (XRS)[1,2], to be flown on the Advanced X-ray Astrophysics Facility (AXAF)[3] is being developed at NASA's Goddard Space Flight Center (GSFC). XRS consists of an array of microcalorimeters and an Adiabatic Demagnetization Refrigerator (ADR)[4]. The ADR provides an operating temperature of 0.065 to 0.100 K. To support extensive development testing of the ADR and the detector array a test dewar has been designed and built at GSFC. Performance specifications say that this dewar must provide an adequate hold time of at least 24 hours with a bath temperature of 1.5 K. The design should minimize microphonic noise to the very sensitive detectors. It also must be flexible and modular to allow quick access to the experiment.

We have designed and built a dewar that surpasses the above requirements. Pumping on the liquid helium bath with a 47 liter per second pump, we have achieved temperatures lower than 1.5 K with a hold time in excess of 72 hours. In addition, the dewar can be operated without liquid nitrogen with a hold time of 36 hours. This feature was incorporated in the design because boiling nitrogen may introduce microphonic noise to the detectors. To further reduce the susceptibility to microphonic vibration a unique suspension system utilizing Kevlar™ fibers was devised which provides both translational and rotational rigidity to the detector mount. Finally, the dewar is very manageable and can be rotated by only one person.

INTRODUCTION

The Advanced X-ray Astrophysics Facility (AXAF) is one of the "Great Observatories" managed by the Marshall Space Flight Center (MSFC). One of the experiments on AXAF is the X-ray Spectrometer (XRS) which is being developed at the Goddard Space Flight Center (GSFC). Within the cryogenic subsystem of XRS is an Adiabatic Demagnetization Refrigerator (ADR) used to cool the microcalorimeter detector array. In order to facilitate extensive development testing of the ADR and the detector array, a laboratory test dewar is required. To meet tight schedule and cost restrictions, the design, fabrication and verification of this test dewar was completed in-house at the GSFC in less than nine months.

The liquid helium test dewar utilizes liquid nitrogen as a secondary cryogen for reducing the heat load to the helium tank. Performance requirements include a need for continuous 24-hour operation at 1.5 K. The sensitivity of the detector array to microphonic vibration requires the dewar have the ability to run without liquid nitrogen. The boiling of liquid nitrogen introduces vibrations which may interfere with development testing. To accommodate this requirement the test dewar is designed with a unique heat exchanger which provides adequate hold time at 1.5 K both with and without liquid nitrogen. The helium tank

Advances in Cryogenic Engineering, Vol. 37, Part B
Edited by R.W. Fast, Plenum Press, New York, 1992

is supported by a suspension system designed to stiffen the detector mount in order to isolate it from microphonic vibration. To support development testing a provision has been included for mounting an X-ray source for controlled introduction of X-rays within the dewar. The modular design provides access to both ends of the helium tank in order to test various types of magnetic shielding. This versatility gives the test conductor the ability to customize the dewar for particular tests.

SYSTEM DESCRIPTION

The AXAF/XRS Test Dewar (Figure 1) consists of a cylindrical stainless steel 55 liter liquid helium tank surrounded by a toroidal aluminum 84 liter liquid nitrogen tank. Both the detector array and the ADR are mounted at the bottom of the helium tank. This end of the helium tank is suspended from the liquid nitrogen tank by a Kevlar™ fiber suspension system. The other end is supported by a stainless steel plate. The stainless steel plate is somewhat isolated from the two tanks by fiberglass washers. The nitrogen tank is suspended within the vacuum shell with a similar Kevlar™ fiber suspension system at the bottom and by four thin-walled stainless steel tubes at the top. Between the top of both cryogen tanks is the heat exchanger which is hard mounted to both and suspended from the vacuum shell by yet another Kevlar™ fiber suspension system. Both cryogen tanks and the heat exchanger are thoroughly blanketed with multi-layer insulation to reduce radiant heat input.

The heat exchanger design (Figure 2) consists of an oxygen free high carbon (OFHC) copper neck assembly bolted to the top of the helium tank with an indium seal. It serves as the vent for the outgoing helium vapor. Mounted to this neck is a deflector and baffle system which directs the vapor through a circuitous route exposing it to the highly conductive copper surface. The current design allows the test conductor to run both with and without liquid nitrogen in the same mechanical configuration. A stainless steel support plate

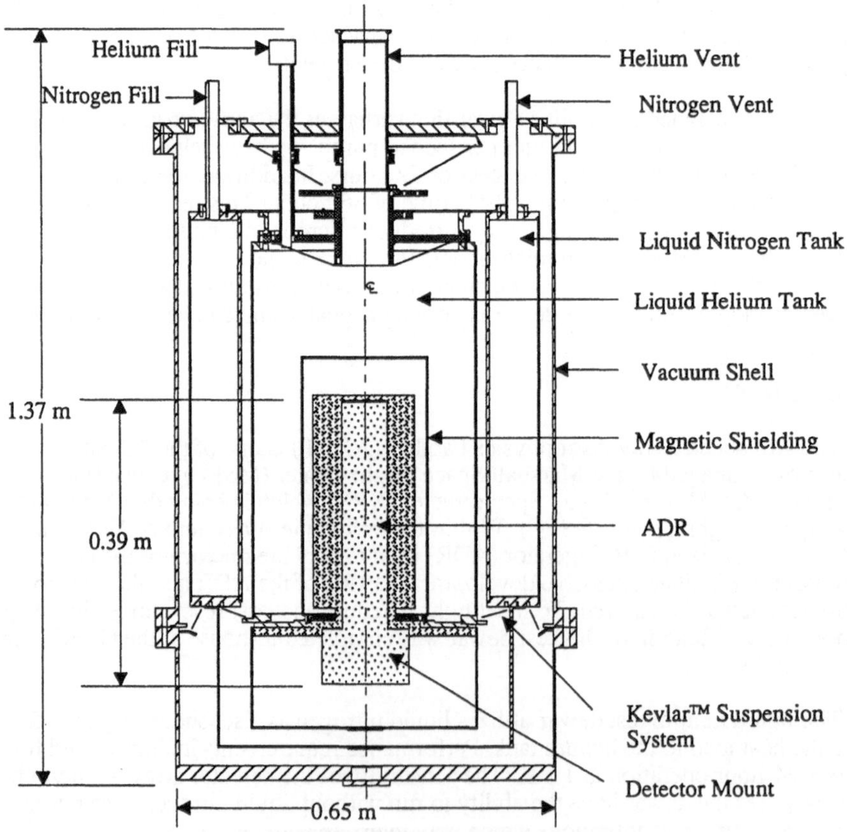

Figure 1. AXAF/XRS Test Dewar.

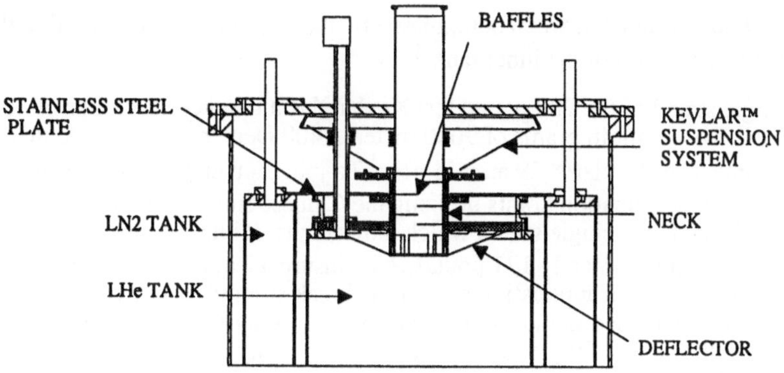

Figure 2. Heat exchanger design.

has been chosen as a compromise between thermally isolating the nitrogen tank when running with liquid nitrogen and thermally anchoring it to the heat exchanger when running with vapor cooling. Incoming radiant heat is intercepted at the bottom by two 1100 aluminum shields mounted on the base of the nitrogen and helium tanks. The walls of the empty nitrogen tank intercept additional incoming radiant heat and conduct it to the outgoing vapor through the stainless steel plate in the neck assembly. In the event that the boiling of the liquid nitrogen interferes with development testing, the stainless steel compromise allows the continuation of cold testing by removing the liquid nitrogen. Liquid nitrogen is transferred to a storage dewar by inserting a vacuum jacketed transfer line into the vent and pressurizing the fill line. This does not interrupt a particular test with the necessary warmup that a mechanical change to the heat exchanger would require.

In order to further reduce microphonic vibrations at the detector mount, a suspension network of Kevlar™ fibers has been incorporated. As shown in Figures 1 & 3, Kevlar™ fibers are attached between the vacuum shell and the nitrogen tank and between the vacuum shell and the heat exchanger, as well as between the nitrogen tank and the helium tank. These fibers are mounted in such a way as to provide radial, axial, and torsional stiffness to the bottom of the helium tank. The thermal contraction of the inner tank contributes to stiffening the support system. Figure 3 details the suspension system between the tanks. The fibers are supported by a fixed yet flexible mount. A standard eyebolt is attached with a stack of #10Belleville™ spring washers, mounted in series and adjusted accordingly. These washers

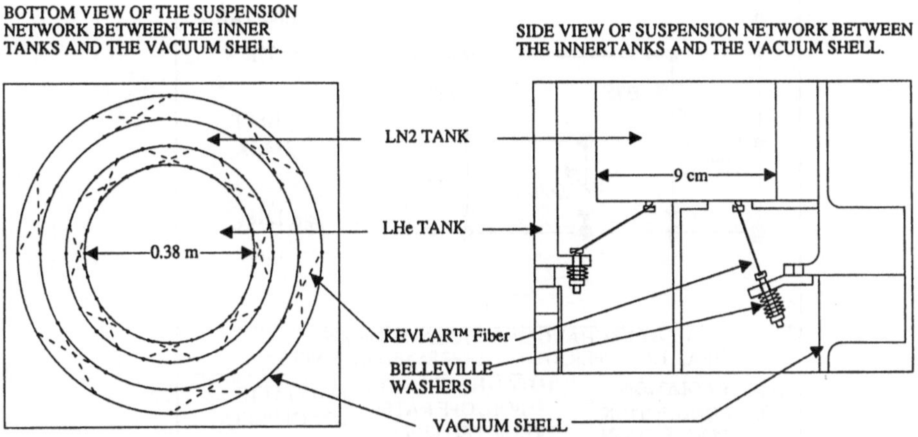

Figure 3. KEVLAR™ suspension network. The length of the fiber between the LHe tank and the LN2 tank is 8 cm and the length of the fiber between the LN2 tank and the vacuum shell is 10 cm.

provide 51-69 pounds tension when compressed by the action of the fiber pulling the eyebolt due to thermal contraction of the inner tank.[5]

A single strand of the 200 pound test Kevlar™ is a braid of 15 groups of 400 Denier yarn.(Denier is the weight in grams of 9000 meters.) 400 Denier yarn is made up of over 250 individual filaments of Kevlar™ 29 aramid fiber. To join each end of a single strand, a long splice is used at the termination points utilizing the Chinese finger grip principle.[6] Pull tests have been performed on a single strand which indicate that failure occurs at the splice when the load surpasses approximately 171 pounds.[7] To insure adequate safety margin, a single strand is doubled up in this application so that the break strength is in excess of 340 pounds. The intent of the tensile load in the suspension system is to stiffen the detector mount and thereby increase the natural frequency to reduce susceptibility to microphonic vibration.

The test dewar design is thoroughly modular with a majority of the the interfaces bolted with indium seals. The helium tank has indium seals at both ends to provide access to the detector mount and heat exchanger. The dewar is mounted in a support dolly that provides 360° access and rotation as well as adjustments to the center of gravity. This makes rotation easier during a variety of experiments.

Routine disassembly of the dewar is possible by one person and can be performed in a single day. The versatility of this modular approach allows for the customization of various components within the test dewar to accommodate a variety of applications. For example, the material used for the support plate in the heat exchanger can be changed from stainless steel to fiberglass in order to better thermally isolate the helium tank and extend hold time if desired.

In order to verify that the design would meet the minimum requirements prior to integration, a simple thermal model was constructed to simulate worst case conditions (as far as the heat load is concerned.) this was run on a Macintosh™ personal computer using the software Excel™ and TK solver™. The dewar thermal model was simplified in order to expedite analysis of the network. Uniform temperature profiles were assumed at the vacuum shell, nitrogen tank, heat exchanger neck assembly and helium tank. The majority of the heat load to the nitrogen tank was assumed to be through radiation. No multi-layer insulation was included nor was there any thermal contact resistance addressed in the model. The thermal network is schematically represented in Figures 4 & 5.

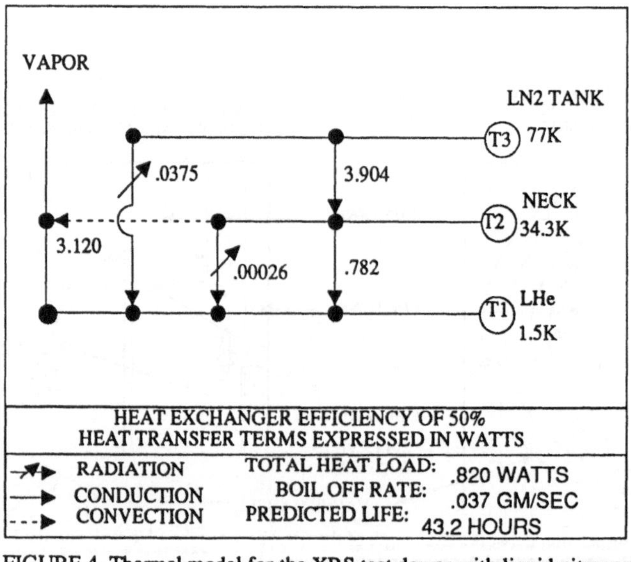

FIGURE 4. Thermal model for the XRS test dewar with liquid nitrogen.

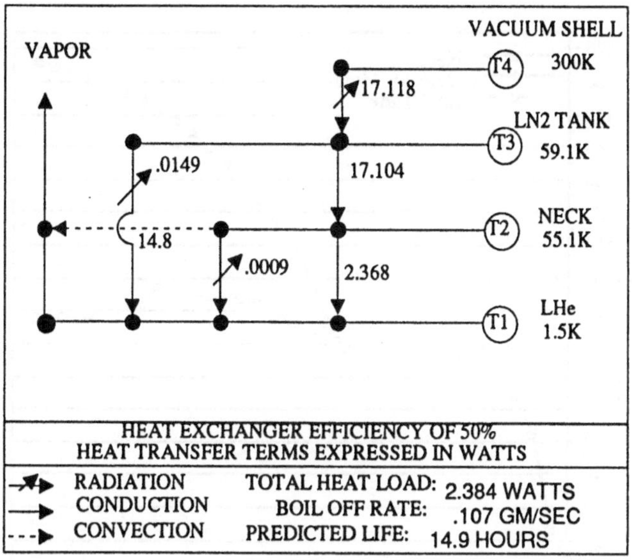

FIGURE 5. Thermal model for the XRS test dewar without liquid nitrogen.

An energy balance was performed on each temperature node. For example, the overall heat flow into the helium tank from conduction and radiation was balanced by boiling the helium. The individual equations used for each radiation and conduction term are summations of the heat flow between two fixed temperature nodes over a simplified geometry of nested cylinders, flat planes and discs. Formulae for these simple forms are available in current heat transfer texts.[8] For each temperature node the specific dewar geometry was simplified to accommodate the these standard shapes. The energy balance yielded an equation for each node. Material properties, such as emissivity, thermal conductivity, specific heat, and latent heat of vaporization were input based on their accepted values as published in current literature.[9,10] The heat exchange of the vapor with the neck is characterized by an efficiency, equal to the ratio of the actual heat transfer rate to the maximum possible. The maximum rate occurs if the vapor temperature rises to the temperature of the heat exchanger.[11] An estimate for one of the unknown temperatures was entered into an Excel™ spread sheet. The spread sheet was customized to solve the various energy balance equations and determine the coefficients of the resulting quartic equation. TK solver™ was then used to solve the quartic equation. With this solution the first temperature estimate could then be revised and the process iterated until convergence. Repeating this process as a function of the heat exchanger efficiency allowed a determination of the efficiency necessary to meet the lifetime requirement.

Given a heat exchanger efficiency of 50% this conservative analysis shows a hold time at 1.5 K of 43 hours with liquid nitrogen and 15 hours when running without. In contrast, Figure 6 shows the temperature profile during an actual performance test at 1.52 K. After 26 hours of running, the liquid nitrogen was removed. The liquid helium that was left lasted 36 hours with vapor cooling. Independent tests run without removing the liquid nitrogen have lasted in excess of 72 hours.

CONCLUSION

The AXAF/XRS test dewar is designed with a versatile approach to meeting several diverse requirements. The heat exchanger is designed in order to run with adequate hold time both with and without liquid nitrogen. The unique suspension system provides both thermal isolation as well as structural stiffness to the detector mount. The modular design provides

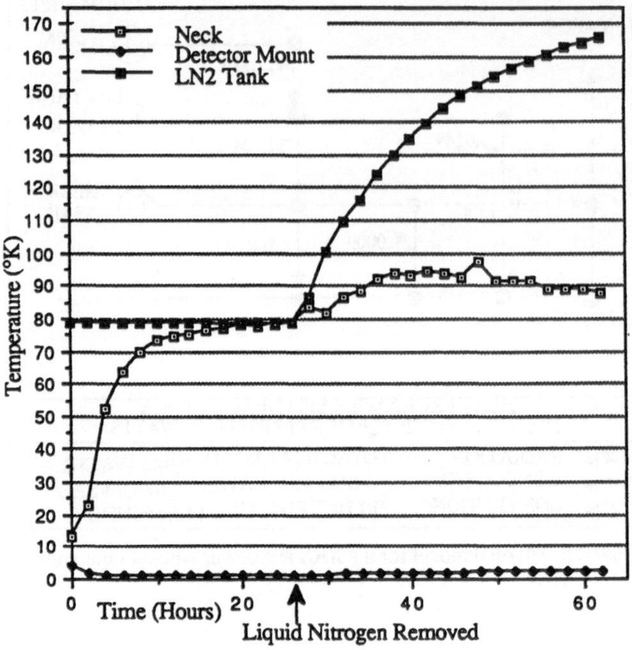

Figure 6. Test dewar temperature profile.

the test conductor with the flexibility of customizing the design to achieve a particular purpose.

Design, fabrication, assembly, and initial performance testing of the AXAF/XRS test dewar was completed entirely in house at the GSFC in nine months. This accelerated schedule has provided the project with a versatile test dewar capable of developing a flightworthy ADR.

REFERENCES

1. S.S. Holt, X-Ray Spectroscopy of AGN with the AXAF 'Microcalorimeter', in: "Astrophysical Letters and Communications," Vol. 26, No. 61, (1987).

2. R.L. Kelley et al., High Resolution X-Ray Spectoscopy Using Microcalorimeters, in: "Proc." SPIE 219 (1988).

3. M.C. Weisskopf, The Advanced X-Ray Astrophysics Facility: An Overview, in: Astrophysical Letters and Communications Vol. 26, No. 1, (1987).

4. A.T. Serlemitsos et al., Adiabatic Demagnetization Refrigerator for Space Use, in: Advances in Cryogenic Engineering, Vol. 35, 1431, (1989).

5. Belleville Spring Washer Specification, Associated Spring, 18 Main Street, Bristol, CT

6. Terminal Splicing Kit, Available from Ashaway Line & Twine MFG. Co., Ashaway, RI

7. NASA/GSFC Internal Memo, Evaluation of Terminal Configurations for Kevlar™ 29 Yarn., M.Viens, Code 313 / Materials Branch, May 11, 1988.

8. J. P. Holman, "Heat Transfer," McGraw Hill, Inc., 1986, p.31, p.407, p. 408.

9. "Handbook on Materials for Superconducting Machinery," Battelle Columbus Laboratories, Columbus, OH, 1977.

10. V. D. Arp and R. D. McCarty, "Thermophysical Properties of Helium-4 From 0.8 to 1500K with Pressures to 2000MPa," NIST Technical Note 1334, Boulder, CO, 1989.

11. William C. Reynolds, "Engineering Thermodynamics," McGraw Hill, Inc., New York 1977, p.574.

5-YEAR LIFETIME HYBRID SUPERFLUID HELIUM DEWAR FOR THE AXAF X-RAY SPECTROMETER (XRS)

Stephen J. Nieczkoski and Richard A. Hopkins
Ball Electro-Optics/Cryogenics Division
Boulder, Colorado

Susan R. Breon
NASA Goddard Space Flight Center
Greenbelt, Maryland

ABSTRACT

The focal plane of the AXAF X-Ray Spectrometer requires an operating temperature of 0.1 K with a mission lifetime of 5 years. This demanding task is accomplished with a hybrid cryogenic subsystem consisting of mechanical coolers, a superfluid helium dewar and an adiabatic demagnetization refrigerator. By using mechanical coolers to remove heat from the dewar outer vapor-cooled shield, a 5-year lifetime is achievable with only a 483-liter tank. This approach takes advantage of flight-proven, high-performance dewar technology and recent success in the development of split, Stirling-cycle mechanical coolers. Although the dewar design principles are similar to those used previously, parasitic heat flow is reduced to a new level by an optimized tension strap support system and careful attention to insulation system details. The benefit of the mechanical coolers is maximized by dewar interface design features that minimize parasitic heating and thermal impedance of the coupling. The dewar design and thermal performance analysis are discussed. Helium lifetime sensitivities and the effects of mechanical cooler failures are predicted.

INTRODUCTION

Launch of the Advanced X-Ray Astrophysics Facility (AXAF) is planned in 1998. The X-Ray Spectrometer (XRS)[1] is one of four first-generation instruments under development. At its focal plane is a calorimeter array that must be maintained at a temperature of 0.1 K. Mission lifetime is 5 years, and the system will be capable of on-orbit helium replenishment. The long mission lifetime requires a new and ambitious level of thermal performance and more demanding levels of reliability and contamination control compared to previously-flown cryogenic systems.

The 0.1 K focal plane temperature is provided by an adiabatic demagnetization refrigerator (ADR)[2] that requires an environmental temperature of less than 1.5 K to operate efficiently. To provide this temperature environment for 5 years within the mass and envelope constraints and with existing technology mandates a hybrid design approach. A systematic optimization process has yielded the best overall cryogenic system architecture for the requirements. The XRS superfluid helium dewar design (Fig. 1) is evolved from technologies proven by the previous IRAS and COBE flight systems.[3,4] Improvements have been made in the designs of the cryogen tank support system and the insulation system to provide improved thermal performance compared to previous systems. Two momentum-compensated pairs of split, Stirling-cycle mechanical coolers are mounted on the dewar and cool the outer vapor-cooled shield (VCS). Without the coolers the helium lifetime would be only about 2 years. The helium lifetime is also enhanced by an external thermal control system that maintains the mainshell at an average temperature of 250 K on orbit.

Advances in Cryogenic Engineering, Vol. 37, Part B
Edited by R.W. Fast, Plenum Press, New York, 1992

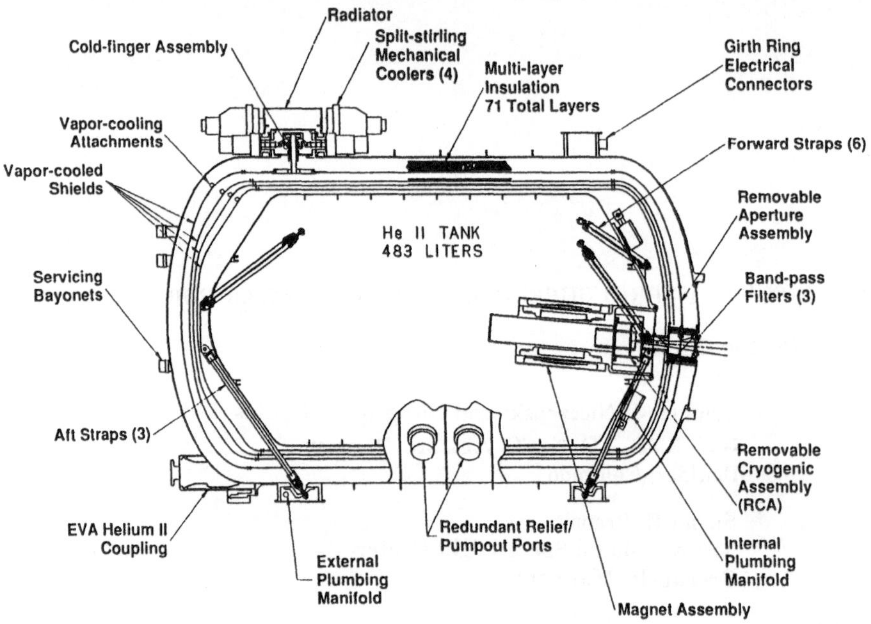

Fig. 1. The 483-liter superfluid helium dewar.

Use of the mechanical coolers raises many issues related to maximizing their benefit to the helium lifetime, assuring a high level of reliability through redundancy and operational strategy, and providing compatibility with the instrument system.[5] The thermal efficiency of the coupling between the coolers and the VCS4 is maximized within the constraints of mechanical considerations. A redundant backup pair of coolers provides a high probability that at least one pair will be operable throughout the mission. The electrical power budget allows all four coolers to operate at once. The lifetime benefit of running one pair is considerably greater than the additional benefit of running the second pair. Therefore, although the predicted lifetime with all four coolers running is 6.7 years, the probability of achieving 5-year lifetime is improved by holding the second pair in reserve until 2 years after launch.

SYSTEM PERFORMANCE REQUIREMENTS AND CONSTRAINTS

Table 1 summarizes the major performance requirements and constraints along with the performance currently predicted. The AXAF will be launched on either the Space Transportation System or a Titan 4.

Table 1. System Performance Requirements and Constraints

ITEM	REQUIREMENT	PREDICTED VALUE
Mass	<351 kg	351 kg
Size	≤168 x 97 cm	168 x 97 cm
Power	<185 W	<185 W
Cryogen Lifetime	4 years with 25 percent analysis margin	5.2 years
Bath Temperature	<1.5 K (orbital operation) <1.9 K (ground operation)	1.3 K 1.6 K (coolers off)
Bath Temperature Stability	±0.05 K over mission duration	±0.04 K (worst case)
Dewar Lockup Capability	6 days without power	9.5 days
Dewar Frequency	≥30 Hz	30 Hz

DEWAR DESIGN

The primary function of the dewar is to cool the ADR and provide an environment for the detectors that is free from interfering disturbances such as contamination and background radiation. The ADR and the detectors are part of the removable cryogenic assembly (RCA), which is bolted to the forward end of the cryogen tank in an arrangement that gives the detectors a line of sight five degrees off the major dewar axis. The RCA is accessed through the removable aperture assembly. The 483-liter cryogen tank provides the heat sink for the RCA. Surrounding the cryogen tank are four aluminum vapor-cooled shields that transfer heat to the helium boiloff in the vent line. VCS4 is also cooled by single-stage, split Stirling-cycle mechanical coolers through a flexible cold finger attachment to the displacer cold tips.

<u>The Dewar Structure</u>

The mainshell provides the structural foundation of the dewar. It supports the cryogen tank and maintains the insulation subsystem in a vacuum environment. On orbit, the mainshell is cooled to an average temperature of 250 K. Both the cryogen tank and dewar mainshell are an all-welded construction from 5083 aluminum. The cryogen tank is supported from the mainshell by nine fiberglass/epoxy tension straps that minimize conducted heat while providing the required strength and stiffness. The support system architecture and the location of the VCS thermal attachments to the straps is thermally optimized.

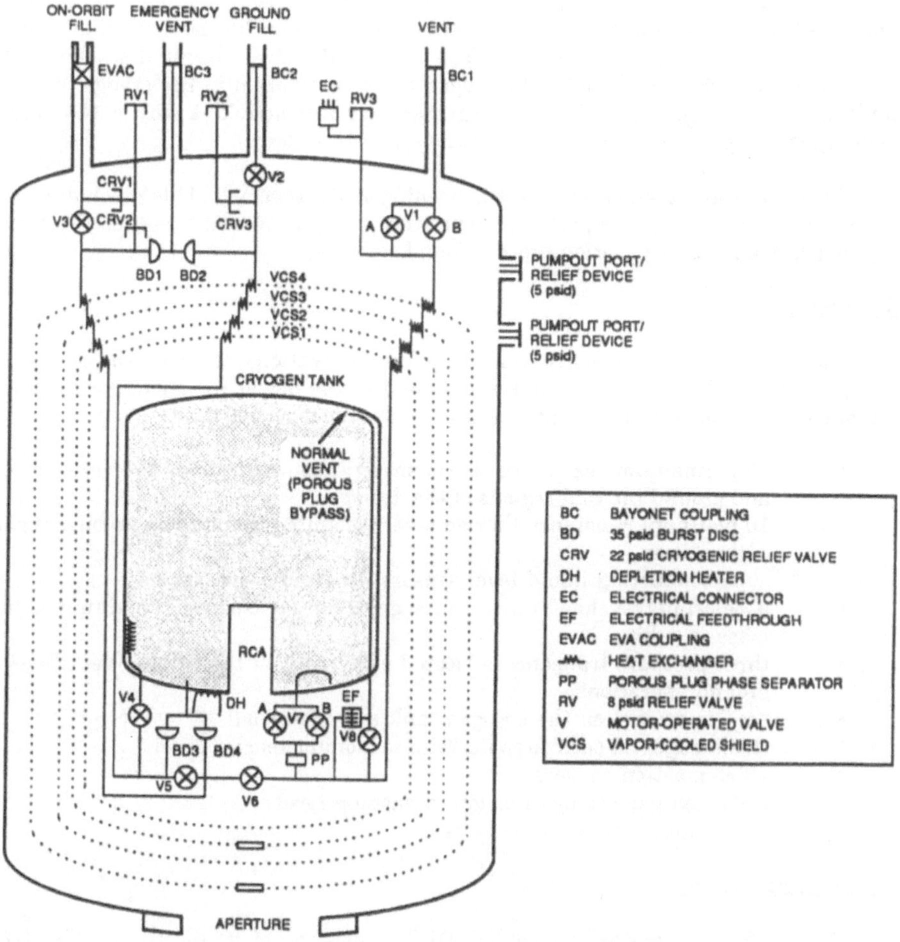

Fig. 2. Dewar fluid management system.

Insulation System

Much work has been done on the performance of insulating materials commonly used in space cryogenic subsystems.[6,7,8] While the performance of these materials is well understood at warm boundary temperatures (i.e., 300-77 K), there is a limited data base and less understanding of their performance in colder regions. Existing data and material properties knowledge suggest that carefully-controlled bare surfaces outperform superinsulations at very low temperatures. Therefore, we use no multilayer insulation inside of VCS2. There are 71 total layers of double aluminized Mylar with Dacron net spacers distributed in three outer MLI blankets between VCS2 and the mainshell. The lifetime benefit of vapor cooling alone is more than a factor of 20. A pair of coolers cooling VCS4 with a heat lift capacity of about 1.5 W at 80 K provides another factor of 2.5.

Fluid Management System

The fluid management system (Figure 2) consists of three lines that penetrate the dewar and internal and external plumbing manifolds. The ground and on-orbit fill lines provide a redundant capability for emergency venting through internal and external burst discs. To provide adequate heat exchange between the VCSs and the helium effluent gas, a 12-in. length of vent line is attached to each shield.

All tubing is 1.27 cm diameter thin-walled stainless steel to minimize conducted heat input. The tubing is connected to the aluminum cryogen tank and mainshell with bi-brazed transition joints. The internal manifold consists of six motor-operated valves, two cold burst discs, and a porous plug phase separator. The external manifold contains four motor-operated valves, two warm burst discs, and relief valves for line sections that could be sealed off while containing cold helium. The baselined valves, burst discs and relief valves are those qualified on the SHOOT program.[9] Vent location in the cryogen tank allows multiple gravity orientation during integration and test activities with a 100 percent fill capability in the prelaunch vertical orientation (aperture down).

Orbital venting commences with the opening of V1, then V7. This precludes liquid helium from breaking through the porous plug before the vent is opened to space, which could prevent the passive phase separation process from beginning.

Instrumentation

Instrumentation is needed to assist in ground operations, to verify the dewar performs as predicted and required during test, to monitor the dewar for anomalous behavior on orbit, and to assist the orbital resupply operation. Sensors consist of:

- 15 germanium resistance thermometers mounted inside the cryogen tank and around the cold regions of the dewar
- 10 platinum resistance thermometers mounted around the warmer regions of the dewar
- superconducting liquid level sensors for ground operations
- a calorimetric, heat pulse mass gage for measuring remaining helium mass on orbit
- three pressure transducers located externally on the fill and vent lines for ground operations
- accelerometers on the cryogen tank and mainshell
- strain gages on each support strap to monitor tension level
- valve position sensors
- a Vac-ion gage to monitor guard vacuum level
- filter rupture detection sensors

Dewar Electrical Cabling

There are seven dewar exit cables (DEC) consisting of miniature stainless-steel conductors in a ribbon form that support the RCA and dewar valves and sensors. The RCA cables are twisted pairs with an overall shield for noise rejection. The DECs are routed into

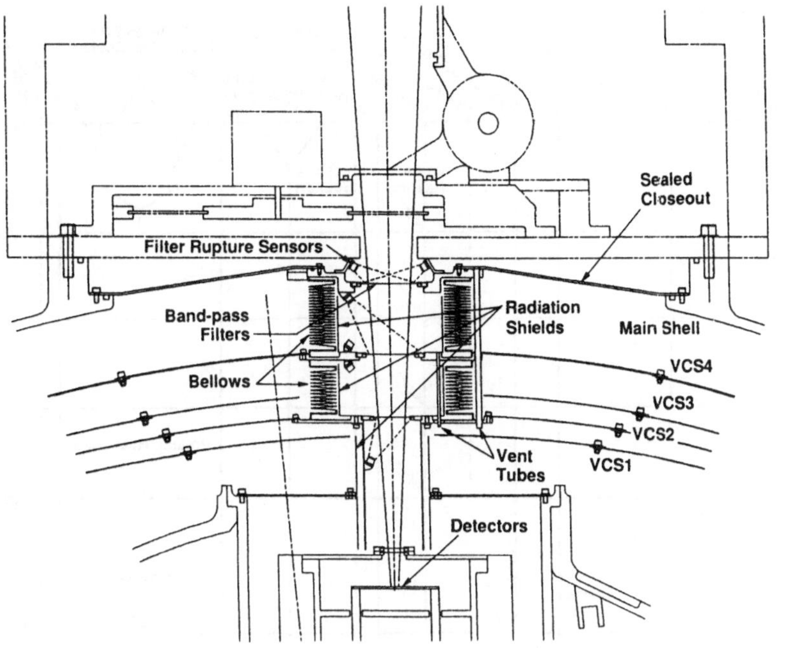

Fig. 3. Aperture design.

the dewar along support straps and bonded to the straps for connection to the VCSs. Power and signal wires to the ADR magnet and sensors inside the cryogen tank are routed through the vent line to avoid the risk of an electrical feedthrough into the cryogen tank from the guard vacuum. The ADR power wires are sized to minimize the combined impact of ohmic and conductive heating.

Aperture

The dewar aperture (Fig. 3) contains very fragile band-pass filters housed within a bellows assembly. The bellows perform the crucial function of preventing water from the multilayer insulation from reaching and condensing on the filters. Since the filters can withstand only 1 torr of pressure differential, careful venting of the bellows cavities along with vacuum pump throttling are needed to avoid filter rupture during guard vacuum acquisition. Also, the venting scheme must not provide a flow path from the warm insulation blanket to the filters. Parasitic heating into the aperture is minimized by 0.005-cm wall stainless steel bellows and radiation baffles.

The Coolers and Their Interface

There are four split Stirling-cycle coolers, compressor and displacer, based on the Oxford design, mounted at the aft girth ring in opposing pairs for momentum compensation. Each cooler pair shares a common cold finger attachment to VCS4. The displacers have a hard mechanical mount to the dewar mainshell to enclose the cold tip within vacuum. The compressors require mechanical isolation to minimize their vibration transmission to the dewar mainshell. Heat is rejected from the coolers by radiation directly to the AXAF cold shroud and conduction to the mainshell.

The cooler-to-dewar interface is designed for maximum thermal efficiency within the mechanical constraints.[5] The substantial motion of the VCS4 relative to the mainshell during launch must be accommodated, and bending of the compliant displacer cold tube must be very limited to avoid regenerator rubbing. Thermal efficiency is maximized by minimizing parasitic heating and interface thermal impedance of the operating coolers, while maximizing interface thermal impedance of non-operating coolers. An active thermal switch based on the principle of differential thermal expansion of dissimilar metals is baselined to thermally disconnect non-operating cooler pairs. A folded fiberglass

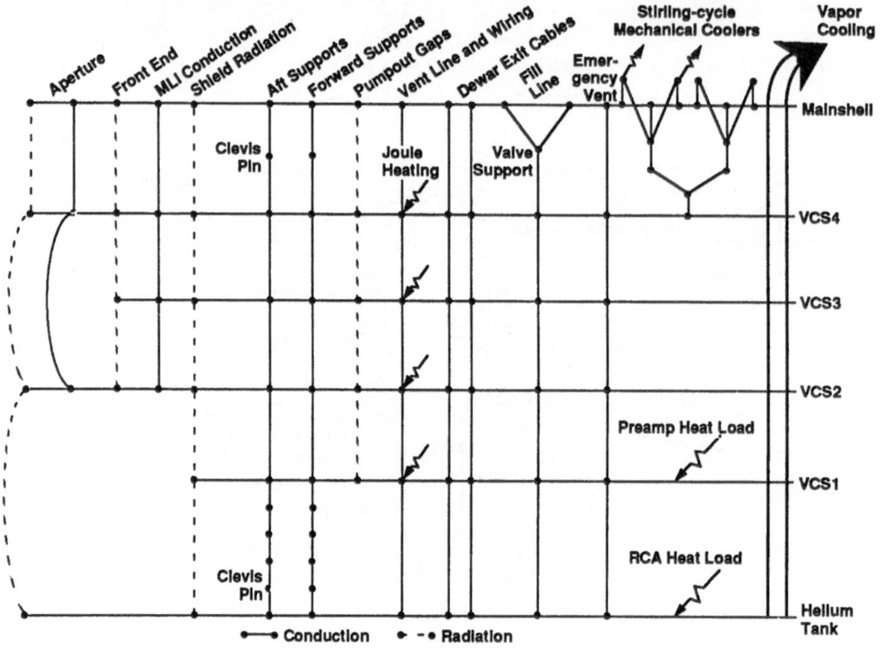

Fig. 4. The dewar thermal network.

tube arrangement provides structural support of the cold finger and thermal switch from the mainshell with minimal heat input. A thermal spreader plate is bolted to VCS4 to minimize the constrictive thermal resistance inherent with point-applied cooling.

THERMAL/MATH MODEL

The dewar design is optimized through studies carried out with the dewar thermal/math model. The methodology and modeling techniques have been taken from the past decade of experience at Ball with the flight-proven IRAS and COBE dewars.

The present model is a thermal network (Fig. 4) of 25 nodes, 70 linear conduction paths, 20 radiation paths (not all shown), and 4 convection paths representing the vapor cooling. Imposed heat loads account for the instrument and preamp power dissipations, ohmic losses in the magnet charge leads, and the heat lift from mechanical coolers. In the case of the coolers, measured performance data at the displacer cold tip is used. Cooler operating temperature dependence is considered. Temperature-dependent material and fluid properties are used. The network is input to the Ball Cryogenic Analyzer Program for steady-state and transient computer solutions. The model uses an algorithm to calculate the vent system pressure drop and accounts for the changing helium thermodynamic conditions for transient solutions.

PREDICTED PERFORMANCE

Final dewar servicing will take place on the launch pad. Analysis supported by COBE dewar experience predicts a 97 percent fill with a 1.3 K bath temperature when the vent valves are closed for prelaunch lockup. Power is not normally available to operate the coolers or valves after this servicing, 5 days prior to launch. Nominally, the vent is opened and a pair of coolers are powered 1 day following launch. Survey lifetime, which is actually an equal combination of observation and hold modes, begins when the bath temperature stabilizes on orbit. Table 2 lists the boundary conditions assumed for thermal performance analysis.

Once in orbit, the helium bath must always remain superfluid if the porous plug phase separator is to function properly. The lockup capability of the dewar is defined as the

Table 2. Performance Analysis Boundary Conditions

PARAMETER	VALUE
Superfluid Helium Launch Mass	67.9 kg
Average Orbital Main Shell Temperature	250 K
Ground Hold Main Shell Temperature	300 K
Dewar Lock-up Duration	6 days
Average RCA Heat Load to CT	2.26 mW
Average JFET Heat Load to VCS1	5.7 mW
Coolers Cooling VCS4	2 (1 pair); type: BAe
Non-operating Coolers at VCS4 (isolated by active thermal switch)	2 (1 pair); type: BAe

amount of time the vent can remain closed before the helium bath goes normal at a temperature of 2.18 K. Thermal masses for each VCS were added to the dewar thermal model to predict transient behavior. Insulation mass was conservatively neglected. Figure 5 shows the predicted lockup capability is 9.5 days without the coolers operating, which is well beyond the required 6 days. At the conclusion of 6 days, the bath temperature will have risen to about 1.8 K. Orbital pumpdown to the 1.3 K steady-state bath temperature will take about 2 weeks and result in a 3 percent helium mass loss. The lockup capability could be extended to 22 days and the mass loss further minimized if power were available to operate a pair of coolers during prelaunch lock-up.

The predicted orbital lifetime of the dewar depends on the chosen operational strategy for the mechanical coolers. Because this cooler technology is just emerging, long-term reliability has yet to be demonstrated. If both cooler pairs were operated without failure, helium lifetime is predicted at 6.7 years. However, since the 5-year lifetime can be achieved by operating just a single cooler pair, the highest probability for a successful mission is to employ a redundant backup cooler-pair strategy. This would require a minimum 2.5-year lifetime from each cooler. For nominal lifetime predictions we are assuming that only one pair of coolers operates at a time over the mission duration. The active thermal switch isolates the non-operating pair so dewar lifetime is not sensitive to whether the pair is in reserve or has failed. The predicted dewar lifetime is 5.2 years (0.40 mg/s helium boiloff rate) and bath temperature is 1.3 K. Table 3 gives the breakdown of heat loads to the cryogen tank. The operational pair of coolers removes a net of 1.5 W of heat from VCS4, cooling it to about 85 K. Additionally, there is between 40 and 50 mW of vapor cooling at each VCS.

If the primary cooler pair does not experience a failure within 2 years on orbit, it would be prudent to switch on the backup pair. With all four coolers running the dewar

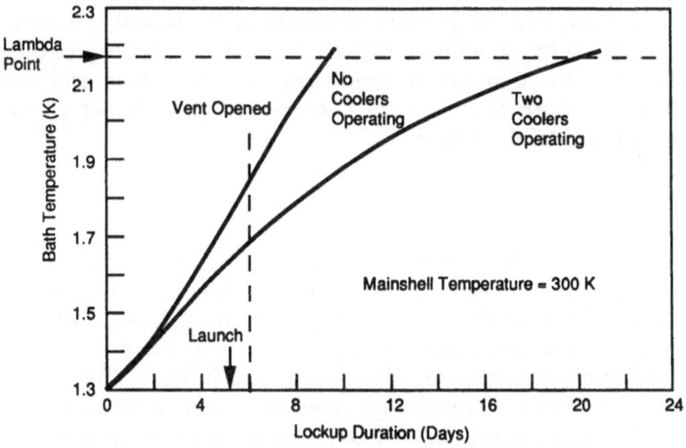

Fig. 5. Dewar non-vented hold capability.

Table 3. Cryogen Tank Heat Loads (with two coolers operating)

SOURCE	HEAT LOAD (mW)	% OF TOTAL
Forward Support Straps	2.43	27.3
Average Instrument Power	2.26	25.4
Aft Support Straps	1.94	21.8
VCS1 Radiation	1.13	12.7
Dewar Exit Cables	0.50	5.6
Fill Lines	0.50	5.6
Aperture Radiation	0.14	1.6
TOTAL	8.9	100.0

Table 4. Major Lifetime Sensitivities (general range of uncertainty is ±10% from nominal)

PARAMETER	NOMINAL VALUE	SENSITIVITY (days/% change from nominal)
Main Shell Temperature	250 K	31.0
Support Strap k A/L	----	6.6
DEC Conductor Area	0.05 cm^2	2.6
Cooler-to-Dewar Interface Thermal Impedance	5.4 K/W	1.6
RCA Heat Load	2.26 mW	1.4
Preamp Heat Load	5.7 mW	0.7
Cooler-to-Dewar Interface Parasitic Heating	200 mW	0.2

boiloff rate is reduced by 25 percent. Using this strategy, the dewar lifetime will be a maximum of 6.2 years if no coolers fail. If both cooler pairs were to become inoperable, the helium boiloff rate would increase by a factor of 2.5, and any remaining lifetime would drop accordingly. If this were to occur just after launch, the predicted lifetime is 2.1 years. Even in this event the mission could be continued because the bath temperature would remain below 1.5 K. All cooler failures are assumed to result in loss of a pair of coolers because the necessary momentum compensation would be lost with the operation of an odd number of coolers.

Table 4 lists the dewar lifetime sensitivities to the major influencing parameters. Excluding cooler operation, lifetime is most sensitive to mainshell temperature, including the effect of compressor mounting temperature on the heat life capacity of the coolers. When the extreme uncertainties in mainshell temperature, RCA heat load and preamp heat load are lumped together, the bath temperature changes by ±0.04 K, which is less than the specified maximum allowable of ±0.05 K.

CONCLUSIONS

Development of the XRS Cryogenic Subsystem is proceeding. Achieving 5-year lifetime within the mass and envelope constraints requires a hybrid approach combining state-of-the-art superfluid helium dewar technology with mechanical coolers to remove heat from the outer region of the dewar insulation system. Use of the mechanical coolers is crucial to making the lifetime requirement. Therefore, reliability of the coolers and planned redundancy are critical issues. The design uses two momentum-compensated pairs of coolers, a primary pair and a backup pair. The lifetime requirement is met if only one pair operates throughout the mission.

ACKNOWLEDGEMENTS

This work was funded by the National Aeronautics and Space Administration under Contract No. NAS5-31251 from the Goddard Space Flight Center. The authors wish to acknowledge the Ball team members who have contributed to the preliminary design effort.

REFERENCES

1. S. R. Breon, R. A. Hopkins, S. J. Nieczkoski, "The X-Ray Spectrometer--A Cryogenic Instrument on the Advanced X-Ray Astrophysics Facility," presented at 1991 Cryogenics Engineering Conference, Huntsville, AL, June 1991.
2. A. T. Serlemitsos et al., "The AXAF/XRS ADR: Engineering Model," presented at 1991 Cryogenics Engineering Conference, Huntsville, AL, June 1991.
3. A. R. Urbach and P. V. Mason, "IRAS Cryogenic System Flight Performance Report," Advances in Cryogenic Engineering, Vol. 29, 1984.
4. R. A. Hopkins, S. J. Nieczkoski and S. M. Volz, "Performance Predictions for Spaceborne, Long-lifetime Helium Dewars Containing Large-aperture Telescopes," Proceedings of SPIE, Vol. 1340, 1990.
5. R. A. Hopkins et al., "Mechanical Cooler-to-Dewar Interfacing in a Long-lifetime Hybrid Stored Cryogen System," presented at 1991 Space Cryogenics Workshop, Cleveland, OH, June 1991.
6. M. S. Bora et al., "Experimental Investigation on Heat Leak Into a Liquid Helium Dewar," Cryogenics, Vol. 30, 1990.
7. I. E. Spradley, T. C. Nast and D. J. Frank, "Experimental Studies of MLI Systems at Very Low Boundary Temperatures," Advances in Cryogenic Engineering, Vol. 35, 1990.
8. E. M. W. Leung, et al., "Techniques for Reducing Radiation Heat Transfer Between 77 and 4.2 K," Advances in Cryogenic Engineering, Vol. 25, 1980.
9. M. J. DiPirro et al., "The SHOOT Cryogenic Components: Testing and Applicability to Other Flight Programs," Proceedings of SPIE, Vol. 1340, 1990.

THERMAL AND MECHANICAL PERFORMANCE OF SUPERFLUID HELIUM DEWAR FOR IRTS

G.Fujii[*], S.Tomoya[*], M.Kyoya[**], M.Hirabayashi[**],
M.Murakami[+], H.Okuda[++], H.Murakami[++] and T.Matsumoto[#]

* Space Development Division, NEC Corp. Yokohama, Japan

** Sumitomo Heavy Industries, Ltd., Toyo, Ehime, Japan

\+ Institute of Engineering Mechanic, University of Tsukuba
Tsukuba, Ibaraki, Japan

++ Institute of Space and Astronautical Science, Sagamihara
Japan

\# Department of Physics, Nagoya University, Nagoya, Japan

ABSTRACT

The Infrared Telescope in Space (IRTS) is a small HeII cooled infrared telescope, launched by the H-II rocket and retrieved by the US Space Shuttle. A 100 L HeII dewar as an engineering model (EM) has been manufactured and tested to evaluate: (1) the heat leak into the cryostat and the temperature distribution, and (2) the mechanical strength and stiffness of the cryostat at both 2 K and 300 K. It is assured from the thermal test that the cryostats meet the thermal requirements for astrophysical observation and the cooling life time. The test results are reflected in the design of the proto-flight model (PFM) currently in progress.

INTRODUCTION

IRTS is a small cryogenically cooled telescope on board the Space Flyer Unit (SFU), which will be launched by the H-II rocket in 1994. The SFU is a reusable free flyer designed as a multipurpose common facility for scientific and engineering experiments and is retrieved by the US Space Shuttle. The configuration and specifications of the SFU are shown in Fig.1. The optical system of IRTS is equipped with a Cassegrain telescope with a primary mirror of 15cm and having a wide field of view suitable for the observation of diffuse IR sources. The major observational objectives extend in wavelength from near infrared to far infrared regions. The characteristic feature of the IRTS is cryogenic cooling down to 1.8 K by stored superfluid helium (HeII) to provide it with both maximum sensitivity and the capability of making photometric observations in far infrared region.

THERMAL DESIGN APPROACH

The observational plan requires that the cooled lifetime should be longer than one month for sufficient sky coverage, the detectors should be colder than 2 K and the highest temperature should be lower than 10 K (30 K for EM) at the forebaffle tip. However, accommodation to the SFU imposes several strict limitations. A minimum attitude control should be

Advances in Cryogenic Engineering, Vol. 37, Part B
Edited by R.W. Fast, Plenum Press, New York, 1992

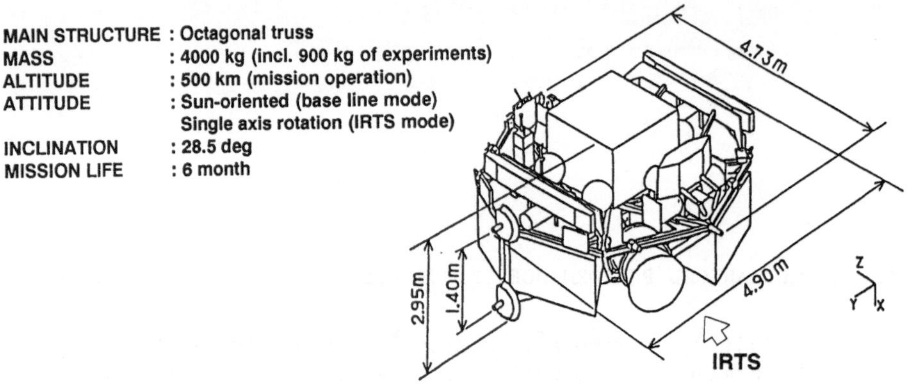

Fig.1 Space flyer unit (SFU) configuration

permitted in order to meet the requirement of Sun and Earth avoidance angles. Limitations of the allowable size, maximum 1.05m in diameter, and weight under 200kg, have great impact on the basic design of the IRTS. The optimum design of the sun shield, the aperture shade and the forebaffle is strictly required to cut the aperture heat load as well as to reduce background IR noise.

The superfluid helium management at the launch site is another key item. The ground operation at the launch site should be well planned, because maximum fill of He II at the moment of the launch is required for maximum lifetime. Vapor evacuation should be continued just prior to the launch, except for short intermissions for the sake of safety during the arming of the rocket. However, "cold launch" condition of the SFU imposes limitationas on the planning of the helium management sequence. The sequence is given in Fig.2. The evacuation pumping must be stopped 12 hours before the launch, because no access to the SFU is allowed during the final count down. Even if the pumping is stopped at the lift-off, vapor evacuation into outer space should be resumed as soon as the atmospheric pressure decreases sufficiently below the vapour pressure in the tank, however, the power for opening the vent valve will not be supplied until 30 minutes after the launch.

CRYOSTAT DESIGN

The major effort in IRTS cryostat design is concentrated on the minimization of heat leak and extremely low background noise. The heat

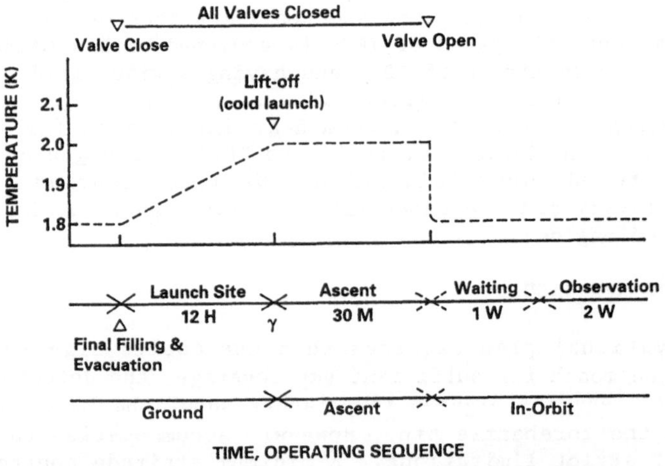

Fig.2 Superfluid He management sequences

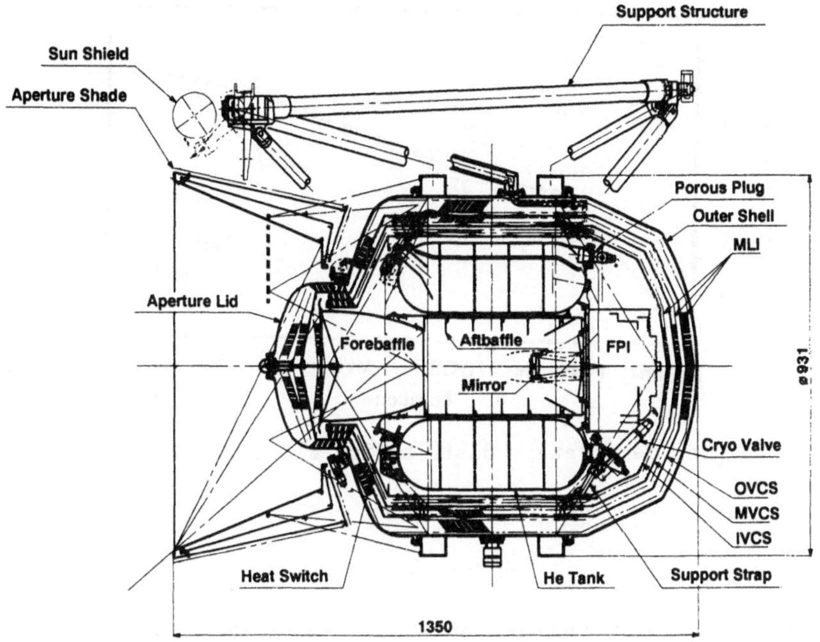

Fig.3 IRTS system schematics

leak to He II gives rise to evaporation, resulting in no temperature rise owing to latent heat of evaporation. The vapour at He II temperature is still capable of absorbing heat and is circulated through a vapour passage around key thermal stages to remove parasitic heat, and is finally exhausted to outer space ideally at the temperature of the outer shell or the highest temperature portion of the system. The design of the optimum distribution of cooling capability of helium gas in the form of sensible heat is found to be of importance for maximum cooled lifetime. It should be noted that this type of open cycle possesses an autonomous thermal force of restitution. That is to say, if the temperature rises, then the evaporation rate increases and the cooling power, in turn, increases resulting in temperature drop.

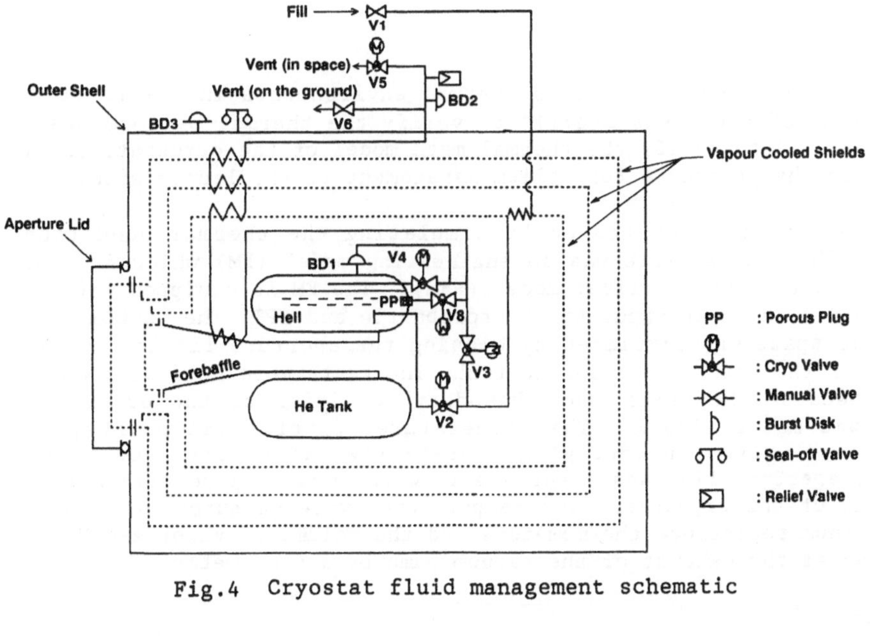

Fig.4 Cryostat fluid management schematic

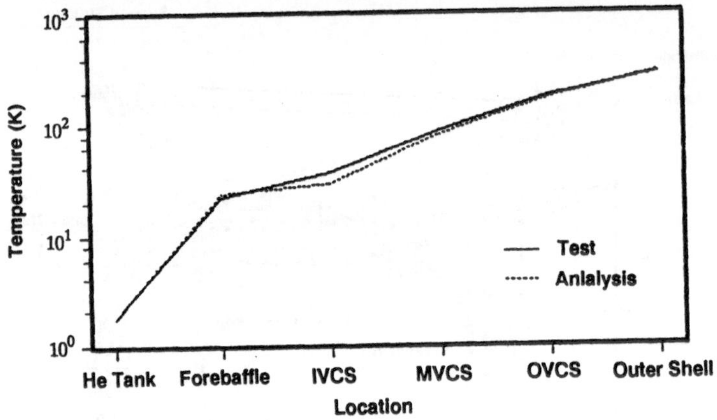

Fig.5 Temperature distribution along He gas flow

By the strong constraints on volume and weight demanded for the system to be accommodated in the SFU, the He II tank volume is limited to 100 L. A cutaway view of the IRTS cryostat is shown in Fig.3, and the fluid management schematic is shown in Fig.4. The cryostat consists of a He tank, an infrared telescope, Focal Plane Instruments (FPI), three vapour cooled radiation shields (IVCS, MVCS, OVCS) and an outer shell. The He tank stores 100 L of superfluid helium of at 1.8 K to cool the FPI to about 2 K. The interior structure composed of the tank, telescope and FPI is suspended by low thermal conductivity support straps from the outer shell to cut conductive heat leak. In order to keep the telescope clean, the aperture lid will be removed just before the observation starts after the environmental conditions have been settled well enough for the cooled telescope to be free from gaseous contamination. Detail design of components of the cryostat are described in the earlier papers[1,2,3]. A design progress from the earlier papers is adding a gas heat switch[4] between the forebaffle and the He tank to cool the forebaffle to below 3 K in case of the best condition of minimum aperture heat load.

THERMAL PERFORMANCE TEST

IRTS will undergo various thermal environments in the mission. The objectives of the tests are:(1) to verify the thermal performance of the cryostat, (2) to verify the thermal math model of the cryostat, and (3) to establish the procedure for helium management at the launch site.

The cryostat was tested by simulating the thermal conditions in Fig.2. The test article was an engineering model (EM) which is thermally equivalent to a proto-flight model (PFM). The EM is equipped with heaters to simulate the heat input to the forebaffle and FPI. The vacuum condition of outer space was simulated by closing the aperture lid and pumping the inside of the outer shell and He tank. The tests were held at room temperature because the temperature boundary condition of the outer shell in orbit was set to 300 K. The temperatures of critical points, i.e. He tank, cryo valves, porous plug, forebaffle, aftbaffle, VCS's, support straps, aperture lid, and outer shell, were measured to obtain the thermal data of the cryostat. The temperatures were measured by using carbon or platinum resistance thermometers and the volume of vaporized He gas was measured at the exhaust of the vacuum pump by a flow meter.

Steady State Thermal Performance

This test was done in order to learn the basic performance of the cryostat. The He tank was filled with HeII and pumping was continued through the open vent until it reached the steady state. The results of the tests and the predicted values are shown in Fig. 5. The results showed that the temperature of the He was 1.8 K, forebaffle was 22.5 K, and the volume of vaporized He gas was 0.93 L/min at 273 K 1 atm. They were in rather good compliance with the analytical results. The required temperatures 1.8 K for He and 30 K for forebaffle were fulfilled. Furthermore, the temperature of the forebaffle may be decreased to 10 K, which is the observational requirement for the proto-flight model (PFM), by changing the material of the vent tube between the forebaffle and IVCS from aluminum to stainless steel and increasing the thermal resistance.

Top-off

It is important to fill the He tank with as much HeII as possible before the launch in order to lengthen the mission lifetime. On the other hand, HeII is obtained by pumping down normal liquid helium (HeI), and in the process the volume of resultant HeII is reduced to 60% of HeI. Thus, in order to increase the final resultant HeII, a procedure called "top-off" was adopted[5]. The procedure is as follows: (1)Pump down both the supply and receiving tanks, and change HeI into HeII. By the operation, helium temperatures become lower than the lambda point (2.17 K) and constant for the whole volume because of the high thermal conductivity of HeII. (2)Pressurize the supply helium tank over the vapour pressure of HeII and transfer the subcooled HeI from the supply tank to the receiving tank. The temperature of the helium in the supply tank is only slightly higher than the lambda point at the bottom while much higher at the surface because of low thermal conductivity of HeI. (3)Pump down the receiving tank to obtain HeII. In the tests, we pumped down the supply tank to 2.1 K, and transferred the helium at various pressure. The results are as follows: (1) 83 L of HeII was filled in one operation. (2)The pressure for supply tank has to be higher than 100 Torr in this system. (3)It is more effective to complete the top-off at pressures higher than 300 Torr in a short time than with lower pressure in a longer time.

Hold Time Test

The vaporized gas in the He tank will be evacuated after topping-off HeII at the launch site to maintain the superfluid state, but the evacuation will be stopped 11 hours before launch and will be restarted 30 minutes after the launch. While evacuation is stopped, the superfluid state must be maintained, even though the temperature of the helium increases owing to heat leak into the tank. In the test, we pumped down the helium in the He tank until it reached steady state at 1.8 K, then closed vent valve V5 and measured the time to reach the lambda point (2.17 K). During the test, only the porous plug vent valve V8 (shown in Fig.4) was kept open, and we observed whether thermal oscillation occurred or not. The volume of HeII was 80 L. The result showed that the superfluid state was able to be kept 20 hours (from 1.8 K to 2.17 K) after stopping of the exhaust. Thermal oscillation did not occur even if V8 was open. This result suggests that the porous plug valve V8 is not essential and may be replaced by V5.

Porous Plug Phase Separation Test

Evacuation of vaporized He gas on orbit is restarted by means of the vacuum of outer space after stopping of the exhaust during the launch.

This test was held after the "hold time test" in the preceding paragraph. The purpose of the test was to certify that the porous plug was able to decrease the temperature of helium to 1.8 K after the temperature had risen. The temperature reached was 1.85 K and the time was 80 hours. The temperature was higher than in the case of exhausting through open vent, but in space there is no pressure drop in the piping from V5 to the vacuum pump, the temperature in the He tank will be reduced to below 1.8 K, and it can fulfill the requirement.

Heat Input Test

The purpose of the test was to estimate the stability of the temperatures in the cryostat for aperture heat loads and FPI heat generation. The aperture heat loads were simulated by supplying 820 mW x 5 min and 33 mW x 85 min per cycle to the forebaffle heater, and 10 mW was supplied continuonsly to the FPI heater, simulating FPI heat generation. Temperature changes in the test are shown in Fig.6. The results showed: (1)The forebaffle suffered maximum temperature change, and the further from the forebaffle the more thermally stable. (2)The temperature change in He was about 10 mK, and the temperature of the forebaffle which is most sensitive, recovered to the steady state in 30 min after giving heat input; thus this system was stable for external turbulence. The mission lifetime which was estimated from the result appeared to be longer than 1 month.

Return to Superfluid Test

Vaporized helium in the He tank is planned to be re-evacuated after stopping the exhaust for 11.5 hours, but there are possibilities of the helium temperature rising over the lambda point before the re-evacuation due to unexpected causes. Thus, it is of great interest if restart of the evacuation in the outer space through the porous plug can decrease the helium temperature and convert to superfluid before observation. The

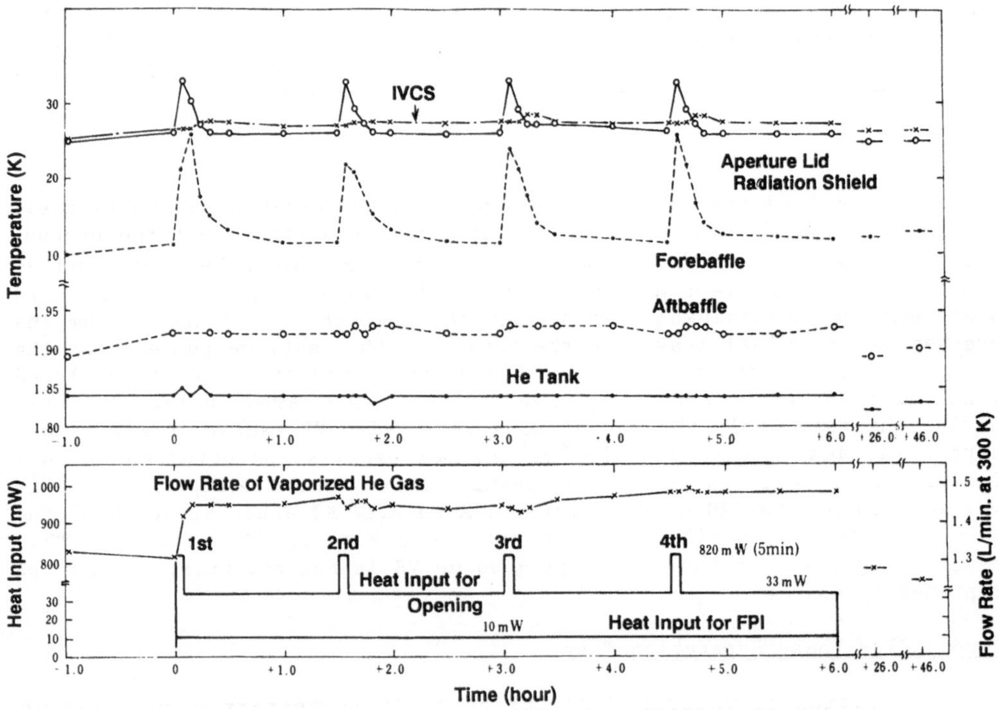

Fig.6 Results of heat input test

Fig.7 Vibration test configuration

tests were held under the following two conditions. (1) The porous plug was set in the gas phase (simulating the pre-launch period). (2) The porous plug was set just under the liquid surface (simulating zero-gravity in space). Under both conditions, after keeping vent valve V5 closed until the temperature of helium rose over 2.2 K, vent valve V5 was opened and exhaust was restarted. The results showed in the case of condition (1) the temperature did not decrease, and in the case of condition (2) the temperature decreased and the helium returned to superfluid state. This means that the vaporized helium took its heat of vaporization from the helium around the porous plug and converted to superfluid, thus the porous plug operated as a phase separator.

MECHANICAL PERFORMANCE TEST

The IRTS cryostat experiences external loads during launch and landing; thus the structural integrity for the loads must be verified. Also the effect of vibration on the thermal performance of the cryostat must be evaluated. The mechanical performance tests for the above purposes were completed successfully.

Static Load Test

The static load test was conducted to verify the strength of the cryostat against the acceleration loads during launch and landing. It is required for the structure that no detrimental deformation occurs by limit loads, and no structural failure occurs by ultimate loads. The tests were conducted for the outer shell, the He tank, and the support structure respectively. The loads were applied by hydraulic jack or tension of the dummy strap. The results were successful and the strength of the cryostat for the external loads was verified.

Vibration Test

The test configuration of the vibration test is shown in Fig.7. In the first stage, ethyl alcohol of the same mass was filled in the He tank instead of HeII. Because of the test jig design, the cryostat was set upside-down. Accelerometers were attached to the components of the cryostat to measure the acceleration level, and the vibration modes were analyzed. The results showed that the cryostat has sufficient strength

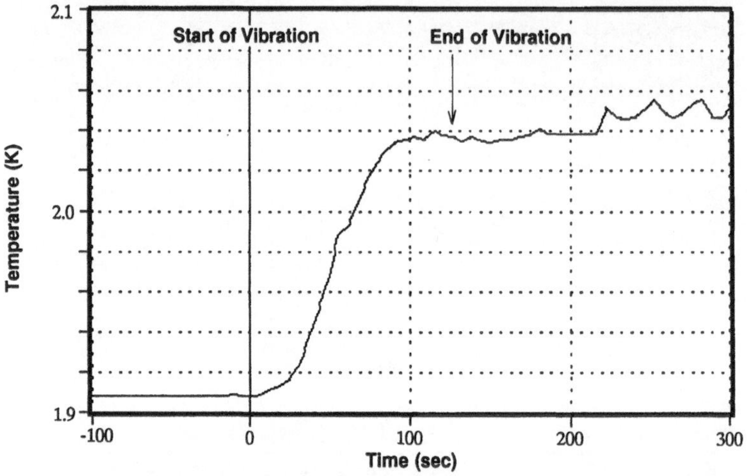

Fig.8 He temperature during vibration

to endure the vibration. The primary natural frequency was 29 Hz and the response amplitude of the He tank was 10 times the added acceleration, which is equal to a damping coefficient ratio of 5%.

After its strength was verified by the preceding vibration test, the cryostat was cooled down to superfluid temperature, and the vibration test was repeated. The purpose of the cooled vibration test is to evaluate the effect of the vibration on the temperature of HeII. The He tank was filled with 50 L of HeII, and the vent valve V5 was closed to simulate the launch condition. The sine vibration was given in X-direction, which is the most critical direction, from 5 to 100 Hz and the maximum acceleration level was 12 G zero to peak at the He tank. As shown in Fig.8, the temperature of HeII rose about 0.13 K by the sine vibration , while no such rises were observed during random vibration tests. The cause of the tem perature increase is thought to be the kinetic energy and it should be taken into account in the helium management plan.

CONCLUSIONS

An EM of the IRTS was designed, manufactured and tested. The results of the tests showed that the cryostat design complies with the required specifications for IRTS. The thermal and structural analyses were verified by the tests. A PFM of the IRTS cryostat has been designed on the basis of the EM and now is being manufactured. Thermal and structural tests for the PFM are planned to be held in 1992.

REFERENCES

1. G. Fujii et al, Thermal Design of a Superfluid Helium Dewar for Infrared Telescope Onboard Space Flyer Unit, ESA SP-288 (1988).
2. M. Murakami et al, Design of Cryogenic System for IRTS, Cryogenics, Vol. 29 (1989), p.553.
3. M. Murakami et al, Thermal design and test of IRTS cryostat,in: "Advances in Cryogenic Engineering", Vol.35, Plenum Press, New York (1990), p.295.
4. L. Duband et al, A Rocket-borne ^{3}He Refrigerator, in: "Advances in Cryogenic Engineering", Vol.35, Plenum Press, New York(1990),p.1447
5. D. Petrac, Top-off Procedure for Space-bound Superfluid Helium Cryostats, in: "Advances in Cryogenic Engineering", Vol.27, Plenum Press, New York (1982), p.1099.

DESIGN OPTIMIZATION OF A VAPOR-COOLED RADIATION SHIELD FOR LHE CRYOSTAT IN SPACE USE

Masahito Yamaguchi, Takao Ohmori

Ishikawajima-Harima Heavy Industries Co. (IHI)
Yokohama, Kanagawa, Japan

Akira Yamamoto

National Laboratory for High Energy Physics (KEK)
Tsukuba, Ibaragi, Japan

ABSTRACT

LHe cryostats for superconducting magnets used in a space environment are required to have superior thermal insulation performance. In order to reduce the heat-in-leakage of radiation from outside the cryostat, vapor-cooled multi-stage radiation shield plates are inserted between the vacuum vessel and the liquid helium vessel, and a number of multi-layer insulation (MLI) layers are wrapped on each surface. The temperature distribution of the radiation shield and the heat-in-leakage are numerically evaluated, and the optimal combination of the shield plates and MLI layers is determined within the allowable weight.

INTRODUCTION

With rapid progress in space science and technology, various superconducting devices, which are to be used in a space environment have been planned for the last several years. Liquid helium is the only useful coolant for space-based superconducting devices at the present time. However, since it can't be supplied very often in space, some measures have to be taken so that liquid helium can be available for a long term mission with extremely low evaporation rate. One of the main sources of heat-in-leakage into the magnet is thermal radiation from outside the cryostat. In order to reduce the heat-in-leakage, multi-stage radiation shield plates, cooled by evaporated helium gas, are inserted between the vacuum vessel and the liquid helium vessel, and several layers of multi-layer insulation (MLI) are wrapped on each surface to improve the thermal insulation characteristics. The greater the number of plates and MLI layers, the more the thermal insulation effect improves. However, we may not employ excessive thermal insulation, because it causes in excessive weight in the cryostat. We have to minimize the heat-in-leakage within the allowable weight.

In this study we try to optimize the number of shield plates and MLI layers considering the thermal insulation performance and the limited weight of the cryostat.

ANALYSIS MODEL

The cryostat of the ASTROMAG (Particle Astrophysics Magnet Facility) test coil which has been developed at KEK[1] is the object of our analysis. ASTROMAG is a large facility for the observation of cosmic rays, and is proposed for the space station. Figure 1 shows the conceptual design of the ASTROMAG test coil cryostat which contains a 2.8 m^3 liquid helium vessel. Multi-stage radiation shield plates are inserted at constant intervals of separation. Every shield plate has a single-pass cooling channel in which helium gas evaporated from the liquid helium vessel flows. Dimple type MLI, which were previously developed by the authors[2], are wrapped on each shield plate surface. The reflective films of this type of MLI have many spherical projections, thereby reducing contact between adjacent layers. Spacers, which are usually necessary in MLI assembly, are therefore not necessary.

Superfluid helium can be adopted for superconducting devices in space use, because the space environment is vacuum, and it is easy to realize the superfluid state. Moreover, cooling by superfluid helium leads to superior characteristics of magnet stability. However, there isn't a remarkable difference between the calculation the heat-in-leakage using either normal or superfluid helium. For this reason, the calculation is performed assuming normal atmospheric-pressure helium (0.1013 MPa).

FUNDAMENTAL EQUATIONS

In this study, the calculation area is divided into control volumes, and the temperature distribution of every radiation shield plate and the heat-in-leakage into the LHe vessel is obtained by calculating the heat balance for every control volume.

Heat transfer of the radiation shield plate

Now, we consider the heat balance for the control volume shown in Fig. 2. For example, the heat flow in the x-direction by conduction, Q_e, can be written as

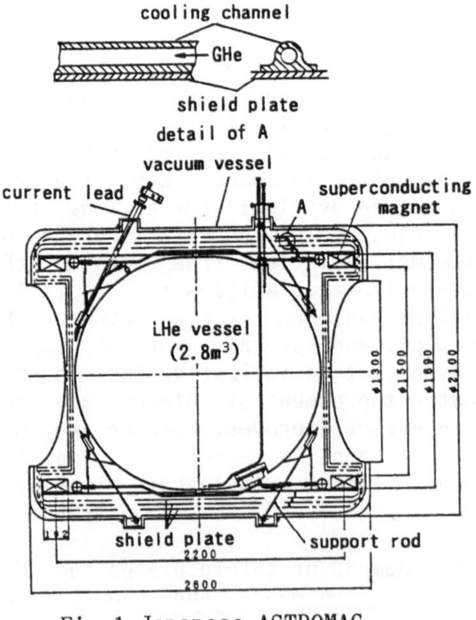

Fig.1 Japanese ASTROMAG test coil cryostat

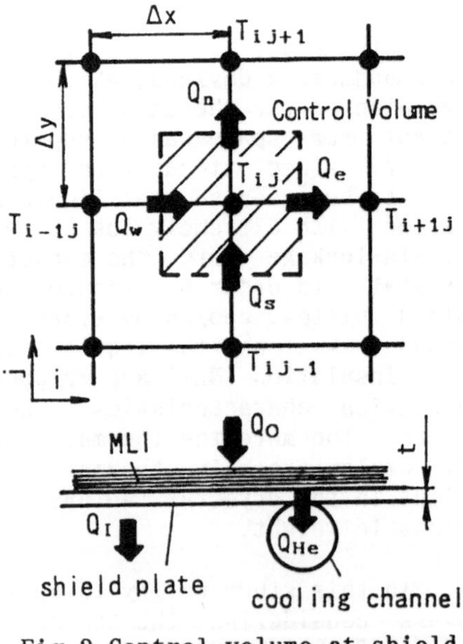

Fig.2 Control volume at shield plate in computation

$$Q_e = \frac{2\lambda_{ij}\lambda_{i+1j}}{\lambda_{ij}+\lambda_{i+1j}} \frac{T_{i+1j}-T_{ij}}{\Delta x} \Delta yt \tag{1}$$

Q_w, Q_n and Q_s are calculated like Q_e. The heat flow by radiation from the outer boundary Q_O, and into the inner boundary Q_I, can be expressed as

$$Q_O = \alpha^O{}_{ij}(T^O{}_{ij} - T_{ij})\Delta x \Delta y$$
$$Q_I = \alpha^I{}_{ij}(T_{ij} - T^I{}_{ij})\Delta x \Delta y \tag{2}$$

where α^O and α^I are the radiation heat transfer coefficients for radiative heat flow from the outer boundary and to the inner boundary, and T^O and T^I means the outer and inner boundary temperature respectively. The heat flow into the helium gas coolant can be calculated from

$$Q_{He} = \alpha^{He}{}_j(T_{ij} - T^{He}{}_j)\pi D \Delta x \tag{3}$$

The heat balance over the control volume leads to the following equation:

$$Q_w - Q_e + Q_s - Q_n + Q_O - Q_I - Q_{He} = 0 \tag{4}$$

Heat transfer of the helium gas coolant

For the helium gas channel, the control volume shown in Fig. 3 is introduced. The heat which is exchanged with the radiation shield plate is given by Eq.(3), and the energy increase of the helium gas can be written as

$$Q_{He} = m(h_j - h_{j-1}) \tag{5}$$

Here, the heat transfer coefficient between the shield plate and the helium gas is given by:

for Re > 2300:

$$\alpha^{He} = 0.023Re^{0.8}Pr^{0.4}\lambda^{He}/D \tag{6}$$

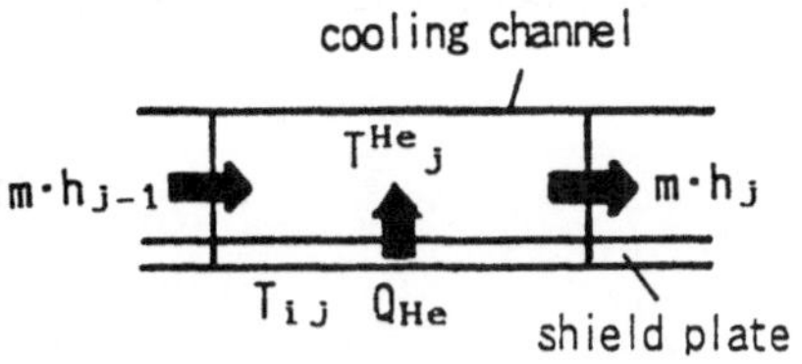

Fig.3 Control volume at cooling channel in computation

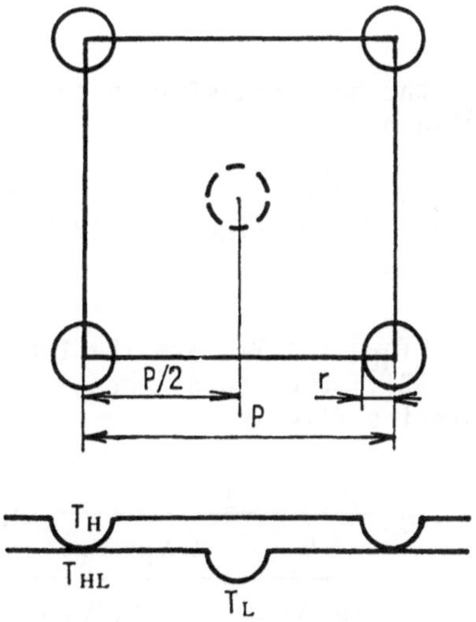

Fig.4 Model of thermal contacting part in dimpled MLI

for Re < 2300:

$$\alpha^{He} = 4.66\ \lambda^{He}/D \tag{7}$$

We can obtain the temperature profiles of the helium gas coolant from Eqs. (3) and (5).

Heat transfer of the multi-layer insulation

Heat transfer of the Super Insulation consists of radiation between the insulating films, contact conduction between films, and conduction in the cross-section of the film. Now we consider the model as shown in Fig. 4, in which two films contact with each other. The radiation heat flow can be expressed as

$$Q_R = \sigma \frac{1}{2/\epsilon_L - 1}(T_H^4 - T_L^4)A \tag{8}$$

Here, T_H and T_L are the hot and cold boundary temperatures. The emissivity ϵ_L is estimated by using the cold boundary temperature, which is obtained from[3]

$$\epsilon = 1.63\times10^{-3} T^{0.495} \tag{9}$$

The contact heat resistance is an important factor for the heat transfer at the contacting surfaces of MLI. In this study, we introduce the contact ratio γ, which is equal to the contact area/non-contact area. The contact area doesn't mean apparent contact area, but means the part in which the heat flow is transferred without any resistance. The heat transfer at the contacting part can therefore be written as

$$Q_C = \frac{2\lambda_H \lambda_L}{(\lambda_H + \lambda_L)t}(T_H - T_L)\ A\gamma \tag{10}$$

here λ_H and λ_L are the thermal conductivities of the cold and hot boundaries respectively. If we assume the heat transfer area to be

$$2\sqrt{2}rt \frac{A}{(P/2)^2} \tag{11}$$

then, the heat transfer in the cross section of the insulating film can be written as

$$Q_C = \frac{\lambda_L}{\sqrt{2}P/2}(T_{HL} - T_L)(2\sqrt{2}rt)\frac{A}{(P/2)^2} \tag{12}$$

where T_{HL} and T_L are the temperatures shown in Fig. 4. From Eqs. (11) and (12), the total heat is transferred by conduction through the insulating film is

$$Q_C = \frac{1}{\dfrac{P}{16r\lambda_L t} + \dfrac{(\lambda_H + \lambda_L)t}{2\lambda_H \lambda_L \gamma}}(T_H - T_L)A \tag{13}$$

Equations (8) and (12) give the total MLI heat flow:

Table 1 Specification of support rod

material	GFRP
outer diameter	6.0 mm
length	0.5 m
total number of rods	16

$$Q_{MLI} = Q_R + Q_C = \alpha_{MLI}(T_H - T_L)A \qquad (14)$$

where

$$\alpha_{MLI} = \frac{\sigma}{2/\epsilon_L - 1}(T_H^2 + T_L^2)(T_H + T_L) + \frac{1}{\dfrac{P}{16r\lambda_L t} + \dfrac{(\lambda_H + \lambda_L)t}{2\lambda_H \lambda_L \gamma}} \qquad (15)$$

From Eq. (14), the total heat transfer coefficient of the MLI, which contains n laminated layers, can be written as

$$1/\alpha\ \text{shield} = \sum_{i}^{n-1} 1/\alpha\ \text{MLIi} \qquad (16)$$

<u>Heat-in-leakage through the support rods</u>

The other source of heat-in-leakage into liquid helium vessel is heat conduction through support rods. The specification of the support rod is shown in Table 1. If we assume the heat-in-leakage per rod to be 1.45 mW, the total heat-in-leakage of all the support rods is

$$Q_{support} = 23.2\ \text{mW}$$

Therefore, heat flow of 23.2 mW leaks into the LHe vessel in addition to the radiation heat transfer.

SELECTION OF CONTACT RATIO AND NUMBER OF MLI LAYERS

It is important to find the appropriate value of the contact ratio with experiment, which was mentioned in the above section with respect to

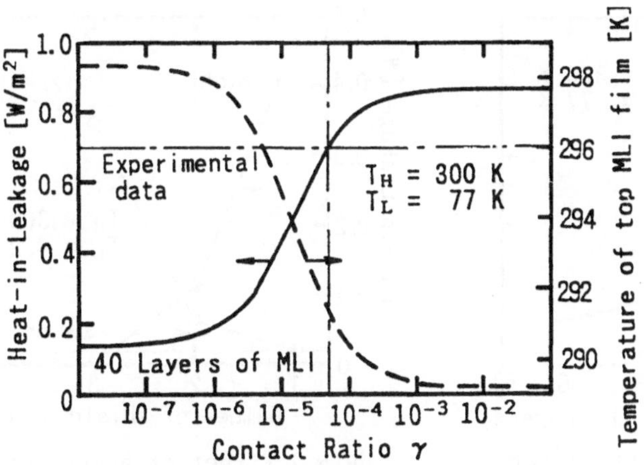

Fig.5 Effect of contact ratio on heat transfer

the MLI heat transfer calculation. The relation between the contact ratio and the calculated heat-in-leakage for 40 layers of MLI is shown in Fig. 5. On the other hand, previous experimental results by the authors[2] and Shu et al.[4] gave a heat-in-leakage of 0.7 W/m^2 for 300-77 K. Therefore we adopt a contact ratio value of 5.0×10^{-5}.

The number of MLI layers is also an important factor for the calculation. Figure 6 shows the heat-in-leakage versus the number of layers of MLI film. We take the number of MLI layers to be 40, because decreases the heat-in-leakage only slowly for more than 40 layers.

CALCULATION RESULTS

The heat-in-leakage for four cases, as a function of number of radiation shield plate, is shown in Fig. 7. In this calculation 40 layers of MLI films are wrapped on every shield. The heat-in-leakage decreases with an increasing number of shield plates, and we can ignore the heat-in-leakage through the shield plates compared with heat conduction through the support rods for more than 3 shield plates. Figure 7 also shows the LHe life time. For 3 shield plates the consumption time is 7 years, and the required condition is satisfied.

Figures 8 and 9 show heat-in-leakage and shield plate temperatures for 5 calculation conditions in which a different number of MLI layers are wrapped on 3 shield plates. The number of MLI layers for the 5 cases is listed in Table 2. From case A to C, the heat-in-leakage and shield plate temperature abruptly decrease, but from C to E they hardly change. It is effective to wrap on the outer and middle shield plates for diminution of heat-in-leakage, but on the other hand it is ineffective to wrap on the inner shield plate and the LHe vessel.

The temperature distribution of the radiation shield plates is shown in Fig. 10 for case C (3 shield plates). Figure 10 shows the developed view of shield plate. The cooling He gas flows from left to right along the top and bottom edges of the rectangular part. The difference between the highest and lowest temperatures in each shield plate is 0.9, 4.5, and 10.8 K for the inner, middle, and outer shield plates respectively.

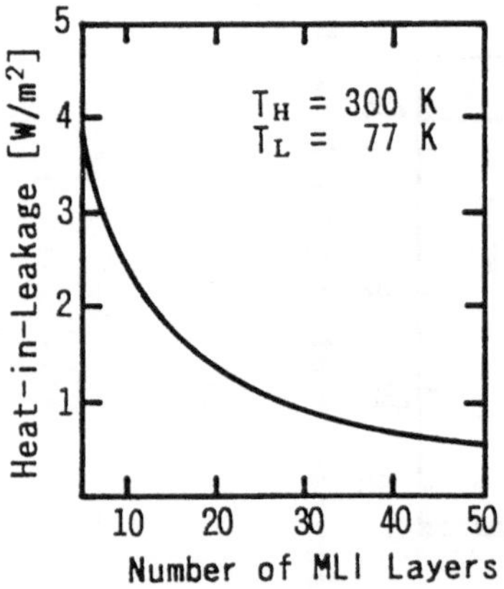

Fig.6 Effect of MLI layers on heat transfer

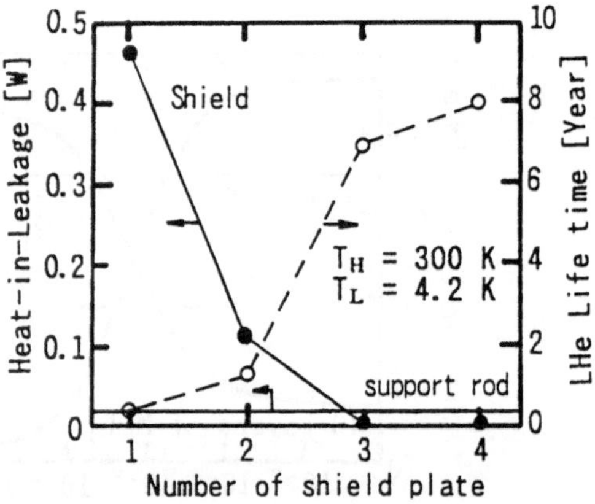

Fig.7 Effect of number of shield plates on heat transfer

Table 2 Method of MLI wrapping

Case	LHe vessel	inner shield	middle shield	outer shield
A	0	0	0	0
B	0	0	0	40
C	0	0	40	40
D	0	40	40	40
E	40	40	40	40

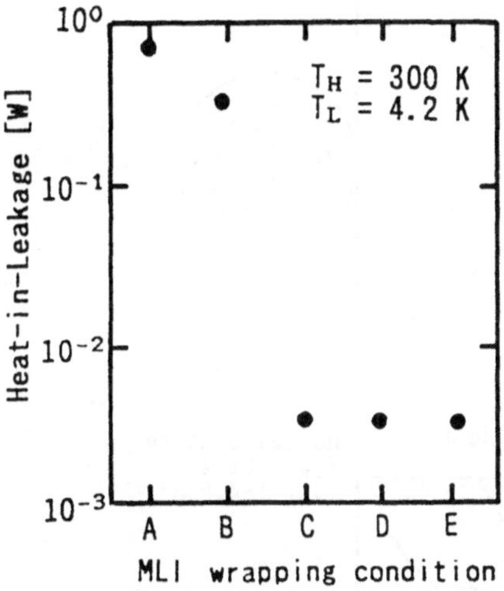

Fig.8 Effect of MLI wrapping method on heat transfer

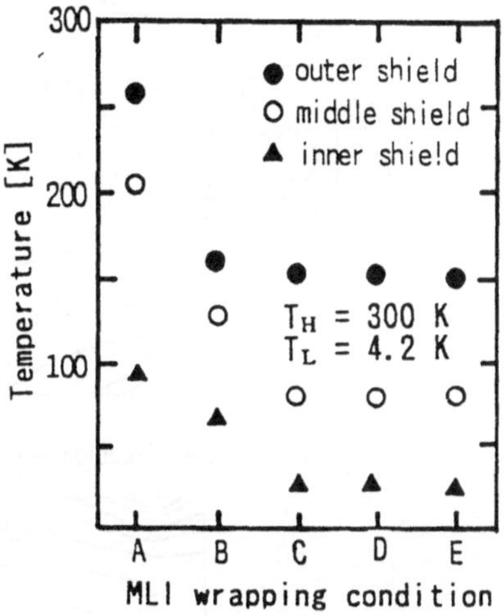

Fig.9 Effect of MLI wrapping method on heat transfer

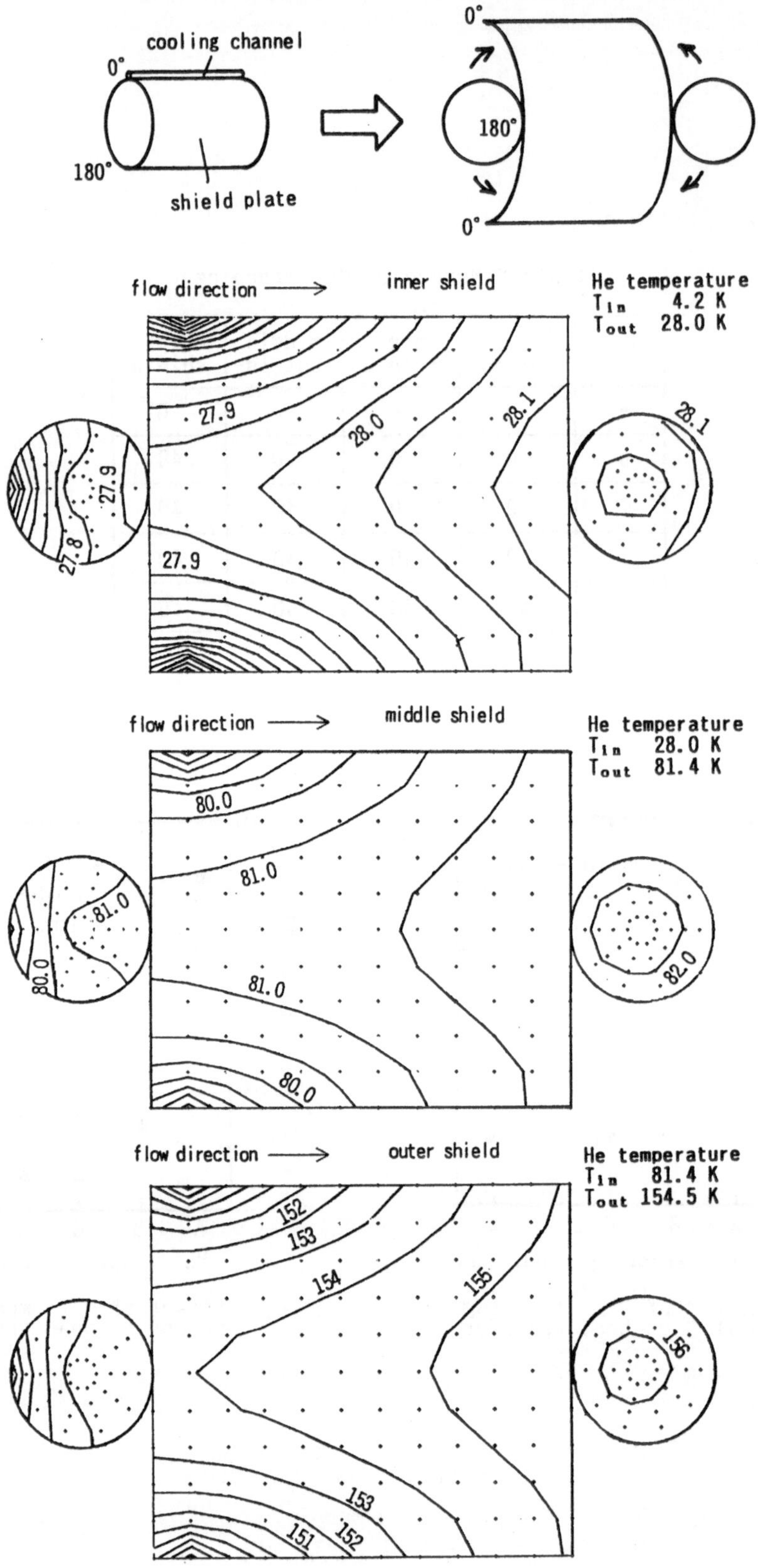

Fig.10 Temperature distribution in shield plates

CONCLUSIONS

The design optimization of a cryostat for space use was achieved considering weight of thermal barrier and heat-in-leakage minimization. We consider that the ASTROMAG test coil cryostat, using three shield plates is the most effective design, and it is not necessary to wrap MLI on all the shield plates. The calculated results shows us that it is sufficient to wrap only on outer and middle shield plate, and LHe is available for 7 years in this design. It is satisfied enough to satisfy the design criterion of the ASTROMAG project.

NOMENCLATURE

A : heat transfer area
D : inner diameter of cooling channel
h : specific enthalpy
m : mass flow rate
P : dimple pitch
Pr : Prandtl number
Q : heat flow
r : radius of dimple
T : shield plate temperature
T^{He} : GHe temperature
t : thickness of shield plate
α : heat transfer coefficient
γ : contact ratio
Δx : distance between grid points in x-direction
Δy : distance between grid points in y-direction
ϵ : emissivity
σ : Stefan-Boltzmann constant (5.67×10^{-8} W/m^2K^4)
λ : thermal conductivity

REFERENCE

1. A. Yamamoto et al. Development of an ASTROMAG Test Coil with Alminum Stabilized Superconductor, IEEE TRANSACTIONS ON MAGNETICS, 27:1944 (1991).
2. T. Ohmori et al. Multilayer Insulation with Aluminized Dimpled Polyester Film, in "Proc. 11th Intl. Cryo. Engr. Conf.,"Butterworth, 567 (1986).
3. S. Tsujimoto and T. Kunitomo, Teion Tasou Dannetuzai no Kaiseki Shuhou ni Tsuite : "Prepr. of 918th Jpn. Soc. Mech. Eng." 80-27 (1980) p.71.
4. Q. S. Shu et al. An Experimental Study of Heat Transfer in Multilayer Insulation System from Room Temperature to 77 K in: "Advances in Cryogenic Engineering" Vol. 31, Plenum, New York 455 (1986).

A MODEL OF THE UNSTEADY BEHAVIOR OF LONG DURATION HELIUM DEWAR

Gerald R. Karr, Lisa B. Jimmerson and Yu-Chung Tsai

University of Alabama in Huntsville
Mechanical Engineering Department
Huntsville, Alabama

ABSTRACT

A numerical model has been developed for simulating the unsteady behavior of a dewar for helium storage. The model includes the dynamic behavior of the vapor cooled shields, MLI insulation, and the effect of parallel conduction and radiation paths. An implicit finite difference program was developed to take into account the time dependent variation of the gas flow through the heat exchanger of each shield, the heat capacity of the shield material, the conductive heat transfer and radiative heat transfer in the system. Initial conditions can be input to any of the shields, the outside temperature, or pumping pressure.

The physical parameters representing a 250 L helium dewar are used to test the model. The dewar is one for which considerable data exists on the temperatures of the shields as a function of time. The data also includes flow rate and liquid level information. It was taken over a period of months on a continuous basis. During the tests the dewar was pumped down to superfluid temperatures and held in that state for long periods of time. One of those tests is used to compare with the model results. The dynamic behavior of the shield temperatures was observed to have time constants of the order days.

INTRODUCTION

The Infrared Telescope (IRT) Experiment was flown on Spacelab 2 in 1985 as a joint project of the Smithsonian Astrophysical Observatory, the University of Arizona and the NASA Marshall Space Flight Center (MSFC). The responsibility at MSFC was for the development of the cryogenic and mechanical systems of the IRT. The cryogenic system provides coolant at near 2 K to the IRT system [1]. Much of the design, development, fabrication and testing of the cryogenic systems of the IRT were performed at the University of Alabama in Huntsville. This paper reports on a computer simulation of the time dependent behavior of the cryogenic system and its correlation to the actual system performance testing.

The system under study (Fig. 1) is comprised of a dewar and a transfer assembly (TA). The dewar consists of an inner tank, temperature shields and multi-layer insulation

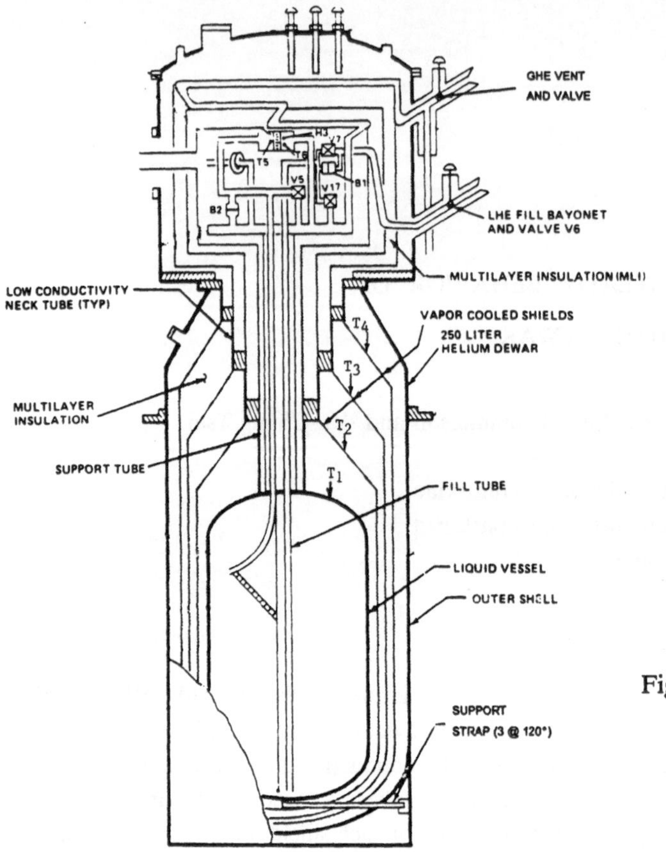

Fig. 1. Diagram of dewar subsystem for IRT experiment with heat stations identified.

(MLI). The inner tank which contains the liquid helium is supported by a neck tube and three support straps at the bottom of the tank. The tank is surrounded by three vapor cooled shields with temperatures of T_2 , T_3 and T_4 where T_2 is the temperature of the innermost shield. MLI is used to reduce heat transfer between the shields.

The helium vapor travels up the neck tube to the TA to be distributed. The TA consists of an outer aluminum case and dome, three nested aluminum vessels, each having a heat exchanger around its top circumference, and each bolting to one of the dewar shield extensions.

ANALYSIS

The computer program was prepared to simulate the performance of the cryogenic system for a wide range of possible operating conditions. The program is based on an energy balance of the heat transfer between the dewar's inner tank, the vapor cooled shields, and other system components. The heat balance on a heat exchanger for an active shield is given by

$$(\rho C_p V)_s \frac{dT_i}{dt} = (\dot{m} C_p)_v \, (T_i - T_{i-1}) + \sum_j C_{ij}(T_j - T_i) + \sum_j R_{ij}(T_j^4 - T_i^4)$$

where ρ is the density, V is the volume, $\dot{m}$ is the helium mass flow rate, C_p is the specific heat with the s and v subscripts referring to solid and vapor, C_{ij} represents the conduction heat

Table 1. Conduction Data

	k (W/m-K)	A (m^2)	L (m)	C12 (W/K)
C12N(Neck Tube)	0.035	2.4580×10^{-3}	0.203	4.2379×10^{-4}
C125(Strap)	0.035	8.4582×10^{-5}	6.5×10^{-4}	4.5544×10^{-3}
C12				4.9782×10^{-3}
C32N(Neck Tube)	0.035	2.4580×10^{-3}	0.0508	1.6935×10^{-3}
C325(Strap)	0.035	8.4582×10^{-5}	0.1268	2.3356×10^{-5}
C32M(MLI10)	1.8×10^{-4}	3.2620	0.0153	3.8376×10^{-2}
C32				4.0093×10^{-2}
C43N(Neck Tube)	0.035	2.4859×10^{-3}	0.0700	1.2430×10^{-3}
C43M(MLI15)	3.0×10^{-4}	3.7091	0.0179	6.2163×10^{-2}
C43				6.3406×10^{-2}
C54N(Neck Tube)	0.035	2.4824×10^{-3}	0.1016	8.5516×10^{-4}
C54M(MLI60)	5.2×10^{-4}	4.2079	0.0255	8.5898×10^{-2}
C54				8.6663×10^{-2}
C53S(Strap)	0.035	8.4546×10^{-5}	0.127	2.3310×10^{-5}
C53				2.3310×10^{-5}

transfer coefficient between the *i* and *j* heat stations, and R_{ij} represents the radiation heat transfer coefficient between the *i* and *j* heat stations [2].

The heat transfer due to conduction has four means of transfer. The fiberglass/epoxy neck tube which supports the inner tank is one mode of conduction. The inner tank is also restrained at the bottom by three fiberglass/epoxy tension straps. This produces a conduction path through the shields and inner tank. The third conduction path is through the MLI between the shields. There is also a small amount of heat transfer due to the residual gas between the shields. The numerical values for the conduction factors are listed in Table 1.

Radiation between the shields was modelled as concentric spheres to simulate the heat leak due to radiation into the inner tank. MLI, which consisted of aluminized mylar with nylon netting, was used between the shields to block radiation. The impact of the MLI was added to the simulation. The number of layers of MLI, the emissivities, and areas of the shields are listed in Table 2. The values for radiation heat transfer coefficients are also listed.

Table 2. Radiation Data

	E	A (m^2)	MLI (No)	R (W/K^4)
R12D(DEWAR)	0.02	2.2996	0	1.3692×10^{-9}
R12T(T.A.)	0.02	0.4149	0	2.7873×10^{-10}
R12				1.7264×10^{-10}
R32D(DEWAR)	0.02	2.8420	10	1.8876×10^{-10}
R32T(T.A.)	0.02	0.4149	10	3.0970×10^{-11}
R32				2.1973×10^{-10}
R43D(DEWAR)	0.02	3.1283	15	1.3414×10^{-10}
R43T(T.A.)	0.02	0.5757	15	2.7025×10^{-11}
R43				1.6117×10^{-10}
R54D(DEWAR)	0.02	3.4309	60	3.5661×10^{-11}
R54T(T.A.)	0.02	0.7727	60	8.4639×10^{-12}
R54				4.4125×10^{-11}

The heat balance for the dewar plus liquid is given by

$$(\rho C_p V)_{s+l}\,\frac{dT_1}{dt} = \dot{m}_e L + \sum_j C_{1j}\,(T_j - T_1) + \sum_j R_{1j}\,(T_j^4 - T_1^4)$$

where $\dot{m}_e$ is the mass rate of evaporation, L is the latent heat of vaporization of liquid helium at temperature T_1, and $s+l$ refers to the solid plus liquid components.

The mass of evaporated helium (m_v) is controlled by a balance of that evaporated, $\dot{m}_e$, and that which is pumped out of the vapor volume, $\dot{m}_p$

$$\frac{dm_v}{dt} = \dot{m}_e - \dot{m}_p$$

The evaporation rate is controlled by the vapor density in the dewar through the expression

$$\dot{m}_e = h_m A(\rho_{sat} - \rho_D)$$

Here A is the surface area of the liquid in the dewar, ρ_{sat} is the saturated vapor density of helium, ρ_D is the density of helium vapor in the dewar, and h_m is the approximate mass transfer coefficient. The ideal gas law is used to express the density differences between the saturated vapor at the liquid surface and the vapor in the dewar which is not at saturated condition.

The mass flow rate through the valve is modeled by the choked flow expression

$$\dot{m}_p = CA^* \frac{P_D}{\sqrt{T_1}}$$

where CA^* represents the resistance of the pumping line, and P_D is the dewar pressure.

PERFORMANCE TESTING OF THE DEWAR

Data is available for the case when the dewar was tested separate from the cryostat. During the test used for model comparison, the dewar was filled and immediately pumped to superfluid. The thermal system was then allowed to stabilize. Within about 8 days, the bath temperature was steady at 1.75K and the shields had reached temperatures of 36K, 100K, and 185K respectively. The mass flow rate stabilized at 8.4 mg/s reflecting a bath heat load of 195 mW. The measured decrease in liquid level of 4.7 liters per day corresponds to 8 mg/s and a dewar lifetime of 42.5 days[3] for a partial fill of 200 liters.

Figures 2 through 7 show the results of the comparison of the numerical simulation with the experimental data. The thermal conductivities of the MLI were adjusted so that the steady state results would agree with the simulation. Notice that the transient results were in general agreement with experiments except for the prediction of the subcooling of the liquid and the lag of the simulated results with respect to the experimental result.

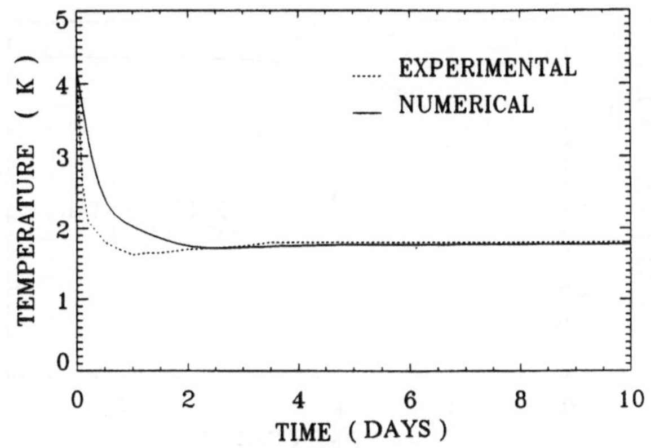

Fig. 2. Comparison of results for T_l, the liquid volume.

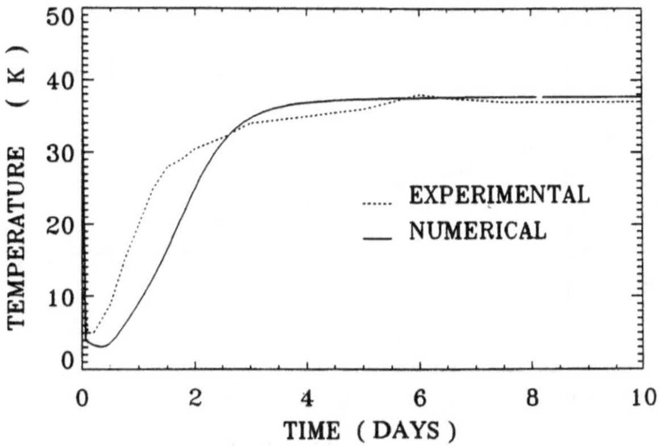

Fig. 3. Comparison of results for T_2, the coldest vapor cooled shield.

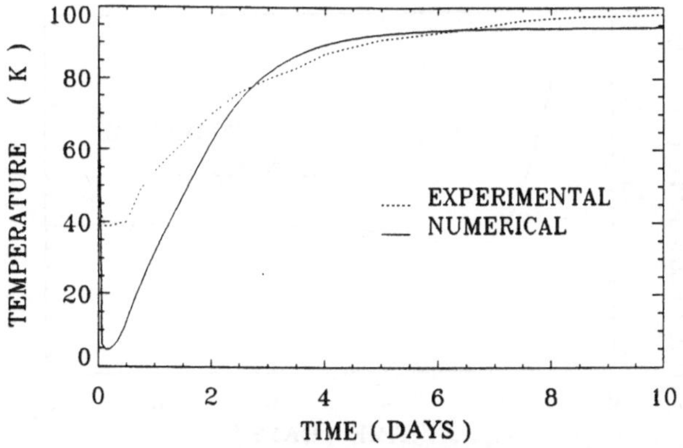

Fig. 4. Comparison of results for the intermediate vapor cooled shield.

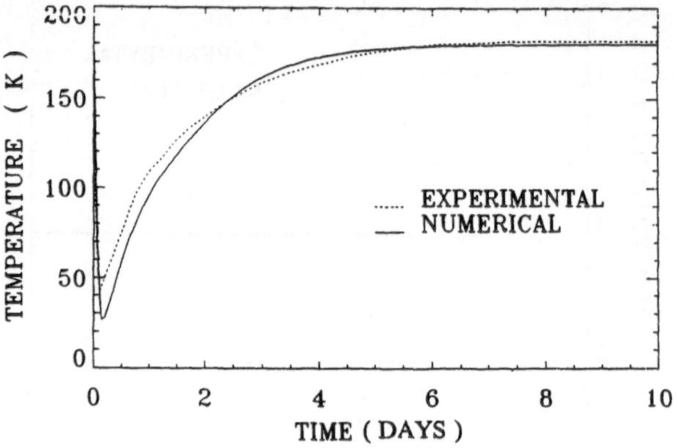

Fig. 5. Comparison of results for T_4, the warmest vapor cooled shield.

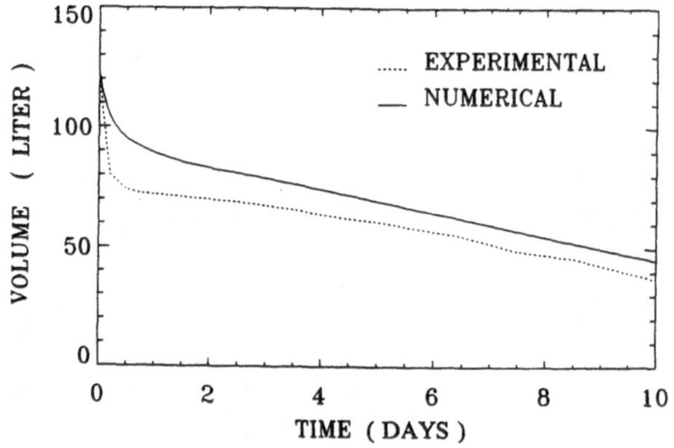

Fig. 6. Comparison of results for the liquid volume.

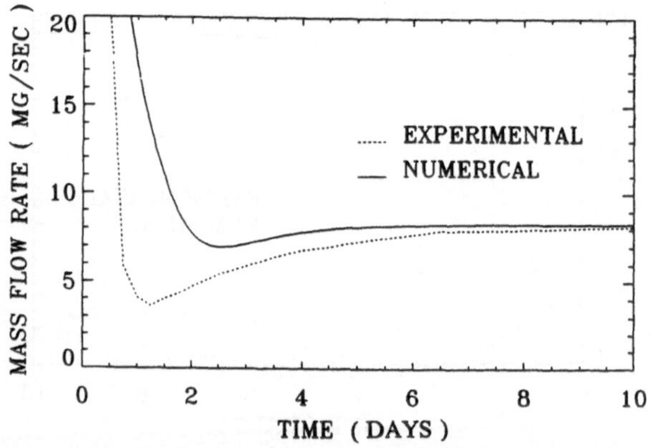

Fig. 7. Comparison of results for the mass flow rate.

CONCLUSIONS

The computer simulation of the time dependent behavior of the cryogenic system correlates well with the actual system. There were several factors which were included in the simulation that had contributing effects. For example, the effect of temperature dependence on specific heat and thermal conductivity were included. The results show general agreement except for the unpredicted subcooling observed in the liquid. The evaporation model is expected to be of major importance in determining the subcooling. Future research will focus on improved treatment of the evaporation model.

REFERENCES

1. L. Katz, J. Hendricks, and G. Karr, Spacelab 2 Infrared Telescope Cryogenic System, in: SPIE, Space Optics, 183:40, (1979).

2. G. Karr, J. Hendricks, E. Urban, L. Katz, D. Ladner, "Cryogenic Sub-System Performance of the Infrared Telescope for Spacelab 2", Proceedings of the Eighth International Cryogenic Engineering Conference, Genova, Italy, June 3-6, 1980, ICEC8.

3. E. Urban, D. Ladner, J. Jolley, M. Armstrong, J. Hendricks, G. Karr, Cryogenic Performance Testing of the Infrared Telescope (IRT) for Spacelab 2, in: Proceedings of 1983 Space Helium Dewar Conference, p. 29-51, UAH Press, Huntsville, AL, 1984.

DESIGN OF THE SOLID CRYOGEN DEWAR FOR THE NEAR-INFRARED CAMERA AND MULTI-OBJECT SPECTROMETER

Rodney L. Oonk

Electro-Optics/Cryogenics Division
Ball Aerospace Systems Group
Boulder, Colorado

ABSTRACT

Ball is developing a multi-purpose, second-generation Hubble Space Telescope (HST) instrument for imaging and spectroscopy in the 1 to 2.5 μm wavelength region. The Near-Infrared Camera Multi-Object Spectrometer (NICMOS) is unique since it is the only HST instrument operating in the near-infrared and cryogenically cooled. The NICMOS detector arrays are cooled to 58 K by a solid-nitrogen (SN_2) dewar with a predicted lifetime of nearly 5 years. To obtain this long lifetime, a hybrid cooling approach using thermo-electric coolers (TECs) is employed to reduce the parasitic heat load on the SN_2. The design features used to promote long life, the predicted lifetime improvements provided by the TECs, and the performance degradation in the event of TEC failure(s) are discussed.

INTRODUCTION

Long-life spaceborne dewars have been developed and flown in support of orbiting instruments for some time. Two examples are the dewars used in the Cosmic Background Explorer[1] and the High Energy Astrophysics Observatories[2]. These dewars have achieved lifetimes of about one year by using state-of-the-art insulation, tank support and low-conductance cabling technology. The next generation spaceborne cryogenically cooled instrument systems are being developed with lifetime requirements of 4 to 5 years; one of which is the Near-Infrared Camera Multi-Object Spectrometer (NICMOS), a second-generation instrument for the Hubble Space Telescope (HST). Mass and envelope constraints imposed by HST mandate the use of some sort of active cooling approach for NICMOS, either in combination with stored cryogens or alone. However, the limited-power and low-vibration requirements preclude the use of the mechanical refrigerators. A hybrid cryogenic system approach, consisting of a high-performance, solid-nitrogen (SN_2) dewar combined with thermo-electric coolers has been baselined for NICMOS. The TECs extend the predicted lifetime of the dewar by nearly a factor of two.

REQUIREMENTS

Table 1 summarizes the NICMOS performance requirements for the cryogenic subsystem. Also shown is the predicted performance for the baseline design.

The interface and constraint requirements for the NICMOS cryogenic subsystem are derived from four sources: (1) the NICMOS instrument, including the other subsystems, (2) the HST, (3) the Space Transportation System (STS, i.e., the shuttle), and (4) integration and testing. Table 2 summarizes the requirements derived from the NICMOS instrument and HST.

Table 1. NICMOS Performance Requirements

ITEM	REQUIREMENT	PREDICTED BASELINE PREFORMANCE
Detector temperature	Cool camera and spectrometer detector arrays to ≤ 62 K	58 K
Detector temperature stability	≤ 200 mK during 1000 second integration; ± 2K over duration of mission	Meets
Cold Stop Temperature	Cool cold stops to ≤ 160 K	155 K
Detector Power Dissipation	< 1 mW during operation, 0 mW during idle	Meets
Lifetime	≥ 4 years	4.9 years

The interfaces and constraints set by the STS and integration and test activities cover the areas of launch loads, safety, venting in the payload bay and ground operations. NICMOS must be designed to withstand the same launch loads as were the first-generation HST instruments. It must meet the safety requirements imposed by the shuttle. STS mission operations require that the cryogenic subsystem can be left unattended for at least five days to cover the final days before launch through launch and deployment.

The driving system requirements for the NICMOS cryogenic subsystem are the lifetime, mass, power and vibration.

SYSTEM TRADES

A trade study was made to select the preferred cryogenic subsystem approach for NICMOS. All possible approaches for the cryogenic subsystem were organized into a design option (or trade) tree. The options were grouped so that approaches that could not meet the performance, interface, or constraint requirements were identified at a high level on the option tree. This allowed rapid convergence on the feasible approaches.

A combination of TECs and radiators would be an ideal approach for NICMOS, since they are low in weight and have virtually no vibration. However, TECs and radiators cannot reach the 58 K detector operating temperature by themselves. TECs are attractive however, since their solid-state construction and the absence of moving parts makes them inherently reliable. Additionally, the use of redundant cooling elements shows promise for further improving TEC reliability[3].

Figure 1 summarizes the results of the system trades. Pure open cycle cryogen systems are too heavy and can only be used in a hybrid approach. Open cycle-TEC/radiator hybrid approaches were therefore traded against closed cycle mechanically driven refrigerator approaches. None of the mechanical refrigerator approaches were found to be viable for NICMOS. Of the numerous mechanical refrigerators currently under development, only the

Table 2. NICMOS and HST Interface and Constraint Requirements

ITEM	REQUIREMENT
Mass	164 kg
Input power available	30 W during operation, 25 W during idle
Maximum vibration	Cannot cause an HST pointing error > 0.0007 arcsec
Thermal	270 ± 1 K main shell temperature 255 K HST radiative sink temperature during operation 232 K HST radiative sink temperature during idle

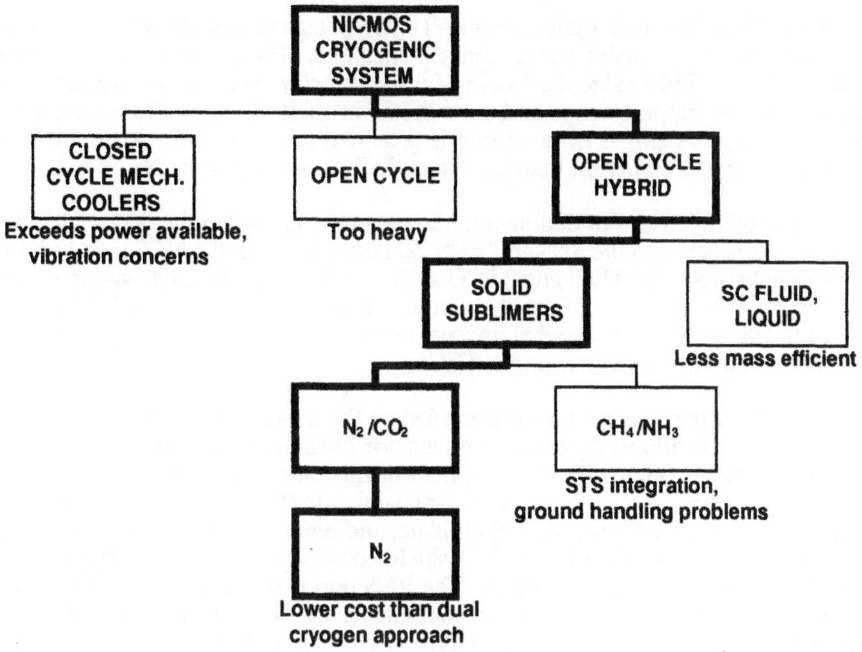

Fig. 1. Trade study results.

British Aerospace (BAe) cooler was considered mature enough. To meet the very stringent induced vibration requirement for HST-borne instruments the BAe cooler would have to be run in an opposed pair configuration, and consequently require about 80 W of input power. This far exceeds the modest (≤30 W) amount of power available to the NICMOS cryogenic subsystem.

Trades were made to select the best combination of cryogens to use for NICMOS. A solid subliming approach was selected because it was more mass efficient than any supercritical or liquid approach. A dual cryogen N_2/CO_2 approach offered nearly the same lifetime as did a CH_4/NH_3 approach and could be more readily operated at 58 K. It avoided significant shuttle integration and ground handling problems associated with hazardous cryogens, and was selected as the baseline approach for NICMOS. Recently an effort was undertaken to reduce the scope and cost of the NICMOS instrument. As part of this effort, the lifetime requirement for the dewar was also reduced from five to four years. This led to selecting the simpler single cryogen N_2 approach (over the dual cryogen N_2/CO_2 approach) as the current baseline for NICMOS.

DESIGN

Figure 2 shows the baseline design of the NICMOS cryogenic subsystem. The 58 K SN_2 stage of 120 liters surrounds a detector chamber containing the three detector arrays. The cryogen tank contains an aluminum foam matrix to retain the SN_2. The high-conductivity foam keeps the SN_2 in good thermal contact with the detector chamber providing a stable detector operating temperature over the entire life of the mission. Surrounding the SN_2 stage is a vapor-cooled shield (VCS). The VCS serves the dual purpose of reducing the parasitic heat leak to the SN_2 and cooling the optical beam filter wheels and cold stops. Surrounding the VCS are two additional shields, each cooled by a pair of thermo-electric coolers (TECs). The four solid-state TECs use the Peltier effect to produce cooling using the 25-30 W of power available. The TECs are soldered to radiators located on the outboard panels of the NICMOS enclosure and connected to the shields via permanently attached flexible cold fingers. They share a common vacuum with the dewar via low-conductivity tubes which minimize heat flow between the radiators and the temperature-controlled main shell. The main shell provides the vacuum close-out for the dewar and mounts to the NICMOS main optical bench. The SN_2 stage is supported from the main shell by six fiberglass/epoxy tension straps, three at each end. These tension straps also support the VCS, TECI and TECO shields. Wiring for the detector arrays and dewar housekeeping is routed along the support straps and exits the main shell via hermetic connectors on the forward girth ring. A single fill/vent line is provided for SN_2 tank. Also provided is an internal heat exchanger to permit freezing and periodic recooling of the N_2.

A total of 124 layers of double-aluminized mylar is used in the dewar, distributed as follows: 40 layers between the SN_2 and VCS, 50 layers between the VCS and the TECI, and 17 layers each between the TECI and TECO shields and the TECO shield and the main shell. Optimizing the thermal design of the dewar resulted in placing most of the MLI in the colder portion of the dewar. This somewhat unusual result is due to the relatively free cooling (in terms of mass impact) available from the TECs.

A multi-purpose access port is provided on the forward dome of the dewar (the left end of Figure 2). It allows access to the detector chamber containing the three detector arrays. It also accomodates the three optical beams and their associated baffling, filter wheels, and cold stops. Access to the detector chamber is obtained by successive removal of the port plates, starting with the main shell plate, and removal of the MLI blankets between each port plate. The port plate at the TECI shield contains a sapphire window to reduce the radiation heat leak into the colder stages. The VCS port plate contains the filter wheels and cold stops. The filter wheels are driven by motors located on the main shell port plate (inside the dewar vacuum space) via low-conductivity drive shafts. One of the filter positions for each optical beam contains a low-emissivity blank-off plate. Besides providing a means to check the detector background noise, the blank-off plate serves as a thermal shutter. During non-observing times (i.e., during idle and while slewing during operation) the blank-off plates will be positioned in the optical beams, reducing the radiation onto the VCS and the SN_2 stages from the warmer stages of the dewar.

Figure 3 shows the arrangement of TE cooling used in the NICMOS dewar. (For clarity only one of each pair of inner and outer TEC/radiator combinations is shown.) The TE-cooled shields substantially reduce the heat load on the VCS. Averaged over the

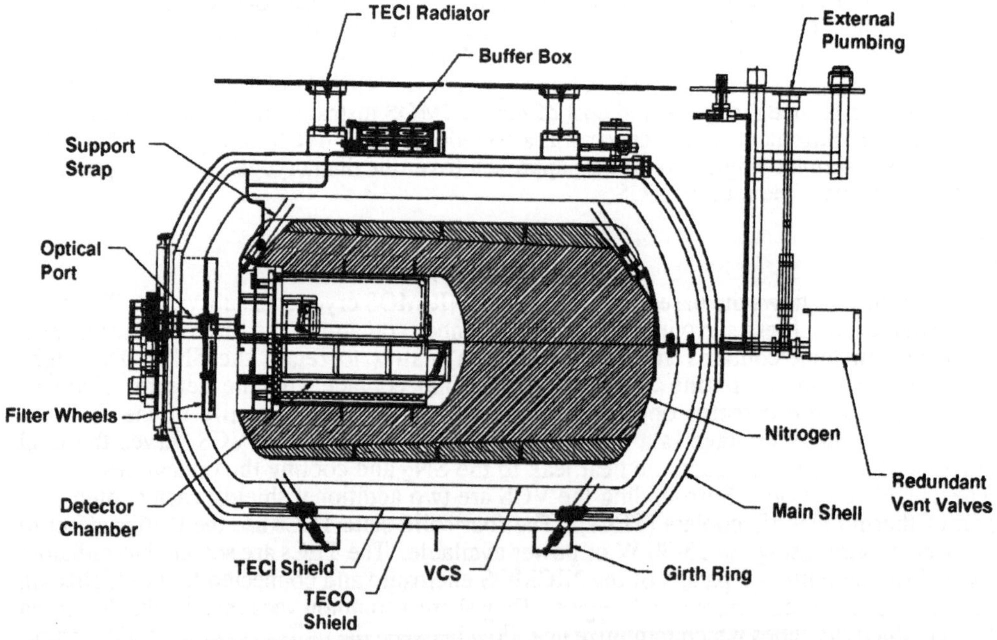

Fig. 2. NICMOS cryogenic subsystem baseline design.

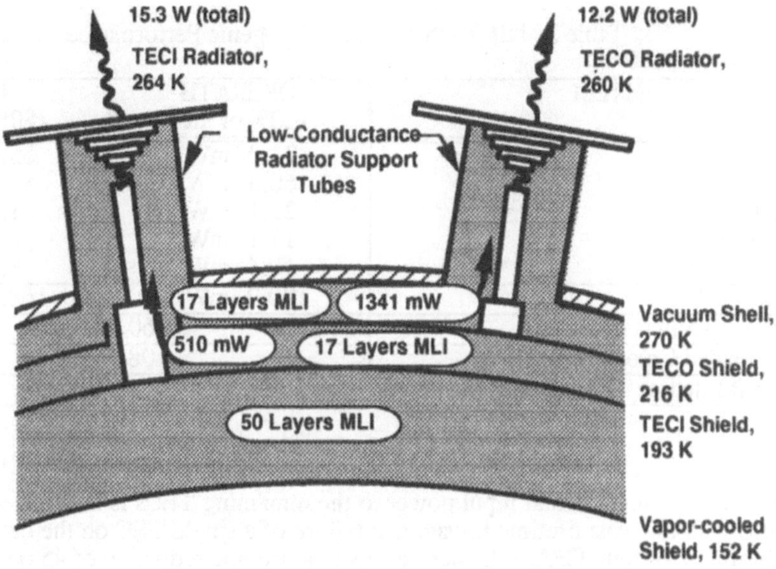

Fig. 3 - Thermo-electric cooling arrangement.

operation and idle modes of NICMOS, the TECs extract a total of 1.85 W from the dewar. This substantially reduces the parasitic heat load on the VCS, and ultimately on the SN_2. Since the cooling capacity of the TECs is non-linear, the temperatures of the TE-cooled shields are not very sensitive to variations in heat loads.

The TEC cold tips are connected to the dewar shields via flexible cold fingers. The designs are nearly identical and consist of copper spindle discs soldered to the TEC cold tips connected in-turn to a second set of discs by fine-wire flexible thermal links. These flexible link assemblies are connected to copper tubes and bolted flanges connected the aluminum shield mounting flanges. The conductive length of the wires soldered to each spindle of the flexible thermal links are 0.5 inch. However, the TEC cold fingers will be installed so that the spindle to spindle distance is 0.4 inches, to allow relative motion between the shields and TEC cold tips in all directions. The design of the flexible thermal link is based on a similar design used for TECs on other Ball programs. Experience has shown this type of link adequately protects the TEC cold tip from side and tensile loads, and can readily be built.

PERFORMANCE

The NICMOS dewar lifetime was calculated using a 26-node thermal network model. This model sizes the dewar to meet the 164 kg mass constraint and while maximizing the lifetime in the process.

The radiative sink viewed by the TEC radiators is warmer during the on-orbit operation than during the on-orbit idle times. This results in higher shield temperatures and a higher N_2 depletion rate during the operate mode. Table 3 shows the predicted on-orbit heat leak to the nitrogen as a function of operating mode. Also shown is the sublimation rate, the loaded SN_2 mass and the on-orbit lifetime. The time-averaged shield temperatures are shown in Figure 3.

A modified version of the dewar thermal model was used to determine the lifetime attainable by a dewar without TE-cooled shields. Holding the overall dewar mass at 164 kg, the predicted lifetime is 2.7 years.

In the event of a TEC failure, Figure 4 shows the lifetime impact of single and multiple TEC failures as a function of time after launch. The analysis assumed a failed TEC

Table 3. NICMOS Baseline Cryogenic Performance

ITEM	OPERATE (20% of life)	IDLE (80% of life)
MLI	71.0 mW	65.4 mW
Supports	60.3 mW	57.0 mW
Wiring	22.1 mW	21.1 mW
Plumbing	17.1 mW	16.2 mW
Optical Ports	10.4 mW	8.9 mW
Total Heat Leak	180.9 mW	168.6 mW
Average Sublimation Rate	60.4 g/day	
Fully-loaded Cryogen Mass	108.5 kg	
On-orbit Lifetime	4.9 years	

provides no cooling and that input power to the remaining TECs is unchanged. The failure scenario with the least lifetime impact is a failure of a single TEC on the outermost shield. The failure of all four TECs at launch results in a lifetime reduction of 45 percent, from 4.9 years to 2.7 years. While this lifetime reduction is severe, many of the science objectives of NICMOS could still be met.

Performance trades showed that for credible TEC failures thermally disconnecting the TEC cold tip provided little performance benefit. For a failure of a single TEC on either shield, the heat leak across the failed (but still connected) device is about 400 to 800 mW. Under these conditions the shield temperature rises, but the non-linear cooling capacity of the remaining TEC limits this temperature rise to between 10 and 20 K. If both TECs on either shield fail, the lifetime impact of not disconnecting the failed TECs is greater but softened by the TEC radiator temperatures being somewhat colder than the main shell. For a complete failure of all four TECs, the lifetime impact of not disconnecting the TECs is less than 0.1 years, again because the radiator temperatures are colder than the main shell.

CONCLUSION

Development is proceeding on NICMOS, a long-life, hybrid cryogenic system for a second-generation Hubble Space Telescope instrument. The hybrid cryogenic system approach of a high-performance, SN_2 dewar combined with thermo-electric coolers is used to cool the NICMOS detector arrays, cold stops and optical filter elements. The TECs take advantage of the power available to the cryogenic system to extend the predicted lifetime of the dewar from 2.7 to 4.9 years. Redundancy is used in the internal design of the TECs to minimize the effects of junction failures on TEC performance. Also, redundant TECs are used in the approach to minimize the lifetime impact of a totally failed TEC. A single failure

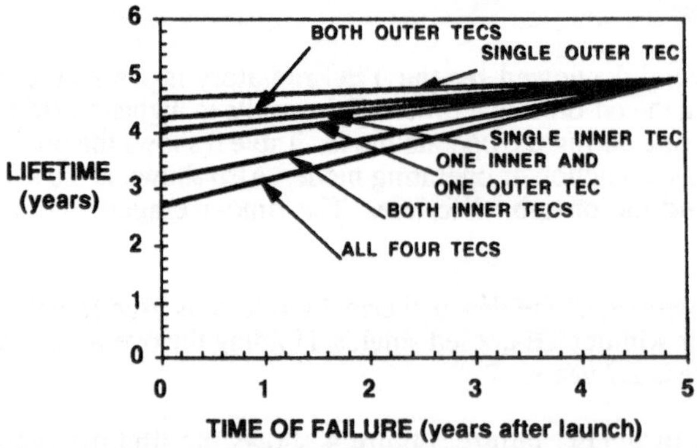

Fig. 4. Lifetime impact of TEC failures.

of one TEC at the start of the mission reduces lifetime by 0.6 years or less while a complete failure of all four TECs at mission start reduces the dewar lifetime to essentially the same as that of a dewar not employing any TECs.

ACKNOWLEDGEMENT

This system discussed in this paper is under development at Ball Aerospace/Electro-Optics and Cryogenics Division. The NASA Goddard Spaceflight Center is funding development of the NICMOS instrument through the University of Arizona under contracts NAS5-30008 and NAS5-31289.

REFERENCES

1. R. A. Hopkins and S. H. Castles, Design of the Superfluid Helium Dewar for the Cosmic Background Explorer, "Proceedings of SPIE," Vol. 509 (1984).

2. W. A. Mahony, J. C. Ling and A. S. Jacobson, The HEAO 3 Gamma-Ray Spectrometer," Nuclear Instruments and Methods," Vol 178 (1980).

3. D. A. Johnson, Improvements in Reliability of Thermoelectric Coolers through Redundant Element Design, in: "Proceedings of SPIE," Vol. 1044 (1989).

DESIGN OF A 2 YEAR LIFETIME SOLID CRYOGEN COOLER

Luo Hui-Yun, C.Z.Wu

LanZhou Institute of Physics
LanZhou, P.R. China

ABSTRACT

Long-lifetime solid cryogen coolers have been utilized in recent years to provide long-term cooling for detector instruments in orbit. In the design of a long-lifetime solid cryogen cooler, optimization studies of the parameters are conducted with the use of a computer program to minimize the weight and volume of the cooler. The lifetime of the cooler is limited by the flow of heat into it through multilayer insulation, support straps, plumbing and detector. Experience has shown that a major source of parasitic heat flow is the multilayer insulation and the support straps. The effect on the total system weight of insulation thickness, apparent thermal conductivity, support strap dimensions, external shell temperature, vent line size and MLI material are investigated. Additional tests showed that the penetration haet leaks through MLI are appreciable if 12 support straps are used. The optimization of the insulation sleeve of the tension strap is performed. The primary specifications of the cooler are a detector operating temperature of 85 K, an optical elements operating temperature of 151 K, a heat load of 80mW for the primary coolant, (methane), 100mW for the secondary coolant, (ammonia), a lifetime of 2 years, and a system weight of 85 kg.

INTRODUCTION

Astrophysics and earth science missions require increasing numbers and types of cryogenic coolers. These coolers have been used to increase the sensitivity and signal to noise ratio of detectors. The lifetime of solid cryogen coolers are limited by the flow of heat into it through MLI, support straps, plumbing and detector. However, for mechanical refrigerators and radiators, variations from the design heat loads result in operating temperature changes, but have no lifetime impact. Solid cryogen coolers, as one of the principal types of cooling systems, have recently been employed in various missions. Six coolers of this type have been flown on spacecraft to provide 60 K cooling for 7 to 11 months and have demonstrated excellent temperature stability, simple orbital operation and high reliability. It is difficult to accurately predict the heat loads for both sensor and cooler and the resulting lifetime when using solid cryogen

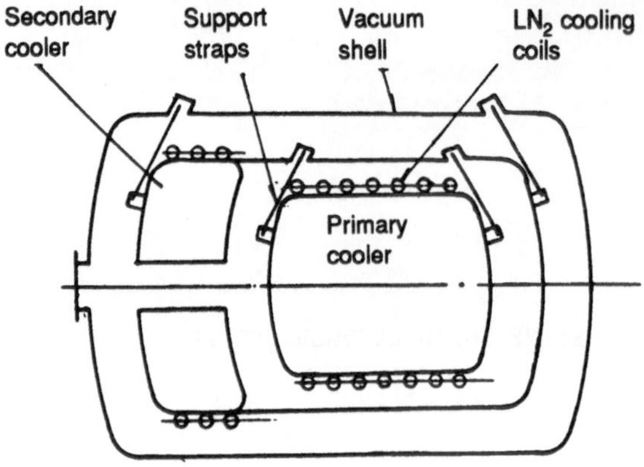

Figure 1. Configuration of a two-stage solid cryogen cooler.

coolers. The ratio of actual life to design life of the six coolers launched are in the range of 0.6 to 0.8[1]. The difference between actual and design life results from errors in the data base (primary thermal conductivity) of multilayer insulation and from degradation of the MLI by penetrating members such as support straps and fill and vent tubes.

This paper describes the design of a 2-year lifetime solid cryogen cooler, the effect of various parameters on cooler weight and the configuration of the cooler. The primary specifications of the cooler are detector operating temperature-85K, detector heat loads-80 mW, heat load of optical elements at 151 K-100 mW, lifetime-2 years, outer shell temperature 293 K, total weight-85 kg, and orbital temperature stability ±2 K.

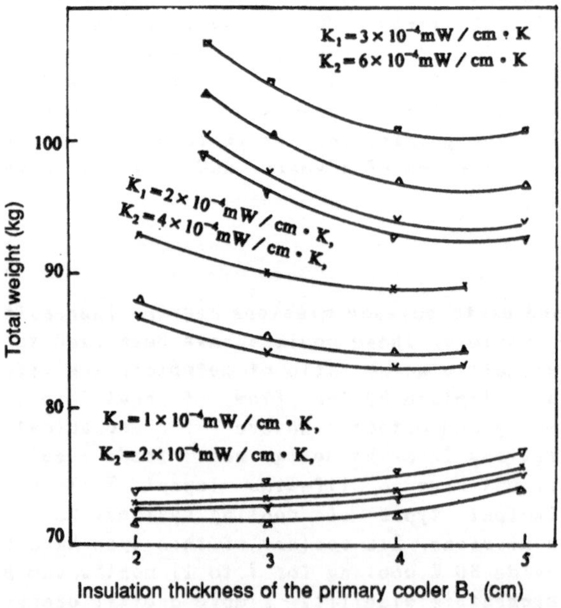

Figure 2. Effect of insulation thickness on cooler weight. ×, □, △, v, ○, ▽ are insulation thicknesses of the secondary cooler, 2 cm, 2.5 cm, 3 cm, 4 cm, 4.5 cm, and 5 cm, respectively.

Table 1 Assumptions for Cooler Tradeoff Studies

Peak accelerations	
axial component	10g
lateral compoonent	10g
Insulation type	Double silk net/Double Aluminized Mylar for primary stage Polyester net/DAM for secondary stage
	20%
Lifetime contingency	80mW for primary stage
Instrument heat loads	100mW for secondary stage

DESIGN APPROACH

In order to satisfy the design requirements, a comparison of the temperature ranges, latent heats of sublimation, and densities of various cryogens suggested a dual-stage cooler utilizing solid methane as the primary cryogen and solid ammonia as the secondary cryogen. A schematic of this cooler is shown in Fig 1. The primary cryogen provides the desired sensor temperature and the secondary cryogen was used as a guard at 151 K to intercept heat from the outer shell and to cool the optical elements.

The trade-off studies of the cooler were conducted utilizing a computer program which involves a deatailed calculation of predicted heat loads to both the primary cryogen and the secondary cryogen and which optimizes the cooler for minimum weight, given the lifetime desired. The sources of heat leak to the cryogen are the MLI, support straps, neck tube radiation, strap sleeve, LN_2 and vent lines. The design parameters are insulation thickness and conductivity, support strap dimensions, ammonia temperature, external shell temperature and the length to diameter ratio

Table 2. Optimum MLI Thickness for Various Conductivity and Lifetime

life(year)	1			2			3		
K_1	$1\cdot10^{-4}$	$2\cdot10^{-4}$	$3\cdot10^{-4}$	$1\cdot10^{-4}$	$2\cdot10^{-4}$	$3\cdot10^{-4}$	$1\cdot10^{-4}$	$2\cdot10^{-4}$	$3\cdot10^{-4}$
K_2 / parameters	$2\cdot10^{-4}$	$4\cdot10^{-4}$	$6\cdot10^{-4}$	$2\cdot10^{-4}$	$4\cdot10^{-4}$	$6\cdot10^{-4}$	$2\cdot10^{-4}$	$4\cdot10^{-4}$	$6\cdot10^{-4}$
MLI thickness on primary cryogen tank(cm)	2.5	2.5	3	3	3.5	4	3.5	4.5	5
MLI thickness on secondary cryogen tank(cm)	2	3	3.5	3	4	5	3.5	5	6
Total weight (kg)	39.80	44.61	49.01	71.67	82.73	92.71	104.46	122.53	138.76

K_1-insulation conductivity of MLI for methane tank(mW/cm.K)
K_2-insulation conductivity of MLI for ammonia tank(mW/cm.K)

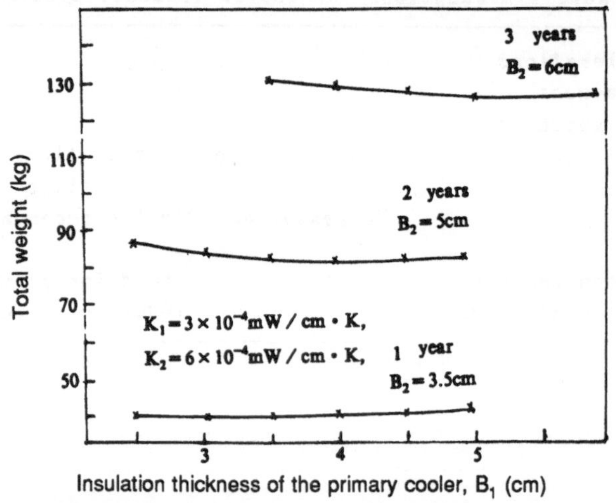

Figure 3. Effect of insulation on cooler weight for various lifetimes. B_2 is the insulation thickness on the secondary cooler.

of the primary coolant vessel. The effect on performance of these parameters was investigated for various lifetimes and cooling loads. In these studies, the major assumptions are summarized in Table 1. The principal results of the tradeoff studies are shown in Figs 2-6. Figures 2 and 3 show the effeot of insulation thickness on the total weight for various insulation conductivity and lifetime. The optimum insulation thickness for various insulation conductivities and lifetimes are summarized in Table 2 where the optimum MLI thickness varied with both the insulation conductivity and lifetime. Fig. 4, Fig. 5 and Fig. 6 indicate that the effect of ammonia temperature, external shell temperature and lifetime on the total weight. The predicted heat load of this cooler, after a number of trade off studies, are given in Table 3.

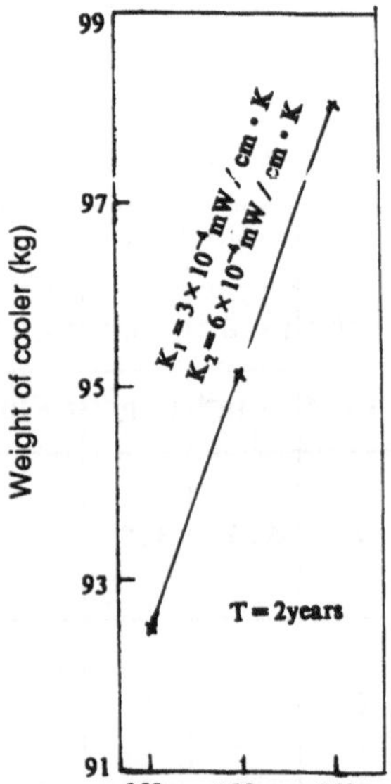

Figure 4. Total weight vs. secondary cryogen temperature.

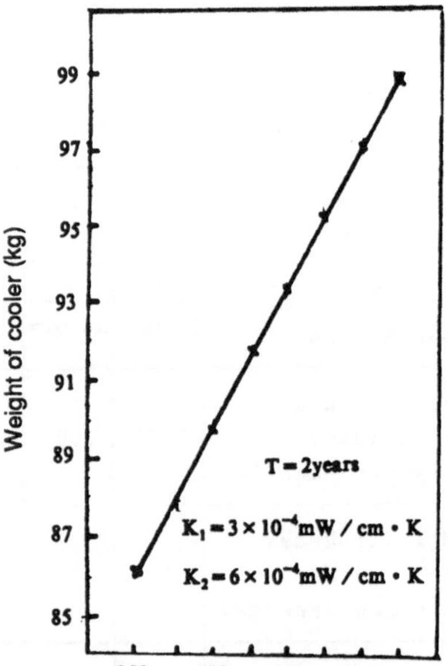

Figure 5. Total weight vs. external shell temperature.

Table 3. Predicted Heat Load Summary and Characterics of Cooler

Source	Heat load (mW)	
	Ammonia	Methane
MLI	288.9	41.6
Support strpas	22.9	12.5
Suppotr strap sleeve	195.3	49.4
Radiation	38.5	36.6
Vent gas	-38	
LN_2 and vent lines	12.5	13.4
Detector, optical elements	100	80
Total	620.1	233.5
Coolant weight(kg)	17.3	27.9
Total weight (kg)	82.7	
Dimensions of cooler (cm)	ϕ64×97	

CONFIGURATION

The components of the cooler are the containers of solid methane and solid ammonia, a cooled shield grounded to the ammonia, vent gas cooled shield, tension straps, multilayer insulation (MLI) around the containers and straps, LN_2 cooling coils, vent lines, outer shell and the shrink-fit assembly.The containers were supported by tension straps from the external

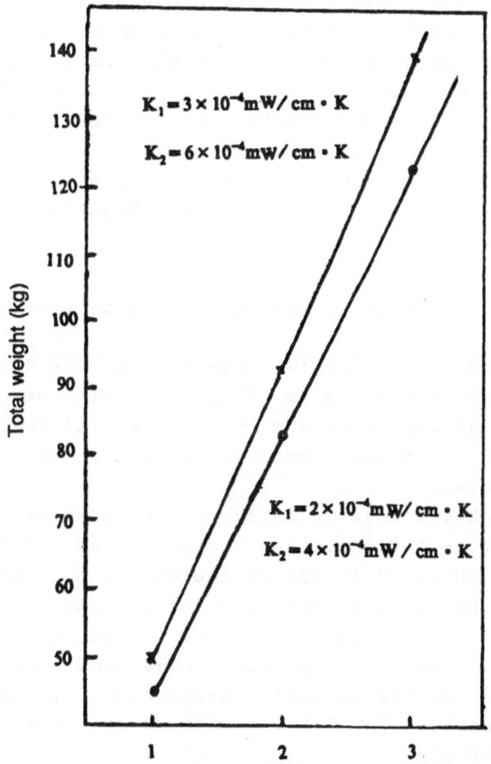

Figure 6. Total weight vs. operating lifetime.

vacuum shell and a cooled shield ground to the ammonia. Internal heat exchangers in the containers were used to achieve solid cryogens during fills and to maintain isothermal conditions during operation. The straps were oriented in a zig-zag pattern at an angle of 35 degrees from the plane normal to the cooler axis. The MLI insulated the methane tank from the ammonia tank and its grounded shield and insulated the ammonia tank and shield from the outer shell.A slitted MLI sleeve was used between the strap and the MLI around the tanks. A trade-off study showed that a 3.5cm MLI thickness for methane tank and a 4cm MLI thickness for ammonia tank were optimum. Vent lines were required to remove the sublimation gases to space. In order to maintain the required temperature, the vent lines must be quite large for ammonia tank. A convoluted Teflon vent line has been used in this cooler to reduce heat leak associated with this line. A vent gas cooled shield between the outer shell and ammonia tank was installed to minimize heat leak to the ammonia tank. Two shrink-fit assmblies connected the detector to the methane tank and the thermal guard and the optics to the ammonia tank.

SUPPORT STRAP DESIGN

The support strap system is one of the key components of the cooler. Since the heat conducted through the support system is a major part of the overall heat load to the cryogen. Therefore, it is important to design the support system so as to minimize heat flow into the cooler subject to constraints on minimum allowable natural frequency of the cooler. maximum allowable stress and no slacking of any strap during launch. Three types of support structure of cooler are high strength fiberglass-epoxy tube, tension strap sysem, and passive orbital disconnect struct. [2,3] The former is used for miniature coolers whose weight is less than 30 kg.The last two have been used for spaceborne dewar in recent years. Experience show that cryogen loss rates are very similar for both systems. The PODS system is more complex than straps and a flight worthy design does not yet exist. The performance of a flight system using straps is more predictable than one using PODS. This cooler used a tension strap system made of fiberglass-epoxy.Based on the test data, the mechanical properties are shown in Table 4.

In design, the following constraint condition must be satisfied:

1) During launch the cooler is subjected to peak accelerations with a 10-g axial component and a 10-g lateral component. The maximum stress seen by any strap due to the inertial reaction of the tanks to the sum of these acceleration components must not exceed a allowable value.
2) During launch,the tension in the straps must be sufficient so that the g-load does not cause any strap to go slack. Therefore, the strain in each strap in the prelaunch state must be larger than the sum of the maximum axial and lateral strains during launch.
3) After filling and solidification with LN_2, a differential contractions between the outer shell and the supported mass occurred and so the support straps must be prestrained at the time of assembly. The total prestress must not exceed a allowable stress of 500 MPa.
4) The natural frequency of the support straps system must be less than 35 Hz.

Table 4. Strap Mechanical Properties

Tensile Strength(MPa)	1400
Modulus of Elasticity(GPa)	36
Fatigue strength(MPa)	700
Creep strength(MPa)	560

The straps shown in Fig 7 zig-zag between the outer shell and the ammonia shield and between the ammonia shield and the methane tank at an angle of 35 degrees from the plane normal to the cooler axis. This angle provides the same stiffness in both the axial and lateral directions. The strains of the methane tank straps and ammonia tank straps due to differential thermal expansion upon filling with cryogen can be calculated according to the formulae:

$$\text{methane tank straps} \quad e_{fill1} = ((\alpha_1(T_a - T_c)\frac{L_{s3}}{2} - \alpha_2(T_a - T_b)\frac{L_{s1}}{2})\sin\gamma$$

$$+ (\alpha_2(T_a - T_b)R_2' \cos\gamma\cos\theta - \alpha_1(T_a - T_c)R_1 \cos\gamma\cos\theta) / L_{eff1}$$

$$\text{ammonia tank straps} \quad e_{fill2} = \alpha_2(T_a - T_b)\,(\frac{L_{s2}}{2}\sin\gamma - R_2\cos\gamma\cos\theta) / L_{eff2}$$

in which α represents coefficient of thermal expansion and L_{eff} is the effective length of strap.

The total prestress in each strap required at the time of assembly is

$$\text{methane tank straps: } \sigma_{pre1} = E_1 e_{fill1} + T_1 / A_1$$

$$\text{ammonia tank straps: } \sigma_{pre2} = E_2 e_{fill2} + T_2 / A_2$$

in which T is the tension in strap. E is the Young's modules. A is strap cross-sectional area, e is the strains in straps.
T_1 and T_2 can be computed from the following equations:

$$T_1 = C \cdot E_1 A_1 (|e_{axial1}| + |e_{lateral1}|)$$

$$T_2 = C \cdot E_2 A_2 (|e_{axial2}| + |e_{lateral2}|)$$

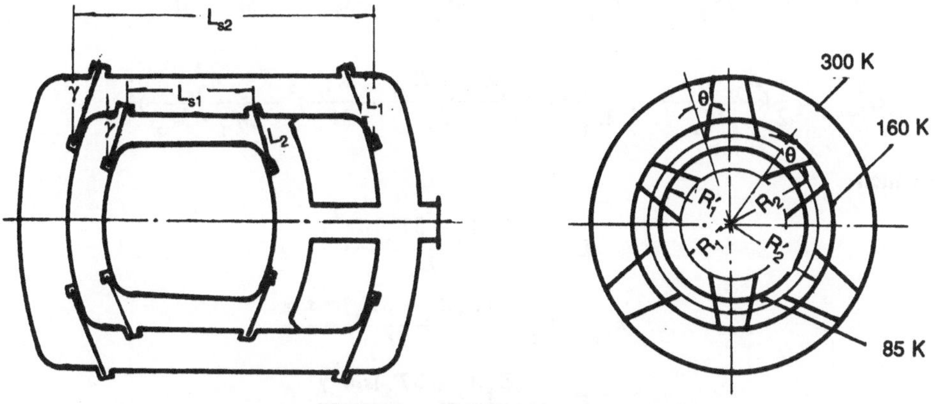

Figure 7. View of arrangement of support straps.

Table 5. Support Strap Configration and Characteristics

Cross-sectional area per strap (cm^2)	Primary tank	Secondary tank
	0.064	0.096
Length of strap(cm)	18	18
Pretension in each stap (kgf)	180	415
Primary resonant frequency (Hz)		
axial	34.8	
lateral	35.2	

in which

$$e_{axial1} = \frac{M_1 \cdot 10g}{12[(E_1A_1) + 3T_1]sin\gamma}$$

$$e_{later\ al1} = \frac{M_1 \cdot 10g}{6(E_1A_1 + 3T_1)cos\gamma}$$

$$e_{axial2} = \frac{(M_1 + M_2)10g}{12(E_2A_2 + 3T_2)sin\gamma}$$

$$e_{lateral2} = \frac{(M_1 + M_2)10g}{6(E_2A_2 + 3T_2)cos\gamma}$$

$$C = 1.1 \sim 1.2$$

The total tension in each strap at the time of assembly is

$$\textit{methane tank straps:}\ F_1 = E_1A_1e_{fill1} + T_1$$
$$\textit{ammonia tank straps:}\ F_2 = E_2A_2e_{fill2} + T_2$$

RESONANT FREQUENCIES

A dual stage solid cryogen cooler can be regarded as dual mass multifreedom spring system. In order to determine the resonant frequency, the axial and lateral stiffness K_x, K_y of each strap are calculated and inserted into the following equations for axial and leteral frequencies.

$$\omega^2_{axial} = \frac{1}{2}(\frac{K_{1x}+K_{2x}}{M_2} + \frac{K_{1x}}{M_1}) - (\frac{1}{4}(\frac{K_{1x}+K_{2x}}{M_2} - \frac{K_{1x}}{M_1})^2 + \frac{K^2_{1x}}{M_1 \cdot M_2})^{\frac{1}{2}}$$

$$\omega^2_{lateral} = \frac{1}{2}(\frac{K_{1y}+K_{2y}}{M_2} + \frac{K_{1y}}{M_1}) - (\frac{1}{4}(\frac{K_{1y}+K_{2y}}{M_2} - \frac{K_{1y}}{M_1})^2 + \frac{K^2_{1y}}{M_1 \cdot M_2})^{\frac{1}{2}}$$

in which

$$K_{1x} = \frac{(E_1A_1 + 3T_1)sin^2\gamma}{L_1}$$

$$K_{1y} = K_{1z} = \frac{(E_1A_1 + 3T_1)cos^2\gamma}{2L_1}$$

$$K_{2x} = \frac{(E_2A_2 + 3T_2)sin^2\gamma}{L_2}$$

$$K_{2y} = K_{2z} = \frac{(E_2 A_2 + 3T_2)\cos^2\gamma}{2L_2}$$

The calculated results of support strap are given in Table 5 for the zig-zag configration.

CONCLUSIONS

Efforts are under way to predict the lifetime of solid cryogen cooler accurately by two techniques. One is to improve the predictability, repeatability and wrap technics of MLI, another is to decrease the penetration heat load to the cooler from the twelve support straps. Trade off studies of mimimum cooler weight showed that the optimum insulation thickness for various lifetime and insulation conductivity is different, however, the insulation conductivity significantiy affected the cooler weight. The design of a 2-year lifetime solid cryogen cooler has been performed and thermal performance predictions of the support system using fiberglass-epoxy straps in zig-zag pattern have been calculated. The engineering model of cooler is being fabricated.

REFERENCES

1. T.C.Nast, Status of solid cryogen coolers, in " Advances in Cryogenic Engineering", Vol. 31, Plenum, New York (1986), P. 835.
2. David Bushwell, Optimum design of dewar supports, J.Spacecraft. 22:432 (1984).
3. R.A. Hopkins, D.A. Payne, Optimized support systems for spaceborne dewars, Cryogenics., 27:209(1987).

VIBRATION TESTING HELIUM CRYOSTAT FOR SPACE QUALIFICATION OF CRYOCOMPONENTS

M. Wanner, J. Wolf, H. Seifers

Linde AG
Process Engineering and Contracting Division
Hoellriegelskreuth, Germany

ABSTRACT

Liquid Helium Space cryostats like the Infrared Space Observatory (ISO) require cryocomponents which have passed qualification tests being more severe than the actual environmental conditions expected during ground operations, the launch and the mission. In order to perform such tests with the ISO components as well as with the focal plane units of the scientific instruments, a vibrational cryostat was built and successfully operated at 4.2 K and at acceleration levels of up to 22,5 g (sine). The specific design allows to vibrate components with up to 12 kg in all three axes without chang- ing their position with respect to the cryostat, i.e. in a single cold cycle.

The cold mounting plate has a standard interface and is mechanically decoupled from the rest of the cryostat structure. Vibration levels are controlled by cryogenic accelerometers mounted on the coldplate. The available cold volume for the test object is 400 mm in diameter and has a height of 400 mm.

LHe boil off can be significantly reduced during standby operation.

INTRODUCTION

The qualification of cryogenic components for space cryostats requires a comprehensive verification of all functional properties under simulated environmental conditions. Therefore the components have to be subject to the cryogenic temperatures and vacuum environment imposed by the space cryostat as well as to the simultaneous dynamic accelerations which are a result of the static and dynamic loads from the launch vehicle. The properties which have to be verified cover amongst others fatique effects, seat tightness of valves and electrical performance.

Table 1 Basic cryostat requirements

Available test volume (diam. x height)	400 x 400 mm
Maximum mass of test object	12 kg
Temperature of test object	4.2 K
Temperature of test object environment	77/4.2 K
Vacuum	$< 10^{-5}$ mbar
Accelerations (3 axes)	
sine	22,5 g
random	6 g RMS
Eigenfrequency of coldplate	> 100 Hz

Although a lot of information can be gained already by tests at LN_2-temperatures, a rigorous approach for LHe-space cryostats requires tests at LHe-temperatures.

This holds especially for the Infrared Space Observatory (ISO), which currently undergoes its qualification tests on payload level /1/. ISO is an ESA project to be flown on Ariane-4 in 1993. Aerospatiale is the prime contractor with MBB/Linde as subcontractor for the cryogenic payload. In order to be able to subject the ISO cryo-components to the required accelerations at LHe temperatures, Linde designed and built a vibration testing helium cryostat.

In addition to the cryocomponents of the ISO cryostat (especially valves, safety devices, the phase separator and the cryo-instrumentation) this cryogenic vibration facility is also capable of testing individual focal plane units of the scientific instruments of the payload.

REQUIREMENTS

The essential performance requirements for the vibration cryostat facility have been derived from the ISO-cryostat specification both for components and for experiments and are summarized in table 1:

Additional requirements resulted from the test conditions of the test objects prior, during and after the cold vibration test. For that purpose feedthroughs for electrical instrumentation and process lines e.g. pressure or vacuum lines have been included. Certain experiments require in addition optical windows for precise optical alignment tests in cold condition before and after the vibration.

A further design driver was the need to operate the cryostat independent from a specific shaker facility. Hence the cryostat should have standard interfaces and should withstand vibration in all three axis. By this way a vibration in different axes requires only a change of the cryostat position relative to the slip table without any impact on the test object. As a consequence the vibration tests in all three axes can be run immediately one after the other in short time.

DESIGN CONCEPT

Starting with these requirements various design concepts have been studied and evaluated. Some of the considerations shall be described in more detail:

The basic design consists of a rigid cryostat with a vacuum vessel, a LHe reservoir, a LN_2-reservoir for shield cooling and support elements which allow to transmit the acceleration forces from the slip table to the cold test object.

Unlike specifically designed space cryostats normal laboratory cryostat in general are not suitable for vibration for various reasons:

- The need to minimize the heat input to the LHe temperature level results in soft structures with a broad spectrum of low frequency eigenmodes. Such soft systems exclude the transmission of defined vibrations.

- Increasing the stiffness of the cryostat structure reduces the hold time of the cryostat significantly and limits the useful test period.

- Liquid motions of the LHe- resp. LN_2-reservoirs superimpose on the eigenmodes of the cryostat and of the test specimen and add by this way unpredictable damping effects to the vibration spectra.

In order to overcome these problems the following characteristics were considered in the design.

- A direct transmission of the vibrations in three axes to the test object is achieved by a stiff connection between the slip table and a so called coldplate. This coldplate represents the thermal and mechanical interface for all test units.

- In order to improve the hold time before and after vibration a standby configuration was considered. In this configuration the mechanical connection between the coldplate and the base plate is loosened and hence the transport of heat is significantly reduced.

- A mechanical decoupling of the remaining cryostat structure from the coldplate avoids crosscoupling of eigenmodes of the cryostat on the vibrations of the coldplate.

- A decoupling of liquid motions from the test object is realized by keeping the liquid volume in the coldplate small. Active cooling is provided by a forced flow from a LHe-reservoir.

- In order to control the actual vibrations of the test object cryogenic accelerometers are placed directly on the coldplate. The signals of these sensors are used as control parameter for the drive of the slip table. Excessive excitations which could lead to an overload of the system are limited by additional sensors on the slip table.

- An actively cooled liquid nitrogen shield which completely encloses the coldplate provides the required 77 K environment and reduces the heat input to the 4.2 K level. As an option the configuration allows to include a second shield at LHe-temperatures.

DESIGN

The configuration of the vibration testing cryostat is sketched in Fig. 1.

The basic structural element is the liquid helium cooled coldplate which is rigidly bolted to the cryostat base plate using glass-fibre reinforced distance rings as thermal barriers. The 12 tension bolts from the coldplate to the base plate extend through the vacuum vessel. Appropriate radial seals allow to adjust the tension forces e.g. after cooldown or prior to vibration from outside the cryostat without affecting the insulation vacuum.

If on the other hand the bolts are loosened additional spring washers lift the coldplate from the distance rings and reduce the mechanical and thermal coupling. Consequently a significant reduction in LHe boiloff is achieved.

The coldplate is made from Al-alloy (Al 5083) and is additionally weight optimized by pocket holes in the lower side. Sufficient stiffness of the structural assembly formed by the coldplate, the support rings and the base plate was verified by FE modelling and was subsequently confirmed by a vibration test of that assembly at ambient temperature prior to cryostat manufacture.

A standard interface between the test object and the coldplate is provided by 8 mm screw holes arranged in a regular square pattern with a distance of 40 mm.

The coldplate is cooled by a small channel at its circumference. Liquid helium supply is done by a small vessel with a capacity of 6 l and a thermosyphon loop. In this way inertia effects that might be caused by sloshing of the liquid volume in the coldplate are kept to a minimum. The

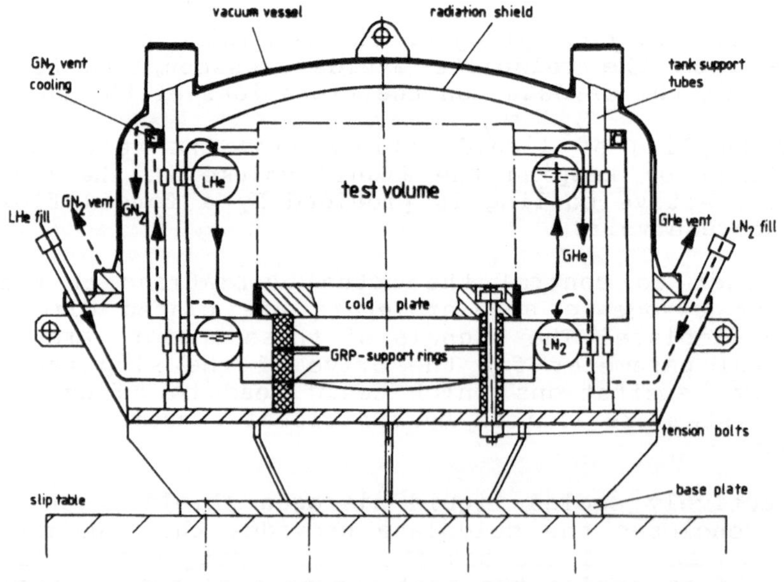

Fig. 1 Schematic configuration of the cryostat

Fig. 2 View into the opened cryostat (vacuum dome at right)

hexagonal cylindrical LHe-vessel is clamped to 12 glass fibre tubes. As can be seen in Fig. 2 these support tubes are arranged in a framework that extends from the bottom plate to fixtures at the top of the vacuum vessel. This construction helps to avoid crosscoupling of the vibrations of the individual parts of the cryostat with the coldplate.

In a similar way a LN_2-vessel of the same shape is also fixed, below the LHe-tank, to the glass fibre support tubes.

A reduction of the heat input to the coldplate and to the LHe-vessel is achieved by an effective routing of the vent gas of the two cryogens.

First the coldplate is completely enclosed by a radiation shield which is actively cooled by the vent gas from the LN_2-vessel. This vent gas is further routed to cool the tension bolts and the glass fibre support rings of the coldplate as well as the glass fibre tubes at appropriate distances and thus intercept the heat flow to the 4.2 K parts.

In addition, the helium boil off is used to precool the fixation points of the He-vessel, the tension bolts and the upper support ring of the coldplate.

The available cold test volume is represented by the diameter of 400 mm of the coldplate and the free height of 400 mm between the coldplate and the covering radiation shield.

Fill and vent lines for the helium and nitrogen supply as well as optional process lines to the test object enter the cryostat from the lower part which is made from stainless steel. In this way, fluid lines do not have to be disconnected during integration of the cryostat and can be left intact. Safety valves protect both tanks which are designed for 5,3 bar max. working pressure.

The vacuum dome is also made from Al-alloy giving a total mass of 260 kg for the whole cryostat. The overall dimensions are approximately 1100 mm diameter and 800 mm height as can be seen by Fig. 3.

Additional ports are arranged at the lower part of the vacuum vessel. These ports are used for evacuation pump and leak detector connections, pressure lines, for optical windows or for electrical feedthroughs.

The mechanical interface between the cryostat and the vibration table is provided by a stainless steel structure with 10 mm screws which are arranged to match the standard square pattern with a distance of 80 mm of most commercial slip tables in Europe.

The cryostat instrumentation consists of several cryogenic temperature sensors and accelerometers /2/ to monitor the coldplate and additional heaters at the coldplate and the cryogen vessels for rapid warm up after tests.

Fig. 3 Overall view of the cryostat together with the housekeeping control electronic

EXPERIENCE

A typical integration and test sequence of a cryocomponent consists of the following steps:

- Integration and mechanical/electrical checkout of the test object and of the cryostat.
- Cooldown of the cryostat with LN_2 and LHe within 4 hours.
- Mechanical/electrical checkout/leak test of the test object and the cryostat at 4.2 K
- Functional tests with the test object
- Transport of the cryostat from the laboratory to the shaker facility.
- Vibration tests including for each axis:
    - sine vibration at constant levels up to 22,5 g (0 - 200 Hz)
    - random vibrations at 6 g RMS (0 - 2000 Hz);
    - before and after each vibration test a resonance search is performed in the range of 0 - 2000 Hz to look for eigenfrequencies and to check for degradations of the test object between vibration runs which would show up in changes of the spectrum.
- Functional tests of the component between vibration runs.
- Mechanical/electrical integrity checkout/leak test of the test object and the cryostat at 4.2 K after vibration
- Transport of the cryostat back to laboratory
- Warm-up and deintegration

After integration and preparation of the cryostat usually all vibration tests can be completed within one working day. Liquid helium and liquid nitrogen refills are done only between the major vibration runs.

The liquid helium hold time amounts to 8 h in standby mode. This time reduces to 40 min. when the tension bolts are tightened and reduces further to appr. 10 min during sine vibration at maximum load. The measured eigenfrequency of the coldplate is 160 Hz and hence is sufficiently above the required 100 Hz.

To date this cryogenic vibration test facility has been successfully in service for 16 complete qualification tests of components and the Short Wavelength Spectrometer (SWS) of the ISO cryostat thereby demonstrating the soundness of the selected concept and the quality of manufacture.

ACKNOWLEDGEMENTS

This work was performed under ESA and MBB contract (MBB contract no. R 3141/3411 R)

REFERENCES

/1/ A. Seidel, Th. Paßvogel
The qualification status of ISO, Adv. Cryog. Eng. (1991)

/2/ Endevco Type 272 or 2272 A

DEVELOPMENT OF A FLIGHT QUALIFIED HEAT SWITCH FOR LN_2 TEMPERATURES

Juan Bascuñán, Terry Nixon, James Maguire

Janis Research Company
Wilmington, Massachusetts

ABSTRACT

A self contained heat switch with no moving parts, has been designed, assembled and tested. The device utilizes concentric cylinders separated by a small gap, filled or emptied of gas to achieve its switching action. The presence or absence of gas is controlled by an adsorption pump containing activated charcoal. During the development of the switch, different geometries were studied - multiple and single gaps, and also different materials - copper, aluminum. Prototypes were built and tested with different gases - helium, nitrogen, argon, neon.

As a result of this study, two gas-gap heat switches with ratios of "ON" to "OFF" conductances of about 600 were manufactured, thermally tested and flight qualified. They will be part of a cryogenic heat pipe experiment, to be conducted in a future shuttle mission.

This paper describes the technical approach and, thermal and structural analyses used in the design of the heat switch, as well as, the results of the thermal and environmental testing.

INTRODUCTION

In low temperature experiments, it is frequently desirable to be able to thermally decouple the instruments from the cryogenic cooler. Some of the advantages are: 1) to reduce close-cycle refrigerator power usage during dormant instruments periods, 2) periodic warm-up of instruments, 3) to reduce parasitic heat loads, 4) to reduce cryogen consumption during "off" periods allowing an extended lifetime.

Various types and variations of cryogenic switches to perform this decoupling have been designed in the past[1,2]. Some operate on the same gas-gap principle as the switch described in this paper. The use of ^4He exchange gas is the oldest and simplest method known[1], with the switching action resulting from the presence or absence of the gas.

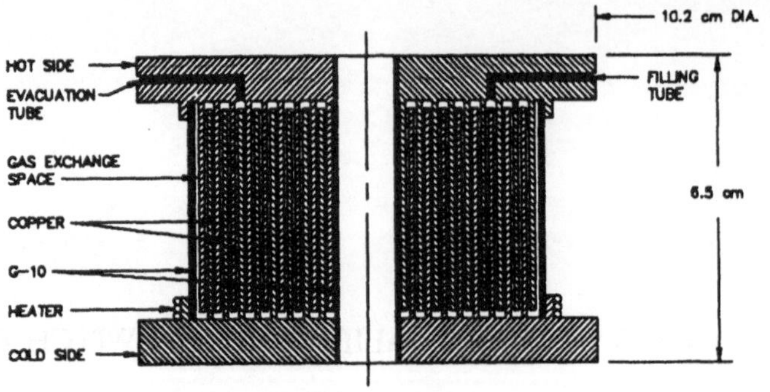

Figure 1. Schematics of the switch.

I. COPPER MULTIFINGERED SWITCH

1. Description

The primary design requirements of the switch are summarized in Table 1. A schematic of the switch is shown in Fig. 1. The "cold" end of the switch is thermally grounded to the cooling system (a LN_2 cooler in this case), while the "hot" end of the switch is thermally grounded to the experiment. The "hot" and "cold" copper ends of the switch are separated by an epoxy fiberglass (G-10 CR) tube, which connect the switch halves, provide alignment, retain the exchange gas and provide mechanical support. The two cylindrical halves of the switch are separated by small gaps across which the exchange gas effects the transfer of energy during operation. The width of these gaps are a key parameter in the switching ratio. The copper surfaces are polished to minimize radiative coupling in the switch.

To determine the number of fins and also the number of gas-gaps, needed to achieve an "ON" conductance of 2.5 W/K, with the given space and for a particular gas, a computer model was developed to optimize the design variables. Based on the analysis, it was decided to use a gas-gap of 0.025 cm and, fins of 4.45 cm height and 0.152 cm thick. The gaps could be filled with gas through the copper capillary tube emerging from the "hot" side and communicating with a gas bottle and a valve system necessary to operate the switch.

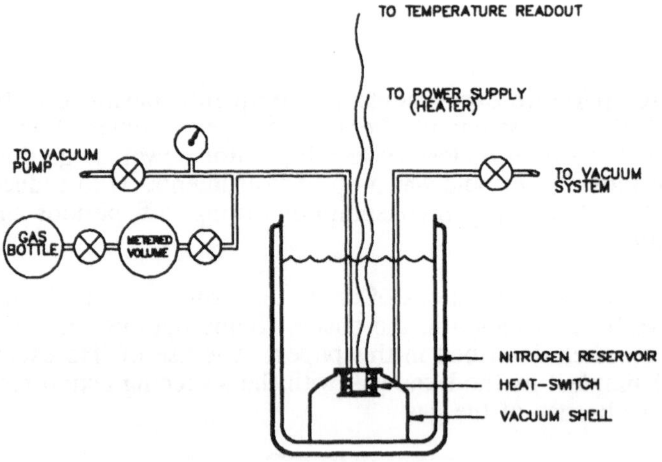

Figure 2. Experimental setup.

Table 1. Primary Design Requirements

Operating Temperature	65 - 77 K Cold End 250 - 300 K Warm End
"ON" Conductance	2.5 W/K
"ON" Heat Flow	20 W maximum
"OFF" Conductance	$\leq$ 0.002 W/K
"OFF" Heat Flow	$\leq$ 0.5 W w/warm end @ 250K $\leq$ 0.75 W w/warm end @ 300K
Weight	1.47 Kg
Diameter	7.62 cm maximum
Overall length	7.62 cm maximum

The final switch geometry yielded a switch weight of 2.3 Kgs and a theoretical "OFF" conductance of 0.009 W/K.

2. Experimental Measurements and Results

The experimental setup is depicted schematically in Fig. 2. The switch was mechanically and thermally anchored to a copper plate of a LN_2 dewar and within its vacuum isolation. A heater and a silicon diode thermometer were attached to the "hot" end of the switch, another silicon diode thermometer was fixed to the "cold" end of the switch.

To determine the switch "ON" conductance, a known volume of exchange gas at room temperature was admitted into the switch through the valve system, the switch being immersed in a LN_2 bath, afterwards power was applied to it and after allowing time for the system to stabilize, the temperature difference across the switch ends was measured. This procedure was repeated for a series of heat inputs and for different exchange gases, namely - Helium, Nitrogen, Neon and Argon.

The steady state "OFF" conductance of the switch was determined by measuring the temperature difference across its ends for each of a series of heat inputs. The measured "OFF" conductance value of the switch was 0.009 W/K.

The performance of the switch in the "ON" position and with the different gases is presented in Fig. 3.

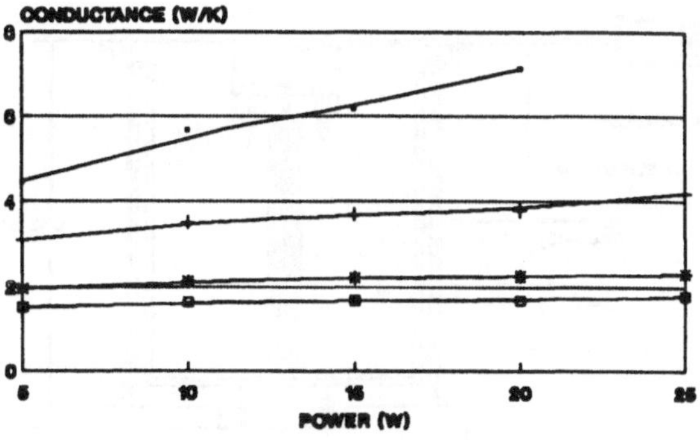

Figure 3. Heat switch performance.

Table 2. Switch/Getter Pump Characteristics

"ON" Conductance	2.2 W/K
"ON" Heat Flow	2.83 W
"OFF" Conductance	0.009 W/K
"OFF" Heat Flow	1.5 W
Annular Gas-Gap	0.025 cm
Weight	2.27 Kg
Exchange Gas	Nitrogen
Getter	2.5 gr of coconut charcoal

Given the lower bound of the switch temperature range and the material intended to be used for gettering, it was decided to use N_2 as the exchange gas. A small getter pump containing the minimum required quantity of activated charcoal needed to remove the nitrogen gas from the switch was designed. The cryopump, with approximately 2.5 grs of charcoal was fabricated from copper and connected to the "hot" side of the switch via a copper capillary tube. The charcoal pump was maintained in weak thermal contact with the bath through an appropriately sized thermal resistance. The heat switch was charged at room temperature with 100 kPa (1 atm) of N_2 gas and then permanently sealed. After cool down, the switch will automatically go into the "OFF" position because its charge of N_2 gas will be adsorbed by the getter pump. To activate the switch to its "ON" position, the charcoal pump is heated using the pump heater. The primary characteristics of the switch/getter pump system are summarized in Table 2.

II. ALUMINUM SINGLE GAP SWITCH

1. Description

Since the multifingered copper switch was too heavy and had an "OFF" conductance value higher than required, new aluminum single gap switches were designed and manufactured. The new switch is shown in Fig. 4. Here, again, the two halves of the switch are connected, by epoxied joints, to a G-10 CR tube which retains the exchange gas and provides alignment, mechanical support and - while the switch is in the "OFF" mode - thermal isolation between the "hot" and "cold" end. A detailed model of the switch was generated and, a finite element

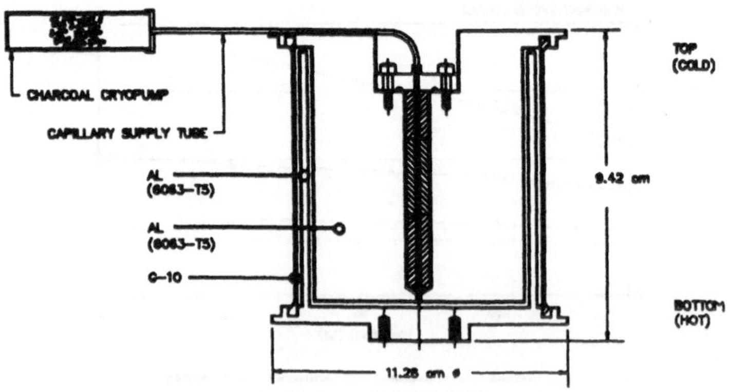

Figure 4. Schematics of heat switch.

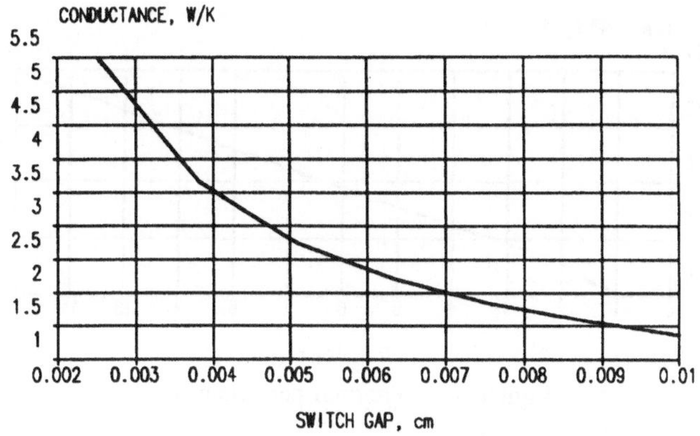

Figure 5. Gap sensitivity analysis.

analysis of the structure was performed. The model, consisting of 882 nodes and 384 elements was analyzed using Images 3D[3] running in a 386 - 25 MHz computer. Part of the switch design process included two "sensitivity" type analyses. One of them, presented in Fig. 5, shows how sensitive is the switch "ON" conductance to its gap width, while the other (Fig. 6) shows the relationship between gas-gap length and the temperature difference across the switch. From these analyses and considering the physical constraints given, a gap width of 0.005 cms was chosen for the new switches. The getter pump, containing about 1 gr of charcoal was connected, via a capillary tube and an indium seal, to the switch. The final weight of the switch was 1.75 Kg, while the cryopump weighed 67 gr. A finite element analysis of the switch indicated that under a static load of 15 gs in any direction, the maximum stress would be found at the G-10 tube and with a magnitude of 15.9 MPa. The maximum displacement of 0.00053 cm was also found to be at the G-10 tube when the 15 gs static load was applied in the x-direction.

2. Tests and Results

In order to satisfy flight qualification requirements, the switches were subjected to radiographic exams, proof pressure tests, leak tests, structural tests and thermal performance tests.

Structural Tests

To determine the ability of the switches to withstand the vibration test conditions as specified, without evidence of damage or deterioration, they were attached to a fixture plate and then securely mounted to an MB/Ling C-150 electrodynamic vibration system. One accelerometer was located on the test

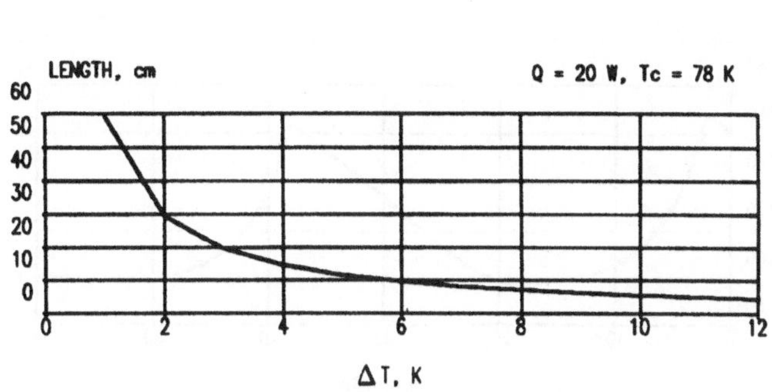

Figure 6. Gap length required to maintain the ΔT.

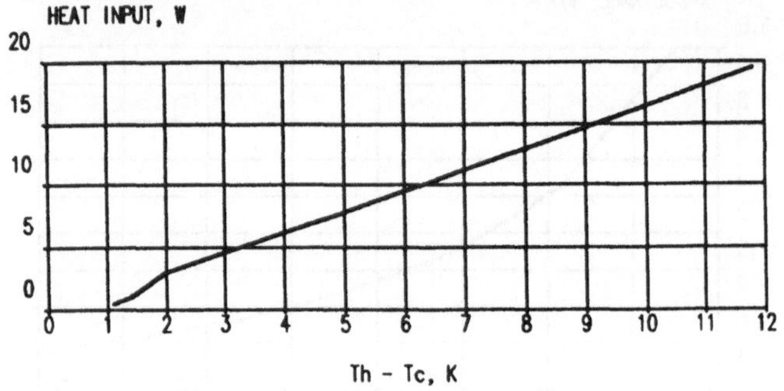

Figure 7. Thermal performance.

fixture while the response accelerometers were attached to the hot side of each switch. The switches were subjected to sine vibration, random vibration, and sine dwells in each of the three mutually perpendicular axes. Vibration was applied at the following levels:

<u>Random Vibration</u>

Flight Level:
20 Hz @ 0.0125 g^2/Hz
50 to 600 Hz @ -.075 g^2/Hz
2000 Hz @ 0.0125 g^2/Hz
Total Grms: 9.1
1 minute/axis
3 axes

Proto-Flight Level:
20 Hz @ 0.025 g^2/Hz
50 to 600 Hz @ 0.15 g^2/Hz
2000 Hz @ 0.025 g^2/Hz
Total Grms: 12.9
1 minute/axis
3 axes

<u>Sine Vibration</u>

Sine Sweep: 20 to 400 Hz @ 0.25 g
2 octaves/minute

Sine Dwell

Flight Level:
20 Hz @ 15 g's
1 second

Proto-Flight Level:
20 Hz @ 18.75 g's
1 second

As a result of the vibration test conditions, there was no evidence of damage or deterioration of the switches, as verified by a post-test radiographic examination.

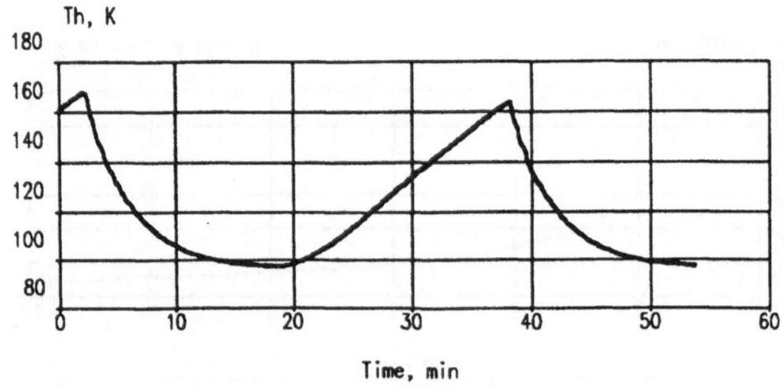

Figure 8. Switch ON/OFF cycling.

Proof Pressure Tests

Each switch was pressurized to 30 psig (2xMEOP) with nitrogen gas. To verify that no changes had occurred in their integrity, the switches were leak checked before and after this test.

Thermal Tests

With the silicon diodes attached to the "hot" and "cold" ends of the switch, as well as, to the cryopump, the assembly was mounted in the vacuum fixture, immersed in the liquid nitrogen bath and allowed to stabilize. Testing began when all sensors indicated less than 80 K. It consisted of applying power to the heater installed on the "hot" side of the switch, first with the device "OFF", then "ON" and, measuring in each case the temperature difference across the switch and therefore its conductance.
Additionally, 3 cooldown tests were performed by warming the "hot" side of the switch to approximately 160 K, with the switch in the "OFF" mode and 20 W of power into the "hot" side, turning the switch "ON" and monitoring the cooldown rate of the "hot" side versus time. The switch was turn "ON" by applying power to the heater wrapped around the cryopump, resulting in the charcoal desorbing the nitrogen gas and fill the switch gap, thus establishing the thermal contact between the "hot" and "cold" sides of the switch. The results of these tests are presented in Figs. 7 and 8. The "OFF" conductance of the switches was measured to be 0.003 W/K.

CONCLUSIONS

A cryogenic self contained gas-gap heat switch, using nitrogen gas as the exchange fluid has been developed, fabricated and tested. Stringent testing, structural as well as thermal demonstrated the adequacy of the design for space flight and the correctness of our analysis. A thermal switching ratio of 600 was attained with switching times of about 15 minutes.

REFERENCES

1. Lounasma, "Experimental Principles & Methods below 1 K," Academic Press, New York (1974).
2. T. Nast, G. Bell, and C. Barnes, Development of Gas Gap Cryogenic Thermal Switch, in: "Advances in Cryogenic Enfineering," vol 27, Plenum Press, New York (1982), p. 1117.
3. Celestial Software Inc., Berkeley, CA.

DEVELOPMENT OF CRYOGENIC RUPTURE DISCS FOR THE SPACE BORNE CRISTA PROJECT

R. Trant, C. Neusser, D. Offermann

University of Wuppertal, Dept. of Physics
Wuppertal, Germany

F. Kesting

Rembe GmbH
Brilon, Germany

ABSTRACT

Space cryostats require safety components to protect the cryogenic system against overpressure. The CRISTA cryostat (Cryogenic Infrared Spectrometers and Telescopes for the Atmosphere), which contains 725 l supercritical helium, will have a three stage safety system. A cryogenic rupture disc mounted directly on the helium tank will be the ultimate safety component. For qualifying cryogenic rupture discs a low temperature test facility was developed. The batch qualification of the cryogenic rupture disc, which is of the reserve buckling type, shows a standard deviation comparable with that at ambient temperature. The design of the rupture disc as well as test program and test results of the successfully performed qualification are described. Furthermore design and performance of the low temperature test facility are treated.

INTRODUCTION

All cryostats have to be equipped with relief devices to protect the cryogenic system against hazardous overpressure. This is also valid for space cryostats to avoid danger for the satellite, the launcher and the personnel. The CRISTA experiment will be launched in 1994 by the Space Shuttle for a 9 day mission. CRISTA will analyse dynamical processes as waves and turbulence in the middle atmosphere by measuring the infrared emission of atmospheric trace gases.[1] The main He tank of the CRISTA cryostat contains 725 l supercritical Helium at a nominal pressure of 3 bar. A second He tank contains 55 l liquid helium at an initial vapor pressure of 0.1 bar.[2] Each He tank has a three stage safety system corresponding to the safety standards of NASA. A cryogenic rupture disc mounted directly on the He tank will be the respective ultimate safety component. The inner diameter of the rupture disc is determined by the maximum vent gas rate due to the most critical failure mode, which is the loss of insulation vacuum.

REQUIREMENTS FOR THE CRYOGENIC RUPTURE DISCS

The safety standards of NASA require that relief devices shall be certified by test to operate at the required conditions of use. The main service conditions for the cryogenic rupture discs are the following:

- operational temperature — $2.5\ k \leq T \leq 10\ K$
- operational pressure — $0.1\ bar \leq p \leq 3.6\ bar$
- endurance — 40 pressure cycles.
  (considering a safety factor of 4)

Furthermore the discs must be capable of full flow at the order of $1 \times 10^3\ g\ s^{-1}$, has to withstand vibrational loads during launch, and must have a helium leak rate of less than $1 \times 10^{-9}\ mbar\ dm^3\ s^{-1}$.

The nominal set pressure of 4.5 bar ± 10% results from the operational pressure of the main He tank and the pressure staging of the three stage safety system. The tolerance band is based on manufacturers recommended standard tolerance of ± 5% for ambient temperature discs. For the cryogenic discs the tolerance band was doubled to take into account low temperature effects.

DESIGN OF THE CRYOGENIC RUPTURE DISCS

The CRISTA cryogenic rupture discs are of the reverse buckling type, which is a highly reliable pressure relief device. This type provides the best tolerance band and allows an operational pressure of up to 90% of the set pressure as well as a high back pressure. The rupture disc consists of a thin nickel membrane welded into a stainless steel housing and a threefold knife made of stainless steel (see Fig. 1). The membrane is prebuckled towards the pressure side. When the set pressure is reached the membrane will flap against the knife, gets cut and the full cross section is released.

A rupture disc is a highly sensitive precision safety element. Therefore special care has to be taken concerning the welding of the membrane. Laser spot welding is used to minimize thermal stresses during manufacturing. In order to avoid mechanical stresses during mounting of the burst disc, there are two grooves in the housing between mounting flange and membrane.

The cross section is calculated in accordance with the German regulation for rupture discs (AD Merkblatt A 1). With respect to the ultimate mass flow rate of about $1 \times 10^3\ g\ s^{-1}$ the inner diameter of 40 mm gives a margin of about 150%.

TEST PROGRAM

A batch qualification had to be performed because an operational test of a rupture disc can only be done in a destructive way. The chosen size of

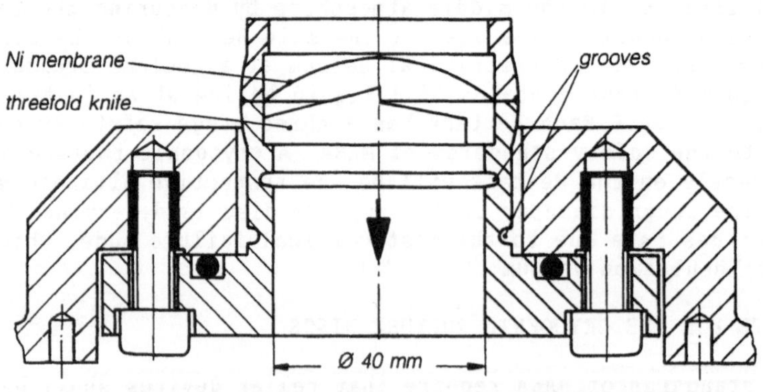

Fig. 1: CRISTA Cryogenic Rupture Disc

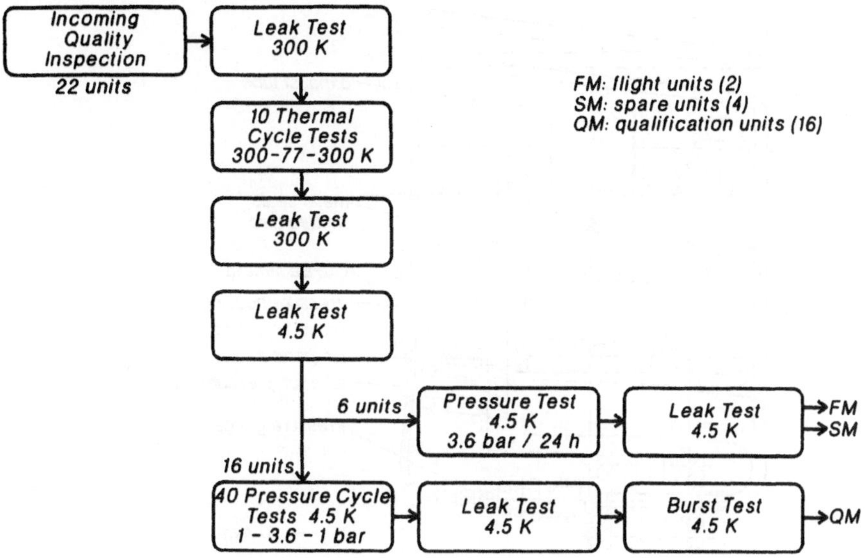

Fig. 2: Test Program

the batch takes into account that there is little experience with cryogenic rupture discs and that the qualification has to be performed at cryogenic temperatures. The whole batch consisted of 22 units, 16 of them were qualification units (QM), two are flight units (FM) and four are spare units (SM).

The main steps of the test program (see Fig. 2) are pressure cycles and burst tests at liquid helium temperature for QM's and a compressive creep test for FM's and SM's. The QM tests shall demonstrate that the rupture discs operate at service conditions within the specifications. The tests of the FM's and SM's shall demonstrate their usability. Before these tests all units undergo thermal cycling in order to age them with respect to low temperature effects. Before and after each step of the test program described above all units are leak tested at the corresponding temperature to show integrity of the unit.

A prebatch of five units was tested at room temperature to verify the manufacturing process and to ascertain the set pressure at room temperature. Another prebatch of six units was tested at 4.5 K to verify the set pressure at operational temperature, which cannot be predicted theoretically. Beyond that the prebatch should demonstrate the feasibility of the testfacility.

LOW TEMPERATURE TEST FACILITY

A test facility was developed which should be able to qualify burst discs at the temperature of liquid helium, liquid hydrogen, or any other cryogenic temperature with high accuracy.

This posed the following requirements:

- temperature range	4.5 K $\leq$ T $\leq$ 300 K
- test unit: - mass	up to 3000 g
- outer diameter	up to 130 mm
- test cycle period	1 test per day at 4.5 K
- temperature stability	< 1 K
- accuracy of pressure reading	< ± 0.05 bar.

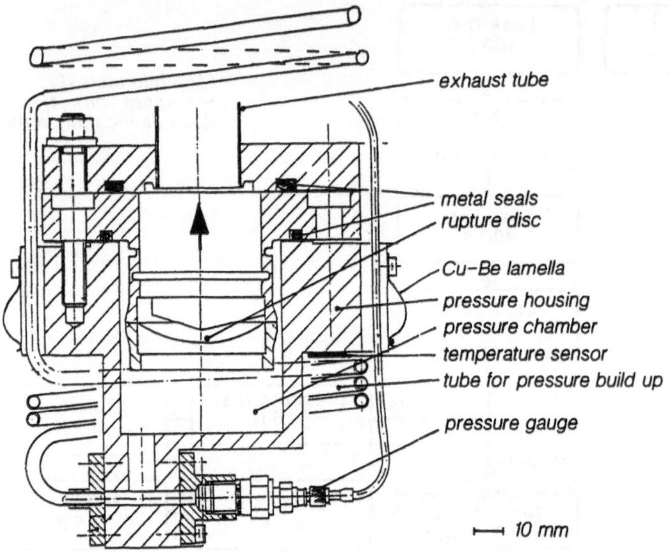

Fig. 3: Probe of the Low Temperature Test Facility

Furthermore the test facility should be operated at the location of the burst disc manufacturer to have short and effective interrelation between manufacturing processes and tests.

A top loading refrigerator cryostat was selected to fulfill these requirements. Three temperature stages of the refrigerator cryostat, together with heating devices at every stage, make it possible to reach any test temperature between 4.5 K and 300 K in an easy and safe manner. Controlled cool-down/warm-up cycles are possible in order to provide reproducible test conditions. The top loading principle has the advantage that a change of the test unit is possible while the refrigerator is still in operation. The test facility consists of three components:

- top loading refrigerator cryostat
- control unit
- probe.[3]

The top loading cryostat combines two refrigerators to assure reliable test conditions at 4.5 K as well as short test cycles. One refrigerator cools the first two temperature stages (80 K and 20 K) as well as the radiation shields of the cryostat. The other refrigerator provides precooling of the helium gas for a Joule Thompson expansion valve which is the third stage (4.5 K). Temperatures in between these stages can be achieved by controlled heating. Thermal contact between the refrigerator cryostat and the probe is provided by helium contact gas inside the top loading tube.

The control unit provides reproducible pressure build up as well as automatic pressure cycle tests. The gas pressure is monitored in three independent ways by a manometer, a pressure transducer inside the control unit, and a pressure gauge at the probe. This enables accurate pressure determination even in the cold.

The probe (see Fig. 3) holds the rupture disc. It is equipped with two cryogenic sensors that measure pressure and temperature. The tube for pressure build up is wound around the probe to ensure that helium gas at test temperature is used for pressure build up . This should exclude any uncontrolled temperature effects to the set pressure. When the set pressure of the relief disc is reached, the gas is released through the exhaust tube. This tube is also used for the leak tests of the relief discs. To improve the thermal contact to the temperature stages of the refrigerator cryostat Cu-Be-lamella are screwed to the pressure housing.

The low temperature test facility achieves the following cool-down times:

- 300 K ⟶ 80 K : 4.0 h
- 80 K ⟶ 20 K : 1.5 h
- 20 K ⟶ 4.5 K : 0.1 h.

The pressure build up for a burst test was divided into two phases. During the first 50 seconds the pressure increased up to 80% of the nominal set pressure. Afterwards the pressure increased by 0.5% of the nominal set pressure per second until bursting. The accuracy of pressure reading including burst pressure determination is ± 0.020 bar. The pressure cycle tests were performed automatically with 9 cycles per hour. The temperature stability during the whole qualification program of one unit is ± 0.1 K at 4.5 K.

TEST RESULTS

The complete test program was performed successfully without any restriction.[3] The average set pressures of the three batches were:

- prebatch at 300 K (5 units) p_{set} = (3.11 ± 0.11) bar
- prebatch at 4.5 K (6 units) p_{set} = (4.32 ± 0.09) bar
- main batch at 4.5 K (16 units) p_{set} = (4.23 ± 0.12) bar

(Standard deviations are given; they correspond to 3.5%, 2.1%, and 2.9% one sigma values).

The essential results are as follows:

- The average set pressure of the main batch is below the specification by 0.27 bar. This does not pose a problem because of the small tolerance band achieved (see Fig. 4).
- The tolerance band is only one half of the one required.
- The standard deviations at 4.5 K are comparable with those at room temperature.
- With respect to the average set pressure at room temperature the one at 4.5 K increases by 36% for the main batch and 39% for the prebatch.

The test results demonstrate that the CRISTA cryogenic rupture discs operate at the required conditions of use.

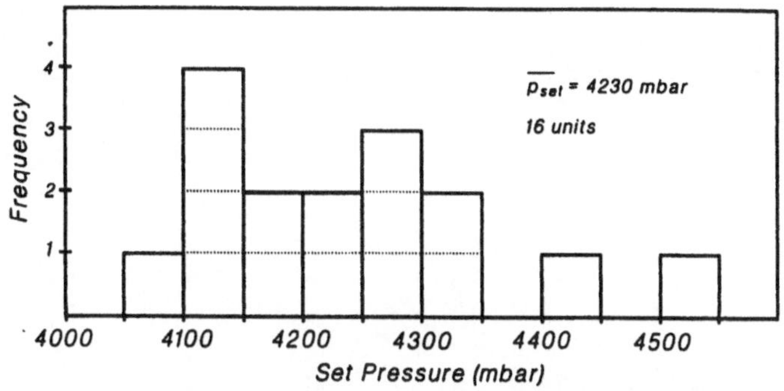

Fig. 4: Set Pressure Histogram of Main Batch

ACKNOWLEDGEMENTS

The CRISTA project is funded by DARA (Deutsche Agentur für Raumfahrtangelegenheiten), Bonn, FRG. The development of the CRISTA cryogenic rupture discs was funded by Ministerium für Wirtschaft, Mittelstand und Technologie des Landes NRW, Düsseldorf, FRG. We acknowledge the fruitful collaboration with Dr. Forth, Leybold AG (Köln, FRG), during the refrigerator development.

REFERENCES

1. Grossmann, K. U. and Offermann, D., The CRISTA experiment on ASTRO-SPAS, in:"Proc. Eigth ESA Symposium on European Rocket and Balloon Programmes and Related Research", ESA SP-270 (1987), p. 411.
2. Trant,R., Grossmann, K.U., Langfermann, M., Offermann, D., Cryogenics of the CRISTA/SPAS Experiment aboard the Space Shuttle, in: "Proc. Thirteenth Intl. Cryo. Engr. Conf.", Butterworth, Guildford, UK (1990), p. 475.
3. Neusser, C., Diplomarbeit, Department of Physics, University of Wuppertal, FRG (1991).

DESIGN AND DEVELOPMENT OF A LEAK TIGHT HELIUM II VALVE WITH LOW THERMAL IMPACT

G. L. Mills

Ball Electro-Optics/Cryogenics Division
Boulder, Colorado

ABSTRACT

The Lambda Point Experiment is a precision measurement of the specific heat of liquid helium near the lambda point phase transition, in the low gravity of the space shuttle. It requires a valve for the helium sample chamber that operates at helium II temperature, has minimal thermal disturbance to the rest of the instrument, and is leak tight to helium II. A valve meeting these and all of the other science and engineering requirements of the mission has been developed by Ball.

Initially, both torque and pressure actuated valve concepts were considered; the final flight design is pressure actuated. The rational for this decision as well as the rest of the valve design are given. The paper also discusses the manufacturing and testing of the prototype and flight vales. Test data is presented and discussed.

INTRODUCTION

The Lambda Point Experiment is a basic science experiment which has a primary goal of performing a very precise measurement of the specific heat of helium near the lambda point in low gravity. Specifically, the goal of the experiment is to measure the heat capacity of a helium sample with a temperature resolution 10^{-10}K over a temperature range of 10^{-6} K on either side of the lambda point transition in the 10^{-3} gravity environment of the space shuttle [1]. This is a test of the renormalization group theory of cooperative transitions [2].

The Lambda Point Experiment instrument is shown schematically in figure 1. The instrument fits into a dewar and is surrounded by Helium II during the experiment. The focus of the instrument is the calorimeter stage, which contains the sample chamber and valve. The helium sample is contained in a copper chamber which has a spherical interior. The chamber is connected by high thermal conductivity straps to high resolution thermometers and is mechanically connected to the rest of the instrument through low thermal conductivity supports. The sample chamber is evacuated and filled through a small tube which is connected to the outside of the instrument. The fill tube must have a valve mounted directly on the chamber itself to allow evacuation of the fill tube after the chamber is filled with helium. In filling the sample chamber, the mass of helium needs to be adjusted so that a small bubble can be created in the helium sample. The sample chamber valve is the subject of this paper.

Advances in Cryogenic Engineering, Vol. 37, Part B
Edited by R.W. Fast, Plenum Press, New York, 1992

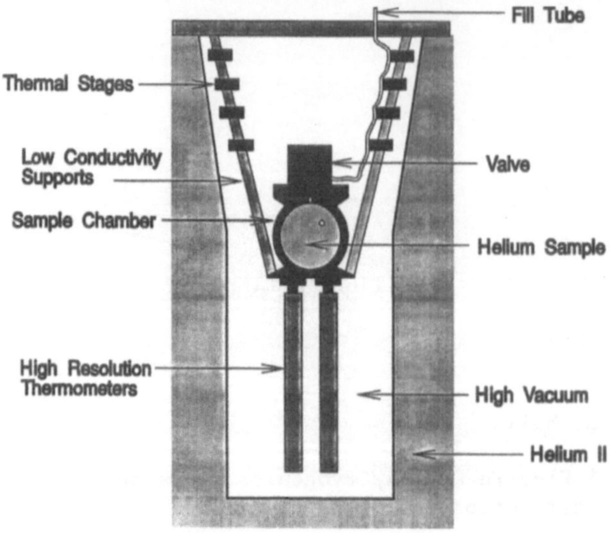

Figure 1: Lambda Point Experiment instrument simplified schematic

VALVE REQUIREMENTS

The requirements for the Lambda point experiment sample chamber valve were established as follows, for any temperature from 300 K to 2.1 K.

1.) Cause no uncontrolled heating or cooling during experiment.
2.) Operation (opening or closing) to cause little or no heating or cooling of valve and sample chamber to allow the helium mass to be precisely adjusted.
3.) Have no greater than a 1 x 10^{-7} scc/s leak across seat after 50 open/close cycles and launch vibration. Includes being leak tight to helium II.
4.) External leaks not greater than 1 x 10^{-9} scc/s.
5.) When closed, be helium leaktight across seat to 1 x 10^{-5} scc/s during launch vibration.
6.) Have sufficient conductance to allow the sample chamber to be filled in a reasonable time period (a few hours or less).
7.) Have a small mass and cause a minimum increase in sample chamber mass.
8.) Contain no magnetic or paramagnetic materials.
9.) Have a minimum impact on the rest of the instrument design
10.) Operate (open and close) with 0 to 1.0 atmosphere external pressure.
11.) Be simple and reliable.

INITIAL CONCEPTS

Several valve concepts were initially considered and were quickly narrowed down to two concepts: pressure and torque actuated. The torque actuated concept was based on a valve which had been successfully used in laboratory experiments at Stanford University. The concept as modified by Ball is shown in figure 2. A simple set screw plunger drives an indium coated diaphragm against a sharp seat. A Belleville spring washer stack is used to make the force more controllable and repeatable. When the set screw is screwed out, the diaphragm is allowed to come off the seat, opening the valve. A drive rod or tube is inserted in the plunger and extends out the dewar through an O-ring seal. It is inserted in the set screw only when the valve is opened or closed to eliminate its thermal effect on the calorimeter when data is being taken.

The pressure actuated concept is similar to the torque actuated concept since a flexible diaphragm is driven against a sharp seat and the

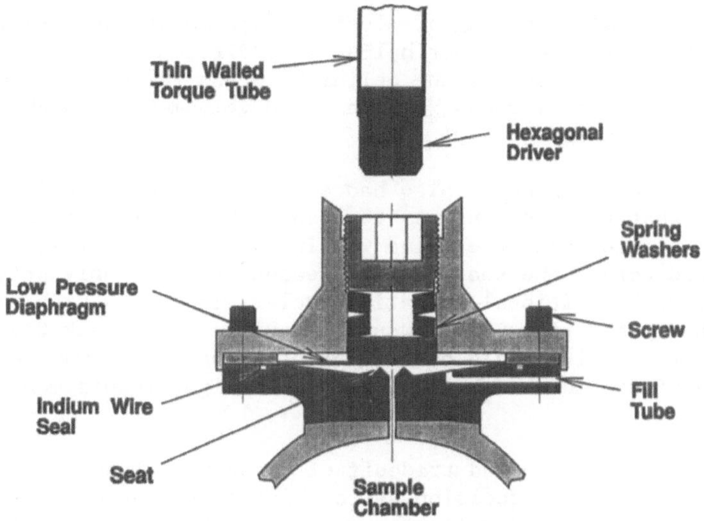

Figure 2: Torque actuated valve concept.

diaphragm is held closed by a stack of Belleville spring washers. The valve is opened by pulling on the valve stem with another, actuating, diaphragm filled with high pressure helium fluid. This allows the low pressure diaphragm to come off the seat, opening the valve.

The decision as to which valve concept to develop into flight design and hardware was difficult. It was recognized that the sample chamber valve represented one of the highest areas of cost and schedule risk to the overall experiment program and had the potential to greatly increase program costs by delaying the overall program. The tradeoff of these concepts are presented in table 1. In general, it appeared that both concepts could probably be made to work and neither concept was clearly superior.

The torque actuated concept had a disadvantage in an important criteria, in that the torque tubes could cause significant heating of the sample

Table 1: Valve concept tradeoff

← increasing importance

criteria / options	No heating during experiment	No thermal effect when valve closes	No internal leakage	Development cost	Minimal effect on inst. design	Small mass
Torque actuated	Actuator pulled out through flaps in shields during exp. GREEN	"Hot poker" effect of actuator difficult to eliminate YELLOW	Seat force can be continously increased until leaktight GREEN	Testing requires rest of the instrument or a mock-up BLUE	Requires free space in middle of entire instrument BLUE	Mass is small GREEN
Pressure actuated	Actuation line and diaphram evacuated during exp. GREEN	Valve closes by reducing actuation pressure GREEN	Constant seat force; any leak cannot be reduced YELLOW	Valve test self-contained GREEN	Almost no effect on rest of instrument design GREEN	Mass larger than torque actuated BLUE

GREEN = MEETS REQUIREMENTS YELLOW = MAJOR DEFICIENCIES
BLUE = MINOR DEFICIENCIES RED = UNACCEPTABLE

container during valve opening or closing. This would make precise adjustment of the sample chamber helium mass difficult. The torque tubes would have to be heat sunk to the helium bath and various controlled temperature stages. This would require the development of high performance sliding thermal anchors.

The pressure actuated concept also had a disadvantage in an important criteria, in that there was no way to change the closing force that provided the sealing once the valve was installed in the instrument. If a leak developed across the seat for any reason it would only stay the same or get worse with additional actuation cycles. To have a high degree of confidence that the flight valve would be sufficiently leak tight at the time of flight, a valve would have to be developed which was several orders of magnitude more leak tight than the flight requirement of less than 10^{-7} scc/s.

It was the identification and tradeoff of these two disadvantages which formed the basis for the decision as to which valve concept was developed. The decision was made to develop the pressure actuated valve. It was thought that while the development of a valve that was leaktight to Helium II to less than 10^{-9} scc/s was difficult, the development of high performance sliding thermal anchors was even more difficult. Ball Electo-Optics/Cryogenics Division (BECD) had recently developed a larger, motor driven valve for the Cosmic Background Explorer program[3,4] (COBE) and this valve was leak tight to Helium II to less than 10^{-9} scc/s. BECD had also recently completed work on thermal contact resistance [5] and we were aware of the difficulty in performing these kinds of measurements. Another consideration was that the development of a leaktight pressure actuated valve was a well contained problem that could be done in parallel with other program activities, but the design of a sliding thermal anchor depended on the design of the rest of the instrument and the final testing of a sliding thermal anchor would require at least a prototype of the entire instrument.

DETAILED DESIGN

The final, flight lambda point sample chamber valve design is shown in figure 3. A photograph of the flight valve and sample chamber is shown in figure 4.

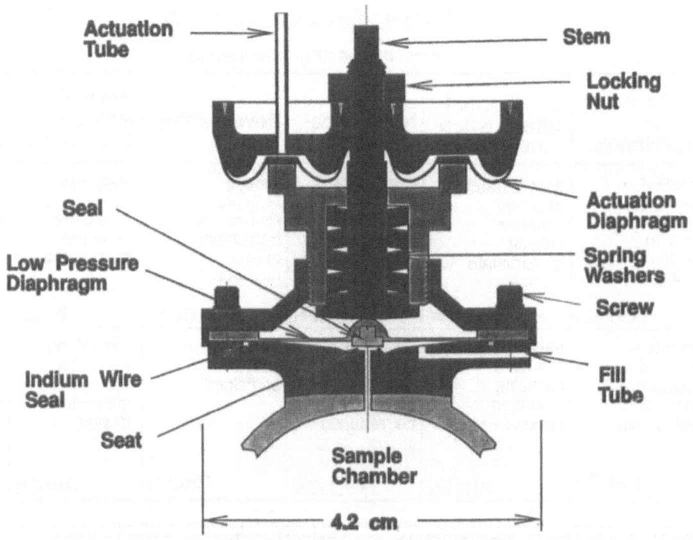

Figure 3: Final pressure actuated valve design

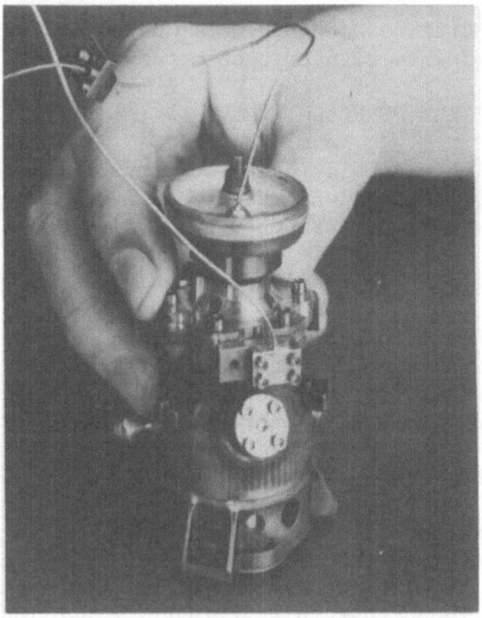

Figure 4: Flight valve and sample chamber

BECD experience with the COBE valve aided greatly in performing the detailed design of the Lambda Point Experiment sample chamber valve. Based on the COBE valve experience, it was decided to make the seat a 110 degree angle with an approximately 0.003 cm. radius on the edge and to machine the seat from hardened beryllium copper. Also similar to the COBE valve was the seal, which was made of vacuum annealed copper 101 (99.99% pure) with a 0.0003 cm. thick gold coating.

There was experimental evidence from the testing of the COBE valve that if the seal moved relative to the seat after the first sealing cycle, much more additional force was required to cause the valve to seal again. For this reason, the seal was soldered into a recess machined into the low pressure diaphragm, so it would not move, but could be (and would have to be) renewed every time the valve was put together. The lambda point valve was designed so that the seal was pushed into the seat with a force of 45 kilograms. This was proportional to the sealing force used on the COBE valve and was based on the relative seat diameters.

The seat diameter for the lambda point valve was chosen to be 0.25 cm. since this was the smallest diameter that could be accurately machined with conventional techniques. The seat diameter of the lambda point valve needed to be large enough to provide a large flow conductance relative to the rest of the system, but since the fill line had an inside diameter of only 0.051 cm., the valve seat could be quite small and still not affect the overall flow conduction. The valve stroke, or distance that the valve seal comes off the seat when open, was set at 0.008 cm. to limit stress on both diaphragms.

The actuation diaphragm was sized to open the valve with an actuation pressure of 1.7 MPa. This actuation pressure was chosen to be well below that of the helium 4 solidification pressure of 2.6 MPa at 1.6 K. Almost all of the valve parts were designed to be built out of beryllium copper. The only parts that were not beryllium copper were the valve seal (pure copper and gold), the titanium screws and the indium static seal. Beryllium copper (copper alloy 172) was chosen because it is very strong in the full hard condition, machines well, welds and solders well, is non-magnetic and is compatible with cryogenic temperatures.

MANUFACTURE AND ASSEMBLY OF FLIGHT VALVE

The lambda point valve was built using conventional machining and assembly processes, but several operations required a large amount of skill and attention to detail to be performed successfully. One side of the actuating diaphragm had very thin (0.015 cm.), curved walls which were difficult to machine without tearing or bending.

The seal was conventionally machined and the sealing surface was polished to a 0.2 micron finish using a fine abrasive. The gold coating was applied using a thin film ion sputtering machine. The seal was then carefully soldered into the low pressure diaphragm using tin-lead solder.

It was recognized that since the valve opened only 0.008 cm., that particles in the helium flow approximately this size or larger would get stuck between the seat and the seal, causing the valve to leak. This created a requirement that the helium fluid system be cleaned so that it had no particles in it larger than 1/10 of the valve stroke or 8 microns. Fortunately, it was found that this stringent cleanliness level could be achieved by carefully following established cleaning and clean room procedures. The final cleaning and assembly of the sample chamber and low pressure diaphragm were done on a class 10,000 flow bench in a class 100,000 clean room.

TEST METHOD

Three prototypes of the sample chamber valve were built and tested before the final, flight valve was built. All of the valves were tested in the same manner. The valve to be tested was thermally and structurally anchored to the outside of the inner tank of a small dewar. When the dewar was assembled and the guard vacuum pumped, this put the valve in a high vacuum as it would be in actual use. Small diameter copper-nickel tubes were attached to the fill and sample chamber sides of the valve seat, and to the high pressure diaphragm using indium sealed flanges. The three tubes were run out to bulkhead fittings in the dewar outside wall. A silicon diode and a germanium thermometer were thermally anchored to the valve body.

The valve was tested for external leaks by attaching a mass spectrometer helium leak detector to the dewar guard vacuum, and pressurizing the three lines with helium. The closed valve was tested for leaks across the seat by attaching the leak detector to the fill tube and pressurizing the sample chamber tube with helium. In both cases, a stable leak detector reading was taken when the tubes were evacuated and after the tubes were pressurized. A leak was an increase from the first to the second reading.

These leak detection tests were done at room temperature, 77 K and 2.1 K with the inner tank of the test dewar filled with liquid nitrogen or helium II as required. If a particular valve assembly was found to leak at 77 K, time and effort was usually not spent cooling the valve to 2.1 K since we have found that if an assembly leaks at 77 K it will be very likely to leak at 2.1 K. Likewise, we have also found that if an assembly is leaktight at 77 K, it is also leaktight below 2.1 K [6]. However, to be safe, all leaktight assemblies were confirmed to be leaktight by testing at 2.1 K.

The conductance of the valve was tested at liquid nitrogen temperature by flowing helium through the fill tube, the open valve and then back out of the dewar with the tube attached to the sample chamber side of the valve.

The gas flow rate was roughly determined by letting it bubble through alcohol and counting the bubbles in a given time period.

TEST RESULTS

The testing of the first and second prototype valves demonstrated the overall feasibility of the valve design. The valve opened and closed, had good conductance and was externally leaktight. The closed valve was initially leaktight across the seat at room temperature and liquid nitrogen temperature, but after being open and closed once, it developed a 10^{-7} std. cubic cm./sec leak. Additional open and close cycles did not significantly affect this leak rate. The valve was disassembled and no obvious cause for the leakage was found. Numerous modifications to the valve seat, seal and assembly procedure were tried with essentially the same results. A launch vibration test of the second prototype decreased the amount of leakage across the seat significantly.

Measurements with a microscope indicated that the impression left by the seat in the seal could be as much as 40% deeper (0.0033 to 0.0055 cm.) on one side than the other. Since the seat and seal were much more parallel than this, it was concluded that the stem force was not centered on the seat and this was affecting sealing. The sealing could be affected if the seat force was moving with respect to the seat during vibration.

A third prototype valve was built which had tighter tolerances than the first, especially with regard to the dimensions that control how well the stem was centered over the seat. In addition, several modifications were incorporated in the design to insure this centering. The valve housing had extensions added to it to which fit closely on the valve base which centered both the housing (which held the stem) and the low pressure diaphragm to the valve base (see figure 3). A small hemisphere was machined into the center of the low pressure diaphragm to insure that the stem could only apply force on its exact center. On the previous valves, this area had been flat.

These modifications proved to be worthwhile, for on the first assembly and test of the prototype valve, it was leaktight to better than 10^{-9} cubic cm./sec across the seat after being open and closed at 2.1 K. Additional testing of the valve showed it to be leaktight to this level after being opened and closed fifty times at liquid nitrogen temperature. The valve was also leaktight to this level during the launch vibration test. This was performed by securely bolting the valve, with the tubes attached, to a vibration test machine. One side of the valve seat was connected to a leak detector, while the other side was pressurized with helium. The valve was then cooled with liquid nitrogen while being vibrated at shuttle launch vibration levels.

This third prototype valve was then disassembled and then reassembled with a new seal and tested. It was also leaktight to the same level. Inspection of the seal from the disassembled valve showed the impression left by the seat to vary in depth less than 6% (.0068 to .0064 cm.).

The flight valve has been assembled on the sample chamber and has been tested for leakage and conductance at 2.1 K. It has a leak rate of less than 10^{-9} std. cubic cm./sec. at 77 and 2.1 K after being opened and closed repeatedly at those temperatures.

The conductance of all of the valves with the fill tube was 1-5 cubic cm./sec of helium at 77 K with 200 torr driving pressure. The conductance of the fill tube alone was found to be approximately the same, therefore the conductance of the valve was concluded to be large compared to the fill line.

The third prototype valve assembly, which was not built into a sample chamber, weighed 182 grams. The flight valve weight is probably very close to this, but it could not be weighed separately from the sample chamber.

CONCLUSIONS

A pressure actuated valve has been built which meets all the requirements of the lambda point experiment calorimeter. Several key features of the design allowed the valve to repeatedly have less than 10^{-9} std. cubic cm./sec helium leakage across the valve seat at 77 K and 2.1 K. These features include: a soft copper and gold seal forced into a hard metal seat, the seal and seat arranged so that they always come together in exactly same spot, and the force on the seal arranged so that it is very uniform and repeatable.

In every case, if a valve was leaktight at 77 K it was also leaktight at 2.1 K. This is consistent with the results of the COBE valve testing [6].

ACKNOWLEDGMENTS

The author wishes to acknowledge Robert Arentz, Dan Payne, and Jerry Siebert - all of BECD - for their significant contributions to this work. This work was supported by Stanford University contract PR4815 and by NASA JPL contract 955057.

REFERENCES

1.) J. A. Lipa, T. C. P. Chui, "Very High Resolution Heat- Capacity Measurements near the Lambda Point of Helium", Physical Review Letters, 51:25, 2291 (1983).

2.) K. G. Wilson, Physical Review B, 4, 3174 (1971).

3.) J. F. Siebert, R. A. Hopkins, H. A. Chameroy, "Development of a Launch-Worthy Motor-Operated Valve for Containment of Superfluid Helium," Proc. of ICEC 9, (1982) 182.

4.) NASA Contract NAS 5-27710.

5.) J. F. Siebert, W.T. Deshler, "Thermal Conductance Measurements of Bolted and Gasketed Aluminum-Aluminum Joints from 1.5-300 K", Proc. of ICEC 10, (1983).

6.) J. F. Siebert, "Experiment Comparison of Helium Leak Rates and Detection Techniques at Temperatures Above and Below the Lambda Point, " Space Cryogenics Workshop, Berlin, Aug. 1984, 89.

A TEMPERATURE CONTROL SYSTEM FOR A VARIABLE TEMPERATURE CRYOSTAT USING AN 80386 PERSONAL COMPUTER

Todd A. Keitel, Lindsay J. Feuling, Evan M. Ludeman

Astronautics Technology Center
Madison, Wisconsin

ABSTRACT

We report on the development of an automatic temperature control system for a variable temperature cryostat operating from 4.2 - 100 K. The large changes that occur in system heat capacity and cooling power over the range of operating temperature have led to the development of a segmented linear control system. The control system has been implemented on an IBM compatible 80386 personal computer which communicates with an external DMM and power supply via IEEE 488 interface. Temperature transducers are carbon glass and platinum resistance thermometers.

Details of the control system, cryostat response and range of operation are included. Application of this system to a vibrating sample magnetometer system for the purpose of obtaining magnetization at constant temperature versus magnetic field, and magnetization at constant field as a function of temperature is discussed.

INTRODUCTION

Interest in the development of magnetic refrigerants for temperature to 80 K has led us to the development of a variable temperature, 0-8 Tesla vibrating sample magnetometer system for three purposes. First, as the refrigerants of primary interest to us above 20 K are ferromagnets, this instrument gives us the capability to rapidly characterize a candidate material interims of ordering temperature and ordering type. Secondly, this instrument yields direct measurement of magnetization as a function of field and temperature which is necessary for calculation of magnetic forces to be expected in a magnetic refrigerator. Finally, this instrument can be used to execute various field/temperature cycles and energy loss due to magnetic hysteresis, if any, can be measured directly.

High quality experimental results from this system are critically dependent upon temperature control. We have achieved satisfactory results in the present system from 4.2 K to beyond 100 K. Details of the system architecture, system response, control system design and performance follow.

Advances in Cryogenic Engineering, Vol. 37, Part B
Edited by R.W. Fast, Plenum Press, New York, 1992

ANALYSIS

Control System Architecture

The Automatic Temperature Control System architecture is shown in Figure 1. The main components of the temperature control loop are the two temperature sensors mounted on the diffuser block, the computer, and the heater. The heater is used to input heat into the system. The diffuser block temperature sensors provide temperature feedback. The computer is used to implement a digital control system that controls the heater element with

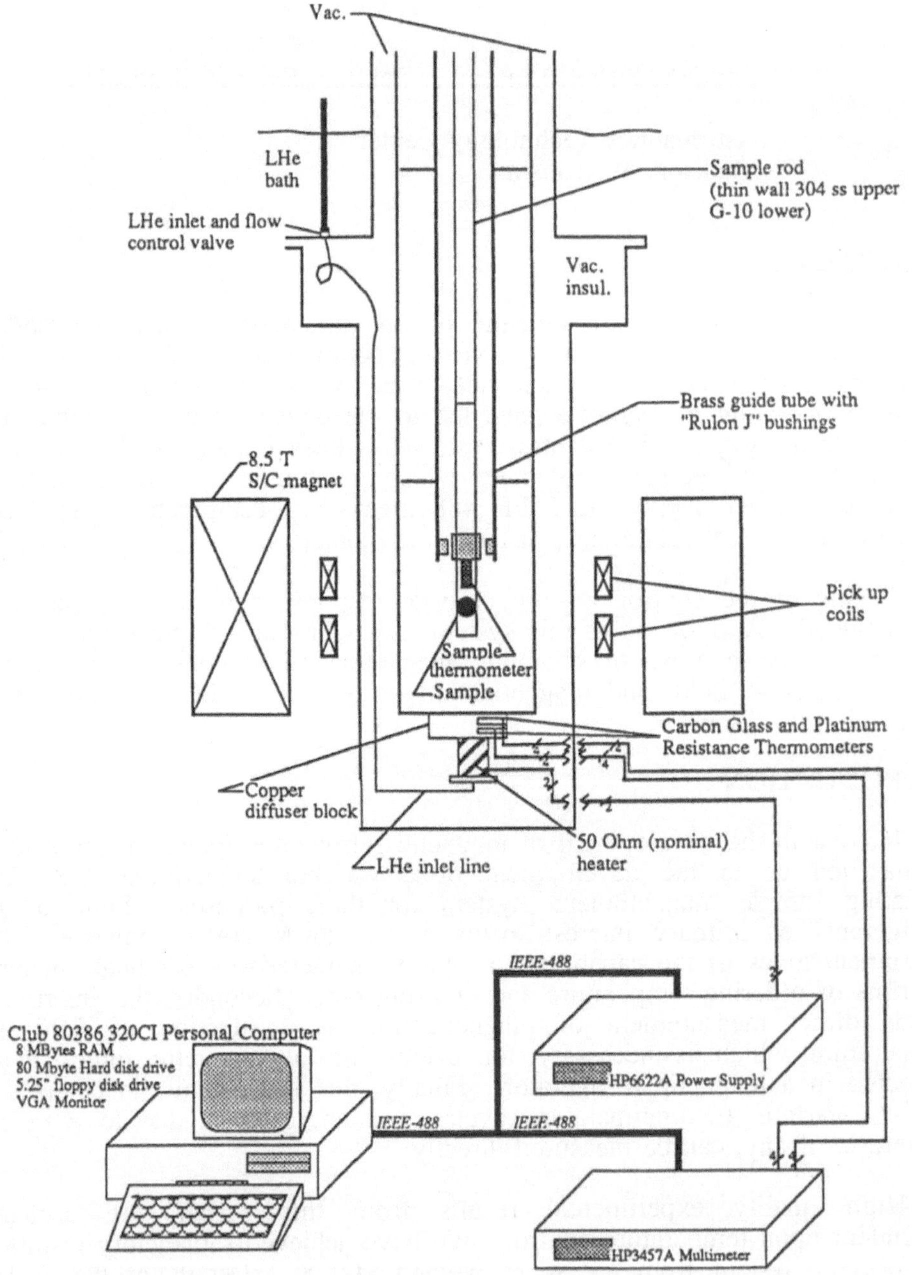

Fig. 1 Automatic temperature control system architecture

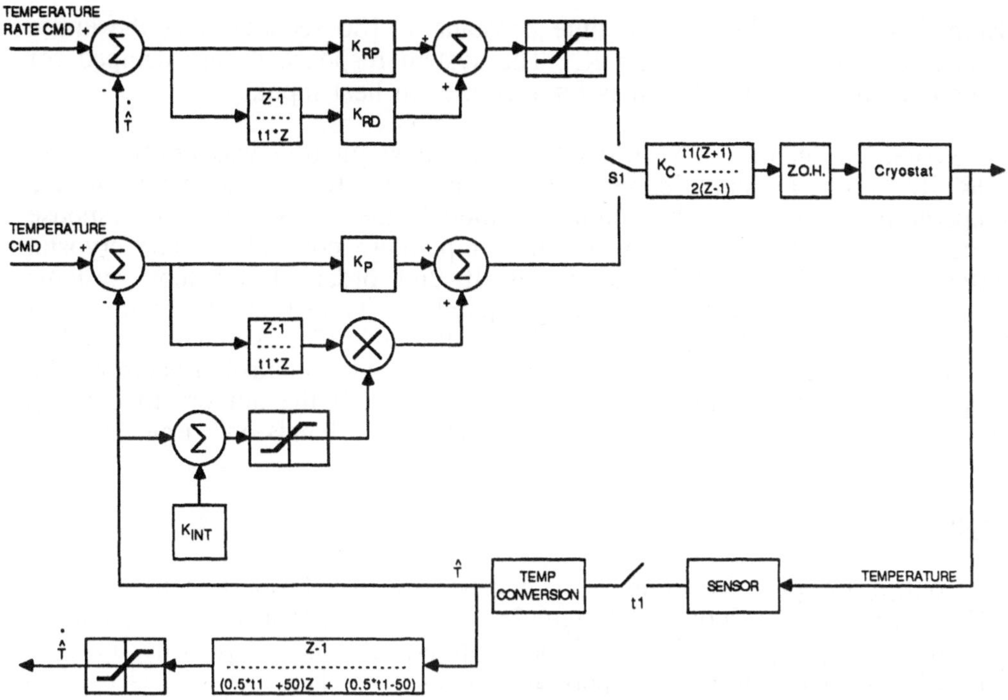

Fig. 2 Automatic temperature control system block diagram

diffuser block temperature feedback. The actual sample temperature follows the diffuser block temperature with some delay.

The software is written in C using the Microsoft QuickC development environment. The software consists of three main components:

1) Control System
2) Graphical User Interface
3) IEEE 488 Interface

The Control System software implements the control algorithms necessary to maintain temperature and temperature rate commands. The Graphical User Interface prompts the user for information and provides strip chart, numeric, and alphanumeric display feedback. The IEEE 488 Interface provides communication to/from the HP3457A Digital Multimeter and the HP6622A Power Supply.

Cryostat System Characterization

System Response. The cryostat system response was characterized by providing heat into the system, via the heater, and observing the diffuser and sample temperatures. The system was characterized over the temperature range of 4.2 - 125 K with varying amounts of heat input. The system cooling capability, which is determined by the manually adjusted gas flow rate, was varied to measure the system response per cooling power.

System Model. The observed system response matches well with a 2nd order polynomial model with temperature/cooling capability dependent damping ratio and natural frequency. The system is characterized as under damped with a nearly constant time constant of 180 seconds over the entire temperature range of 4.2 - 125 K. The system response is less damped for reductions in heat input than it is for increases in heat input.

System Simulation. A closed loop digital simulation was developed, in FORTRAN on a Sun Sparcstation, to aid in the development of the temperature controller. The simulation models the cryostat system response to heat input and the operation of the temperature controller. The cryostat system response is modeled as a set of 2nd order closed loop systems dependent on temperature and cooling capability. Transitions between system models is performed using linear interpolation. The simulation models timing, Analog/Digital, Digital/Analog, and computer control aspects of the temperature control system. The simulation enabled the authors to develop the temperature controller with a minimum of "hands-on" time on the cryostat.

Control System

Control Loops. The controller consists of two Proportional/Derivative (PD) closed loops as shown in Figure 2. The position loop is used to control a specific temperature. The position loop is designed to provide an over damped system response to temperature commands. The rate loop is used to control a specific temperature rate of change. The rate loop is designed to provide an under-damped system response to a temperature rate (ramp) command. The control system switches from position loop to rate loop per real-time request from the operator. The control system automatically switches from the rate loop to the position loop when it reaches an operator-specified temperature ramp end point.

Integrator Voltage Control. Both control loops are designed to function independently of the specific cryostat cooling capability - as set manually via gas flow. The output of the active control loop is feed into an integrator. The output of the integrator is used to command a voltage to be placed across the heater element. The
integrator enables the control system to adapt to changes in the cryostat cooling capability.

Temperature Variant Gain. The position loop derivative gain is a function of temperature to compensate for the changes in the cryostat system response per changes in temperature. The derivative gain, Kp, is computed as three simple linear segments as described in equation (1).

$$Kp = 30.0 + stemp, \quad 50.0 <= drgain <= 130.0 \qquad (1)$$

Where stemp is equal to measured temperature in K. The gain, Kp, is increased by 20% if the temperature command is less than the present temperature minus two degrees. This increase in damping factor, for negative change in temperature commands, is necessary due to the asymmetrical response of the cryostat.

Aperiodic Control Implementation. The controller is designed to operate aperiodically to achieve the highest possible bandwidth while allowing for significant requirement changes during the development phase. Control system filters are implemented using sampling period dependent filter coefficients to accommodate variances in the sampling period. This dynamic

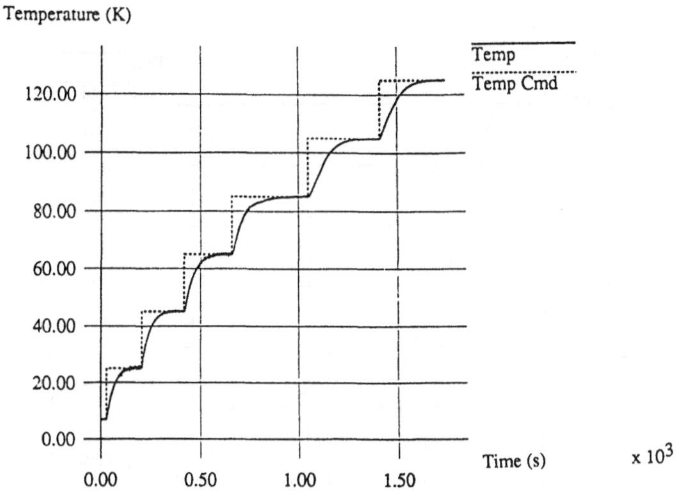

Fig. 3 Temperature loop results - 20 K temperature step commands

filtering method is stable over small variations in period for systems of 4th order or less.

Temperature Measurement. The temperature sensors are sampled by the HP3457A multimeter over a period of one or more power line periods. The operator is given the capability to choose the number of power line periods over which the multimeter samples and integrates. The controller reads the resultant sensor resistance and converts it to temperature using a cubic spline interpolation method from an off-line developed data table. An on-screen graphical strip chart displays real-time temperature values. The controller automatically switches between using a Carbon Glass resistor thermometer below 78 K and a Platinum resistor thermometer above 78 K. Sufficient hysteresis is built into the switch to avoid bouncing. The controller optimally sets the multimeter's range based on temperature.

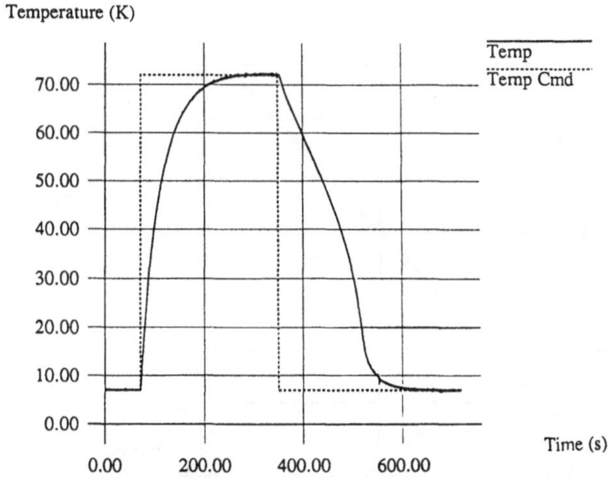

Fig. 4 Temperature loop results - 65 K step commands

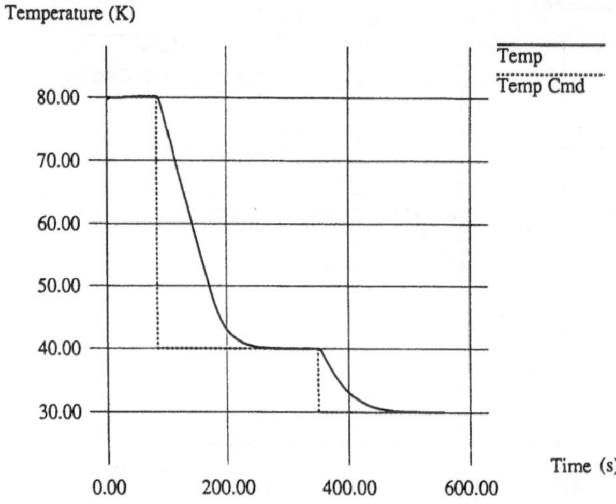

Fig. 5 Temperature loop results - 40 & 10 K step commands

Control System Results

The controller temperature loop maintains constant temperatures with a measured error of less than 0.2% of commanded value (less than 1% of commanded change in temperature). The time constant of the controlled system is dependent on the temperature-variant derivative gain. The resulting time constants range from 50-125 seconds over the temperature range of 7 - 125 K. See Figures 3, 4 and 5. The controller properly responds to any temperature command occurring at any time independent of whether the system has reached a steady state temperature from initial conditions or a previous command.

The controller temperature rate loop maintains a steady rate of temperature change for commanded rate changes of 0.3 to 5.0 K per minute. The controller transitions into the temperature loop at the operator-specified ramp end point. The temperature loop maintains an over-damped system response at the ramp end point. See Figures 6 and 7.

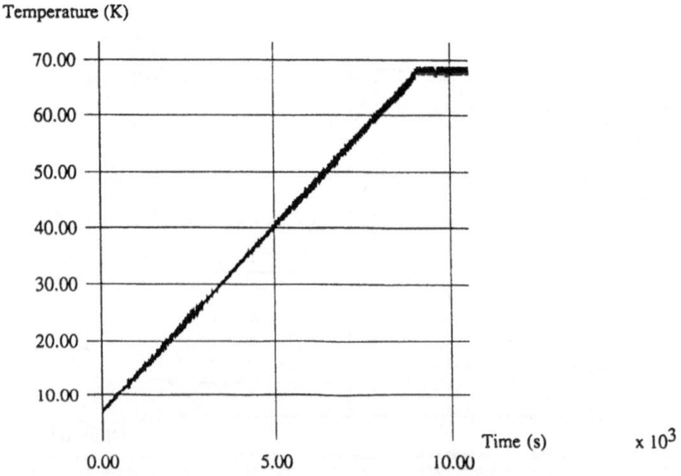

Fig. 6 Temperature rate loop results - 0.4 K/min ramp command

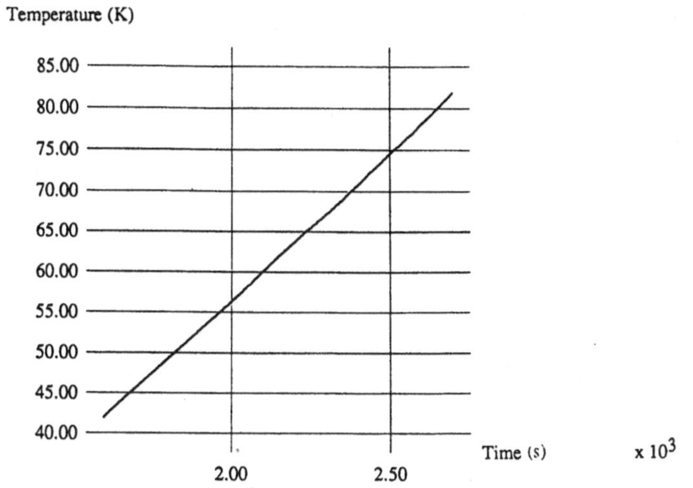

Fig. 7 Temperature rate loop results - 2.0 K/min ramp command

Application to a Variable Temperature Magnetometer

The temperature control system described has been developed specifically for use in an automated variable temperature Vibrating Sample magnetometer (VSM). This system is being used for the characterization of magnetic materials for refrigeration applications and as such, is used in three principle areas: 1) screening of new materials of interest for magnetic transition temperature(s) and type of ordering, 2) measurement of temperature dependent magnetization at high fields for the purpose of calculating the magnetic forces that will act between a magnetic refrigerant and its associated magnet system, and 3) careful study of the magnetic reversibility of refrigerants over the temperature range of application. Operations 1 and 2 both involve maintaining a constant magnetic field during a temperature ramp typically 2 K/min.). In the former case, the applied field is generally small, typically 0.1 Tesla for the study of ferromagnets. In the case of a more general characterization (#2, above), fields up to 8 Tesla may be applied. In the case of hysteresis studies, a constant temperature is maintained as nearly as possible, while the magnetic field is swept through a predetermined cycle, lost energy being determined from the area enclosed in the M-H loop.

CONCLUSIONS

The automatic temperature control system provides accurate temperature and temperature rate control of a commercial helium gas flow cryostat. The broad range of the temperature and temperature rate control leads to the use of this controller in cryostats for thermometer characterization as well as other possible uses. The 80386 PC implementation of the automatic temperature control system provides a good base from which to implement a variable temperature magnetometer with correlated data acquisition and temperature control.

A DEVELOPMENT OF A THIN WIRE RESISTANCE THERMOMETER

WITH ISOTROPIC MAGNETO-RESISTANCE

Koichi Nara, Hideyuki Kato and Masahiro Okaji

National Research Laboratory of Metrology
1-4, Umezono 1-chome, Tsukuba, Ibaraki, Japan

ABSTRACT

A thin wire thermometer with isotropic magneto-resistance can be made by adjusting a pitch angle θ_p in the winding of the sensing wire to a magic angle, $\arctan(1/\sqrt{2})$. Several sensors with different pitch angles were examined and a systematic relationship between their pitch angles and the magneto-resistance was observed. In case of the best sensor, the anisotropy of the magneto-resistance is about 4%, which is reduced from 35% for a commercial platinum resistance sensor.

INTRODUCTION

Thermometry under magnetic fields is essential to the reliable measurement of many physical properties at low temperatures. Several kinds of thermometers, such as capacitance thermometer[1-2], carbon glass resistance thermometer[3] and ZrN film resistance thermometer[4] have been reported to be suitable for the use under strong magnetic fields because of their small field dependence. Some small platinum resistance thermometers have been also reported to be applicable for the same purpose, if their magneto-resistance is properly corrected.[5-9] From a practical point of view, however, it is an unfavorable character that their magneto-resistance depends not only on field strength but also on field direction.[10,11]

Recently, a new design for a thin wire resistance thermometer with a small anisotropy was proposed by the present authors.[12] The isotropic character was achieved by adjusting the pitch angle θ_p in the winding of the sensing wire to a magic angle, $\arctan(1/\sqrt{2})$. In this report, several sensors were made with different pitch angles to assess the validity of the designing principle. The measurement of the magneto-resistance clearly supports the predicted relationship[12] between the magneto-resistance and the pitch angle.

PITCH ANGLE DEPENDENCE OF THE MAGNETO-RESISTANCE

The basic design of the sensors is illustrated in Fig.1, where a thin wire is wound with a constant pitch p around a cylindrical glass core of a diameter of d. The relative increment of line resistance per unit length $\rho(T,B)$ is assumed as shown by equation (1).[12]

$$\rho(T,B)/\rho(T,0)-1=e_{\perp}B_{\perp}^2+e_{\parallel}B_{\parallel}^2 \qquad (1)$$

Advances in Cryogenic Engineering, Vol. 37, Part B
Edited by R.W. Fast, Plenum Press, New York, 1992

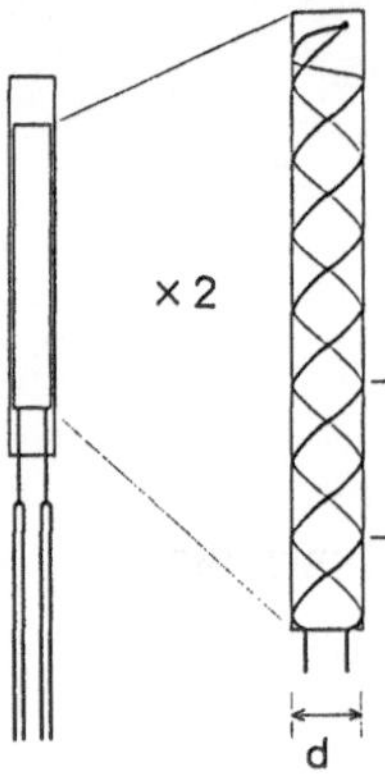

Fig.1 Schematic view of a thin wire resistance thermometer

where $B_\perp$ and $B_{/\!/}$ are the components of the magnetic field perpendicular and parallel to the wire element, respectively. Then the total resistance of the sensor, R(T,B), can be readily calculated by integrating Eq.1 along the wire and the magneto-resistance is given as follows. In the calculation, Z-axis is chosen to be parallel to the axis of the sensor core.

$$R(T,B)/R(T,0)-1=e_\perp((1-\frac{\cos^2\theta_p}{2})B_{x-y}^2+\cos^2\theta_p\,B_z^2)+e_{/\!/}(\frac{\cos^2\theta_p}{2}B_{x-y}^2+\sin^2\theta_p\,B_z^2) \quad (2)$$

where the pitch angle, θ_p, is defined as $\arctan(p/(\pi d))$. By replacing θ_p in Eq.(2) with a magic angle, $\theta_m = \arctan(1/\sqrt{2})$, Eq.(2) is simplified to have no dependence on the field direction.

Following the design described above, several sensors were made from the same thin wire as used for the new-Japanese industrial standard platinum resistance thermometer (new-JIS PRT hereafter), GR0705, which has been reported to be highly stable.[8] When the winding pitch is measured, the sensors are dipped in the clove oil to avoid the lens effect by the glass core. Fig. 2 and 3 show the photographs of the new-JIS PRT and a PRT of the present design with the best isotropic character, respectively.

The magneto-resistance of the sensors are measured under the magnetic fields of up to 8T. The measurement system is the same as one reported earlier.[8] The magnetic field is applied parallel or perpendicular to the axis of the sensors. As the best example, Fig. 4 shows the ratio of the magneto-resistance under the field perpendicular to the axis of the sensor against that with the field parallel to the axis. In this case, the anisotropy is about 4%, which can be compared with 35% for the commercial new-JIS PRT, GR0705. At 40K, the ratio is about 0.96 and independent of the magnetic field. Below 30K the ratio is close to 0.96 at lower fields but decreases

Fig.2 Commercial PRT, GR0705

Fig.3 Sensor of the present design

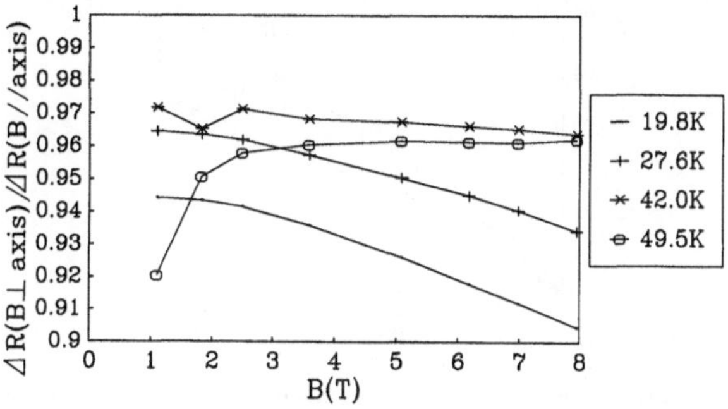

Fig.4 Field dependence of the ratio of the magneto-resistance with the field perpendicular to the axis of the sensor against that with the parallel field.

as the field increases. The dependence of the ratio on the fields or temperatures indicates that the assumption in Eq.(1) is no longer valid at low temperatures under high magnetic fields.

ASSESSMENT OF THE DESIGN

The experimental setup above includes two extreme conditions that the field is parallel or perpendicular to the axis of the sensors. From Eq.(2) the magneto-resistance under these conditions can be derived as follows.

$$R(T,B)/R(T,0)-1=e_{\perp}\cos^2\theta_p\, B^2+e_{/\!/}\sin^2\theta_p\, B^2 \qquad B /\!/ \text{ Axis of the sensor} \quad (3)$$

$$=e_{\perp}(1-\frac{\cos^2\theta_p}{2})B^2+e_{/\!/}\frac{\cos^2\theta_p}{2}B^2 \qquad B \perp \text{ Axis of the sensor} \quad (4)$$

As is clearly seen from Fig. 2, the pitch angle for the commercial new-JIS PRT is very small and $\cos\theta_p$ is 0.9992. Then $e_{\perp}B^2$ and $e_{/\!/}B^2$ at 42K can be determined. This leads us to calculate the theoretical magneto-resistance for a wide range of θ_p by use of Eqs.(3) and (4). Figure 5 compares these theoretical lines with experimental results at 42.0K under 8T. The theoretical lines give good fits at large θ_p but give only poor fits at smaller θ_p. The reason for this disagreement is not clearly known yet.

COMPARISON AMONG THE SENSORS

The data on the magneto-resistance for platinum thermometers can be summarized from a view point of its field direction dependence.

Figure 6 shows the magneto-resistance and its anisotropy for the commercial new-JIS PRT, GR0705. The data are characterized by its small variation of magneto-resistance among the sensors and a large anisotropy. Its

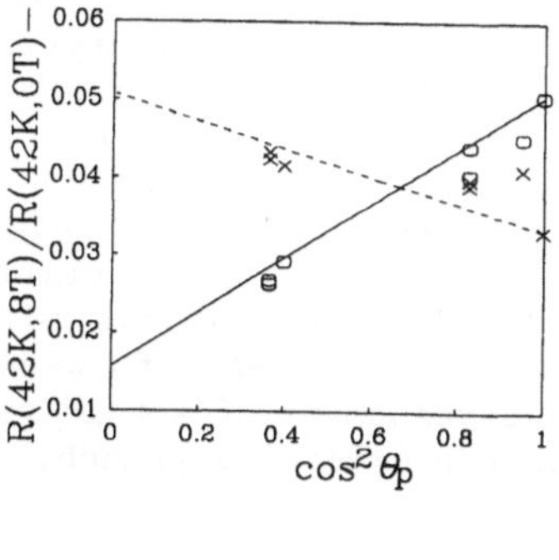

Fig.5 Pitch angle dependence of the relative resistance increase at 42K : solid line(calculation), o(experiment); the direction of the field is parallel to the axis of the sensor , dotted line(calculation), ×(experiment); perpendicular to the axis

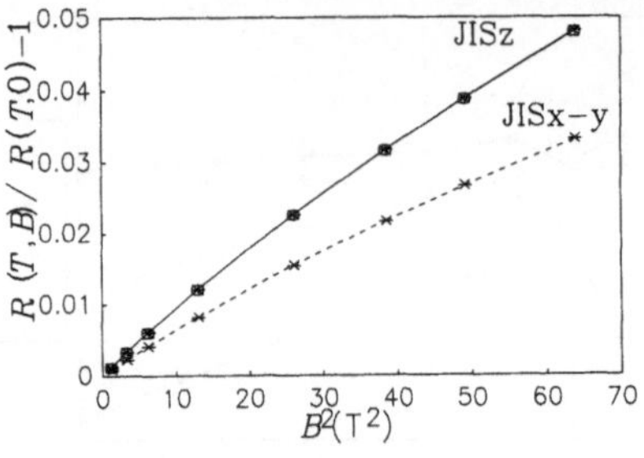

Fig.6 Relative resistance increase under magnetic field for the commercial new-JIS thermometer GR0705 at 42.0K : three solid lines ; the axis of the sensor is parallel to the field, dotted line ; perpendicular to the field. Three solid lines fall on a single line because of the small sensor dependence.

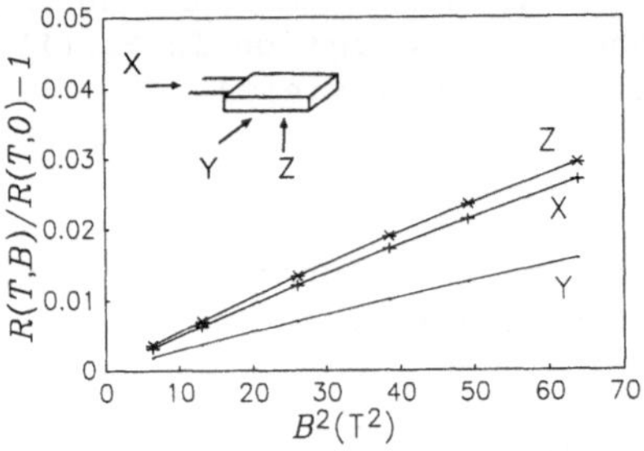

Fig.7 Relative resistance change under magnetic field for commercial thin film platinum thermometer, PTF-7, at 42.0K : the inset shows the direction of the magnetic field.

small variation is explained by its small pitch angle θ_p in Eq. 3. As $\cos\theta_p$ is insensitive to a change in θ_p for a small θ_p, a variation of 10% in the pitch of the winding only induces a change of 0.004% to the magneto-resistance.

Figure 7 shows another result for a commercial thin film platinum resistance thermometer, PTF-7.[13] Its magneto-resistance is largest when the field is applied perpendicular to the plane of the sensing film. It shows the smallest magneto-resistance when the field is applied along the y-axis. The origin of its large anisotropy could be explained if necessary information on the film pattern were provided. It would be difficult to develop a thin film thermometer with a small anisotropy if the film is deposited on a single flat plane. If it is possible to form a thin film on a curved surface, a similar design discussed above will be effective.

CONCLUSION

A sensor with isotropic magneto-resistance is developed. The typical design parameters of the sensor are shown in table 1. The best sensor showed an anisotropy of as small as 4%, which is reduced from 35% for the commercial PRT, GR0705. The designing assumption was proved to be satisfied above 40K under the magnetic fields of up to 8T. On the other hand, the sensing wire should be wound with as small pitch as possible to achieve a small variation of magneto-resistance among the sensors, from the viewpoint of its pitch angle dependence. From a technical point of view, it is not easy to wind a thin wire with a very precise pitch. Therefore, a development of a thin film thermometer with an accurate pitch would be another promising approach.

Table 1 Typical design parameters of the present sensors

Diameter of the platinum wire	15μm (new-JIS Pt100)
Diameter of the glass core	0.7mm
Pitch of the winding	1.55mm
Outer diameter of the sensor	0.8mm
Outer length of the sensor	8mm

ACKNOWLEDGEMENTS

The authors thank Hayashi Denko, Co. Japan for the production of the sensors. They also thank Mr. M. Tamura for his useful advice in the determination of the pitch angle of the sensors.

REFERENCES

1. W. N. Lawless, Aging phenomena in a low-temperature glass-ceramic capacitance thermometer, Rev. Sci. Instrum. 46:625(1975).
2. J. B. Hartmann and T. F. McNelly, NaF:OH and KCl:OH magnetic field-independent capacitance thermometers, Rev. Sci. Instrum. 48:1072(1977).
3. H. H. Sample, B. L. Brandt and L. G. Rubin, Low-temperature thermometry in high magnetic fields. V. Carbon-glass resistors, Rev. Sci. Instrum. 53:1129(1982).
4. T. Yotsuya, M. Yoshitake and J. Yamamoto, New type cryogenic thermometer using sputtered Zr-N films, Appl. Phys. Lett. 51:235(1987).
5. B. L. Brandt, L. G. Rubin and H. H. Sample,Low-temperature thermometry in high magnetic fields. VI. Industrial-grade Pt resistors above 66K ; Rh-Fe and Au-Mn resistors above 40K, Rev. Sci. Instrum. 59:642(1988).
6. T. Haruyama and R. Yoshizaki, Thin-film platinum resistance thermometer for use at low temperatures and in high magnetic fields, Cryogenics 26:536(1986).
7. Y. Iye, Small low-cost platinum resistance thermometers for thermometry in magnetic fields, Cryogenics 28:164(1988).
8. K. Nara, H. Kato and M. Okaji, Magneto-resistance of a highly stable industrial-grade platinum resistance thermometer between 20 and 240K, Cryogenics 31:16(1991).
9. K. Nara, H. Kato and M. Okaji, A Derivation of a Universal Correction Function for the Magneto resistance of a Newly Introduced Japanese Industrial Standard Platinum Resistance Thermometer GR0705, Teionkogaku 25:394(1990).(in Japanese, synopsis is available in English)
10. L. J. Neuringer, A. J. Perlman, L. G. Rubin and Y. Shapira, Low Temperature Thermometry in High Magnetic Fields. II. Germanium and Platinum Resistors, Rev. Sci. Instrum. 42:9(1971).
11. H. Alms, R. Tillmanns and S. Roth, Magnetic-field-induced temperature error of some low-temperature thermometers, J. Phys. E 12:62(1979).
12. K. Nara, H. Kato and M. Okaji, Possible design for a thin wire resistance thermometer with isotropic magneto-resistance, to be published in Cryogenics.
13. K. Nara, H. Kato and M. Okaji, Development of a heat-capacity measurement system under magnetic fields. (1) Magneto-resistance and its anisotropy of a highly stable thin film platinum resistance thermometer, PTF-7, submitted to Teionkogaku.

TEMPERATURE COMPENSATION FOR PIEZORESISTIVE PRESSURE TRANSDUCERS AT CRYOGENIC TEMPERATURES

D. L. Clark

Martin Marietta Astronautics Group
Denver, CO

ABSTRACT

By thermally-coupling the pressure transducer to the cryogenic vessel or transfer line, the measurement uncertainty can be significantly reduced as well as the heat leak associated with sensing lines extending to ambient temperatures. Temperature compensation algorithms are presented along with experimental results using piezoresistive pressure transducers which are operated at the cryogen temperature.

Pressure measurements with accuracies better than 0.5% of full scale can be obtained with piezoresistive transducers at any temperature from 70 K up to 300 K using appropriate algorithms. The resulting reduction in heat leak minimizes the effects of Thermo-Acoustic-Oscillations (TAO) and reduces measurement uncertainty.

Data is presented from testing with liquid nitrogen and liquid hydrogen as well as implementation considerations.

INTRODUCTION

Cryogenic pressure measurements have always challenged the system designer. Traditional methods of mounting the transducer in the ambient environment generate heat leak through the sensing lines and are subject to Thermo-Acoustic-Oscillations (TAO). These effects can induce fluctuations that exceed the level of the measurement itself. Previous testing[1,2,4] has demonstrated the value of thermally-coupling the pressure transducer to the cryogenic fluid. These methods eliminate the large temperature differential across the sensing lines, reducing the heat leak and the associated induced pressure fluctuations. Typical systems installations attempt to reduce TAO by using restrictions in the sensing lines or large damping volumes. These methods reduce TAO but also eliminate any dynamic response in the measurement.

Pressure transducers which are constructed to operate at cryogenic temperatures have recently been developed[3]. These transducers can withstand extreme cold shock while providing

Advances in Cryogenic Engineering, Vol. 37, Part B
Edited by R.W. Fast, Plenum Press, New York, 1992

repeatable, predictable outputs. The sensing diaphragm of these transducers are constructed with a silicon strain sensing bridge. The bridge output is temperature dependant with increasing sensitivity at cryogenic temperatures. The piezoresistive bridges exhibit similar response trends which can be used to generate a response versus temperature algorithm over the entire temperature range. These algorithms can be used to provide pressure measurements at cryogenic temperatures comparable to ambient pressure sensors.

TEMPERATURE COMPENSATION

Two components of the temperature response must be considered with the piezoresistive sensor. For purposes of this analysis, we separate the zero-shift due to temperature from the sensitivity effects. A differential pressure transducer with a range of 0 to 202 kPa was use for this testing.

The test setup for determining the temperature characteristics of the sensor used a calibrated reference transducer mounted in the ambient environment with common sensor port connections for the thermally-coupled transducer. The test transducer was mounted in a large thermal mass to reduce thermal transients. A type "E" thermocouple mounted to the transducer body provided the reference temperature. The test transducer was placed in a cryogenic container and covered with liquid nitrogen. The test pressure was applied and the transducer was allowed to slowly warm up after draining the liquid nitrogen.

The zero-shift results of the test transducer are shown in Figure 1 where the zero-pressure output is plotted versus the sensor temperature. The figure includes the zero shift from two similar

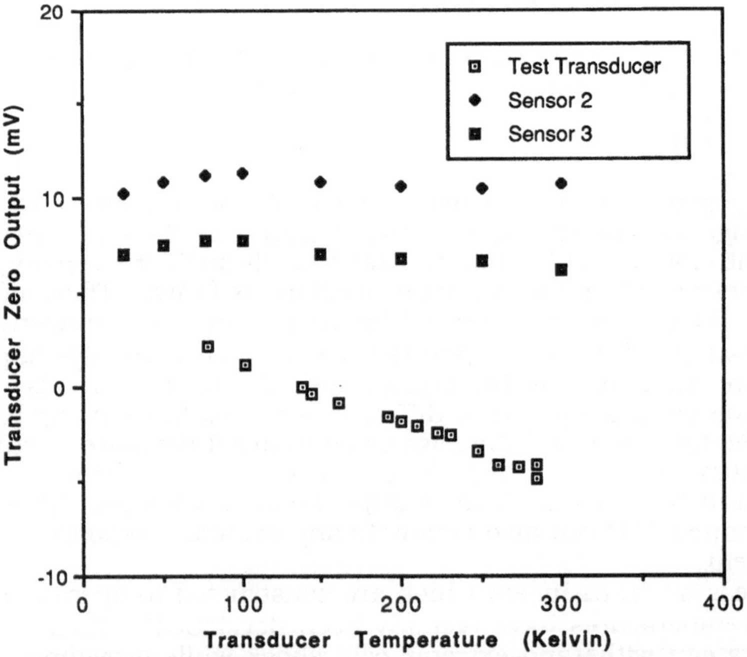

Figure 1 - Transducer Zero Output vs. Temperature

transducers for comparison. The zero-shift is modelled with the following linear equation:

$$Z(t) = 4.21 - .03051(T) \tag{1}$$

In this equation T is the transducer temperature in Kelvin and Z(t) is the zero offset in millivolts. The worst-case value for the test transducer was found to be 0.43 millivolts which corresponds to 0.35% of full scale output.

The sensitivity shift is of greater magnitude than the zero-shift and requires a more complicated model. The output sensitivity the test sensor is plotted versus temperature in Figure 2. The values shown are the output of the device in kPa per millivolt with the zero-pressure output subtracted.

Again the sensitivities of two similar transducers are included in the figure. These values were modeled using a third-order polynomial with the following equation:

$$S(t) = 0.835369 - 8.8577e^{-4} * T + 2.9961e^{-5} * T^2 - 5.6536e^{-8} * T^3 \tag{2}$$

In this equation, S(t) is the transducer sensitivity in kPa per millivolt. The actual pressure is then calculated by determining the zero offset from equation (1) and multiplying the difference of the raw output voltage and the zero offset by the sensitivity calculated from equation (2).

$$P = (mV - Z(t)) * S(t) \tag{3}$$

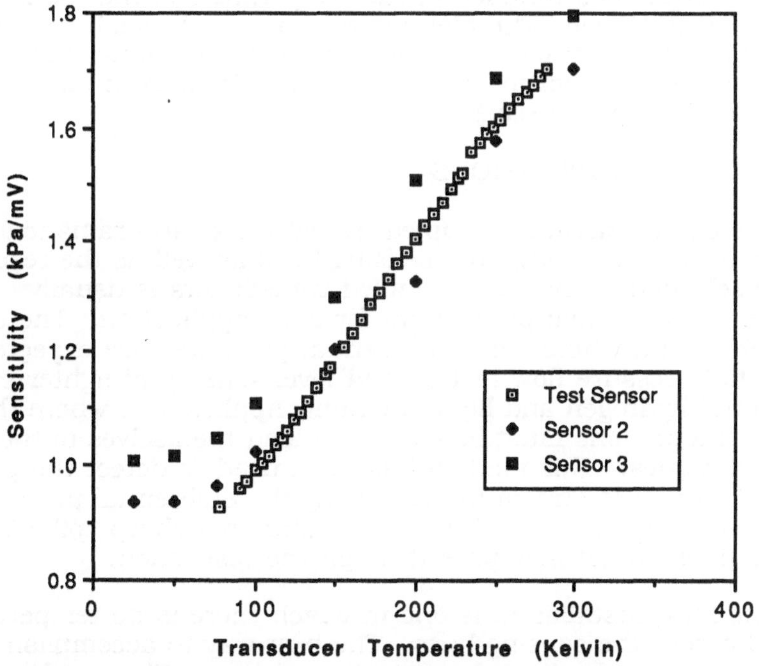

Figure 2 - Transducer Sensitivity vs. Temperature

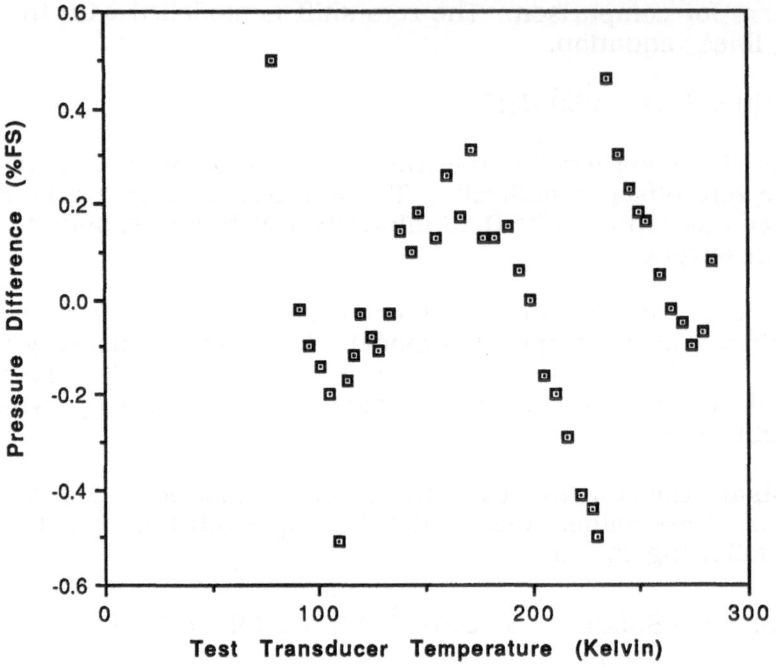

Figure 3 - Operational Transducer Comparison

An operational test was performed to test the temperature compensation algorithm. The test used an externally mounted pressure transducer connected to the ullage space on a cryogenic tank.

The test transducer was thermally-coupled to the tank wall and also measured the tank ullage pressure. A comparison of the two measurements is shown in Figure 3 where the difference between the readings is plotted as a function of the temperature of the thermally-coupled test transducer. The plot shows that the thermally-coupled transducer accurately tracked the pressure over the entire temperature range from ambient down to liquid nitrogen. The maximum difference was less than 0.5% of full scale.

APPLICATION CONSIDERATIONS

The use of thermally-coupled transducers can dramatically reduce the heat leak through the sensing lines as well as the resulting TAO. The selection of thermally-coupled transducers is usually dependant on the system parameters for each application. The concept is extremely useful where small differential pressures are expected such as differential pressure flow meters and level sensing of lighter cryogens such as liquid hydrogen and liquid helium. Applications where the actual flow fluctuations must be measured lend themselves to thermally-coupled techniques. This method has been used to detect the presence of bubbles in the flow stream by measuring the differential pressure across a flow restriction. The bubbles register as a sharp spike in the differential pressure as they pass through the restriction.

The ideal installation is one in which there is no temperature differential across the sensing lines. The best way to accomplish this is to directly immerse the transducer in the cryogen. This method provides the most constant temperature but may not be possible for all applications. Additionally, the sensor leads must be brought out of the vessel, requiring hermetic feedthroughs. This method can be used to completely eliminate TAO effects and provide very accurate measure measurements. Since there are no induced oscillations, the transducer can track very small flow fluctuations. This method has been used to measure pressure drop across capillary retention devices and provides an indication of screen breakdown when bubbles are ingested into the device. Other applications include level sensing[4] and in-line vapor bubble detection.

A method for measuring pressures in flow lines mounts the transducer on the flow line inside the vacuum jacket. A saddle between the transducer and the line insures good thermal contact. The sensing lines are also thermally-coupled to the flow line to reduce vapor bubbles in the sensing lines. The entire sensing system is covered with MLI to reduce heat leak and insure optimum thermal transfer to the transducer.

An example of the flow measurement is shown in Figure 4 where the output of a thermally-coupled transducer is shown during a liquid hydrogen no-vent fill test. The transducer is measuring the pressure drop across a venturi meter in the primary fill line. The measurement shows large fluctuations in the initial flow due to two-phase conditions as the line cools. The fluctuations then drop as the sub-cooled liquid hydrogen enters the tank. The overall flow rate decreases steadily as the pressure in the unvented tank increases with the rising liquid level. Finally, the flow stops completely as the maximum level is reached and the pressure fluctuations in the flowmeter increase due to the generation of vapor bubbles in the transfer line. This application used a pre-calibrated transducer since the flow temperature was known. This method provided accurate data for liquid hydrogen but was not valid for other temperatures. The differential pressure scale in the figure is only approximate since a more complicated algorithm is required to determine the true pressure.

CONCLUSIONS

The use of thermally-coupled pressure sensing techniques can greatly improve certain measurements in cryogenic systems. By eliminating the large temperature differential in the sensing lines, the heat leak and induced oscillations are greatly reduced. Sensitive measurements of low density fluids can be made. Actual flow fluctuations can be measured including entrained vapor bubbles.

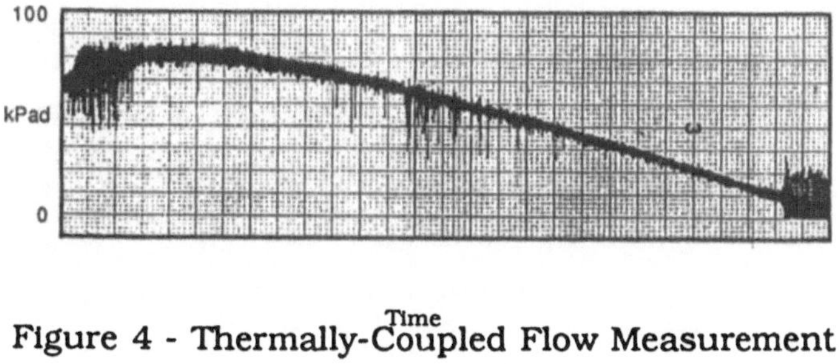

Figure 4 - Thermally-Coupled Flow Measurement

Piezoresistive cryogenic transducers can provide measurements over the entire temperature range with accuracies approaching those of conventional transducers. With the elimination of TAO, the uncertainty is reduced for many applications. This paper investigated one particular transducer as an example of the possible uses for thermally-coupled sensing techniques. Data from the manufacturer is included for two similar sensors for comparison. This particular approach was used to provide a transducer for a wide operating temperature range. The sensors can achieve much higher accuracies at a fixed, known temperature with appropriate calibrations.

REFERENCES

1. P. L. Walstrom and J. R. Maddox, Use of Siemens KPY Pressure Transducers at Liquid Helium Temperatures, Cryogenics, 27:439 (1987).
2. A. Kashani, A. L. Spivak, R. A. Wilcox and C. E. Woodhouse, Performance of Validyne Pressure Transducers in Liquid Helium, Proc. 12th Int. Cryo. Eng. Conf., Southhampton, U.K. (1988)
3. C. Boyd, D. Juanarena and M. G. Rao, Cryogenic Pressure Sensor Calibration Facility, in; Advances In Cryogenic Engineering, Vol. 35, R. W. Fast,ed., Plenum Publishing Corp., New York (1990), pp 1573-1581
4. D. L. Clark, Thermally-Coupled Cryogenic Pressure Sensing, in; Applications Of Cryogenic Technology, Vol. 10, J. Patrick Kelley,ed.,Plenum Publishing Corp., New York (1991)
5. D. B. Juanarena, M. G. Rao, Integrated Cryogenic Sensors, in: "Proc. 37th Int. Inst. Symp.,"San Diego, CA (1991), p.741-753

ACOUSTIC COMPOSITION SENSOR FOR CRYOGENIC GAS MIXTURES

P. Shakkottai, E.Y. Kwack, T.S. Luchik and L.H. Back

Jet Propulsion Laboratory
4800 Oak Grove Drive
Pasadena, California

ABSTRACT

An acoustic sensor useful for the determination of the composition of a gaseous binary mixture in cryogenic liquid spills has been characterized. One version of the instrument traps a known mixture of helium and nitrogen at ambient temperature in a tube which is interrogated by sonic pulses to determine the speed of sound and hence the composition. Experimental data shows that this sensor is quite accurate. The second version uses two unconfined microphones which sense sound pulses. Experimental data acquired during mixing when liquid nitrogen is poured into a vessel of gaseous helium is presented. Data during transient cooling of the tubular sensor containing nitrogen when the sensor is dipped into liquid nitrogen and during transient warm-up when the sensor is withdrawn are also presented. This sensor is being developed for use in the mixing of liquid cryogens with gas evolution in the simulation of liquid hydrogen/liquid oxygen explosion hazards.

INTRODUCTION

The Radioisotope Thermoelectric Generator (RTG) safety program at the Jet Propulsion Laboratory (JPL) had a need to experimentally determine the rate of gas evolution resulting from an accidental spill of liquid hydrogen (LH_2) and liquid oxygen (LO_2). The evolution of the gases would be due to two effects: flash vaporization since the two fluids were in equilibrium above one atmosphere and boiling due to heat transfer since the fluids would spill out on to a nearly room temperature surface. These two effects result in a complex measurement since the time scales for flashing are significantly shorter than that for heat transfer. Since LO_2 and LH_2 are reactive, LN_2 was to be substituted for LO_2 in the planned inert tests. LN_2 has been shown to be a good thermodynamic substitute for LO_2[1].

Since each of the cryogens would be evaporating simultaneously, it is necessary to know the temperature, velocity and composition of the mixture in order to determine the rate of evolution of each gas. Since the measurements were to be made in a region where the gases were "well-mixed", each gas component in the mixture could be treated

Advances in Cryogenic Engineering, Vol. 37, Part B
Edited by R.W. Fast, Plenum Press, New York, 1992

as having the same temperature and velocity. These two quantities can easily be measured using thermocouples and pitot-static tubes, respectively. The measurement of mixture composition is somewhat more difficult.

The rate of gas evolution in this particular problem is rapid, due to the flashing and occurs over a rather long period, due to the film and pool boiling. Gas chromatography for the determination of composition is not suited to this problem because it is too slow and would miss the initial transient. Mass spectrometry is not practical for this problem since mass spectrometers do not recognize very light gases like helium and hydrogen. This was the motivation to develop a technique and sensor which could be used to determine mixture composition in mixtures of hydrogen and nitrogen.

The speed of sound in a mixture of gases is a function of temperature and composition. Therefore, for known gas mixtures, the measurement of speed of sound can be used to accurately determine temperature. An acoustic technique for measurement of temperature is described by Venkateshan et al[2]. For binary gas mixtures of gases, theoretically the mixture ratio can be determined accurately if the temperature of the mixture is known and the gases are well mixed.

In this paper, we address the development of the acoustic composition sensor which can be used to determine the composition of a binary gas mixture. Two sensors are described, each having one quality that the other lacks. Results of transient heating and cooling of nitrogen-helium gas mixtures are discussed.

EXPERIMENTAL SETUP

Two types of acoustic sensors were used. The first is a tubular sensor made of a stainless steel tube of internal diameter 1.27 cm and of length 116.1 cm to which a

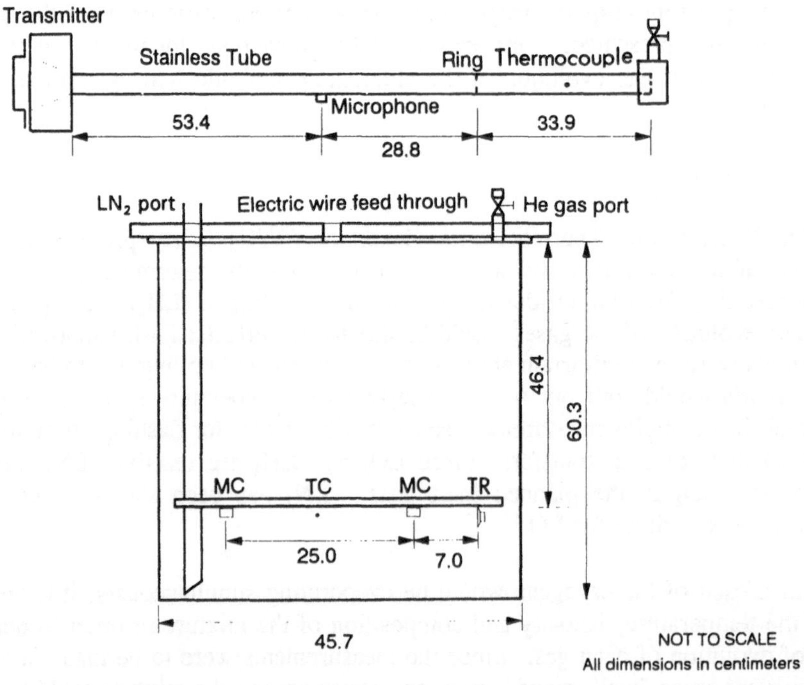

Figure 1. Schematic of single-microphone tubular sensor (top) and dual-microphone unconfined sensor (bottom).

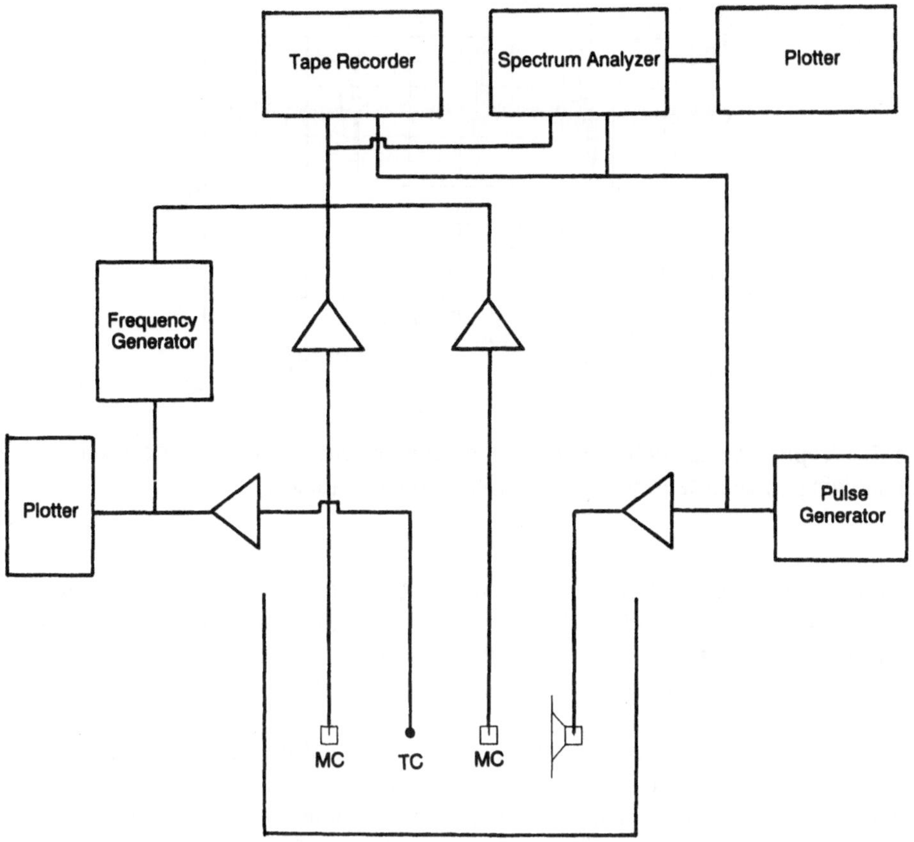

Figure 2. Block diagram of signal processing system.

transition piece of lucite was fixed at one end as shown in Fig. 1. A brass end fitting with a valve closed the other end. This allowed the entry of chosen gases into the tube. At the transition piece, a piezoelectric loud speaker (tweeter) was attached and driven by a pulse generator to produce sonic pulses in the tubular sensor.

A microphone located 53.4 cm from the transition piece was used to sense the acoustic wave motion in the tube. The sound pulses reflected by a ring located 28.8 cm from the microphone and the reflections from the end wall, located 33.9 cm from the ring, were used to determine the speed of sound in the region between the ring and the end wall as shown in Fig. 1. A thermocouple measured the temperature at the mid point of this region. Multiple reflections can also be used to improve time resolution, if necessary.

If the region between the ring and end wall is made porous by drilling with many holes of small size, the tubular sensor can be used to measure the speed of sound in the same region for the case where the gases outside the tube diffuse or convect to the inside of the tube. This time constant was too long in some cases of accidental spills of the cryogenic liquids and was therefore not employed here.

The second version of the sensor consisted of two-microphones and a transmitter mounted in a straight line. This arrangement was located in a large stainless steel vessel (Fig. 1) of diameter equal to 45.7 cm and height 60.3 cm. This vessel was closed by a lid of lucite (1.27cm thick). A gas port allowed helium to enter the chamber and another port allows liquid N_2 to be poured in through the tube that extended to the bottom of the vessel. In Fig. 1, the locations of the microphones and the transmitter are given. The apparatus

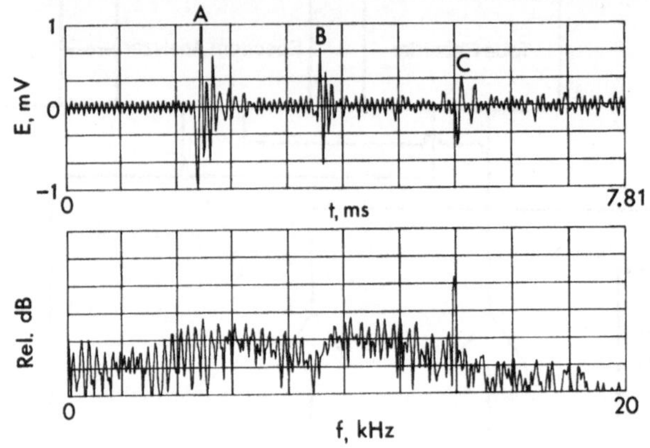

Figure 3. Typical signal obtained from the tubluar sensor (above), spectrum analyzer output showing spike at thermocouple carrier frequency (below).

was deliberately made to allow leaks at the lid so that it was possible to pour liquid nitrogen through the port quickly. To prevent all the helium initially in the vessel from being displaced immediately at the start of the experiment, a small flow of helium gas was allowed and continued throughout the experiment.

Fig. 2 shows the signal processing system. A pulse generator produced pulses repeated approximately once a second. These pulses were amplified and applied to a transmitter. The signals received by the two microphones were amplified and added together. Also added to these is the frequency modulated signal from the thermocouple. This signal was obtained by D.C. amplifying the thermocouple signal and using a voltage to frequency converter. The combined signal shows a sine wave of low amplitude representing the temperature as well as pulses associated with signals detected by one microphone or two microphones, as the case may be. The composite signal was recorded on tape and played back later. The original pulse was also recorded to aid in triggering the signal analyzer. The output of the signal analyzer was plotted by using a digital plotter.

EXPERIMENTAL DATA AND DISCUSSION

A typical signal from the tubular sensor is shown in the upper plot of Fig. 3. In the Figure, A is the initial pulse sensed by the microphone, B is the signal due to reflection

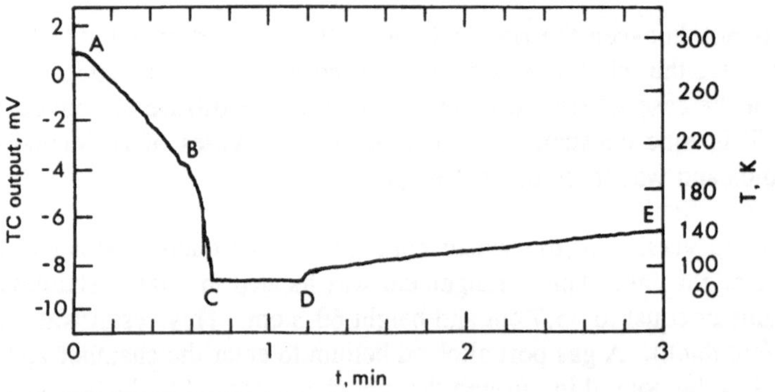

Figure 4. Thermocouple trace during LN_2 cool-down and subsequent warm-up.

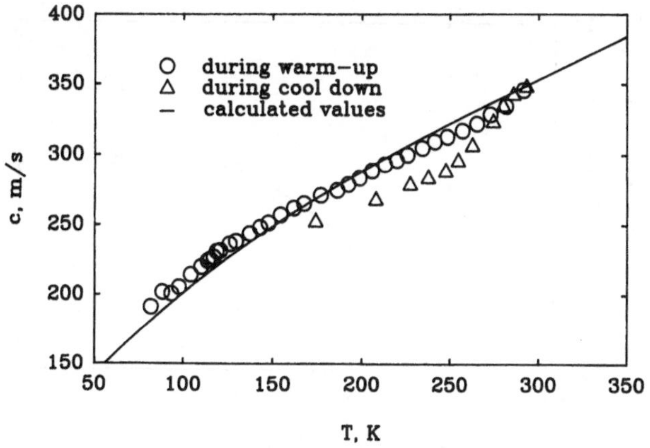

Figure 5. Sonic speed measured during cool-down and warm-up.

from the ring and C is the reflection from the end wall. The sine wave occurring as a small baseline has the frequency corresponding to the thermocouple signal. This frequency is easily read from the strong peak in the spectrum shown in the lower part of Fig. 3. The temperature scale is linear in frequency and varies from 15 kHz at 296 K to 6.4 kHz at 81 K. The speed of sound in the test section is determined by dividing 33.9 cm by half of the time interval from B to C which is 1.98 ms in Fig. 3.

The tubular sensor was plunged up to the location of the microphone into a dewar containing liquid Nitrogen in a test to record the transient temperatures. The signal from the thermocouple, shown in Fig. 4, decreases linearly for approximately 30 sec from A to B and then decreases more rapidly along BC indicating more rapid heat transfer. It is possible that a small amount of liquid nitrogen leaked in at the bottom through the valve to cause this rapid cooling inside. The minimum temperature is reached at C rather abruptly and remains almost constant along CD. Near D, the sensor was pulled out and was allowed to warm up in air. This warming process along DE is much slower.

Data were recorded on tape all through the test and were reduced by playing the tape back. In Fig. 5 the measured values of the speed of sound, c, are plotted vs. temperature, T, and compared with the known speed of sound of nitrogen. The speed of sound measured corresponds to an average in the region between the end wall and the ring whereas the calculated value corresponds to the speed of sound at the temperature measured by the thermocouple.

During the transient period, cooling and warming, the temperature at the center of the tube, measured by the thermocouple, differs from the average temperature of the gas. Therefore, during cooling, the measured speeds of sound are lower than the values at the thermocouple, as expected. Since heat transfer is more rapid during cooling than warming, values of c measured during cooling are further from the values calculated using the measured temperature than those during warming. From 80K to 180K they are slightly higher indicating that the thermocouple is colder than the average which is to be expected. But in the region from 180K to 300K, the thermocouple is warmer than the average which was not expected. This may be due to axial temperature gradients caused by the large thermal mass (brass block) at the end of the test section. Errors due to these axial gradients could be minimized by using more thermocouples located axially along the sensor. During the cooling period it was not possible to get data between 180K and 80K because the recorded signals were too noisy. It is possible that liquid nitrogen entered the tube and caused a hissing noise due to boiling inside the tube.

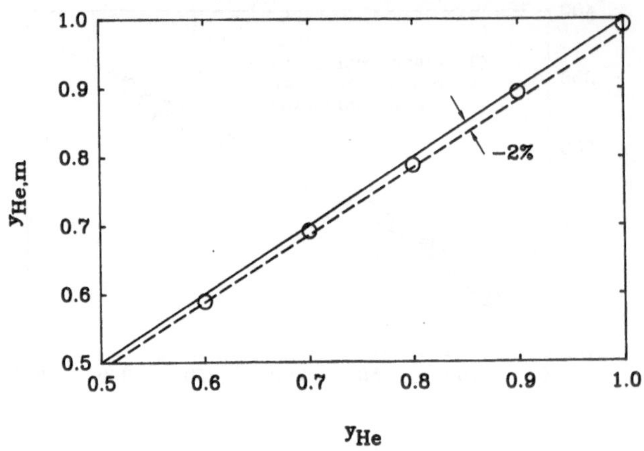

Figure 6. Comparison of known and acoustically measured volume fraction of helium-nitrogen mixtures at room temperature.

Tests with Gas Mixtures

Tests were conducted by measuring the speed of sound of known mixtures of helium and nitrogen at ambient temperature. A similar tubular sensor was employed and data were acquired at two values of gage lengths between the ring and the wall at 30.9 cm and 78.6 cm. The detailed description of these sensors and results are reported elsewhere[3]. A plot of measured composition vs. the known composition is shown in Fig. 6. The measured values of volume fraction of helium are lower than the known values by less than 2% for y_{He} from 0.6 to 1.0. The differences are most likely due to air leaks into helium. The speed of sound measurement is quite satisfactory to measure the composition.

To simulate the mixing of two cryogenic liquids, a vessel was filled with He gas, and LN_2 was poured into the vessel. For fast time response, the speed of sound was measured using the two microphone sensor technique (Fig. 1). Figure 7 shows a typical signal measured in air at room temperature which show multiple reflections from the vessel wall in addition to the main pulses A and B which were detected by the two microphones. When the vessel was filled with helium, the time scale was shortened to approximately one third the values in Fig. 7. Only the peaks A and B were used to determine the speeds of sound. The tape recorder signal contained a baseline sine wave as in Fig. 3, which is not shown in Fig. 7.

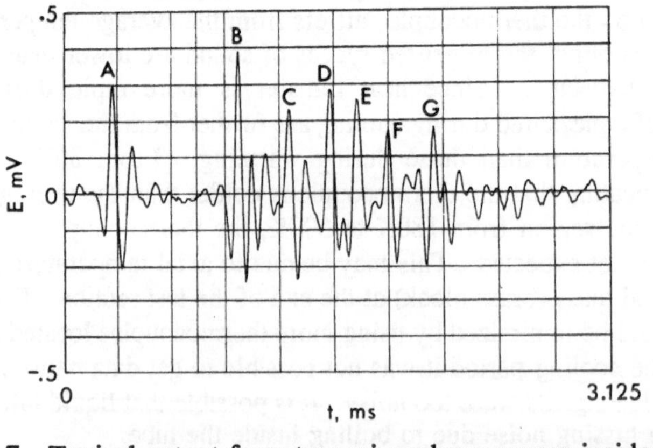

Figure 7. Typical output from two-microphone sensor inside vessel.

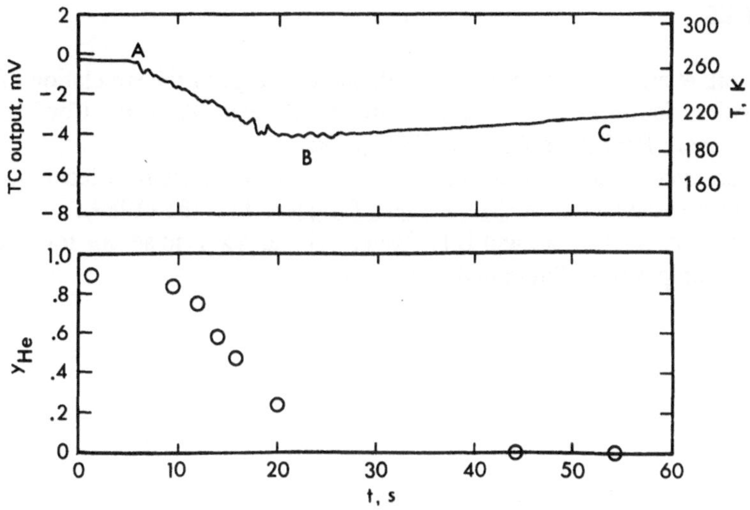

Figure 8. Thermocouple output during LN_2 addition (above) and corresponding measured volume fraction of helium.

In Fig. 8, the variation of temperature with time is shown in the experiment where liquid N_2 was poured into the vessel containing helium. The temperature dropped from an initial temperature at A to B rapidly during which time liquid nitrogen was boiling at the bottom below the microphone. The temperature trace was unsteady in this region. From B to C the gas mixture warmed up slowly after all the liquid nitrogen had boiled away. The minimum temperature reached in this experiment was only 200 K.

In the lower plot of Fig. 8, values of the volume fraction of helium, y_{He}, calculated from the measured speed of sound are plotted against time. It is seen that helium was displaced by the nitrogen in 30 sec. For times greater than 30 sec., up to 210 sec. (the duration of the test), the region near the microphones was pure nitrogen in spite of the fact that helium was continuously flowed into the vessel.

CONCLUSION

Transient measurements of temperature and composition have been made in an experiment with a time scale of 30 sec. It is clear that much more rapid phenomena can also be monitored since measurements can be obtained from every pulse, which can be repeated several times per second, if desired.

The tubular sensor is less sensitive to external noise and if the phenomena is slow enough, it is the better choice of sensors. The porous tubular sensor had a time constant of a minute or so. If the phenomena is more rapid, bare microphones must be used and under noisy conditions some data will be unusable. The smallest separation distance for the two microphone configuration depends upon the speed of sound in the gas mixture. It is about 30 cm for the case of the speed of sound equal to 1000 m/s. Generally, long separation distances are better for improved resolution.

ACKNOWLEDGEMENT

The research carried out in this paper was carried out in the Applied Technologies Section of the Jet Propulsion Laboratory, California Institute of Technology, under contract to the National Aeronautics and Space Administration. The work was internally funded through the Flight Projects Office as part of the RTG Safety Task.

REFERENCES

1. T.S. Luchik et al, Cryogenic mixing with phase change in the simulation of LO_2/LH_2 explosion hazards, in: "Proc. Ninth Intl. Heat Transfer Conf.", Vol. 1. Hemisphere Publishing Corp. (1990), p. 365.
2. S.P. Venkateshan et al, Acoustic temperature profile measurement technique for large combustion chambers, ASME J. Heat Transfer, 111:461 (1989).
3. E.Y. Kwack, P. Shakkottai, and L.H. Back, Tubular acoustic sensor for measurement of gas composition, Submitted to Rev. Sci. Inst.

MONITORING OF CRYOGENIC FLOWS: REALIZATION OF

THE RADIO FREQUENCY METHOD

Yu. P. Filippov, A. I. Alexeyev, N.I. Lebedev,
I. S. Mamedov, S. V. Romanov

Joint Institute for Nuclear Research
Particle Physics Laboratory
Dubna, USSR

ABSTRACT

The report presents the sensors, instruments, and methodology to control such characteristics of single-phase and two-phase flows as average temperature T, mean density ρ, void fraction φ, quality x and flow rate G. Radio frequency method is used to achieve a high accuracy. The hydrodynamics features specific to channels of round and annular cross-sections are taken into account while working out each type of the sensor. Intelligent controllers of two modifications: a CAMAC module and an independent device equipped with the RS 232 interface are described. Application to fast determination of vapor-liquid expander efficiency, helium liquefaction capacity, etc. is demonstrated.

INTRODUCTION

"Over 90% of all measurements are temperature, pressure, level and flow rate measurements".[1] This also applies to cryogenics, where the correct choice of cryogen parameters determines safety and efficiency of research and industrial installations. Cryogens are quite often used in a two-phase state or in the form of a 2-component or 3-component mixture; so, besides the parameters mentioned, it is necessary to determine phase ratio characteristics (local and average vapor content, concentrations and flow patterns). It should also be mentioned that some specific features of cryogenic systems and cryogens impose significant restrictions on the measuring equipment.

Besides low temperature, the specific features affecting the choice of the principle of operation and the design of the sensor for cryogenic systems are: 1. Relatively difficult replacement and size limits. So the sensitive element of the sensor must be small in size and sufficiently reliable, which is achieved by a simple design without moving parts; 2. Possible transitions from two-phase to single-phase flows. In this connection, one should choose at least one quantity carrying information about both single-phase and two-phase states of a fluid as a measured (signal) characteristic of a flow; 3. Significant difference between the physical properties of cryogens and the properties of phases, which is usually observed in practice. Helium is a typical example: two-phase HeI[2] and especially HeII gas flows.[3]

A helpful tool for diagnostics of both two-phase flows and single-phase flows close to saturation is a resonance radio frequency sensor.[4] Its signal characteristic depends on the dielectric constant of the medium ε. In a two-phase vapor-liquid medium the dielectric constants of the phases ε_l and ε_g depend on the saturation pressure P_s. So, measuring P_s (or T_s) and ε one can obtain information on the qualitative ratios of phases in the flow, both local $\varphi_{loc}=\varphi(\vec{r})$ and average $\varphi=\langle\varphi(\vec{r})\rangle=A_g/(A_g+A_l)$. As it is shown below, the void fraction φ and the quality $x=G_g/(G_g+G_l)$ are connected to each other through simple relations. It allows determination of mean flow-rate thermodynamic characteristics: enthalpy $i=i_g x+i_l(1-x)$, entropy $s=s_g x+s_l(1-x)$, etc. In a single-phase region the quantities P and ε unambiguously identify the thermodynamic state of a fluid and, like P and T, they allow determination of other thermophysical characteristics. The analysis shows that RF sensors can comply with the above-mentioned specific features of cryogenic measurements. This paper deals with cryogenic applications of intelligent RF sensors alone and in combination with other devices for determination of qualitative and quantitative characteristics of cryogens.

SENSITIVE ELEMENTS

The output parameter of an RF sensor is the resonance frequency f, which is defined in a general case as[4] $f^2=f_0^2\left(1+\int \varepsilon(\vec{r})\vec{E}^2(\vec{r})d\vec{r}/\int \vec{E}^2(\vec{r})d\vec{r}\right)$ where f_0 and $\vec{E}$ are the resonance frequency and the electric field strength of the sensor "filled" with vacuum. Thus, the quantity f depends not only on the mean integral dielectric constant of the flow but also on the distribution of phases over the channel cross section. So, by exciting oscillations with different configurations of the electric field $\vec{E}(\vec{r})$, one can both determine the mean characteristics and study the fine structure of the flows. Let us consider two examples involving measuring resonators with annular or circular channels.

Sensor with an annular channel

If there is an annular channel, it seems reasonable to use a sensor in the form of a short-circuited coaxial resonator with inductive excitation of oscillations (Fig. 1). When the first harmonic of class TEM oscillations is excited in the resonator, the electric field strength between the channel walls varies logarithmically, and at a sufficiently small value of δ/d_{in} the field is close to the uniform one. The length l_s of the sensitive element must satisfy the resonance conditions and be relatively large ($l_s \ll d_{in}$) to reduce nonlinear edge effects. Uniformity

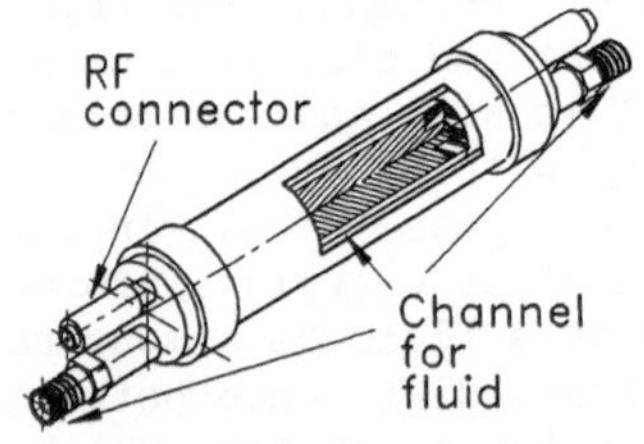

Fig. 1. Sensor with channel of annular cross-section.

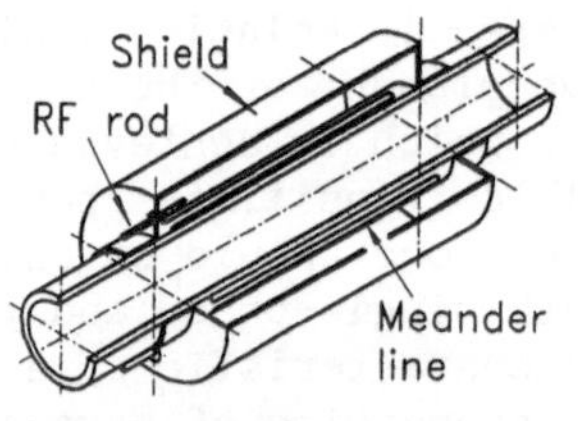

Fig. 2. Sensor with channel of round cross-section.

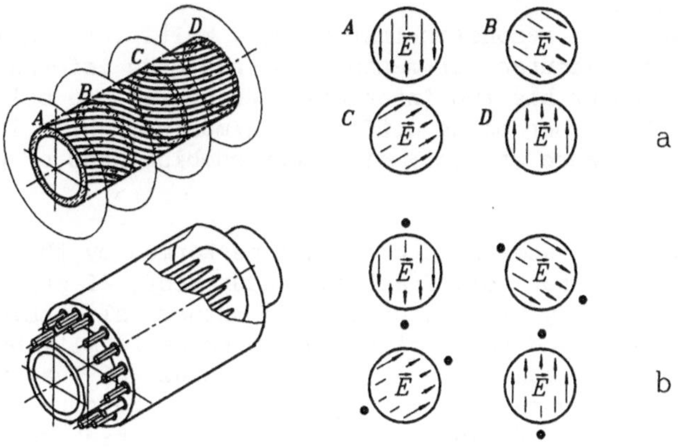

Fig. 3. Sensors with a twisted meander and several pairs of rods.

of the field allows measurement of the cross-section mean dielectric constant and thus the void fraction of flow φ. By exciting oscillations of higher harmonics in the resonator, one can obtain more complicated configurations of the electric field. Analyzing a series of resonance frequencies of the first and higher harmonics f_0, f_1, ..., f_n, one can obtain information on the phase distribution structure in the flow on the basis of the relevant algorithms.

Sensor with a round channel

A sensor with a round channel is an open resonator with capacitive excitation of oscillations. A long meander-type line is applied to the external surface of a dielectric tube that serves as a measuring channel.[4] When a resonance of the first-mode oscillations (of the dipole type) is excited in the line, the potential distribution over the tube surface is close to the harmonic one. If the sensor is sufficiently long and the meander pitch is sufficiently small, there is a highly uniform field in the measuring volume. So, as in the previous case, the measurement at the first harmonic of TEM oscillations allows one to determine the mean dielectric constant of the flow. Yet, there are some difficulties arising from the fact that even an ideal uniform field is distorted by a nonuniform dielectric, and the resonance frequency f thus depends on orientation of the phase boundaries with respect to the direction of $\vec{E}$. It is best manifested in stratified flows or flows close to stratified ones.

There are two ways to decrease the corresponding error. 1- Using a twisted meander. In a sensor of this design (see Fig. 3a) much of the error is automatically corrected. Yet, the twisted meander sensor can be mainly used to determine integral characteristics. 2- Using several pairs of communication rods (see Fig. 3b). The resulting value of φ is calculated as a mean $\varphi=\varphi(\varphi_1,\varphi_2,\ldots,\varphi_n)$ of the values of φ_i obtained for each pair of rods. A sensitive element with several pairs of exciting and detecting rods allows development of a device for investigations of the internal flow patterns in a round channel. Analyzing the first harmonic frequencies $f_{01},f_{02},\ldots,f_{0n}$ of the resonator for different pairs of rods and similar frequency sets $f_{11},f_{12},\ldots,f_{1n}$; $f_{m1},f_{m2},\ldots,f_{mn}$ of higher harmonics, one can reconstruct the two-dimensional pattern of the phase distribution in the flow in some approximation, using mathematical methods of tomography. This design of a "dielectric tomographer" resembles the "flow image" capacitive device presented in reference.[5]

However, the use of a resonator instead of a multiplate capacitor[5] allows to obtain a higher accuracy of measurement and more information, since a resonator tomographer has two information channels, one being different pairs of rods, the other being different harmonics, while in the capacitive device there are only different combinations of pairs of plates.

The calculations and the experiments performed by the authors[6] show that RF sensors used for void fraction measurements of cryogen flows allow in principle a very high accuracy, the methodical (unavoidable) error related to the field and phase distribution nonuniformity being ≤2.5%. Yet, the actual error can be larger mainly because of inaccurate evaluation of φ by the resonance frequency f and pressure P, i.e. because of calibration errors. Besides, in many cases the measurements are aimed at finding the quality x and not φ, the interrelation of these quantities being determined by the flow hydrodynamics. These two important problems of measurement analysis are dealt with in the next section.

INTERPRETATION OF MEASUREMENT

Calibration

The calibration of a radio frequency VF sensor with a uniform energy distribution of the electric field consists in determining the dependence of the resonance frequency f on the dielectric constant ε of the medium in the sensor. In a wide variation range of ε this problem can probably be solved only experimentally or, in some cases, by numerical simulation. Yet, in the practically important pressure interval the difference in vapor and liquid dielectric constants of different cryogenic media is comparatively small. For example, at P=100 kPa we have $(\varepsilon_l-\varepsilon_g)/\varepsilon_l \cong 0.04$ for helium, $(\varepsilon_l-\varepsilon_g)/\varepsilon_l \cong 0.18$ for hydrogen, $(\varepsilon_l-\varepsilon_g)/\varepsilon_l \cong 0.29$ for nitrogen (here ε_l and ε_g are the dielectric constants of the liquid and gaseous cryogens on the saturation line). Our measurements showed that the relative variation of the resonance frequency $(f_l-f_g)/f_l$ in the φ interval from 0 to 1 is not large either.* On the other hand, to choose an approximation, which is valid with the prescribed degree of accuracy in a small variation interval of the argument ε and the function f, is quite a solvable problem. It can be done by fixing the form of the function with several coefficients that can be found from the experimental data. One can solve the function $f(\varepsilon)$, for example, as a power series in $\Delta\varepsilon = \bar{\varepsilon}-\varepsilon_l$: $f(\bar{\varepsilon}) = f(\varepsilon_l+\Delta\varepsilon)=f_l+a_1\Delta\varepsilon+a_2(\Delta\varepsilon)^2+a_3(\Delta\varepsilon)^3+\dots$. However, there seems to be a more preferable way when the resonator is replaced by an equivalent circuit with the concentrated parameters, which allows a physically substantiated form of the function $f(\varepsilon) = F_{phy}(\varepsilon, K_1, K_2, \dots, K_n)$ with a set of experimentally determinate coefficients $K_1, K_2, \dots, K_n$. The coefficients K_i are determined by the measurements of resonance frequencies f_i while one fills the sensor with media of known dielectric constants ε_i. Compared with the previous approach, this one must ensure the prescribed accuracy at a much smaller number of coefficients, which is proved by the experiments with helium and nitrogen. In Fig. 4 there is the simplest equivalent circuit of an annular sensor and the corresponding calibration curve $f(\varepsilon)=K_1/(\varepsilon+K_2)^{0.5}$

---

*As pointed out in reference,[6] the value of $(f_l-f_g)/f_l$ for helium is ≅ 0.016, which makes the requirements for the measuring equipment more stringent.

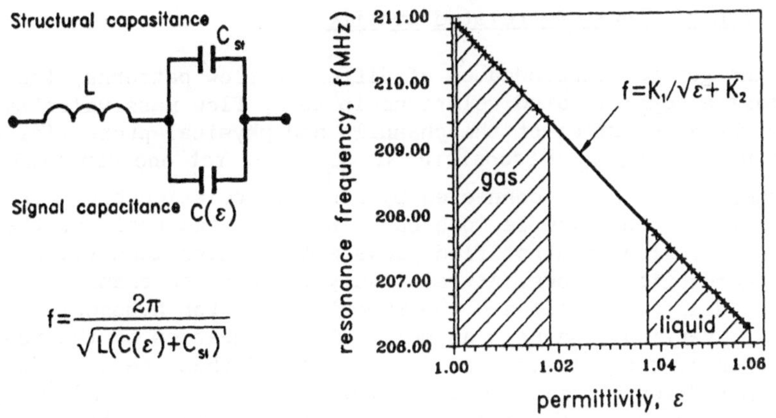

Fig. 4. Equivalent circuit and calibration curve of annular sensor for helium.

The coefficients K_1 and K_2 were calculated after processing a set of the experimental data obtained while the sensor was filled with liquid and gaseous helium on the saturation line at pressures 50 kPa $\le P \le$ 210 kPa. The variation interval of ε, "directly" measured during the calibration, appears to be 2/3 of the whole signal range of the measurements. The adequacy of the calibration function in the whole range is proved by the experiments, when the quantity φ was determined by the saturated helium and nitrogen filling level of the sensor.

Proceeding from the calibration function of the sensor filled with a single-phase fluid $f=f(\varepsilon)$ to the functions $\varphi=\varphi(f)$ for a two-phase flow, we assume that $\varepsilon=\varphi\varepsilon_g+(1-\varphi)\varepsilon_l$, which is a fairly accurate assumption for cryogenic fluids $(\varepsilon_l-\varepsilon_g)/\varepsilon_l \ll 1$. With this assumption, the calibration function $\varphi(f)$ for an annular sensor takes on the form $\varphi = (\varepsilon_l-(K_1^2/f^2-K_2))/(\varepsilon_l-\varepsilon_g)$. Fig. 5 shows the calibration functions for three cryogens. One can see that the functions $\varphi(f)$ are close to linear ones. More accurate calibration characteristics can be obtained through numerical calculation of the resonators. The methods and software for this calculation are well developed.

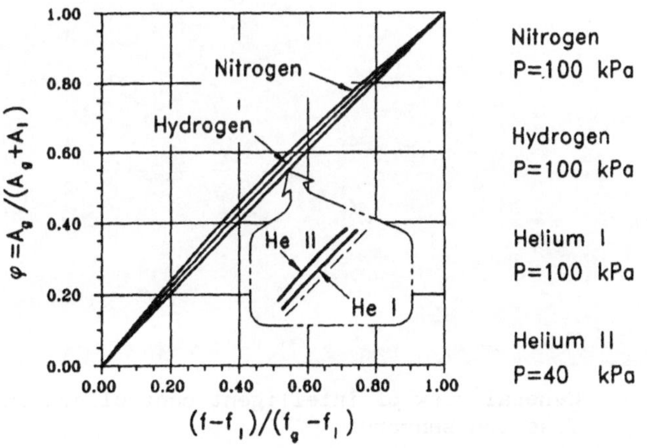

Fig. 5. Calibration functions for some cryogens.

Determination of flow-rate related quality

In the general case, because of different flow patterns, the quantity x bears a sophisticated relationship to φ, flow mass velocity m, geometric characteristics of the channels and physical properties of the phases: densities ρ_l, ρ_g, viscosities μ_l, μ_g.... Yet one can find the regions where x and φ are connected by a simple relationship. In horizontal flows, for example, one can single out two regions: the low mass velocity region of stratified flows and the high mass velocity region of homogenized flows. At relatively large m the channel orientation does not matter and the function $x(\varphi)$ for homogenized flows can apply to vertical channels as well. The boundaries of the regions and the types of the relations $x=x(\varphi)$ in each of them are considered, for example, in reference,[7] and for two-phase helium flows in reference.[2] For example, for narrow rectangular or annular channels with helium $x(\varphi)$ functions are determined as follows: $x_s=\varphi/((\rho_l/\rho_g)^{4/7}(\mu_g/\mu_l)^{1/7}(1-\varphi)+\varphi)$ in the case of stratified flows, and $x_n=\varphi/((\rho_l/\rho_g)(1-\varphi)+\varphi)$ in the case of homogenized flows. In reference[6] it is shown that the above functions are in good agreement with the experimental results. Thus, when the void fraction φ is known, the quantity x for hydrodynamically stabilized flows can be unambiguously determined by simple formulae, if the flows are definitely stratified or homogenized. However, none of these formulae can be absolutely accurate, which increases the error of determination of x as compared with the error $\delta\varphi$. As a result, δx may amount, e.g. for helium, to ≅5-7%.

As it seen from the above, the RF sensor signal processing requires electronic computer means. Suitable for this procedure are specially designed intelligent controllers with standard interfaces for connection to data acquisition and processing systems.

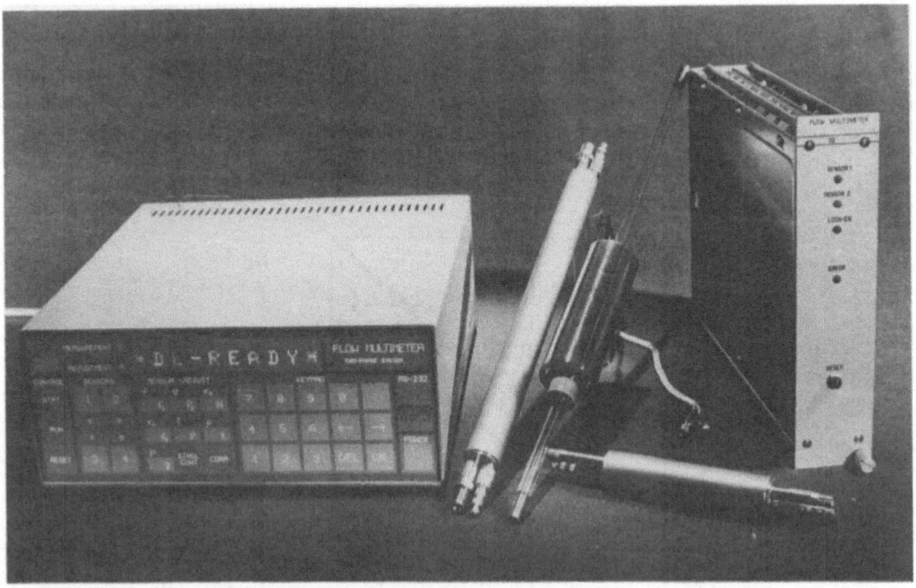

Fig. 6. General view of intelligent controllers and void fraction sensors.

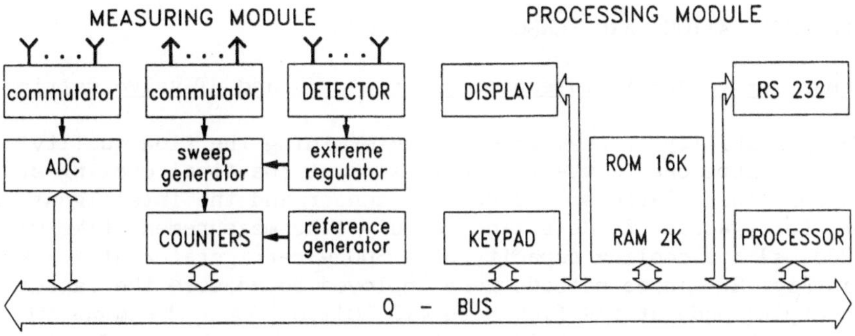

Fig. 7. The block-diagram of the independent controller.

INTELLIGENT CONTROLLERS

A general view of a controller as an independent block is given in Fig. 6. The block-diagram of this device is shown in Fig. 7. The device is structurally divided into two modules - a measuring and a processing one. The measuring module consists of a sweep generator (*SG*), an extreme regulator (*ER*), *RF* commutators, a commutator of analogue signals, a 10-bit *ADC* (for pressure measurement). To determine the resonance frequency, the *SG* sends a signal to the sensor. From the sensor the frequency signal goes to the extreme regulator, which controls the sweep generator and maintains the resonance frequency *f*. From the second output of the *SG* the signal is applied to a frequency meter consisting of two counters (f and f_{ref}). To achieve a higher count accuracy, the frequency variation period is chosen to be a multiple of the *ER* gating period.

Another version of the intelligent controller is made to the CAMAC standard (Fig. 6). Its structure is similar to that of the independent controller considered. Control, data input and output are carried out through the CAMAC bus. Two resonators are connected to this controller. This block calculates fewer characteristics than the independent one because further analysis of the data can be performed by the computer controlling the crate.

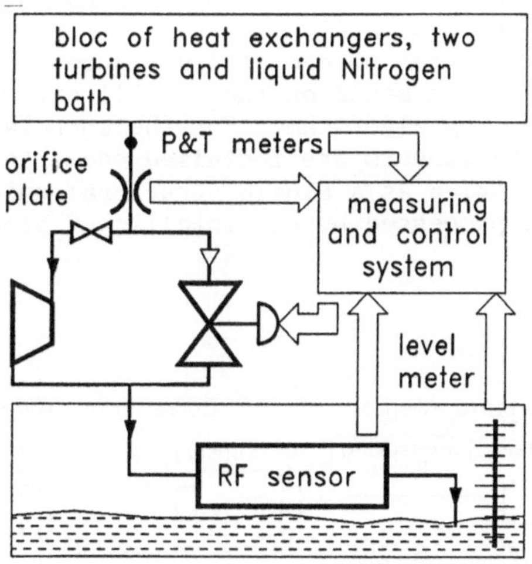

Fig. 8. Applications of void fraction sensor in refrigerator.

Measurement of vapor-liquid expander efficiency and JT-valve setting

The possibility of unambiguously determining the flow quality naturally prompts one to use the RF sensors at the final cooling stages of cryogenic installations. We use the sensor and the intelligent controller, for example, to determine rapidly (the measurement time was from 20 to 200 ms) the cooling capacity of a 300 W refrigerator at 4.5 K, as shown in Fig. 8. Three methods were employed to measure the capacity in the liquefying mode at the flow rate $(13\text{-}20)\cdot 10^{-3}$ kg/s: by means of a level meter, by means of a void fraction sensor, and by *P* & *T* values in front of the *JT*-valve. The results obtained are in good agreement with each other. The discrepancy is within 2-9%, which is due to the orifice plate and the RF sensor calibration errors and to possible partial entrainment of liquid helium in the reverse flow. Also the void fraction sensor is used to determine the thermodynamic efficiency of a vapor-liquid turbine where it seems to be indispensable (the turbine was manufactured at the SKIF factory, Moscow).

Our experience indicates that *RF* sensors are valuable measuring devices for automatic regulation systems in cryogenic installations, producing a relatively fast effect on a regulator after a possible disturbance.

Flow rate measurement

When the single-phase flow rate is determined by the pressure difference ΔP in a narrowing device $G=\xi(\rho\ \Delta P)^{0.5}$ or by the number of turbine revolutions n $G=\xi \rho n$, the *RF* sensor can serve as mean density meter. So one of the necessary parameters determined by the intelligent device is the mean density ρ.

The two-phase flow rate can also be determined by the pressure difference at the orifice plate[8] or by the number of turbine revolutions n.[7] In this case, however, the flow rate G is involved in sophisticated relations $G = G(\Delta P,\ x,$ thermophysical properties) or $G=G(n, x,$ thermophysical properties) and the *RF* sensor is to be used for estimation of x. The correlation and tagging methods or their combination seem to be more preferable for determination of the two-phase flow rate (see Fig. 9). When we used these methods with *RF* sensors, the tags were heat pulses applied to the flow by a heater placed at a distance from the sensor and in front of it. The device is based on the intelligent controller with an additional data processing algorithm. For the correlation measurements of the flow rate two *RF* sensors are installed one after another in the flow (they can be assembled as a single structural unit). Their frequency signals are processed by one intelligent block, which computes

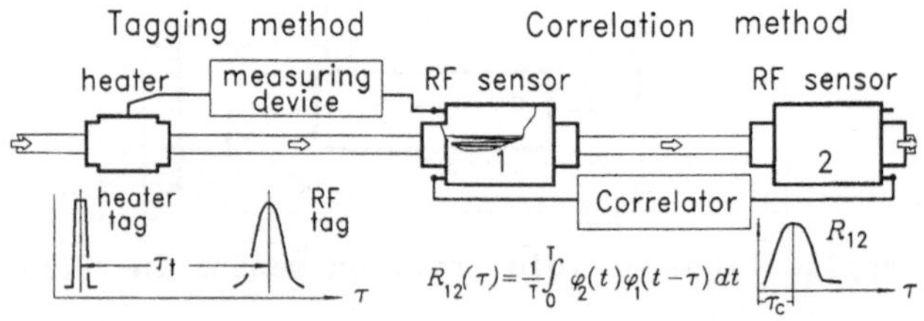

Fig. 9. Determination of the two-phase flow rate.

the mutual correlation function $R_{12}(\tau) = \frac{1}{T}\int_0^T \varphi_2(t)\varphi_1(t-\tau)dt$. The flow rate is calculated as $G=\beta\rho SL/\tau_c$, where β is the correction coefficient, S is the channel cross section, L is the distance between the sensors, τ_c is the time for which R_{12} has the maximum value. A flowmeter of this type on the basis of the independent controller described in the previous section is also being developed now. For any pair (in the set of 4 sensors) there is a possibility of calculating the mutual correlation function and the mass flow rate. The preliminary experiments showed that the accuracy of the two-phase helium flow rate measurement by the tagging and correlation methods is noticeably higher as compared with the calorimetric measurement[6] or measurement with a flowmeter based on an orifice plate and a void fraction sensor.

Determination of concentrations of binary and triple mixtures

The dielectric constant of a binary mixture ε_{bin} depends on the concentration of components ζ and on their dielectric constants ε_1 and ε_2: $\varepsilon_{bin}=\varepsilon_{bin}(\zeta,\varepsilon_1,\varepsilon_2)$. In their turn, the dielectric constants are state functions $\varepsilon_1=\varepsilon_1(P,T)$, $\varepsilon_2=\varepsilon_2(P,T)$. Thus, measuring the pressure P, the temperature T and the dielectric constant of the flow with RF sensors, one can find the dielectric constant of a binary mixture and, consequently, its concentration. In cryogenics, for example, a $^3He+^4He$ solution is used to obtain very low temperatures. Miniature *RF* sensors can be a suitable tool for determining concentrations of components. To avoid heat input through *RF* cables a capacitive, i.e. completely contactless, connection can be used between the resonator and the measuring equipment.

To analyze a triple mixture, one needs one more parameter, for example, the medium density ρ_{trip}. The density of a cryogenic mixture flow can be measured with a density meter operating on the basis of the high-temperature superconductor levitation effect. It is small in size, which is important for placing it inside a pipeline, and highly accurate ($\delta\rho\leq 0.06\%$). Concentrations of components ζ_1, ζ_2, ζ_3 are found by solving the set of three equations

$$\begin{cases} \varepsilon_{trip} = \varepsilon_{trip}(\zeta_1,\zeta_2,\zeta_3,\varepsilon_1(P,T),\varepsilon_2(P,T),\varepsilon_3(P,T)) \\ \rho_{trip} = \rho_{trip}(\zeta_1,\zeta_2,\zeta_3,\rho_1(P,T),\rho_2(P,T),\rho_3(P,T)) \\ \zeta_1+\zeta_2+\zeta_3=1 \end{cases}$$

Although at first glance this way of analysis seems to be sophisticated sometimes it can be much simpler and cheaper than traditional chromatography.

CONCLUSION

To monitor cryogen flows, it is convenient to use a radio frequency method, which involves a signal capacitor connected to an oscillator circuit whose resonance frequency depends on the dielectric constants of cryogens. This method has the following advantages:

- high accuracy due to a specific design and low-temperature calibration;
- allowance for specific features of hydrodynamics in round and annular sensitive elements;
- adaptability to any dielectric media.

Using intelligent radio frequency sensors, one can determine different parameters of cryogens: in the two-phase region they are the void fraction φ, the quality x, and φ- and x-dependent mean-volume and mean-flow-rate thermodynamic characteristics - enthalpy, density, entropy, etc.; in the single-phase region they are the mean-volume thermodynamic quantities - temperature, density, enthalpy, etc., and in two-/three-component mixtures they are concentrations of components. Combining these sensors with orifice plates, turbines, pulsed heat sources, or using them in pairs, one can measure flow rates of both two-phase and single-phase media.

It is reasonable to use intelligent *RF* sensors for diagnostics of cryostabilization systems in superconducting devices, for fast determination of the liquefaction capacity and vapor-liquid expander efficiency, in cryogenic propellant filling systems, in automatic regulation systems of cryogenic installations and for science research.

REFERENCES

1. E. H. Higman, R. Fell, A. Aiaya, J. Meas. and Contr. 19:47 (1986).
2. A. I. Alexeyev, Yu. P. Filippov, I. S. Mamedov, Flow Pattern of Two-Phase Helium in Horizontal Channels, Cryogenics 31:330 (1991).
3. G. Brianti et al, Chapter 6, Cryogenics, in: "Design Study of the Large Hadron Collider (LHC)," CERN/AC/DI/FA/90-06 (1990), p. 109.
4. V. A. Viktorov, B. V Lunkin, A. S. Sovlukov, RF-Measurements of Technologic Process Parameters, Energoatomizdat, Moscow (1989).
5. S. M. Huang, A. B. Plaskowski, C. G. Xie, M. S. Beck, J. Phys. E.: Sci. Instrum. 3:173 (1989).
6. V. V. Danilov, Yu. P. Filippov, I. S. Mamedov, in: "Advances in Cryogenic Engineering," Vol. 35, Plenum Press, New York (1990), p. 745.
7. "Encyclopedia of Fluid Mechanics," Vol. 3, Gas-Liquid Flow, Ed. N. P. Cheremisinoff, Gulf Publishing Company Book Division, Houston (1985)
8. D. Chisholm, in: "Proc. 13th Intl.Conf. Inst. Rerfig. Congr.," 2:781 (1980).

CALIBRATION SCENARIO OF LEVEL GAUGE FOR PRESSURIZED CRYOGENIC VESSELS

Burt X. Zhang and Alex R. Varghese

Air Products and Chemicals, Inc.
Gardner Cryogenics Department
Lehigh Valley, Pennsylvania

ABSTRACT

Conventional pressure differential gauges have been used for the evaluation of the mass quantity of cryogen contained in pressurized cryogenic vessels. The pressure sensing ports of the gauge are connected to the top and the bottom of the inner vessel of the container through two spool piping circuits (SP circuits) in the annular space between the inner and the outer vessels of the container. Due to the gravitational force, there exists a pressure variation through each SP circuit. This study investigates the magnitude of the pressure variation through the SP circuits and its impact on the correlation of mass quantity and gauge reading. The temperature and pressure distributions along the SP circuits are numerically evaluated. The mass-gauge reading correlation has to be calibrated for each combination of vessel geometry and operating pressure. Calibration curves corresponding to saturation and supercritical states are analyzed and compared to the ones excluding the SP circuits' pressure variation effect. This pressure variation effect manifests itself at higher operating pressure and becomes significant as the pressure rises above the critical point of the contained cryogen. The geometrical dependence of the mass quantity for a given level interval will then disappear, and the mass-PDG reading correlation appears to be linear. For helium, the error due to the omission of the pressure variation through the SP circuits in terms of percentage of mass quantity is approximately 2% near the critical point. In the supercritical state, the error can be as high as 11% at $p = 791$ kPa.

INTRODUCTION

For a cryogenic vessel, it is highly desirable to be able to monitor the mass quantity of the contained cryogen. Such requirement becomes more important when a cryogenic vessel is used as a storage tank. In this case, the cryogen will be withdrawn from time to time at required quantity. The quantity of the remaining cryogen should be known at any time. This objective can be easily achieved, if the cryogen is saturated, by detecting the liquid level. Electronic devices have been developed to detect the location of the liquid/vapor interface[1]. The liquid level may be determined using resistance or capacitance level gauges. In recent years, several kinds of level probes have been developed using superconducting materials. These devices work on the basis that liquid and vapor have different heat-dissipating capabilities. Properly calibrated under specified pressure, these devices can offer a level accuracy within millimeters. The correlation between the liquid level and the remaining mass quantity can be easily calibrated for the saturation state. However, this type of liquid level detectors fails to function when the contained cryogen is in a single phase, which frequently occurs in applications. This work involves the calibration and analysis of a pressure differential gauge (PDG) which is used to monitor the mass quantity within a cryogenic container. The PDG senses the pressure differential between its two sensing ports. These two sensing ports are connected to the top and the bottom of the inner vessel of the container respectively. This configuration is schematically shown in Figure 1. Another pressure gauge is used to register the pressure on the top of the inner vessel. A scheme is developed to interpret the two pressure readings from these gauges, to predict the available mass quantity within the container.

The existence of the SP circuits which connects the PDG to the inner vessel has an impact on the gauge reading accuracy. This impact is due to the gravitational force which acts on the mass contained in the SP circuits. As a result, the pressure difference registered by the PDG differs from the pressure differential between the top and the bottom of a level container, and consequently, the quantity of mass interpreted from the gauge readings deviates from the actual quantity of mass in the container. Therefore, the magnitudes of the pressure variations over the upper and lower sections of the SP circuits and their influences on the mass-gauge reading correlation have to be investigated. The actual operating condition is numerically simulated and the fluid properties are evaluated by using a commercial program GASPAK developed by Arp and McCarty[2].

THEORY

Ideally, the quantity of mass contained in a vessel can be evaluated by interpreting the pressure differential between the top and the bottom of the vessel using the hydrostatic equation

$$\Delta p = \rho g \Delta h \tag{1}$$

where p is the pressure, ρ the fluid density, g the gravitational acceleration, and h the elevation. Once the density is known, the total mass can be calculated by integration through the total volume. As the thermodynamic state of the cryogen varies during transportation or storage, different approaches to evaluate the mass quantity have to be adopted accordingly.

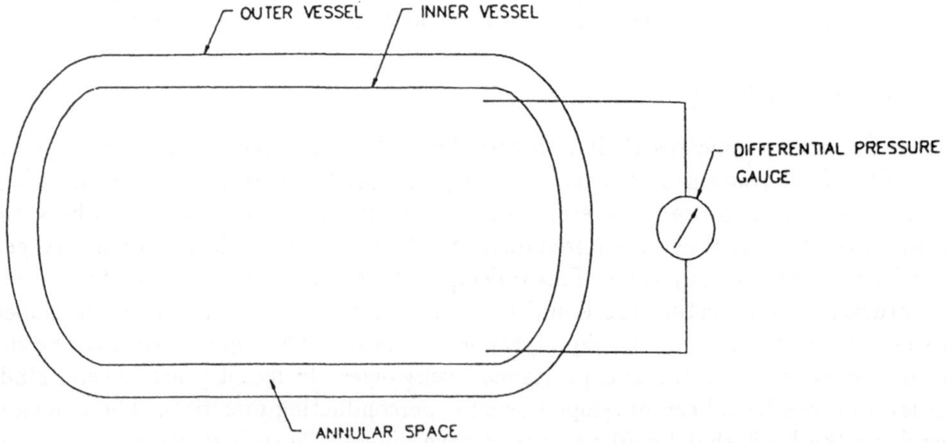

Figure 1: Schematic of level gauge for a cryogenic container

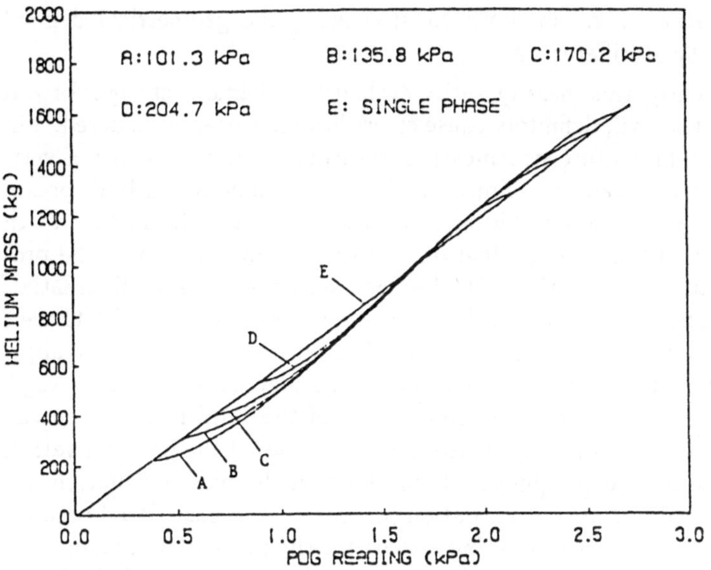

Figure 2: Ideal calibration curves for 11,355 liter LHe container

Saturation

When the contained cryogen is saturated, Equation 1 becomes

$$\Delta p = \sum_{i=1}^{2} \rho_i g \Delta h_i \tag{2}$$

In this case, the pressure heads due to liquid and vapor have to be evaluated separately and added together, and two integration processes will be involved to evaluate the liquid and vapor mass quantities separately. The relative elevation of the liquid-vapor interface with respect to a reference line will be determined and the volumes occupied by liquid and vapor can be obtained through numerical integration. The parameters used to determine the thermodynamic state, and hence to obtain the densities will be saturation and pressure. Finally, the total mass contained in the vessel can be calculated by multiplying the liquid and vapor densities to the volumes they occupy respectively, and adding them together to obtain

$$m = \sum_{i=1}^{2} \rho_i V_i \tag{3}$$

Single Phase

The process of mass quantity evaluation is simplified when the cryogen is in a single phase. In this case, the cryogen could be liquid, vapor, or in the supercritical state. The density will be determined using the hydrostatic equation of the form

$$\Delta p = \rho g H \tag{4}$$

where H is the elevation differential between the two nozzles for the upper and the lower SP circuits. The total mass will be determined using this density and the total volume. Calibration curves produced by using this scheme at different pressures are presented in Figure 2. As illustrated in this figure, the single phase curve indicates a nearly linear mass-PDG reading relationship. The curves corresponding to the saturation states bend on both ends, showing the geometrical dependence of the mass quantity at different levels.

Apparently, this mass quantity evaluation scheme is straightforward. However, two factors that will definitely cause errors have not been considered. First, the gravity and the temperature stratification effect dictate the existence of vertical pressure and temperature gradients. Since the density is a function of both pressure and temperature, it will be a variable instead of a constant. The impact of this factor on the accuracy of the mass evaluation and the method to resolve the problem will be discussed later. Second, the PDG does not sense the pressure differential between the top and the bottom of the vessel. The two sensing ports of the PDG are connected to the nozzles of the upper and the lower SP circuits. The cryogen entrapped in each SP circuit leads to an undesirable pressure variation across the circuit. Negligence of this pressure variation causes an underestimate of the total mass in the container. The objective of this work is to evaluate the magnitudes of the pressure variations through the SP circuits, and compensate their effect on the mass evaluation accuracy. Such effort is carried out in the calibration process for the mass-PDG reading correlation.

In the process of generating a chart for the mass-PDG reading correlation, one has to evaluate the pressure variations through the upper and the lower SP circuits. First, the thermal boundary conditions for the SP circuits have to be identified, and then, the temperature distribution along each of the SP circuits will be determined. Once the temperature distribution is known, the pressure distribution along each SP circuit can be computed through an iteration process. The SP circuits are devided into elements at arbitrarily small intervals. From the nozzle end of each SP circuit, the vessel pressure and the local temperature are used to determine the local fluid density. Once the local fluid density value is obtained, Equation 1 can be applied to calculate the local pressure at the next element. For each SP circuit, this process is continued until the other end of the SP circuit, which connects to the sensing port of the PDG, is reached. The sum of the absolute magnitudes of the pressure variations in the upper and the lower SP circuits represents the deviation of the registered pressure difference from the actual pressure differential between the top and the bottom of the inner vessel.

COMPUTATIONAL ALGORITHM

Thermal Properties of Cryogen

A computer program GASPAK[2] is used to generate the thermal properties of the cryogen. The input parameters are pressure and temperature, and the thermal properties are evaluated along the SP circuits at each iteration and constantly updated.

Volume Computations

In the saturation state, at a certain liquid level, the total volume occupied by liquid is integrated as follows

$$V_l = V_h + V_s \tag{5}$$

$$V_h = \int_0^Z \frac{B}{A}\pi \left[R^2 - (R-Z)^2\right] dZ \tag{6}$$

$$V_s = \int_0^Z 2L_s\sqrt{R^2 - (Z-R)^2}dZ \tag{7}$$

where A and B are the semiaxes b of the ellipsoidal heads, and R and L_s are the

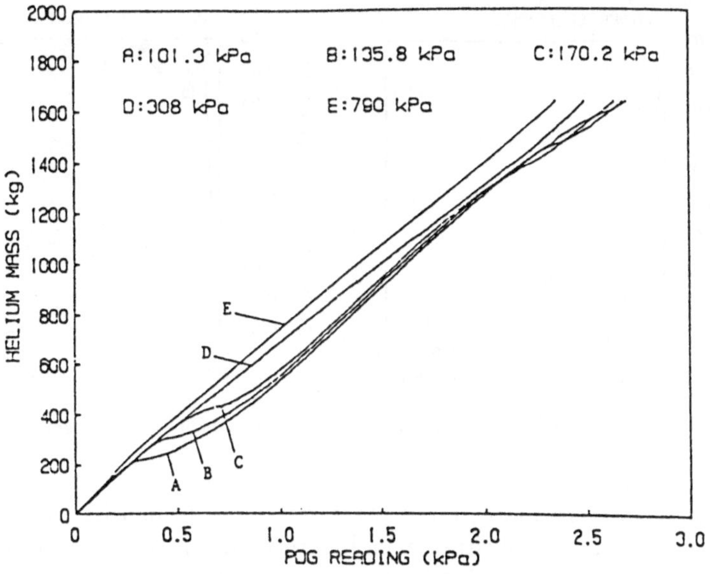

Figure 3: Calibration curves for 11,355 liter LHe container

radius and the length of the inner shell. V_l here denotes the liquid volume and V_h, V_s represent the head and the shell volumes respectively. The integrations are carried out in the vertically upward direction and the axis variable Z is bounded by $0 \leq Z \leq 2R$. The vapor volume can be easily obtained by subtracting the liquid volume from the total volume. Theoretically, the exact total volume is given by

$$V_{et} = \pi R^2 L_s + \frac{4}{3}\pi ABC \tag{8}$$

where C is another semiaxis of the ellipsoidal heads and in this case we have $C = A$. The first term in this equation denotes the shell volume, and the second term in this equation represents the volume of both heads. Only one integration process (total volume) is required if the contained fluid is in a single phase as the density does not have an abrupt change in this situation.

Temperature Boundary Conditions

The two ends of each SP circuit are assumed to have the vessel temperature and the ambient temperature respectively. Intermediate temperature boundary conditions along each SP circuit are also specified at two locations where the SP circuit is thermally attached to the first and second thermal shields.

RESULTS

The computed calibration curves for the 11,355 liter LHe portable customer station are shown in Figure 3. These mass-PDG reading curves may be compared with those presented in Figure 1 to show the the SP circuit pressure variation effect. Evidently, the PFG reading differences between the two groups of calibration curves due to the SP circuit pressure variation effect indicate errors in the evaluation of the mass contained in the vessel under different situations. To demonstrate the error due to the existence of the SP circuits, a set of curves representing the pressure variations through the upper and the lower SP circuits is presented in Figure 4. The magnitude of the pressure variation through each SP circuit depends on the cryogen density distribution along the circuit. Since the temperature distribution along the

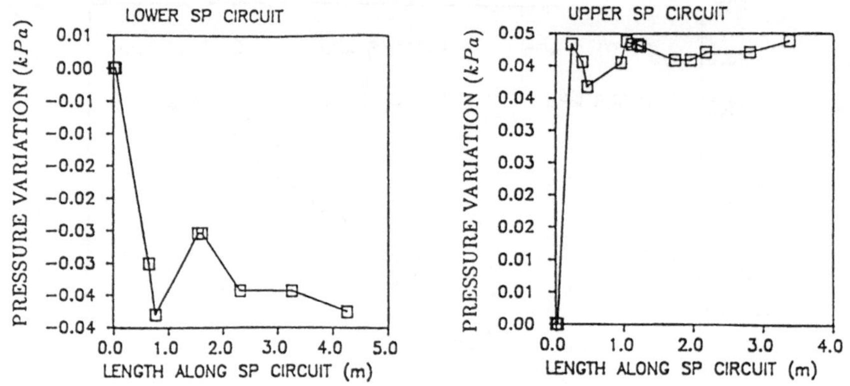

Figure 4: Pressure variations through upper and lower SP circuits

SP circuits does not vary much, the cryogen density is primarily controlled by the pressure in the vessel. As illustrated in Figure 5, the magnitude of the pressure variation through a SP circuit increases as the pressure in the vessel rises. This behavior can be quantitatively explained by examination of the relationships among the relevant thermodynamic properties of the contained cryogen. Because of the fixed temperature boundary conditions, the temperature distribution along the majority portion of each SP circuit reaches an equilibrium. The only portion of the SP circuit along which the temperature distribution slightly varies in time is the one between the nozzle and the first thermal anchor. This variation is dictated by the temperature in the vessel, which however, does not vary much. Under this kind of circumstance, the density of the cryogen entrapped in the SP circuits will be higher corresponding to a higher vessel pressure, hence causing higher pressure variations through the SP circuits.

CONCLUSION

The numerical results clearly indicate that the omission of the SP circuits pressure variation effect causes erroneous evaluation of the total mass of cryogen contained in a vessel under different situations. The SP circuit pressure variation

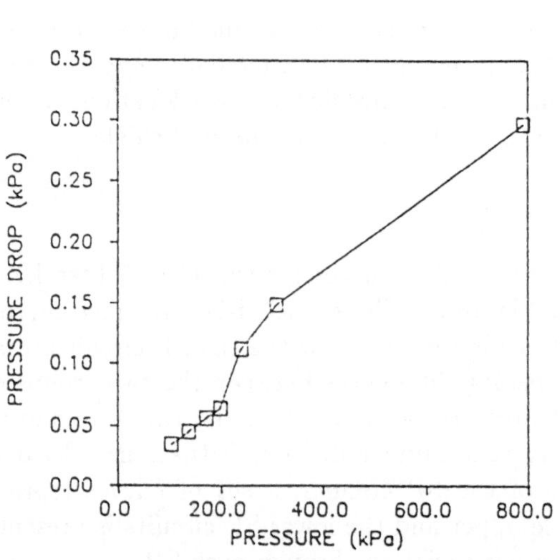

Figure 5: Pressure dependence of pressure drop through SP circuit

effect manifests itself at higher vessel pressures and becomes significant as the vessel pressure rises above the critical pressure of the contained fluid. In the saturation state, when the pressure approaches the critical point, the error due to the omission of the SP circuit pressure variation in terms of percentage of mass is approximately 2% for a 11,355 liter container. In the supercritical state however, the error can be as high as 11% at $P = 791$ kPa.

The result of this study is applicable to different pressurized vessels. Adjustments have to be made for different configuration of the SP circuits and container geometry. Future improvement of this mass quantity evaluation scheme could be the inclusion of the temperature stratification effect which occurs under certain circumstances. Depending on the significance of such effect, it may impose a considerable impact on the accuracy of the current mass quantity evaluation scheme. There is a restriction for the application of this mass quantity evaluation scheme as far as the contained cryogen is concerned. GASPAK only provides thermodynamic properties for fifteen (15) fluids. This scheme cannot be applied to other fluids until their thermodynamic properties are available and preferably in the form of computer program subroutines.

REFERENCE

1. M.J. DiPirro and A.T. Serlemitsos, *Discrete Liquid/Vapor Detectors for Use in Liquid Helium*, Adv. Cryo. Eng. 33:1617 (1990).

2. V.D. Arp and R.D. McCarty, *GASPAK* user's guide, version 2.2, Cryodata, Niwot, Colorado (1989).

DESIGN CONSIDERATIONS FOR AN ULTRAHIGH-PRESSURE SYSTEM TO DELIVER CRYOGENS TO CRYOGENIC PUMP TEST CELLS

Anthony A. Cassano and Glenn E. Kinard

Process Systems Group
Air Products and Chemicals, Inc.
Allentown, Pennsylvania

ABSTRACT

A Component Test Facility (CTF) is presently under construction at NASA's Stennis Space Center (SSC) in Mississippi to test high-pressure turbopump assemblies. These high-pressure pumps will deliver fuel and oxidant to the Advanced Launch System (ALS) main and booster engines to provide greater thrust and thus deliver heavier payloads into orbit. Liquid hydrogen and liquid oxygen pumps will be tested in the CTF for 25 seconds "steady state" duration.

The turbine-driven test pumps will deliver the cryogens in the 41368 kPa (6000 psi) to 55158 kPa (8000 psi) range. Three test cells, one for oxygen and two for hydrogen turbopumps, each contain a gas generator, a turbine, and a pump . Ultrahigh-pressure (UHP) gas systems at 103420 kPa (15000 psi) will be used to pressurize and move the cryogenic fluids to the test cells. While the fluids to the test cells must flow at defined steady state conditions for the short duration of the test, the delivery system is in a state of change throughout a test run. The challenge was to provide one set of controls and equipment to handle the unsteady state fluids and deliver steady state fluids to the test cell.

INTRODUCTION

The turbopumps are a critical part of the ALS main and booster engines. The pumps, with the exception of valves, contain the the only moving parts of the engine. All of the propellant flow is provided by the pumps. Figure 1 is a simplified flow scheme for an engine indicating the turbopump role. Liquid hydrogen is boosted in pressure by the booster pump to ensure no cavitation will occur in the turbopump. The turbopump provides sufficient fluid pressure to both drive the turbines and supply the thrust required for the engine. A portion of the high pressure hydrogen is diverted to channels in the thrust chamber walls to cool the chamber cone and prevent damage from the intense heat released by the thruster burners. Part way up the chamber hydrogen is withdrawn and combined with the remaining hydrogen turbopump effluent and used as fuel to the burners of the hydrogen and oxygen turbopumps. The fuel is combusted with liquid oxygen and the generated hot gas drives the turbopump turbines. Near the top of the chamber the remaining coolant hydrogen is withdrawn and used to drive the liquid hydrogen and liquid oxygen booster pump turbines. The hydrogen gas from the booster pump turbines is combined with the hot gas, still mostly hydrogen, from the turbopumps turbine exhaust and used as the fuel in the thruster burner (gas generator). The hydrogen-rich fuel combusts with liquid oxygen in the gas generator giving a high pressure hot gas which provides thrust to the engine. Liquid oxygen is boosted in

Advances in Cryogenic Engineering, Vol. 37, Part B
Edited by R.W. Fast, Plenum Press, New York, 1992

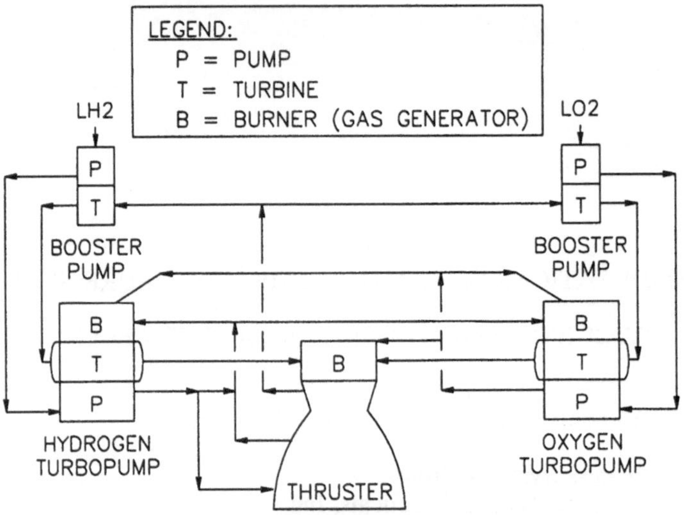

Fig. 1. Simplified engine flow scheme

pressure in the booster pump to ensure no cavitation in the turbopump. The oxygen turbopump provides sufficient fluid pressure to supply the oxidant to the hydrogen turbopump, the oxygen turbopump, and the thruster burner.

THE SSC CTF (COMPONENT TEST FACILITY)

The CTF is comprised of five major areas (see Figure 2): test cells, a delivery system, a fill system, a storage system, and a disposal system. Each of the three test cells contains a facilities gas generator, and the test turbine and turbopump. Only one turbopump will be tested at any time. The delivery system contains the necessary run tanks and pressurant gas bottles to deliver the cryogenic fluids to the test cells,and will be discussed in more detail below. Liquid hydrogen and oxygen to fill the run tanks are pressure transferred from their respective liquid storage tanks. The fill system converts liquids to high-pressure gases for charging the ultrahigh-pressure (UHP) and high-pressure (HP) gas storage bottles between test runs, and includes a liquid hydrogen pump, a liquid nitrogen pump, and vaporizers for each fluid. Liquid hydrogen and nitrogen for the fill system are pressure transferred to their respective pumps and pumped to 106300 kPa (15000 psi). The fluids are then vaporized and used to fill their respective gas bottles. The storage system consists of liquid storage tanks for hydrogen, nitrogen, and oxygen along with their appropriate loading stations to receive liquid from trailers. The disposal system includes flares for pumped liquid hydrogen and turbine exhaust, and a LOX catch tank and vent stack. It is designed to handle normal, upset, and emergency conditions.

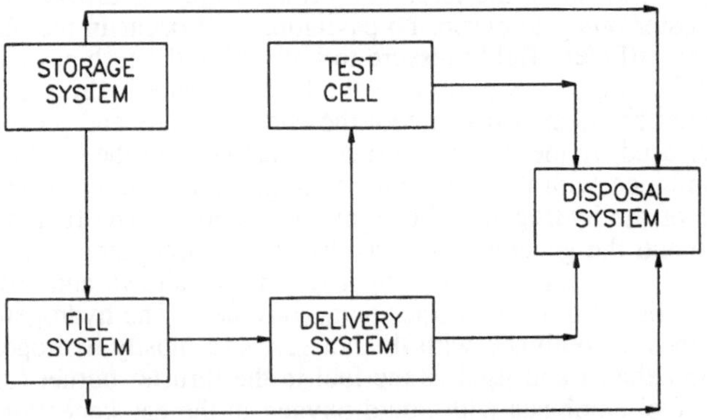

Fig. 2. Simplified block diagram of the SSC facility

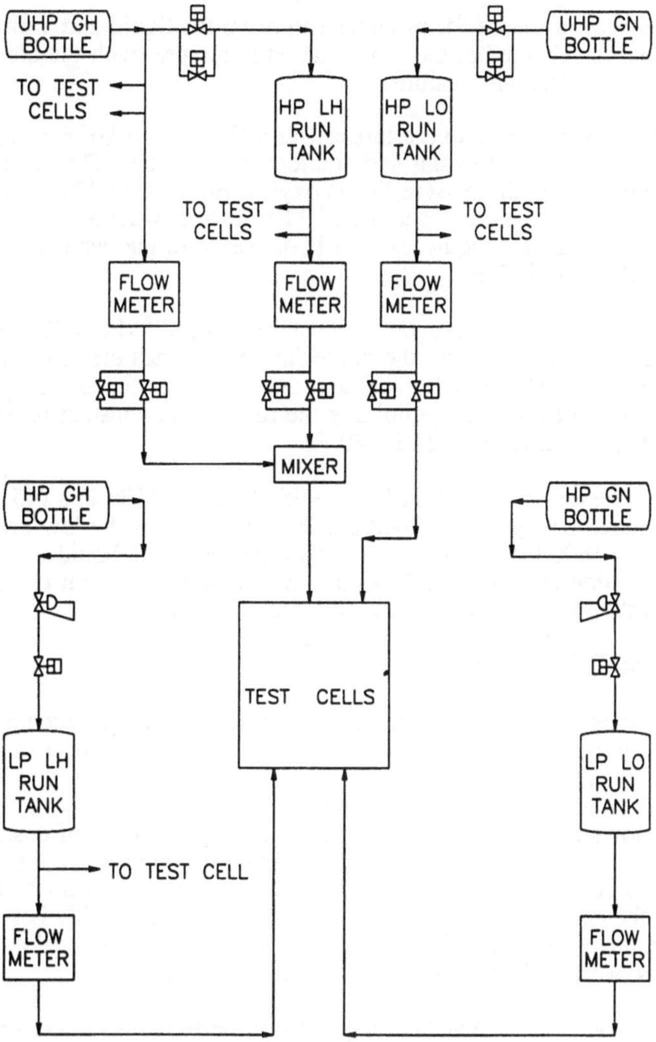

Fig. 3. Simplified delivery system PFD

DELIVERY SYSTEM

Figure 3 is a simplified PFD of this system. Note that the UHP gas bottles provide pressurant gases to the HP run tanks; the HP gas bottles do the same for the low-presure (LP) run tanks. The fuel and oxidant required for the gas generator are pressure transferred from the HP run tanks by gas from the UHP gas bottles. The hydrogen circuit is designed to deliver fuel as a gas, a liquid, or combinations of both. As the cryogen being pumped is drawn from the low-pressure run tanks by the test turbopump, gas from the HP gas bottle maintains a constant pressure to the turbopump suction. Pressures for both HP and LP run tanks are sensed at the bottom of the run tank to eliminate the variation in liquid static head as the tanks empty.

At the start of a test run, the ultra-high hydrogen gas bottle is at 93180 kPa (13515 psia) and ambient temperature. If a liquid fuel is required, the gas from the bottle is used to pressurize the high pressure liquid hydrogen run tank to 37230 kPa (5400 psia) and then propel the liquid hydrogen at a rate of 16.3 kg/sec while holding the pressure constant in the HP run tank. The liquid fuel is delivered to the test cell at 28640 kPa (4154 psia) and 36.11 K (65 R). If a gaseous fuel is used, the fuel is supplied directly from the gas bottle at the same rate and pressure, but at slightly lower than ambient temperature. For a combination requirement both liquid and gas are mixed in a mixer to attain uniform temperature to the test cell. The hydrogen UHP gas bottle provides two services: a source for propellant to drive the liquid and a direct source for the fuel. The fuel is delivered at the same rate and pressure, with the blend ratio varied to deliver the required test temperature.

The oxidant is delivered to the test cell as a liquid. The UHP nitrogen gas bottle initially is at 93180 kPa (13515 psia) and ambient temperature. The gas from the bottle is used to pressurize the high pressure liquid oxygen run tank to 37230 kPa (5400 psia) and then propel the liquid oxygen at a rate of 11.8 Kg/sec while holding the pressure constant in the run tank. The liquid oxidant is delivered to the test cell at 28640 kPa (4154 psia) and 104 K (187.2 R).

The high pressure gas hydrogen bottle starts at 41470 kPa (6015 psia) and ambient temperature. The gas from the bottle first pressurizes the low pressure liquid hydrogen run tank to 1550 kPa (225 psia) and then displaces the liquid hydrogen to the test cell at a rate of 93 kg/sec, again holding the run tank at constant pressure. The liquid to the turbopump inlet is at 22 K (40 R).

The high pressure gas nitrogen bottle initially is at 31130 kPa (4515 psia) and ambient temperature. The gas from the bottle first pressurizes the low pressure liquid oxygen run tank to 1900 kPa (275 psia) and then displaces the liquid oxygen to the test cell turbopump suction at a rate of 527 kg/sec, again holding the run tank at constant pressure. The turbopump feed to the test cell is at 93 K (167.4 R).

CONTROL CONSIDERATIONS

The total elapsed time for the test run is approximately 40 seconds. Some controls must be pre-programmed because control loops may not react in time for reliable control. Certain of the valves demanding fast response times and high power require hydraulic actuators. The test procedure is to increase the flows to 20% of design and hold those flows for 10 seconds, then ramp flows up to 100% of design and run for 25 seconds, then ramp the flows down to zero. Figure 4 indicates the transient nature of the pressurant gas conditions in the gas bottle, and the pressurant flow to the run tank in order to follow the desired flow profile of liquid to the test cell. This is typical for the two UHP gas bottles serving the fuel and oxidant HP run tanks as well as the lower pressure systems of the HP gas bottles and the LP run tanks.

Gases from the bottles displace liquids from the run tanks at constant pressure as the liquids flow to the test cell. As gas is withdrawn from the gas bottle the fluid is isentropically expanded in the bottle, cooling the gas left in the bottle. The pressurant gases entering the run tanks are constantly changing in temperature during the test run, and consequently the gas flow required to displace the liquid at constant flow is constantly increasing. Two other design factors need to be considered - the collapse

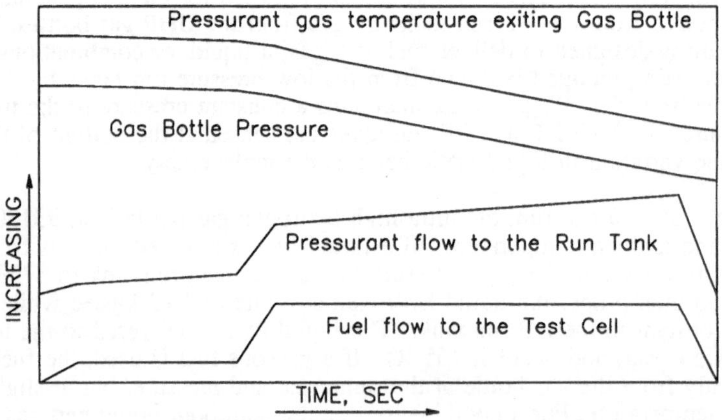

Fig. 4. Relationship between fuel flow and gas bottle

factor and rangeability of control valves. The gas entering the run tank will be cooled by its cold environment, causing the gas introduced to the run tank to "collapse" or fall in pressure. This "collapse" will be discussed in a later section. Flexibility was built into the gas system. The test run usually will start with the ultra-high gas bottles at 93180 kPa (13515 psia) but the customer also wanted to know the minimum pressure required in the ultra-high pressure gas bottle in order to complete a test run. This meant the gas flow had to be controlled with the gas bottle pressure ranging from 93180 kPa (13515 psia) to approximately 40000 kPa (5800 psia). The ultra-high pressure circuits contain two control valves in parallel, a "big" and "little" valve. These valves control the gas flow to hold the pressure of the run tank. The "little" valve opens to pressurize the tank and maintain its pressure as the flow ramps up to 20% of design flow. The "big" valve then comes in control to increase the gas flow to the run tank as liquid flow ramps to 100% design flow and continues to open to maintain the run tank pressure as the liquid leaves the run tank.

The liquid flow rates from the high pressure run tanks are measured with venturi meters. The design of these meters is critical due to the high pressure fluids being measured. In order to get reliable signals from the meters, constraints need to be observed. The maximum allowable pressure drop was set at 20680 kPa (3000 psi), the minimum allowable drop was set at 5% of the inlet pressure, and the throat diameter to pipe diameter ratio was set at 0.3 minimum. Two control valves in parallel as described above are used to control the fluid flow to the test cell.

The flow of ultra-high pressure gaseous fuel to the test cell is also measured by a venturi meter. Two parallel control valves control the flow rate required at the test cell. The flow rate is constant, but the pressure and temperature to the valves from the gas bottles change throughout the test run.

The high pressure gas flow from the HP gas bottle to the LP run tank is controlled using two control valves in series. Critical pressure drop was taken across a pressure regulator with the operating Cv set for the required flow rate. The second control valve then was used to control the fluid pressure at the bottom of the low pressure run tank. The flow rates of liquids from the LP run tanks to the test turbopump intakes are measured in mass (turbine type) flowmeters.

COLLAPSE FACTOR

As mentioned above, the quantity of gas required to pressurize the HP run tank and to displace "liquid" from it during the test run depends on how much it cools while in the HP run tank during the test run. Heat transfer from the pressurizing gas to the walls of the HP run tank and to the liquid, as well as mixing of the pressurizing gas with the liquid, reduce the pressurant gas temperature and tend to "collapse" (reduce) the gas pressure, thereby requiring more pressurant gas for the test run. The collapse factor is defined here as the ratio of the actual mass of UHP (or HP) gas used for pressurization and liquid expulsion during the test run after accounting for heat and mass transfer, relative to the mass of UHP (or HP) gas required for an adiabatic displacement. The collapse factor is critical to the proper sizing of the volume of the UHP or HP gas bottles to ensure sufficient pressurizing gas for the desired test duration, and for proper sizing of lines and valves.

The quantity of gas required for adiabatic displacement neglects any heat transfer to or from the pressurant gas and any mixing of it with the liquid in the run tank. The UHP "H2 gas on H2 liquid" HP run tank is chosen as an example. The initial pressure in the UHP bottles is 93,180 kPa (13,515 psia). To establish the minimum design volume for the UHP gas bottle, its pressure is assumed to drop from the maximum starting pressure (set by the maximum allowable working pressure of the UHP gas bottle as a pressure vessel) to the minimum final pressure, set by the required operating pressure of the HP run tank plus the pressure drop allocated to the piping and control valves from the UHP gas bottle to the HP run tank. Rangeability of the control valves must be considered in selecting this pressure drop, considering the variation in flows and gas bottle pressure. The HP run tank initially contains gas and liquid saturated at 108 kPa (15 psia), with the liquid occupying 80% of the tank volume. At the end of the test run, the UHP bottle pressure has fallen to 39,090 kPa (5670 psia) and the run tank is at 37,230 kPa (5400 psia). The final liquid in the run tank at the end of the test run is the initial liquid less the liquid withdrawn for the test run. Neither the initial vapor nor pressurant gas added during the test leaves the HP run tank.

The gas remaining in the UHP gas bottle at any moment during the test run has expanded from its initial pressure and done work to expel the gas which has left it and entered the HP run tank. Consequently the gas inside the UHP gas bottle decompresses along an adiabatic and reversible (isentropic) path. As a result of this isentropic expansion the gas leaving the UHP bottle drops in temperature as the bottle pressure falls during the test run. The gas exiting the UHP gas bottle is throttled across a control valve at constant enthalpy to the pressure of the HP run tank. Across that control valve its temperature rises if conditions are outside the region bounded by its Joule-Thomson inversion curve. This occurs during the entire test run for the hydrogen going to the hydrogen HP run tank. (For nitrogen to the HP oxygen run tank, this occurs during most of the test run, but by the end of it, the nitrogen conditions cross the inversion curve and its temperature drops across the throttle valve.) As a net result of (1) the temperature drop due to the isentropic expansion in the gas bottle and (2) the temperature change across the throttle valve as the gas drops to the run tank pressure, the temperature of the pressurant gas entering the HP run tank drops during the test run. The low-pressure vapor and liquid originally in the HP run tank undergo isentropic compression by the pressurant gas to the high pressure of the test run, and both their volumes shrink during pressurization.

The heat transfer from the walls of the UHP gas bottle and the gas contained in it occurs at a low rate by natural convection, and is neglected for calculations both with and without heat transfer in the liquid run tank. This is slighty conservative, as any heat transferred to the UHP gas from the UHP gas bottle wall will tend to reduce the required quantity of UHP gas. High gas velocities occurring locally at the gas nozzles of the UHP and HP gas bottles, together with the cold temperature of the gas due to the isentropic expansion, can lead to temperatures at the nozzles which are lower than allowed by the vessel metallurgy. Stainless steel "sleeve" inserts extending into the UHP vessel were recommended at these nozzles to essentially insulate the carbon steel vessel wall near the nozzles from the cold gas.

The design volume of the HP run tank is determined by the quantity of liquid withdrawn during the test run, and the volumes of the initial gas ullage and final "heel" of liquid. To determine the minimum design volume of the UHP gas bottle needed to provide sufficient pressurant gas to complete the desired test run duration, the average density of the pressurant gas in the HP run tank at the end of the test run must first be found. From the isentropic nature of the gas expansion in the UHP gas bottle and the isenthalpic pressure drop across the control valve, the average temperature of the gas entering the HP run tank can be determined. Assuming no heat transfer in the run tank, this establishes the average temperature of the pressurant gas in the HP run tank and hence its density at the given run tank pressure. The volumes occupied by the final HP run tank liquid and the residual (initially low pressure saturated) vapor are determined by their masses and their densities after isentropic compression by the pressurant gas. These volumes subtracted from the entire volume of the HP run tank give the volume occupied by the pressurant gas which, along with the previously determined average density of the pressurant gas, establishes the mass of gas taken from the UHP gas bottle into the run tank. The initial UHP gas bottle pressure and temperature, the final pressure, and the isentropic expansion path of the gas in the UHP gas bottle determine the initial and final densities of the UHP bottle gas. This change in density and the required mass to be taken from the UHP gas bottle into the run tank determine the minimum design volume of the UHP gas bottle, based on no heat transfer or mixing of pressurant with liquid in the HP run tank (i.e., with a collapse factor of 1.0).

A computer program was developed to model the actual heat transfer from the pressurant gas to the walls and liquid, and to calculate the collapse factor. The program divides the run tank gas and liquid into control volumes, each of which is assumed to be well mixed and of (different) uniform temperatures. Transient mass and energy balances are written for each control volume including heat transfer between the gas and the HP run tank wall. The heat transfer coefficient was calculated as the sum of a natural convection coefficient plus a forced convective coefficient which decays with distance from the run tank inlet gas distributor [1,2] . Due to the thick walls of the HP run tank and the short test duration, only the inner portion of that vessel wall approaches the pressurant gas temperature. The wall was modelled as slabs at uniform temperatures of finite thickness equal to 70% of the thermal penetration thickness. The thermal penetration thickness is the distance into a semi-infinite slab at which the temperature change is less than 1% of a step change in temperature imposed on its surface; it is a function of the wall's thermal diffusivity and the test run time. The 70% was chosen to give heat transfer from the gas to the wall identical to it as to a semi-infinite slab [3,4].

The warm pressurant fluid element just above the interface and the cold "liquid" element just below it also exchange heat. The heat transfer coefficient across the interface included a component to account for mixing that occurs between the two fluids. Even though the pressures in the HP run tanks are well above the critical pressures of liquid oxygen and hydrogen it is assumed that an interface is identifiable, or at least conceptually valid. Due to the low velocities of gas from the run tank gas distributor, the gas penetration into the liquid is under one inch, and the pressurant is assumed to turn and flow parallel to the "liquid" interface. This free jet entrains some liquid which cools the pressurant. As a worst case, it is assumed that the jet velocity along the surface is equal to the impingement velocity for the incoming gas, i.e. the pressurant gas turns at the interface without loss of velocity. Classical free jet models often express the entrainment as a velocity across the boundary surface of the jet[5] . The entrainment velocity was taken at 8% of the jet velocity relative to the "liquid" velocity, times the square root of the jet density relative to the entrained fliud density. This pressurant cooling effect due to mixing was encorporated as a pseudo heat transfer coefficient across the interface. Values of 0.42 and 0.295 watts/cm2-K (740 and 520 BTU/hr-ft2-F) were calculated for the hydrogen and oxygen HP run tanks respectively. When added to the computer model, this effect increased the pressurant gas useage for both HP run tanks by about 10%. The H2 on H2 collapse factor increased from 1.11 to 1.24 when the mixing effect was added to the heat transfer effect. These values neglect any "hold" period after pressurization which would result in further cooling of the pressurant gas. A minimum collapse factor of 1.4 was recommended, and was used to calculate the required gas flows for sizing lines and control valves from the UHP and HP gas bottles to their respective run tanks. Higher collapse factors (near 2.0) were used to establish the final UHP bottle volumes, giving an allownace for test "hold" periods after pressurization before start of flow.

CONCLUSION

Several factors were important in the process design aspects of the CTF delivery system. The transient nature of the process, the short run time, and the requirement for the fuel to be liquid or gas, all place extreme demands on the rangeability of flow meters and control valves, often requiring multiple valves to cover the range of flows demanded by the test flow profile. The isentropic expansion of the gas in the UHP and HP bottles leads to falling temperatures of the gas from those vessels as their pressures fall during the test run. Insulating insert "sleeves" were required in the UHP gas bottles in the area of the nozzles to avoid excessively low temperature of the UHP vessel walls near them. The pressurant gas experiences an isenthalpic pressure drop across the control valve into the run tank, causing a temperature change which may be an increase or decrease depending on the conditions relative to the pressurant gas's Joule-Thomson inversion curve. Cooling of the pressurant gas by heat transfer to the run tank walls and by mixing with the "liquid" leads to increased gas requirement relative to an adiabatic case. To account for these cooling effects, a minimum collapse factor of 1.4 was recommended for sizing lines and control valves.

ACKNOWLEDGEMENT

Air Products and Chemicals, Inc. wishes to thank the personnel of the Stennis Space Center and Bechtel National, Inc. for their counsel and cooperation during design of the CTF.

REFERENCES

1. Epstein,M. and H.K.Georgius,"A Generalized Propellant Tank-Pressurization Analysis",Int.Adv.Cryo.Eng.,Vol 10-B, K.D.Timmerhaus, ed., Plenum Press, N.Y.(1965), pp. 290-302.
2. Nein, M.E. and J.F. Thompson,"Experimental and Analytical Studies of Cryogenic Propellant Tank Pressurant Requirements", NASA TN D-3177, Washington, D.C.,(Feb.1966),88 pp.
3. Bird, R.B., W.E. Stewart, and E.N. Lightfoot, "Tranport Phenomena", John Wiley & Sons, N.Y.(1960),p.354.
4. Carslaw,H.S. and J.C. Jaeger,"Conduction of Heat in Solids", 2nd ed., Oxford at the Claredon Press (1959).
5. Albertson et al, Proc.Am.Civil Engr.,74,pp. 1751-96(1948).

CAPACITY REQUIREMENTS FOR PRESSURE RELIEF DEVICES ON CRYOGENIC CONTAINERS

Alex P. Varghese and Burt X. Zhang

Air Products and Chemicals, Inc.
Gardner Cryogenics Department
Lehigh Valley, Pennsylvania

ABSTRACT

An equation is derived for the evaluation of the capacity requirements for pressure relief devices on cryogenic containers. It comprises terms such as the heat transfer rate into the container and the specific heat input corresponding to a given situation. This equation can be used to calculate the capacity requirements of pressure relief devices on cryogenic containers irrespective of the thermodynamic state of the cryogen. This equation is developed on the premise that all the vapor generated by the heat input need not be released to maintain a constant pressure inside the container.

The equation is then adapted for the prevalent cases such as liquid full-liquid discharging, saturated liquid and vapor-vapor discharging, saturated liquid and vapor-liquid discharging and supercritical fluid discharging.

In adverse situations, such as an overturned cryogenic container on fire, connections to the pressure relief devices may be covered by liquid. For this reason, discharge rates for vapor and liquid, as well as supercritical cryogens are discussed and design guidelines are provided.

INTRODUCTION

Pressure relief devices are used to keep the pressure at or below a safe limit in all process, transfer, transportation, and storage systems involving fluids subject to pressure rises due to system energy increases from changes in process or system boundary conditions. The pressure relief devices limit the pressure increases by discharging the fluids from the system.

The majority of the equations in use for determining the capacity requirements of pressure relief devices are based on the basic requirement of discharging all the vapor generated by the heat input from a fire or other emergency situations. While this approach is conservative, it imposes a heavy financial burden when the set pressure of a relief device is close to the critical pressure of the fluid.

A single equation is developed in this paper for calculating true capacity requirements of pressure relief devices for a fluid irrespective of its thermodynamic state. This equation is developed on the premise that all the vapor generated by the heat input need not be released to maintain a constant pressure inside the container.

Methods for Calculating the mass discharge requirements for fluid systems with design pressures far lower than the critical pressures of the fluids concerned are covered in publications of American Petroleum Institute (API) and Compressed Gas Association (CGA). In such literature, the mass discharge rate requirements are dependent on the rate of heat transfer into the fluid or rate of energy generation within the fluid, and the properties of the fluid. To keep the discussion simple and general, mathematical expressions for the discharge requirements will be derived for a case of heat transfer into the fluid. In the CGA and API expressions for mass discharge rate requirements, the heat transfer into the fluid system causes a certain mass of fluid to change phase and this mass of fluid has to be discharged to keep the pressure from rising. In this formulation, the density of vapor is considered negligible compared to the density of liquid. This is true only when the pressure of the fluid is far lower than the critical pressure of the fluid which is generally the case for most fluid systems. However, fluids like helium and hydrogen have critical pressures low enough to fall in the working pressure ranges of their cryogenic systems. The expressions developed here are applicable to fluids involved in both single phase or change of phase processes.

THEORY

API-CGA Formula

The API-CGA mass discharge rate formula for equipment involved in a fire, abnormal process heat input or breakdown of insulation, essentially has its basis in the relation

$$\dot{m} = \frac{\dot{Q}}{L} \tag{1}$$

It is evident from thermal property tables of fluids that the latent heat of vaporization decreases as the pressure approaches its critical value, and goes to zero at pressures equal to or greater than the critical pressure. This will give extremely large mass discharge rates at sub-critical pressures near the critical pressure if the above basic relation is used. It will also be convenient to have a general expression that could be simplified to suit the conditions at hand.

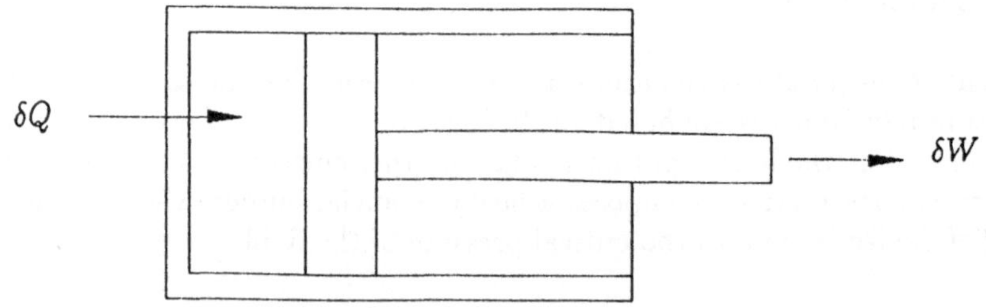

Figure 1: Illustration of first law of thermodynamics.

Specific Heat Input

It is easy to see that the API-CGA basic formula requires all the vapor generated by the heat flow into the fluid be discharged. When heat is added to a fluid system at constant pressure, the fluid needs additional space to expand to maintain a constant pressure. The mass of fluid in this additional space has to be vented to maintain the pressure constant.

Consider a mass of fluid m at pressure p and temperature T in a cylinder with a piston as illustrated in Figure 1, The first law of thermodynamics for the system yields,

$$\begin{aligned} \delta Q &= dU + \delta W && (2)\\ \delta W &= pdV && (3)\\ H &= U + pV && (4)\\ dH &= dU + pdV + Vdp. && (5) \end{aligned}$$

For a constant pressure process,

$$dp = 0 \tag{6}$$

and therefore

$$dH = \delta Q \tag{7}$$

For a heat input of δQ to a fluid system at constant pressure, let the increase in volume of the fluid be dV. A fluid system that has to be protected by a pressure relief device almost always has a fixed volume. In a constant volume fluid system, the relief devices should relieve the mass

$$\begin{aligned} m_l &= \frac{dV}{v} \\ &= m\frac{dv}{v} && (8) \end{aligned}$$

to keep the pressure constant for a heat input of

$$\delta Q = mdh \tag{9}$$

Thus the mass expulsion needed per unit of heat transfer to maintain constant pressure is

$$\begin{aligned} \frac{m_l}{\delta Q} &= m\frac{dv}{vmdh} \\ &= \frac{1}{v\left(\frac{dh}{dv}\right)} \\ &= \frac{1}{v\left(\frac{\partial h}{\partial v}\right)_p} && (10) \end{aligned}$$

as a constant pressure process was assumed. Values of the specific heat input

$$\theta = v\left(\frac{\partial h}{\partial v}\right)_p \tag{11}$$

are given in tables of thermophysical properties of various fluids from National Institute of Standards and Technologies (NIST). They can also be generated by using computer programs[5].

Therefore, for a heat transfer rate of $\dot{Q}$, the mass expulsion rate of fluid required to maintain a constant pressure is given by

$$\dot{m} = \frac{\dot{Q}}{\theta} \tag{12}$$

APPLICATION TO FLUID SYSTEMS

The expression for mass expulsion rate derived above in terms of the specific heat input θ can now be adapted for fluid systems involving different phases.

Liquid Full System–Liquid Discharging

If values of the specific heat input θ are calculated at different temperatures for a liquid under constant pressure p, the minimum value of the specific heat input θ_{min} will be found at the saturation temperature. For this θ_{min}, a mass expulsion rate calculated using Equation 12 will be considered as the minimum mass rate of liquid flow requirement. This system will eventually reach a condition when it will contain both liquid and vapor phases. When values of θ are not available, the equivalent equation in terms of specific heat and cubical expansion coefficient given as formula C-9 in API-520 may be useful.

Fluid System Containing Saturated Liquid and Vapor

In this case, any heat transfer into the fluid system at constant pressure will result in evaporation of the saturated liquid. The minimum mass discharge rate required for this case will depend on whether the pressure relief devices are connected to the vapor or liquid side. Generally, pressure relief devices are connected to the vapor side. However, under accident conditions like overturned fluid transport tanks, connections to the pressure relief devices could end up covered by liquid. These two cases have to be considered separately. For an evaporation process, the differential quantity in the specific heat input term $v(\partial h/\partial v)_p$ can be replaced by difference quantity to give

$$v\frac{\Delta h}{\Delta v} = v\frac{L}{v_{fg}}. \tag{13}$$

For the cases where heat addition to the vapor is not negligible or when the system eventually becomes vapor-full, it shall be noted that the minimum mass discharge rate requirement shall be evaluated using Equation 17. Variations, if any, in heat transfer rates to liquid and vapor phases may be included in these evaluations.

a. Vapor Discharge: If pressure relief devices are connected to the vapor side, the mass discharge will be in vapor form and the minimum mass discharge rate requirement expression for heat addition to the liquid phase only, will become

$$\dot{m} = \dot{Q}\left(v_g\frac{L}{v_{fg}}\right)^{-1} \tag{14}$$

b. Liquid Discharge: In cases where relief device connections are covered by liquid resulting in liquid discharge, the minimum mass discharge rate requirement equation for heat addition to the liquid phase only, at constant pressure becomes

$$\dot{m} = \dot{Q}\left(v_f\frac{L}{v_{fg}}\right)^{-1} \tag{15}$$

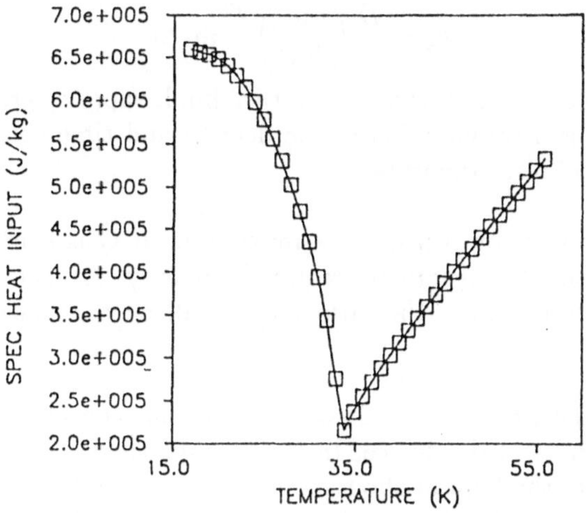

Figure 2: Specific heat input as a function of temperature for hydrogen at 1,379 kPa.

Minimum mass discharge rate requirement for the heat added to the vapor phase can be evaluated by the expression

$$\dot{m}_v = \frac{\dot{Q}_v v}{\theta_{min} v_f} \tag{16}$$

where v is the specific volume of vapor at the temperature where θ_{min} occurs for the specified pressure and $\dot{Q}_v$ is the heat addition rate to the vapor phase. Obviously, this can yield a very high relief device capacity requirement compared to other cases if the pressure in the system is far lower than the critical pressure. It may be less expensive to eliminate the possibility of occurrence of this condition by altering the design of the system rather than providing high capacity pressure relief devices.

Vapor Full or Super Critical System

For single phase, constant pressure systems, the specific heat input θ is temperature dependent. Variation of θ with respect to temperature for hydrogen at 1,379 kPa is shown in Figure 2. It is evident from the above curve that the specific heat input for a fluid has its minimum value at a particular temperature for a specified pressure. Thus, the mass discharge rate required to maintain pressure below a certain value is given by

$$\dot{m} = \frac{\dot{Q}}{\theta_{min}} \tag{17}$$

SIZING OF RELIEF DEVICES FOR SUPERCRITICAL FLUID, VAPOR OR GAS FLOW

Capacities of relief devices for supercritical fluid, vapor or gas flow are generally stated by the manufacturers in Standard Cubic Feet per Minute (Standard Cubic Meter per Hour) of air at the set pressure of the device. The formula for the required flow capacity in Standard Cubic Feet per Minute (Standard Cubic Meter per Hour) of air is obtained by combining the equations developed in this paper with the formula given in ASME Code Section VIII, Division 1. The formula for required capacity becomes

$$\dot{F}_a = \frac{D}{C}\left(\frac{p}{RM}\right)^{1/2} \dot{m}\sqrt{v} \tag{18}$$

where v is the specific volume of the supercritical fluid, vapor or gas at the inlet of the relief device and D is a combination of the density and time conversion multipliers dependent on units of measurements.

$\dot{F}_a$ will be a maximum if $\dot{m}\sqrt{v}$ is a maximum. If $\dot{Q}$ is constant, $\sqrt{v}/\theta$ has to be a maximum to yield the maximum required capacity. At constant pressure, as v and θ varies with temperature, the ratio $\sqrt{v}/\theta$ shall be maximized with respect to temperature.

Locating the relief device away from the cryogenic container can raise the fluid temperature at the inlet of the device. In such a case, the ratio of the square root of the specific volume of the fluid at the temperature at the inlet of the relief device to the specific heat input at the temperature of the fluid inside the container shall be maximized.

CONCLUSION

Realistic mass flow rate requirements are provided by the expressions developed in this work for fluid systems frequently encountered in practice. In designing pressure relief devices for fluid system, evaluation should be made of the probabilities of occurrences of each of the cases discussed here or any special cases that may develop. The minimum mass discharge rate requirements for each and every case and any probable combinations of cases shall be computed and pressure relief devices capable of meeting the highest flow requirement shall be provided.

NOMENCLATURE

C – constant for gas or vapor related to ratio of specific heats, given in ASME Code
$\dot{F}$ – volume flow rate
H – enthalpy
h – enthalpy of unit mass
L – latent heat of vaporization
M – molecular weight of fluid
$\dot{m}$ – mass discharge rate required
m – mass
p – pressure
Q– quantity of heat absorbed by system
$\dot{Q}$ – heat transfer rate
R – particular gas constant
T – temperature
U – internal energy
u – internal energy of unit mass
V – volume
v – specific volume
W – work
θ – specific heat input

Subscripts

a – air equivalent
f – liquid phase, saturated

g – vapor phase, saturated
fg – difference in saturation property between liquid and vapor
l – expulsion required
p – under constant pressure
v – vapor phase

REFERENCES

1. API recommended Practice 520, Part I-Design, 4th Edition (1976).
2. *Pressure Relief Device Standards*, Compressed Gas Association Inc., (1980).
3. R.D. McCarty and L.A. Weber, *Thermophysical Properties of Parahydrogen*, NBS Technical Note 617.
4. H.M. Long and P.E. Loveday, *Safe and Efficient Use of Liquid Helium*, Technology of Liquid Helium, NBS.
5. V.D. Arp and R.D. McCarty, *GASPAK* user's guide, version 2.2, Cryodata, Niwot, Colorado (1989).

DESIGN AND PERFORMANCE OF LIQUID HYDROGEN TARGET SYSTEMS FOR THE FERMILAB FIXED TARGET PROGRAM*

D. H. Allspach, J. F. Danes, J. A. Peifer and R. P. Stanek

Fermi National Accelerator Laboratory
Batavia, Illinois

ABSTRACT

The Fermilab 1990-1991 Fixed Target Program featured six experiments utilizing liquid hydrogen or liquid deuterium targets as part of their apparatus. Each design was optimized to the criteria of the experiment, resulting in variations of material selection, methods of refrigeration and secondary containment. Collectively, the targets were run for a total of 14,184 hours with an average operational efficiency of 97.6%. The safe and reliable operation of these targets was complemented by an increased degree of documentation and component testing. This operation was also aided by several key upgrades. All the systems were designed and fabricated under a set of written guidelines that blend analytical calculations and empirical guidance drawn from over twenty years of target fabrication experience.

INTRODUCTION

Fermilab, the world's highest energy accelerator, has been in operation since 1974. During that time, the accelerator has transformed from 400 GeV fixed target operation to 800 GeV fixed target operation and finally to 900 GeV collider operation. It currently cycles between the latter two modes.[1] Hydrogen targets have played a key role in the fixed target program since its inception. In all, approximately 50 target systems have been installed, with sizes ranging from 0.2 L to 300 L and diameters up to 10 cm. Although the majority of targets have used hydrogen, deuterium targets have also shared the work load. The physics objectives have been quite varied, as witnessed by the experiment titles for the 1990 run (see Table 1); however, the requirements for a tightly controlled, safe and reliable target system have been very consistent.[2]

The general design criteria has been to provide a "clean" interaction region of either hydrogen or deuterium with as little additional material surrounding the fluid as possible. This translates to fabricating the cryogenic and vacuum vessels from low Z materials, such as Mylar, beryllium or Rohacell,† a closed cell polymethacrylimide. All support equipment including the refrigeration system must be kept out of the particle trajectories to avoid unnecessary irradiation

*This work is sponsored by the Universities Research Association, under contract with the U.S. Department of Energy.

†Registered trademark of Rohm Tech, Inc., Malden, Massachusetts.

Table 1. Target Operating Data

EXPERIMENT	EXPERIMENT TITLE	OPERATIONAL EFFICIENCY	TOTAL OPERATING TIME (hrs.)	STATIC HEAT LOAD (W)
665-Deuterium	Muon Scattering with Hadron Detection	99.90%	2736	4
665-Hydrogen	Muon Scattering with Hadron Detection	93.60%	3360	3
683	Photoproduction of High Pt Jets	First Run Pending	0	5
687	Photoproduction of Charm and B	95.40%	4608	6
690	Study of Charm and Bottom Production	99.96%	1248	3
704	Experiments with the Polarized Beam Facility	99.20%	2232	4
706	A Comprehensive Study of Direct Photon Production in Hadron Induced Collisions	First Run Pending	0	4

and halo effects. Controlling the target density to a few tenths of a percent is essential for experiments trying to measure their interaction cross sections.

TARGET SYSTEM COMPONENTS

A generalized schematic of the target system components is shown in Figure 1. The target flask material chosen for a particular system varies with the expected amount of radiation energy absorbed by the target from the beam. The flask is designed for a maximum allowable working pressure of at least 172.3 kPa differential, while taking into account pressure, liquid

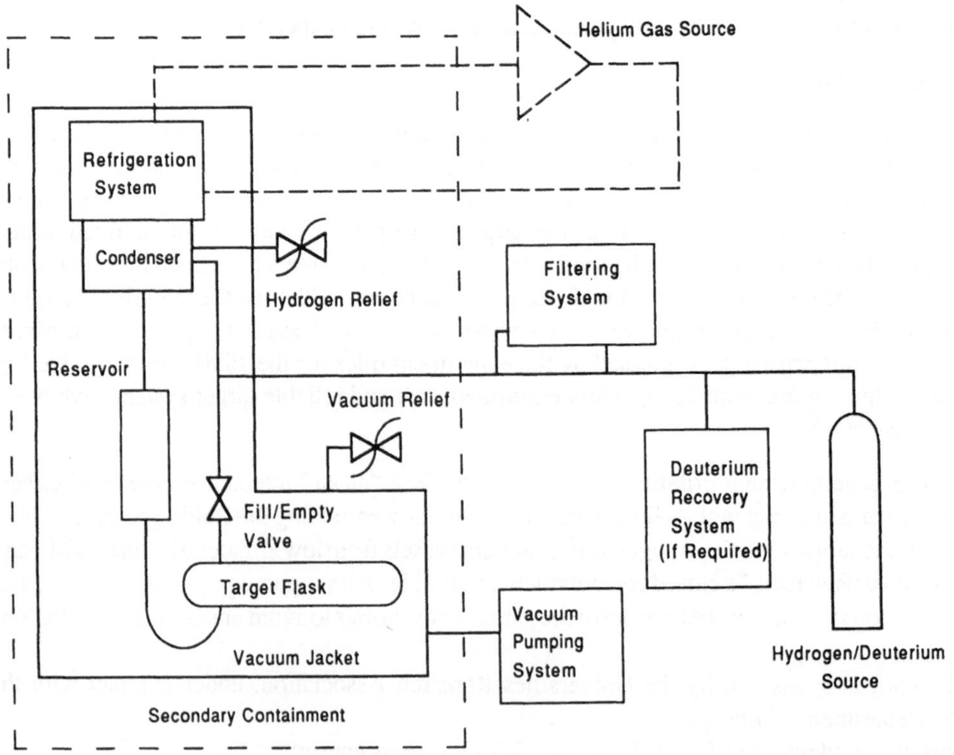

Figure 1. General Target System Components

head and cooldown loadings. The flask, along with the rest of the hydrogen circuit, is protected with a safety relief valve set at the maximum allowable working pressure.

A vacuum jacket surrounds the flask as well as the refrigeration unit(s). The vacuum volume is sized to contain the liquid hydrogen in the target flask as cold vapor at atmospheric pressure. This reduces the chance of subsequently damaging the vacuum jacket in the case of a target flask failure. The vacuum system is supplied with relief devices set to relieve slightly above atmospheric pressure. The evacuation system consists of a roughing pump and a diffusion pump which is in series with a fore pump. Cryostat vacuum pressure after cooldown of the target is approximately 7.0 x 10^{-8} Pa. Continuous pumping of the vacuum space is required due to the permeability of gases through Mylar. The roughing pump is used to prepump the cryostat vacuum jacket and evacuate the hydrogen circuit during the pump and purge cycles.

Refrigeration is supplied to a plate in contact with the hydrogen vapor space in the condensing pot. Hydrogen circulates in the system using a thermal siphon effect. The liquid from the condensing pot passes through the reservoir and into a tube connected to the bottom of the flask. The hydrogen vaporizing in the flask, bubbles out the top of the flask through a tube to the top of the condensing pot and then recondenses back into the reservoir. The rate at which the liquid condenses is controlled by a heater to maintain a constant pressure in the hydrogen circuit, such that the liquid hydrogen density varies less than 0.2%. The hydrogen pressure, in most cases, is controlled at 103.4 kPa. Hydrogen is supplied during a target fill from high pressure cylinders. In the event that a recovery system is installed for deuterium use, the deuterium is supplied from recovery tanks rather than from cylinders. Refrigeration, when supplied by a mechanical cryocooler incorporates an APD Cryogenics Inc. Gifford-McMahon unit. Fermilab targets use both 10 W and 50 W models. Actual cooling capacity has been measured for the "10 W" unit at 12 W and for the "50 W" unit at 36 W for optimum conditions. Multiple refrigeration units are used at some target locations to provide adequate cooling capacity. In each case a heater is attached to the condensing plate which regulates the excess refrigeration.

The compressor used to supply high pressure helium to the 10 W APD refrigeration unit is the APD model HC-8. The compressor discharge pressure is 2306 kPa and the suction pressure is 514.8 kPa. The helium flowrate is approximately 3.2 g/sec for each 10 W unit. The 50 W units are supplied with helium from a high pressure header maintained at 1962 kPa. Suction pressure for this unit is at 108.2 kPa. The suction and discharge headers also service other helium systems at Fermilab. The helium flow through a 50 W unit is approximately 11.1 g/sec.

The liquid hydrogen target systems each have some form of secondary hydrogen containment. Its purpose is to contain and control the release of hydrogen gas from the hydrogen target system in the event of a failure. In some cases this secondary containment is in the form of a tent constructed with flame retardant materials. However, when the experimental apparatus dictates a more compact design, a vacuum buffer volume is used. It is sized to create a total vacuum volume large enough to contain the equivalent amount of gas generated by the liquid at room temperature and subatmospheric pressure.

DESCRIPTION OF 1990-1991 TARGET SYSTEMS

Design Parameters

The following paragraphs describe the design criteria and solutions for each of the liquid hydrogen targets built for use in the 1990-1991 Fermilab Fixed Target Run. Six different experiments requested liquid hydrogen targets for this run (see Table 2).

Table 2. Target Design Parameters

EXPERIMENT	FLUID	TARGET VOLUME (L)	TARGET DIAMETER (cm)	TARGET LENGTH (cm)
665	Hydrogen	8	10	100
	Deuterium	8	10	100
683	Hydrogen or Deuterium	2.2	7.62	50.8
687	Deuterium	7	5.08	340
	Hydrogen	8	Annular, 7.62 o.d., 5.08 i.d.	330.2
690	Hydrogen	0.2	3.81	14.27
704	Hydrogen	2.9	6	100.1
706	Hydrogen	0.5	6.35	15.24

Experiment 665 - This experiment consists of one 8 L liquid hydrogen target and one 8 L liquid deuterium target. The two targets are independent so far as insulating vacuum and refrigeration are concerned. Each of these targets is cooled with a 50 W APD refrigerator. The target flask is 10.0 cm in diameter and made with 0.254 mm thick Mylar. The target vacuum shell is constructed of 2.54 cm thick Rohacell foam with a 0.127 mm thick layer of Mylar covering it. The liquid hydrogen target insulating vacuum space is common with an evacuated Mylar target adjacent to it, in order to provide experimenters with a method of running a calibration with an "empty target". These targets, along with a solid target wheel capable of holding seven types of solid targets, are supported on a moveable table. The table movement is computer controlled so as to position a different target in the beam for each spill, thus reducing the systematic error of the experiment. All of these targets are included inside the secondary containment volume. Secondary containment is provided in the form of a tent with dimensions of 4.9 m height, 3.7 m width and 1.4 m depth. The tent material is Herculite,* an antistatic and flame retardant material. It is used on all but one side of the tent. The fourth side of the tent is covered with Lexan polycarbonate sheet. This is a transparent material which has been included on the tent as a means to assure that no personnel are left inside the tent when interlocking the radiation area. The top of the tent is a sheet metal hood which attaches to ducting leading outdoors. In the case of a hydrogen release into the tent, a hydrogen detector triggers an exhaust fan which vents hydrogen outdoors through the ducting. Intake air is supplied at the base of the tent. The deuterium target includes a recovery system to minimize losses. The 8 L of liquid deuterium when warmed to room temperature is contained in four 3785 L tanks at a pressure below the primary deuterium circuit safety relief device set pressure. When refilling the target, the deuterium is routed through the filtering system before it re-enters the target for reliquefaction.

Experiment 683 - The target flask built for this experiment has a volume of 2.2 L. The experiment will run with liquid hydrogen in the target flask during some portions of the Fixed Target Run and with liquid deuterium during the remaining periods. The flask is constructed of 0.254 mm thick Mylar. The target vacuum shell is made from aluminum type 6061-T6 with Mylar beam windows. The upstream beam window diameter is 8.26 cm and uses 0.178 mm thick Mylar. The downstream window has a 22.9 cm diameter and uses 0.381 mm thick Mylar. The target is cooled with two 10 W APD refrigerators which provide a shorter cooldown time and redundancy during operation. It is supported on a moveable table which is again, computer controlled. A solid target wheel is directly adjacent to the hydrogen target which is capable of holding eight types of solid targets. All of the targets are located inside of a tent having dimensions of 4.3 m height, 3.4 m width and 1.4 m depth. The construction is similar to that of the tent used at experiment 665.

---

*Registered trademark of Herculite Products, Inc., New York, New York.

Experiment 687 - This target is different in concept than any of the others built for the current fixed target run. Beam travels through a flask containing approximately 7 L of subcooled liquid deuterium. The flask containing the liquid deuterium is fabricated from aluminum type 6061-T6 and has a diameter of 5.04 cm. The deuterium flask is surrounded by 8 L of liquid hydrogen in the annular space between the outer wall of the deuterium vessel and another aluminum tube with a 7.62 cm diameter. The hydrogen is cooled by three 10 W APD refrigerators. The liquid hydrogen in turn subcools the deuterium which has a higher saturation temperature at our chosen operating pressures. The deuterium pressure is maintained at 322.0 kPa, while the hydrogen pressure is maintained at 103.4 kPa. The system operating temperature is 20.3 K resulting in 8.0 K of subcooling to the deuterium. A recovery system is used to minimize deuterium losses. The liquid deuterium when warmed to room temperature is contained in one 3785 L tank at a pressure below the primary deuterium circuit safety relief device set pressure of 827 kPa. The vacuum jacket material used for this target is stainless steel type 304 with aluminum type 6061-T6 beam windows of 0.762 mm thickness. Using stainless steel and aluminum as target materials results in higher maximum allowable working pressures, and thus, higher safety relief device set pressures for this target as compared to others built for this fixed target run. The hydrogen circuit is relieved at 274 kPa while the vacuum circuit relieves at 205 kPa. The target system uses no tent or vacuum buffer volume. It is located inside a beam enclosure and is very well protected with concrete blocks positioned primarily for radiation shielding purposes. The exhausts of all safety relief devices for each circuit (deuterium, hydrogen and vacuum) are vented outdoors. A hydrogen detector is located in the enclosure which triggers an exhaust fan in the case of a release of either the hydrogen or deuterium. The gas is then safely vented outdoors. Special procedures are also in effect for accessing the enclosure in which this target is located. The main reason for using a metallic target flask is that this system is located in a beam position which receives 4×10^{12} protons per pulse at primary (800 GeV) energy. This results in a radiation exposure which prohibits the use of Mylar material. The cooling method chosen assures a subcooled deuterium target even under peak beam loading. Other target systems have solved this problem by utilizing a forced flow concept coupled with direct heat exchange to the target fluid.[3]

Experiment 690 - The target flask in use at this experiment, holding on the order of 0.2 L of liquid hydrogen, is the smallest flask built for any experiment at Fermilab. The flask material is Mylar at a thickness of 0.127 mm. The beam enters the target insulating vacuum space through a 0.178 mm thick Mylar window. The vacuum shell surrounding the target flask was constructed with 5.1 mm Rohacell wrapped with 0.076 mm thick woven fiber glass cloth saturated with epoxy, thus bonding it to the Rohacell vacuum jacket. The target is located inside a helium purged gas chamber which is part of a series of helium and flammable gas chambers. The hydrogen for this target is cooled with one 10 W APD refrigerator. The target support stand may be manually moved upstream of its operating position allowing experimenters to access experimental apparatus otherwise difficult to reach. Secondary containment is in the form of a vacuum buffer volume. Safety relief valves on both the hydrogen and the vacuum circuits vent through ducting to the outdoors.

Experiment 704 - The target flask has a 2.9 L volume. The flask has a 6.0 cm diameter with a 100.1 cm length. It is fabricated with Mylar of thickness 0.178 mm. The target vacuum shell is made with 1.91 cm Rohacell with 0.127 mm Mylar fixed on its outer surface. The upstream vacuum window is a 0.127 mm Mylar window with a diameter of 6.99 cm. The target is cooled with two 10 W APD refrigerators. The target support stand may be moved manually such that other fixed target experiments may be moved into the beamline. The target uses a standard design tent as secondary containment. Its dimensions are 4.0 m in height, 1.6 m in width and 1.1 m in depth.

Table 3. Major Causes of Down-Time

PROBLEM	TYPE	DOWN-TIME (hrs.)
Refrigerator	Loss of Refrigeration	220
Relief Valve	Damaged O-rings	130
Compressor Systems	Various Trips	50
Fill/Empty Valve	Control Instability	25
Heater Controller	Controller Malfunctions	19

Experiment 706 - This is a small target holding approximately 0.5 L of liquid hydrogen. The flask is built of 0.178 mm thick Mylar. Its diameter is 6.35 cm and its length is 15.24 cm. The vacuum shell is made of stainless steel and is cylindrical in shape. The beam windows for the vacuum jacket are made of beryllium plate and are secured using circular stainless steel flanges. The upstream beryllium beam window is 6.35 cm in diameter and the downstream window is 7.62 cm in diameter. Both windows are approximately 0.254 cm thick. The target is supported inside a box as are silicon strip detectors positioned directly adjacent to the target on both its upstream and downstream ends. The box is located on a table which may be manually moved out of the beam line. It is critical for this experiment that the silicon strip detectors sense no vibrations; therefore, an alternative hydrogen refrigeration system was chosen. Helium is transferred from commercial liquid helium dewars through a vacuum insulated transfer line to a heat exchanger that cools the hydrogen. After the helium passes through the heat exchanger a portion of the remaining cooling capacity is used to intercept the heat load by connecting the tube to a radiation shield inside the vacuum can. The helium is then vented outdoors. The secondary containment and hydrogen venting system are similar to that used for experiment 690.

Operational Data

The targets that were operated during the first portion of the 1990-1991 Fixed Target Run totaled over 14,000 hours of running time. Operating efficiencies of these targets ranged from 93.6% to 99.96%. The static heat load of each target was measured and ranges from 3 W to 6 W. See Table 1 for individual target operating data. The major cause for downtime during the first portion of this fixed target run was refrigerator related problems. In each case refrigerator efficiency was sufficiently reduced, thereby failing to provide the cooling capacity required to maintain a stable target. Approximately 220 hours of downtime were accumulated due to this problem. This represents 1.6% of the total operating time for all the targets. See Table 3 for other major causes of downtime during the first portion of this fixed target run.

SYSTEM UPGRADES

The years of experience associated with the target systems operated at Fermilab has led to some key component upgrades which have increased reliability and decreased the need for operator intervention.

Refrigeration System

The compressors used with the 10 W refrigerators have been upgraded to APD model HC-8, rotary type which has proven to be much more reliable than the previously used reciprocating machines. An interesting by-product of this compressor upgrade was that the new compression system with associated filtration had an extremely low oil carryover. The intake/exhaust valve of the Gifford McMahon refrigerator was left with no lubrication and started to fail prematurely. As a result, the graphite filled Teflon valve was replaced with a valve made from molybdenum disulfide filled Vespel. This material is self lubricating and has a lifetime of over 2000 hours. Instrumentation used in analyzing the performance of the refrigerator includes a Hastings mass flowmeter on the helium circuit and a transducer made by PCB Piezotronics, Inc. installed on the first stage of the refrigeration unit. The transducer allows monitoring of the pressure change during a refrigeration cycle in the first stage of the refrigeration unit. The refrigerator may be tuned to optimize its performance with the pressure curve of the refrigeration cycle as an indicator.

Target System

The hydrogen circuit of each target system includes two safety relief valves. The primary safety relief has been upgraded to an Anderson Greenwood Series 80 valve. This valve is fully open within 110% of set pressure and does not exhibit a problem with "sticking" as did the previously used modulating valves. The safety relief device used on the cryostat vacuum space has also been changed. A parallel plate relief device designed at Fermilab has replaced the Kapton rupture disc.

TARGET GUIDELINES

Over the many years of fabrication and operation of hydrogen targets, several standard practices have been developed. These have been compiled with some detailed rules for safety analysis in a document entitled, "Guidelines for the Design, Fabrication, Testing, Installation and Operation of LH_2 Targets." These rules are intended to complement the official Fermilab Safety Standards and will eventually be incorporated into the Safety Manual. An abbreviated table of contents is shown in Table 4.

The guidelines prohibit the use of hydrogen batch filling of the target flask and require that all targets be refrigerated either by mechanical cryocoolers or direct heat exchange with liquid helium. Recommended materials for the target flask are polyester film (Mylar), polyimide film (Kapton) and metals, such as stainless steel type 304 or aluminum type 6061-T6. All target flasks are designed, tested and relieved for a maximum allowable working pressure of at least 172.3 kPa differential. Allowable stresses are set at 63% of yield strength for plastic films and the lesser of 66% of yield or 25% of ultimate strength for metals. If available, the ASME code allowable stresses are referenced. Because of concerns with polyester film degradation, strict controls are placed on initial and periodic testing of material samples. A total radiation exposure limit of 10^8 rads of absorbed energy is imposed for Mylar flasks. High radiation exposure targets are fabricated with an acceptable metal.

Table 4. Table of Contents for LH_2 Guidelines

I.	Scope
II.	Design Fabrication and Testing:
	Refrigeration System
	H_2 Reservoir
	Target Flask
	Instrumentation and Control System
	Vacuum Vessel
	Secondary Containment
III.	System Testing and Installation
IV.	Safety Analysis and Review
V.	Operation
VI.	Target Safety Review Documentation

The successful fabrication of Mylar target flasks has developed over time to a point where both materials and techniques can be standardized. All joints are designed to place the epoxy bond in shear when the flask is pressurized. Dimensions for joint overlaps, surface preparation, the epoxy curing procedures and clean up are all carefully controlled. The recommended adhesives are Shell Epon 838 or 815 with a curing agent of V40 or V25. To decrease flask distortion during cooldown, longer targets incorporate artificial seams to compensate for material added due to joints. Connections to the hydrogen supply and vent tubes are made using a Vespel ring, which also joins the two parts of the cylindrical flask. Plastic flasks are relieved at 170.2 kPa by a dual relief system that is sized for the worst case air condensation heat flux.

The guidelines also set requirements for the secondary containment system that contains and safely vents any release of hydrogen gas from the target system due to failure. Controlling the environment of the target has proven crucial for the safe operation of these systems. By definition, these targets are part of an experimental apparatus, and are therefore in close proximity to equipment requiring personnel access. Each target vacuum box is sized to allow for the initial release of the entire liquid hydrogen volume. As the hydrogen warms, the gas is vented to a position that will not endanger personnel or equipment.

Rigorous testing of target system components is required by the guidelines. In all cases where applicable, the ASME code pressure testing procedures are referenced. Operating and emergency procedures, valve lists and failure mode analyses are developed for each system. The process of authorizing target cooldown requires a written permit signed by the safety panel chairperson and the respective division head.

CONCLUSIONS

Although the criteria for each experiment varied, we were successful in building target systems that met the expectations of both the experimenters and the target guidelines. The liquid hydrogen targets proved to be both safe and reliable during the first period of the 1990-1991 Fixed Target Run.

ACKNOWLEDGEMENTS

The authors would like to acknowledge the assistance of J. R. Kilmer in the design and upgrades of these systems. We would also like to acknowledge the mechanical and electrical technicians led by J. A. Peifer for the construction and installation of each system.

REFERENCES

1. J. C. Theilacker, Current operating experience and upgrade plans of the Tevatron cryogenic system, in: "Advances in Cryogenic Engineering," Vol. 35B, Plenum Press, New York (1990), p. 917.
2. R. Rubinstein, "1990 Fermilab Research Program Workbook," internal publication (1990).
3. J. W. Mark, A 650 mm long liquid hydrogen target for use in a high intensity electron beam, in: "Advances in Cryogenic Engineering," Vol. 29, Plenum Press, New York (1984), p. 1033.

A SOLID HYDROGEN TARGET FOR NUCLEAR PHYSICS

A.F. Zeller, R. Fontus, H. Laumer, and B. Sherrill

National Superconducting Cyclotron Laboratory
Michigan State University
East Lansing, Michigan

ABSTRACT

As part of a program to study the nuclear structure of exotic nuclear beams produced in heavy ion fragmentation reactions a thin hydrogen target is being constructed. A pure hydrogen target is superior to hydrogenous plastic targets because all of the mass is in the element of interest. The target will be attached to a small liquid helium dewar which is mounted on a flange to allow use in various locations. The enthalpy of the helium gas is used to cool an intermediate shield, so no liquid nitrogen is required.

INTRODUCTION

The National Superconducting Cyclotron Lab's K1200 cyclotron, combined with the A1200 Spectrometer is capable of producing beams of radioactive nuclei[1], with intensities of 10's to 10^4 particles per second. Some of the fundamental properties of these nuclei that will be measured are the atomic mass, and the properties of the single particle states. For example, the mass of ^{22}Al can be measured by the reaction ^{1}H(^{23}Al,^{22}Al)^{2}H. Ordinarily, spectroscopic information about nuclear levels, such as the neutron hole states populated by removing a neutron from ^{11}Li, would be done by using a ^{11}Li target and a proton beam, looking at the outgoing deuterons. However, making targets with a half life of 9 msec is very difficult, so the inverse reaction, p(^{11}Li,^{10}Li)d, is done with a radioactive beam. This requires a hydrogen target, which can be one of four types: a solid material which has a significant hydrogen content, or a target cell with windows which contain either solid, liquid, or gaseous hydrogen.

The standard hydrogenous material is polyethylene which consists of many chains of ($-CH_2-CH_2-$). The drawback of this material is that there is six times more areal density of carbon than hydrogen, resulting in large energy losses and high counting rates from contamination reactions from the vastly more abundant carbon.

Gas cells provide more freedom from background reactions, but suffer from relatively low areal densities, or from having to go to an extended target length to gain the required density. Therefore, the natural choice is for a liquid or solid target, with the drawback that cryogenics are now required. The choice of liquid or solid targets hinges on whether the power density of the beam is large or on the availability of cooling to 20 K or 4 K. When the energy deposition is high, such as from high energy electron beams, then a circulating liquid target is necessary. This, of course, is very expensive and very complicated. However, radioactive beams are of low intensity, with power depositions on the order of micro- or milli- watts. At low beam power density solid hydrogen has an advantage over liquid targets because the density[2] of

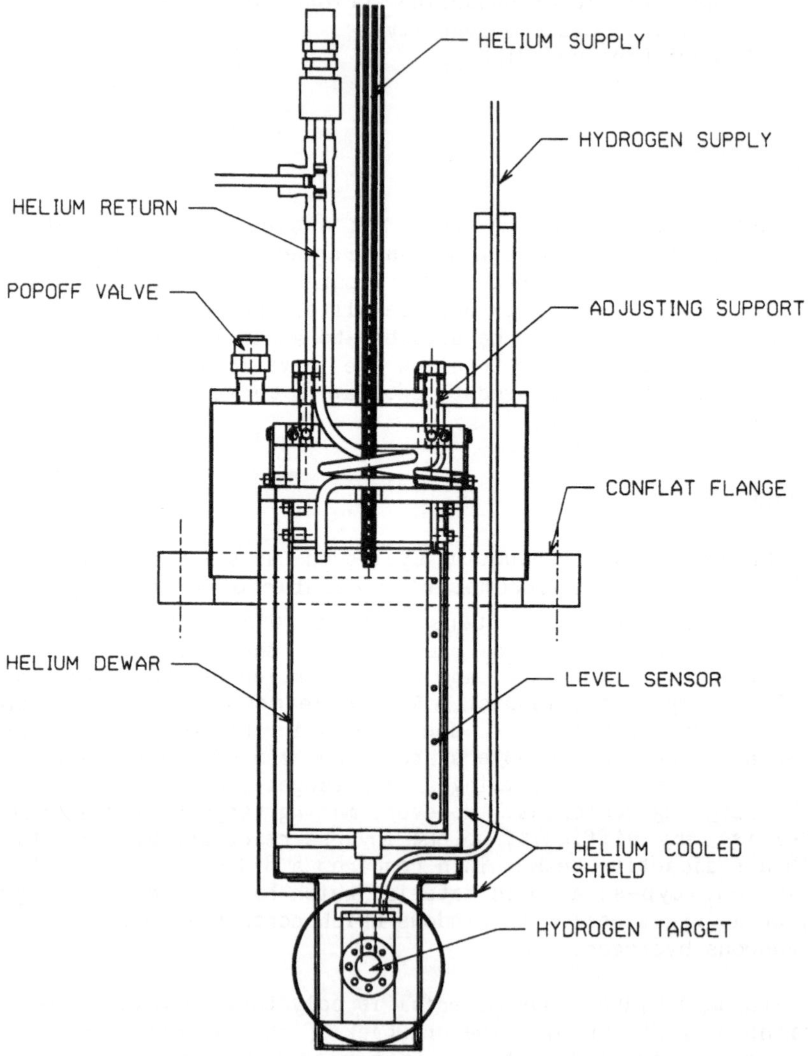

Fig. 1. Schematic view of the target cell and associated hardware. The cell and shields are copper, and the helium reservoir and hydrogen feed line are stainless steel. Not shown are the heater wires, temperature sensors, nor the solid angle limiting cones.

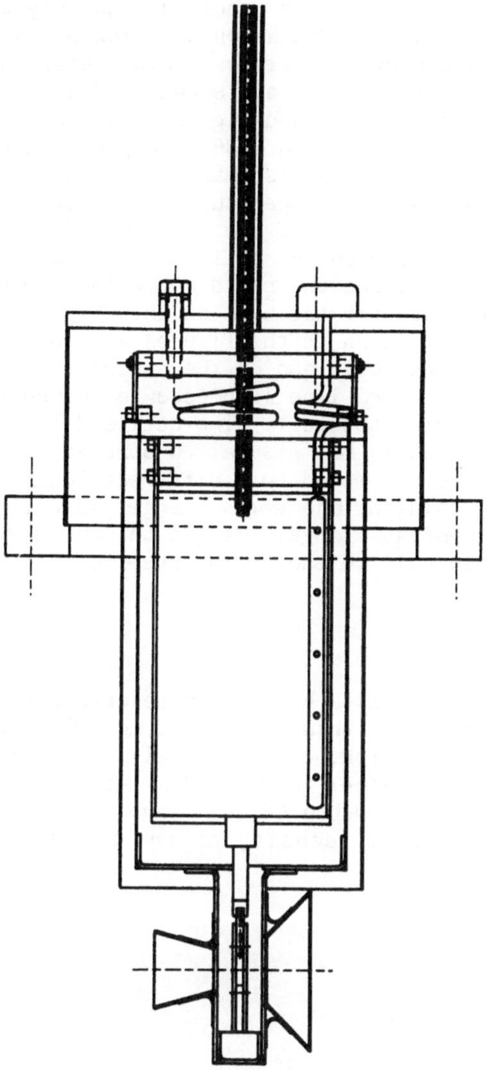

Fig. 2. The same as Fig. 1, but seen from the side to show the solid angle limiting cones and the cell thickness.

solid hydrogen at 4 K is 15% greater than liquid at 20 K. Additionally, the density is very constant over a several degree temperature range around 4 K, while the density of the liquid is a strong function of the temperature. The thermal conductivity of the solid and liquid differ significantly: The thermal conductivity of para hydrogen (<0.5% ortho hydrogen) at 4 K is 1.4 W/cm K while that of liquid at 20 K is 0.001 W/cm K - a difference of 1000. However, if the solid is formed by condensing out from a warm gas the ratio of ortho- to -para cannot be predicted. If it approaches the composition of room temperature gas, then the thermal conductivity may be as low[3] as 0.0025 W/cm K, which is still larger than the liquid.

TARGET DESIGN

The availability of liquid helium leads to a design which is readily transportable to any target location, and which uses liquid from a portable dewar. No liquid nitrogen is required, and only warm helium gas is returned to the refrigerator. Operation with low pressure differentials allow the use of thin beryllium windows which keeps the background to a minimum. Schematics of the target are shown in Fig. 1, which shows the target from the beam direction, and Fig. 2 which shows a side view. The target is cooled by conduction by a 6.35 mm copper rod extending from the helium reservoir. The cell is also made from copper to insure that the hydrogen remains at a constant temperature. The double heat shield is used to insure the maximum use of the enthalpy of the helium gas and to reduce the radiant heat load on the cell. The entrance and exit cones, shown in Fig. 2, also reduce the solid angle for 300 K radiation, while providing sufficient space for the beam to enter and exit with a non-zero scattering angle. Not shown in the figures are the temperature sensors or the heater wires on the cell and on the stainless steel hydrogen feed line. The heater on the feed line is needed to insure that hydrogen does not freeze before entering the cell. The cell and the heat shields are supported by G10 links, which also allow some adjustment of the target position when the target is moved to a different location. To facilitate use in different locations the whole target assembly is attached to a standard 20 cm (8") Conflat flange. Target thickness can be changed by simply attaching a new cell to the helium reservoir.

OPERATIONAL SCENARIO

The target cell and attached helium reservoir are shown in Fig. 3. The target areal density for this cell is 10 mg/cm². Two thin beryllium windows, each 0.00125 cm thick, close the cell. The windows are attached to the cell with the two keeper plates also shown in the figure. The windows are also attached with GE 7031 varnish to assure that they are leak tight. The total areal density of the beryllium windows is 4.7

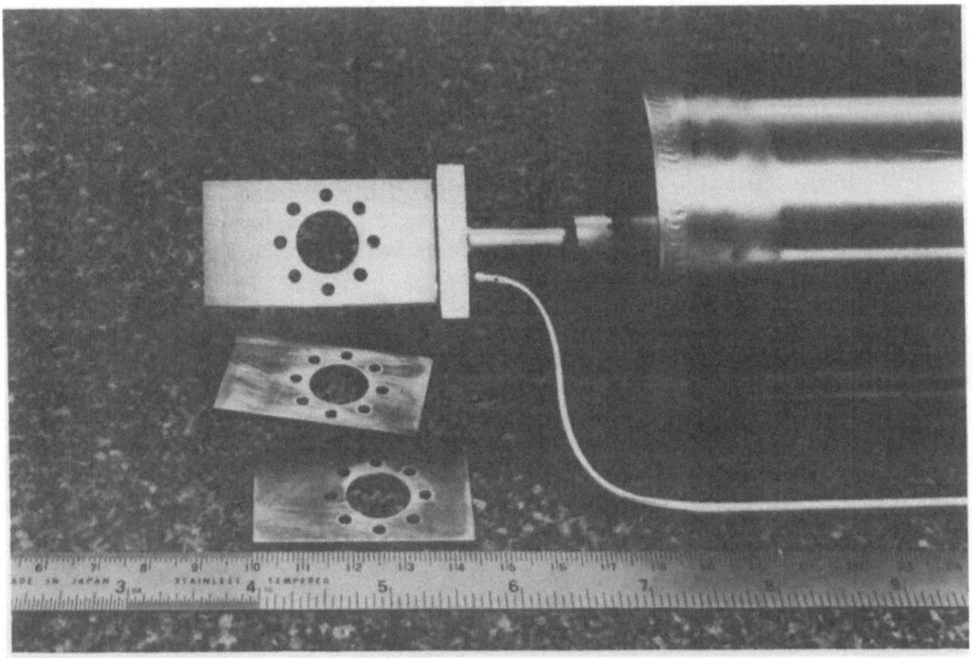

Fig. 3. Photograph of the target cell and attached liquid helium reservoir. Also shown are the window keeper plates.

mg/cm², less than half that of the hydrogen. It should be noted that the number of hydrogen target atoms is twenty times the number of beryllium atoms. We are presently trying to determine the minimum window thickness that can be produced. It is likely that an areal density of 1 mg/cm² will be possible, before porosity of the foils is encountered. Additionally, beryllium is mono-isotopic, so contamination peaks will be easier to interpret.

To use the target, the cell is first pumped out and the liquid helium reservoir is filled. The heater on the hydrogen fill line is turned on to keep the temperature in the line above 13 K, and a known amount of hydrogen gas is admitted. After the cell has been filled, the cell heater is used to raise the temperature to above the melting point of hydrogen, to insure that the target is uniform upon refreezing. Initial calibration of the temperature monitors and gas handling system will be done before the system is installed. Additionally, transparent windows will be used in testing to verify that the target is uniform. During the experiment the background peaks can be observed by removing the hydrogen and running only on the empty cell. The calculated hold time with a 100 L helium supply dewar is two days, providing a convenient time for background runs while the dewar is being refilled.

CONCLUSIONS

An inexpensive solid hydrogen target with thin windows has been built and, because it requires only liquid helium from a portable dewar, is readily transferable to any target location. The target is presently being assembled for final testing. It will be used when accelerator time is available.

ACKNOWLEDGMENTS

This work supported in part by the National Science Foundation under Grant No. PHY86-11210.

REFERENCES

1. B. M. Sherrill et al, The NSCL radioactive beam facility, in: "Proc. First Intl. Conf. on Radioactive Nuclear Beams", World Scientific Singapore, (1990), p. 72.
2. R. B. Scott, Properties of cryogenic fluids, in: "Cryogenic Engineering", Van Nostrand, Princeton, NJ (1959), p. 268.
3. G. E. Childs, L. J. Ericks and R. L. Powell, "Thermal Conductivity of Solids at Room Temperature and Below", NBS Monograph 131, National Bureau of Standards, Boulder, CO (1973).

LNG VAPORIZER UTILIZING VACUUM STEAM CONDENSING

Y. Miyata, M. Hanamure and H. Kujirai
Production Engineering Development Center
Tokyo Gas Co.,Ltd.
Yokohama, Japan

Y. Sato, H. Shohtani and Y. Ikeda
Heat Exchangers Engineering & Development Dept.
Sumitomo Precision Products Co.,Ltd.
Hyogo, Japan

ABSTRACT

This report concerns the field test results of a new type of peak-shaving LNG vaporizer (VSV) whose heat source is vacuum steam. The VSV utilizes the condensation latent heat of vacuum steam to vaporize and superheat LNG within heat transfer tubes. Prior to the field test, water and liquified nitrogen (LN_2) were used as the medium for the heated fluid in order to confirm the basic heat transfer characteristices of vacuum steam and the influence of non-condensable gasses. In the field tests, the heat transfer performance, controlability and a transient response were confirmed in a scaled-down model device using external steam as the heat source.

As a result, it was confirmed that : (1) a high heat transfer coefficient is obtained on the outer surfaces of the heat transfer tubes where the vacuum steam is condensing. (2) quick load change is possible and the outlet temperature drop rate is gradual when the heat source is cut off. (3) non-condensable gasses on the outer tube surfaces which obstructs heat transfer could be easily removed. These results indicate that this vaporizer responds flexibly to operating conditions and is suitable for a wide range of applications.

INTRODUCTION

At present, Tokyo Gas Co., Ltd. is mainly using the Open Rack Vaporizer (ORV) for base load and the Submerged Combustion Vaporizer (SCV) for peak-shaving and emergency applications. However, the new emphasis on energy saving and the intensifying competition between energy sources has necessitated the development of a more reliable and economical vaporizer for peak shaving. As a result, we, Tokyo Gas Co.,Ltd. and Sumitomo Precision Products Co.,Ltd. decided to jointly develop a new vaporizer, named the Vacuum Steam Vaporizer (VSV), utilizing a vacuum boiler which produces steam inside of a drum with sub-atmospheric pressure.

Advances in Cryogenic Engineering, Vol. 37, Part B
Edited by R.W. Fast, Plenum Press, New York, 1992

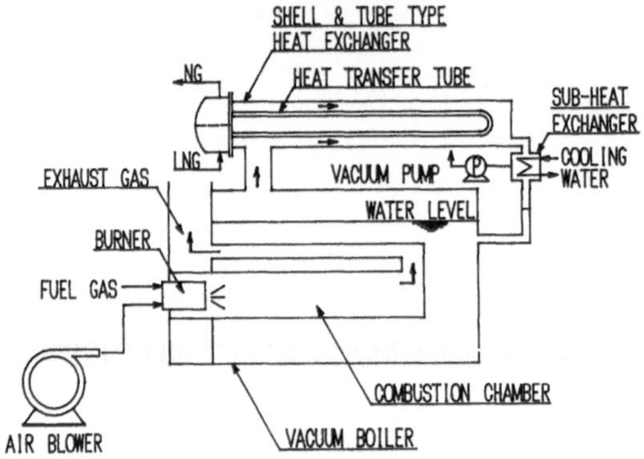

Fig. 1. Schematic diagram of VSV

The VSV is a steam-intermediate fluid type vaporizer which takes advantage of the high heat transfer coefficient of steam condensation. As shown in Fig. 1, the VSV is basically composed of a vacuum boiler and shell-and-tube-type LNG heat exchanger. The condensation latent heat of vacuum steam, which is produced in the vacuum boiler, vaporizes and superheats LNG in the heat transfer tubes which are located in the heat exchanger. The heating medium (vacuum steam) circulates from the drum to the heat exchanger and back to the drum in cycles as water, vacuum steam, and water again. In contrast, the SCV (see Fig. 2) is composed of heat transfer tubes in a water bath and a submerged combustion device as shown in Fig. 2. The exhaust gas from the combustion device is emitted into the water, stirring the water as it rises and vaporizes LNG in the process. In this vaporizer, a minimum exhaust gas flow rate must be maintained in order to sufficiently stir the water, regardless of the combustion load.

The advantages of the VSV are as follows:

1. As the combustion process is not submerged and the air supply can be adjusted in proportion to the combustion load, the cost of electric power required for the air blower can be reduced in comparison with the SCV.
2. The high heat transfer coefficient of steam condensation and high temperature of vacuum steam provide heat transfer surface area savings in comparison with the SCV.
3. An auxiliary external heat source such as excess steam from the gas turbine can be easily applied to the VSV, so as to realize significant running cost savings.
4. The VSV has leeway time in which it is possible to restart before final shut down in cases where the heat source is cut off due to misfires because the heat transfer process continues for some time after the heat source is cut off. This increases the reliability of the system.

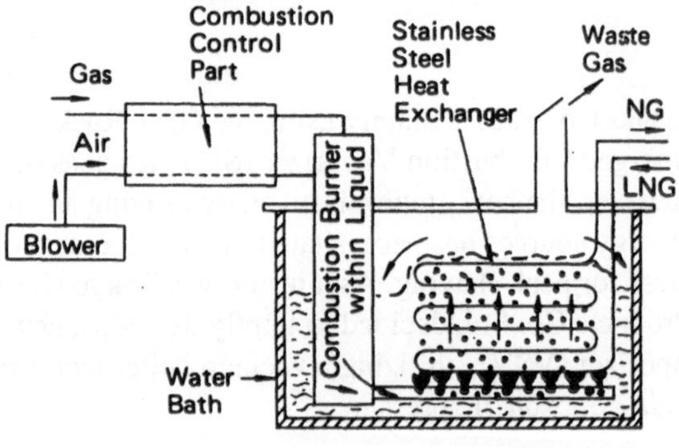

Fig. 2. Schematic diagram of SCV

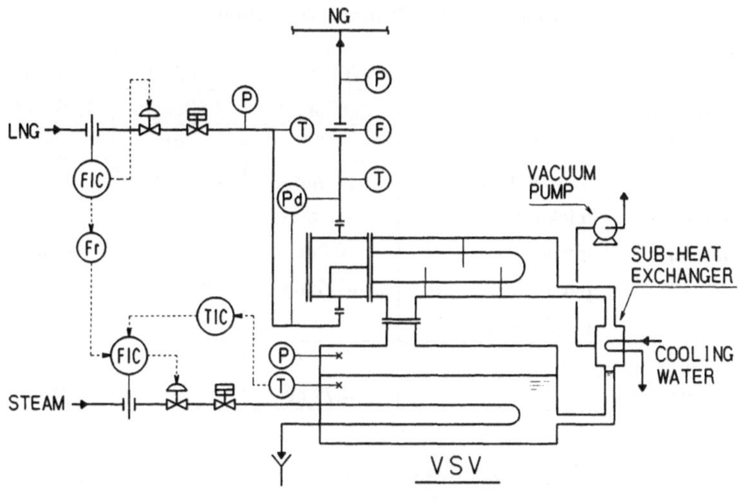

Fig. 3. Flow sheet of test equipment

As the first stage of development, basic experiments were conducted using water and LN_2 as the heated fluid in order to confirm the basic heat transfer characteristics of vacuum steam and the influence of non-condensable gasses. Following these experiments, field tests, where the vacuum steam was produced by external steam, were carried out at the Tokyo Gas Negishi Works using three kinds of heat exchangers in order to confirm heat transfer performance, controlability and transite response. The results of both the basic experiments and field tests are described under RESULT AND EVALUATION.

TEST EQUIPMENT

Fig. 3 and 4 show the flow sheet and configuration for the field tests and equipment, respectively. The upper portion of Fig. 4 is an LNG heat exchanger while the lower portion is a vacuum boiler producing vacuum steam. The three types of heat exchangers (type I, II, and III) shown in Table 1 were used in the field tests in order to estimate the heat flow and steam flow in an industrial-size plant. Heat transfer tubes for actual use were distributed in type I, while tubes with a diameter half the size were used in type II and III. The performance of type II and III were analyzed and compared to evaluate the effect of tube bundles.

The LNG was heated as it flowed through the heat transfer tubes where it was finally vaporized and superheated. A subsidiary heat exchanger was installed for the purpose of removing non-condensable gasses from the surface of heat transfer tubes in the heat exchanger. A condenser using industrial water as a cooling source was used as a subsidiary

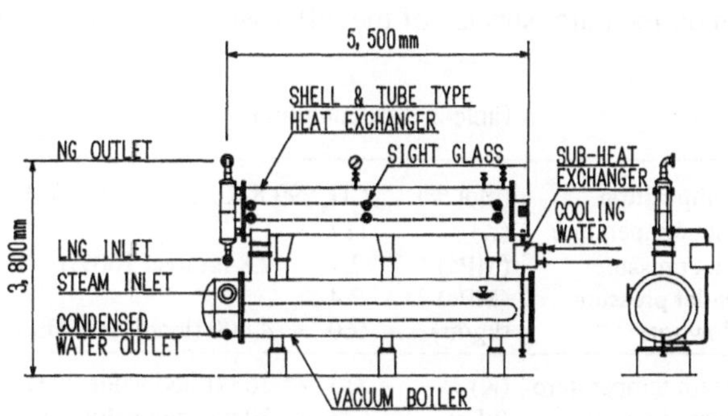

Fig. 4. Outline of test equipment

Table 1. Heat exchanger specifications

		TYPE I	TYPE II	TYPE III
LNG Flow rate (total)		4,600kg/h	4,600kg/h	1,200kg/h
Number of tubes		8	30	8
Dimension of heat transfer tubes	outside diameter	19.0mm	10.5mm	10.5mm
	thickness	1.6mm	1.2mm	1.2mm
	length	10.0m	4.0m	4.0m
	pitch	47.5mm	26.3mm	26.3mm
	number of passes	2	2	2
Material of heat transfer tubes		SUS304	SUS304	SUS304
Heat transfer area		4.79 m^2	4.00 m^2	1.07 m^2

heat exchanger. A vacuum pump was installed in order to initially decompress the drum and to periodically discharge non-condensable gasses.

External steam was used as the heat source in the field tests. In order to keep the drum temperature at test conditions, the inlet steam flow rate was controlled according to the LNG flow rate (feed forward factor) and the drum water temperature (feed back factor).

TEST METHOD

MEASURING ITEMS

As shown in Fig. 3, the flow rate, temperature and pressure of both inlet LNG and outlet natural gas (NG), the flow rate of heat source steam, the temperature and pressure of steam within the vacuum boiler, and the pressure loss in the steam drum and the tubes of the LNG heat exchanger were measured as conditions of operation. In addition, the temperature of the fluid in the heat transfer tubes as well as the temperature of the outer surfaces were measured using thermo couples and recorded with a high-speed data logger.

TEST CONDITION

The flow rate of LNG and drum water temperature were changed as shown in Table 2.

RESULT AND EVALUATION

HEAT TRANSFER COEFFICIENT OF CONDENSATION

In the initial basic experiment with water as the fluid medium, The heat transfer coefficient of condensation on the outer surface of the tubes was calculated by the difference in

Table 2. Test condition

LNG	Composition	(mol %)	CH_4 : 99.8, C_2H_6 : 0.1, N_2 : 0.1
	Inlet temperature	(K)	117 ~ 122
	Inlet pressure	(MPa)	2.7 ~ 2.8 (at load 100%)
	Outlet pressure	(MPa)	2.4
	Flow rate	(kg/hr)	460 ~ 4,600 (load 10~100%)
	Vacuum steam temperature	(K)	313 ~ 363 (rated value 343)
	Vacuum steam pressure	(kPa)	7.33 ~ 70.6 (rated value 30.9)

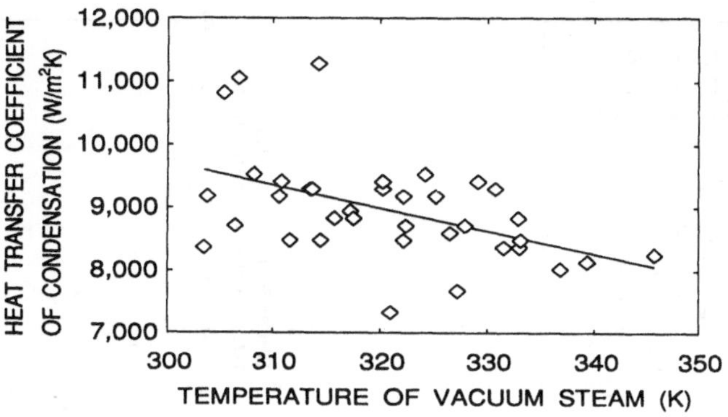

Fig. 5. Mean heat transfer coefficient of steam condensation

water temperature at inlet and outlet. Fig. 5 shows the results plotted in relation to mean vacuum steam temperature. The heat transfer coefficient was 8,000 to 9,000 W/m^2K and tended to increase as the steam temperature decreased. This tendency is consistent with the Nusselt formula concerning filmwise condensation on a horizontal tube.

The heat transfer coefficient of the tube interior (αi) was calculated using Equation (1).

$$\alpha i = Gm \cdot jw \cdot Cpw / Pr^{2/3} \qquad (1)$$

Next the outer tube surface condensation heat transfer coefficient (αo) was derived using Equation (2)

$$1/\alpha o = 1/K - do/(di \cdot \alpha i) - 0.5 \cdot do/\lambda m \cdot \ln(do/di) \qquad (2)$$

where

Gm = mass flow rate per unit cross sectional area (kg/m^2 s)
jw = Colburn's j factor of inside water (experimental value) (–)
Cpw = specific heat at constant pressure of inside water (J/kg K)
Pr = Prandtl number (–)
K = mean overall heat transfer coefficient (W/m^2 K)
do = outer diameter of a tube (m)
di = inner diameter of a tube (m)
λm = thermal conductivity of tube metal (W/m K)

STATIC CHARACTERISTIC

Fig. 6 shows the measured temperature of outlet gas according to LNG load using the drum water temperature as a parameter. We can see that when the drum water temperature is 353K (80°C) the VSV has 3 times the vaporization rate of the SCV, where the vaporization rate is defined as the ratio of the LNG flow rate to unit heat transfer surface area. We can also see that when the drum water temperature is kept constant, the temperature of outlet gas increases remarkably as the LNG load is reduced. For energy saving purposes, the drum water temperature was controlled according to LNG load so as to avoid the excess increase of NG temperature. Type II results have clarified the fact that a bundle with a larger tube number decreases the heat transfer coefficient on the outer surfaces due to non-condensable gasses. The test results show that the heat transfer coefficient in type II is 10 % less than that of type III.

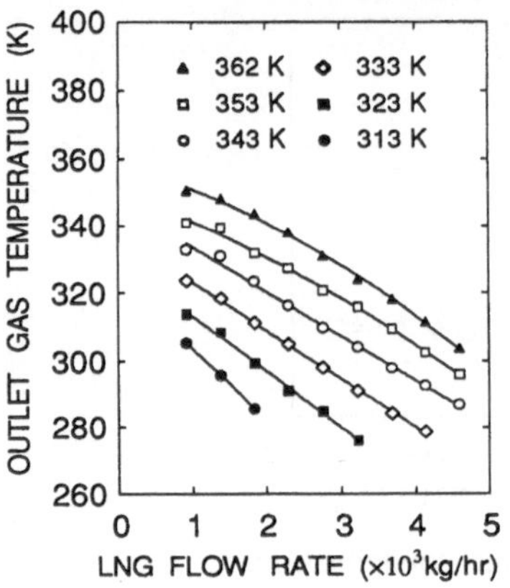

Fig. 6. Outlet gas temperature according to LNG load

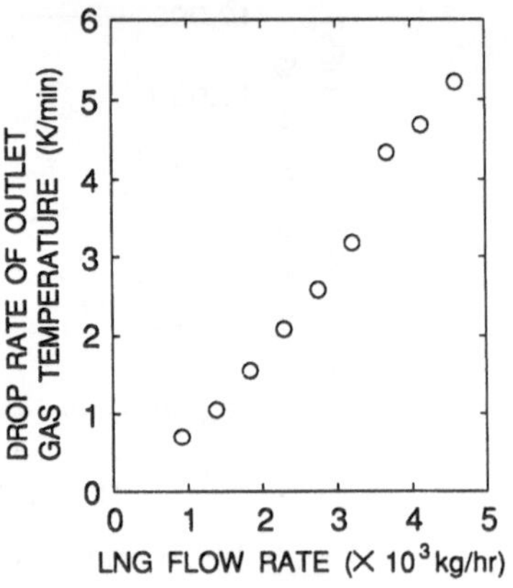

Fig. 7. Drop rate of outlet gas temperature in the case that heat source is cut off

DYNAMIC CHARACTERISTIC

In cases with both a constant and changing load (maximum change in speed of 30%/min), there was little change in the outflow rate of gas which implies stable vaporization. In order to clarify the transient response when the heat source is cut off, the heat-source steam was quickly turned off and the drop rate of outlet gas temperature was measured, under the condition of keeping the LNG load constant. Fig. 7 shows the drop rate of outlet gas temperature relative to LNG load. Heat is conserved in the drum water which continues to supply heat to the LNG tubes, thereby giving the VSV leeway time before final shutdown in cases where the heat source is suddenly cut off. This provides an opportunity to restart the heat source and is a useful feature for increasing reliability.

INFLUENCE OF NON-CONDENSABLE GAS

When non-condensable gasses such as leak air enter the vacuum steam drum, heat transfer performance is significantly decreased. In cases such as LNG vaporization where the fluid medium has an extremely low temperature, an ice layer may form on the outer tube surfaces. In the field test equipment, the subsidiary heat exchanger was installed as a steam condenser in the return pipe in the boiler in order to remove non-condensable gasses from the outer tube surfaces. This setup caused the non-condensable gasses to remain in the subsidiary heat exchanger.

The amount of non-condensable gasses increased with time and finally ice formed around bends in the heat transfer tubes where non-condensable gasses concentrate most. The ice was melted immediately each time non-condensable gasses were discharged by the vacuum pump operation. During long-run tests the vacuum pump was activated on a cycle of once every several hours. The heat loss attributed to the subsidiary heat exchanger was less than 1%.

CONCLUSION

The following points were confirmed in these experiments.

1. The heat transfer area of the VSV can be reduced to 1/3 that of the SCV due to high temperature of the vacuum steam and high heat transfer coefficient of condensation.
2. The VSV quickly and stably responds to a rapid load change.

3. The leeway time before final shutdown in cases where heat source is cut off is approximately 4 times that of SCV
4. Non-condensable gasses can be easily removed from the heat transfer surfaces by installing a subsidiary heat exchanger.
5. Because external waste heat can easily be incorporated into the system as a heat source, additional heat sources such as gas turbine exhaust gas can be used to economize operations.

ACKNOWLEDGEMENT

The authors would like to thank Prof. T. Fujii and Prof. H. Honda of Kyushu University for their sophisticated technical advice on the development of the prototype.

REFERENCE

1. T. Fujii et al, "Heat transfer and flow resistance of low pressure steam flowing through tube banks," Int. J. Heat Mass Transfer, Vol. 15, p. 247(1972).

DESIGN OF A 12 K SUPERCONDUCTING MAGNET SYSTEM, COOLED VIA THERMAL CONDUCTION BY CRYOCOOLERS

M.T.G. van der Laan, R.B. Tax, H.H.J. ten Kate,
L.J.M. van de Klundert

University of Twente, Applied Superconductivity Centre
Dept. of Appl. Physics, Enschede, The Netherlands

ABSTRACT

The operation of superconducting magnets is usually involved with the handling of cryogenic liquids and the associated aspects of availability and safety. In order to improve this a magnet system is developed which is cooled by small cryocoolers in the absence of liquids. The dry magnet having Nb_3Sn windings is mounted in a vacuum vessel with only one intermediate thermal shield. Three small model coils were fabricated, one made with the react and wind technique, the other two were first wound and then reacted. case specific conduction cooled leads were designed and tested. Also a dummy coil at full scale was cooled down to 11 K for which the shield temperature had values between 41 K and 46 K.

INTRODUCTION

Medium sized magnets are used in a numerous variety of machines and apparatus like magnetic resonance imaging, balloon borne particle detectors, ship propulsion, magnetic separation and maglev trains. For this type of magnets there exists the possibility for cooling them by means of cryocoolers. A few systems of this type has been described in ref. [1], [2] and [3]. In order to investigate the system behaviour and its properties, we are developing a 1:3 scale model MRI magnet. This magnet has an inner diameter of 400 mm, a length of 600 mm, and can generate a central field in the warm bore of 1 T with an energizing current of 100 A. The maximum field occurs in the windings equals to 1.9 T. The magnet will operate at 12 K while the intermediate thermal shield has to remain below 50 K. Cooling is provided by two double-stage Gifford-McMahon cryocoolers. The cooling power is brought to the magnet windings, leads and suspension by solid heat conductors only. This implies that, besides the helium gas in the closed circuit of the cryocoolers, no liquid or gaseous helium is present in the system. A major goal is to create a magnet system which is reliable, user friendly, economically feasible and having a good field quality and experimental possibilities.

Advances in Cryogenic Engineering, Vol. 37, Part B
Edited by R.W. Fast, Plenum Press, New York, 1992

DESIGN OF THE MAGNET SYSTEM

The magnet is build up as a stack of six circular coils and five aluminium spacing rings. The aluminium rings serve as buckling cylinders as well as for centering purposes. They also add to a better heat conductivity of the magnet. The coils are wound with 0.6 mm diameter Nb_3Sn wire, insulated with a glass braid, on a stainless steel coil former and are being reacted after winding. At this stage of the production process the coil is encapsulated with an aluminium strip and then vacuum impregnated. The aluminium strip has 12 radially arranged M6 threaded tubes. The channels in the tubes are used as an entry for the epoxy resin and the thread on the outside is used for connecting the heat conductors aluminium strips to the magnet. The coils and spacers are then stacked and held together with four axial tension rods. Cooling is provided by a system of aluminium and copper conductors which collect the heat and connect the coil system to the cryocoolers. At the outside of the magnet there are 12 aluminium strips bolted to the circular coils and spacing rings. These strips come together on an aluminium ring which is in turn bolted and connected to the cold head via a flexible copper braid. This is shown in Fig. 1 and in Fig. 5. The shield is connected to each cold head with 16 small copper braids. Other aspects of the system were treated in refs. [4] and [5]. The complete system is enclosed by a stainless steel vacuum vessel. One end of the vessel has a larger diameter than that part which surrounds the magnet. This creates more volume and flexibility facilitating experiments with the system as to include and test heat switches. In the case of an operational magnet the vessel can be cylindrically shaped of course and the cryocoolers, heat switches and other cryogenic parts can be put in a separate cold box.

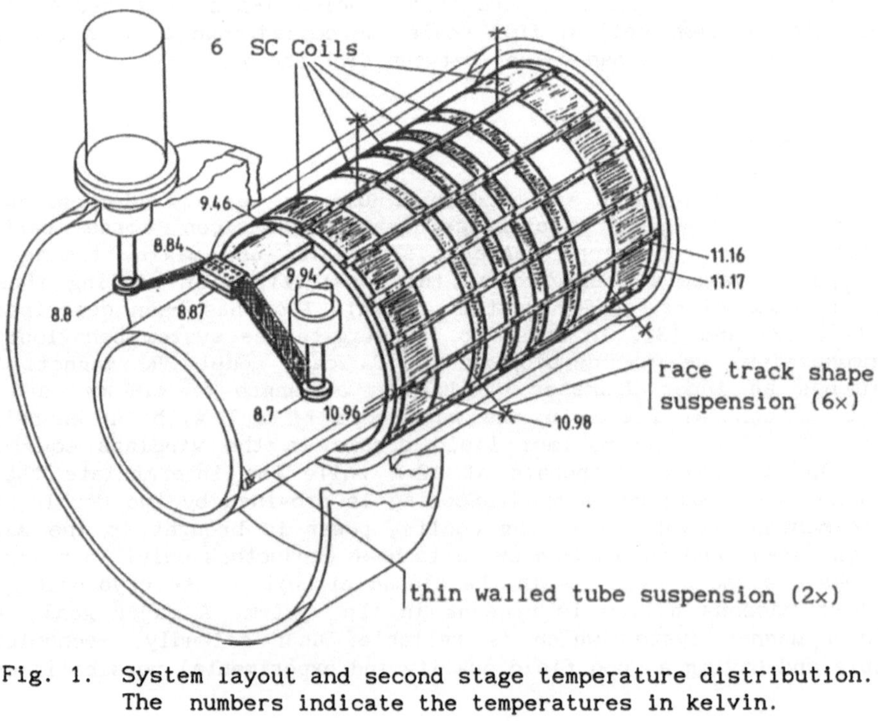

Fig. 1. System layout and second stage temperature distribution. The numbers indicate the temperatures in kelvin.

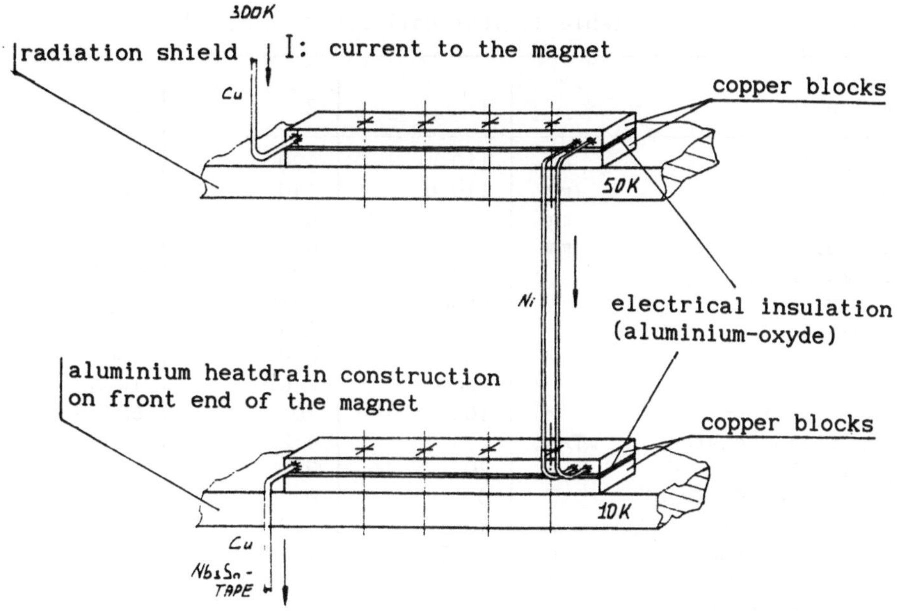

Fig. 2. Current lead construction.

The suspension

The magnet suspension system consists of six race track shaped belts and two thin walled stainless steel tubes. The former is present for counterbalancing the radial forces and the latter for the axial forces. The belts are 115 mm long and they have at the end a radius of curvature of 4.5 mm. They contain windings of aramid fiber and they are impregnated with epoxy in a mould. The tensile strength of this belt is 780 N while the fracture appeared in the mid section at 25 mm from the end of the belt. The heat leak through one belt is 1.5 mW from 300 K to the magnet. The belts are fitted between these temperature levels and are not thermally anchored to the first stage. The connection with the outer wall of the cryostat is made via spiral springs for centering purposes and for compensating the thermal contraction of the magnet.

Leads and thermal anchoring

The current leads consist of three parts. The first part is made of copper wire and thermally anchored to R.T. and the radiation shield. The second part is a nickel wire of 99.999 % purity and it is connected to the first and second stage. The connection between the second stage and the magnet terminals is a 5 mm × 50 μm Nb_3Sn tape conductor with a copper stabilizer. The leads are optimized for a current of 100 A following the standard current lead theory [7], while the heat produced is removed from the system by thermal conduction. A brief outline of the construction is given in Fig. 2. The thermal anchoring is a stack of two copper strips with a 0.6 mm Al_2O_3 strip in between. Indium foil between each layer improves the thermal conductivity of the stack. The calculated heat loads on the system at 100 A are: 0.78 W at the second stage and 9.2 W at the first stage.

Table 1. Test coil data

coil number		1	2	3
inner diam.	mm	110	130	40
outer diam.	mm	110.6	140	70.8
length	mm	50	8	24.3
wire diam.	mm	0.3	0.55	0.6
filaments		4536	192	~1530
matrix		bronze	copper	copper
insulation		none	glass braid	glass braid
process		R&W	W&R	W&R
supplier		VAC□	ECN*	TWCA+
number of turns		100	70	627
max. field constant	mT/A	2.4	4.2	17.9
I_q at 4.2 K	A	80	331	200
background field	T	0	0	10
I_q on cold head	A	25	100	150
at coil temp.	K	10.3	14.6	11.7
at lead temp.	K	-	16.7	12.8
background field	T	0	0	0

VAC□ = Vacuumschmelze
ECN* = the Netherlands Energy Research Foundation
TWCA+ = Teledyne Wah Chang Albany

Model coils

Before starting to wind the large magnet, three small test coils were made to learn about our winding and impregnating techniques. The coil and wire dimensions and the quench current measurements are summarized in Table 1. The first coil is wet wound using prereacted wire. This is a very risky process because the wire can easily be damaged and this was just what happened at every attempt to wind with the thin wire. This has resulted in relatively low quench currents.

The second and third coil are wound on a stainless steel coil form and then reacted and vacuum impregnated. In the second coil the lead temperature was 2 K higher than the coil temperature. An improved heat drain construction for the leads in coil 3 gave a measured lead/coil temperature difference of 1.1 K.

Test of a dummy magnet

The cooling down of the complete system should be finished within a reasonable time. In order to simulate a system cool down we have installed the radiation shield and a dummy magnet in the cryostat. In this case there are no current leads present. The dummy magnet is a steel cylinder with a diameter of 400 mm and a length of 600 mm. The aluminium shield weighs 54 kg and the steel dummy has a mass of 42 kg. The total magnet cold mass will be 65 kg. This mass is larger then the 42 kg it can have because of experimental and financial reasons. The thermal connection system is the same as that shown in Fig. 1. The temperature versus time characteristics are shown in Fig. 3. The dummy is cold within 41 hours. When the thermal switches couple the first and second stage of the cold

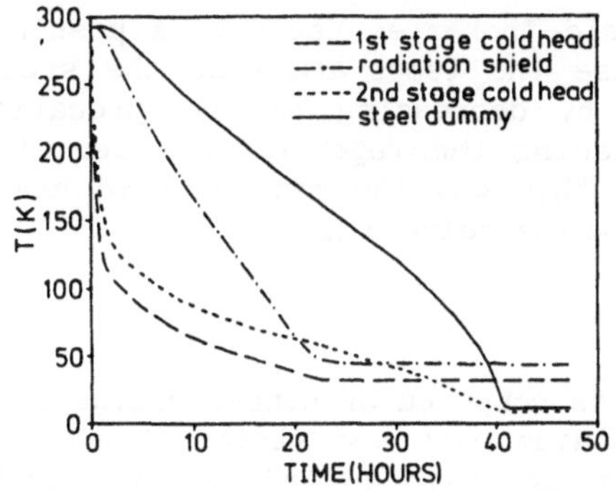

Fig. 3. Cool down curves for the dummy magnet.

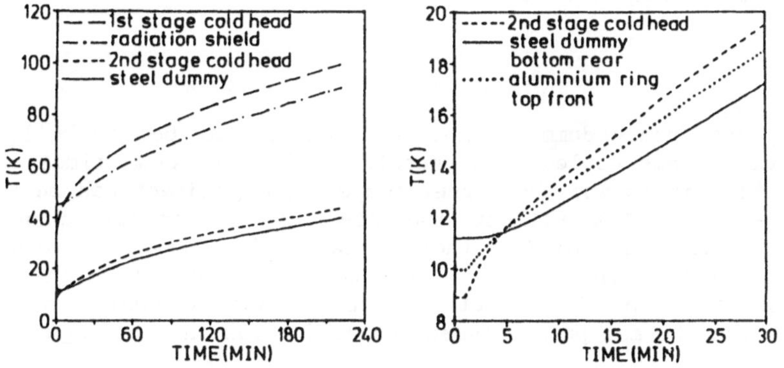

Fig. 4 Warm up curves for the dummy magnet.

heads are installed and working correctly it can be expected that the cooldown time will be reduced to less than 25 hours. The warm up curves are measured after switching off the compressors of the cold heads. They are shown in Fig. 4. It is shown that cryocooler failure during a short time will not directly cause an important temperature raise of the magnet.

Temperature distribution

The cooling down process has been observed using numerous GaAlAs temperature sensors. The stationary temperature distribution after the system was cold is shown partly in Fig. 1 in the case of the second stage temperatures and in Fig. 5 for the first stage temperatures. The maximum temperature gradients during the cool down of the system are shown in Fig. 3 where it is noticed that less than 15 % of the gradient is over the dummy. The indicated temperatures are of the cold head and of the bottom rear of the dummy.

In the stationary situation the maximum gradient measured over the dummy magnet is 1.2 K. This value can be reduced to 0.5 K when Al 1100 instead of Al 6063 is used as material for the heat collecting ring. The connection between the cold heads and the ring is a copper braid with a measured heat conduction coefficient of 1740 W/m/K at 10 K. When the cold heads are placed at a larger distance from the magnet, which results in less vibration and noise in the vicinity of the magnet, the required quantity of copper braid increases rapidly. A possibility to reduce this quantity is to decrease the temperature of the second stage cold head. This can be achieved by decreasing the reciprocating frequency of the displacer and by replacing the regenerator material Pb by Er_3Ni as was shown in ref. [8]. In this way the minimum cold head temperature can be lowered from 8 K to a value below 4 K.

Thermal impedances

The thermal path is composed of copper braids soldered to OFHC copper terminals and of aluminium bars and strips. The connections are made by screwing the parts together with an indium layer in between. The thermal impedances of these contacts have been measured at their normal working temperatures. An M6 Cu-In-Cu screw contact has a thermal impedance of 0.075 K/W at 32 K while a Cu-In-Al version has a value of 1.85 K/W at 37.5 K. At a temperature level of 9 K the latter type of contact has 0.2 K/W for ten parallel contacts.

CONCLUSIONS

It appears from a dummy magnet test that it may be to build medium sized magnet systems cooled by cryocoolers. The cool down times could be in reasonable limits and the magnet temperature gradient can be adjusted at a value below 1 K without the need of an extreme quantity of construction materials in the thermal path. When Er_3Ni is used as a regenerator material instead of Pb the cold heads could be fitted at a larger distance from the magnet with a relatively small increase of material in the thermal path between the cold heads and the magnet.

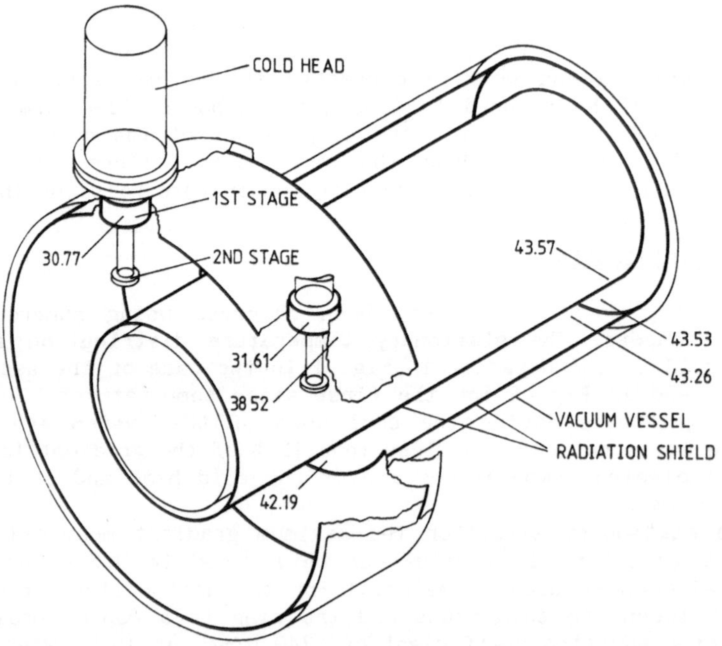

Fig. 5. System lay out and first stage temperature distribution. The numbers indicate the temperatures in kelvin.

ACKNOWLEDGEMENTS

This research in the program of the Foundation for Fundamental Research on Matter (FOM) has been supported in part by the Netherlands Technology Foundation (STW).

REFERENCES

1. M.O. Hoenig, Design concepts for a mechanically refrigerated 13 K superconducting magnet system, IEEE Trans. Magn., Vol. MAG-19 (1983), p. 880.
2. GEC, Superconductive magnetic resonance magnet, European patent application, publication nr. 0 350 267 A1, date 10.01.90.
3. GEC, Coupling a cryogenic cooler to a body to be cooled, European patent application, publication nr. 0 350 266 A2, date 10.01.90.
4. M.T.G. van der Laan, R.B. Tax, H.H.J. ten Kate, L.J.M. van de Klundert, A mechanically driven switch for decoupling cryocoolers, in: "Advances in Cryogenic Eng.," Vol. 35, p. 1457.
5. M.T.G. van der Laan, R.B. Tax, H.H.J. ten Kate, L.J.M. van de Klundert, The cryogenic system of a conduction cooled 12 K superconducting magnet. Presented at ICEC-13, Beijing, China, 1990.
6. M. Furuyama, H. Yamamoto, M. Tanaka, M. Kaneko, Y. Matsubara, T. Ogasawara, Performance of Nb_3Sn tape magnet cooled by indirect conduction method, in: "Advances in Cryogenic Engineering," Vol. 35, Plenum Press, New York (1990).
7. Yu.L. Buyanov, A.B. Fradkov, I.Yu. Shebalin, A review of current leads for cryogenic devices, Cryogenics 15: 193 (1975).
8. T. Kuriyama, M. Takahashi, H. Nakagome, H. Seshake, T. Eda, T. Hashimoto, Two-stage GM refrigerator with Er_3Ni regenerator for helium liquefaction. Paper presented at the 6th International Cryocooler Conference (1990).

AUTHOR INDEX

SUBJECT INDEX